STATISTICAL QUALITY CONTROL

McGraw-Hill Series in Industrial Engineering and Management Science

CONSULTING EDITORS

Kenneth E. Case, *Department of Industrial Engineering and Management, Oklahoma State University*

Philip M. Wolfe, *Department of Industrial and Management Systems Engineering, Arizona State University*

Barnes: *Statistical Analysis for Engineers and Scientists: A Computer-Based Approach*
Bedworth, Henderson, and Wolfe: *Computer-Integrated Design and Manufacturing*
Black: *The Design of the Factory with a Future*
Blank: *Engineering Economy*
Blank: *Statistical Procedures for Engineering, Management, and Science*
Bridger: *Introduction to Ergonomics*
Denton: *Safety Management: Improving Performance*
Grant and Leavenworth: *Statistical Quality Control*
Hicks: *Industrial Engineering and Management: A New Perspective*
Hillier and Lieberman: *Introduction to Mathematical Programming*
Hillier and Lieberman: *Introduction to Operations Research*
Huchingson: *New Horizons for Human Factors in Design*
Juran and Gryna: *Quality Planning and Analysis: from Product Development through Use*
Khoshnevis: *Discrete Systems Simulation*
Kolarik: *Creating Quality: Concepts, Systems, Strategies, and Tools*
Law and Kelton: *Simulation Modeling and Analysis*
Marshall and Oliver: *Decision Making and Forecasting*
Moen, Nolan, and Provost: *Improving Quality through Planned Experimentation*
Nash and Sofer: *Linear and Nonlinear Programming*
Nelson: *Stochastic Modeling: Analysis and Simulation*
Niebel, Draper, and Wysk: *Modern Manufacturing Process Engineering*
Pegden: *Introduction to Simulation Using SIMAN*
Riggs, Bedworth, and Randhawa: *Engineering Economics*
Taguchi, Elsayed, and Hsiang: *Quality Engineering in Production Systems*
Wu and Coppins: *Linear Programming and Extensions*

STATISTICAL QUALITY CONTROL

SEVENTH EDITION

Eugene L. Grant
Professor of Economics of Engineering, Emeritus
Stanford University

Richard S. Leavenworth
Professor of Industrial and Systems Engineering, Emeritus
University of Florida

THE McGRAW-HILL COMPANIES, INC.
New York St. Louis San Francisco Auckland Bogotá Caracas Lisbon
London Madrid Mexico City Milan Montreal New Delhi
San Juan Singapore Sydney Tokyo Toronto

McGraw-Hill

A Division of The McGraw-Hill Companies

STATISTICAL QUALITY CONTROL

This book is printed on acid-free paper.

1 2 3 4 5 6 7 8 9 0 FGR FGR 9 0 9 8 7 6 5

P/N 0-07-024162-7
PART OF
ISBN 0-07-844354-7

This book was set in Times Roman by The Clarinda Company.
The editors were Eric M. Munson, Kiran V. Kimbell, and James W. Bradley;
the production supervisor was Leroy A. Young.
The cover was designed by Amy Barovick.
Quebecor Printing/Fairfield was printer and binder.

Library of Congress Cataloging-in-Publication Data

Grant, Eugene Lodewick, (date)
Statistical quality control / Eugene L. Grant, Richard S. Leavenworth. —7th ed.
p. cm. —(McGraw-Hill series in industrial engineering and management science)
Includes bibliographical references and index.
ISBN 0-07-024162-7
1. Quality control—Statistical methods. I. Leavenworth, Richard S. II. Title. III. Series.
TS156.G7 1996
658.5'62' 015195—dc20 95-41429

THE INTERNATIONAL EDITION

When ordering this title, use ISBN 0-07-114248-7.

ABOUT THE AUTHORS

EUGENE L. GRANT is Professor of Economics of Engineering, Emeritus, at Stanford University. He holds B.S. and C.E. degrees from the University of Wisconsin, an M.A. in Economics from Columbia, and an honorary Doctorate of Engineering from Montana State University. After serving in the U.S. Navy in World War I and with the U.S. Geological Survey thereafter, he joined the faculty of Montana State University in 1920 and left there with the rank of Professor in 1930 to become Associate Professor of Civil Engineering at Stanford. Prior to becoming emeritus in 1962, he served as Executive Head of the Civil Engineering Department (1947–1956) and Chairman of the Industrial Engineering Committee (1946–1952).

From 1941 through 1944, he directed the Engineering, Science, and Management War Training (ESMWT) program at Stanford where, in conjunction with Holbrook Working, he helped develop an intensive short course in Quality Control by Statistical Methods for key personnel of war industries. This became the model for a nationwide program of similar courses sponsored jointly by the War Production Board and the United States Office of Education; alumni of these courses formed the American Society for Quality Control in 1946.

He received the 1952 Shewhart Medal from ASQC and was awarded honorary membership in 1968. In 1966 ASQC established an annual E. L. Grant Award for distinguished contributions to quality control education. He is one of three honorary academicians of the International Academy for Quality. In 1965, the Engineering Economy Division of the American Society for Engineering Education established an annual E. L. Grant Award for the best paper in each volume of *The Engineering Economist.* Professor Grant received the Founders Award from the American Institute of Industrial Engineers in 1965 and the Wellington Award in 1979. He received a Distinguished Service Citation from the University of Wisconsin College of Engineering in 1964. In 1987 he was elected to the National Academy of Engineering.

In addition to "Statistical Quality Control," his books are "Principles of Engineering Economy," now in its eighth edition (1990), published by John Wiley & Sons and now coauthored with W. G. Ireson and R. S. Leavenworth; "Depreciation," published by The Ronald Press Company (1949), coauthored with P. T. Norton, Jr.; and "Basic Accounting and Cost Accounting," published by McGraw-Hill

Book Company (1956 and 1964), the second edition coauthored with L. F. Bell. He was coeditor with W. G. Ireson of "Handbook of Industrial Engineering and Management," published by Prentice-Hall (1955 and 1971).

RICHARD S. LEAVENWORTH received his Ph.D. degree in Industrial Engineering from Stanford University in 1964. After two years at Virginia Polytechnic Institute and State University, he joined the faculty of the University of Florida, where he is now Professor Emeritus of Industrial and Systems Engineering. Over the years, he has won teaching awards and has served the University as Acting Chairman of the Department, and as Assistant Dean for Planning and Analysis.

Dr. Leavenworth's research, sponsored in part by the Office of Naval Research, has resulted in numerous technical reports and publications in the *Journal of Quality Technology, Naval Research Logistics Quarterly,* and *Transactions of the Institute of Industrial Engineers*. He has developed a number of courses for in-plant training and presented seminars and short courses nationally and internationally. His consulting and training activities have included such organizations as the U.S. Department of Commerce, General Electric Company, Manhattan Industries, Florida Department of Transportation, U.S.D.A. Food Safety Inspection Service, Harris Corporation, the Tennessee Valley Authority, the Naval Aviation Depot, Jacksonville, Florida, Blue Cross/Blue Shield of Florida, George Washington University, and Technology Training, Inc.

He is coauthor with E. L. Grant and W. G. Ireson of "Principles of Engineering Economy" (8th ed., 1990) published by John Wiley & Sons, New York, and with E. L. Grant of "Statistical Quality Control" published by McGraw-Hill, New York, since its fourth edition (1972). Both texts have been translated into Spanish and are widely distributed through International Students Editions.

He has served the Institute of Industrial Engineers (IIE) nationally as Region Vice President and Vice President for International Operations, has held numerous offices in local chapters, and has served as editor, *The Engineering Economist*. In 1984 he was presented the IIE Quality Control and Reliability Engineering Division Award of Excellence. He is a Fellow of IIE and a Senior Member of the American Society for Quality Control and a member of the American Society for Engineering Education and the American Statistical Association and is a Registered Professional Engineer in California and Florida. In 1991, he served on the Board of Examiners for the Malcolm Baldrige National Quality Award and, in 1993, on the Board of Examiners for the Florida Governor's Sterling Award for quality.

TO THE MEMORY OF
John Charles Lounsbury Fish

CONTENTS

LIST OF EXAMPLES

PREFACE

This book is a practical working manual. It deals primarily with various types of control charts and acceptance sampling systems and procedures. These are simple but powerful techniques that have been widely used in many industries and in many countries throughout the world to improve product and service quality and to reduce costs. The most effective use of these techniques depends upon their being understood by production and inspection operators and their supervisors, by engineers, and by middle and upper level management.

The objective has been to write a book that might be immediately useful to all of these groups. No attempt has been made to write for the professional statistician or the mathematician. The aim has been to give just enough theory to supply practical working rules that will enable one to recognize the limitations of the methods as well as their many uses.

A special feature of this book is the liberal use of descriptions of actual cases from a number of economic sectors. Each example has been selected to bring out one or more important points. These examples reflect the general viewpoint of the book that the statistical techniques described should be studied primarily as a means to various ends desired by a cost-conscious management. A number of examples deal not only with the behavior of random variables but also with the behavior of people in various industrial and business situations.

The book retains its intuitive approach to probability and statistics. The authors believe that this approach is very important to engineering and business students who may have very little industrial experience but may have some responsibility for teaching SQC courses in-house to operators and inspectors. Just as in previous editions, the changes from the preceding edition have been made in part to improve the presentation of fundamental principles and in part to try to keep the treatment of various topics up-to-date. Some of the changes are as follows:

1. The book is now divided into four parts. Part One is a short introduction to the topics covered in the other three parts. Part Two treats topics concerning statistical process control (SPC), Part Three discusses scientific sampling, and Part Four deals with the economics of quality as well as some management and teaching aspects of quality control.

2. The treatment of simple $\overline{X}$ and R charts (Chap. 2) begins Part Two. This change is intended to emphasize the idea that control needs to be achieved before histograms and estimates of process centering and dispersion (introduced in Chaps. 3 and 4) can be relied on to predict future results.
3. The material on rational subgrouping (Chap. 8) has been separated from that on process capability (Chap. 9) to emphasize the importance of establishing purpose when setting up sampling procedures and methods of recording data. The addition of error of measurement material and gage repeatability and reproducibility (R&R) studies at this point emphasizes the importance of purpose in determining rational subgrouping.
4. Process capability analysis and aspects of design and inspection specifications and design tolerances have been combined into Chap. 9 to emphasize their importance in process improvement.
5. Chapter 10 on special process control procedures now contains comments on the so-called precontrol technique (10.2.8), the use of box plots (10.3.4), exponentially weighted moving average (EWMA) charts (10.6.3), and the special problems introduced by extremely small production runs (10.9) and extremely high-quality production (10.10). The chapter concludes with CuSum charts for averages.
6. The scope of Part Three has been somewhat reduced without deleting material important to U.S. students. All standard measures of effectiveness of sampling procedures are now covered in Chap. 11. The emphasis has shifted to that of analyzing any sampling process, rather than simply sampling for product acceptance purposes, because nearly all quality inspection involves sampling, either from an unknown and possibly shifting universe, or from a lot. Standard plans therefore are used more as a vehicle for understanding the implications of various kinds of sampling.
7. Chapter 17 on economic aspects of quality control has been expanded to include the ASQC categories of quality costs (17.3.1) and Taguchi's loss function and its related index, C_{pm} (17.4).
8. A new chapter (Chap. 18) describes some of the history and evolution of statistical quality control and the dramatic impact that Walter Shewhart, Harold Dodge, and W. Edwards Deming have had on quality control in the twentieth century.
9. Another new chapter (Chap. 19) introduces and compares the two most influential quality management models that are evolving in this last quarter of the twentieth century: *(a)* the total quality model best exemplified by the criteria and guidelines for the U.S. Malcolm Baldrige National Quality Award and *(b)*, the registration requirements of the ISO 9000 set of quality standards. The difficulty of integrating and implementing these management models in an operational setting, such as a factory floor or a service facility, is presented through the use of problem solving models, the "seven basic tools," and storyboard models.
10. A computer disk accompanies the book. It contains simple SPC software, some easily identifiable data tables for examples and problems in the book,

and some probability calculation programs written in BASIC. TXT files on that disk contain instructions for running the programs. Instructions are also included in the Instructor's Solutions Manual that accompanies this book and is available to instructors from the publisher.

This edition follows the general pattern of the six earlier editions, which were greatly influenced by the viewpoint and philosophy of Dr. W. Edwards Deming. The authors also wish to acknowledge their debt to Harold F. Dodge, whose extensive comments had a great influence on the writing of the second and third editions. Particular mention should be made of Ms. Bonnie Small for her review and suggestions for the fourth edition and of the staff of the University of Tennessee Center for Productivity through Quality for their review of the fifth edition. Our thanks are also extended to a number of users of the sixth edition for their suggestions for changes and improvements.

Reviewers of the manuscript of this edition provided much valuable input in its modification. These reviewers included Suraj Alexander, Roger Berger, Richard Buhman, Kenneth Case, who also is a Consulting Editor in the field for McGraw-Hill, Owen Miller, Joel Nachlas, Ahmad Seifoddini, Jill Swift, and Gary Wasserman. Their approaches to the subject, which ranged from strong management to strong statistics orientations, contributed many insights. Needless to say, not all of their suggestions could be incorporated while still maintaining the essential character of the book as previously described. Further acknowledgment should be made of the reviews by Professors Swift and Wasserman of the accompanying software developed by Mr. Mark Shewhart.

In addition, our thanks are offered to Mr. Seymour Selig for his assistance with material on military standards and to Dr. John F. Mahoney for his comments and help in the development of new tables of factors for control charts. Nevertheless, as in all technical books, the final responsibility for the selection and treatment of material must fall on the shoulders of the authors, and they should receive the blame for any deficiencies.

Eugene L. Grant

Richard S. Leavenworth

PART ONE

INTRODUCTION

1

INTRODUCTION AND OVERVIEW

The long-range contribution of statistics depends not so much upon getting a lot of highly trained statisticians into industry as it does in creating a statistically minded generation of physicists, chemists, engineers, and others who will in any way have a hand in developing and directing the production processes of tomorrow.

—W. A. Shewhart *and* W. E. Deming[†]

1.1 THE MEANING OF QUALITY

How do you define *quality?* A curt response might be, "I really can't define it, but I sure do know it when I see it." A more reflective observer might say, "Quality is in the eye of the beholder." Assertions such as these frequently are offered as flippant responses to the question. Nevertheless they do offer some insight into the true nature of "quality" because they focus directly on the respondent rather than on a thing or an action being judged. In the popular sense, identifying quality is purely a judgment call. It is entirely dependent on the perceptions of the individuals, or collection of individuals, making the determination. You and I may agree on the quality of something, or we may completely disagree. If the good, service, activity, or whatever *satisfies* us, we judge it to be something of quality. If we are not satisfied, we judge it to lack quality. In the popular sense, we judge the *degree of excellence* of a good or service by comparing it explicitly or implicitly to something else and in the context of our life experiences.

[†]W. A. Shewhart (edited by W. E. Deming), *Statistical Method from the Viewpoint of Quality* ~ p. 49, The Graduate School, Department of Agriculture, Washington, D.C., 1939.

The people who make the satisfaction judgment calls may be described as *customers.* Service professionals, such as lawyers or engineers, may prefer to call them clients; computer service companies may call them users or subscribers; doctors use the term patients; educators prefer the term students; the customers of live entertainment performances may be called audiences or fans. It is a matter of complete indifference what term is used; in the broadest sense they all are customers and they are the ones that must be satisfied. Quality, then, must be judged in terms of *customer satisfaction.*

Webster's International Dictionary offers the following as its first two definitions of the noun *quality:*

> **1a.** peculiar or essential character; **1b.** a distinctive inherent feature; property, virtue. **2a.** degree of excellence; degree of conformance to standard; **2b.** inherent or intrinsic excellence of character or type; superiority in kind.[†]

In the context of the goods and services provided by business, industry, and government, Webster's definitions suggest that it is the features, properties, attributes, characteristics, etc., of a good or service that define its quality. They go on to imply, if not state, that we must judge quality excellence or quality conformance using some metric. If some good or service is judged to be superior, it can be so only in relation to some absolute standard or relative to another good or service that is judged to be inferior. This implies the need to establish measurement systems, specifications, standards of acceptability, procedures for provision or production and test, and schemes for assessment and verification of results.

As engineers, production or service managers, or businesspeople, this customer satisfaction concept is not sufficient to translate needs and desires into products and services. There is a specification-production-inspection cycle that requires more careful definition than has been given so far. Thus we may consider four quite different technical meanings of the word *quality.*

The first meaning is purely descriptive.[‡] It defines some characteristic of the good or service. An electric heater element has resistance, which is measured in ohms. A steel wire has weight and tensile strength as two of its characteristics. Light bulbs have average lives. Doctors' offices have waiting times, and airports have takeoff and landing delays. These are characteristics. There is no implication of degree of excellence, just the potential for recording measurable fact.

The other three technical meanings of quality pertain to *degree of excellence* in design, specification, and production or implementation. The second meaning bears on the *quality of design.* It addresses the question, "How well does the design meet its objectives?" These design objectives must be enunciated by management. They are derived from the perceived needs and desires of customers. Much interaction between provider and customer may be required before design

[†] *Webster's Third International Dictionary*, Merriam Webster Company, 1971.

[‡] This discussion of quality follows that of E. L. Grant and L. F. Bell, "Some Comments on the Seman-[illegible]uality and Reliability," *Industrial Quality Control,* Vol. 17, No. 11, May 1961, pp. 14–17, American [illegible] Quality Control (ASQC), Milwaukee, Wis.

objectives are clarified to the point where all agree that the design is satisfactory in terms of its degree of excellence.

The third meaning relates to *quality of specification.* It addresses questions about the effectiveness of design specifications in realizing stated or implied objectives. This is the stage at which concepts are turned into drawings, procedures, dimensions, bills of material, processes, timetables, and so on.

The fourth meaning relates to *quality of conformance.* It addresses questions about degree of excellence of conformance of the service or manufactured product to the design specifications. Some writers on quality control say that quality means conformance to specifications and nothing more. As we have indicated, quality must be built into design, specification, and conformance. Conformance to improper specifications or to specifications that do not derive from an excellent design will not lead to excellent services and products.

At a number of places in this book we address problems encountered in manufacturing and services that relate to these different meanings and to combinations of them. It is important, therefore, for the student to realize that these different meanings do exist and to be able to identify which meaning of quality is being discussed when people, even those in technical fields, speak of quality. The problem is further compounded when such persons make unconscious and unidentified shifts from one technical meaning to another.

1.2 THE FOCUS OF THIS BOOK

This book deals primarily with various types of Shewhart control charts and with various types of acceptance sampling systems and procedures. Developed in the 1920s, Shewhart control charts got their name from their inventor, Walter A. Shewhart, when he was with Bell Telephone Laboratories.[†] Also with Bell Labs at that time was Harold F. Dodge, who is best remembered for his many developments in the use of tables and procedures for product acceptance based on the mathematics of probability. While the tools were developed in a manufacturing environment, they have proved to be equally useful in nonmanufacturing areas. A brief discussion of the impact on manufacturing of these two intellectual giants is given in Chap. 18.

1.3 THE CONTROL-CHART VIEWPOINT

One purpose of this book is to explain the control-chart point of view in some detail. Part Two, entitled "Statistical Process Control," is primarily devoted to this exposition. In spite of the apparent simplicity of the control chart, many people involved in trying to manage and improve quality find that its use calls for an entirely new point of view. Briefly stated, the main principles are these:

[†]See W. A. Shewhart, *Economic Control of Quality of Manufactured Product,* D. Van Nostrand, Inc., Princeton, N.J., 1931. This important work was published again in 1980 as the Fiftieth Anniversary Commemorative Reissue, with dedication by W. E. Deming, and is available from ASQC.

1. Measured quality of manufactured product is always subject to a certain amount of variation as the result of chance.
2. Some "constant system of chance causes" is inherent in any particular scheme of production and inspection.
3. Variation within this stable pattern is inevitable.
4. The reasons for variation outside this stable pattern may be discovered and corrected.

These simple principles provide the foundation of statistical process control.

The *constant system of chance causes* referred to by Shewhart also has been called a *stable system* or defined as *statistical stability.* W. Edwards Deming (see Chap. 18) referred to these constant or stable systems as *common causes of variation,* constantly operating within a system. Variation outside this stable system, called *assignable cause* variation by Shewhart and *special cause* variation by Deming, may be detected using control charts and its cause or causes eliminated through analysis and corrective action. Often these actions result in substantial improvement in the quality of products and services and reductions in spoilage, rework, and error rates. Moreover, by identifying certain quality variations as inevitable chance variations, the control chart tells when to leave a process alone and thus prevents unnecessarily frequent adjustments that tend to increase variability rather than decrease it.

Through its disclosure of the natural capabilities of a production process, the control-chart technique permits better decisions on engineering tolerances and better comparisons between alternative designs and between alternative production or service methods.

1.3.1 An Aid to Cooperation

A common complaint among production personnel is that engineers responsible for specifications do not understand production problems. Inspection personnel often complain not only about the poor quality of manufactured product but also about the unreasonableness of specified tolerances. In fact, very frequently inspection practices develop that substitute the inspector's views regarding proper tolerances for those actually specified by the engineers. In many organizations, there is evident need for a basis on which designers, production personnel, and inspectors can understand each other's problems. The development of Quality Circles in Japan was an approach to solving this problem.

In the past many arguments among these three groups have been carried on with more heat than light because of the absence of facts in a form that would provide a basis for agreement. In many cases these facts can be provided by the use of statistical quality control techniques. In fact, statistical quality control provides a common language that may be used by all three groups in arriving at a rational solution of mutual problems.

1.3.2 Some Simple Statistical Tools

Many of the techniques developed by mathematical statisticians for the analysis of data may be used in the control of product quality. The expression *statistical quality control* may be used to cover all uses of statistical techniques for this purpose. However, it often relates particularly to four separate but related techniques that constitute the most common working statistical tools in quality control. These tools are:

1. The Shewhart control charts for measurable quality characteristics. In the technical language of the subject, these are described as charts for variables, or as charts for $\overline{X}$ and R (average and range) and charts for $\overline{X}$ and s (sample average and standard deviation).[†]
2. The Shewhart control chart for fraction rejected, or p chart.
3. The Shewhart control chart for number of nonconformities, or c chart.
4. The portion of sampling theory that deals with the quality protection given by any specified sampling acceptance procedure.

Part Two of this book deals with (1), (2), and (3) and also introduces some other statistical techniques for comparing processes. Part Three includes a number of procedures for sampling acceptance, many of which may be used as an aid in sampling for process control purposes. In the use of statistical methods to control product quality, these are the tools for cost reduction and quality improvement that are the most widely applied.

1.3.3 Variables and Attributes Data

An important distinction in the technical language of statistics is that between *variables* and *attributes.* When a record is made of an actual measured quality characteristic, such as a dimension expressed in hundredths of a millimeter, the quality is said to be expressed by variables. When a record shows only the number of articles conforming and the number failing to conform to any specified requirements, it is said to be a record by attributes. Variables data are measured on a continuous scale with individual measurements rounded to some desired number of decimal points. Attributes data involve counts of articles or counts of events.

All manufactured product and many service activities must meet certain requirements, either express or implied. Many of these requirements, or specifications, may be stated as variables. Examples are dimensions, hardness, operating temperatures, tensile strengths, weights, or percentage of a particular impurity in a chemical compound. Many specifications of variables give both an upper (designated U or USL_x) limit and a lower (designated L or LSL_x) limit for the measured

[†]The symbol $\overline{X}$ is read as "X bar" or as "bar X." The bar over any symbol always indicates an average. Thus $\overline{X}$ means an average of the X's. R and s are alternative measures of the dispersion of a set of data. Their calculation and meaning are explained in Chaps. 2 and 3.

value. Some, such as the percentage of a particular impurity in a chemical compound, may have an upper limit only; others, such as strength, may have a lower limit only.

Attributes data may be used in either of two ways. Many requirements are necessarily stated in terms of attributes rather than variables. This applies, for example, to many things that may be judged only by visual examination. The glass cover of a pressure gage is either cracked or not cracked. A lithographed label either has a certain desired color or does not. The surface finish of a piece of furniture either presents a satisfactory appearance or does not. A glass lens contains no more than a certain number of air bubbles not exceeding a specified size, or it does not. In general, the thing examined either conforms or does not conform to specifications. On the other hand, attributes data may provide the count of the number of times a particular event occurs. Examples are a count of the number of surface imperfections on the hood of an automobile, or the number of rings before a telephone is answered, or the number of nonconforming solder connections on an electronic circuit card.

In addition to numerous quality characteristics that are specified without reference to dimensions, many characteristics that are specified as measurable variables are inspected merely as conforming or nonconforming to specifications. This applies, for example, to gaging of dimensions of machine parts by go and not-go gages.

1.3.4 Some Benefits to Be Expected from the Use of Control Charts for Variables

Trouble is a common state of affairs in manufacturing. Whenever the trouble consists of difficulty in meeting quality specifications that are expressed in terms of variables, the Shewhart control charts for $\bar{X}$ and R are indispensable tools in the hands of the troubleshooter. They provide information on three matters, all of which need to be known as a basis for action. These are:

1. Basic variability of the quality characteristic
2. Consistency of performance
3. Average level of the quality characteristic

No production process is good enough to produce all items of product exactly alike. Some variability is unavoidable; the amount of this basic variability will depend on various characteristics of the production process, such as the machines, the materials, the operators. Where both upper and lower values are specified for a quality characteristic, as in the case of dimensional tolerances, one important question is whether the basic variability of the process is so great that it is impossible to make all the product within the specification limits. When the control chart shows that this is true and when the specifications cannot be changed, the alternatives are either to make a fundamental change in the production process that will reduce its basic variability or to face the fact that it will always be necessary to sort out the acceptable product. Sometimes, however, when the control chart shows so much basic variability that some product is sure to be made outside the tolerances, a review of the situation will show that the tolerances are tighter than necessary for the functioning of the product. Here the appropriate action is to change the specifications to widen the tolerances.

Variability of the quality characteristic may follow a chance pattern, or it may behave erratically because of the occasional presence of assignable causes that can be discovered and eliminated. The control limits on the chart are so placed as to disclose the presence or absence of these assignable causes. Although their actual elimination is usually an engineering job, the control chart tells when, and in some instances suggests where, to look. As previously mentioned, the action of operators in trying to *correct* a process may actually be an assignable cause of quality variation. A merit of the control chart is that it tells when to leave a process alone as well as when to take action to correct trouble. The elimination of assignable causes of erratic fluctuation is described as bringing a process *under control* and is responsible for many of the cost savings resulting from statistical quality control.

Even though the basic variability of a process is such that the *natural tolerance range* is narrower than the specified tolerance range, and even though the process is under control, showing a consistent pattern of variability, the product may be unsatisfactory because the average level of the quality characteristic is too low or too high. This also will be disclosed by the control chart. In some cases the correction of the average level may be a simple matter, such as changing a machine setting; in other situations, such as increasing an average level of strength, it may call for a program of research and development work.

Once the control chart shows that a process is brought under control at a satisfactory level and with satisfactory limits of variability, one may feel confident that the product meets specifications. This suggests the possibility of basing acceptance procedures on the control chart, using it to determine whether this happy state of affairs is continuing. Under these favorable circumstances substantial savings are often possible in costs related to inspection. Where inspection consists of destructive tests, it may be possible to reduce the number of items tested, thus saving both in testing cost and in the cost of the product destroyed.

1.3.5 Some Benefits to Be Expected from the Use of Control Charts for Fraction Rejected

Most routine inspection of manufactured products is inspection by attributes, classifying each item inspected as either accepted or rejected (with possibly a further division of rejects into spoilage and rework). This statement applies to both 100% inspection and sampling inspection. In such inspection it is common practice to make a record of the number of items rejected.

The practice of recording at the same time the number of items inspected is not so universal. However, if quality performance at one time is to be compared with that at another time, the record of total number inspected is just as necessary as the record of number rejected. The ratio of the number of items rejected to the number of items inspected is the *fraction rejected.*†

Thus the Shewhart control chart for fraction rejected generally makes use of data that either are already available for other purposes or can readily be made available. Simple statistical calculations provide control limits that tell whether assignable causes of variation appear to be present or whether the variations from day to day (or lot to lot, supplier to supplier, or whatever the classification basis may be) are explainable on chance grounds.

It will be shown in later chapters that this control chart for attributes (the p chart) is somewhat less sensitive than the charts for variables ($\bar{X}$ and R charts) and does not have as great a diagnostic value. Nevertheless, it is an extremely useful aid to production supervision in giving information as to when and where to exert pressure for quality improvement. It is a common experience for the introduction of a p chart to be responsible for substantial reductions in the average fraction rejected. In some instances the p chart will disclose erratic fluctuations in the quality of inspection, and its use may result in improvement in inspection practices and inspection standards. Moreover, the p chart often serves to point out those situations needing diagnosis of trouble by the control chart for variables.

In addition to its use in process control, the p chart may be of great value in dealing with outside suppliers. Suppliers may differ both in the quality level submitted and in the variability of that quality level. It is particularly desirable to know whether the quality of product submitted by a supplier today is a reliable indication of what may be expected to be submitted next month. The p chart gives useful guidance on this point.

1.3.6 Some Benefits to Be Expected from the Use of Control Charts for Nonconformities

This type of control chart applies to two rather specialized situations. One is the case where a count is made of the number of nonconformities of such type as blemishes in a painted or plated surface of a given area, weak spots in the insulation of rubber-covered wire of a given length, or imperfections in a bolt of cloth. The other is the case of inspection of fairly complex assembled units in which there are a great many opportunities for occurrences of nonconformities of various types, and the total number of nonconformities of all types found is recorded for each unit.

As in other types of control charts, the control limits are set in a way to detect the presence or absence of assignable causes of variation, and they therefore tell when to take action on the process and when not to do so. Experience indicates that erratic variation in inspection standards and inspection practices seems particularly likely to exist in this type of inspection and that the control chart for nonconformities generally proves helpful in standardizing inspection methods.

†It is commonly expressed as a decimal fraction, such as 0.023. The decimal fraction may be multiplied by 100 to convert it into percent rejected, such as 2.3%.

1.4 SCIENTIFIC SAMPLING

Acceptance inspection is a necessary part of manufacturing and may be applied to incoming materials, to partially finished product at various intermediate stages of the manufacturing process, and to final products. Acceptance inspection may also be carried out by the purchaser of manufactured products. Equivalent statements may be made about services and service processes simply by substituting the appropriate words for *manufacturing* and *product.*

The introduction of Just-In-Time (JIT) inventory control makes formal acceptance sampling procedures by the customer impractical except for quality audit purposes. The supplier is required to perform any sampling inspection and to provide statistical evidence of process control and/or acceptable product with each lot shipped to the customer. This evidence may take the form of control charts, results of inspections, and/or quality indexes.

Much inspection, either for process control or for product acceptance, is by sampling. Often 100% inspection turns out to be impracticable or clearly uneconomical. Moreover, the quality of the product accepted may actually be better with scientific acceptance sampling procedures than would be the case if the same product were subjected to 100% inspection. Sampling inspection has a number of psychological advantages over 100% inspection, where inspectors' fatigue on repetitive operations may be a serious obstacle.

It is common knowledge that on many types of inspection, even several 100% inspections will not eliminate all the nonconforming product from a stream of product, a portion of which does not conform to specifications. The best protection against the acceptance of nonconforming product is, of course, having the product made right in the first place. Good sampling acceptance procedures may often contribute to this objective through more effective pressure for quality improvement than can be exerted with 100% inspection. Some sampling schemes also provide a better basis for diagnosis of quality troubles than is common with 100% inspection.

It should be recognized that although statistical sampling acceptance procedures are generally superior to traditional sampling methods established without reference to the laws of probability, anyone who uses acceptance sampling must face the fact that whenever a portion of the stream of products submitted for acceptance does not conform to specifications, some nonconforming items are likely to be passed by any sampling acceptance scheme. The statistical approach to acceptance sampling frankly faces this fact. It attempts to evaluate the risk assumed with alternative sampling procedures and to make a decision as to the degree of protection needed in any instance. It is then possible to choose a sampling acceptance scheme that gives a desired degree of protection with due consideration for the various costs involved.

1.5 USE OF EXAMPLES

Throughout this book there are many descriptions of cases involving use of statistical quality control methods. For convenience in reference, these are numbered consecutively throughout each chapter.

Before proceeding in Parts Two and Three with a detailed description of the tools of statistical quality control, it seems advisable to give a general perspective on the subject by illustrating the use of some of the more common tools. Examples 1.1 and 1.2 illustrate the control chart for variables; Example 1.3 illustrates the use of elementary probability theory to evaluate the quality protection given by a common sampling plan.

Example 1.1

$\bar{X}$ and R charts. Experimental control charts for process control

Facts of the case In the kit of tools provided by statistical quality control, one of the most potent for the diagnosis of production problems is the Shewhart control chart for variables. A course in statistical quality control often starts with a brief introduction of this control chart. It was the topic for discussion in the third 2-h lecture in an evening course given in the subject.

One of the members of the class was a production supervisor in a small department in a plant that had never before used statistical quality control methods. After hearing the 2-h lecture, this supervisor, in order to become familiar with the control chart, made an experimental application to one of the operations in the department.

This operation consisted of thread grinding a fitting for an aircraft hydraulic system. The pitch diameter of the threads was specified as 0.4037 ± 0.0013 in. All these fittings were later subject to inspection of this dimension by go and not-go thread ring gages. This inspection usually took place several days after production. In order to minimize gage wear in this inspection operation, it was the practice of the production department to aim at an average value a little below the nominal dimension of 0.4037 in.

To make actual measurements of pitch diameter to the nearest ten-thousandth of an inch, the supervisor borrowed a visual comparator that had been used for other purposes. Approximately once every hour the pitch diameter of five fittings that had just been produced was measured. For each sample of five the average and the range (largest value in sample minus smallest value) were computed. The figures obtained are shown in Table 1.1.

Two charts that are not control charts If these measurements had been made without benefit of the supervisor's introduction to the control-chart technique, the averages for each sample might well have been calculated but probably not the ranges. Figure 1.1 shows two types of charts that are *not* control charts but that sometimes are made from information of this type. These charts may be of interest to production supervision, but they do not give the definite basis for action that the control chart supplies.

Figure 1.1*a* shows individual measurements plotted for each sample. It also shows the nominal dimension and upper and lower tolerance limits. With the exception of one fitting in sample 8, all the fittings examined met the specified tolerances.

TABLE 1.1 MEASUREMENTS OF PITCH DIAMETER OF THREADS ON AIRCRAFT FITTINGS

Values are expressed in units of 0.0001 inch in excess of 0.4000 in. Dimension is specified as 0.4037 ± 0.0013 in

Sample number	Measurement on each item of five items per hour					Average $\bar{X}$	Range R
1	36	35	34	33	32	34.0	4
2	31	31	34	32	30	31.6	4
3	30	30	32	30	32	30.8	2
4	32	33	33	32	35	33.0	3
5	32	34	37	37	35	35.0	5
6	32	32	31	33	33	32.2	2
7	33	33	36	32	31	33.0	5
8	23	33	36	35	36	32.6	13
9	43	36	35	24	31	33.8	19
10	36	35	36	41	41	37.8	6
11	34	38	35	34	38	35.8	4
12	36	38	39	39	40	38.4	4
13	36	40	35	26	33	34.0	14
14	36	35	37	34	33	35.0	4
15	30	37	33	34	35	33.8	7
16	28	31	33	33	33	31.6	5
17	33	30	34	33	35	33.0	5
18	27	28	29	27	30	28.2	3
19	35	36	29	27	32	31.8	9
20	33	35	35	39	36	35.6	6
Totals	. . .	. . .	. . .	. . .	. . .	671.0	124

Figure 1.1*b* shows the averages of these samples. A chart of this type may be useful to show trends more clearly than one of the type of Fig. 1.1*a*. However, without the limits provided by the Shewhart technique, it does not indicate whether the process shows lack of control in the statistical sense of the meaning of *control.*†

†The word *control* has a special technical meaning in the language of statistical quality control. A process is described as *in control* when a stable system of chance causes seems to be operating. This meaning is developed more fully in Part Two of this book. However, the word is often misused and misinterpreted, particularly by those who have been briefly exposed to the jargon of statistical quality control without having had a chance to learn its principles.

For example, a government inspection officer in a certain plant was shown a chart that gave average values of an important quality characteristic of successive lots of a certain item manufactured in this plant. This chart, prepared by one of the inspectors, was similar to Fig. 1.1*b* in showing averages but no control limits. "Any fool could see this process is in control," the officer exclaimed. Later an assistant calculated control limits and plotted them on the chart, showing many points out of control. A few minutes later the inspection officer was overheard giving this advice to the plant manager: "Any fool can see that this process is out of control."

As a matter of fact no one—fool or otherwise—can tell by inspection of a chart such as Fig. 1.1*b* whether or not the process is in statistical control.

It should be noted that because Fig. 1.1*b* shows averages rather than individual values, it would have been misleading to indicate the tolerance limits on this chart. It is the individual article that has to meet the tolerances, not the average of a sample. Averages of samples often fall within tolerance limits even though some of the individual articles in the sample are outside the limits. This was true in sample 8, in which the average was close to the nominal dimension

FIGURE 1.1 Pitch diameter of threads of fitting for aircraft hydraulic system: (*a*) individual measurements; (*b*) averages of samples of five.

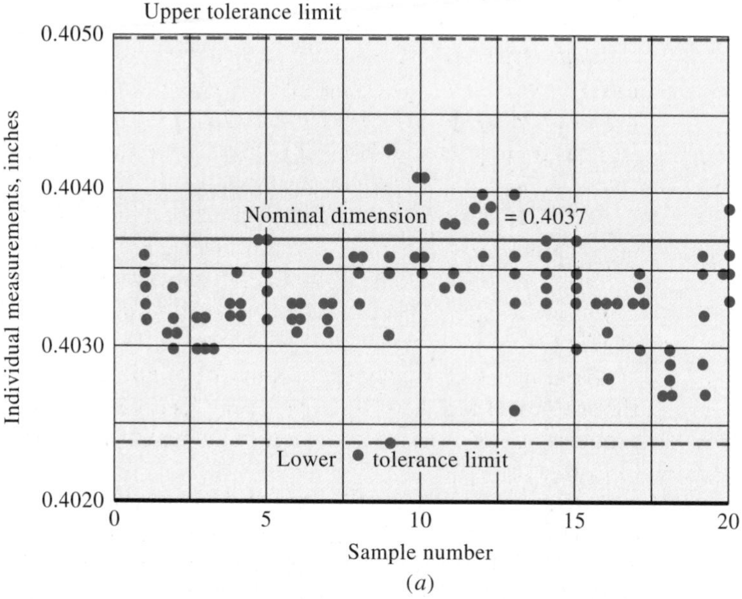

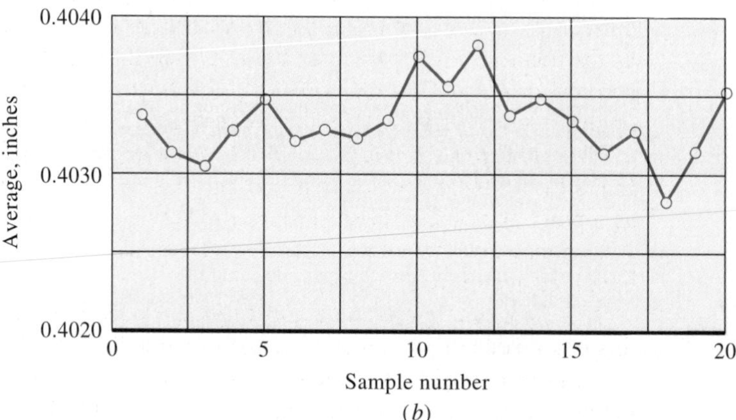

even though one item was below the minimum limit. Therefore, a chart for averages that shows tolerance limits tends to give a false sense of security on the question of whether tolerances have been met.

Two control charts Figure 1.2*a* shows the control chart for averages $\bar{X}$. It will be noted that this is Fig. 1.1*b* with the addition of control limits and with the elimination of the irregular line connecting the points. Figure 1.1*b* shows the control chart for range *R*.

Each of these control charts has a solid line to indicate the average value of the statistic that is plotted. The grand average $\bar{\bar{X}}$ (i.e., the average of the averages) is 33.6 (measured—as in Table 1.1—in units of 0.0001 inch in excess of 0.4000 in). This is the sum of the averages, 671.0, divided by the number of samples, 20. The average of the ranges is 6.2. This is the sum of the ranges, 124, divided by the number of samples, 20.

Each chart also shows two dashed lines marked *upper control limit* and *lower control limit.* The distances of the control limits from the line showing the average value on each chart depend on the subgroup size and the average range $\bar{R}$. The methods for calculating such distances are explained in Chap. 2.

On the $\bar{X}$ chart of Fig. 1.2 this distance is 3.6 for both control limits. The upper control limit is 37.2, 3.6 above the grand average of 33.6, and the lower control limit is 30.0, 3.6 below the grand average.

On the *R* chart, the control limits are not equidistant from the average range of 6.2; the upper limit is 13.1, and the lower limit is 0.

These limits shown on Fig. 1.2 are what are described in Chap. 2 as *trial limits.* Before projecting them into the future (i.e., past sample 20) to control future production, they need to be slightly modified by methods that are there explained.

The charts show lack of control Three points (samples 10, 12, and 18) are outside the control limits on the chart for averages. Two points (samples 9 and 13) are outside the control limits on the chart for ranges. This indicates the presence of assignable causes of variation in the manufacturing process, i.e., factors contributing to the variation in quality that it should be possible to identify and correct. Of course not much could be done about these past assignable causes, as the control limits were not established until the end of the 20-h record. The control charts in Fig. 1.2 merely give evidence that there should be a good opportunity to reduce the variability of the process.

The dividends from the control chart come in the application of the control limits to future production. The prompt hunting for assignable causes as soon as a point goes out of control gives opportunity for their immediate discovery. Action may be taken not only to correct them at once but in many cases to prevent their recurrence.

This lack of control was evident to the production supervisor who prepared this chart. A continuation of the chart permitted the identification of the assignable causes of variation in the average—mostly related to machine setting—and the assignable causes of variation in the range—mostly related to carelessness

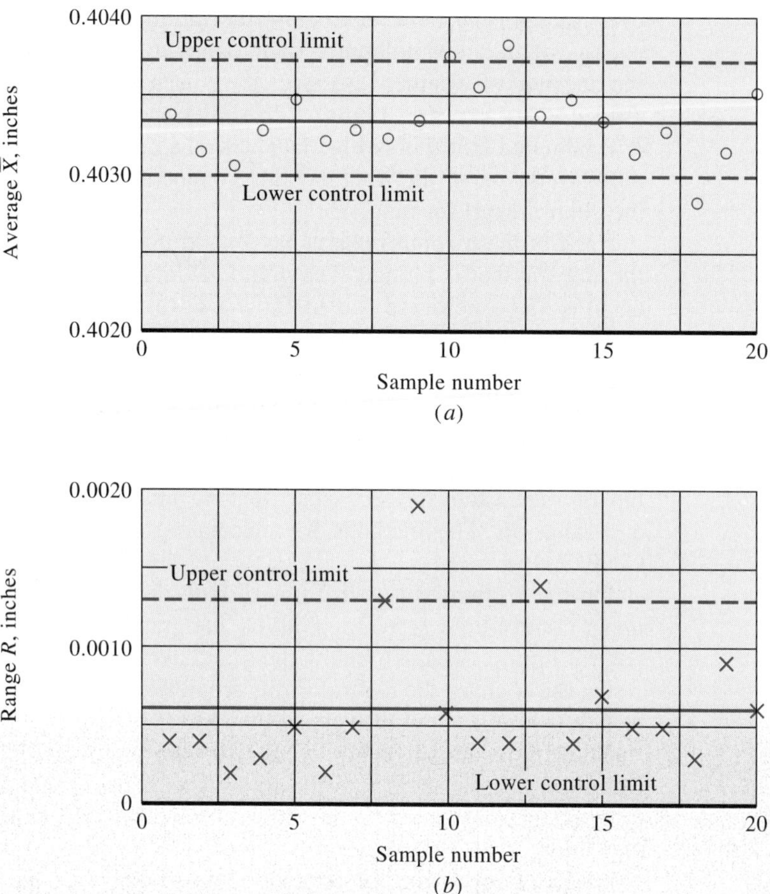

FIGURE 1.2 Pitch diameter of threads of fitting for aircraft hydraulic system: (*a*) control chart for averages $\overline{X}$; (*b*) control chart for ranges *R*.

of a particular operator. An effort to prevent their frequent recurrence resulted in a substantial improvement in product uniformity.

Other conclusions from this control chart The simplicity of the control chart is evidenced by the fact that this alert supervisor was able to use it to advantage after receiving only 6 h instruction in statistical quality control with only two of those hours devoted to construction of the control charts for $\overline{X}$ and *R*. This successful application was to process control, i.e., to the detection and elimination of assignable causes of quality variation. However, several conclusions that might have been suggested by the control-chart analysis were not evident to the supervisor, although they doubtless would have been evident to someone with additional training and experience in this subject. These conclusions, the basis for which will be developed in later chapters, are as follows:

1. If control (in the statistical sense) can be maintained, the natural tolerances of this process appear to be about ± 0.0006 in. Thus, by maintaining statistical control there should be no difficulty in making all the product well within the specified tolerances of ±0.0013 in.
2. As the practice has been to center the process at a dimension somewhat below the nominal dimension of 0.4037 in to minimize gage wear on 100% inspection with go and not-go thread ring gages, the question arises at what level the process ought to be centered. If statistical control can be maintained, this level must be not less than 0.4030 in to ensure that practically all product be within the specifications. Actually the level shown by the 20-h record was about 0.4034 in. This would be definitely on the safe side if the process could be kept within statistical control. But with the process out of control, there is always danger of nonconforming product regardless of the level.
3. Whenever natural tolerances are within the specification tolerances, consideration should always be given to the advisability of eliminating 100% inspection and substituting sampling inspection with the use of the control chart. In this case, five measurements of the actual dimension at specified intervals might replace 100% inspection with the go and not-go gages, except where the control chart showed lack of control. Such a change should not be made until the control chart had been maintained on this operation for some time with all points falling inside the control limits. Once this change was made, the motive of reducing wear on the ring gages would be eliminated and the nominal dimension of 0.4037 in should be aimed at.

The methods by which one may obtain these conclusions, and similar conclusions of economic importance in other cases, are developed in the following chapters.

Some Comments on Example 1.1

Because management's preconceived ideas on statistical quality control sometimes include a number of misconceptions, it is worthwhile to make several comments on Example 1.1 directed at these common misconceptions.

1. One misconception is that because the methods are *statistical,* they can be applied only where there is a long period of record. Often this is the basis for the feeling that it is not worth applying them to new operations that are to be continued for only a few months.

   It should be noted that the necessary data for a successful application in this case were obtained in 20 h. As a matter of fact, as past records are seldom in a form for the most effective use of the control chart for variables, it is usually necessary to start securing the required data *after* the decision is made to use this technique. The time needed to get enough information to supply a basis for action depends on how long it will take to manufacture enough units for a suitable chart.
2. Another misconception is that the methods are necessarily highly mathematical. This application involved only simple arithmetic.

3. A common misconception is that the methods are so complicated that they cannot be operated by ordinary production and inspection employees.

   As already pointed out, this particular supervisor made a useful application after only 2 h instruction in the particular technique. This is not to suggest that 2 h is sufficient time to devote to explanation of the control chart for variables; it merely is evidence that the essential features of this method can be explained simply in a short time. As also pointed out, this supervisor missed some of the possible useful conclusions that might have been drawn from the data. The more persons holding positions of responsibility in an organization who understand statistical quality control and the better they understand it, the better the chance for all the possible cost-saving applications to be made.
4. Another misconception is that the techniques are good to use only when you are conscious that you are in trouble.

   It is true that the places where one is conscious of trouble are likely to provide the best opportunities for saving costs. Nevertheless, the techniques often turn up cost-saving opportunities in places where there is no particular awareness of trouble. The possibility in Example 1.1 of substituting sampling inspection by variables for 100% inspection by attributes is an example. The supervisor chose this particular operation somewhat at random in order to provide an opportunity to experiment with the techniques.
5. Still another serious misconception is that effective use of the techniques of statistical quality control may be obtained by applying them only in one department.

In Example 1.1 we have noted that the data secured by the *production* supervisor point to a possibility of a saving in *inspection* costs. Budgetary control systems in industry generally are operated in a way that gives a supervisor credit for a cost saving in the supervisor's own department but not for savings in other departments. In this particular case, if the decision had been made to adopt acceptance sampling by variables at the point of production, with a control chart kept in the production department to be used for process control as well as for acceptance, this decision would have involved the question of whether the measurements were to be made by production or inspection personnel. This question could have been answered only at a management level above both production and inspection. If—as might well have been the case—the decision had been that it was advisable to have the measurements made by production personnel, it would have superficially appeared to *increase* costs in the production department. That is, it would have seemed to increase costs for which the production supervisor was held responsible and to decrease costs in a place where—as a matter of budgetary routine—the supervisor was given no credit.

Without a full understanding by top management, production supervision, and inspection supervision as to the basis of some of the cost savings that may be effected by statistical quality control, it is evident that the routine operations of a budgetary control system may actually prove an obstacle to securing the savings. This topic is explored at greater length in Chap. 17.

Example 1.2

$\bar{X}$ and R charts. Revision of tolerances

Facts of the case A rheostat knob, produced by plastic molding, contained a metal insert purchased from a supplier. A particular dimension determined the fit of this knob in its assembly. This dimension, which was influenced by the size of the metal insert as well as by the molding operation, was specified by the engineering department as 0.140 ± 0.003 in. Many molded knobs were rejected on 100% inspection with a go and not-go gage for failure to meet the specified tolerances.

A special gage was designed and built to permit quick measurement of the actual value of this dimension. Five knobs from each hour's production were measured with this gage. Table 1.2 shows the measurements obtained on the first 2 days after they were started.

TABLE 1.2 MEASUREMENTS OF DISTANCE FROM BACK OF RHEOSTAT KNOB TO FAR SIDE OF PINHOLE

Values are expressed in units of 0.001 in. Dimension is specified as 0.140 ± 0.003 in

Sample number	Measurement on each item of five items per hour					Average $\bar{X}$	Range R
1	140	143	137	134	135	137.8	9
2	138	143	143	145	146	143.0	8
3	139	133	147	148	139	141.2	15
4	143	141	137	138	140	139.8	6
5	142	142	145	135	136	140.0	10
6	136	144	143	136	137	139.2	8
7	142	147	137	142	138	141.2	10
8	143	137	145	137	138	140.0	8
9	141	142	147	140	140	142.0	7
10	142	137	145	140	132	139.2	13
11	137	147	142	137	135	139.6	12
12	137	146	142	142	140	141.4	9
13	142	142	139	141	142	141.2	3
14	137	145	144	137	140	140.6	8
15	144	142	143	135	144	141.6	9
16	140	132	144	145	141	140.4	13
17	137	137	142	143	141	140.0	6
18	137	142	142	145	143	141.8	8
19	142	142	143	140	135	140.4	8
20	136	142	140	139	137	138.8	6
21	142	144	140	138	143	141.4	6
22	139	146	143	140	139	141.4	7
23	140	145	142	139	137	140.6	8
24	134	147	143	141	142	141.4	13
25	138	145	141	137	141	140.4	8
26	140	145	143	144	138	142.0	7
27	145	145	137	138	140	141.0	8
Totals	. . .	. . .	. . .	. . .	. . .	3,797.4	233

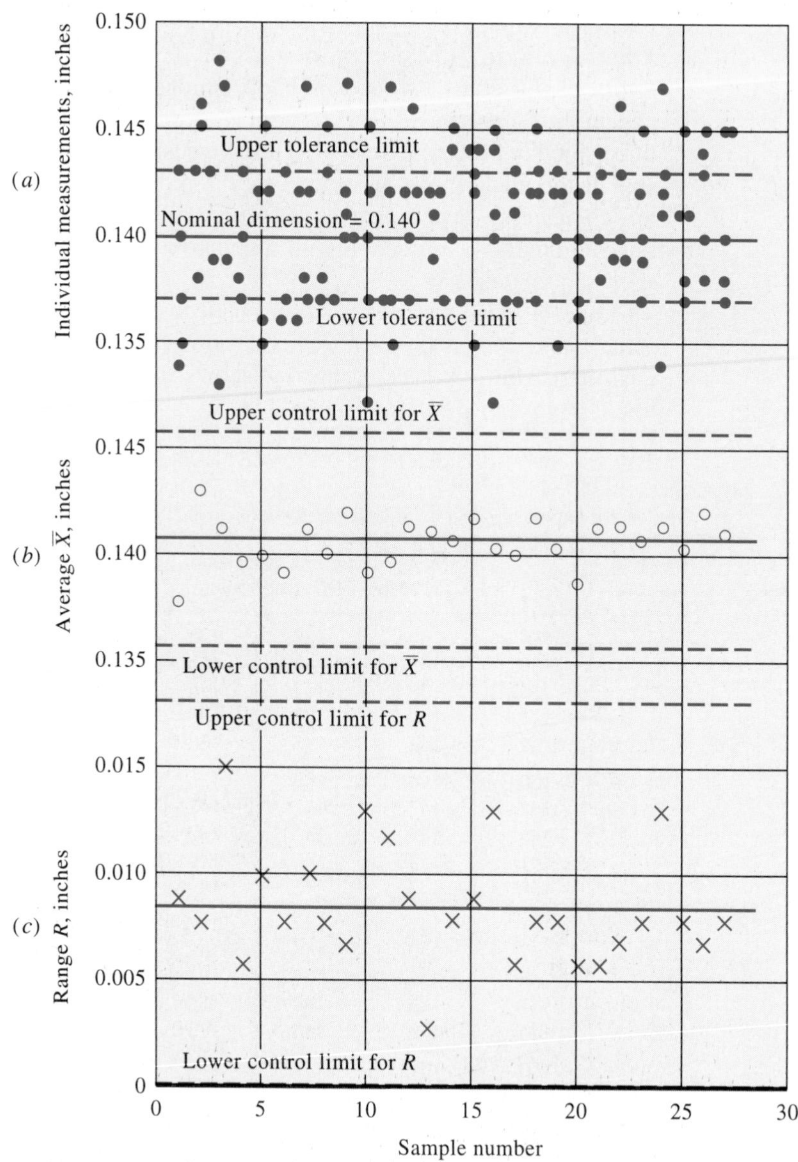

FIGURE 1.3 Measurements of dimension on rheostat knob: (*a*) individual measurements; (*b*) control chart for averages $\overline{X}$; (*c*) control chart for ranges *R*.

Analysis of the facts by the control-chart technique Figures 1.3*b* and 1.3 *c* are control charts for $\overline{X}$ and *R*, respectively, for the 27 samples taken in these 2 days. Figure 1.3*a* is not a control chart, but—like Fig. 1.1*a* in Example 1.1—it shows individual measurements.

If we count all the points outside the tolerance limits that are shown on Fig. 1.3*a*, we find that 42 of the 135 knobs measured failed to meet the specifications for this dimension. This is approximately 31% nonconforming, a very high figure.

At the same time it is evident that all the points on the $\overline{X}$ chart (chart for averages) and the R chart (chart for ranges) are inside the control limits. The values of averages and control limits are as follows:

$$\overline{\overline{X}} \text{ (grand average)} = \frac{3{,}797.4}{27}$$

$$= 140.6 \text{ (expressed in units of 0.001 in as in Table 1.2)}$$

$$\overline{R} \text{ (average range)} = \frac{233}{27} = 8.6$$

$UCL_{\overline{X}}$ (upper control limit for averages) = 145.6
$LCL_{\overline{X}}$ (lower control limit for averages) = 135.6
UCL_R (upper control limit for ranges) = 18.2
LCL_R (lower control limit for ranges) = 0

This process was evidently in statistical control, even though it was in control with a spread that was unsatisfactory from the standpoint of the specified tolerances of ±0.003 in. No assignable causes of variability were indicated. The variations from hour to hour were chance variations; they could not be reduced by hunting for changes that took place from one hour to the next. In a situation of this type, improvement is not likely to be obtained by the plant superintendent's bringing pressure on the department supervisor, or by the supervisor's bringing pressure on the machine operator.

Action based on the control-chart analysis When the quality control engineer in this plant studied the situation, it was discovered that although part of the spread in values of this dimension was due to the inherent variability of the plastic molding operation, the most serious factor was the variability of the metal insert from the outside supplier. Because the part was essential for an important contract and available suppliers were scarce, nothing could be done immediately about the variability of the metal insert.

The immediate alternatives were either to continue to eliminate many parts by 100% inspection or to widen the tolerances. The quality control engineer requested the engineering department to review the tolerances. This review showed that the specified tolerances of ±0.003 in were much narrower than necessary for a satisfactory functioning of the rheostat knob as part of its assembly. After trying knobs with different values of this dimension and judging when the fit was satisfactory, the tolerances were changed to

$$\begin{cases} +0.010 \\ -0.015 \end{cases}$$

It may be noted from inspection of Table 1.2 or Fig. 1.3*a* that these revised tolerances would have permitted the acceptance of all parts measured in the 27 samples of 5 there shown. With the change in tolerances and the process shown to be in statistical control, it then became possible to use the control chart as a substitute for 100% inspection of this dimension.

Subsequent work with the supplier resulted in some reduction in the variability of the metal insert. At a later date, assignable causes of variation developed in the plastic molding process; these were promptly detected by the control chart and the conditions corrected. Later it proved satisfactory to reduce the number of parts measured from five every hour to five every 4 h; subsequently this was cut down to five every 8 h.

This plastic molding operation was carried on intermittently with runs of 2 or 3 days spaced several weeks apart. The control-chart technique is particularly helpful in process control on this type of operation.

Some Comments on Example 1.2

This example affords opportunity for some contrasts with the situation described in Example 1.1 and gives further illustration of some of the uses of the control chart for variables:

1. In Example 1.2 the quality control engineer used the control chart as the natural weapon to adopt for the diagnosis of trouble. Without the control chart, and with 31% of nonconforming items produced, the most common action is for management to "get tough," bringing pressure on production supervision to do better. The control chart has been described as the substitution of "get smart" for "get tough" in managerial policy in dealing with quality troubles, because it provides supervisors with clues to the causes of correctable day-to-day troubles. In this case, however, the first message of the control chart was to tell management what *not* to do. The chart said, "It's no use to get tough," and it also said, "It's no use to hunt for causes of hour-to-hour or day-to-day variation."

   This is the type of situation in which the alternatives presented are (*a*) to make a fundamental change in the process, (*b*) to change the specifications, or (*c*) to resign yourself to the necessity of continuing to try to separate out the acceptable product by 100% inspection. In this case it happened that the specified tolerances had been set arbitrarily without consideration of the particular needs of this assembly; therefore, the appropriate action was to change the specifications.
2. In the situation described in Example 1.2, an active statistical quality control program was in operation. It was run by a quality control engineer who reported directly to the works manager. Where the information secured from the analysis of control charts points to action by more than one department, such a plan of organization is often advantageous.
3. The contrast between Examples 1.1 and 1.2 emphasizes the point that *statistical control* is an expression that describes the pattern of variability of the

process rather than the past performance of the process in meeting specifications. In Example 1.1, 25% of the samples showed lack of control, whereas only 1% of the product examined failed to meet specifications. In Example 1.2, although none of the samples showed lack of control, 31% of the product failed to meet specifications.

4. The point is made in Example 1.1 that it is always misleading to show tolerance limits on a chart for averages ($\bar{X}$ chart). Example 1.2 emphasizes this point. Even though many of the individual rheostat knobs measured had dimensions greater than 0.143 or less than 0.137 in, no average of a sample of five was outside these tolerance limits. Someone who made a comparison between the tolerance limits and the charted values of $\bar{X}$ might well have reached the incorrect conclusion that all the product examined met the specification of 0.140 ± 0.003 in.

1.6 MEANINGS AND USAGE OF THE WORDS *DEFECTIVE* AND *DEFECT*

The technical meanings of the words *defective* and *defect* as used in the manufacturing industries differ from the common meanings of these words as used in everyday speech. This difference between the technical and popular meanings of these words has been a source of misunderstanding in litigation, particularly in product liability suits.

Among other matters, Example 1.2 illustrates the technical meaning of the word *defective* as used in manufacturing. In this sense, a manufactured part or article is defective if it fails to conform to specifications in some respect. Similarly, a *defect* is a failure of the part or article to conform to some one specification. A manufactured item that is defective may contain only one defect or it may contain two or more defects.

When the measurements shown in Table 1.2 were made, the rheostat knobs appeared to be 31% defective because a certain dimension on 31% of the knobs did not conform to the specification of 0.140 ± 0.003 in. Shortly thereafter, with no change whatsoever in the product, the knobs were 0% defective because the tolerances had been widened to

$$\begin{cases} +0.010 \\ -0.015 \end{cases}$$

In contrast, when the word *defective* is used in its popular sense as applied to a manufactured article, it means that the article is unsatisfactory with reference to its intended purpose. None of the rheostat knobs in Example 1.2 were defective in the popular sense either before or after the relaxation of tolerances. In the popular sense of the word, a *defect* of a manufactured article is some characteristic of the article that makes it unsatisfactory for its intended purpose.

In his Edwards Medal address in 1974 before the American Society for Quality Control, the distinguished past president of the Society, Arthur Bender, Jr., suggested that the best way to avoid such misunderstandings would be to abandon the technical usage of the words.[†] His admonition was widely adopted in the United States in the 1980s.

Obviously it is impossible for manufacturers to take any action that will change the popular meanings of *defective* and *defect.* The one way the confusion between the technical and popular meanings can be eliminated is to substitute other words or phrases when the technical meanings are intended. An article that does not conform to specifications in some respect can be described as *nonconforming* rather than as *defective.* A failure to conform to a particular specification can be called a *nonconformity* or *nonconformance* rather than a *defect.* In describing the results of acceptance inspection, it is possible to say *percent rejected* or *nonconforming* rather than *percent defective.*

Part Two of this book deals with various aspects of process control, with particular emphasis on the contributions that can be made by the different types of control charts. In this portion of the book, it has been possible to eliminate the words *defective* and *defect* in nearly all cases. However, a few such usages remain in quotations and in the reproduction of certain forms and charts.

Part Three of this book describes a number of acceptance sampling systems that use the words *defective* and *defect* in their technical sense. While some of the U.S. commercial versions of standards for acceptance sampling have replaced them, the more frequently used military versions have retained them. As long as these words continue to be used in the official documents describing such systems, writers explaining the systems will also need to use the words.

In all technical subjects, terminology ought to be as clear as possible. Thus a clarification of the usage of *defective* and *defect* would have been desirable even if it had not been motivated by litigation about product liability. Chapter 17 contains some comments on the economic aspects of product liability in relation to the subject matter of this book.

Example 1.3

Weakness of a common acceptance sampling scheme

Should sampling be used for acceptance? Two opposite attitudes toward sampling are common in industrial inspection. They are:

1. An uncritical suspicion of all sampling. This is based in part on a belief that a lot may be very different from the sample taken from it.
2. An uncritical approval of all sampling. This is based on a belief that a lot will be like the sample taken from it.

The truth is somewhere between these two viewpoints. Samples may give a misleading idea of lot quality. On the other hand, it is possible to devise sampling procedures that do provide a desired quality protection. The basis for distinguishing bad sampling schemes from good ones is developed throughout this

[†]Arthur Bender, Jr., "Don't Say THAT!," *Quality Progress,* vol. 7, no. 7, p. 8, July 1974.

entire book. At this point it seems worthwhile to stimulate curiosity about good sampling schemes by a critical examination of a common acceptance sampling procedure that is bad.

This common procedure calls for inspecting 5 articles from each lot of 50. If every article in this 10% sample conforms to specifications, the lot is accepted. If one or more nonconforming articles are found in the sample of 5, the lot is rejected.

This acceptance procedure gives adequate protection if the articles in a lot are either all acceptable or all rejectable. (In fact, if the items are either all acceptable or all rejectable, a sample of one is sufficient.) It gives little protection against product with a moderate percentage of nonconforming articles.

The truth of this latter statement may be demonstrated by assuming that the product submitted for inspection is, on the average, 4% nonconforming; i.e., on the average there will be 2 nonconforming articles in a lot of 50. Some lots will contain more; some less. Column *B* of Table 1.3 shows the number of lots that would be expected to have each number of nonconforming articles according to the laws of chance; it is assumed that 1,000 lots are submitted. It will be noted that the lots submitted vary from 0 to 12% nonconforming. In 1,000 lots with 50 articles in each lot, 50,000 articles are submitted for acceptance.

Column *D* gives for each lot quality the relative frequency of having at least one nonconforming article in a random sample of 5 taken from a lot of 50. The first figure, 0.00, is obvious; there can be no nonconforming article in the sample if there is none in the lot. The second figure, 0.10, merely states the reasonable conclusion that if there is only one nonconforming article in a lot of 50, that article will appear in only 10% of the samples of 5 taken from the lot. The source of the remaining figures in column *D* is not so obvious. They can be calculated by simple probability mathematics as explained in Chap. 5.

It is evident that if 10% of the lots are rejected, 90% will be accepted; if 19% are rejected, 81% will be accepted, etc. That is, the sum of the decimals in columns *D* and *E* must be 1.00. Columns *F, G, H,* and *I* are derived from the preceding columns as indicated in the table.

The relationship between the proportion of submitted lots that will be accepted in the long run (usually referred to as the *probability of acceptance*) and the percentage of nonconforming articles in the submitted lots is shown graphically by the curve in Fig. 1.4. In the language of statistics, such a curve is called an *operating characteristic curve* or, more briefly, an *OC curve.* In the discussion of acceptance sampling in Part Three of this book, we shall have occasion to examine many such curves.

The practical conclusions from the table are obtained from an analysis of the totals of columns *F, G, H,* and *I.* Column *G* tells that 815 of the lots are accepted; this is a total of 40,750 articles. The significant figure desired is the quality of these accepted lots. Column *I* indicates that they contain 1,468 nonconforming articles. The ratio of 1,468 to 40,750 is 0.036; that is, there are 3.6% of nonconforming articles in the accepted lots.

TABLE 1.3 ANALYSIS OF ACCEPTANCE SAMPLING PLAN

Take random sample of 5 from lot of 50. Accept lot if no nonconforming articles are found in sample; reject if one or more nonconforming articles are found. Examine 1,000 lots. Assumed process average is 4% nonconforming

Number nonconforming in lot	Number of lots submitted	Number of nonconforming items, $A \times B$	Proportion of lots submitted that will be		Number of lots rejected $D \times B$	Number of lots accepted $E \times B$	Number of nonconforming articles in rejected lots $A \times F$	Number of nonconforming articles in accepted lots $A \times G$
			Rejected	Accepted				
(A)	*(B)*	*(C)*†	*(D)*	*(E)*	*(F)*	*(G)*	*(H)*	*(I)*
0	130	0	0.00	1.00	0	130	0	0
1	270	270	0.10	0.90	27	243	27	243
2	275	550	0.19	0.81	52	223	104	446
3	185	555	0.28	0.72	52	133	156	399
4	90	360	0.35	0.65	32	58	128	232
5	35	175	0.42	0.58	15	20	75	100
6	15	90	0.49	0.51	7	8	42	48
Totals	1,000	2,000	. . .	. . .	185	815	532	1,468

†Column *C* gives the total number of nonconforming articles in all the lots in column *B* for each specified number nonconforming. The total number of nonconforming articles submitted is 2,000. This is 4% of the 50,000 submitted.

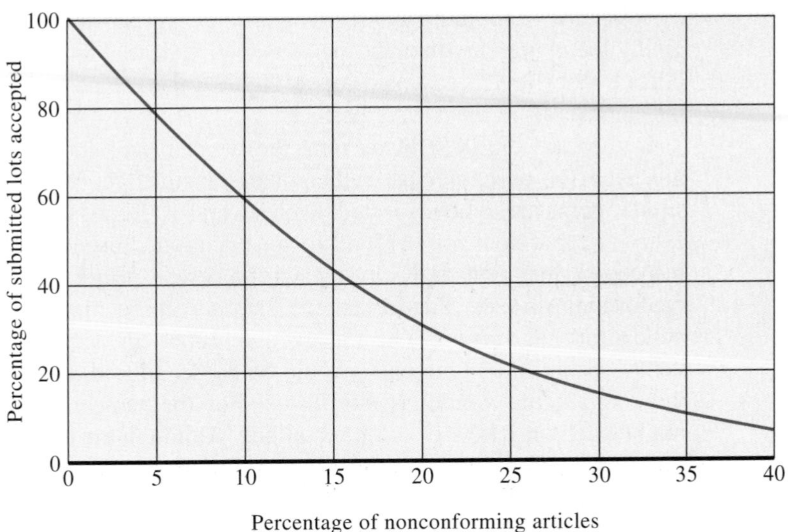

FIGURE 1.4 Operating characteristic (OC) curve for the acceptance sampling plan of Example 1.3.

It is evident that this acceptance procedure has effected a negligible improvement in quality. The incoming quality submitted was 4% nonconforming, and the outgoing quality after inspection is 3.6% nonconforming. To accomplish this small improvement, 185 lots out of the 1,000 submitted were rejected.

The quality of these rejected lots also is a matter of interest. They are, on the average, 5.75% nonconforming; they contain 532 nonconforming articles among their 9,250 articles. In them, each sample of 5 contained at least one nonconforming article; calculations based on probability mathematics indicate that 14 of them may be expected to have 2 nonconforming articles in the sample of 5. This is a total of 199 nonconforming articles found in the samples. If these are eliminated from the 185 lots, 333 nonconforming articles will remain among a total of 9,051 articles; this is 3.68% nonconforming. In other words, once the nonconforming articles found in the samples are removed, the quality of the rejected lots is practically as good as that of the accepted lots.

Some Comments on Example 1.3

A serious obstacle to sensible and economic acceptance plans is the illusion that perfection (in the sense of 100% conformity to specifications) is possible as the result of *any* inspection plan. This illusion often is cherished both by those who reject all sampling plans in favor of 100% inspection, and by those who adopt sampling plans that accept a lot if a sample is perfect and reject it if the sample contains one or more nonconforming articles. Perhaps without being conscious of giving the matter any thought, persons who favor the latter view would have confidence that the 815 accepted lots of Table 1.3 were perfect or nearly so because the samples from them were perfect.

In the case of 100% inspection, the belief in perfection through inspection is mistaken for psychological rather than statistical reasons. Where a great many similar items are to be inspected, even several 100% inspections may not weed out every one that does not conform to specifications. Inspection fatigue on repetitive inspection operations may cause even the best inspector to miss some of the nonconforming articles. For instance, a first screening might eliminate 95% of the nonconforming pieces. On the second 100% inspection, the nonconforming items—being less numerous—would be harder to find, and possibly only 90% of those remaining would be eliminated. For the same reason, a third inspection might catch only 80% of those remaining. Although the percentages missed in any given case will depend on the difficulty of the inspection operation and on the skill and diligence of the inspectors, the general principle is sound that if some nonconforming product is submitted for screening inspection, a portion of it is likely to be passed even with several 100% inspections.

Errors in distinguishing between acceptable and rejectable articles are somewhat less likely to occur in sampling inspection than in 100% inspection. There are a number of reasons why this is true. The inspector in sampling inspection obviously has a more responsible assignment; a mistake in classifying an article as satisfactory or unsatisfactory may decide the acceptance or rejection of an entire lot rather than only one article. This greater responsibility tends to influence inspectors to do a better job. Moreover, the greater variety of sampling inspection as compared to 100% inspection tends to cause less inspection fatigue. It is more monotonous and fatiguing to spend all day inspecting one or two kinds of articles than to inspect many different kinds of articles. Because fewer inspectors are needed for sampling inspection, it may be possible to secure better inspectors.

Nevertheless, in spite of these possible advantages over 100% inspection, sampling inspection cannot ensure that all accepted product will conform to specifications. If nonconforming product is included in lots submitted for acceptance under sampling inspection, then, according to the laws of chance, some lots that contain such product are likely to be accepted.

Most sampling plans for lot-by-lot acceptance in small lots that base the decision on each small lot on the evidence of a small sample from that lot are likely to have the same weakness as the plan examined in Table 1.3. Under ordinary conditions the product passed by such a scheme will not be substantially better than the product submitted for examination.

Before such acceptance plans, which often delude the user with a promise of perfection they cannot fulfill, can be discarded in favor of better ones, there must be a recognition of the fact that there is a limit to the degree of perfection to be expected. The question of just what this limit is in any case, i.e., just what is an acceptable quality level, is basically a question of economy. Some methods of getting as good an assurance as possible of this acceptable quality level are explained in Part Three. They require as background some understanding of the principles of the Shewhart control chart, which are developed in Part Two.

1.7 MANY ECONOMY STUDIES CALL FOR THE VIEWPOINT OF STATISTICAL QUALITY CONTROL

Many decisions on matters related to quality are called for in manufacturing. In making such decisions it is desirable to examine the relative economy of the alternatives under consideration. The techniques of statistical quality control may make a useful contribution to such economy studies.

Consider, for example, the question of the filling of containers—a problem for the food industries and all other industries that package their products. Suppose that government regulations require that all or some specified large percentage of the packages contain at least a certain stipulated weight.[†] Where a conscientious attempt is made to meet such a specification, it is usually done by overfilling enough to be on the safe side.

But the question always remains, "How much overfill is necessary?" This question is often answered on a practical basis of, "Make it enough so that we avoid trouble with the government inspectors." If these inspectors really do a critical job, the resulting average overfill is likely to be too much for maximum economy.

This is not to say that the economic answer is to be gained by increasing the amount of trouble with the inspectors. It is rather a matter of finding the facts about the variability of weights and analyzing these facts. This can be done by the Shewhart control chart for variables better than by any other known technique.

Can the variability of the process be reduced without any change in the physical methods being used to fill the containers? In other words, does the process show lack of control (with the word *control* used in its statistical sense)? If so, what are the reasons for the out-of-control points? Perhaps they may be corrected in such a way that they will be unlikely to recur. Or possibly their correction may call for the maintenance of a control chart continuously (rather than briefly as an information-getting and troubleshooting device) with some attendant cost for measurement, clerical labor, and supervision. It may even be necessary to maintain several such charts (for instance, on weights of filled containers, weights of empty containers, density of filling substance).

If out-of-control points can be eliminated, it is obviously possible to work closer to the minimum specification limit and thus reduce the cost of the overfill. This is a matter of balancing the cost of maintaining control against the cost of the extra overfill. The control-chart data giving the basic variability of the process will

[†]In the sense used in statistical quality control, a quality of a manufactured product may be any characteristic of that product. Thus the *quantity* of material in a container is a *quality* of the filled container.

provide the information for decision as to the average level to be aimed at in order to meet specifications and will thus provide an estimate of the money saving possible from better control.

If different methods of filling containers are proposed (for example, improved automatic controls vs. controls that are only partially automatic), one element in the cost comparison is the possible difference in the cost of overfill with each method. This requires the use of the Shewhart control chart to estimate the basic variability of each method.

If the specification of minimum filling weights has not been made either by government regulation or by a customer contracting for the product but must be made by the producer, the economic decision becomes even more complicated. Here the question may be the effect on consumer acceptance in a competitive market of occasional shortages below the amount stated on a package. Although it may be difficult or impossible to place a definite money value on this effect, it is extremely helpful to have an idea of the costs associated with various degrees of consumer protection. This information is provided by the control-chart technique.

1.8 STATISTICAL QUALITY CONTROL MAY HAVE USEFUL BY-PRODUCTS

The techniques of statistical quality control bring certain desirable results that cannot be achieved as well in any other way. These might be described as the direct benefits of statistical quality control. In addition, the introduction of these techniques into any business often causes certain desirable changes that might be described as *by-products.*

One such by-product may be the establishment or improvement of inspection standards, with the preparation of definite instructions for each inspection procedure. Another may be the periodic evaluation of departmental performance in quality terms. Still another may be the evaluation of different suppliers' quality performance in terms of average fraction rejected, with choice of future suppliers based on these findings.

Sometimes an important by-product of statistical quality control may be the establishment of effective process inspection where none has previously existed. In some manufacturing concerns there is little or no process inspection; inspection takes place some days—or even weeks or months—after production with no chance to associate any rejected product with possible causes in the production departments. Statistical quality control, with its emphasis (explained in later chapters) on keeping track of the order of production, tends to call for inspection close to the point of production.

Although the introduction of process inspection is sometimes a by-product, it should be noted that a direct object of statistical quality control is to provide a new tool that makes process inspection more effective. The information obtained by process inspection—either conducted by roving inspectors or by machine operators themselves—is often misused to make too frequent machine adjustments. As already pointed out under the discussion of the Shewhart control chart for variables, these too frequent adjustments have the opposite effect from that intended: they increase rather than decrease the variability of the process. In the introduction of statistical quality control, reports such as these are common: "After a week of the control chart we accomplished substantial improvement in product quality by persuading the operator to let the machine run itself rather than changing the settings whenever a critical dimension gets close to the specification limit." "We found the roving inspector was shutting down the machine for resetting three times as often as necessary."

1.9 REASON FOR USE OF THE ADJECTIVE *STATISTICAL*

Statistics is a word with two quite different meanings. In one sense, it refers to any facts stated in terms of numbers; in this sense, it is a plural noun. Thus one may say "Statistics *are* kept in the sales department regarding all branch office sales." In the other sense, it refers to a body of methods by which useful conclusions can be drawn from numerical data. In this sense, it is a singular noun. Thus one may say "Statistics *is* based in large part on the law of large numbers and the mathematical theory of probability." It is in this second sense that the adjective *statistical* is accurately used in the expression *statistical quality control.*

The control of quality of manufactured product is a function that existed long before statistical methods were applied to the analysis of quality data and that exists today whether or not statistical techniques are used. Properly used, the expression *quality control* applies to a function much broader than the expression *statistical quality control.* The use of *quality control,* or *Q.C.,* in the sense of *statistical quality control* inevitably leads to confusion as to the meaning of the expression.

In the long run this confusion is likely to be more serious than any troubles introduced by the use of an extra word. For this reason, throughout this book the expression *quality control* is always used in the broader sense of the control of quality of product by whatever methods may be used and the adjective *statistical* is always employed where the control of product quality by statistical methods is referred to.

1.10 FOUR DIFFERENT LEVELS OF UNDERSTANDING STATISTICAL QUALITY CONTROL

In any manufacturing company, government procurement agency, or other organization in which substantial statistical quality control applications are to be made, experience indicates that appropriately there may be four levels of understanding of the subject.

One is the level of understanding the mathematics on which are based the control charts and sampling tables and their relationship to the many other tools for the analysis of data that have been developed by mathematical statisticians. A person on this level should be able to read the literature of mathematical statistics without great difficulty and should have reasonable familiarity with this literature.

Persons on this level should be available in any comprehensive statistical quality control program.

The second level is that of general understanding of the principles underlying the various types of control charts and sampling tables. It calls for understanding why these methods work, how to interpret their results, and how to decide which method to use in any particular case.

A third level is that of a broad understanding of the objectives and possible uses of statistical quality control, even though this understanding is not sufficiently detailed and precise to permit close supervision of statistical quality control work. This type of understanding is particularly helpful at higher management levels.

The fourth level calls merely for use of one or more of the techniques on a rule of thumb basis. In any plant in which many applications are made, there will doubtless be a number of inspectors, machine operators, and possibly some clerks on this level.

This book is aimed at the second level. The success of any statistical quality control program is likely to depend on the number of people in an organization who are competent on this level and on the distribution of these people among various departments. The cooperation among departments necessary for the fullest benefits from statistical quality control has been mentioned and is emphasized throughout succeeding chapters. The more persons in inspection supervision, production supervision, methods engineering, tool engineering, engineering design, and top management who understand the basic principles of statistical quality control, the better the opportunity for effective use of these techniques.

This chapter, although intended primarily as an introduction for people interested in the second level, is also intended to be helpful to those who are interested only in the third level. Chapter 18 is also suggested as reading for individuals primarily interested in the third level.

1.11 NONMANUFACTURING APPLICATIONS OF STATISTICAL QUALITY CONTROL TECHNIQUES

Although control charts and statistical types of acceptance sampling procedures were originally developed for use in mass production manufacturing, these techniques are applicable to most other types of activities in all sectors of the economy including service business, government, education, and health care. Figure 19.9 in Chap. 19 illustrates an application from the health care sector, more specifically, a postanesthesia care unit problem. The quality improvement activity follows a particular model referred to as a *storyboard* or *story book* model.

Applications to business processes and in areas outside manufacturing are discussed in many places throughout this book. Some of the examples are drawn from the service and public sectors of the economy. For instance, certain acceptance sampling systems are well suited for checking errors in clerical work and verifying the accuracy of inventory counts. Control charts may be applied to many business variables to discover their average values, the range of variation that can be expected as a matter of chance, and the presence or absence of special cause variation. These efforts may lead to uncovering potential improvement projects and, when completed, will help assess the success of the effort.

Those whose interests are primarily in the nonmanufacturing applications should review the problems for solution at the end of each chapter. Many of the problems refer to nonmanufacturing situations.

1.12 TOPICS COVERED IN PART FOUR OF THIS BOOK

As stated in Sec. 1.2, the main focus of this book is on the use, and sometimes misuse, of control-chart techniques to control all kinds of processes and of various types of acceptance sampling systems. Naturally, the economic use of any or all of these techniques depends on an evaluation of the costs and benefits involved. Chapter 17 discusses a number of aspects of this determination.

It is impossible to plan where we are going if we don't know where we have been. Chapter 18 offers a brief history of statistical quality control from its beginnings in the early part of the twentieth century to the present (1995). It will help the student understand where these tools fit in the current world of local, state, and national quality awards and will provide some background on international standards for what constitutes a "quality system."

Chapter 19 discusses the primary models of quality management: the criteria of the U.S. Malcolm Baldrige National Quality Award and the quality manual–based model of the ISO 9000 series of quality standards. These management models are then used as the basis for discussing models for quality assurance, process control, and quality improvement.

Chapter 20 discusses and illustrates some statistical sampling demonstrations useful in classroom presentations to help students understand random phenomena. Except for the sampling demonstrations, which may best be introduced at various stages in the course, each topic is best understood after the student has had broad exposure to the statistical tools presented in Parts Two and Three.

PART TWO

STATISTICAL PROCESS CONTROL

2

DIRECTIONS FOR SIMPLE $\overline{X}$ AND R CHARTS

The purpose of analysis is insight, rather than numbers. Therefore, the best analysis is the simplest analysis which provides the needed insight. Given its simplicity, robustness, sensitivity, and versatility, there is no one technique which can successfully compete with Shewhart's control charts.

—Donald J. Wheeler[†]

2.1 SETTING UP AND OPERATING CONTROL CHARTS FOR $\overline{X}$ AND R

Example 1.1, and the figures accompanying it, explained the essential difference between data plots in general and control-chart plots. In particular, Fig. 1.1*b* showed a plot of the averages of five measurements that constituted a subgroup. It was stated that this chart didn't tell much of a story beyond the possible detection of drifts or runs in the sample average values over time. Such charts frequently are called *run charts* for that reason. Run charts, often used by management, show what has been happening over some specified period of time. Because of their great popularity as a tool of historical narration, persons not well schooled in the use of statistical process control sometimes refer to them as control charts. Control charts, such as those shown in Figs. 1.2 and 1.3, have control limits and central lines on them and are accompanied by decision rules to signal to the user when action should be taken on a process. Run charts offer no scientific basis for making decisions about when corrective action should or should not be taken.

[†]Donald J. Wheeler, *Myths, Foundations, and Competitors for Shewhart's Control Charts,* 1991, p. 9, Statistical Process Controls, Inc., Knoxville, Tenn.

Sections 1.3 and 1.4 introduced control charts for variables data and offered two examples of applications of $\bar{X}$ and R charts. The $\bar{X}$ chart is used to analyze the *average level* of a process; the R chart is used to analyze its *spread* or *dispersion.* Taken together, they indicate whether or not the process is stable, that is, whether a *constant system of chance causes* appears to be operating. Since this pair of charts is the most frequently used for monitoring variables data, we will address it first.

This chapter discusses the mechanics of setting up and running control charts for $\bar{X}$ and R, provides a checklist of steps in using the charts, and introduces some process capability concepts in relating specifications to process measurements.

2.2 CHECKLIST OF NECESSARY STEPS IN USING $\bar{X}$ AND R CHARTS

It is helpful to visualize the decisions and calculations that must be made and the actions that must be taken as occurring in a sequence somewhat as follows:

- I. Decisions preparatory to the control charts
  - A. Some possible objectives of the charts
  - B. Choice of the variable
  - C. Decision on the basis of subgrouping
  - D. Decision on the size and frequency of subgroups
  - E. Setting up the forms for recording the data
  - F. Determining the method of measurement
- II. Starting the control charts
  - A. Making and recording the measurements and recording other relevant data
  - B. Calculating the average $\bar{X}$ and range R for each subgroup
  - C. Plotting the $\bar{X}$ and R charts
- III. Determining the trial control limits
  - A. Decision on required number of subgroups before control limits are calculated
  - B. Calculation of trial control limits
  - C. Plotting the central lines and limits on the charts
- IV. Drawing preliminary conclusions from the charts
  - A. Indication of control or lack of control
  - B. Interpretation of processes in control
  - C. Relationships between processes out of control and specification limits
  - D. Suggestions about preliminary conclusions and actions based on the control charts
- V. Continuing to use the charts
  - A. Revision of central lines and control limits
  - B. Use of the control chart as a basis for sorting sublots into homogeneous grand lots
  - C. Use of the control charts for action on the process
  - D. Use of the control charts for acceptance inspection
  - E. Use of the control charts for action on the specifications

2.2.1 Decisions Preparatory to the Control Charts

A. *Some Possible Objectives of the Charts*

In general, where control charts for variables, either $\overline{X}$ and R or $\overline{X}$ and s, are undertaken, some or all of the following purposes are present:

1. To analyze a process with a view to one or more of the following objectives:
   a. To secure information to be used in establishing or changing specifications or in determining whether a given process can meet specifications. This was illustrated in Example 1.2.
   b. To secure information to be used in establishing or changing production procedures. Such changes may be either elimination of assignable causes (special causes) of variation or fundamental changes in production methods (common causes) that may be called for whenever the control chart makes it clear that specifications cannot be met with present methods. Production changes of one or the other of these types are referred to in Examples 1.1, 2.1 to 2.3, 4.1, and 4.2.
   c. To secure information to be used in establishing or changing inspection procedures or acceptance procedures, or both. This objective is referred to in Examples 1.1, 1.2, 2.1 to 2.3, and 4.2.
2. To provide a basis for current decisions during production as to when to hunt for causes of variation and take action intended to correct them, and when to leave a process alone. This is nearly always one of the purposes of any control chart for variables.
3. To provide a basis for current decisions on acceptance or rejection of manufactured or purchased product. This is illustrated in Examples 2.1 and 4.1 and discussed further in Chap. 15. Sometimes, as in Example 4.1, the control chart is undertaken primarily for this purpose. Often when it is undertaken for other purposes, there is a hope that, as time goes on and the other purposes are accomplished, it will ultimately be possible to reduce inspection costs by using the control chart for variables for acceptance. This is almost essential to meet the conditions of Just-In-Time inventory management. It was actually accomplished in Example 1.2 and was suggested as a possibility in the comments on Example 1.1.
4. To familiarize personnel with the use of control charts. Although this would seem to be a legitimate purpose only in the early stages of the use of statistical quality control techniques in any organization, control charts undertaken for this purpose often disclose opportunities for cost savings.

B. *Choice of the Variable*

The variable chosen for control charts for $\overline{X}$ and R must be something that can be measured and expressed in numbers, such as a dimension, hardness number, tensile strength, weight, etc. The real basis of choice is always the prospect of reducing or preventing costs. From the standpoint of the possibility of reducing *production* costs, a candidate for a control chart is any quality characteristic that is causing rejections or rework involving substantial costs. From the *inspection and acceptance* standpoints, destructive testing always suggests an opportunity to use the control chart to reduce costs. Expensive or lengthy analytical procedures also suggest the possibility of reducing inspection costs with the control charts.

In general, if acceptance is on a sampling basis and the quality tested can be expressed as a measured variable, it is likely to be worthwhile to examine inspection costs to form a basis for judgment as to the possibilities of reducing these costs by basing acceptance on the control chart for variables. Where 100% inspection takes place on an attributes basis, as with go and not-go gages, the chances for savings in costs depend on considerations outlined in Chap. 18.

Frequently the best chances to save costs are in places that would not be suggested by an examination either of costs of spoilage and rework or of inspection costs. These depend on the use of the control chart to analyze a process. The discussion in Chap. 1 of the problem of filling containers illustrates this type of opportunity. Concealed costs often exist that are not apparent in any cost statement.

In the introduction of the control-chart technique in any organization, the choice of the right variables is often troublesome (see Example 9.5). Occasionally the large number of possible variables is a source of confusion. One large manufacturing plant counted several hundred thousand specified dimensions on the many parts going into its products. Obviously only a very small fraction of these were legitimate candidates for control charts for $\bar{X}$ and R. In selecting variables for initial application of the control-chart technique, it may be important not only to choose those with opportunities for cost savings but to select carefully a type of saving that everyone in a supervisory or managerial capacity will readily accept as being a real saving. This usually—although not always—suggests starting where the spoilage and rework costs are high.

C. Decision on the Basis of Subgrouping

The key idea in the Shewhart method is the division of observations into what Shewhart called *rational subgroups.* The success of the Shewhart technique depends in large measure on the discrimination used in the selection of these subgroups.

Generally speaking, subgroups should be selected in a way that makes each subgroup as homogeneous as possible and that gives the maximum opportunity for variation from one subgroup to another. As applied to control charts on production, this means that it is of vital importance not to lose track of the order of production. Particularly if the primary purpose of keeping the charts is to detect shifts in the process average, one subgroup should consist of items produced as nearly as possible at one time; the next subgroup should consist of items all produced at a single later time; and so forth.

This basis of subgrouping was illustrated in Examples 1.1 and 1.2 in which five items constituting one subgroup were produced in succession at about 8 o'clock; five items constituting the next subgroup were produced in succession at about 9 o'clock; none of the items produced between the 8 and 9 o'clock subgroups were measured for purposes of the control charts. With this scheme of subgrouping, it sometimes is desirable that the exact time of choosing a sample should vary a bit one way or the other from the stipulated time and that this variation should not be predictable by the operator. For instance, the 9 o'clock sample might be taken one day at 9:10 and the next day at 8:45. It is better if the operator is not able to be sure in advance just which items are to be selected as the sample for inspection.

This desirable scheme of subgrouping to make each sample as homogeneous as possible may sometimes need to be modified either because of practical difficulties in taking homogeneous samples or because the control chart is intended to serve several different purposes. For instance, if one of the purposes of a control chart is to provide a basis for acceptance, it may be desirable to have each subgroup as nearly representative as possible of the production over a given period of time. For this purpose it would be better to inspect five items selected at random from the production in some given period of time than to inspect five items produced in succession at the beginning or end of the period. Or, if each item takes a considerable period of time to produce, it may be more convenient for the items inspected for each subgroup to be spaced approximately uniformly over the production of a given period. This is illustrated in Example 2.2.

D. Decision on the Size and Frequency of Subgroups

Shewhart suggested four as the ideal subgroup size. In the industrial use of the control chart, five seems to be the most common size. Because the essential idea of the control chart is to select subgroups in a way that gives minimum opportunity for variation *within* a subgroup, it is desirable that subgroups be as small as possible. On the other hand, a size of four is better than three or two on statistical grounds; the distribution of $\bar{X}$ is nearly normal for subgroups of four or more even though the samples are taken from a nonnormal universe; this fact is helpful in the interpretation of control-chart limits. A reason sometimes advanced for the use of five is ease of computation of the average, which can be obtained by multiplying the sum by 2 and moving the decimal point one place to the left.

Subgroups of two or three may often be used to good advantage, particularly where the cost of measurements is so high as to veto the use of larger subgroups.

Larger subgroups such as 10 or 20 are sometimes advantageous where it is desired to make the control chart sensitive to small variations in the process average. The larger the subgroup size, the narrower the control limits on charts for $\bar{X}$ and the easier it is to detect small variations; this is true only if subgroups are selected in a way that these variations in process average occur between and not within subgroups. Generally speaking, the larger the subgroup size, the more desirable it is to use standard deviation rather than range as a measure of subgroup dispersion. A practical working rule is to use $\bar{X}$ and s charts rather than $\bar{X}$ and R charts whenever the subgroup size is greater than 15.

In certain problems of process control, the issue of subgroup size may be viewed as an economic problem that is closely related to the issue of the tightness of control limits, which is examined briefly in Chap. 4.

In the introduction of the control-chart technique in any organization, charts for variables are often applied to data already collected for some other purpose. Often it is wise to start the control charts with no change in the method of collecting data, putting off any changes that might improve the control charts until such time as the charts have proved their usefulness to management. In such cases the subgroup size is likely to be determined by the way in which the data are collected. Sometimes this involves variable subgroup sizes; methods of determining limits and plotting charts for these are explained in Chap. 10.

Where the control-chart analysis is applied to past data already tabulated, this may necessitate much larger subgroups than would ordinarily be selected. This is illustrated in Chap. 10.

No general rules may be laid down for frequency of subgroups. Each case must be decided on its own merits, considering both the cost of taking and analyzing measurements and the benefits to be derived from action based on the control charts. In the initial use of a control chart for analyzing a process, it may be desirable to arrive at conclusions quickly by taking frequent samples. Later on, after the troubles have been diagnosed and corrected and the function of the control chart has become the maintenance of process control on current production, it may be advisable to reduce the frequency of sampling. This was illustrated in Example 1.2, in which the frequency of sampling was ultimately reduced from one subgroup every hour to one subgroup every 8 h.

The frequency of taking a subgroup may be expressed either in terms of time, such as once an hour (as in Examples 1.1 and 1.2), or as a proportion of the items produced, such as 5 out of each 100 (as in Example 2.3).

E. Setting Up the Forms for Recording the Data

Although the exact details of the forms used will vary from one organization to another, two general types of forms are in common use. Both types are shown in Example 2.3 near the end of this chapter. The type shown in Fig. 2.7, in which the successive measurements in each subgroup are recorded one below the other, avoids the necessity of mental arithmetic (or calculations on scratch paper) in figuring averages.

Figure 2.8 shows the other common type of form, in which each line contains all the measurements in one subgroup. This type avoids the necessity of copying the $\bar{X}$ and R values to compute the grand averages and generally allows room for more subgroups on a page. Where only the final two digits of the measurement are subject to variation, the mental arithmetic involved in obtaining averages is not complicated.

All forms need spaces for indicating the item measured, the unit of measurement, and other relevant information. It is particularly important to provide opportunity for remarks regarding any production changes (for example, changed machine setting, changed operator, tool sharpened, etc.), inspection changes, or other matters observed that might give clues to the causes of any out-of-control points.

F. *Determining the Method of Measurement*

Decisions must be made as to the measuring instruments to be used and the way in which the measurements are to be made. Usually it is desirable to prepare definite written instructions on this point.

2.2.2 Starting the Control Charts

A. *Making and Recording the Measurements and Recording Other Relevant Data*

The actual work of the control charts starts with the first measurements. It should always be remembered that the information given by the control chart is influenced by variations in measurement as well as by variations in the quality characteristic being measured. Any method of measurement will have its own inherent variability; it is important that this not be increased by mistakes in reading measuring instruments or errors in recording data. It is also important that notes be made about any occurrences that, if the control chart later shows lack of control, might provide help on the investigation of assignable causes of variation.

B. *Calculating the Average and Range for Each Subgroup*

The average for each subgroup i is calculated by summing the individual measurements in the subgroup and dividing by the number of measurements. The central line for the $\overline{X}$ chart is found by summing the averages and dividing by the number of subgroups. Where n is the subgroup size and k the number of subgroups, the formulas for these calculations are

$$\overline{X}_i = \sum_{j=i}^{n} \frac{X_{ij}}{n}$$

$$\overline{\overline{X}} = \sum_{i=1}^{k} \frac{\overline{X}_i}{k}$$

To find the range for each subgroup, the highest and lowest numbers in the subgroup must first be identified. With large subgroups it is helpful to mark the highest value with the letter H and the lowest with the letter L. The subgroup range is computed by subtracting the lowest value from the highest. The central line for the R chart is found by summing the subgroup range values and dividing by the number of subgroups. Formulas for these calculations are

$$R_i = X_{i,\ \max} - X_{i,\ \min}$$

$$\overline{R} = \sum_{i=1}^{k} \frac{R_i}{k}$$

C. Plotting the $\bar{X}$ and R Charts

$\bar{X}$ and R charts have already been illustrated in the preceding chapter, and many others are given throughout this book. Such charts are often plotted on rectangular cross-section paper having 8 or 10 rulings to the inch. Profile paper may also be used. Some special forms developed for this purpose have used rulings similar to profile paper, with vertical lines spaced $\frac{1}{6}$ in apart and horizontal lines spaced $\frac{1}{10}$ in apart.

The vertical scale at the left is used for the statistical measures $\bar{X}$ and R. The horizontal scale is used for subgroup numbers. Dates, hours, or lot numbers may also be indicated on the horizontal scale. Each point may be indicated on the chart by a dot, circle, or cross. In this book the general practice is to use a circle for points on $\bar{X}$ charts and a cross for points on R charts; there are some exceptions, however. There is no standard practice in industry in this regard. Points on control charts may or may not be connected. In this book they are not connected except where the connecting line serves some definite purpose or where computer-generated or published charts having connecting lines are being reproduced.

Points on both $\bar{X}$ and R charts should be kept up to date. This is particularly important where the charts are posted in the shop and are used by machine operators, setters, and supervisor.

2.2.3 Determining the Trial Control Limits

A. Decision on Required Number of Subgroups before Control Limits Are Calculated

The determination of the minimum number of subgroups required before control limits are calculated is a compromise between a desire to obtain the guidance given by averages and control limits as soon as possible after the start of collecting data and a desire that the guidance be as reliable as possible. The fewer the subgroups used, the sooner the information thus obtained will provide a basis for action but the less the assurance that this basis for action is sound.

On statistical grounds it is desirable that control limits be based on at least 25 subgroups. Moreover, experience indicates that the first few subgroups obtained when a control chart is initiated may not be representative of what is measured later; the mere act of taking and recording measurements is sometimes responsible for a change in the pattern of variation.

For these reasons, if 25 subgroups can be obtained in a short time, it is desirable to wait for 25 or more subgroups; this would be true, for example, where a new subgroup was measured every hour.

However, where subgroups are obtained slowly, there is a natural desire on the part of those who initiated the control charts to draw some conclusions from them within a reasonable time. This impatience for an answer frequently leads to the policy of making preliminary calculations of control limits from the first 8 or 10 subgroups, with subsequent modification of limits as more subgroups are obtained.

B. Calculation of Trial Control Limits

When the grand average of the subgroups $\bar{\bar{X}}$ and the average of the ranges $\bar{R}$ have been calculated, the factors contained in Table *D,* App. 3, are used to calculate control limits. These formulas are

$$UCL_{\bar{X}} = \bar{\bar{X}} + A_2\bar{R}$$

$$LCL_{\bar{X}} = \bar{\bar{X}} - A_2\bar{R}$$

$$UCL_R = D_4\bar{R}$$

$$LCL_R = D_3\bar{R}$$

The trial limits thus obtained are appropriate for analyzing the past data that were used in their calculation. They may require modification before extending them to apply to future production.

The preceding formulas for limits for $\bar{X}$ and R assume a constant subgroup size. Chapter 10 explains the necessary calculations where the subgroup size is variable.

C. Plotting the Central Lines and Limits on the Charts

The central line on the R chart should be drawn as a solid horizontal line at $\bar{R}$. The upper control limit should be drawn as a dashed horizontal line at the computed value of UCL_R. If the subgroup size is seven or more, the lower control limit should be drawn as a dashed horizontal line at LCL_R. If the subgroup size is six or less, the lower control limit for R does not exist.

The central line on the $\bar{X}$ chart should be drawn as a solid horizontal line at $\bar{\bar{X}}$. The upper and lower control limits for $\bar{X}$ should be drawn as dashed horizontal lines at the computed values.

2.2.4 Drawing Preliminary Conclusions from the Charts

A. Indication of Control or Lack of Control

Lack of control is indicated by points falling outside the control limits on either the $\bar{X}$ or the R charts. Some users of the control charts identify such out-of-control points by a special symbol. For instance, if each point is represented by a circle, a cross is made in the circles designating out-of-control points; if each point is represented by a dot, a circle may be drawn around the dot for out-of-control points.

When, because points fall outside the control limits, we say that a process is "out of control," this is equivalent to saying, "Assignable causes of variation are present; this is not a constant-cause system." As explained in Chap. 4, we can make this statement with considerable confidence that it is correct; a constant-cause system will seldom be responsible for points falling outside control limits set as indicated.

In contrast to this, when all points fall inside the control limits, we cannot say with the same assurance, "*No* assignable causes of variation are present; this *is* a constant-cause system." No statistical test can give us this positive assurance. When we say, "This process is in control," the statement really means, "For practical purposes, it pays to act as if no assignable causes of variation were present."

Moreover, even in the best manufacturing processes, occasional errors occur that constitute assignable causes of variation but that may not constitute a basis for action. This fact may lead to various practical working rules on the relationship between satisfactory control and the number of points falling outside limits. One such rule is to consider not more than 1 out of 35, or 2 out of 100, points outside control limits as evidence of control.[†]

Even though all points fall within control limits, lack of control may be indicated by runs of seven or more points in succession on the same side of the central line, or by the presence of other extreme runs such as those listed in Chap. 4.

The actions suggested by the evidence of the control chart depend on the relationship between what the process is doing and what it is supposed to do. That is, the apparent universe pattern of variation implied by the control chart needs to be compared with the specifications. This comparison is simplest and most meaningful if the process appears to be in control.

B. *Interpretation of Processes in Control*

With evidence from the control chart that a process is in control, we are in a position to judge what is necessary to permit the manufacture of product that meets the specifications for the quality characteristic charted. The control-chart data give us estimates of:

1. The centering of the process (μ may be estimated as $\overline{\overline{X}}$)
2. The dispersion of the process (σ may be estimated as $\overline{R}/d_2$)[‡]

The discussion in Chap. 3 will make it clear that estimates of μ and σ are subject to sampling errors. For this reason, any conclusions obtained from a short period (such as 25 points on the control chart) must be regarded as tentative, subject to confirmation or change as the control chart is continued and as more evidence becomes available. However, the reasonable thing to do at any stage in the proceedings is to make the best interpretation possible of the data already available. Once it is evident what actions are suggested by this interpretation, a decision can be made whether action should be taken at once or whether it

[†]"Control Chart Method of Controlling Quality during Production, ANSI Standard Z1.3—1969" (Reaffirmed 1975), p. 18, American National Standards Institute, New York, 1975; and "ASQC Standard B3-1985," American Society for Quality Control, Milwaukee, Wis.

[‡]Sample statistics used to estimate true universe parameters are discussed at some length in Chap. 4. The Greek lowercase letter mu (μ) frequently is used to designate the true centering; the lowercase Greek letter sigma (σ) is used to designate the standard deviation measure of dispersion.

should await more data. In applying the following analysis of processes in control, the limitations of the current estimates of μ and σ should always be kept in mind.

In Chap. 3 it is pointed out that practically all (all but 0.27%) of a normal distribution falls within limits of $\mu \pm 3\sigma$; that is, for practical purposes the spread of the distribution may be thought of as approximately 6σ. In the preliminary analysis of control-chart data, there is hardly enough evidence to permit judgment as to whether the distribution is approximately normal. Certainly the evidence is seldom sufficient to tell whether 0.27% or 0.6% or 1.1% or some other small percentage of the distribution will fall outside $\mu \pm 3\sigma$. Evidence of the exact form of the frequency distribution and the percentage of the distribution outside these limits may be obtained only after a long accumulation of data under control. In the meantime a satisfactory rough guide to judgment is provided by the assumption that 6σ is a measure of the spread of the process.

Actions based on the relationship between the specifications and the centering and dispersion of a controlled process depend somewhat on whether there are two specification limits, a maximum or upper limit USL_x and a minimum or lower limit LSL_x, as is always true of dimensions, or only one specification limit, either USL_x or LSL_x, as might be true of a specified minimum tensile strength, or minimum weight of the contents of a container, or maximum percentage of a particular chemical impurity. Some of the various possible situations that may exist with two limits are shown in Figs. 2.1 to 2.3. Some situations that may exist with a single limit are shown in Figs. 2.4 to 2.6.

FIGURE 2.1 Some cases where the spread of a process is less than the difference between specifications.

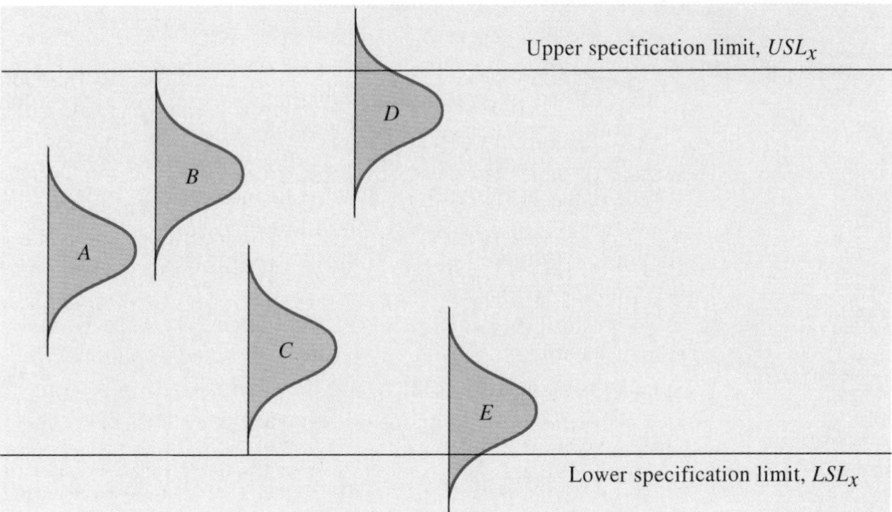

Possible Relationships of a Process in Control to Upper and Lower Specification Limits

When a controlled process must meet two specification limits on individual values, USL_x and LSL_x, all possible situations may be grouped into three general classes as follows:

1. The spread of the process (6σ) is appreciably less than the difference between the specification limits ($USL_x - LSL_x$).
2. The spread of the process (6σ) is approximately equal to the difference between the specification limits ($USL_x - LSL_x$).
3. The spread of the process (6σ) is appreciably greater than the difference between the specification limits ($USL_x - LSL_x$).

The first situation is illustrated in Fig. 2.1, in which the specification limits are shown by the upper and lower horizontal lines. Frequency curves *A, B, C, D,* and *E* indicate various positions in which the process might be centered. With any of the positions *A, B,* or *C,* practically all the product manufactured will meet the specifications as long as the process stays in control.

In general, the conditions represented in Fig. 2.1 *A, B,* and *C* represent the ideal manufacturing situation. When the control chart shows that one of these conditions exists, many different possible actions may be considered; the choice among the various actions is a matter of relative economy.

For example, it may be considered economically advisable to permit X to go out of control if it does not go too far; that is, the distribution may be allowed to move between positions *B* and *C.* This may avoid the cost of frequent machine setups and of delays due to hunting for assignable causes of variation that will not be responsible for unsatisfactory product. In Chap. 10, the use of so-called modified control limits for this purpose is explained.

Or, where acceptance has been based on 100% inspection, it may be economical to substitute acceptance based on the control chart for $\bar{X}$ and R.

Or, if there is an economic advantage to be gained by tightening the specification limits, such action may be considered.

If none of these things are to be done, it may be economical to discontinue the use of the control chart, or at least to increase the time interval between control-chart inspections. The larger the ratio of $USL_x - LSL_x$ to the process spread (6σ), the more favorable the situation is to getting good product without assistance from any control chart.

With the process in position *D* of Fig. 2.1, some product will fall above the upper specification limit; in position *E* some product will fall below the lower specification limit. In either case the obvious action is to try to change the centering of the process, bringing it closer to position *A.* Once this has been done, consideration may be given to the various actions just enumerated.

The second type of situation is illustrated in Fig. 2.2. Only if the process is exactly centered between the specification limits, as in position *A,* will practically all the product conform to the specifications. If the distribution shifts away from

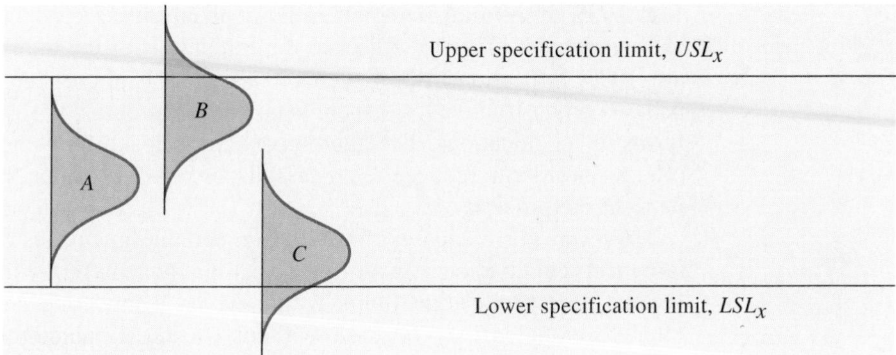

FIGURE 2.2 Some cases where the spread of a process is approximately equal to the difference between specification limits.

this exact centering, as in *B* or *C,* it is apparent that some of the product will fall outside the specification limits.

Here the obvious action is to take all steps possible to maintain the centering of the process. This usually calls for continuous use of the control charts for $\overline{X}$ and R with subgroups at frequent intervals and immediate attention to points out of control. If fundamental changes can be made that reduce process dispersion, they will ease the pressure. Consideration also should be given to the question of whether the tolerances are tighter than really necessary.

The third type of situation is illustrated in Fig. 2.3. Here the specification limits are so tight that even with the process in control and perfectly centered as in position *A,* some nonconforming product will be made. This calls for a review of tolerances, as was illustrated in Example 1.2. It also calls for effort to make fundamental changes that will reduce process dispersion. It is still important to maintain the centering of the process; the curves in positions *B* and *C* show how a shift in process average will increase the nonconforming percentage.

FIGURE 2.3 Some cases where the spread of a process is greater than the difference between specification limits.

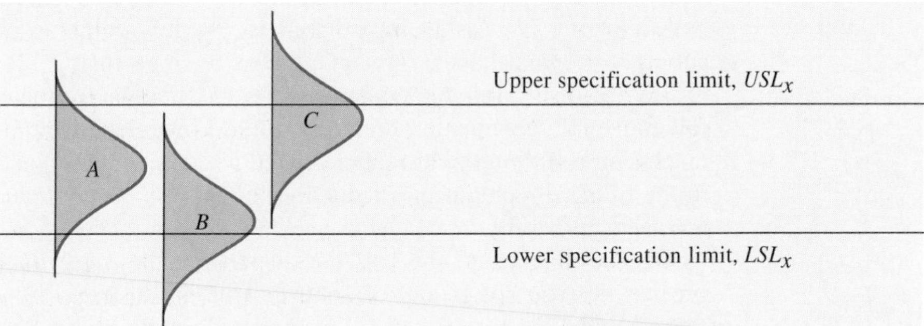

If 100% inspection is possible, the nonconforming product may be sorted out and eliminated. (This is subject to the limitations of the human error involved in any 100% inspection.) But if such 100% sorting is impossible because acceptance is based on destructive tests, there is no chance to obtain product all of which conforms to specifications. The alternatives are to make fundamental changes in the process in order to reduce the process dispersion or to widen the specification limits to fit the process.

However, in some cases (for example, certain dimensions of mating parts, certain electrical characteristics of electrical components) a twilight zone exists just beyond the specification limits within which a moderate percentage of out-of-tolerance articles may be used without causing trouble; this situation usually calls for continued action to maintain the centering of the process.

Possible Relationships of a Process in Control to a Single Specification Limit

The key to the most useful classification of the situations involving two specification limits was the process dispersion. A similar key to the classification of situations involving a minimum limit is the position of $(\mu - 3\sigma)$ with respect to LSL_x. Again three situations may be considered:

1. The low value of the process distribution $(\mu - 3\sigma)$ is appreciably above LSL_x. This is illustrated in Fig. 2.4
2. The low value of the process distribution $(\mu - 3\sigma)$ is approximately at LSL_x. This is illustrated in Fig. 2.5.
3. The low value of the process distribution $(\mu - 3\sigma)$ is appreciably below LSL_x. This is illustrated in Fig. 2.6.

The first situation is one in which there is a margin of safety, the second is one in which the specification is just barely met as long as the process stays in control, and the third is one in which some nonconforming product is inevitable unless a fundamental change is made in the process. The fundamental change may be either a decrease in process dispersion or an increase in the process average.

Three curves—*A*, *B*, and *C*—with different dispersions have been shown in Figs. 2.4 to 2.6 to emphasize the interrelationship of process average, process dispersion, and specification minimum limit. All three distributions—*A*, *B*, and *C*—have the same low value. However, distribution *B* with a greater dispersion must have a greater process average than *A* for the low points to be at the same level; similarly, *C* must have a greater process average than *B*. It is evident that the greater the dispersion, the higher the average must be for the entire distribution to fall above the specification minimum limit. This relationship between average and dispersion is an important matter related to costs in many instances; for example, in the filling of containers a reduction in dispersion may reduce cost by reducing the average overfill.

On the other hand, the less the dispersion, the more important it is that the process average not go out of control. This is illustrated by comparing distributions *A* and *C* in Figs. 2.5 and 2.6. In Fig. 2.6 both process averages have shifted an equal amount below their position in Fig. 2.5. However, the proportion of bad

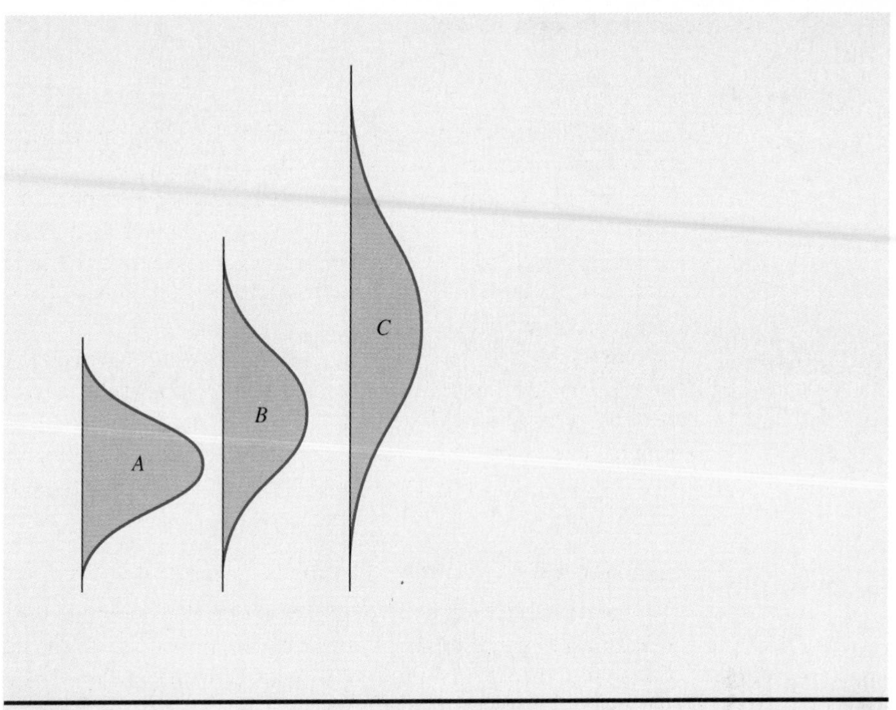

FIGURE 2.4 Some cases where the low value of the process distribution is above the specification minimum.

FIGURE 2.5 Some cases where the low value of the process distribution is approximately at the specification minimum.

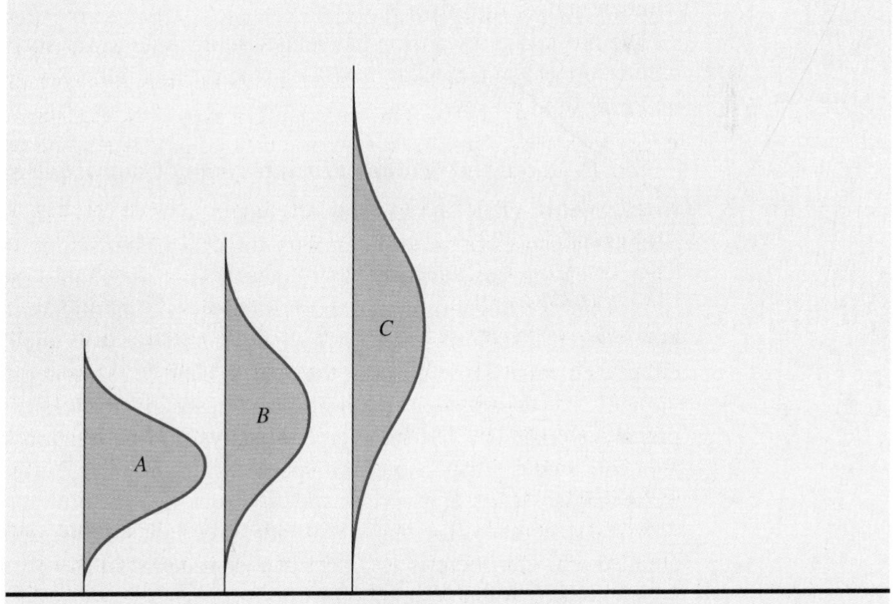

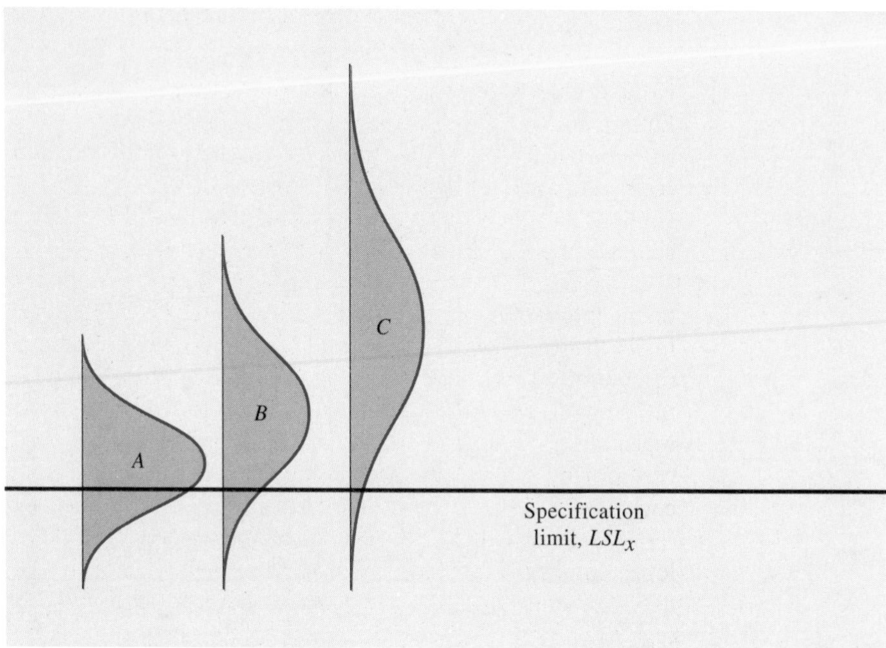

FIGURE 2.6 Some cases where the low value of the process distribution is below the specification minimum.

product, as indicated by the area of the distribution below the specification limit, is much greater in *A* than in *C*.

The preceding discussion has related to a single lower specification limit LSL_x, with no upper limit. Similar reasoning would apply if there were an upper limit with no lower limit.

C. Relationships between Processes out of Control and Specification Limits

If the control chart shows lack of control, one obvious step is to hunt for the assignable causes of variation and try to correct them. Some guidance is given in Chap. 4 on the statistical aspects of detection of assignable causes.

It is also pertinent to inquire into the process centering and dispersion that may be expected if control is obtained. If the out-of-control points on the *R* chart are eliminated, what evidence does the control chart give about process dispersion? In view of this process dispersion and of the specification limits, where should the process average be? The same type of analysis of the relationship between process average, process dispersion, and specification limits that was discussed for controlled processes is appropriate for processes not yet brought under control. The practical differences are that it may not be certain whether control can actually be attained and that there is less basis for confidence in the estimates of μ and σ at which control might be attained.

D. *Suggestions about Preliminary Conclusions and Actions Based on the Control Charts*

If the R chart shows control, estimate σ as $\bar{R}/d_2$. If there is an upper specification limit, compare $(\bar{\bar{X}} + 3\sigma)$ with USL_x. If there is a lower specification limit, compare $(\bar{\bar{X}} - 3\sigma)$ with LSL_x. If there are two specification limits, also compare 6σ with $(USL_x - LSL_x)$. Determine from these comparisons which type of condition is applicable of the various types illustrated in Figs. 2.1 to 2.6. Consider the possible actions suggested in our discussion of these figures.

If the charts show lack of control, try to judge from the charts and all other pertinent information what may be the assignable causes and whether or not it is likely that they can be eliminated. After removing the subgroups that showed lack of control on the R chart, calculate a revised $\bar{R}$ and new control limits for R. If these limits show additional subgroups out of control, remove those subgroups and repeat the calculation. Estimate σ as $\bar{R}/d_2$, using the final revised $\bar{R}$ after all out-of-control values of R have been eliminated. Consider this as a value of σ that might conceivably be obtained if the process were brought into control.

Consider whether or not the process average can readily be set at any desired level. (In some cases, as in dimensions resulting from machining operations, this may be merely a matter of machine setting; in other cases, as in tensile strengths or percentages of chemical impurities, this may be a difficult matter involving costly changes in the process.) If so, decide on the desired value of $\bar{X}_0$ at which to aim, assuming that control could be maintained with σ as estimated. If not, in the light of the known value of $\bar{\bar{X}}$, decide on an attainable $\bar{X}_0$. Use the values of $\bar{X}_0$ and σ thus obtained as a basis for an analysis of the relationship of the specification limits to the process average and dispersion.

Decisions as to whether to take action now on the basis of the control-chart data already at hand, or to defer action until more data are obtained, should give weight to the uncertainties in the estimates of μ and σ and to the costs and possible consequences of the proposed action.

Example 2.1 describes a case in which control charts on product received from two suppliers were used to diagnose entirely different sources of trouble in the production processes of the different suppliers.

Example 2.1

Use of control charts by a purchaser to help suppliers improve their processes

Facts of the case A manufacturer of electronic devices had trouble with the cracking of a certain small cross-shaped ceramic insulator used in the device. The cracking generally took place after the manufacturing operations were nearly completed and did so in a way that made it impossible to salvage the unit. Hence the costs resulting from each cracked insulator were many times the price of an insulator. Moreover, the cracking of some insulators during manufacturing operations suggested that others might be likely to crack under service conditions.

In an effort to improve the situation, all incoming insulators of this type were given 100% inspection. This inspection included a proof load test in which the inspector applied a more or less standardized finger pressure in an effort to break each insulator. This 100% inspection failed to decrease the percentage of units rendered defective by cracked insulators.

A simple testing device was then constructed to measure the actual strength in flexure by testing insulators to destruction. From each incoming lot of insulators, 25 were tested. As the insulators came from two suppliers, control charts for $\overline{X}$ and R were maintained for each supplier. The tests showed that both suppliers had approximately the same percentage of defective insulators. However, the control charts indicated that the explanations for the defectives were totally different in the two cases.

Supplier A had high average strength but complete lack of anything resembling statistical control. Supplier B had excellent statistical control but at a level such that an appreciable part of the frequency distribution was below the required minimum strength.

This diagnosis of the situation was brought to the attention of both suppliers. They were encouraged to exchange information about production methods. Certain techniques used by supplier A (largely related to mixing and molding the clay) were adopted by supplier B to try to raise average strength. Certain techniques used by supplier B (largely related to control of temperature and humidity during firing) were adopted by supplier A to try to bring the process into control. Both suppliers established control charts to help in the maintenance of control. As a result of these actions, both suppliers brought their product into control at a satisfactory level and the trouble with the cracking of the insulators was eliminated.

2.2.5 Continuing to Use the Charts

A. *Revision of Central Lines and Control Limits*

The trial control limits served the purpose of determining whether past operations were in control. The continuing use of the control chart, with each out-of-control point used as a possible basis for hunting for an assignable cause of variation and taking action to eliminate that cause, may require revised limits.

If the process has been in control with both average and dispersion satisfactory from the standpoint of the specification limits, the trial control limits should be extended to apply to future production. As more data accumulate, the limits may be reviewed from time to time and revised whenever necessary. It is desirable to establish regular periods for this review, such as once every week, once every month, or once every 25, 50, or 100 subgroups.

If the dispersion has been in control, as evidenced by the R chart, but the average $\overline{X}$ has been out of control, the trial central line and trial limits on the R chart should continue to be used, and control limits on the $\overline{X}$ chart should continue to be a distance of $A_2\overline{R}$ on either side of the central line. The location of the revised central line on the $\overline{X}$ chart calls for a decision whether this should be at an aimed-at level (sometimes called a standard level), designated as $\overline{X}_0$, or whether it should be based on the past data, a revision of $\overline{\overline{X}}$. When making this choice it should be recognized that wherever the central line (that is, the assumed process average for the purposes of the control chart) is placed, a point outside control limits is interpreted as meaning. "This variation is more than would be expected as a matter of chance if the assumed process average is maintained."

If the process average may be changed by a fairly simple adjustment of the manufacturing process, such as a machine adjustment influencing a dimension, it is nearly always desirable to decide on a standard or aimed-at value $\overline{X}_0$. This aimed-at process average should be determined by consideration of the relationship between process average, process dispersion, and specification limits, along the lines already discussed. Where there are upper and lower specification limits and product falling outside one limit may be reworked whereas product falling outside the other limit must be scrapped, the aimed-at average should be chosen with due consideration for the difference between costs of spoilage and rework.

Unless the process average can be changed by a definite adjustment, $\overline{\overline{X}}$ is likely to be the result of a complex set of factors, with the influence of each factor not clearly known. In such instances, a control chart based on an aimed-at value of $\overline{X}_0$ that was different from the past $\overline{\overline{X}}$ might show many points out of control; these out-of-control points would simply indicate that the aim had been poor without giving much help in correcting it. For this reason it is more sensible in such cases to use the past $\overline{\overline{X}}$ as the new central line, or possibly a revised $\overline{\overline{X}}$ corrected by the elimination of the past out-of-control points. Then if different changes in the process are undertaken one after another with the purpose of finding a way to bring the process average to a new desired level, the control chart based on the past process average will give evidence whether each change constitutes an assignable cause of variation from past performance. If these assignable causes move the process average closer to the desired average, they should be continued rather than eliminated. Each change that is continued may call for a new central line on the $\overline{X}$ chart based on the changed performance.

Where the R chart shows that the process dispersion is out of control, it is desirable to estimate the value of $\overline{R}$ and σ that might be attained if the dispersion were brought into control. This estimate is necessary even though it cannot be made with great assurance that it is correct. One possible method is to eliminate the values of R above the control limits and make a new calculation of $\overline{R}$. If new limits calculated from this $\overline{R}$ throw more points above the control limits, the calculation of a revised $\overline{R}$ may be repeated again. This method is illustrated in Example 2.3.

This revised $\overline{R}$ may be used as the new central line on the R chart, as the basis for calculating the new limits on the R chart and for calculating $A_2\overline{R}$ to get the distance of the control limits from the central line on the $\overline{X}$ chart. This has the effect of tightening the limits on both the R and the $\overline{X}$ charts, making them consistent with a σ that may be estimated from the revised $\overline{R}$ as $\overline{R}/d_2$. This σ may also serve as a basis for a preliminary analysis of the relationship among process average, process dispersion, and specification limits, as outlined in our discussion of Figs. 2.1 to 2.6.

Where both $\overline{R}$ and $\overline{X}$ are out of control and new limits on the $\overline{X}$ chart are to be based on an aimed-at $\overline{X}_0$, the calculation of the revised $\overline{R}$ should be made before making a decision regarding the value of $\overline{X}_0$ to be used. This is illustrated in Example 2.3.

Revised limits should be reviewed from time to time as additional data are accumulated.

B. *Use of the Control Chart as a Basis for Sorting Sublots into Homogeneous Grand Lots*

If some quality of a manufactured product shows unsatisfactory variability and if the control chart shows that lack of control is responsible for this variability, the natural action seems to be to make a strong effort to bring about a state of control. However, it may turn out that for some reason it is impracticable to achieve control. Sometimes a satisfactory alternative may exist in the sorting of sublots of product into grand lots that seem to be statistically homogeneous. The variability from one grand lot to another may then become the basis for different correction factors somehow applied in the use of each grand lot. Conceivably, even though control could actually be obtained at a cost, it might be more economical to carry out this sorting.

This sorting is a possible procedure, for example, in the case of artillery ammunition. Experience shows that the variation in ballistic properties from one lot to another may be greater than the variation within any lot. (In other words, the averages show lack of statistical control.) Successive lots, the samples from which, under ballistic tests, all fall within control-chart limits, may be treated as sublots of one grand lot. Different corrections in the powder charge, or in the firing tables used by the artillery operator, could then be made for each grand lot.

In the case of chemical or bacteriological tests that different laboratories make of the same product, experience shows that it is often impracticable to get control. The variation from one laboratory to another will be greater than the variation in successive tests made by any one laboratory. Here also, correction factors might be applied to the results given by any laboratory that did not seem to be part of the grand lot.[†]

Although the control chart provides the best way to sort sublots into homogeneous grand lots, it sometimes happens that people who do not fully understand the principles behind the control chart use it for sorting in ways that are incorrect. This point is illustrated in Example 2.2.

Example 2.2

An unsound proposal for sorting product into classes

Facts of the case This example is adapted from an actual case, with slight modifications intended to disguise the source and to simplify the situ-

[†]See Example 8.1 for an ingenious application of this correction factor technique to a manufacturing problem.

ation in a way that will concentrate attention on the important principle involved.

A certain munitions product, which for purposes of this story will be called the *XYZ*-77, was subject to destructive testing in order to determine a particular quality characteristic. Four of these *XYZ*-77s from each lot of 500 were tested to establish a basis for acceptance. Control charts were plotted, using each set of four tests as a subgroup. These exhibited good control with only an infrequent point falling outside control limits.

The specifications stated that the average value of this quality characteristic should be 540. Tolerances were ±20. It proved possible to hold the average (that is, $\bar{\bar{X}}$ on the control chart) at 540. The upper control limit on the $\bar{X}$ chart was 549; the lower control limit was 531. This indicated that 3-sigma limits on individual values were 558 and 522. From this it seemed reasonable to believe that, if the distribution were normal or nearly so, as long as control was maintained, practically all the product would fall within the specification limits of 560 and 520. As the process showed good control, this was a very satisfactory state of affairs.

Suddenly there was a demand for some *XYZ*-77s that would hold the much closer tolerances of 540 ± 6. Not all the product was required to meet these specifications, but it was desired that a substantial portion of it do so.

One of the team who had been working with the control chart had a suggestion that was thought would solve the problem. It called for no change in production methods, and for a continuance of destructive testing of four out of each lot of 500 with the control chart carried on as before. However, it was proposed to add two new inner limits to the control chart. These were to be set at a distance from $\bar{\bar{X}}$ equal to the tolerance limits divided by $\sqrt{n}$. This meant inner limits at $540 \pm 6/\sqrt{4}$, that is, at 543 and 537.

It was further proposed that whenever the average of a sample plotted within these inner limits, the lot was to be classified as meeting the new specifications of 540 ± 6. Whenever the average of the sample plotted within control limits but above 543 or below 537, the lot was to be classified as meeting the original specifications of 540 ± 20.

Comment on Error in Reasoning Involved in Example 2.2

This suggestion could not have been made except by someone who had completely misunderstood the correct interpretation of the control chart.

When $\bar{X}$ and R charts for some quality characteristic of a manufactured product exhibit control over a long period, this means that the successive subgroups behave like random samples drawn from a single bowl. The existence of control indicates that for all practical purposes the bowl was unchanged throughout the period. If, as in Example 2.2, each subgroup is a small sample from a large lot, the evidence is that the large lots are practically alike. The variation from one subgroup to another is simply a chance variation to be expected in random sampling.

With all points on the control chart for $\bar{X}$ falling between control limits of 549 and 531, the fact that the sample from one lot has an $\bar{X}$ of 545 and the sample from the next lot has an $\bar{X}$ of 540 is an indication that the lots are *alike,* not that they are different. It is only in the case of lack of control that the control chart can be used to sort lots into grand lots having different dispersions or different averages.

In Example 2.2, the control chart gave conclusive evidence that the manufacturing process, as it was then carried on, could not meet the proposed new tolerance of ± 6. If 100% inspection were possible, the product meeting these close tolerances (about two-thirds of the total product if the distribution were approximately normal) could be sorted from the product not meeting the tolerances. As the test was destructive, this was not possible. It was necessary, therefore, either to make some fundamental change in the manufacturing process that would reduce the variability of the product or to devise some nondestructive test that gave substantially the same information as the destructive test and would thus permit 100% inspection.

C. Use of the Control Charts for Action on the Process

In continuing the use of the control charts, there may be three different kinds of action on the process, as follows:

1. Action to remove assignable causes of variation that are brought to attention by out-of-control points
2. Action to establish the process average
3. Action to establish the process dispersion

Once a process is brought into control with a satisfactory average and dispersion, an important purpose of the control chart is to help continue this happy state of affairs. The most common routine use of the control chart for variables is for this purpose. This involves simply leaving a process alone as long as it stays in control, and hunting for and removing assignable causes of variation whenever the control charts show lack of control.

Actions to establish a process average at some desired level follow the lines that we have suggested. It should be emphasized that whenever an aimed-at value $\bar{X}_0$ is used for the central line on the control chart, there should be an actual effort made to aim at that value in the process itself. It does no good to put such a line on the control chart unless this forms a basis for action by those who carry out the operations.

Actions to reduce process dispersion often call for fundamental changes in machine or methods. Action may also be undertaken to match various process dispersions with different jobs to be undertaken. The information given by control charts about the natural tolerances that will be held by various machines or various production methods may make it possible to fit the process dispersion to the job in hand. Operations calling for close tolerances may be assigned to the machines that will hold the close tolerances, and operations on which wide tolerances are satisfactory may be assigned to those machines that will hold only wide tolerances. This use of the control chart in production planning may be its chief contribution in some jobbing shops.

D. Use of the Control Charts for Acceptance Inspection

Our discussion of the interesting and important subjects of the strategy and tactics of acceptance inspection and sampling is deferred until Part Three of this book.

We shall see there that one useful concept in acceptance sampling is that an overall decision with respect to the acceptance or rejection of a manufacturing process often is superior to a series of unrelated decisions regarding acceptance or rejection of separate lots of product. Control charts may give evidence that a process is in statistical control with satisfactory centering and dispersion and thus provide a rational basis for the acceptance of the process.

In Part Three we shall see that another useful concept is that the best way to obtain satisfactory product is to have it made right in the first place—that it usually is difficult or impossible to "inspect quality into a product." We have already noted a case in Example 2.1 in which the feedback of information obtained from control charts played an important part in the diagnosis of troubles that were interfering with the making of satisfactory product.

E. Use of the Control Charts for Action on the Specifications

The control charts for variables may influence specifications in two ways.

They may be used to determine the capabilities of a manufacturing process before the specification limits are set. This is often a sensible procedure and is discussed in Chap. 10.

They may also be used to give evidence that, because of the inability of a manufacturing process to meet existing specification limits even when it is in control, a review of specification limits is called for.

A general principle applicable here is that the basis of all specification limits should be the prospective use of the part or product for which the limits are specified. Ideally, all specification limits should be exactly right from the standpoint of what is really needed. Actually, as illustrated in Example 1.2, many specification limits are made tighter than really necessary, often because no time or effort has been given to finding out what is necessary. Moreover, in most cases there is no one right value of specification limits that can be settled independently of cost factors involved; these cost factors cannot be properly judged without information regarding the capabilities of the manufacturing process such as is given by the control chart. For these reasons, many cases exist in which the appropriate conclusion from the control chart is to change the specifications.

Example 2.3

Milling a slot in an aircraft terminal block. An example to illustrate the steps in the use of $\bar{X}$ and R charts on a manufacturing operation

Decisions preparatory to the control chart High percentages of rejections for many of the parts made in the machine shop of an aircraft company indicated the need for examination of the reasons for trouble. As most of the rejections were for failure to meet dimensional tolerances, it was decided to try to find the causes of trouble by the use of $\bar{X}$ and R charts.

These charts, which of course required actual measurement of dimensions, were to be used only for those dimensions that were causing numerous rejections. Among many such dimensions, the ones selected for control charts were those having high costs of spoilage and rework and those on which rejections were responsible for delays in assembly operations. Although the initial purpose of all the $\bar{X}$ and R charts was to diagnose causes of trouble, it was anticipated that some of the charts would be continued for routine process control and possibly for acceptance inspection.

This example deals with one of these dimensions, the width of a slot on a duralumin forging used as a terminal block at the end of an airplane wing spar. The final machining of this slot width was a milling operation. The width of the slot was specified as $0.8750 \left\{ \dfrac{+0.0050}{-0.0000} \right.$ in. The designing engineers had specified this dimension with a unilateral tolerance because of the fit requirements of the terminal block; it was essential that the slot width be at least 0.8750 in and desirable that it be as close to 0.8750 as possible.

Most of the aircraft parts produced in this machine shop were large parts fabricated in lots whose size varied from a few hundred to several thousand. It was felt that practical considerations called for a single decision as to the method of subgrouping and the size and frequency of sample to apply to all the $\bar{X}$ and R charts to be used. One limiting factor was the small number of available personnel for the control-chart inspection in relation to the number of control charts it was desired to keep. On this basis it was decided that for each chart the sample inspected would be approximately 5% of the total production of the part in question. Because of the many general considerations favoring five as the subgroup size, this size was adopted. It was considered essential that, wherever possible, all measurements be made at the point of production. As lots of five of these large parts did not accumulate at the machine, it was decided that one part would be measured out of approximately every 20 produced, and that a subgroup would consist of five such measurements.

The type of form used for recording the data is illustrated in Fig. 2.7. It was chosen as a result of the decision to measure many of the dimensions to the nearest ten-thousandth of an inch: it was believed that with so many significant figures, delays and errors would be introduced by any type of form calling for much mental arithmetic. If measurements had been made only to thousandths of an inch, the other type of form would have been appropriate. This is illustrated in Fig. 2.8, in which the same measurements as in Fig. 2.7 have been recorded to the nearest thousandth of an inch.

The method of inspection to secure data for each $\bar{X}$ and R chart was stated in written instructions. In the case of the slot width of the terminal block, this was to measure the width with a micrometer at two specified positions in the slot. The recorded slot width was the average of these two measurements.

$\bar{X}$ AND R CONTROL CHART DATA SHEET

Product *Terminal block* Dept No. *78* Order No. *54321*

Characteristic *Width of slot* Specified limits { *0.8800 in.* Max. / *0.8750 in.* Min. }

Unit of measurement *0.0001 in. over 0.8000*

Subgroup No.	1	2	3	4	5	6	7
a	772	756	756	744	802	783	747
b	804	787	773	780	726	807	766
c	779	733	722	754	748	791	753
d	719	742	760	774	758	762	758
e	777	734	745	774	744	757	767
Total	3851	3752	3756	3826	3778	3900	3791
Average, $\bar{X}$	770	750	751	765	756	780	758
Range, R	85	54	51	36	76	50	20
Date or time	3/7	3/7	3/7	3/8	3/8	3/8	3/9

Subgroup No.	8	9	10	11	12	13	14
a	788	757	713	716	746	749	771
b	750	747	730	730	727	762	767
c	784	741	710	752	763	778	785
d	769	746	705	735	734	787	772
e	762	747	727	751	730	771	765
Total	3853	3738	3585	3684	3700	3847	3860
Average, $\bar{X}$	771	748	717	737	740	769	772
Range, R	38	16	25	36	36	38	20
Date or time	3/9	3/9	3/10	3/10	3/10	4/2	4/2

Subgroup No.	15	16					
a	771	767					
b	758	769					
c	769	770					
d	770	794					
e	771	786					
Total	3839	3886					
Average, $\bar{X}$	768	777					
Range, R	13	27					
Date or time	4/3	4/3					

	$\bar{X}$	R
1	770	85
2	750	54
3	751	51
4	765	36
5	756	76
6	780	50
7	758	20
8	771	38
9	748	16
10	717	25
11	737	36
12	740	36
13	769	38
14	772	20
15	768	13
16	777	27
	12,129	621

Calculation of limits

$\bar{\bar{X}} = 12{,}129 \div 16 = 758$

$\bar{R} = 621 \div 16 = 39$

$A_2\bar{R} = .58\ (39) = 23$

$UCL_{\bar{X}} = \bar{\bar{X}} + A_2\bar{R}$

$= 758 + 23 = 781$

$LCL_{\bar{X}} = \bar{\bar{X}} - A_2\bar{R}$

$= 758 - 23 = 735$

$UCL_R = D_4\bar{R}$

$= 2.11\ (39) = 82$

$LCL_R = D_3\bar{R} = 0$

FIGURE 2.7 $\bar{X}$ and R data sheet for Example 2.3.

Starting the control charts The actual measurements for the first 16 subgroups are shown in Fig. 2.7. This number of subgroups corresponds to a production order for 1,600 of these terminal blocks. Averages and ranges were calculated as shown in Fig. 2.7 and were plotted as shown in Fig. 2.9.

At the time of the twelfth subgroup, before the completion of this production order and before the calculation of central line or control limits, the quality control inspector noticed that the machine operator was occasionally checking performance by a micrometer measurement on width of slot on a terminal block

RECORD SHEET FOR $\overline{X}$ & R CHART

Material or part name Terminal block Part No. 1-2345
Characteristic measured Width of slot Plant 6 Dept. 78
Unit of measurement 0.001 in. over 0.800 Recorded by J.S.

Series No.	Date produced	A	B	C	D	E	$\overline{X}$ Av. of items	R Range of items	Record of inspection
1	3/7	77	80	78	72	78	77.0	8	
2		76	79	73	74	73	75.0	6	
3		76	77	72	76	74	75.0	5	
4	3/8	74	78	75	77	77	76.2	4	
5		80	73	75	76	74	75.6	7	
6		78	81	79	76	76	78.0	5	
7	3/9	75	77	75	76	77	76.0	2	
8		79	75	78	77	76	77.0	4	
9		76	75	74	75	75	75.0	2	
10	3/10	71	73	71	70	73	71.6	3	
11		72	73	75	74	75	73.8	3	
12		75	73	76	73	73	74.0	3	
13	4/2	75	76	78	79	77	77.0	4	Operator's check measure-
14		77	77	78	77	76	77.0	2	ments have been made
15	4/3	77	76	77	77	77	76.8	1	on hot part.
16		77	77	77	79	79	77.8	2	Instructed to wait until part
17									has cooled before making
18									check measurement, and to
19									center process at 0.8775
20									inches.
	Totals						1212.8	61	

(Measurements on each of five items in series: A–E)

$\overline{\overline{X}} = \frac{1212.9}{16} = 75.8$

$\overline{R} = \frac{61}{16} = 3.8$

$A_2\overline{R} = 0.58\,(3.8) = 2.2$

$D_4\overline{R} = 2.11\,(3.8) = 8.0$

$UCL_{\overline{X}} = 75.8 + 2.2 = 78.0$

$LCL_{\overline{X}} = 75.8 - 2.2 = 73.6$

$UCL_R = 8.0$

$LCL_R = 0$

FIGURE 2.8 An alternative form of $\overline{X}$ and R data sheet.

that had just come off the machine. As the block was still hot from the milling operation, this dimension as measured by the operator was too high because of the expansion of the metal as a result of the temperature. Moreover, the operator was influenced by the unilateral tolerance to aim at a dimension at or very slightly above the nominal dimension of 0.8750 in.

Even without a central line or control limits, it was evident from the chart and the data sheet that this was producing many slots that were too narrow. After the twelfth subgroup the operator was instructed to make check measurements on parts that had cooled to room temperature and to aim at a dimension of 0.8775, halfway between the upper and lower tolerance limits. This was reflected in the results in subgroups 13 to 16.

Determining the trial control limits Calculation of trial control limits was made after the first 16 subgroups which completed the production order. As shown in Fig. 2.7, this was done using the A_2 and D_4 factors and formulas from Table *D*, App. 3. These trial control limits are shown for the first 16 subgroups in the control charts in Fig. 2.9.

Drawing preliminary conclusions from the charts Subgroup 1 is above the upper control limit on the *R* chart. Subgroup 10 is below the lower control limit on the $\bar{X}$ chart. Moreover, the last 10 of the 16 points on the *R* chart all fall below the central line. It is evident that the measurements obtained are not the result of a constant system of chance causes.

If subgroup 1 is eliminated from consideration, $\bar{R}$ for the remaining 15 subgroups is $\frac{536}{15} = 36$. This gives as the revised upper control limit

$$D_4\bar{R} = 2.11(36) = 76$$

Subgroup 5 falls exactly on the control limit.

A common experience on hand-operated machines, where the dispersion of a controlled process is dependent in part on the care taken by the operator, is

FIGURE 2.9 $\bar{X}$ and R control charts for Example 2.3.

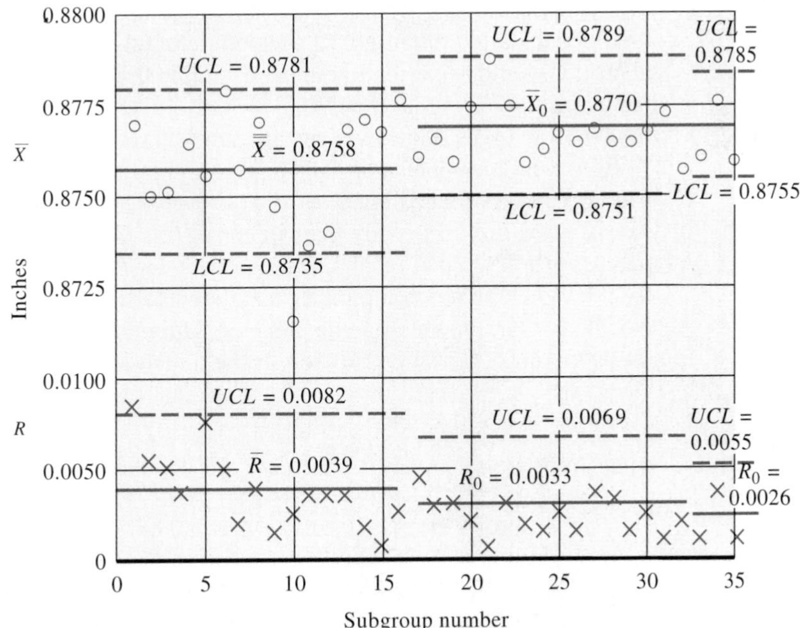

that the introduction of the control chart increases the care taken by the operator and thus reduces process dispersion. For this reason the ranges of the first few subgroups may not be representative of what may be expected as time goes on. The general appearance of this R chart with its run of the last 10 points below $\bar{R}$ suggests this as probably true of the slot width. Hence a second revision of $\bar{R}$, with subgroup 5 eliminated, seems reasonable. This gives $\bar{R} = \frac{460}{14} =$ 33; that is, $\bar{R} = 0.0033$ in.

From this second revision of $\bar{R}$ an estimate may be made of σ, the process standard deviation that might be anticipated if the process were controlled in the future. This estimate $\sigma = \bar{R}/d_2 = 0.0033/2.326 = 0.0014$ in. If this should be the value of σ, the natural tolerance or spread of the controlled process, 6σ, will be

$$6(0.0014) = 0.0084$$

This spread may be compared with the tolerance spread:

$$U - L = 0.8800 - 0.8750 = 0.0050$$

It is evident that the natural tolerance of this process is considerably greater than the specified tolerance. Unless the process dispersion can be reduced, it is evident that even though the process can be brought into control, a high percentage of nonconforming product will be produced.

This situation is like the one shown in curve B of Fig. 2.3; namely, the dispersion is too wide and the process average ($\bar{\bar{X}} = 0.8758$) is too low. It is evident that the process average is capable of adjustment; the instructions given to the operator after subgroup 12 seemed definitely to raise the average. At first glance, it would seem that the most desirable thing to do is to aim at minimum rejections by trying to hold the process in control at an average midway between the upper and lower specification limits, namely, at 0.8775.

However, this conclusion fails to give weight to the fact that a slot that is too narrow can be widened but a slot that is too wide cannot be narrowed. In other words, rework is less costly than spoilage. It is therefore desirable to center the process at a level that results in few slots over the upper specification limit of 0.8800 in, even though a number may be under the lower specification limit of 0.8570 in. The upper 3-sigma limit on individual values might be placed at 0.8800 to make a trial calculation of the aimed-at $\bar{X}_0$. If this is done,

$$\begin{aligned} \bar{X}_0 + 3\sigma &= 0.8800 \\ \bar{X}_0 + 3(0.0014) &= 0.8800 \\ \bar{X}_0 &= 0.8758 \end{aligned}$$

By chance, this is exactly the $\bar{\bar{X}}$ of the first 16 subgroups. It is evident that such a centering will continue to result in considerable rework.

Experience on similar jobs indicates that it is reasonable to expect that process dispersion may be further reduced. Hence it seems wise to center the process somewhat above 0.8758. Just how much above depends on how much improvement is expected and on the relative costs of spoilage and rework. A figure of 0.8770 was selected.

Continuing to use the charts For the continuation of the control chart for the next production order, which started several weeks later, the central line was set as $\bar{X}_0 = 0.8770$. The 3-sigma control limits were based on assuming $\sigma = 0.0014$. Using Table F of App. 3,

$$UCL_{\bar{X}} = \bar{X}_0 + A\sigma = 0.8770 + 1.34(0.0014) = 0.8789$$

$$LCL_{\bar{X}} = \bar{X}_0 - A\sigma = 0.8770 - 1.34(0.0014) = 0.8751$$

$$UCL_R = D_2\sigma = 4.92(0.0014) = 0.0069$$

$$\text{Central line}_R = R_0 = d_2\sigma = 2.326(0.0014) = 0.0033$$

$$LCL_R = D_1\sigma = 0$$

(As σ was estimated from an $\bar{R}$ of 0.0033, the same limits would have been obtained using the factors and formulas from Table *D* of App. 3, with an $\bar{R}$ of 0.0033.)

These limits are shown for subgroups 17 to 32 on the control charts of Fig. 2.9. The averages and ranges of these subgroups were as follows (as in the data sheet shown in Fig. 2.7, $\bar{X}$ is in units of 0.0001 in above 0.8000 and R is in units of 0.0001 in):

Subgroup number	$\bar{X}$	R
17	761	47
18	766	31
19	760	32
20	775	22
21	788	7
22	775	32
23	760	21
24	763	18
25	768	27
26	766	17
27	769	38
28	766	35
29	766	17
30	769	26
31	774	14
32	758	24
Totals	12,284	408

None of the pieces inspected for control-chart inspection in subgroups 17 to 32 fell outside the specification limits. The average values for these 16 subgroups are

$$\overline{\overline{X}} = \frac{12{,}284}{16} = 768 \qquad \text{(that is, 0.8768 in)}$$

$$\overline{R} = \frac{408}{16} = 26 \qquad \text{(that is, 0.0026 in)}$$

It is evident that there has been a further narrowing of the process dispersion. This should be recognized by a revision of control limits starting with subgroup 33. As there seems to be no reason for a change in the aimed-at average, these revised control limits should be computed from an $\overline{X}_0$ of 0.8770 and an $\overline{R}$ of 0.0026, using the factors from Table *D* of App. 3.

$$UCL_{\overline{X}} = \overline{X}_0 + A_2\overline{R} = 0.8770 + 0.58(0.0026) = 0.8785$$

$$LCL_{\overline{X}} = \overline{X}_0 - A_2\overline{R} = 0.8770 - 0.58(0.0026) = 0.8755$$

$$UCL_R = D_4\overline{R} = 2.11(0.0026) = 0.0055$$

$$LCL_R = D_3\overline{R} = 0$$

These limits are shown on the control chart of Fig. 2.9 as applying to subgroups 33, 34, and 35. With *R* reduced to 0.0026, the estimate of σ is now

$$\frac{0.0026}{2.326} = 0.0011 \text{ in}$$

If control can be maintained at this level, $\mu + 3\sigma = 0.8803$ and $\mu - 3\sigma = 0.8737$. This indicates that a small amount of spoilage and a moderate amount of rework will still be produced; however, the situation is greatly improved as compared to that which existed before the start of the control chart. As time went on, it proved possible to maintain control and to decrease $\overline{R}$ (and σ) further to the point where nearly all the product fell within specification limits.

In situations where the specification limits are as tight as this in relation to the process dispersion, it is not appropriate to use the control chart for acceptance as a substitute for 100% inspection. Neither was this a situation in which the tolerance limits could be widened; the required fit of the part properly controlled the specifications despite the fact that the natural tolerance of the process seemed to be wider than the specified tolerance.

Comment on Example 2.3

The reader should be warned against the inference that a reduction in universe dispersion can always be readily attained, even though it was actually attained in this case. Although such reduction in σ is possible in certain machine shop operations in which the skill and care of the operator are controlling factors, it is not possible in many operations on automatic machines, where the process dispersion is almost entirely a matter of what the machine will do and of the variability of the materials being machined. Neither is it possible in operations on many quality characteristics other than dimensions.

Example 2.3 has been explained at some length. The purpose of this full explanation has been not only to show forms and computations involved in the simple control charts for $\bar{X}$ and R but also to show the way in which judgment enters into interpretation of these charts and into the action based on this interpretation.

A point to be emphasized is that no fixed rules may be laid down regarding the appropriate action based on interpretation of a control chart. The person who makes decisions about action—whether a quality control engineer, supervisor, methods engineer, or machine operator—must understand both the process being analyzed and the general principles underlying the control-chart analysis. Even though all situations may be grouped into a few simple classes from the statistical viewpoint, each actual case is somewhat different from all others; decisions regarding action are economic decisions that should be based on all the facts of each particular case.

It is recommended that all users of $\bar{X}$ and R charts examine the American National Standards Institute pamphlets on control charts referred to in Chap. 3. The presentation of the subject in this and the two succeeding chapters has been greatly influenced by these standards.

2.3 SOME COMMENTS ON COMPUTER SOFTWARE FOR STATISTICAL PROCESS CONTROL

The number of computer programs available for process control and statistical analysis grew at a rapid pace during the early and mid-1980s. Much of this growth has resulted from the development and wide acceptance of the microcomputer (personal computer). Since 1984, *Quality Progress* magazine annually has published a directory of quality assurance/quality control software.[†] The March 1994 issue indicated 95 sources dealing with process capability, 111 with process control, and 115 with general statistical analysis, among others. *Quality* magazine (Hitchcock Publishing Company, Wheaton, Ill.) also publishes lists of software in a convenient matrix format showing features, prices, and sources.

Prices for these software packages range from less than \$100 to many thousands. For microcomputers, prices tend to be in the lower range typically from \$100 to \$500. Those for local area networks (LANs), minicomputers, and mainframes tend to be in the higher range.

Selection of appropriate software is difficult at best. However, there are some software features that the authors believe to be necessary for stand-alone operation on a microcomputer or multiterminal minicomputer system.

[†]See, for example, *Quality Progress*, pp. 60–128, March 1994, American Society for Quality Control, Milwaukee, Wis.

1. First and foremost, it must be "user friendly." If subprofessionals are to use the system, options should be limited and error correction opportunities frequent. Directions must be clear and offer many on-line "help" instructions.
2. Video displays of $\bar{X}$ and R charts should provide at least the results from 40 subgroups, usually the most recent ones.
3. Printed-chart output should be able to handle a complete file or any range from the file desired by the user.
4. Files should be of sufficient size to hold at least 300 subgroups of typical size (four or five measurements). The user should be able to add to and delete from files. As a file reaches capacity, there should be a provision for dumping the oldest records, preferably to hard-copy output for historical purposes.
5. The user should be able to direct calculation of $\bar{\bar{X}}$ and $\bar{R}$ from any subset of the data file and have control limits calculated automatically from those data. Their numerical values also should be displayed and/or printed. The elimination of out-of-control points in making these calculations should be at the user's discretion. The user should have the option to input control limits and central lines directly.
6. It should be possible to develop a histogram from all or a selected portion of the data, with or without out-of-control point data eliminated, and to view the result either on the video display or printed or both. These outputs may or may not show specification limits and provide statistics of nonconforming proportions and capability indexes (discussed in Chap. 9).

The desirability of output in color (either video or printed), the ability to interface directly with automated inspection equipment (software that interfaces with standard spreadsheet programs), and other features offered by some software packages depend on the application and user needs and desires. Those in a position to influence these purchase decisions need to keep in mind the skill levels of those who will operate the software, the equipment needs for each package, and the extent of training required to make the system functional.

PROBLEMS

2.1. Control charts for $\bar{X}$ and R are maintained on a certain dimension of a manufactured part, measured in inches. The subgroup size is 4. The values of $\bar{X}$ and R are computed for each subgroup. After 20 subgroups, $\Sigma\bar{X} = 41.340$ and $\Sigma R = 0.320$. Compute the values of the 3-sigma limits for the $\bar{X}$ and R charts, and estimate the value of σ on the assumption that the process is in statistical control.
Answer: 2.079, 2.055; 0.0360, 0; 0.0078.

2.2. The dimension referred to in Prob. 2.1 is specified as 2.050 ± 0.020 in. Compare the natural tolerances of this process, 6σ, with the specification range, $USL_x - LSL_x$. Refer to Fig. 2.1 through 2.3, and comment on the ability of this process to meet specifications.

2.3. Control charts for $\bar{X}$ and R are maintained on the shear strength in pounds of test spot welds. The subgroup size is 3. The values of $\bar{X}$ and R are computed for each subgroup. After 30 subgroups, $\Sigma\bar{X} = 12{,}930$ and $\Sigma R = 1{,}230$. Compute the values of the 3-sigma limits for the $\overline{X}$ and R charts, and estimate the value of σ on the assumption that the process is in statistical control.

Answer: 472.8, 389.2; 105.4, 0; 24.2.

2.4. In Prob. 2.3 the specified minimum strength for a weld is 370 lb. If the process is in statistical control, what can you say about its capability to meet this specification and about its current performance? Use Fig. 2.4 through 2.6 in forming your answer.

2.5. Example 1.1 discussed the use of experimental control charts for process control for measurements of pitch diameter of threads on certain aircraft fittings. Table 1.1 shows the data for 20 subgroups of 5 measurements that were used to prepare the control charts of Fig. 1.2. Use the data to verify the location of the central lines and control limits for the $\overline{X}$ and R charts.

2.6. In Prob. 2.5 the specified nominal dimension and tolerances for the pitch diameter were 0.4037 ± 0.0013 in. The process showed definite lack of control with several points beyond the control limits on both the $\overline{X}$ and R charts. Thus it is imprudent to make any definitive statements about the ability of the process to meet specifications. Nevertheless one might want to speculate as to the potential capability if the process is brought under control. Of the six figures, 2.1 through 2.6, which in your opinion most nearly fits the situation as described in Example 1.1? Discuss in general terms what might be expected from this process once it is brought under control.

2.7. Example 1.2 discussed a critical measurement in the production of a rheostat knob. Table 1.2 shows the data for 27 subgroups of 5 measurements that were used to prepare the control charts of Fig. 1.3*b* and *c*. Use the data to verify the location of the central lines and control limits for the $\overline{X}$ and R charts.

2.8. In Prob. 2.7 the specified nominal dimension and tolerances for the measurement were given as 0.140 ± 0.003 in. While the process indicated statistical control, a substantial proportion of knobs failed to meet specifications on this dimension. Of the six figures, 2.1 through 2.6, which in your opinion most nearly fits the situation described in Example 1.2? Discuss in general terms what options might be available for both the consumer and producer.

2.9. Control charts for $\overline{X}$ and R are maintained on a critical dimension in a certain manufacturing process. The subgroup size is 5. After 25 subgroups, $\Sigma\overline{X} = 562.5$ mm and $\Sigma R = 90.0$ mm. Compute the values of the central lines and 3-sigma control limits for this process. Estimate the value of σ on the assumption that the process was operating in a state of statistical control.

Answer: 20.41, 22.5, 24.59; 0, 3.60, 7.60; 1.55.

2.10. The diameter of one end of a gyro drive shaft is subject to statistical control using $\overline{X}$ and R charts. After 30 subgroups of 5 shafts each have been examined, $\Sigma\overline{X} = 34{,}290$ and $\Sigma R = 330$. Calculate control limits for the $\overline{X}$ and R charts. Estimate the mean μ and standard deviation σ of the process assuming that it is in statistical control. What are the natural tolerances of this process?

Answer: 1136.6, 1149.4, 0, 23.2; 1143.0, 4.73; 1128.8, 1157.2.

2.11. Control charts for $\overline{X}$ and R are maintained on the braking strength in pounds in a certain destructive test of a particular type of ceramic insulator used in vacuum tubes. The subgroup size is 4. After 30 subgroups, $\Sigma\overline{X} = 1950.0$ and $\Sigma R = 123.5$. Compute the values of the central lines and 3-sigma control limits for $\overline{X}$ and R

charts, and estimate the value of σ on the assumption that the process is in statistical control.

2.12. $\overline{X}$ and R control charts are to be run to establish control over the manufacture of a drive shaft. After 30 subgroups have been drawn and inspected, and the diameters recorded, $\Sigma\overline{X} = 22.5150$ in and $\Sigma R = 0.1410$ in. The subgroup size is 5. Calculate the values of the central lines and control limits for the $\overline{X}$ and R charts, and estimate the standard deviation σ of the process assuming it is in statistical control.

2.13. Calculate the central lines and control limits for $\overline{X}$ and R for the data of Prob. 3.10. Test the data for control by checking for any values of $\overline{X}$ or R that fall beyond these limits.

2.14. Calculate the central lines and control limits for $\overline{X}$ and R for the data of Prob. 3.19. Test the data for control by checking for any values of $\overline{X}$ or R that fall beyond these limits.

3

WHY THE CONTROL CHART WORKS; SOME STATISTICAL CONCEPTS

There is no such thing as constancy in real life. There is, however, such a thing as a constant-cause system. The results produced by a constant-cause system vary, and in fact may vary over a wide band or a narrow band. They vary, but they exhibit an important feature called stability. Why apply the terms constant and stability to a cause system that produces results that vary? Because the same percentage of these varying results continues to fall between any given pair of limits hour after hour, day after day, so long as the constant-cause system continues to operate. It is the distribution of results that is constant or stable. When a manufacturing process behaves like a constant-cause system, producing inspection results that exhibit stability, it is said to be in statistical control. The control chart will tell you whether your process is in statistical control.

—*W. E. Deming*[†]

3.1 THE NEED FOR UNDERSTANDING STATISTICAL PRINCIPLES

This chapter and the one that follows deal with the principles behind control charts for variables. Although much good work in statistical quality control has been done by individuals who had only a vague notion of the principles behind the control chart, experience shows that the lack of a clear understanding of these principles may lead to certain costly mistakes that otherwise could be avoided. How-

[†]W. E. Deming, "Some Principles of the Shewhart Methods of Quality Control," *Mechanical Engineering,* vol. 66, pp. 173–177, March 1944.

ever, few of the many engineers, production personnel, and inspection supervisors who will make use of statistical quality control techniques will be called upon to advance the frontiers of theoretical knowledge. The theory developed in this and subsequent chapters, therefore, will emphasize an *intuitive* approach rather than a purely *mathematical* approach. Much of the theory presented uses ordinary arithmetic with the occasional assistance of algebra. The reader should be aware, however, that the mathematical proofs that back up these intuitive explanations are frequently highly complex and sophisticated.

The methods, definitions, and examples that follow are intended to portray the use of statistical theory in quality control work and to stimulate the reader's imagination by suggesting parallel opportunities in the reader's own industry.

3.2 DESCRIPTION OF PATTERNS OF VARIATION

As already stated in Chap. 1, variation seems inevitable in nature. Manufacturing processes are no exception to this. Whether one is attempting to control a dimension of a part that is to go into a precision assembly, the resistance of a relay, the acidity of a solution used for dyeing textiles, the weight of the contents of a container, or any other quality of a product or service, it is certain that the quality will vary.

It follows that it is necessary to have some simple methods of describing patterns of variation. Statisticians have developed such methods. One useful method involves a *frequency distribution.* Another involves the finding of a measure of the central tendency of a distribution (that is, an *average*) combined with some measure of the *dispersion,* or spread, of the distribution.

This chapter explains and illustrates these methods of describing patterns of variation. It explains that there may be both stable and unstable patterns of variation; it points out that the practical conclusions to be drawn from frequency distributions, averages, and measures of dispersion depend on the stability of the pattern of variation. The control chart, which is a test for this stability, is illustrated.

3.2.1 Counting the Frequencies of Different Observations

Table 3.1 gives the results of measurements of drained weights[†] in the canning of tomatoes. Several times during a shift in this particular cannery, a company inspector took from the production line five cans that had been filled and sealed. The inspector opened each can, emptied out and weighed the solid contents, and

[†]These are drained weights immediately after filling the cans and before processing them. From a control viewpoint, it is advantageous to determine drained weight at this time to provide an immediate basis for action whenever the measurements disclose the existence of trouble. Actual specifications for finished product ordinarily relate to drained weights after processing. A necessary part of the control procedure that is not discussed here is a determination of the relationship between drained weights before and after processing.

TABLE 3.1 DRAINED WEIGHT AFTER FILLING OF CONTENTS OF SIZE NO. $2\frac{1}{2}$ CANS OF STANDARD GRADE TOMATOES IN PURÉE

Weight given in ounces

Subgroup no.	Date	Hour	Measurement on each can of five cans per subgroup					Average $\bar{X}$	Range R
1	Sept. 21	9:30	22.0	22.5	22.5	24.0	23.5	22.9	2.0
2		10:50	20.5	22.5	22.5	23.0	21.5	22.0	2.5
3		11:45	20.0	20.5	23.0	22.0	21.5	21.4	3.0
4		2:30	21.0	22.0	22.0	23.0	22.0	22.0	2.0
5		5:25	22.5	19.5	22.5	22.0	21.0	21.5	3.0
6	Sept. 22	10:00	23.0	23.5	21.0	22.0	20.0	21.9	3.5
7		1:15	19.0	20.0	22.0	20.5	22.5	20.8	3.5
8		5:00	21.5	20.5	19.0	19.5	19.5	20.0	2.5
9	Sept. 23	9:30	21.0	22.5	20.0	22.0	22.0	21.5	2.5
10		1:15	21.5	23.0	22.0	23.0	18.5	21.6	4.5
11		1:45	20.0	19.5	21.0	20.0	20.5	20.2	1.5
12		3:30	19.0	21.0	21.0	21.0	20.5	20.5	2.0
13	Sept. 25	8:00	19.5	20.5	21.0	20.5	21.0	20.5	1.5
14		10:25	20.0	21.5	24.0	23.0	20.0	21.7	4.0
15		11:30	22.5	19.5	21.0	21.5	21.0	21.1	3.0
16		2:30	21.5	20.5	22.0	21.5	23.5	21.8	3.0
17		3:15	19.0	21.5	23.0	21.0	23.5	21.6	4.5
18		5:30	21.0	20.5	19.5	22.0	21.0	20.8	2.5
19	Sept. 26	2:00	20.0	23.5	24.0	20.5	21.5	21.9	4.0
20		3:00	22.0	20.5	21.0	22.5	20.0	21.2	2.5
21		4:45	19.0	20.5	21.0	20.5	22.5	20.7	3.5
22	Sept. 27	7:30	21.5	25.0	21.0	19.0	21.0	21.5	6.0
23		8:35	22.5	22.0	23.0	22.0	23.5	22.6	1.5
24		10:40	22.5	22.0	22.0	19.5	20.5	21.3	3.0
25		1:45	18.5	22.0	22.5	21.0	21.5	21.1	4.0
26		3:30	21.5	20.5	20.5	16.5	21.5	20.1	5.0
27		4:00	24.0	22.0	17.5	21.0	22.5	21.4	6.5
28		4:40	19.5	22.5	15.5	20.0	22.5	20.0	7.0
29	Sept. 28	7:15	22.0	17.5	21.0	22.0	23.5	21.2	6.0
30		7:45	22.0	20.0	20.5	24.0	21.5	21.6	4.0
31		10:00	22.5	21.0	19.5	21.5	22.5	21.4	3.0
32		1:15	20.0	22.0	20.0	21.5	20.0	20.7	2.0
33		3:30	21.0	19.5	22.0	20.0	20.0	20.5	2.5
34	Sept. 29	9:00	22.5	21.5	21.0	21.5	23.5	22.0	2.5
35		10:50	22.0	21.0	21.0	20.5	21.0	21.1	1.5
36		1:15	25.0	20.0	20.0	20.5	22.5	21.6	5.0
37		2:30	20.5	21.0	21.0	19.0	21.0	20.5	2.0
38		4:10	21.5	22.0	22.0	20.0	21.0	21.3	2.0
39		5:20	21.5	22.0	21.5	20.5	22.5	21.6	2.0
40	Sept. 30	9:30	22.5	24.5	25.5	20.0	21.0	22.7	5.5

(continued)

TABLE 3.1 *(continued)*

Subgroup no.	Date	Hour	Measurement on each can of five cans per subgroup					Average $\bar{X}$	Range R
41		11:15	21.5	24.0	21.5	21.5	22.5	22.2	2.5
42		2:10	23.0	23.5	21.0	21.5	21.5	22.1	2.5
43		3:30	22.5	19.5	21.5	20.5	20.0	20.8	3.0
44	Oct. 2	8:20	23.5	23.0	24.5	21.5	20.5	22.6	4.0
45		2:30	21.0	21.0	24.5	23.0	22.5	22.4	3.5
46		3:30	24.5	21.5	21.5	22.5	22.5	22.5	3.0
47		5:00	24.0	21.0	24.0	22.0	20.5	22.3	3.5
48	Oct. 3	9:15	23.5	22.5	20.0	20.0	21.0	21.4	3.5
49		10:00	22.0	20.5	21.0	22.5	23.0	21.8	2.5
50		1:00	22.0	23.5	24.0	22.0	22.0	22.7	2.0
51		3:00	23.5	21.0	23.5	21.5	23.0	22.5	2.5
52		4:30	24.5	21.5	21.0	24.5	22.5	22.8	3.5

recorded the results to the nearest half ounce. Table 3.1 shows such measurements covering 11 days of operation on the day shift, a total of 260 measurements.

One way to organize such figures to show their pattern of variation is to count the number of times each value occurs. This may be done conveniently on a check sheet such as Fig. 3.1. The results of such a count are called a *frequency distribution.*

3.2.2 Definitions Relative to Frequency Distributions†

- *A grouped frequency distribution* of a set of observations is an arrangement that shows the frequency of occurrence of the values of the variable in ordered classes.
- The interval, along the scale of measurement, of each ordered class is termed a *cell.*
- The *frequency* for any cell is the number of observations in that cell.
- The *relative frequency* for any cell is the frequency for that cell divided by the total number of observations.

3.2.3 Cells and Cell Boundaries

In Table 3.1, each measured weight was recorded to the nearest half ounce, as no greater precision of measurement was required. However, the weight was actually

†Taken by permission from "Manual on Quality Control of Materials," American Society for Testing and Materials, Philadelphia, Pa. The latest version is "ASTM Manual on Presentation of Data and Control Chart Analysis," STP 15D, 1976. This ASTM manual contained an excellent detailed exposition of matters treated briefly in this chapter.

what is called a *continuous variable.* For example, the contents of a given can recorded by this weighing process as weighing 21.0 oz does not necessarily weigh exactly 21 oz. If more precise methods of measurement were used, it might be found to weigh 20.897 or 21.204 oz or any other value that is nearer to 21.0 than to 20.5 or 21.5. The statement in the frequency distribution that 41 cans had a drained weight of 21.0 oz really means that the weight in each of the 41 cans was somewhere between 20.75 and 21.25 oz. The figure of 21.0 oz is the *midpoint* of a cell whose boundaries are 20.75 and 21.25.

In grouping data into a frequency distribution, the questions always arise as to how many cells there should be, and where the cell boundaries should be placed. A rough working rule used by statisticians is to aim to have about 20 cells. This is often subject to exceptions dictated by other considerations. For example, if there are only 100 measurements rather than 260, as in Table 3.1, about 10 cells would be a more appropriate figure. The distribution of Fig. 3.1 has 21 cells. Some writers suggest that the number of cells approximate the square root of the number of observations.

Cell boundaries should be chosen halfway between two possible observations. Cell intervals should be equal. In grouping the data of Table 3.1, the smallest possible cell width is 0.5 oz as measurements were not recorded to any smaller unit.

FIGURE 3.1 Check sheet to determine frequency distribution from data of Table 3.1.

25.5	/
25.0	//
24.5	𝍸 /
24.0	𝍸 ////
23.5	𝍸 𝍸 ////
23.0	𝍸 𝍸 ////
22.5	𝍸 𝍸 𝍸 𝍸 𝍸 𝍸 /
22.0	𝍸 𝍸 𝍸 𝍸 𝍸 𝍸 ////
21.5	𝍸 𝍸 𝍸 𝍸 𝍸 𝍸 ///
21.0	𝍸 𝍸 𝍸 𝍸 𝍸 𝍸 𝍸 𝍸 /
20.5	𝍸 𝍸 𝍸 𝍸 𝍸 /
20.0	𝍸 𝍸 𝍸 𝍸 ////
19.5	𝍸 𝍸 //
19.0	𝍸 //
18.5	//
18.0	
17.5	//
17.0	
16.5	/
16.0	
15.5	/

However, if measurements had been made to the nearest 0.1 oz, it might still have been advantageous to use a cell width of 0.5 oz.

Table 3.2 illustrates an appropriate way to present a frequency distribution showing cell midpoints and cell boundaries. It illustrates also the effect of a coarser grouping, using the cell width as 1.0 oz instead of 0.5 oz. With measurements made to the nearest 0.5 oz, no cell width between 0.5 and 1.0 is possible. Thus figures must be grouped into either 21 cells or 11 cells.

3.3 GRAPHIC REPRESENTATION OF A FREQUENCY DISTRIBUTION

Three common ways of graphic presentation of frequency distributions are shown in Fig. 3.2. Of these, the *frequency histogram* of Fig. 3.2*a* is in some respects the best. In this graph the sides of the columns represent the upper and lower cell boundaries, and their heights (and areas) are proportional to the frequencies within the cells. Figure 3.2*b*, the *frequency bar chart,* uses bars centered on the midpoints of the cells; the heights of the bars are proportional to the frequencies in the respective cells. Figure 3.2*c*, the *frequency polygon,* consists of a series of straight lines joining small circles that are plotted at cell midpoints with a height proportional to cell frequencies.

3.3.1 Cumulative Frequency Distributions

It is sometimes advantageous to tabulate the frequencies of values less than or greater than the respective cell boundaries. Table 3.3 illustrates this. It shows both the number of cans and the percentage of cans having less than a given weight.

Such a *cumulative frequency distribution* may be presented graphically in the manner shown in Fig. 3.3*a*. This type of graph is called an *ogive* because of its similarity to the ogee curve of the architect and the dam designer.

The plotting of relative frequency on a probability scale,[†] as in Fig. 3.3*b,* tends to smooth the ogive to something closer to a straight line. It also concentrates somewhat more attention on the extreme variations.

3.3.2 Frequency Distributions Often Supply a Basis for Action

A frequency distribution relative to any quality of a manufactured product supplies a useful picture of the way in which that quality has varied in the past. Through-

[†]Normal probability paper is the invention of an engineer, the late Allen Hazen. It can be purchased from most manufacturers of graph paper. The probability scale is so designed that the ogive of a normal curve (explained later in this chapter) will plot on it as a straight line.

TABLE 3.2 EXAMPLES OF GROUPED FREQUENCY DISTRIBUTIONS SHOWING CELL MIDPOINTS AND CELL BOUNDARIES

Data of Table 3.1 on drained weights of contents of size No. $2\frac{1}{2}$ cans of tomatoes

a. Finer grouping			*b.* Coarser grouping		
Cell midpoints	Cell boundaries	Observed frequency	Cell midpoints	Cell boundaries	Observed frequency
	25.75	—		25.75	—
25.5		1			
	25.25	—	25.25		3
25.0		2			
	24.75	—		24.75	—
24.5		6			
	24.25	—	24.25		15
24.0		9			
	23.75	—		23.75	—
23.5		14			
	23.25	—	23.25		28
23.0		14			
	22.75	—		22.75	—
22.5		31			
	22.25	—	22.25		65
22.0		34			
	21.75	—		21.75	—
21.5		33			
	21.25	—	21.25		74
21.0		41			
	20.75	—		20.75	—
20.5		26			
	20.25	—	20.25		50
20.0		24			
	19.75	—		19.75	—
19.5		12			
	19.25	—	19.25		19
19.0		7			
	18.75	—		18.75	—
18.5		2			
	18.25	—	18.25		2
18.0		0			
	17.75	—		17.75	—
17.5		2			
	17.25	—	17.25		2
17.0		0			
	16.75	—		16.75	—
16.5		1			
	16.25	—	16.25		1
16.0		0			
	15.75	—		15.75	—
15.5		1			
	15.25		15.25		1
				14.75	
Total	.....	260	Total	.....	260

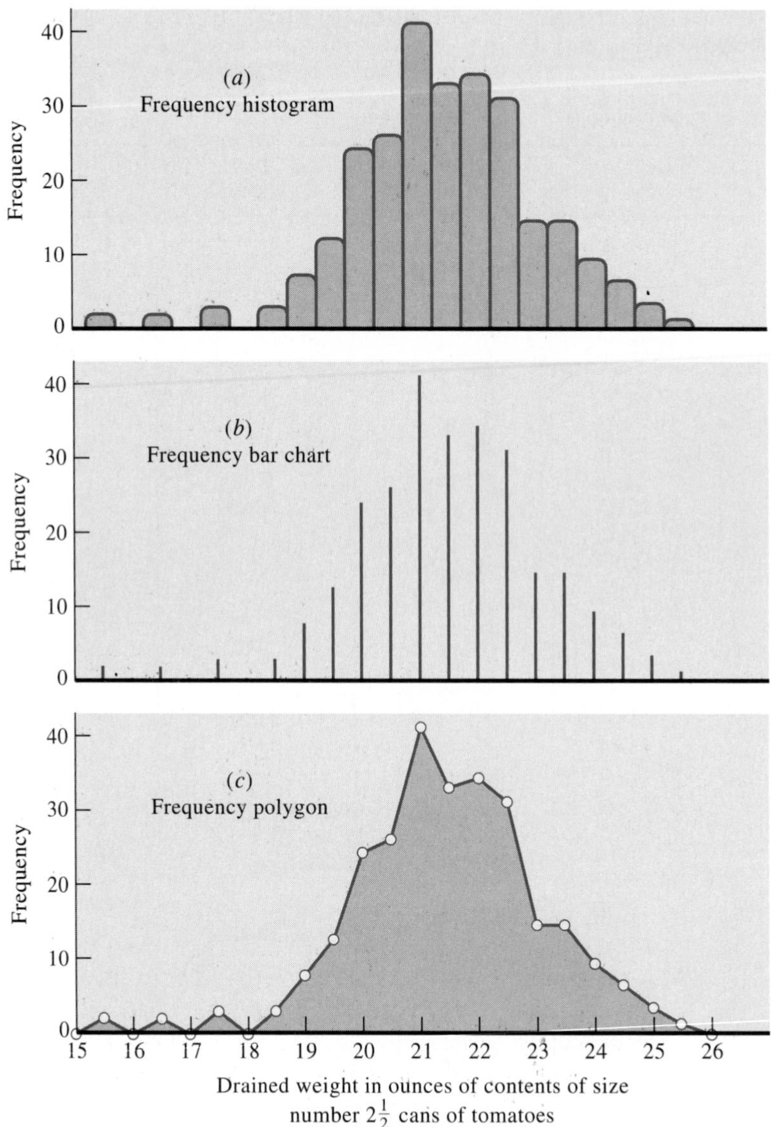

FIGURE 3.2 Three methods of graphical presentation of a frequency distribution—data of Table 3.2.

out this book, there are many illustrations of actions that may be taken more intelligently with such a picture than they are likely to be taken without it. Some of these relate to specification of the quality characteristic and its tolerance limits; others, to action on the process; still others, to the planning of inspection and acceptance procedures.

TABLE 3.3 EXAMPLES OF CUMULATIVE FREQUENCY DISTRIBUTIONS
Data of Table 3.1 on drained weights of contents of size No. $2\frac{1}{2}$ cans of tomatoes

Weight, oz	Number of cans having less than given weight	Percentage of cans having less than given weight
25.75	260	100.0
25.25	259	99.6
24.75	257	98.8
24.25	251	96.5
23.75	242	92.3
23.25	228	87.7
22.75	214	82.3
22.25	183	70.4
21.75	149	57.3
21.25	116	44.6
20.75	75	28.8
20.25	49	18.8
19.75	25	9.6
19.25	13	5.0
18.75	6	2.3
18.25	4	1.5
17.75	4	1.5
17.25	2	0.8
16.75	2	0.8
16.25	1	0.4
15.75	1	0.4
15.25	0	0.0

Example 3.1

A use of a frequency distribution to accomplish a cost saving

Facts of the case In 100% inspection of completed electronic devices, a certain critical electrical characteristic was measured on a meter. This meter had a dial gage on which the value of the electrical characteristic was indicated by a pointer. Two fixed red lines were set on the dial, one at the minimum and the other at the maximum specified value. This permitted the inspector to tell at a glance whether or not each device met the specifications, so that the inspector could rapidly sort the nonconforming product from the acceptable product.

Because this characteristic was responsible for a number of rejections and it was desired to study its pattern of variation, the inspector was asked to read the value registered on the dial gage for every device and to make a line on a frequency distribution check sheet indicating this value. One check sheet was made for each shift and sent immediately to the quality control engineer of the plant. These sheets generally looked something like Fig. 3.4.

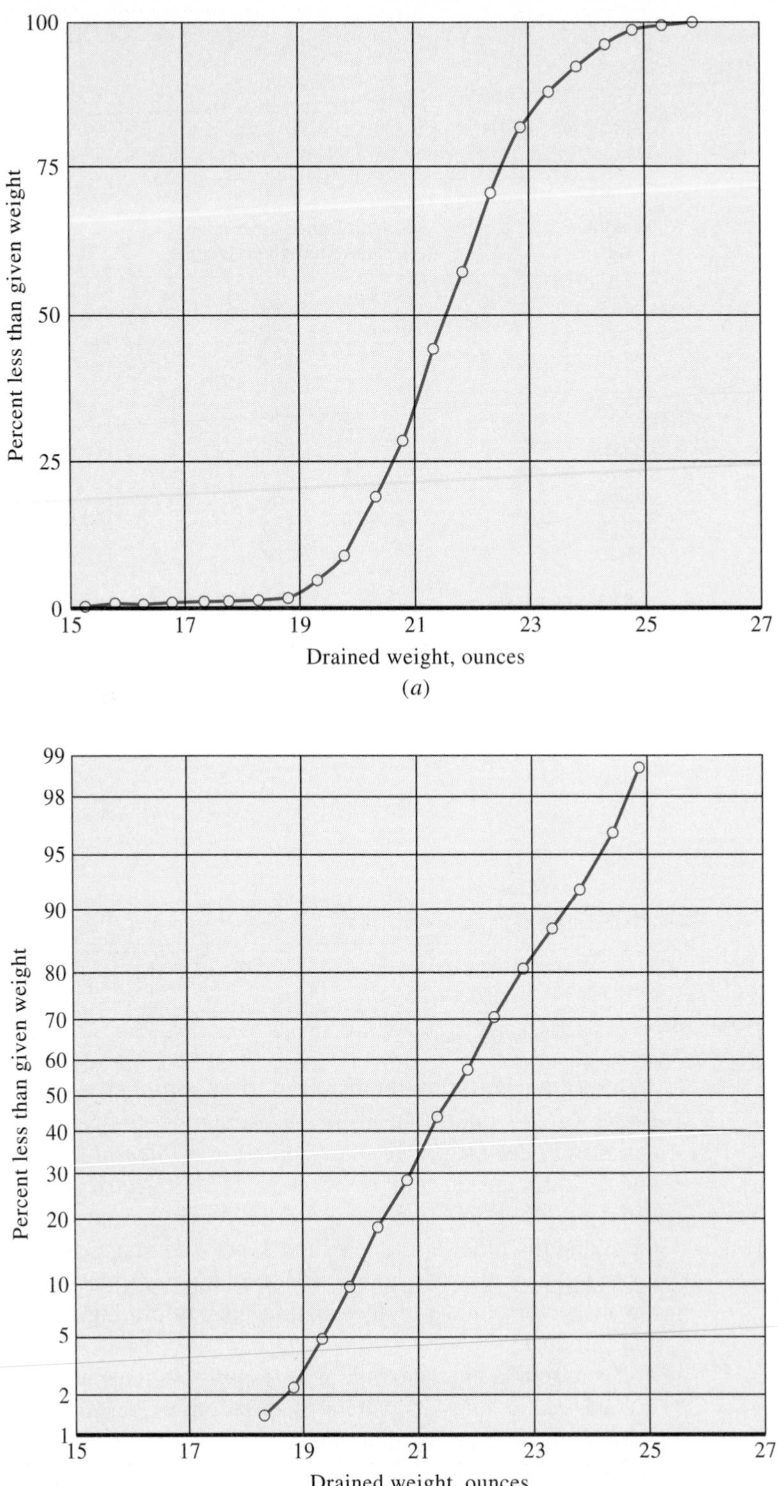

FIGURE 3.3 Graphical presentation of a cumulative frequency distribution—data of Table 3.3; *(a)* rectangular coordinate ruling; *(b)* probability ruling.

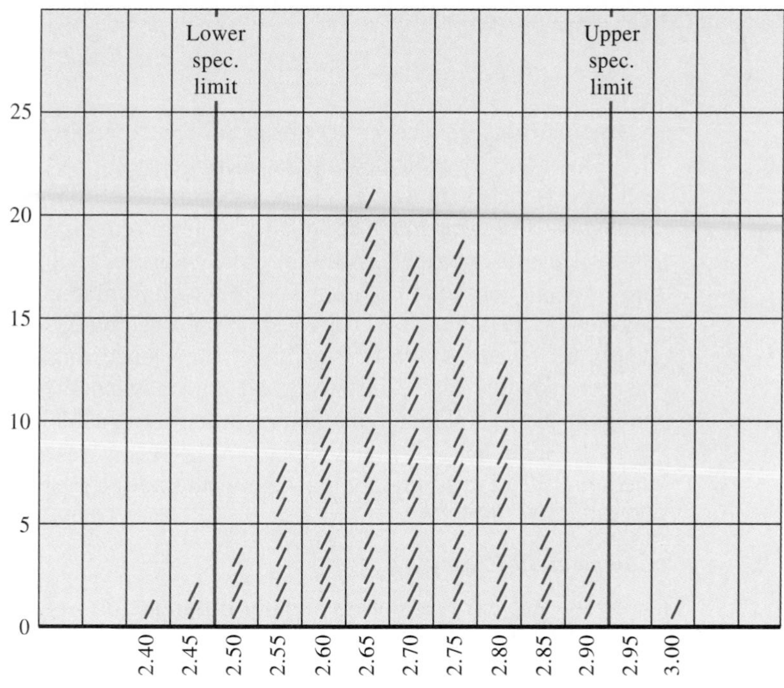

FIGURE 3.4 Typical check sheet showing frequency distribution of electrical characteristic of electronic device.

FIGURE 3.5 This check sheet called for investigation of the reasons for the unusual distribution pattern.

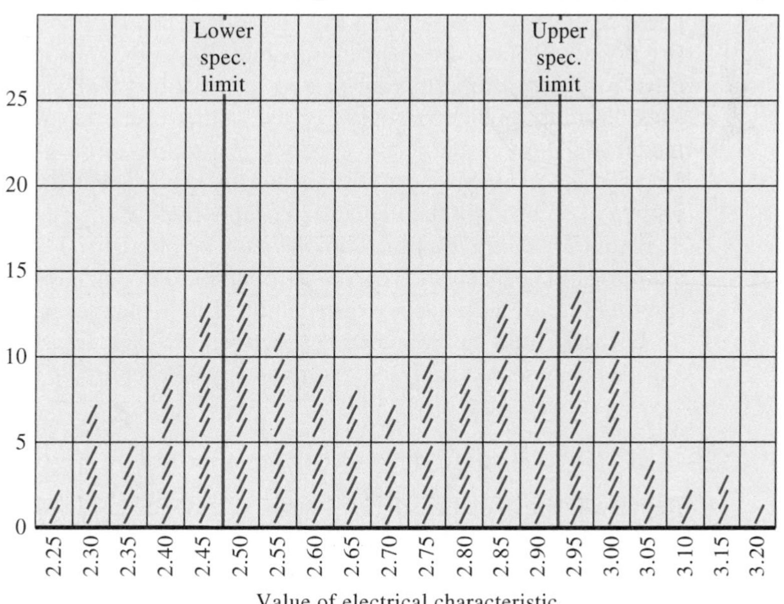

One morning the check sheet from the preceding night shift looked like Fig. 3.5. This not only indicated a great increase in the proportion of nonconforming units but also showed an unusual distribution pattern.

Analysis and action The quality control engineer, suspecting that this unusual distribution pattern might indicate an inspection error, went immediately to the inspection station and rescued from the scrap bin the devices rejected on the night shift. They were all retested by the day shift inspector, and nearly all proved to be satisfactory. It developed that there had been a new inspector on the night shift who had not understood how to operate the test equipment. Several hundred dollars worth of acceptable product was thus saved from the scrap heap. The new night shift inspector was instructed in the correct method of testing so that the mistake would not be repeated.

Comment on Example 3.1

Figures 3.4 and 3.5 represent a slightly different form of check sheet from that shown in Fig. 3.1. They are arranged to give a graphic picture of the pattern of variation somewhat like that given by a *frequency histogram* (Fig. 3.2*a*).

In this case the inspector made a record of the frequency distribution with a minimum of clerical labor. The inspector did not actually write down any figures for the measured value of the electrical characteristic but merely made a single line in the appropriate cell on the check sheet each time a measurement was made.

It often happens that a person with some understanding of statistics and with a thorough technical knowledge of a process can make a quick and accurate guess at a reason for trouble from looking at the behavior pattern shown on either a frequency distribution or a control chart. The quality control engineer made such a guess in this case, recognizing Fig. 3.5 as a *bimodal distribution,* that is, one with two cells with high frequencies separated by cells with lower frequencies. This often results from the mixing of two distributions with different modes or averages. From technical knowledge of the testing of electronic devices, the engineer recognized how a particular type of inspection error would have this effect. Guesses made in this way are not always correct, but they have a much better chance of being right than hunches unsupported by any statistical evidence.

Frequency distributions of samples may be plotted and examined to guide decisions regarding acceptance of lots of product from which the samples are taken. An informal use of a frequency distribution in this manner is described in Example 15.1. The Shainin lot plot method, described in Chap. 15, is a formal procedure for acceptance based on analysis of a frequency distribution of a sample of 50 articles.

3.4 AVERAGES AND MEASURES OF DISPERSION

Representation of sample data by means of a frequency distribution is always bulky, frequently time-consuming, and sometimes misleading. One way to simplify the information contained in a frequency distribution is to use two or more numbers that describe its characteristics. Two essential numbers, or *statistics,* are those that measure the *central tendency* of the data and measure their spread or *dispersion.* Such statistics *summarize* the information contained in the frequency distribution and therefore may suppress some important information contained in it.

The most commonly used measures of central tendency are the *median,* the *mode,* and the *mean* or, more correctly, the *arithmetic mean.* In popular language, the word *average* usually implies the technical arithmetic mean. Such usage is also common in the literature of quality control. In this book the word *average* means *arithmetic mean* unless otherwise qualified.

The arithmetic mean of a set of n numbers is the sum of the numbers divided by n. Expressed algebraically,

$$\overline{X} = \frac{X_1 + X_2 + X_3 + \cdots + X_n}{n} = \frac{1}{n} \sum_{j=1}^{n} X_j \tag{3.1}$$

where the symbol $\overline{X}$ represents the arithmetic mean; X_1, X_2, and so forth, represent specific numbers; and the Greek capital letter Σ (sigma) is the mathematical symbol for the taking of a sum.

The other important measures of central tendency mentioned previously have found relatively less use in quality control. Control charts based on the median, which is the magnitude of the middle case when the numbers that make up the sample are arranged by increasing order of magnitude, are discussed in Chap. 10. The mode is the magnitude of the case that occurs most frequently, the value corresponding to the high point on a frequency histogram.

Two measures of dispersion that are extremely useful in statistical quality control are the *range* R† and the *sample standard deviation* s. Use of the range as a measure of dispersion, and its importance in control charts for variables, was introduced in Chap. 2. The range is almost never used for large subgroups, that is, for subgroups of more than 10 or 15 items. Expressed algebraically,

$$R = X_{max} - X_{min} \tag{3.2}$$

where X_{max} is the largest number and X_{min} is the smallest number in the set.

The sample standard deviation of a set of numbers from their arithmetic mean is designated in this book by s. Expressed algebraically,

$$\begin{aligned} s &= \sqrt{\frac{(X_1 - \overline{X})^2 + (X_2 - \overline{X})^2 + \cdots + (X_n - \overline{X})^2}{n-1}} \\ &= \sqrt{\frac{\Sigma(X_j - \overline{X})^2}{n-1}} \end{aligned} \tag{3.3}$$

†R is the symbol commonly used in industry in the United States. An alternative symbol for range is w, used in certain British literature on quality control.

While the sample standard deviation is less frequently used as a measure of dispersion of subgroups in control charts for variables, it is often used in measuring the dispersion of a frequency distribution such as might be pictured in a histogram. This is true particularly when the data have not been recorded in subgroups in order of production. The mathematical formula for s is applicable whether we are talking about a subgroup drawn for control-chart purposes or any other group of data.

Still another measure of dispersion is the *sample variance.* This measure is the sum of the squares of the deviations from the arithmetic mean divided by $n - 1$. In other words, the sample standard deviation is the square root of the sample variance.[†] In many of the derivations in mathematical statistics that deal with the relationship between samples and the universes from which samples are drawn, there are mathematical advantages in the use of the variance as the measure of dispersion.

A possible source of confusion to persons who read two or more books that deal with principles or applications of statistics may be the different names, symbols, and formulas that writers use in connection with measures of mean-square deviation. This somewhat troublesome topic involves the relationship between the dispersion of samples of a finite size and the dispersion of the "universe" from which, in concept, samples are assumed to have been drawn. At this point in our discussion, we have not yet dealt with sampling theory and are merely defining the sample standard deviation of a set of numbers as given in Eq. (3.3).

3.4.1 Calculation of a Subgroup Sample Average, Standard Deviation, and Range

Computation of the statistics for a small number of observations may be illustrated by reference to the five weights on the first line of Table 3.1 (subgroup no. 1).

Using Equation (3.1) to find the sample mean,

$$\overline{X} = \frac{1}{n} \Sigma X_j$$

$$= \tfrac{1}{5}(22.0 + 22.5 + 22.5 + 24.0 + 23.5) = 22.9$$

Using Equation (3.2) to find the range of the sample,

$$R = (X_{\text{max}} - X_{\text{min}}) = (24.0 - 22.0) = 2.0$$

[†]In the technical language of statistics, the sample standard deviation is the square root of an unbiased estimator of the variance. It will be seen in Chap. 4 that s is not an unbiased estimator of the universe standard deviation σ.

Using Equation (3.3) to find the sample standard deviation,

$$s = \sqrt{\frac{(22.0 - 22.9)^2 + (22.5 - 22.9)^2 + (22.5 - 22.9)^2 + (24.0 - 22.9)^2 + (23.5 - 22.9)^2}{4}}$$

$$= \sqrt{\frac{0.81 + 0.16 + 0.16 + 1.21 + 0.36}{4}}$$

$$= \sqrt{0.675} = 0.822$$

Note the simplicity in the calculation of R in comparison with that of s.

A useful alternative form for Equation (3.3) is

$$s = \sqrt{\frac{X_1^2 + X_2^2 + X_3^2 + \cdots + X_n^2 - n\bar{X}^2}{n - 1}}$$

$$= \sqrt{\frac{\Sigma X_j^2 - n\bar{X}^2}{n - 1}} \qquad (3.4)$$

The example calculation using this form for s is as follows:

$$s = \sqrt{\frac{(22.0)^2 + (22.5)^2 + (22.5)^2 + (24.0)^2 + (23.5)^2 - (5)(22.9)^2}{4}}$$

$$= \sqrt{\frac{2{,}624.75 - 2{,}622.05}{4}} = \sqrt{\frac{2.70}{4}}$$

$$= 0.822$$

This form of the equation may prove more convenient, especially where a calculator or computer is available. It should be noted, however, that more significant figures must be carried along during the calculation. In this example, the fifth and sixth significant figures were the important ones. Many small desktop and pocket calculators provide for the automatic calculation of $\bar{X}$ and s based on the entry of a set of numbers.

3.4.2 Shifting the Origin to Simplify the Arithmetic

One possible trick useful in simplifying the arithmetic and also in reducing the number of significant figures that must be carried throughout the calculations is to shift the measuring scale to a new zero point. For example, a new origin of 20.0 oz could have been chosen for the variable in the preceding calculations. The new variable of interest would be the excess weight above 20 oz. Thus the calculations of $\overline{X}$ and s would have proceeded as follows:

$$\overline{X}\,(\textit{above the origin 20.0}) = \frac{2.0 + 2.5 + 2.5 + 4.0 + 3.5}{5} = 2.9$$

$$\overline{X}\,(\textit{above the origin 0}) = 2.9 + 20.0 = 22.9$$

Had an assumed origin of 23.0 been used,

$$\overline{X}\,(\textit{above the origin 23.0}) = \frac{-1.0 - 0.5 - 0.5 + 1.0 + 0.5}{5} = -0.1$$

$$\overline{X}\,(\textit{above the origin 0}) = -\,0.1 + 23.0 = 22.9$$

Likewise, the addition or subtraction of a constant from all of any set of numbers does not change s. Assuming an origin of 20.0, the calculation of s becomes

$$s = \sqrt{\frac{2^2 + 2.5^2 + 2.5^2 + 4^2 + 3.5^2 - (5)(2.9)^2}{4}}$$

$$= \sqrt{0.675} = 0.822$$

The reader will note a considerable saving in the number of significant figures required in this calculation compared with previous calculations of s.

3.4.3 Calculation of Sample Average and Standard Deviation of Grouped Frequency Distributions—Long Method

In Table 3.2*a* the data of Table 3.1 were rearranged in a grouped frequency distribution based on dividing the entire set of data into 21 equally spaced cells of 0.5 oz. These results were then used to prepare, among other things, the frequency histogram in Fig. 3.2*a*. In so doing, the subgrouping order of production was completely lost for the 52 subgroups taken over the 11-day period. Whatever pattern of variability results from analyzing the grouped frequency distribution will include any changes in μ and σ that occurred during this period. Thus, if the process lacks evidence of statistical control, estimates of μ will have no meaning beyond the articles inspected, and σ will *(a)* be too large to reflect the capability of the process and *(b)* not be useful to predict the future.

When the order of production is not known, the analysis of grouped frequency distributions can prove beneficial in many ways, as discussed previously and in many places throughout this book, including obtaining rough estimates of μ and σ.

Although Eqs. (3.1) and (3.3) might be applied to the actual measured weight in computing the average and standard deviation of the 260 observed weights of Table 3.1, the computation would be time-consuming, and the chances of an error in calculations would be great. To get the average, one would add the 260 individual values:

$$\overline{X} = \frac{\Sigma X_j}{n} = \frac{5{,}589.5}{260} = 21.498 \text{ oz}$$

The first step in calculating the standard deviation would be the squaring of each of the 260 numbers. This long calculation is not illustrated here.

The same results could be obtained with less effort by considering the numbers grouped into a frequency distribution. This is illustrated in Table 3.4.

TABLE 3.4 COMPUTATION OF SAMPLE AVERAGE AND STANDARD DEVIATION OF FREQUENCY DISTRIBUTION—LONG METHOD

Data of Tables 3.1 and 3.2*a* on drained weights of contents of size No. $2\frac{1}{2}$ cans of tomatoes

Weight, oz (midpoint of cell) X	Frequency f	Total weight fX	Squared weight X^2	Sum of squares for cell fX^2
25.5	1	25.5	650.25	650.25
25.0	2	50.0	625.00	1,250.00
24.5	6	147.0	600.25	3,601.50
24.0	9	216.0	576.00	5,184.00
23.5	14	329.0	552.25	7,731.50
23.0	14	322.0	529.00	7,406.00
22.5	31	697.5	506.25	15,693.75
22.0	34	748.0	484.00	16,456.00
21.5	33	709.5	462.25	15,254.25
21.0	41	861.0	441.00	18,081.00
20.5	26	533.0	420.25	10,926.50
20.0	24	480.0	400.00	9,600.00
19.5	12	234.0	380.25	4,563.00
19.0	7	133.0	361.00	2,527.00
18.5	2	37.0	342.25	684.50
18.0	0	0.0	324.00	0.00
17.5	2	35.0	306.25	612.50
17.0	0	0.0	289.00	0.00
16.5	1	16.5	272.25	272.25
16.0	0	0.0	256.00	0.00
15.5	1	15.5	240.25	240.25
Totals	260	5,589.5	......	120,734.25

$$\overline{X} = \frac{\Sigma fX}{n} = \frac{5{,}589.5}{260} = 21.498 \text{ oz}$$

$$s = \sqrt{\frac{\Sigma fX^2 - n\overline{X}^2}{n-1}} = \sqrt{\frac{120{,}734.25 - 120{,}162.64}{259}}$$

$$= \sqrt{2.207} = 1.485 \text{ oz}$$

The importance of carrying extra significant figures in the calculation of s by the use of Eq. (3.3) is again evident here. For any practical conclusions drawn from the average $\bar{X}$, there would be no need to carry it out to three decimal places as 21.498. However, for correct calculation of s these three decimal places are necessary. If $\bar{X}$ were rounded off to 21.50 for this calculation, the effect would be to make a 2% error in s, as s would then appear to be 1.45 instead of the correct figure of 1.485.

An inspection of Table 3.4 will show that 5,589.5 is necessarily the same sum that would be obtained if the 260 numbers were added one by one, and 120,734.25 is necessarily the same sum that would be obtained if each of the 260 numbers were separately squared and squares were then added. Consider the numbers in any cell of the frequency distribution. For example, consider the cell corresponding to a weight of 20.0 oz. The 24 observations of 20.0 oz, if added one by one, would contribute $24 \times 20.0 = 480$ to the sum of the numbers. As the square of 20 is 400, these 24 observations would contribute $24 \times 400 = 9{,}600$ to the sum of the squares.

Although less laborious than calculations from the same data without grouping, calculations of the type of Table 3.4 are still wasteful of time. Table 3.4 has been introduced at this point primarily to provide a transition to the common shortcut calculations for frequency distributions from the direct calculations that use the basic formulas defining average and standard deviation. These shortcut calculations are illustrated in Table 3.5. If the reader fully understands Table 3.4, no difficulty should be experienced in understanding Table 3.5.

3.4.4 Calculation of Sample Average and Standard Deviation of Grouped Frequency Distributions—Short Method

Just as the shift of origin simplified the arithmetic in the calculation from ungrouped data, so also may a shift of origin greatly simplify calculations of average and standard deviation of frequency distributions. Another change in the direction of simplification is a change of units; one cell is equal to one unit throughout the calculations. At the end of the calculation in cell units from the assumed origin, the results for average and standard deviation are converted into units of the original measurements by multiplying by the value of one cell unit. The average is then converted to its original origin by adding the value of the assumed zero point. This is illustrated in Table 3.5.

A further shortcut, which may be used when a rapid calculation is more important than accuracy, may be obtained by grouping the frequencies into larger cells. This could be accomplished by using the grouping of Table 3.2*b*, which has 1 oz rather than 0.5 oz as the cell width.

In all calculations from grouped data, the assumption is made that the true values of the variable are concentrated at the center of the cell in which they fall. In the coarse grouping of Table 3.2*b*, this assumption does not agree with the known facts. For example, in the cell with the midpoint 21.25 it is known that

TABLE 3.5 COMPUTATION OF SAMPLE AVERAGE AND STANDARD DEVIATION OF FREQUENCY DISTRIBUTION—SHORT METHOD
Data of Tables 3.1 and 3.2*a* on drained weights of contents of size No. $2\frac{1}{2}$ cans of tomatoes

Weight, oz (midpoint of cell) X	Frequency f	Deviation in cells from assumed origin d	fd	fd^2
25.5	1	9	9	81
25.0	2	8	16	128
24.5	6	7	42	294
24.0	9	6	54	324
23.5	14	5	70	350
23.0	14	4	56	224
22.5	31	3	93	279
22.0	34	2	68	136
21.5	33	1	33	33
21.0	41	0	0	0
20.5	26	−1	−26	26
20.0	24	−2	−48	96
19.5	12	−3	−36	108
19.0	7	−4	−28	112
18.5	2	−5	−10	50
18.0	0	−6	0	0
17.5	2	−7	−14	98
17.0	0	−8	0	0
16.5	1	−9	−9	81
16.0	0	−10	0	0
15.5	1	−11	−11	121
Totals	260	. . . . .	259	2,541

$$\bar{X} \text{ (in cells from assumed origin of 21.0)} = \frac{\Sigma fd}{n} = \frac{259}{260} = 0.996$$

$$\bar{X} \text{ (in original units from true origin)} = \text{assumed origin} + \frac{\Sigma fd}{n} \text{ (cell width)}$$

$$= 21.0 + 0.996(0.5) = 21.498 \text{ oz}$$

$$s \text{ (in cell units)} = \sqrt{\frac{\Sigma fd^2 - (\Sigma fd)^2/n}{n-1}} = \sqrt{\frac{2{,}541 - 258}{259}}$$

$$= \sqrt{8.815} = 2.969$$

$$s \text{ (in original units)} = (s \text{ in cell units}) \text{ (cell width)} = 2.97(0.5) = 1.485 \text{ oz}$$

there are 41 occurrences in the lower half of the cell and only 33 in the upper half. In most cases some error will be introduced whenever a coarser grouping is used to shorten computations. In general, it is desirable that the least count of the measuring instrument be such as to permit grouping the data into at least 20 cells.

3.5 SAMPLING STATISTICS AND UNIVERSE PARAMETERS

Many of the useful actions that may be taken in manufacturing are related to future production rather than completed lots of items. The knowledge that must be gained from the sampling procedure is knowledge of the pattern of variation of the production process from which the sample was drawn.

Statisticians use particular words and phrases to define the unknown pattern of variation from which the known sample has been drawn. It is frequently called the *universe, parent distribution,* or *population.* Some of the more recent texts in mathematical statistics refer to it as the *sample space* or *description space.* The true, but unknown, numerical values that describe the universe are called parameters.

In order to draw conclusions about an unknown universe, it is necessary to rely on numerical values derived from samples drawn from that universe. Such numerical values—which include the sample mean, median, RMS or standard deviation, range, variance, and so forth—summarize the information contained in the sample data. Each is referred to as a *statistic* of the sample and may be used to estimate the corresponding *parameter* of the unknown universe.

While statistics literature has adopted no uniform standard for notation, Greek letters have been widely used to identify the parameters of a distribution. Most often the lowercase Greek letter mu (μ) is used to identify the parametric value of the mean and the lowercase sigma (σ) to identify the parametric value of the standard deviation.

In the development of the techniques of statistical quality control in the Bell Telephone Laboratories in the 1920s and 1930s, the prime ($'$) symbol was adopted to identify a parameter of a universe. Thus the symbols $\bar{X}'$ (read "X bar prime") and σ' (read "sigma prime") identified the parametric values of the arithmetic mean and standard deviation, respectively, for a given universe. These symbols, and similar ones for attributes data, have been in use in the United States for many years. They—and other symbols, terms, and definitions—were adopted in the "ASTM Manual on the Presentation of Data" in 1937 and in 1941 in the American War Standards on Quality Control. The earliest standards in quality control, promulgated by the then American Standards Association (now American National Standards Institute), incorporated them as did the American Society for Quality Control.[†] Thus for over 50 years, as well as in the first five editions of this book, they were standard symbols in the literature of quality control.

The 1978 revision of ANSI/ASQC Z1.5 changed to the use of the Greek notation for the mean and standard deviation. However, the other standards have

[†]The principal ANSI/ASQC standards are ANSI Z1.1-1958, "Guide for Quality Control"; ANSI Z1.2-1958, "Control Chart Method of Analyzing Data"; ANSI Z1.3-1958, "Control Chart Method of Controlling Quality during Production" (ASQC designation for these standards is B-1, B-2, and B-3, respectively; all three were reaffirmed in 1985); and ANSI Z1.5-1978, "Definitions, Symbols, Formulas and Tables for Control Charts" (ASQC designation A-1). They are available from either the American National Standards Institute, New York, or the American Society for Quality Control, Milwaukee, Wis. The principal ASTM publication is "ASTM Manual on Presentation of Data and Quality Control Chart Analysis," STP 15D, American Society for Testing and Materials, Philadelphia, Pa., 1976.

retained the use of the prime notation. Certain other changes introduced by this standard and by the 1976 revision of the ASTM Manual are discussed in Chap. 4.

The parametric symbols may be applied to a known distribution, as in the bowl drawing experiments discussed later in this chapter, or when analyzing data from all items in a production lot. More commonly, parameters are estimated from one or more samples drawn from an unknown universe. Thus $\bar{X}$, the average of a sample of size *n*, may be used to estimate the parametric mean μ. Often a "caret" or "hat" (^) symbol is used above the Greek letter to differentiate more clearly the estimator from the parametric value. Thus an estimate of the parametric mean obtained from $\bar{X}$ will be designated by $\hat{\mu}$.

3.5.1 Using Samples to Make Estimates about a Universe

Assume that measurements are made on some variable and that it is proposed to base some actions on these measurements. In considering the merits of various proposals for action, it usually is helpful to think of the measurements as one or more samples taken from a universe (or perhaps from several universes). If the variable is some quality characteristic of a manufactured product (such as a dimension, weight, or electrical resistance), the universe may be thought of as the potentially unlimited output of the manufacturing process in question. Or, in some instances, the universe may be viewed as a particular lot of manufactured articles from which one or more samples have been taken.

It is appropriate to consider the following types of questions before making decisions about proposals for action:

1. Does it appear that we have only one universe? Or, more realistically, is it good enough for practical purposes to take action as if we had a single universe?
2. If the answer to the first question is "Yes," how much do we know about the universe? More specifically, what do we know about its centering and dispersion? What are our best estimates of its parameters μ and σ, and how satisfactory are these estimates as a basis for action?

3.5.2 Inferences about Pattern of Variation to Be Drawn from Sample Average and Standard Deviation

It clearly would be advantageous if substantially all the information contained in a frequency distribution could be packed into two figures, such as a measure of central tendency and a measure of the spread or dispersion. Under favorable circumstances it is possible to come close enough to this ideal for many purposes. A practical problem is to judge whether the circumstances are favorable.

Mathematical statisticians have discovered that the best measure of central tendency for this purpose is the *arithmetic mean* (referred to in this book as the *average*). A useful measure of dispersion for this purpose is the *standard deviation.*

The ideal would be to be able to say, given any average and standard deviation, just what proportion of the measurements fell within any specified limits. For example, if μ were 21.60 mm for some dimension of a manufactured part and σ were 0.05 mm, it would then be possible to tell what proportion of the parts fell between the limits $\mu \pm 2\sigma$, that is, between 21.50 and 21.70 mm. Or it would be possible to tell the proportion that fell between $\mu - 4\sigma$ and $\mu - 2\sigma$, that is, between 21.40 and 21.50 mm, or between any other desired limits. If this could be done, one could completely reconstruct a frequency distribution provided one knew only two figures, its average and its standard deviation.

For this to be possible, the equation of a frequency curve that described the pattern of variation of the particular quality characteristic would need to be fully defined once the average and standard deviation were known. These two numbers are, in fact, sufficient to define the so-called *normal curve.* Many observed frequency distributions of measured qualities of manufactured product, and many other frequency distributions found in nature, do correspond roughly to this normal curve.

3.6 THE NORMAL CURVE

Figure 3.2*a* showed a graphical representation of a frequency distribution by means of a frequency histogram. The area contained in each column of the histogram is proportional to the frequency within its cell. If there were enough observed numbers in the frequency distribution, the number of cells might be increased more and more and the width of a cell made smaller and smaller.[†] The series of steps that constitutes the top of the histogram would then approach a smooth curve. The height of the curve at any point would be proportional to the frequency at that point, and the area under it between any two limits would be proportional to the frequency of occurrence within those limits. Such a curve is called a *frequency curve* or *reference distribution.*

One can imagine frequency curves of many different shapes. Mathematicians have developed equations that exactly define many common types. The most useful of these curves is the *normal curve.* This curve is variously called the normal law, the normal curve of error, the probability curve, the gaussian curve, the laplacian curve, and the normal distribution curve. Its bell-shaped symmetrical form is illustrated in Fig. 3.9*a.* Although most of the area under it is included within the limits $\mu \pm 3\sigma$, the curve extends from $-\infty$ (minus infinity) to $+\infty$ (plus infinity). The curve is fully defined by μ and σ.

The most commonly quoted limits in connection with the normal curve are as follows:

Limits	Percent of total area within specified limits
$\mu \pm 0.6745\sigma$	50.00
$\mu \pm \sigma$	68.26
$\mu + 2\sigma$	95.45
$\mu \pm 3\sigma$	99.73

[†]See Fig. 3.6 for an example of a histogram with 61 cells.

This means that in those distributions that roughly approximate the normal curve, about two-thirds of the occurrences fall within one standard deviation on either side of the average, all but about 5% within two standard deviations, and practically all fall within three standard deviations.

Table *A*, App. 3, gives to four decimal places the proportion of the total area under the normal curve that occurs between $-\infty$ and any chosen point expressed in terms of multiples of σ on either side of μ. It can be used to find the area between any two chosen points. For example, the area between $\mu + 2.00\sigma$ and $\mu - 1.75\sigma$ is found as follows:

$$\begin{aligned} \text{Table } A \text{ reading for } +2.00 &= 0.9773 \\ \text{Table } A \text{ reading for } -1.75 &= \underline{0.0401} \\ \text{Area enclosed} &= 0.9372 \end{aligned}$$

Mathematically, Table *A* is described by the formula

$$F(z) = \frac{1}{\sqrt{2\pi}} \int_{-\infty}^{z} e^{-x^2/2}\, dx$$

where z is a *normalized* statistic found by subtracting from some designated value the arithmetic mean μ and dividing this difference by σ. Further discussion of the normal curve and limitations on its use is deferred to Chap. 5.

Suppose it is desired to compare the sum of the frequencies in the cells from 19.5 to 23.0, inclusive, in Table 3.2*a* with the frequencies expected for the same region from a normal curve having the same μ and σ. The upper limit of this group of cells is 23.25 (the upper limit of the cell with the midpoint 23.0) and the lower limit is 19.25. These points must be expressed in terms of multiples of σ from μ. The estimators of μ and σ are $\overline{X}$ (the average of the sample of size $n = 260$) and s (the sample standard deviation).†

From either Table 3.4 or 3.5,

$$\hat{\mu} = \overline{X} = 21.498 \text{ oz} \qquad \text{and} \qquad \hat{\sigma} = s = 1.485 \text{ oz}$$

Thus, for the lower limit,

$$z_1 = \frac{19.25 - 21.498}{1.485} = -1.51$$

and for the upper limit,

†Those familiar with statistical inference know that, when the standard deviation is estimated from the sample, Student's *t* distribution theoretically is more correct than the normal distribution when the samples are drawn from a normally distributed universe. However, as *n* gets large, say, 100 or more, Student's *t* converges to the normal distribution. Hence, it is reasonable to use the normal in this and similar cases. See Chap. 5.

$$z_2 = \frac{23.25 - 21.498}{1.485} = +\,1.18$$

Table A reading for + 1.18 = 0.8810
Table A reading for – 1.51 = 0.0655
Area included = 0.8155

The total frequency in Table 3.2a is 260: this multiplied by the figure of 0.8155 gives a frequency of 212 cases (that is, occurrences of the variable) to be expected in cells 19.5 to 23.0 if the distribution were normal. This compares favorably with an actual frequency count of

$$14 + 31 + 34 + 33 + 41 + 26 + 24 + 12 = 215$$

cases observed in these cells.

3.6.1 Use of Bowl Drawing to Illustrate the Relationship between a Known Universe and a Series of Samples under a Constant-Cause System

In actual situations, the two questions posed on p. 91 are relevant because the universes are unknown. However, a good way to obtain a feeling for the rational basis underlying the procedures and formulas used in answering such questions is to conduct an experiment using a known universe.

For example, a number is written on each of a group of physically similar chips. The chips are placed in a bowl and mixed thoroughly, and one chip is drawn at random. The number written on this chip is recorded, and the chip is replaced in the bowl. The chips are mixed again, and a chip is again drawn and the number is recorded. By repeating this process enough times, a series of numbers may be obtained that, if the chips are stirred well enough between drawings, will have been drawn in a chance or purely random manner. By examining the behavior of this variable that seems to be influenced only by a constant system of chance causes, it is possible to understand the way in which all constant-cause systems operate.

This type of demonstration frequently is made in courses in statistical quality control. If the chips are all alike, so that each one has as good a chance as any other to be drawn, it is possible to illustrate the meaning of many of the formulas used in control-chart work. These formulas, developed by mathematical statisticians, deal with the relationship between a universe and samples taken from it.

It is appropriate that bowl drawings should be part of explanations of how to use the control chart, as Shewhart used them in the early stages of his work on the applications of statistical methods in manufacturing. In his original treatise on the subject, he gives the data from 4,000 drawings of chips from each of three bowls, one containing a normal universe, one a rectangular universe, and one a triangular universe. The following discussion makes use of the first 400 drawings from Shewhart's normal bowl. (His rectangular and triangular universes are discussed and the results of his drawings from them are briefly summarized later in this chapter.)

3.6.2 Contents of Shewhart's Normal Bowl

Table 3.6 gives the distribution of the markings on the 998 chips in this bowl. Figure 3.6*a* shows the histogram for this frequency distribution. Figure 3.6*b* shows a more conventional histogram for the same distribution with a coarse grouping into 13 cells rather than 61; it illustrates the effect of coarse grouping on the histogram of a normal distribution.

The average of this symmetrical distribution obviously is 30. The standard deviation may be calculated to be 9.954 or, in round numbers, 10.

In referring to the distribution in this bowl as a normal distribution, we are taking some slight liberties with the normal curve. A true normal curve is continuous,

TABLE 3.6 MARKING ON CHIPS IN SHEWHART'S NORMAL BOWL[†]

Marking on chip X	Number of chips	Marking on chip X	Number of chips	Marking on chip X	Number of chips
60	1	39	27	19	22
59	1	38	29	18	19
58	1	37	31	17	17
57	1	36	33	16	15
56	1	35	35	15	13
55	2	34	37	14	11
54	2	33	38	13	9
53	3	32	39	12	8
52	4	31	40	11	7
51	4	30	40	10	5
50	5	29	40	9	4
49	7	28	39	8	4
48	8	27	38	7	3
47	9	26	37	6	2
46	11	25	35	5	2
45	13	24	33	4	1
44	15	23	31	3	1
43	17	22	29	2	1
42	19	21	27	1	1
41	22	20	24	0	1
40	24				

[†]W. A. Shewhart, *Economic Control of Quality of Manufactured Product,* app. II, tables *A, B,* and *C,* Van Nostrand Reinhold Co., Princeton, N.J., 1931. These bowl drawings from table *A* are used here by permission of the author and publishers. Shewhart's normal bowl contained numbers from −3.0 to +3.0. Numbers from 0 to 60 have been substituted here to simplify the presentation by eliminating negative numbers and decimals.

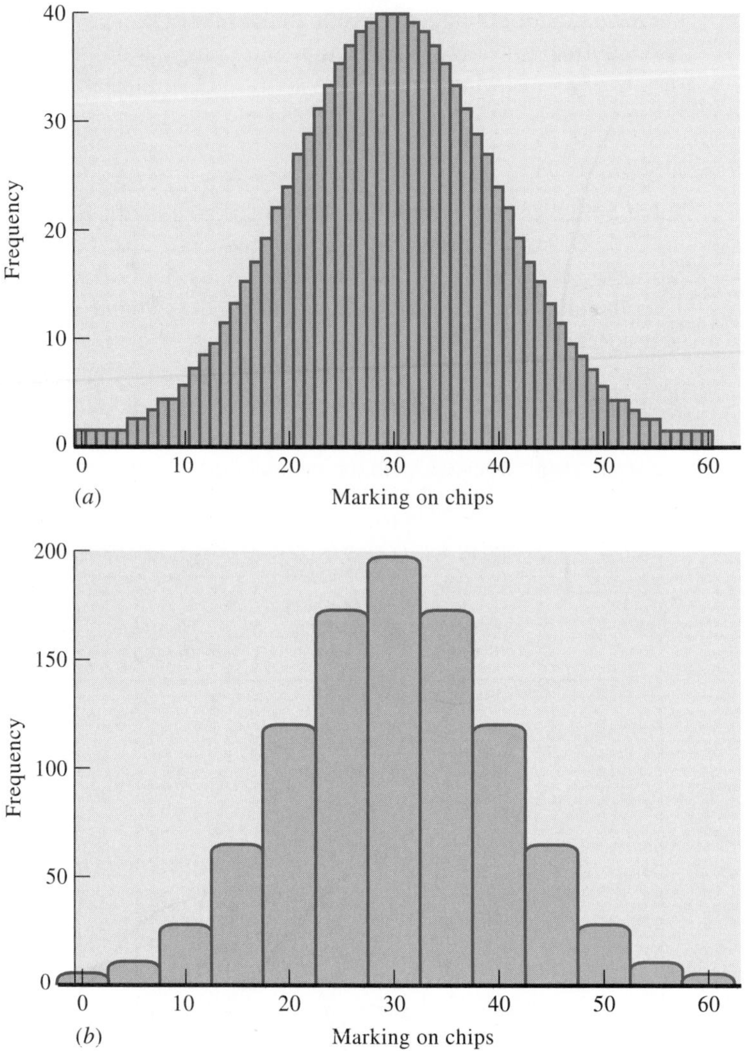

FIGURE 3.6 Histograms showing distribution of chips in Shewhart's normal bowl.

whereas this or any other bowl distribution is discontinuous (or, in the language of statistics, *discrete*). A distribution is discontinuous or discrete if there are intermediate values of the variable that cannot occur. (For example, a chip marked 31 or 32 may be drawn from the bowl but not one marked 31.94279651.) Moreover, the normal curve extends from $-\infty$ to $+\infty$ even though the percentage of area under the curve outside the 3-sigma limits is very small; a bowl would require several million chips to describe accurately the normal curve out to, say, 5-sigma limits. What can be said about the distribution in Shewhart's normal bowl is that it is as close to a normal distribution as it is possible to get with 998 chips and that its departure from normality is of no practical significance from the standpoint of the uses to be made of the bowl drawings.

3.6.3 Drawings from Shewhart's Normal Bowl

Table 3.7 gives the results of 400 drawings from Shewhart's bowl of 998 chips. Each time a chip was drawn, it was replaced and the chips stirred before the next

TABLE 3.7 400 DRAWINGS FROM SHEWHART'S NORMAL BOWL ARRANGED INTO SUBGROUPS OF FOUR

Number of drawings	Markings on chips in subgroup				Average $\bar{X}$	Range R	Standard deviation s
1–4	47	32	44	35	39.50	15	7.1
5–8	33	33	34	34	33.50	1	0.6
9–12	34	34	31	34	33.25	3	1.5
13–16	12	21	24	47	26.00	35	14.9
17–20	35	23	38	40	34.00	17	7.6
21–24	19	37	31	27	28.50	18	7.6
25–28	23	45	26	37	32.75	22	10.1
29–32	33	12	29	43	29.25	31	12.9
33–36	25	22	37	33	29.25	15	7.0
37–40	29	32	30	13	26.00	19	8.8
41–44	40	18	30	11	24.75	29	12.8
45–48	21	18	36	34	27.25	18	9.1
49–52	26	35	31	29	30.25	9	3.8
53–56	52	29	21	18	30.00	34	15.4
57–60	26	20	30	20	24.00	10	4.9
61–64	19	1	30	30	20.00	29	13.7
65–68	28	34	39	17	29.50	22	9.5
69–72	29	25	24	30	27.00	6	2.9
73–76	21	37	32	25	28.75	16	7.1
77–80	24	22	16	35	24.25	19	7.9
81–84	28	39	23	21	27.75	18	8.1
85–88	41	32	46	12	32.75	34	15.0
89–92	14	23	41	42	30.00	28	13.8
93–96	32	28	46	27	33.25	19	8.8
97–100	42	34	22	34	33.00	20	8.3
101–104	20	38	27	32	29.25	18	7.6
105–108	30	14	37	43	31.00	29	12.5
109–112	28	29	32	35	31.00	7	3.2
113–116	35	30	37	26	32.00	11	5.0
117–120	51	13	45	55	41.00	42	19.1
121–124	34	19	11	16	20.00	23	9.9
125–128	32	28	41	40	35.25	13	6.3
129–132	14	31	20	35	25.00	21	9.7
133–136	25	44	29	27	31.25	19	8.7
137–140	18	22	20	33	23.25	15	6.7

(continued)

drawing. For purposes of analysis, the drawings are shown in subgroups of four. Table 3.7 gives the average $\overline{X}$, the range R, and the sample standard deviation s for each subgroup.

It is helpful to think of these figures as representing the variation of some quality characteristic of a manufactured product. They might be the last two digits of the measurements of a dimension measured to 0.001 mm. Or they might be the final two digits, showing grams and tenths of a gram, of filling weights of a con-

TABLE 3.7 *(continued)*

Numbers of drawings	Markings on chips in subgroup				Average $\overline{X}$	Range R	Standard deviation s
141–144	21	31	39	25	29.00	18	7.8
145–148	17	44	54	13	32.00	41	20.1
149–152	36	48	19	41	36.00	29	12.4
153–156	25	31	38	30	31.00	13	5.4
157–160	35	21	20	34	27.50	15	8.1
161–164	21	22	44	19	26.50	25	11.7
165–168	39	22	24	29	28.50	17	7.6
169–172	40	44	24	18	31.50	26	12.5
173–176	23	25	46	29	30.75	23	10.5
177–180	23	37	44	34	34.50	21	8.7
181–184	36	52	30	28	36.50	24	10.9
185–188	35	23	11	5	18.50	30	13.3
189–192	33	15	40	29	29.25	25	10.5
193–196	18	30	22	25	23.75	12	5.1
197–200	23	30	20	19	23.00	11	5.0
201–204	7	32	36	38	28.25	31	14.4
205–208	29	30	39	31	32.25	10	4.6
209–212	36	12	34	25	26.75	24	10.9
213–216	36	37	39	32	36.00	7	2.9
217–220	38	9	25	39	27.75	30	14.0
221–224	11	44	29	29	28.25	33	13.5
225–228	31	18	31	25	26.25	13	6.2
229–232	22	47	12	27	27.00	35	14.7
233–236	29	24	32	44	32.25	20	8.5
237–240	42	26	32	27	31.75	16	7.3
241–244	29	40	43	29	35.25	14	7.3
245–248	23	22	23	39	26.75	17	8.2
249–252	34	27	52	28	35.25	25	11.6
253–256	27	40	23	24	28.50	17	7.9
257–260	34	38	16	28	29.00	22	9.6
261–264	39	19	39	32	32.25	20	9.4
265–268	42	25	25	42	33.50	17	9.8
269–272	30	25	38	39	33.00	14	6.7
273–276	43	22	10	28	25.75	33	13.7
277–280	17	31	10	16	18.50	21	8.9

(continued)

tainer. Or they might be any other characteristic, such as hardness, tensile strength, electrical resistance, or temperature.

3.6.4 Relationship between μ, σ, and the Values of $\overline{X}$

In statistics, the bar above any symbol means an average. Thus $\overline{X}$ is an average of the values of X. In the control chart for variables, each subgroup has its $\overline{X}$. The symbol $\overline{\overline{X}}$ thus refers to an average of the $\overline{X}$ values, that is, an average of the averages of values of X.

TABLE 3.7 *(continued)*

Numbers of drawings	Markings on chips in subgroup				Average $\overline{X}$	Range R	Standard deviation s
281–284	40	49	38	37	41.00	12	5.5
285–288	22	39	26	18	26.25	21	9.1
289–292	30	36	34	18	29.50	18	8.1
293–296	41	37	27	32	34.25	14	6.1
297–300	5	20	43	26	23.50	38	15.7
301–304	38	26	38	25	31.75	13	7.2
305–308	27	38	40	33	34.50	13	5.8
309–312	20	23	28	35	26.50	15	6.6
313–316	29	29	34	29	30.25	5	2.5
317–320	25	35	37	42	34.75	17	7.1
321–324	42	59	38	28	41.75	31	12.9
325–328	24	32	22	22	25.00	10	4.8
329–332	38	40	31	52	40.25	21	8.7
333–336	22	52	33	27	33.50	30	13.1
337–340	46	32	20	50	37.00	30	13.7
341–344	27	29	24	15	23.75	14	6.2
345–348	31	26	34	35	31.50	9	4.0
349–352	32	46	30	32	35.00	16	7.4
353–356	35	20	34	46	33.75	26	10.7
357–360	55	25	33	54	41.75	30	15.1
361–364	22	46	52	42	40.50	30	13.0
365–368	14	24	2	43	20.75	41	17.4
369–372	36	52	19	50	39.25	33	15.3
373–376	29	21	17	9	19.00	20	8.3
377–380	33	31	32	18	28.50	15	7.1
381–384	52	34	17	5	27.00	47	20.5
385–388	23	41	21	29	28.50	20	9.0
389–392	28	22	45	21	29.00	24	11.1
393–396	32	27	16	30	26.25	16	7.1
397–400	23	23	27	36	27.25	13	6.1
Totals	. .	. .	. .	. .	3,007.50	2,076	932.8

For the 100 subgroups of Table 3.7, $\overline{\overline{X}} = 3{,}007.50/100 = 30.08$. This is very close to 30.00, μ, the average of the distribution in the bowl. (In fact it is considerably closer to the universe average than would ordinarily be obtained in 400 drawings.) The average of sets of 80 drawings (20 subgroups of 4) may be shown to vary from 28.89 to 31.46. The averages of subgroups of 4 vary from 18.50 to 41.75, and the individual drawings run from 1 to 59.

The more drawings averaged, the more likely it is that their average will be close to the average in the bowl. Or, stated in more general terms, the larger the sample taken from any universe, the more likely it is that the average of the sample will be close to the average of the universe. This would seem to be an acceptable proposition on commonsense grounds even without the support given by statistical theory or by experimental bowl drawings.

Table 3.8 uses an analysis of the 400 drawings from Shewhart's bowl to show how the spread of the averages depends on sample size. These 400 drawings were

TABLE 3.8 RELATIVE FREQUENCIES OF $\overline{X}$ VALUES IN SAMPLES OF VARIOUS SIZES FROM 400 DRAWINGS FROM SHEWHART'S NORMAL BOWL
All frequencies expressed as percentages of total

Cell boundaries†	Distribution in bowl	$\overline{X}$						
		$n = 2$	$n = 4$	$n = 8$	$n = 16$	$n = 40$	$n = 80$	$n = 400$
58.31–61.31	0.2							
55.31–58.31	0.3							
52.31–55.31	0.8							
49.31–52.31	1.3	1.0						
46.31–49.31	2.4	0.5						
43.31–46.31	3.9	2.0						
40.31–43.31	5.8	3.5	5					
37.31–40.31	8.0	9.0	3	2				
34.31–37.31	9.9	11.5	11	8	4			
31.31–34.31	11.3	17.0	21	22	32	20	20	
28.31–31.31	12.0	18.5	23	34	36	60	80	100
25.31–28.31	11.3	12.0	21	26	28	20		
22.31–25.31	9.9	11.0	10	4				
19.31–22.31	8.0	8.5	3	4				
16.31–19.31	5.8	2.0	3					
13.31–16.31	3.9	0.5						
10.31–13.31	2.4	2.0						
7.31–10.31	1.3	1.0						
4.31– 7.31	0.8							
1.31– 4.31	0.3							
−1.69–+1.31	0.2							

†Following the rule that cell boundaries should always be located halfway between two possible values, the cell boundaries for the distribution in the bowl should properly be 58.5 to 61.5, 55.5 to 58.5, etc., rather than 58.31 to 61.31, etc. However, in making frequency distributions of the averages, it was necessary to choose the cells in such a way that no average fell on a cell boundary. In choosing one set of cell boundaries to apply to all the frequency distributions compared in Table 3.8, it was necessary to violate the rule stated.

divided into 200 subgroups of 2, the averages calculated for each subgroup, and a frequency distribution made of the averages. The same was done for subgroups of 4, 8, 16, 40, and 80. To permit easy comparisons of these distributions having different total frequencies, Table 3.8 shows them all expressed in *percentage* of total frequency.

Table 3.8 suggests certain ideas about averages of samples. It is evident that if many random samples of any given size n are taken from a universe, the averages ($\bar{X}$ values) of the samples will themselves form a frequency distribution. This frequency distribution of $\bar{X}$ values is similar to all frequency distributions in having its own central tendency and dispersion or spread, which might be expressed in terms of average and standard deviation. The average $\bar{\bar{X}}$ of such a frequency distribution of $\bar{X}$ values apparently tends to be near μ, the average of the universe. The spread of this frequency distribution of $\bar{X}$ values seems to depend not only on the spread of the universe but also on the sample size n; the larger the value of n, the less the spread of the $\bar{X}$ values.

Statistical theory gives us some very definite information about these matters. It tells us that in the long run the average of the $\bar{X}$ values will be the same as μ, the average of the universe. And in the long run, the standard deviation of the frequency distribution of $\bar{X}$ values will be $\sigma/\sqrt{n}$, that is, the standard deviation of the universe divided by the square root of the sample size. Thus if $n = 4$, the standard deviation of the frequency distribution of the $\bar{X}$ values will tend to be only half as great as the standard deviation of the universe. If $n = 16$, the spread of the frequency distribution of the $\bar{X}$ values will be one-fourth as great as that of the universe. If $n = 400$, it will be one-twentieth as great. A picture of how this works in a sampling experiment may be obtained by comparing the distributions of $\bar{X}$ values given in Table 3.8 with the distribution in the bowl.

This standard deviation of the expected frequency distribution of the averages is represented by the symbol $\sigma_{\bar{X}}$. It is referred to in textbooks on statistics as the *standard error of the mean,* or standard error of the average.

Regardless of the form of the universe, whether normal or otherwise, it is true that the expected $\sigma_{\bar{X}} = \sigma/\sqrt{n}$ and that the expected $\bar{\bar{X}} = \mu$. If the universe is normal, statistical theory tells us that the expected frequency distribution of the $\bar{X}$ values also will be normal. This is reinforced by experimental evidence from bowl drawings. Figure 3.7, taken from Shewhart's book, shows the observed frequencies of averages of 1,000 samples of four from his normal bowl and indicates the excellent fit of this distribution to the normal curve.

As previously stated, any distribution is completely specified if it is known to be normal and its average and standard deviation are known. This means that in sampling from a normal distribution that has a known average and standard deviation, statistical theory gives a complete picture of the expected pattern of variation of the averages of samples of any given size.

3.6.5 Importance of the Normal Curve in Sampling Theory

Even though the distribution in the universe is not normal, the distribution of the $\bar{X}$ values tends to be close to normal. The larger the sample size and the more

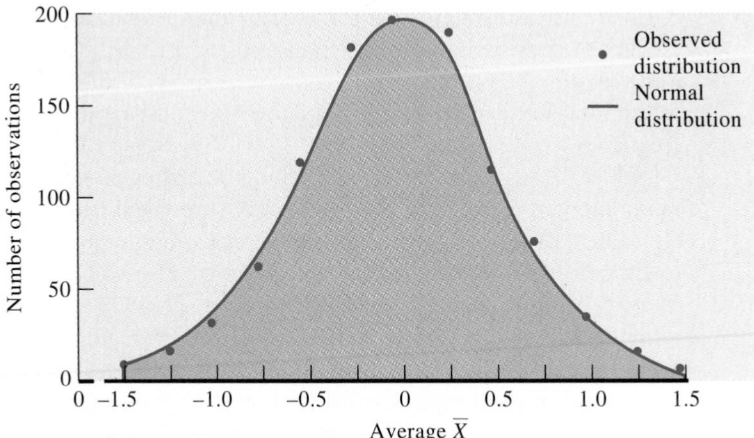

FIGURE. 3.7 The distribution of $\overline{X}$ values from 1,000 samples of 4 from Shewhart's normal bowl was a very close fit to the normal curve. (*Reproduced by permission from "Economic Control of Quality of Manufactured Product" by W. A. Shewhart, 1931, p. 181, Van Nostrand Reinhold Co., Litton Educational Publishing, Inc.)*

nearly normal the universe, the closer the frequency distribution of averages approaches the normal curve.

However, even if n is as small as four and the universe is far from normal, the distribution of the averages of samples will be very close to normal. Shewhart illustrates this by showing the distributions of averages of 1,000 samples of four from each of two bowls of chips, one containing a rectangular and the other a triangular distribution. Figure 3.8, taken from Shewhart,[†] compares these universes, neither of which even faintly resembles the normal curve, with the close fit of the normal curve to the distribution of averages of samples of four.

It has already been stated that many observed distributions of industrial characteristics do correspond roughly to the normal curve. Nevertheless, many others do

[†]The rectangular universe contained 122 chips, with two chips marked −3.0, two marked −2.9, two marked −2.8, and so on, to two marked +3.0. The triangular universe contained 820 chips with 40 chips marked −1.3, 39 chips marked −1.2, 38 marked −1.1, and so on, to one chip marked +2.6. (They are, in effect, "normalized" statistics.)

The main point to be noted here from Fig. 3.8 is that even with a great departure from normality in the universe, the distribution of $\overline{X}$ values with $n = 4$ is approximately normal; in sampling from most distributions found in nature and industry, the distribution of $\overline{X}$ values will be even closer to normal. However, it is of interest to observe that distributions similar to the rectangular and triangular distributions sometimes are found in industry. Although they seldom occur as a result of production alone, they may be found as a result of production followed by 100% inspection. For example, if a production operation gives a distribution on a certain dimension that is roughly normal with a standard deviation of 0.001 and the specified tolerances on the dimension are ±0.001 cm, it is obvious that only about 68% of the product will meet the specifications. If the production operation accurately centers the dimension at its specified nominal value, about 16% of the product will be rejected by the go gage and another 16% by the not-go gage. The distribution of the accepted product will not be far from rectangular. There will be two distributions something like the triangular, one of the product rejected by the go gage, and the other of the product rejected by the not-go gage.

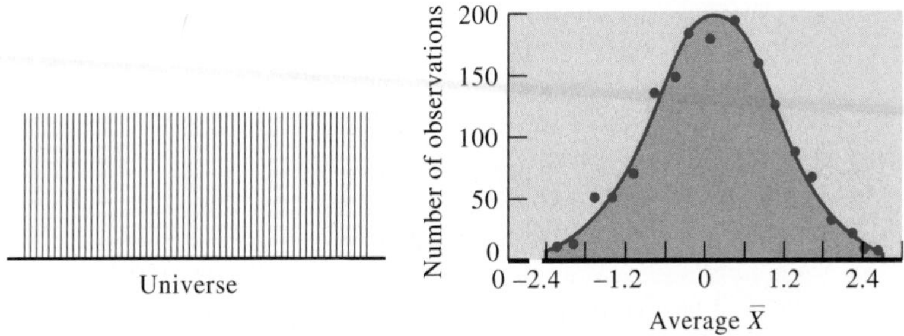

RECTANGULAR UNIVERSE

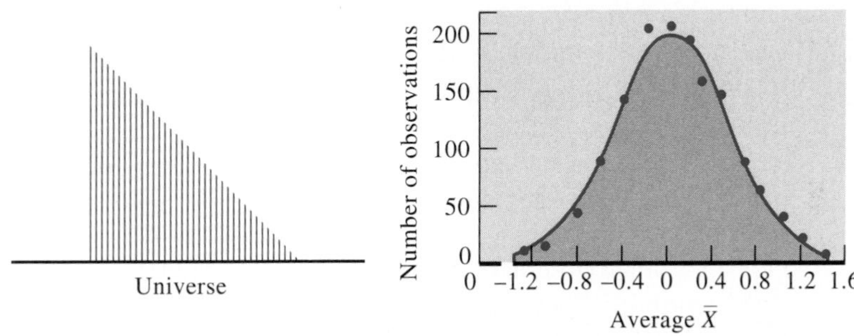

RIGHT TRIANGULAR UNIVERSE

FIGURE 3.8 Even from rectangular and triangular universes, the distribution of $\overline{X}$ values from samples of four is approximately normal. *(Reproduced by permission from Shewhart, op. cit., p. 182.)*

not. Serious mistakes often are made when it is assumed that the distribution of an industrial quality characteristic is necessarily normal.

The great practical importance of the normal curve arises even more from its uses in sampling theory than from the fact that some observed distributions are described by it well enough for practical purposes. Of great practical significance is the fact that distributions of averages of samples tend to be approximately normal even though the samples are drawn from nonnormal universes. This very important result of the theory of probability is discussed in greater detail in Chap. 5.

3.6.6 Relationship between σ and $\overline{s}$

In Table 3.7, the s of each subgroup was computed by the method explained. The s values are similar to the $\overline{X}$ values in that they differ greatly from one subgroup to the next; the smallest s observed was 0.6, and the largest was 20.5. However, one difference between the $\overline{X}$ and s distributions is evident. Whereas the $\overline{X}$ values tend to be centered at 30, the universe average, the s values seem to be centered at a somewhat lower figure than the universe standard deviation of 9.954.

In the long run, the sample standard deviations of samples of any size from a normal universe will follow a chance pattern that can be predicted by mathematics. Statistical theory also predicts the ratio between $\bar{s}$, the average of the s deviations of samples of any given size (such as the subgroups of four), and σ, the standard deviation of the universe from which the samples are taken. This ratio, represented by the symbol c_4,[†] is given in Table *C* of App. 3.

The value of c_4 for samples of four is 0.9213. This factor may be used to estimate an unknown universe standard deviation σ from $\bar{s}$, the average observed sample standard deviation of any given set of subgroups. From the 100 subgroups of Table 3.7, $\bar{s} = 932.8/100 = 9.328$. This gives an estimate of σ of

$$\hat{\sigma} = \frac{\bar{s}}{c_4} = \frac{9.328}{0.9213} = 10.12$$

Because, in this case, the frequency distribution in the bowl is known, it is possible to compare this estimated σ of 10.12 with the known σ of 9.954. The error in estimate is a little less than 2%.

It is of interest to observe that, even with the use of this c_4 factor, the s of an individual small subgroup gives no reliable information about σ. For instance, the estimate σ from the first subgroup (drawings 1–4) is 7.1/0.9213 = 7.8; from the second subgroup it is 0.6/0.9213 = 0.7; from the third it is 1.5/0.9213 = 1.6; from the fourth it is 14.9/0.9213 = 16.2; and so forth.

If the 100 subgroups are divided into 5 sets of 20 subgroups each, the resulting estimates of σ are as follows:

Set of subgroups	$\bar{s}$	Estimate of σ
1–20 (drawings 1–80)	8.262	8.97
21–40 (drawings 81–160)	9.80	10.64
41–60 (drawings 161–240)	9.65	10.48
61–80 (drawings 241–320)	8.34	9.06
81–100 (drawings 321–400)	10.58	11.48

Whereas some of the estimates from individual subgroups missed the true σ by as much as 100%, none of the estimates from the sets of 20 subgroups missed it by more than 20%. The use of all 100 subgroups gave an estimate within 2%. It seems evident that the larger the number of subgroups included in the calculation of $\bar{s}$, the greater the confidence in the estimate from $\bar{s}$ of the unknown standard deviation of the universe.

[†]Insofar as possible, the symbols used in this book are those generally used in statistical quality control work in the United States. Unfortunately, the lowercase form of the letter c is commonly used with three different meanings. c_4 is used for the ratio of $\bar{s}$ to σ; c is used to represent number of nonconformities in the control chart for nonconformities per unit; c, c_1, and c_2 are used to indicate acceptance numbers in the Dodge-Romig single and double sampling tables described in Chap. 13. The first two uses have the authority of the American National Standards Institute and the American Society for Quality Control; the third is common in the literature and usage of sampling tables. As the possible confusion arising from departure from standard usage seems more serious than the confusion from the use of the same symbol in different ways, c is used in all three senses in this book. The three uses relate to such different situations that the context should always make clear which meaning of c is intended.

3.6.7 Relationship between σ and $\overline{R}$

Given n, the sample or subgroup size, statistical theory gives the expected value of the ratio between $\bar{s}$ and σ in random sampling from a normal universe. Similarly, it gives the expected value of the ratio between the average range $\overline{R}$ and σ. This latter ratio, designated as d_2, also is given in Table C, App. 3.

One practical use of this d_2 factor is to provide an alternative method of estimating an unknown universe standard deviation σ from a series of samples or subgroups. It is of interest to compare the estimates of σ obtained from $\overline{R}$ with those obtained from $\bar{s}$ in the bowl drawings of Table 3.7.

For the 100 subgroups of 4, $\overline{R}$ = 2,076/100 = 20.76. The d_2 factor given for $n = 4$ in Table C is 2.059. Hence the estimate of σ is

$$\hat{\sigma} = \frac{\overline{R}}{d_2} = \frac{20.76}{2.059} = 10.08$$

If the 100 subgroups are divided into 5 sets of 20 as was done in the estimates from $\bar{s}$, the estimates of σ are as follows:

Set of subgroups	$\overline{R}$	Estimate of σ
1–20 (drawings 1– 80)	18.40	8.94
21–40 (drawings 81–160)	21.65	10.51
41–60 (drawings 161–240)	21.65	10.51
61–80 (drawings 241–320)	18.30	8.89
81–100 (drawings 321–400)	23.80	11.56

With ranges distributed all the way from 1 to 47, it is evident that estimates of σ made from a single range could vary all the way from 0.5 to 22.8. The comments regarding the unreliability of estimates of σ from the s of a single small sample apply with even greater force to estimates of σ from a single value of R. The range of one small subgroup gives little information about the standard deviation of the universe.

A comparison of the two estimates of σ from each set of 20 subgroups shows a close agreement between the estimate based on $\bar{s}$ and the one based on $\overline{R}$.

	Estimate of σ from $\bar{s}$	Estimate of σ from $\overline{R}$
First set	8.97	8.94
Second set	10.64	10.51
Third set	10.48	10.51
Fourth set	9.06	8.89
Fifth set	11.48	11.56

In only one of the five cases is the difference between the two estimates as great as 1% of σ. In four of the five cases, the estimate of σ from $\bar{s}$ is a little closer to the true value of σ (9.954) than the estimate from R. On the other hand, the estimate from the $\overline{R}$ of the 100 subgroups, 10.08, happens to be a little closer to the true value than the estimate 10.12 from the $\bar{s}$ of 100 subgroups. In all cases the differences seem negligible for practical purposes.

A practical point is that it often is much easier[†] to compute R for a subgroup by making a single subtraction than it is to compute s by calculating several squares and a square root. In control-chart work, this ease of calculation of R usually is much more important than any slight theoretical advantage that might come from the use of s as a measure of dispersion of subgroups. However, in some cases where the measurements themselves are costly (for example, destructive tests of valuable items) and it is necessary that the inferences from a limited number of tests be as reliable as possible, the extra cost of calculating s for each subgroup may be justified.

3.7 OTHER FREQUENCY CURVES

The normal curve is useful in many ways as an external reference distribution in solving practical problems in industry. Some of these ways will be developed in succeeding chapters. Its general pattern, with a concentration of frequencies about a midpoint and with small numbers of occurrences at the extreme values, repeats itself again and again. Nevertheless, the normal curve is frequently misused; it is not safe to assume that unknown distributions are necessarily normal.

There are many other frequency patterns that are like the normal in that frequencies decrease continuously from the center to the extreme values but—unlike the normal—are not symmetrical. Extreme values occur more frequently in one direction from the center than in the other. Two such distributions are illustrated in Fig. 3.9*b* and *c*. In the technical language of statistics, these lopsided frequency curves are known as *skewed* curves. Figure 3.9*b* is skewed to the right; that is, extreme variations occur more frequently above the mode than below it. Figure 3.9*c* is skewed to the left. The distribution of weights shown in the graphs of Fig. 3.2 also appears to be somewhat skewed to the left.

3.8 WHAT THE AVERAGE AND STANDARD DEVIATION OF A SET OF NUMBERS REALLY TELL

It has been pointed out that it would be convenient if the combination of sample average and standard deviation could tell us just what proportion of a set of numbers fell within any specified limits. It has also been pointed out that this combi-

[†]The statement that it is easier to compute R than s for a subgroup calls for two reservations. In this statement, "easier" means that the arithmetic is less laborious and, hence, less time-consuming and less costly. It does not mean that there is any less likelihood of mathematical error in computing R than in computing s. To compute R, the largest and smallest numbers in a subgroup must first be identified. Whenever an error is made in picking either the largest or the smallest number, the calculated value of R is less than the correct value. Such errors are common in the rapid calculations made for control-chart purposes in industrial plants. As there is no reason for compensating errors making R greater than its true value, computed values of R are sometimes in error on the low side.

A second reservation is brought about by technological developments in calculators and computers. Many of these provide for the automatic calculation of $\bar{X}$ and s simply by the input of the individual readings. Still others can be programmed by the user to perform these calculations. It is basically easier for such machines to calculate s for a subgroup than to calculate R.

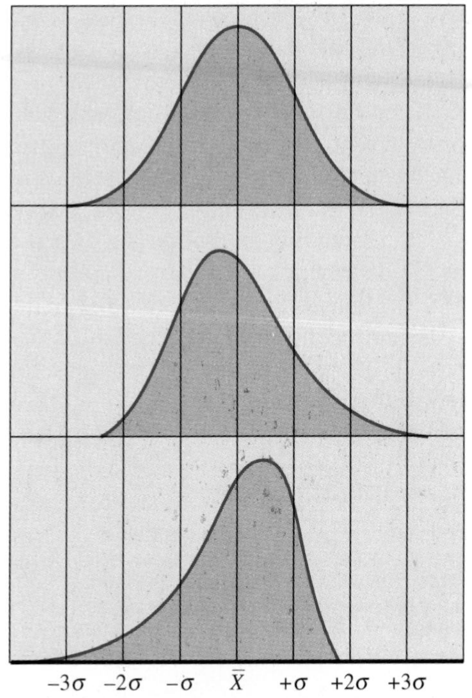

FIGURE 3.9 *(a)* Normal frequency curve; *(b)* and *(c)* typical skewed frequency curves.

nation does supply such information in those cases where it is known that the numbers are distributed according to the normal curve.

What do $\overline{X}$ and s tell if nothing whatever is known about the pattern of variation? One answer to this question is given by *Tchebycheff's inequality.* This mathematical theorem states that more than $1 - (1/t^2)$ of *any* set of finite numbers must fall within the closed range $\mu \pm t\sigma$ (where t is not less than 1). For example, if $t = 3$, this means that more than $1 - (1/3^2)$, or $\frac{8}{9}$, of any set of numbers must fall within the limits $\mu \pm 3\sigma$, where μ and σ have been computed from the numbers themselves. In other words, less than $\frac{1}{9}$ of the numbers can fall outside these limits. In this case, μ is calculated using the formula for $\overline{X}$, Eq. (3.1). The formula for σ is similar to Eq. (3.3) for s except that n is used in the denominator rather than $n - 1$. The reason for this is that the mean of the set of numbers is known if the entire set is known. Further discussion of this matter is left for Chap. 5.

The actual fraction falling outside $\mu \pm 3\sigma$ may be much less than $\frac{1}{9}$; the Tchebycheff inequality simply states that it cannot be as much as $\frac{1}{9}$. It should be emphasized that this limit of $\frac{1}{9}$ applies to the *sum* of the fractions above an upper limit of $\mu + 3\sigma$ and below a lower limit of $\mu - 3\sigma$.

Frequently it is the case that the set of numbers constitutes a sample taken from some unknown parent distribution (or more than one such distribution). In this case, the same formulas apply. However, calculations regarding this sample apply to the sample only. That is, the calculations of μ and σ can tell only about the *sample* and not about the underlying distribution from which the sample was taken. If $\overline{X}$ and s have been calculated for this sample, $\mu = \overline{X}$ and $\sigma = \sqrt{(n-1)/n}\, s$.

Usually it is desired to have $\overline{X}$ and σ tell more about a distribution than can be determined from Tchebycheff's inequality. An adaptation of the inequality by Camp and Meidell states that under certain circumstances more than $1 - (1/2.25t^2)$ of any distribution will fall within the closed range $\overline{X} \pm t\sigma$. These circumstances are that the distribution must have only one mode, that the mode must be the same as the arithmetic mean, and that the frequencies must decline continuously on both sides of the mode. Many distributions that are not normal do actually come close enough to meeting these conditions for the Camp-Meidell inequality to be applied with confidence.

The percentages indicated by the normal curve and by the two inequalities are compared as follows:

Limits	If the distribution is roughly normal, *approximately* the following percentages of cases will be *outside* the limits	If the Camp-Meidell conditions apply, *less than* the following percentages of cases will be *outside* the limits	Under any and all circumstances, *less than* the following percentages of cases will be *outside* the limits
$\overline{X} \pm 2\sigma$	4.55	11.1	25.0
$\overline{X} \pm 2.5\sigma$	1.24	7.1	16.0
$\overline{X} \pm 3\sigma$	0.27	4.9	11.1
$\overline{X} \pm 3.5\sigma$	0.05	3.6	8.2
$\overline{X} \pm 4\sigma$	0.006	2.8	6.3
$\overline{X} \pm 4.5\sigma$	0.0007	2.2	4.9
$\overline{X} \pm 5\sigma$	0.00006	1.8	4.0

The rules summarized in the foregoing tabulation are given in many books on statistics. It sometimes happens that the readers of such books misinterpret these rules as pertaining to the relationship between the statistics obtained from samples and the parameters of the universe from which the samples are drawn. It should be emphasized that the rules apply merely to any set of numbers, whether those numbers are viewed as constituting a sample or a universe. For example, Tchebycheff's inequality tells us that it is impossible to construct any set of numbers in which 4.0% or more of the numbers are more than 5σ away from the arithmetic mean.

PROBLEMS

The problems in this section marked with a star (☆) are well suited to solution on a PC. Your instructor will advise you about which data sets you may enter into a spreadsheet or, if you are using special SPC software, into that data bank. While SPC software will eliminate the necessity for the student to enter control-chart limit formulas, the student should verify the accuracy of the results. It is important to learn how to make these calculations.

3.1. In the production of an electrical device operated by a thermostatic control, five control switches were tested each hour to determine the "on" temperature at which the thermostat actually operated under a given setting. Results of the test over a 3-day production period were as follows:

Date	Subgroup number	"On" temperature at which thermostatic switch operates (temperature units not specified)				
		a	*b*	*c*	*d*	*e*
Apr. 25	1	54	56	56	56	55
	2	51	52	54	56	49
	3	54	52	50	57	55
	4	56	55	56	53	50
	5	53	54	57	56	52
	6	53	47	58	55	54
	7	52	55	54	55	56
	8	56	53	53	54	55
	9	55	52	53	56	55
	10	50	54	53	55	55
Apr. 26	11	57	54	53	52	53
	12	52	52	54	53	55
	13	54	53	55	52	52
	14	54	55	54	53	55
	15	56	53	57	56	54
Apr. 27	16	58	57	56	54	54
	17	55	55	55	56	53
	18	54	57	54	55	54
	19	54	53	56	53	55
	20	53	53	57	54	53
	21	53	55	57	56	55
	22	59	54	53	54	55
	23	54	55	58	55	54
	24	56	53	51	55	59
	25	56	55	55	55	55

Using a check sheet similar to Fig. 3.1, make a tally of these 125 measurements to arrange them in a frequency distribution. From this check sheet prepare a table similar to Table 3.2*a* showing cell midpoints, cell boundaries, and observed frequencies.

☆ 3.2. Plot the frequency distribution of Prob. 3.1 as a frequency histogram similar to Fig. 3.2*a*.

☆ 3.3. Arrange the frequency distribution of Prob. 3.1 into a cumulative frequency distribution similar to the one shown in Table 3.3. Show both "number less than" and "percentage less than" for each cell boundary. Plot the ogive on rectangular coordinate paper.

3.4. Find the arithmetic mean and sample standard deviation of subgroup 2 of Prob. 3.1. Compute the standard deviation by both Eqs. (3.3) and (3.4). Which method would be more useful for machine calculation?

3.5. Find the arithmetic mean and sample standard deviation for subgroup 3 of Prob. 3.1. Compute the standard deviation by both Eqs. (3.3) and (3.4). Which method would be more useful for machine calculation? Why?

☆ **3.6.** Using the short method illustrated in Table 3.5, compute the arithmetic mean and sample standard deviation of the frequency distribution data in Prob. 3.1.
Answer: $\overline{X} = 54.23$; $s = 1.93$.

3.7. Using the mean and standard deviation found in Prob. 3.6, determine what proportion of the original data actually fall outside $\overline{X} \pm 3s$. Comment on the differences between this percentage and the three percentages for $\overline{X} \pm 3\sigma$ given in the table on page 108.
Answer: 0.8%.

3.8. Using the $\overline{X}$ and s calculated in Prob. 3.6, what percentage of the distribution would fall between the limits 51.5 and 57.5 assuming that the underlying distribution is approximately normal? Compare this result with the actual results tabulated in Prob. 3.1.
Answer: 87.8%, 90.4%.

3.9. A normal curve has the same average and standard deviation as those calculated in Prob. 3.6. What percentage of the area under the curve will fall between limits of 52.5 and 56.5? Compare this to the percentage of cases observed between these limits in the frequency distribution of Prob. 3.1.

3.10. A small electronic device is designed to emit a timing signal of 200 milliseconds (ms) duration. In the production of this device, subgroups of five units are taken at periodic intervals and tested. $\overline{X}$ and R are calculated for each subgroup and used to plot control charts. The results of inspection of 125 of these devices are shown in the following table.

Using a check sheet similar to Fig. 3.1, make a tally of these 125 measurements to arrange them in a frequency distribution. Use cell midpoints of 189.5, 191.5, 193.5, and so on. From this check sheet prepare a table similar to Table 3.2*a* showing cell midpoints, cell boundaries, and observed frequencies.

	Duration of automatic signal, ms						
	Sample letter						
Subgroup number	a	b	c	d	e	Average $\overline{X}$	Range R
1	195	201	194	201	205	199.2	11
2	204	190	199	195	202	198.0	14
3	195	197	205	201	195	198.6	10
4	211	198	193	199	204	201.0	18
5	204	193	197	200	194	197.6	11
6	200	202	195	200	197	198.8	7
7	196	198	197	196	196	196.6	2
8	201	197	206	207	197	201.6	10
9	200	202	204	192	201	199.8	12
10	203	201	209	192	198	200.6	17

Subgroup number	Duration of automatic signal, ms					Average $\bar{X}$	Range R
	Sample letter						
	a	*b*	*c*	*d*	*e*		
11	195	198	196	204	201	198.8	9
12	193	203	197	198	201	198.4	10
13	200	206	208	199	200	202.6	9
14	199	199	197	204	202	200.2	7
15	189	199	205	197	199	197.8	16
16	198	196	199	205	197	199.0	9
17	198	201	201	206	206	202.4	8
18	206	200	190	202	196	198.8	16
19	197	198	198	195	201	197.8	6
20	196	199	197	198	204	198.8	8
21	196	207	203	193	197	199.2	14
22	202	202	206	209	202	204.2	7
23	200	213	196	193	199	200.2	20
24	204	192	198	205	199	199.6	13
25	199	201	194	205	207	201.2	13
Σ	...	...	...	...	...	4,990.8	277

☆ **3.11.** Plot the frequency distribution of Prob. 3.10 as a frequency histogram similar to Fig. 3.2*a*.

3.12. Find the sample standard deviation of subgroup 6 of Prob. 3.10 by both Eqs. (3.3) and (3.4). Which method would be more useful for machine calculation? Why?

3.13. Find the sample standard deviation of subgroups 16 through 20 in Prob. 3.10. Use Eq. (3.3) to calculate the standard deviations of subgroups 16 and 17 and Eq. (3.4) for subgroups 18 through 20. Which method would be preferable for machine calculation? Why?

3.14. Using the long method illustrated in Table 3.4, compute the arithmetic mean and sample standard deviation of the frequency distribution obtained in Prob. 3.10.

3.15. Using the short method illustrated in Table 3.5, compute the arithmetic mean and sample standard deviation of the frequency distribution obtained in Prob. 3.10.

3.16. Using the results of Prob. 3.10, plot a frequency histogram and a cumulative frequency curve. Overlay the two curves, one on the other, using rectangular coordinate paper with the ordinate scale for the histogram on the left side and the ordinate scale for the ogive on the right.

3.17. A normal curve has the same average and standard deviation as those calculated in Prob. 3.14 or 3.15. What percentage of the area under the curve will fall between limits of 190.5 and 210.5? Compare this with the percentage of cases observed between these limits in the frequency distribution of Prob. 3.10.

3.18. A normal curve has the same average and standard deviation as those calculated in Prob. 3.14 or 3.15. What percentage of the area under the curve will fall between limits of 194.5 and 208.5? Compare this with the percentage of cases observed between these limits in the frequency distribution of Prob. 3.10.

3.19. The high-voltage output of a certain power supply used in a copy machine is specified as 350 ± 5 V dc at 20 milliamps (mA). Subgroups of four power supply units are drawn from the process and inspected approximately every half-hour. The data from 25 subgroups are shown in the following table along with the average and range for each subgroup.

Using a check sheet similar to Fig. 3.1, make a tally of these 100 measurements to arrange them in a frequency distribution. Use cell midpoints of 347.0, 347.5, 348.0, 348.5, and so on. From this check sheet prepare a table similar to Table 3.2*a* showing cell midpoints, cell boundaries, and observed frequencies.

	DC-voltage output at 20 mA					
	Sample letter					
Subgroup number	*a*	*b*	*c*	*d*	Average $\overline{X}$	Range *R*
1	348.5	350.2	348.3	350.3	349.3	2.0
2	351.3	351.2	347.1	349.7	349.8	4.2
3	348.5	350.5	348.5	349.0	349.1	2.0
4	351.4	350.4	348.6	353.2	350.9	4.6
5	349.4	348.0	349.6	351.1	349.5	3.1
6	351.1	348.1	349.2	350.1	349.7	3.0
7	348.3	349.9	350.7	348.5	349.4	2.4
8	349.9	349.1	349.0	349.6	349.4	0.9
9	349.2	348.7	348.8	350.3	349.3	1.6
10	349.2	351.6	351.9	349.2	350.5	2.7
11	350.1	350.5	351.2	347.9	349.9	3.3
12	350.4	350.8	350.3	352.6	351.0	2.3
13	347.7	349.6	348.6	349.3	348.8	1.9
14	349.0	351.1	350.2	348.0	349.6	3.1
15	350.7	349.3	349.3	350.2	349.9	1.4
16	350.0	351.8	352.3	349.8	351.0	2.5
17	350.1	349.8	349.6	349.2	349.7	0.9
18	351.1	350.6	346.9	349.8	349.6	4.2
19	351.4	349.3	349.7	349.6	350.0	2.1
20	348.8	349.6	351.3	349.2	349.7	2.5
21	349.4	350.2	350.2	351.8	350.4	2.4
22	351.7	351.6	349.9	347.1	350.1	4.6
23	350.4	349.0	349.2	349.6	349.6	1.4
24	349.4	348.7	350.3	348.8	349.3	1.6
25	349.6	349.1	349.6	351.2	349.9	2.1
Σ	.....	.....	......	.....	8,745.2	62.8

☆ **3.20.** Plot the frequency distribution of Prob. 3.19 as a frequency histogram.

3.21. Find the sample standard deviation of subgroup 15 of Prob. 3.19 by both Eqs. (3.3) and (3.4). Which method would be preferable for machine calculation? Why?

3.22. Find the sample standard deviation of subgroup 22 of Prob. 3.19 by both Eqs. (3.3) and (3.4). Which method would be preferable for machine calculation? Why?

3.23. Find the sample standard deviations of subgroups 6 through 10 of Prob. 3.19. Use Eq. (3.3) to calculate the standard deviations of subgroups 6 and 7 and Eq. (3.4) for subgroups 8 through 10. Which method would be preferable for machine calculation? Why?

3.24. Using the long method illustrated in Table 3.4, compute the arithmetic mean and sample standard deviation of the frequency distribution obtained in Prob. 3.19.

3.25. Using the short method illustrated in Table 3.5, compute the arithmetic mean and sample standard deviation of the frequency distribution obtained in Prob. 3.19.

☆ **3.26.** Using the results of Prob. 3.19, plot a frequency histogram and a cumulative frequency curve. Overlay the two curves, one on the other, using rectangular coordinate paper with the ordinate scale for the histogram on the left side and the ordinate scale for the ogive on the right.

3.27. A normal curve has the same average and standard deviation as those calculated in Prob. 3.24 or 3.25. What percentage of the area under the curve will fall between limits of 347.25 and 352.75? Compare this with the actual percentage of cases observed between these limits in the frequency distribution of Prob. 3.19.

3.28. A manufacturer of electrical products purchases many parts from outside suppliers. A lot of 20,000 of a certain small component is received from a new supplier. The receiving inspection department for the manufacturer has taken a random sample of 200 components from this lot and measured the resistance of each component. These resistances in ohms have been arranged into the following frequency distribution:

Cell boundaries, ohms	Frequency
88.5–86.5	2
86.5–84.5	5
84.5–82.5	16
82.5–80.5	24
80.5–78.5	40
78.5–76.5	44
76.5–74.5	25
74.5–72.5	22
72.5–70.5	13
70.5–68.5	7
68.5–66.5	2

(a) Compute the average and sample standard deviation of this frequency distribution.
(b) What percentage of a normal distribution having your computed estimates of μ and σ would fall outside the specification limits 75 ± 10 ohms?
(c) If you make the arbitrary assumption that resistances are distributed uniformly throughout each cell, what percentage of the actual distribution fell outside these limits?

3.29. The contained weight of a certain dry product is labeled as 500 gm. Periodically a sample is taken from the packaging line and the contents weighed. After 250 samples were weighed and the weight recorded, the following frequency distribution was formed:

Cell boundaries	Frequency	Cell boundaries	Frequency
505.5–506.0	1	501.5–502.0	34
505.0–505.5	2	501.0–501.5	25
504.5–505.0	7	500.5–501.0	17
504.0–504.5	12	500.0–500.5	13
503.5–504.0	25	499.5–500.0	7
503.0–503.5	29	499.0–499.5	4
502.5–503.0	41	498.5–499.0	2
502.0–502.5	30	498.0–498.5	1

(a) Compute $\overline{X}$ and s of this frequency distribution.
(b) If a normal distribution had these values for μ and σ, what percentage of the distribution would fall below the label weight of 500 gm?
(c) What proportion of the actual frequency distribution fell below the label weight?
(d) What conclusions, if any, can you reach on the question of whether or not the producer was maintaining good statistical control of this quality characteristic? Explain your answer.

3.30. Data have been collected on the cycles to failure of a certain relay device. The so-called short method has been used to develop the following information. The origin is at 575 cycles, and the cell interval is 50 cycles.

$n = \Sigma f = 200 \quad \Sigma fd = -63 \quad \Sigma fd^2 = 1{,}795$

(a) Compute the values of $\overline{X}$ and s for this distribution.
(b) If this process is assumed to generate a normal distribution, what proportion of the product meets a minimum specification of 400 cycles?

3.31. Some 200 samples of a certain nylon fiber have been tested to breakage for tensile strength. From these data, a frequency histogram has been prepared using a cell width of 25 gm. Using the short method and with an origin at 1,000 gm,

$n = \Sigma f = 200 \quad \Sigma fd = 340 \quad \Sigma fd^2 = 1{,}200$

(a) Compute the values of $\overline{X}$ and s for this frequency distribution.
(b) Using these values as estimates of μ and σ, respectively, find the proportion of a normal curve that would fall below a minimum tensile strength specification of 1,000 gm.

3.32. A normal curve has the same average and standard deviation as those calculated in Prob. 3.4. What percentage of the area under the curve will fall between limits of 48.5 and 58.5? Compare this with the actual percentage of cases observed between these limits in the frequency distribution of Prob. 3.1.

3.33. The mean value of the modulus of rupture of a large number of test specimens of green Sitka spruce has been found to be 5,600 lb/in^2.

(*a*) If the standard deviation is 840 lb/in^2 and the distribution is approximately normal, the modulus of rupture will fall between 5,000 and 6,200 for what percentage of the specimens?

(*b*) For what percentage will it be above 4,000?

(*c*) Below 3,500?

Answer: (*a*) 52.5%; (*b*) 97.2%; (*c*) 0.62%.

3.34. Tests of the stiffness of a number of aluminum alloy channels gave the following frequency distribution. Stiffness was measured in "effective *EI* in lb/in^2."

Stiffness	Frequency	Stiffness	Frequency	Stiffness	Frequency
2,640	1	2,440	33	2,280	14
2,600	2	2,400	41	2,240	5
2,560	7	2,360	35	2,200	3
2,520	11	2,320	22	2,160	1
2,480	25				

(*a*) Compute $\overline{X}$ and s.

(*b*) If a normal distribution had these values for μ and σ, what percentage of the distribution would fall below 2,150?

Answer: (*a*) $\overline{X} = 2{,}399.6$; $s = 83.6$; (*b*) 0.14%.

3.35. What proportion of a frequency distribution would you expect to fall outside $\overline{X} \pm 2.3\sigma$ limits:

(*a*) If it is known to be approximately normal?

(*b*) If it is known only that it satisfies the conditions of the Camp-Meidell inequality?

(*c*) If nothing is known about the form of the distribution?

3.36. What proportion of a frequency distribution would you expect to fall outside $\overline{X} \pm 1.8\sigma$ limits:

(*a*) If it is known to be approximately normal?

(*b*) If it is known only that it satisfies the conditions of the Camp-Meidell inequality?

(*c*) If nothing is known about the form of the distribution?

3.37. The dimension referred to in Prob. 2.1 is specified as 2.050 ± 0.020. If the dimension falls above USL_X, rework is required; if below LSL_X, the part must be scrapped. If the process is in statistical control and normally distributed, what can you conclude regarding its ability to meet specifications? Can you make any suggestions for improvement?

3.38. In Prob. 2.3 the specified minimum strength for a weld is 370 lb. If the process is in statistical control and normally distributed, what can you conclude regarding its ability to meet this specification?

3.39. Control charts for $\overline{X}$ and R are maintained on dissolved iron content of a certain solution in parts per million (ppm). After 125 hourly samples have been drawn and analyzed, the data are organized into 25 subgroups of 5 measurements each, maintaining the time order of sampling. From these data, $\Sigma\overline{X} = 390.8$ and $\Sigma R = 84$. Find the values of 3-sigma control limits for $\overline{X}$ and R, and estimate the value of σ for this process under the assumption that the process is in control.

3.40. The specification on the process described in Prob. 3.39 calls for no more than 18 ppm dissolved iron in the solution. Assuming that a normal distribution underlies the process and that the process continues to be in statistical control with no change in average or dispersion, what proportion of the sample measurements may be expected to exceed this specification?

4

WHY THE CONTROL CHART WORKS; SOME EXAMPLES

Statistical methods serve as landmarks which point to further improvement beyond that deemed obtainable by experienced manufacturing men. Hence after all obvious correctives have been exhausted and all normal logic indicates no further gain is to be made, statistical methods still point toward a reasonable chance for yet further gains; thereby giving the man who is doing trouble shooting sufficient courage of his convictions to cause him to continue to the ultimate gain, in spite of expressed opinion on all sides that no such gain exists.

—G. J. Meyers, Jr.[†]

4.1 THE USE OF CONTROL CHARTS TO JUDGE WHETHER OR NOT A CONSTANT SYSTEM OF CHANCE CAUSES IS PRESENT

Control charts for $\overline{X}$, R, and s supply a basis for judgment on a major question of practical importance. This question might be phrased in different ways, such as, "Were all these samples drawn from the same bowl?" or "Is there one universe from which these samples appear to come?" or "Do these figures indicate a stable pattern of variation?" or "Is this variation the result of a constant-cause system?" or merely "Do these measurements show statistical control?"

Any rule that might be established for providing a definite "Yes" or "No" answer to these questions is bound to give the wrong answer part of the time. The

[†]G. J. Meyers, Jr., Discussion of E. G. Olds, "On Some of the Essentials of the Control Chart Analysis," *Transactions, American Society of Mechanical Engineers,* vol. 64, pp. 521–527, July 1942.

decision where to draw the line between a "Yes" and a "No" answer must be based on the expected action to be taken if each answer is given.

In quality control, the answer "No, this is not a constant-cause system" leads to a hunt for an assignable cause of variation and an attempt to remove it, if possible. The answer "Yes, this is a constant-cause system" leads to leaving the process alone, making no effort to hunt for causes of variation. The rule for establishing the control limits that will determine the "Yes" or "No" answer in any case should strike an economic balance between the costs resulting from two kinds of errors—the error of hunting for trouble when it is absent (whenever the "No" answer is incorrect) and the error of leaving a process alone because of not hunting for trouble when it really is present (whenever the "Yes" answer is incorrect).

Any rule for establishing control limits should be a practical one based on this point of economic balance. In most of this book, we follow the common practice in the United States of using so-called 3-sigma limits. Experience indicates that in most cases 3-sigma limits do actually strike a satisfactory economic balance between these two types of errors. (Although, in principle, the choice of limits is a problem of minimizing the sum of certain costs, there are great practical difficulties in making good estimates of the relevant unit costs.)

4.1.1 Stable and Unstable Patterns of Variation

Figure 3.1 presented a frequency distribution of the 260 drained weights of contents of cans of tomatoes obtained over an 11-day period. At first glance, it might appear that this distribution presents a satisfactory picture of a pattern of variation that existed throughout that 11 days.

Such a statement is not necessarily true. It may be that the pattern of variation has gone through several changes during the period. The *first* question that must be answered is whether there was a stable pattern. If there is evidence of a lack of stability, it may be true that the 260 weights represent samples from a number of different universes that existed at different times. The resulting frequency distribution, in this case, would represent a weighted average of the various universes that existed at the times the samples were taken.

Whenever a frequency distribution is to be used as a basis for estimating the capabilities of a process, making a proposed change in the process, or reviewing specifications, the control chart becomes an important tool in establishing stability. Once stability has been established, the analyst may feel relatively confident that one universe is being measured and not a weighted average of many. *The order in which the measurements were made should always be preserved in recording data for a frequency distribution.*

4.2 USE OF THE CONTROL CHART IN INTERPRETATION OF A FREQUENCY DISTRIBUTION

Figure 4.1 shows the control charts for $\overline{X}$ and R for the basic data on drained weights contained in Table 3.1. None of the $\overline{X}$ points is outside the control lim-

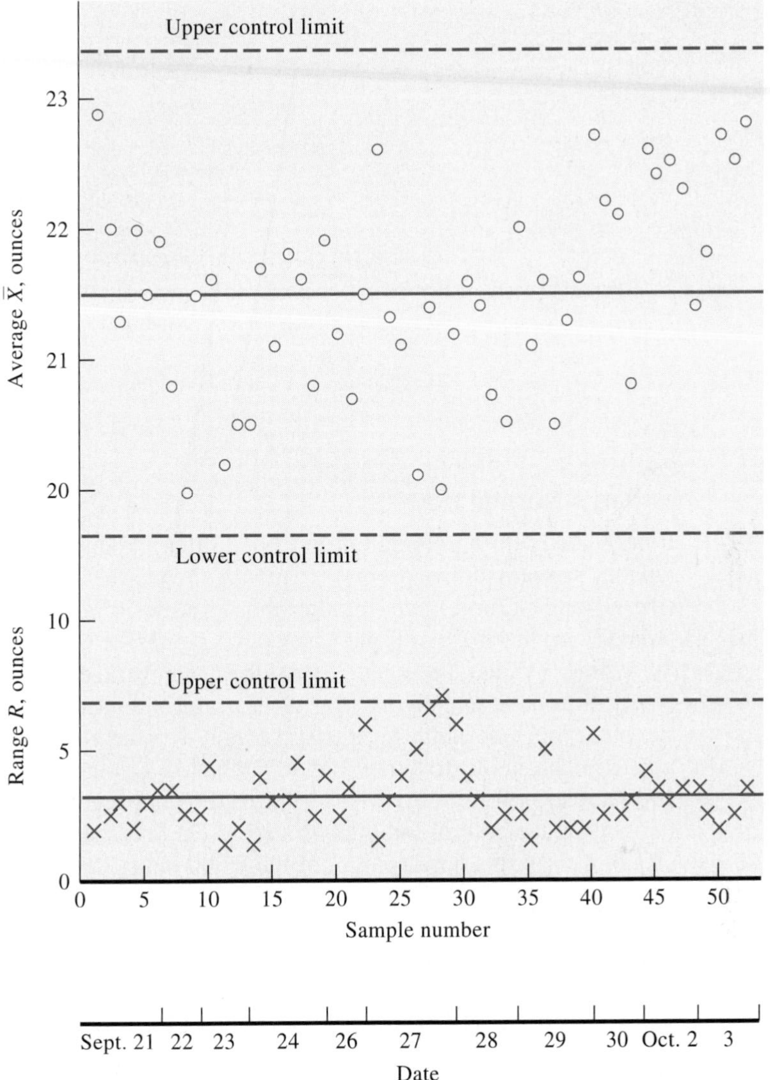

FIGURE 4.1 $\overline{X}$ and R control charts for drained weights of contents of cans of tomatoes—data of Table 3.1.

its; one of the 52 points for R is outside the limits. Application of secondary rules based on the theory of runs, discussed later in this chapter and in Chap. 5, indicates some slight shifts during the 11-day period. While usage of these data to estimate μ and σ might be subject to question, it does not necessarily follow that this much departure from a constant pattern of variation is unsatisfactory from a practical standpoint. Many production processes never do as well.

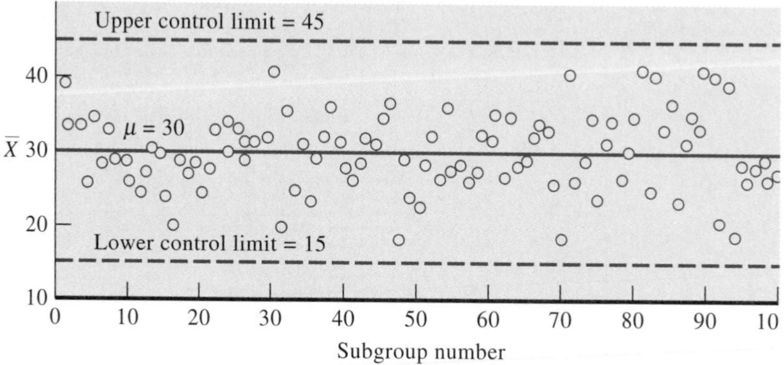

FIGURE 4.2 Control chart for $\overline{X}$ for 100 subgroups of 4 drawn from Shewhart's normal bowl.

4.2.1 The $\overline{X}$ Chart for Drawings from Shewhart's Normal Bowl

It has already been pointed out in Chap. 3 that the distribution of the $\overline{X}$ values of random samples drawn from one universe tends to be normal when the sample size is four or more, even though the universe is not normal. It has also been pointed out that nearly all the cases (all but 0.27%) in a normal distribution will fall within 3-sigma limits on either side of the average.

It follows that as long as a series of samples (or subgroups) are really random samples from one universe, their averages will nearly always fall within limits $\mu \pm 3\sigma_{\overline{X}}$. This is illustrated by Fig. 4.2, which is a control chart for the averages of 100 samples of 4 from Shewhart's normal bowl, using data of Table 3.7. No points fall outside the control limits of 45 and 15. In Shewhart's 1,000 drawings of samples of 4 from this bowl, only 2 of the 1,000 points fell outside these control limits.[†]

In Fig. 4.2, the central line on the chart could be set at 30, that is, at μ, the known average of the universe. The 3-sigma limits could be based on a calculation from the known value of σ, the standard deviation in the bowl, which in round numbers was 10. The standard error of the average is

$$\sigma_{\overline{X}} = \frac{\sigma}{\sqrt{n}} = \frac{10}{\sqrt{4}} = 5$$

The 3-sigma limits were therefore $3\sigma_{\overline{X}} = 3(5) = 15$ on either side of the average of 30. This placed the upper control limit at 45 and the lower control limit at 15.

[†]It is of interest to examine the distribution of markings on chips as given in Table 3.6. These markings extended from 0 to 60, with 60 of the 998 chips below 15 and another 60 above 45. Obviously in any random series of drawings, occasional subgroups of 4 will contain chips with numbers low enough for the average to be below 15; others will have the average above 45. The point to be emphasized is that if the chips are drawn in a random manner, the laws of chance operate in a way that these averages below 15 or above 45 will be infrequent. Whenever one of these infrequent events occurs, the control chart will seem to say, "The universe has changed; look for trouble," when in reality the composition of the universe is unchanged and no trouble can be found.

It is not strictly accurate to say that in the long run 3-sigma limits on an $\overline{X}$ chart will show points out of control only 27 times in 10,000 (that is, 0.27% of the time) provided the universe is really unchanged. This would be strictly true only if the distribution of the $\overline{X}$ values were exactly normal and the control limits were based on known values of μ and σ. Actually, although the distribution of $\overline{X}$ values is roughly normal, it is not exactly so unless the universe is normal; the 3-sigma limits are necessarily calculated from the observed data rather than from parameters of the universe. Hence 3-sigma limits may give false indications of lack of control somewhat more often than indicated by the normal curve.

Nevertheless, such false indications of lack of control will be infrequent. The 3-sigma limits seldom make the error of indicating trouble (that is, indicating an assignable cause of variation) when there is no trouble to be found. If points on the $\overline{X}$ chart fall outside 3-sigma limits, there is good reason for confidence that they point to some factor contributing to quality variation that can be identified.

4.2.2 Calculation of 3-Sigma Limits on Control Charts for $\overline{X}$

The use of Tables *C, D,* and *E* of App. 3 for the calculation of control limits may be illustrated by using the first 20 subgroups of the bowl drawings of Table 3.7.

After calculating the averages and ranges of subgroups, the next step in the calculation of limits is to find $\overline{\overline{X}}$ and $\overline{R}$. For the first 20 subgroups, these are

$$\hat{\mu} = \overline{\overline{X}} = \frac{\Sigma \overline{X}_i}{20} = \frac{577.75}{20} = 28.9$$

$$\overline{R} = \frac{\Sigma R_i}{20} = \frac{368}{20} = 18.4$$

If Table *C* is to be used, the next step is to estimate σ. For this it is necessary to find in Table *C* the d_2 factor for the subgroup size. In this case $n = 4$, and Table *C* gives $d_2 = 2.059$.

$$\text{Estimate of } \sigma, \hat{\sigma} = \frac{\overline{R}}{d_2} = \frac{18.4}{2.059} = 8.94$$

Then $3\sigma_{\overline{X}}$ can be calculated from the relationship $\sigma_{\overline{X}} = \hat{\sigma}/\sqrt{n}$:

$$3\sigma_{\overline{X}} = \frac{3\hat{\sigma}}{\sqrt{n}} = \frac{3(8.94)}{\sqrt{4}} = 13.4$$

$$\text{Upper control limit}_{\overline{X}} = \overline{\overline{X}} + 3\sigma_{\overline{X}} = 28.9 + 13.4 = 42.3$$
$$\text{Lower control limit}_{\overline{X}} = \overline{\overline{X}} - 3\sigma_{\overline{X}} = 28.9 - 13.4 = 15.5$$

The two steps in the calculation of $3\sigma_{\overline{X}}$ might be consolidated as

$$3\sigma_{\overline{X}} = \frac{3\overline{R}}{d_2\sqrt{n}} = \frac{3}{2.059\sqrt{4}}\overline{R} = 0.73\overline{R} = 0.73(18.4) = 13.4$$

To shorten the calculation of control limits from $\overline{R}$, this factor $3/d_2\sqrt{n}$, the multiplier of $\overline{R}$ in the preceding calculation, has been computed for each value of n from 2 to 20 and tabulated in Table D of App. 3. This factor is designated A_2. The formulas for 3-sigma control limits on charts for $\overline{X}$ then become

$$UCL_{\overline{X}} = \overline{\overline{X}} + A_2\overline{R}$$
$$LCL_{\overline{X}} = \overline{\overline{X}} - A_2\overline{R}$$

If control limits are to be calculated from $\overline{s}$ rather than from $\overline{R}$, the calculations for the first 20 subgroups of Table 3.7 are as follows:

$$\overline{\overline{X}} = 28.9$$

$$\overline{s} = \frac{\Sigma s_i}{20} = \frac{165.2}{20} = 8.26$$

Using the c_4 factor from Table C to estimate σ,

$$\text{Estimate of } \sigma, \hat{\sigma} = \frac{\overline{s}}{c_4} = \frac{8.26}{0.9213} = 8.97 \quad 3\sigma_{\overline{X}} = \frac{3\hat{\sigma}}{\sqrt{n}} = \frac{3(8.97)}{\sqrt{4}} = 13.5$$

$$UCL_{\overline{X}} = 28.9 + 13.5 = 42.4$$
$$LCL_{\overline{X}} = 28.9 - 13.5 = 15.4$$

As in the calculation from $\overline{R}$, the two steps in the calculation of $3\sigma_{\overline{X}}$ can be consolidated as

$$3\sigma_{\overline{X}} = \frac{3}{c_4\sqrt{n}}\,\overline{s} = 1.63(8.26) = 13.5$$

To shorten the calculations for control limits from $\overline{s}$, this factor $3/c_4\sqrt{n}$, the multiplier of $\overline{s}$ in the above calculation, has been computed for each value of n from 2 to 25, then by 5's to 100, and tabulated in Table E of App. 3. This factor is designated A_3. The formulas for 3-sigma control limits using this factor are

$$UCL_{\overline{X}} = \overline{\overline{X}} + A_3\overline{s}$$
$$LCL_{\overline{X}} = \overline{\overline{X}} - A_3\overline{s}$$

For those situations where it is desired to calculate control limits directly from known or standard values of σ and μ, the factor $3/\sqrt{n}$ has been computed and tabulated in Table F, App. 3. This factor is designated as A. The formulas for 3-sigma control limits using this factor are

$$UCL_{\overline{X}} = \mu + A\sigma \text{ or } \overline{X}_0 + A\sigma_0$$
$$LCL_{\overline{X}} = \mu - A\sigma \text{ or } \overline{X}_0 - A\sigma_0$$

As applied to the control chart of Fig. 4.2, using the known values $\mu = 30$ and $\sigma = 10$ and the value $A = 1.50$ given by Table F for a subgroup size of 4,

$$UCL_{\overline{X}} = 30 + 1.50(10) = 45.0$$

$$LCL_{\overline{X}} = 30 - 1.50(10) = 15.0$$

The various equations for central lines and 3-sigma limits on control charts for $\overline{X}$, R, and s are assembled for convenient reference in Table 4.1. The factors (such as A, A_2, and so on) referred to are given in Tables C to F of App. 3. The reader will note that the spread of the limits on $\overline{X}$ charts as well as on R or s charts depends on the process dispersion. Limits on all the charts may be calculated directly from a known or assumed σ by estimating σ from either $\overline{R}$ or $\overline{s}$. In most cases in industrial practice, limits are computed from $\overline{R}$.

4.2.3 Control Charts for Range and Sample Standard Deviation

A general formula for the control limits on the $\overline{X}$ chart is $\mu + 3\sigma_{\overline{X}}$. Similarly, general formulas for control charts for measures of subgroup dispersion are:

Standard, aimed-at, or target values of μ ($\overline{X}_0$) may be substituted for μ or $\overline{\overline{X}}$; likewise, targets (R_0 or s_0) may be substituted for $\overline{R}$ or $\overline{s}$.

1. For the R chart: $\overline{R} \pm 3\sigma_R$
2. For the s chart: $\overline{s} \pm 3\sigma_s$

However, when these three formulas are applied to the calculation of lower control limits, such limits will turn out to be less than 0 where n is 6 or less for the

TABLE 4.1 EQUATIONS FOR COMPUTING 3-SIGMA LIMITS ON SHEWHART CONTROL CHARTS FOR VARIABLES

Table references for required factors pertain to App. 3. CL = central line

Method	$\overline{X}$ chart	R chart	s chart
μ and σ known or assumed	$CL = \overline{\overline{X}} = \mu$ $UCL_{\overline{X}} = \mu + A\sigma$ $LCL_{\overline{X}} = \mu - A\sigma$ Table F	$CL = \overline{R} = d_2\sigma$ $UCL_R = D_2\sigma$ $LCL_R = D_1\sigma$ Tables C and F	$CL = \overline{s} = c_4\sigma$ $UCL_s = B_6\sigma$ $LCL_s = B_5\sigma$ Tables C and F
μ and σ estimated from $\overline{\overline{X}}$ and $\overline{R}$	$CL = \overline{\overline{X}}$ $UCL_{\overline{X}} = \overline{\overline{X}} + A_2\overline{R}$ $LCL_{\overline{X}} = \overline{\overline{X}} - A_2\overline{R}$ Table D	$CL = \overline{R}$ $UCL_R = D_4\overline{R}$ $LCL_R = D_3\overline{R}$ Table D	
μ and σ estimated from $\overline{\overline{X}}$ and $\overline{s}$	$CL = \overline{\overline{X}}$ $UCL_{\overline{X}} = \overline{\overline{X}} + A_3\overline{s}$ $LCL_{\overline{X}} = \overline{\overline{X}} - A_3\overline{s}$ Table E		$CL = \overline{s}$ $UCL_s = B_4\overline{s}$ $LCL_s = B_3\overline{s}$ Table E

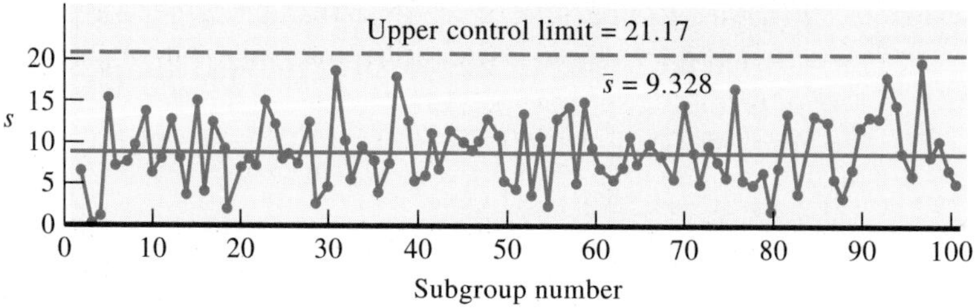

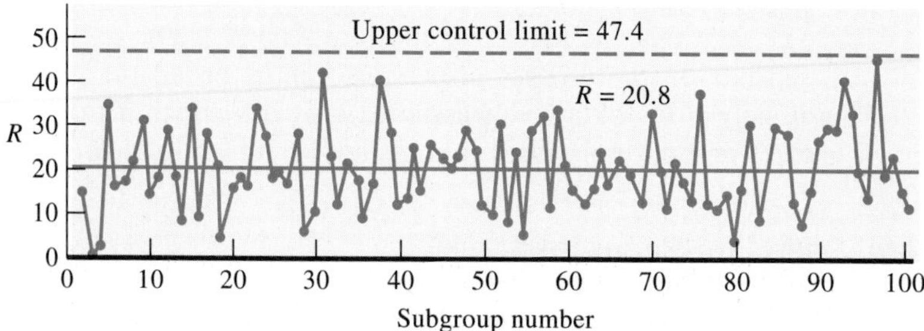

FIGURE 4.3 Control charts for sample standard deviation and range—drawings from Shewhart's normal bowl.

R chart and where n is 5 or less for the s chart. Because R and s cannot be less than 0, the lower control limit is not used in these cases.

Figure 4.3 shows control charts for s and R for the 100 subgroups of 4 given in Table 3.7. Limits on these charts were calculated by using the respective values of $\bar{s}$ and $\bar{R}$ found in Table 3.7 and by the equations indicated in Table 4.1.

$$UCL_s = B_4\bar{s} = 2.27(9.328) = 21.17$$

$$LCL_s = B_3\bar{s} = 0(9.328) = 0$$

$$UCL_R = D_4\bar{R} = 2.28(20.8) = 47.4$$

$$LCL_R = D_3\bar{R} = 0(20.8) = 0$$

The similarity between the variation from subgroup to subgroup shown in the s and R charts is emphasized here by the use of lines[†] connecting the successive points. It seems clear that these two charts tell practically the same story. Either one may be used in any instance; there is no need to use both.

[†]In practical control-chart work, the points on $\bar{X}$ and R charts are sometimes connected and sometimes not. In this book points on the charts are not connected except in special cases such as Fig. 4.3. Generally speaking, experience shows that the connection of points on control charts for variables is likely to lead to their misinterpretation, particularly by many people who are not familiar with the principles behind the control chart. It is usually advantageous if control charts for $\bar{X}$ and R do not look like ordinary trend charts.

In subsequent illustrations throughout this book, only one measure of dispersion will be calculated. Usually this will be R, although occasionally it may be s. The purpose of calculating both R and s for the subgroups in Table 3.7 was to illustrate that R and s were alternative measures of the same thing, that they led to similar estimates of σ, similar control limits on $\overline{X}$ charts, and similar control charts showing subgroup dispersion.

In practical control-chart work in industry, R rather than s should nearly always be used as a measure of subgroup dispersion. R is easier to explain; almost everyone can understand range, whereas people with little background in statistics have difficulty understanding standard deviation.

4.2.4 Estimates of σ from $\overline{R}$ and $\overline{s}$ from Various Subgroup Sizes in a Controlled Process

The preceding discussion of $\overline{X}$ charts makes it clear that the width of the band between upper and lower control limits depends entirely on the variability within the subgroups, which we have measured by either the average range $\overline{R}$ or the average sample standard deviation $\overline{s}$. Both $\overline{R}$ and $\overline{s}$ lead to estimates of σ, the standard deviation of the universe. It is of interest to examine data from a controlled process to see whether subgroup size seems to influence these estimates of σ.

As the chips were drawn one by one from Shewhart's bowl and as each chip drawn was replaced and the chips stirred before the next drawing, there is no natural subgroup size; it is permissible to group the drawings into subgroups of any size. To illustrate the effect of different sizes of subgroups, the 400 drawings have been divided into subgroups of 2 and 8 as well as 4, and the values of R and s have been computed for each subgroup. (The individual R and s values for subgroups of 2 and 8 are not shown here.) $\overline{R}$ and $\overline{s}$ for subgroup sizes of 2 and 8 have been computed for each set of 80 drawings as well as for the entire 400 drawings. σ has been estimated from each value of $\overline{R}$ and $\overline{s}$, using the appropriate d_2 and c_4 factors from Table C.

Table 4.2 gives the estimates of σ for each set of 80 drawings and for the entire 400 drawings, using $\overline{R}$ and $\overline{s}$ as estimators and using the three different subgroup sizes. The fairly close agreement among the different estimates of σ from any given set of drawings is striking. The variation in estimates of σ from one set of 80 drawings to another is evidently much greater than the variation among the different estimates from any set. It would seem that, at least from the standpoint of estimating the dispersion of the universe, many different subgroup sizes are acceptable. (Incidentally, it will be noted that $\overline{R}$ and $\overline{s}$ give identical estimates of σ for the subgroup size of 2. The range of a subgroup of 2 is $\sqrt{2}$ times the sample standard deviation. d_2 is therefore $\sqrt{2}$ times c_4, and $\overline{R}$ is $\sqrt{2}$ times $\overline{s}$.)

Although satisfactory estimates of σ may be made with various subgroup sizes, often there are good reasons why some one particular subgroup size may be the best to use in any given instance. Various considerations entering into the choice of subgroup size are discussed in Chap. 8.

TABLE 4.2 COMPARISON OF ESTIMATES OF UNIVERSE STANDARD DEVIATION σ BASED ON SUBGROUP SIZES OF 2, 4, AND 8
Known value of $\sigma = 9.95$

	Estimates of σ					
	Subgroup size of 2		Subgroup size of 4		Subgroup size of 8	
Drawings	From $\bar{R}$	From $\bar{s}$	From $\bar{R}$	From $\bar{s}$	From $\bar{R}$	From $\bar{s}$
1–80	8.62	8.62	8.94	8.97	9.24	8.98
81–160	10.75	10.75	10.51	10.64	10.50	10.58
161–240	9.73	9.73	10.51	10.48	9.76	9.89
241–320	8.86	8.86	8.89	9.06	8.85	9.02
321–400	11.68	11.68	11.56	11.48	11.98	12.17
1–400	9.93	9.93	10.08	10.12	10.07	10.13

4.2.5 The Distribution of the Standard Deviation

Universes seem to give nourishment to many other distributions that have less spread, such as distributions of averages, standard deviations, and ranges. And just as each universe has its average and standard deviation, so also does each distribution of averages, standard deviations, or ranges have its own average and standard deviation.

Unfortunately, statistical theory cannot give us such useful generalizations about the distribution of s as it can about the distribution of $\bar{X}$. In the case of the distribution of $\bar{X}$, theory gave the expected average μ and the expected standard deviation $\sigma/\sqrt{n}$, both of which were independent of the form of the universe. Moreover, theory told us that if the universe distribution were normal, the distribution of $\bar{X}$ values would be normal regardless of sample size, and that even if the universe distribution were not normal, the distribution of $\bar{X}$ values would approach normality as sample size increased.

However, if the universe is normal, statistical theory can tell us the expected average and the expected standard deviation of the distribution of s. As already stated, in samples from a normal universe the expected average $\bar{s}$ is $c_4\sigma$.[†] A commonly used approximate estimate of σ_s, the expected standard deviation of the distribution of s for samples from a normal universe, is $\sigma/\sqrt{2(n-1)}$. It also is known that, as n increases, the distribution of s becomes closer and closer to a symmetrical distribution.

[†]Where

$$c_4 = \sqrt{\frac{2}{n-1}}\,\frac{[(n-2)/2]!}{[(n-3)/2]!} \quad \text{and} \quad \left(\frac{1}{2}\right)! = \frac{\sqrt{\pi}}{2}$$

In this formula, the symbol ! means "factorial." See Chap. 5 for an explanation of factorials.

Theoretical knowledge of the distribution of s in samples from a normal universe is the basis for 3-sigma limits on the control chart for s. The central line on the control chart is set at $\bar{s}$. The limits are set at $\bar{s} \pm 3\sigma_s$.

The approximate value of σ_s for a normal universe is

$$\sigma_s = \frac{\sigma}{\sqrt{2(n-1)}} \tag{4.1}$$

Modern statistical theory gives the exact value as[†]

$$\sigma_s = \sqrt{1 - c_4^2}\,\sigma \tag{4.2}$$

When n is large, the difference between (4.1) and (4.2) is negligible. Equation (4.2) is the basis for the control limit factors where n is 25 or less; Eq. (4.1) is used where n exceeds 25.

When 3-sigma limits on a chart for s are calculated from an observed $\bar{s}$, they are

$$UCL_s = \bar{s} + 3\sigma_s = B_4\bar{s}$$

$$LCL_s = \bar{s} - 3\sigma_s = B_3\bar{s}$$

When limits are based on a known or assumed value of universe standard deviation σ, they are

$$UCL_s = c_4\sigma + 3\sigma_s = B_6\sigma$$

$$LCL_s = c_4\sigma - 3\sigma_s = B_5\sigma$$

In computing σ_s for the B_4 and B_3 factors given in Table *E* of App. 3, σ is assumed to be $\bar{s}/c_4$. The B_6 and B_5 factors are given in Table *F*.

4.2.6 The Distribution of the Range

Although no simple formula gives either the expected average range R or the standard deviation of the range σ_R, statistical theory does give the ratio of these figures to universe standard deviation σ in sampling from a normal universe. Theory also fully defines the expected distribution of R in sampling from a normal universe.[‡]

[†]See "ASTM Manual on Presentation of Data and Control Chart Analysis," STP 15D, p. 138, American Society for Testing and Materials, Philadelphia, Pa., 1976. Also see Frederick Mosteller, "On Some Useful 'Inefficient' Statistics," *The Annals of Mathematical Statistics,* vol. 17, pp. 377–408, December 1946.

[‡]See Simon, op. cit., p. 204; E. S. Pearson, "The Probability Integral of the Range in Samples of n Observations from a Normal Population," *Biometrika,* vol. 32, 1942; and L. C. H. Tippett, "On the Extreme Individuals and the Range of Samples Taken from a Normal Population," *Biometrika,* vol. 17, pp. 364–387, 1925. See also N. L. Johnson and Fred Leone, *Statistics and Experimental Design,* vol. I, John Wiley & Sons, Inc., New York, 1964.

When 3-sigma limits are calculated from an observed R, they are

$$UCL_R = \overline{R} + 3\sigma_R = \overline{R} + 3\frac{d_3}{d_2}\overline{R} = D_4\overline{R}$$

$$LCL_R = \overline{R} - 3\sigma_R = \overline{R} - 3\frac{d_3}{d_2}\overline{R} = D_3\overline{R}$$

When limits are based on a known or assumed value of universe standard deviation σ, they are

$$UCL_R = d_2\sigma + 3\sigma_R = d_2\sigma + 3d_3\sigma = D_2\sigma$$

$$LCL_R = d_2\sigma - 3\sigma_R = d_2\sigma - 3d_3\sigma = D_1\sigma$$

In these formulas, the d_2 factor expresses numerically the expected value of $\overline{R}/\sigma$ and the d_3 factor expresses the standard deviation of this relationship. Values of d_2 and d_3 are given in Table *C* of App. 3. The factors necessary to calculate control limits appear in Tables *D* and *F.*

4.2.7 A Modification of d_2 When Relatively Few Subgroups Are Available

In estimating σ from $\overline{R}$, we have used the fraction $\overline{R}/d_2$. The mathematical theory underlying the d_2 factor assumes that sampling has been from a normal universe. The d_2 factor depends on the subgroup size. For example, it is 2.326 for a subgroup size of 5.

Strictly speaking, the validity of the exact value of the d_2 factor assumes that the ranges have been averaged for a fair number of subgroups, say, 20 or more. Where only a few subgroups are available, a better estimate of σ is obtained by using a factor that writers on statistics have designated as d_2^* (read as "dee-sub-two-star"). Table 4.3, adapted from Military Standard 414 of the U.S. Department of Defense, illustrates the dependence of this factor on the number of subgroups in the special case where the subgroup size is 5.

TABLE 4.3 RATIO d_2^* OF EXPECTED $\overline{R}$ TO σ IN AVERAGING RANGES OF VARIOUS NUMBERS OF SUBGROUPS OF 5 FROM A NORMAL UNIVERSE

No. of subgroups of 5	d_2^*	No. of subgroups of 5	d_2^*
1	2.474	8	2.346
2	2.405	10	2.342
3	2.379	12	2.339
5	2.358	20	2.334
6	2.353	∞	2.326

In the ordinary construction of control charts for quality control in industry, it is good enough for practical purposes to use factors based on d_2 rather than on d_2^*. However, in certain other statistical applications it is desirable to use d_2^*. The only such application included in this book is in Chap. 15 in our discussion of Military Standard 414 for acceptance sampling by variables.

Acheson Duncan has tabulated values of d_2^* for subgroup sizes from 2 to 15 and for numbers of subgroups from 1 to 15.[†]

4.3 CONTRIBUTION OF THE CONTROL CHART TO ELIMINATION OF CAUSES OF TROUBLE

As pointed out and briefly illustrated in Chap. 1, actions based on the control chart for variables are of many kinds. Some of these actions, particularly those related to specifications and tolerances and to acceptance procedures, need to start from evidence that a process is in control. For such actions, it is satisfactory to understand the behavior of constant-cause systems.

Other useful actions start from evidence of the control chart that a process is out of control. Troubleshooting in manufacturing is a particularly important example of this. In this type of control-chart application, the control chart sometimes says, "Leave this process alone," and at other times it says, "Hunt for trouble and try to correct it."

A major virtue of the control chart is that it tells—within reasonably satisfactory limits—*when* to hunt for the cause of variation. It is always helpful to know *when;* sometimes this may be sufficient to indicate *where* to look. Nevertheless, there is often a fair amount of hard work between the decision to hunt for trouble and the actual discovery and correction of the root cause of the trouble. This fact is responsible for H. F. Dodge's frequently quoted statement that "statistical quality control is 90% engineering and only 10% statistics."[‡]

The control chart unaided cannot put its finger on exactly *where* the cause of trouble can be found. Nevertheless, users of the control-chart technique sometimes develop an ability to diagnose causes of process troubles with surprising accuracy. This ability usually depends on a combination of an understanding of the principles of the control chart and an intimate knowledge of the particular processes to which the control chart is applied.

No general book on statistical quality control can supply the necessary knowledge of various processes. However, some guidance may be given on the statisti-

[†]A. J. Duncan, *Quality Control and Industrial Statistics,* 4th ed., p. 950, Richard D. Irwin, Inc., Homewood, Ill., 1974.

[‡]This statement has sometimes been misinterpreted to belittle the practical contributions of statistics in industrial quality control.

When statistics—by means of the control chart or otherwise—points to the need to hunt for trouble, it often is true that 90% of the hard work remains to be done; the tough engineering job of hunting for causes and eliminating them is still ahead. In cases where this troubleshooting has led to spectacular reductions in scrap and rework, these cost reductions would not have been made without the hard engineering work. It is sometimes forgotten that neither would they have been made without the statistics.

A fair analogy is to think of the use of the control chart in troubleshooting as a chain with ten links. One of them is the statistical link; the other nine are engineering links. From the standpoint of the strength of the chain, weakness in the statistical link is just as serious as weakness in one of the engineering links.

cal aspects of interpretation of the control chart for purposes of troubleshooting. We shall try to provide this guidance by examining several general ways in which lack of control may occur and noting the effect of each on the appearance of control charts for $\bar{X}$ and R.

4.3.1 Lack of Statistical Control Implies a Shift in the Universe

Drawing chips from a bowl is a helpful analogy in clarifying what really happens when a manufacturing process shows lack of control. The system of chance causes in operation at any particular moment corresponds to the universe, that is, to the distribution of chips in the bowl. The items actually manufactured at that moment correspond to a sample drawn from that bowl or universe. When points fall outside the limits on the control charts, this is evidence that the universe has changed; it is as if samples were being drawn from a different bowl.

Usually the items *produced* in any period constitute a much larger sample from the universe than the items actually *measured* for control-chart purposes during the same period. Thus the control chart gives evidence not only regarding the universe (that is, the chance cause system operating) but also regarding the items produced that were not measured.

4.3.2 A Classification of Ways in Which Lack of Control May Occur

Because lack of control corresponds to the substitution of a new bowl, a classification of different types of lack of control may be thought of as a classification of ways in which two bowls may differ in their distribution of chips. It is helpful to give separate consideration to three ways in which universes may differ, as follows:

1. They may differ in average only.
2. They may differ in dispersion only.
3. They may differ in both average and dispersion.

Shifts in universe average influence the control charts for $\bar{X}$ and R in one way; shifts in universe dispersion influence them in another way.

Shifts in the universe may be sustained shifts over a period of time, as if many subgroups were drawn from one bowl and then many more drawn from another bowl. Or shifts may be frequent and irregular as if there were a great many bowls with the drawings from each bowl continued for periods of different lengths. Or shifts may be gradual and systematic.

In Fig. 4.4, the frequency curves are intended to represent the universe or bowl, that is, the chance cause system in operation at any moment. Figure 4.4*a* shows a situation in which the universe continued for a while at one average, then shifted for a while to a higher average, and finally was brought back to the original average. Figure 4.4*b* shows a situation in which the universe average varies erratically

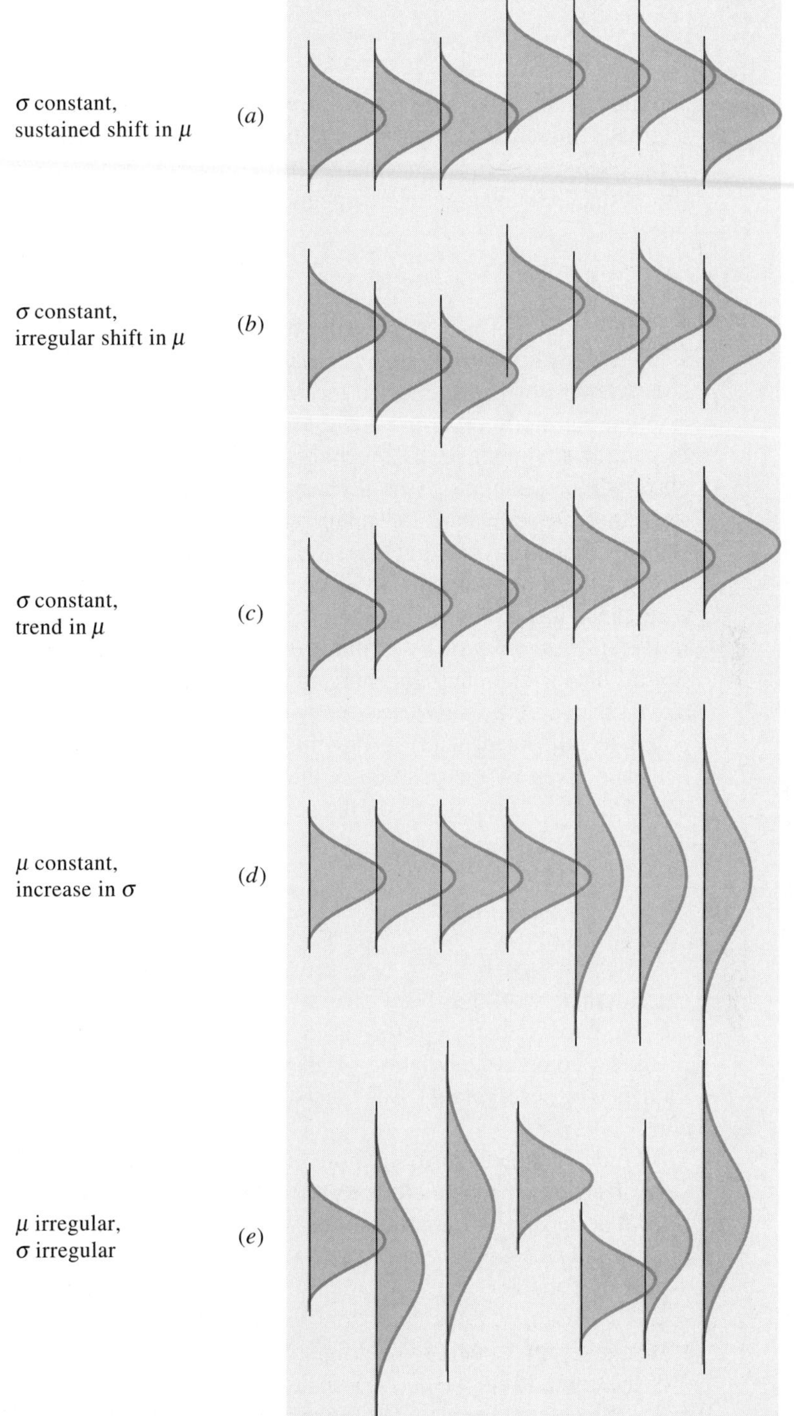

FIGURE 4.4 Chance cause systems (represented here by frequency curves) may change in different ways: *(a)* temporary jump shift in universe average with constant spread; *(b)* irregular shifts in universe average with constant spread; *(c)* steady trend in universe average with constant spread; *(d)* change in universe spread with no change in average; *(e)* irregular changes in both average and spread.

and universe dispersion remains constant. Figure 4.4*c* shows the universe average gradually increasing. Figure 4.4*d* shows a situation in which, although the universe average remained unchanged, the universe dispersion doubled. Figure 4.4*e* shows universe average and dispersion both varying erratically.

4.4 CHANGES IN UNIVERSE AVERAGE

A common type of lack of control observed in manufacturing is a shift in universe average with little or no change in universe dispersion. In such cases, the control chart is often of great value to the machine setter, helping the setter to center the machine setting in order to produce at a desired process average. This type of lack of control is shown on the $\overline{X}$ chart; unless the changes in universe average take place within a subgroup, the R chart will show control.

In those cases where the main reason for keeping a control chart is to detect changes in the universe average, the appropriate scheme of selection of subgroups differs from those cases in which the control chart has several purposes, including acceptance inspection.

Because, as previously explained, control limits are set far enough from the central line on the chart for there to be very few points outside the limits without a real change in the universe, small shifts in universe average will not cause many points to fall out of control. For this reason, it is often useful to supplement the evidence given by the position of the points relative to the control limits with evidence given by tests based on the statistical theory of runs or sequences.

4.4.1 Some Tests for Lack of Control Based on Runs of Points above or below the Central Line on the Control Chart

Considerable work has been done by mathematicians on the development of various types of statistical tests based on the theory of runs. Many of these tests involve a complete tabulation of all the runs, long and short alike, in any definite sequence of observations. Such tests provide useful tools for the study of research data.

In order to detect shifts in a universe parameter in the common applications of control charts in manufacturing, the most practical plan is to use a few simple rules that depend only on the extreme runs. The following are suggested:

Consider that grounds exist for suspicion that the universe parameter has shifted:

- Whenever, in 7 successive points on the control chart, all are on the same side of the central line
- Whenever, in 11 successive points on the control chart, at least 10 are on the same side of the central line
- Whenever, in 14 successive points on the control chart, at least 12 are on the same side of the central line

- Whenever, in 17 successive points on the control chart, at least 14 are on the same side of the central line
- Whenever, in 20 successive points on the control chart, at least 16 are on the same side of the central line

The theoretical basis for these rules is discussed in Chap. 5.

The sequences mentioned in each of these rules will occur as a matter of chance, with no change in the universe, more frequently than a point outside of 3-sigma control limits. (In fact, such sequences did occur in the 400 drawings from Shewhart's normal bowl.) For this reason they provide a less reliable basis for hunting for trouble than the occurrence of a point outside of control limits.

Moreover, if all these different rules are used to judge whether the universe parameter has shifted, the chances of the false indication of a shift are greater than if only one of the rules were used.[†]

Sequences of points on one side of a central line can be useful in relation to possible action on the control chart even where they are not used for hunting for trouble in the process. They are particularly helpful in the special case where the lower control limit is zero. For instance, a reduction in the value of σ calls for a change in the location of the upper and lower control limits on the $\bar{X}$ chart and a change in the center line and limits on the R chart. But in all R charts for which the subgroup size is not more than six, it is impossible for the change in σ to be shown by a point falling below the lower control limit of zero. However, an indication of the decrease in process dispersion can be obtained by applying the theory of runs.

Certain other applications of the theory of runs to the interpretation of control charts are discussed in Chap. 10.

4.4.2 Sustained Shift in Universe Average

In the control chart for $\bar{X}$ given in Fig. 4.5, the universe varied in the manner indicated in Fig. 4.4*a*. The universe average for the first 40 subgroups was 30; for the next 40 subgroups it was 40; for the final 20 subgroups it went back to 30. The universe standard deviation was 10 throughout the entire period.

Figure 4.5 was made by adding 10 to each drawing from the 161st to the 320th in Table 3.7. This has the same effect as if subgroups 41 to 80 were drawn from a bowl in which each chip was marked with a number 10 higher than a corresponding chip in the distribution described in Table 3.6. Of course, subgroups 1 to 40 and 81 to 100 were drawn from the bowl described in Table 3.6. Figure 4.5 is similar to Fig. 4.2 except that points 41 to 80 have been raised 10 units. However, the control limits are drawn on Fig. 4.5 as if they were established by the $\bar{X}$ and R of the first 40 subgroups.

During the period in which the universe average has changed, namely, subgroups 41 to 80, occasional points are above the upper control limit. Tests based

[†]See S. W. Roberts, "Properties of Control Chart Zone Tests," *Bell System Technical Journal,* vol. 37, pp. 83–114, January 1958.

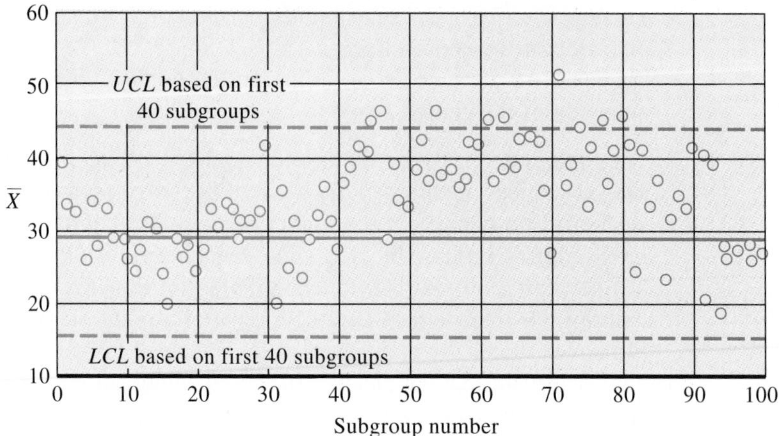

FIGURE 4.5 $\overline{X}$ chart showing sustained shift in universe average.

on the theory of runs also give clear evidence that the universe average has shifted.

This shift in average was equal to one standard deviation of the universe. It is obvious that the greater this shift in the average, the stronger the evidence of lack of control, and the sooner the shift is likely to be detected by the control chart.

When sustained shifts of process average occur *after* the control limits have been established, they result in all the out-of-control points falling outside *one* control limit, and all suspicious runs occurring on the *same* side of the central line. However, if the shift occurs *during* the period from which the control limits were established, the evidences of lack of control shift from one side to the other at the time of shift in process average.

No *R* chart has been given to accompany the $\overline{X}$ chart of Fig. 4.5. With this type of shift in process average, the *R* chart gives no indication of lack of control. It is evident that without a change of universe dispersion, the dispersion within the subgroups will not be affected.

Examples 4.1 and 4.2 illustrate situations involving shifts in process average.

Example 4.1

Shift in process average of acidity of dye liquor

Facts of the case In the dyeing of woolen yarns, it is desirable to control the acidity of the dye liquor. Unless the dye liquor is sufficiently acid, the penetration of color is unsatisfactory; on the other hand, a too acid liquor affects the durability of the products made from the yarn. Acidity is conveniently measured as pH (hydrogen ion concentration). A low pH corresponds to high acidity, and vice versa. In any dyeing operation there is a band of pH values within which the best results as to both color penetration and durability are obtained. A control chart for pH is helpful in maintaining acidity within the desired band.

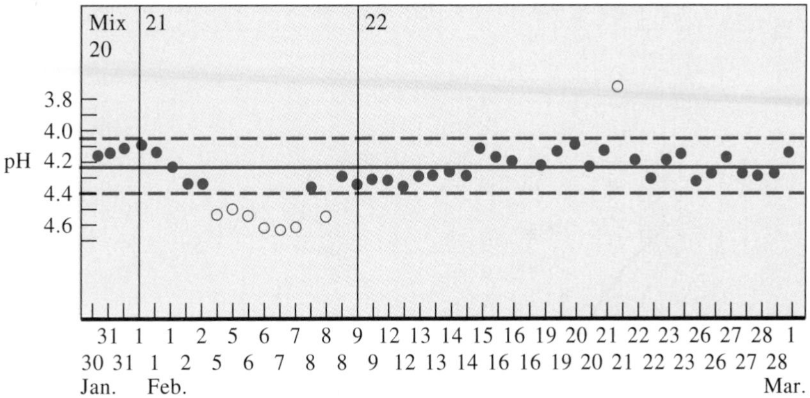

FIGURE 4.6 Control chart for average $\overline{X}$ for pH of dye kettles.

Such a control chart is shown in Fig. 4.6. On this chart are plotted the average X values of pH of dye liquor from five Hussong kettles used for the dyeing of blanket wool. Table 4.4 gives the actual $\overline{X}$ and R values for the period of approximately 5 weeks covered by this chart. Generally, two determinations of pH were made from each kettle every day, although a few days show one or three determinations.

The central line on the chart at $\overline{X}_0 = 4.22$, and the control limits of 4.05 and 4.39 were established by previous data. This process average and dispersion had proved to be satisfactory from the standpoint of the desired characteristics of the dye liquor. The chart is plotted with high acidity (that is, low pH) at the top of the graph.

Analysis and action The acidity of the dyeing solution depends not only on the constituents put into the dye liquor but also on the characteristics of the wool being dyed. From time to time it is necessary to use wools from sources that have different characteristics. Although blends of wools from various sources are made, successive blends will differ somewhat from one another.

On February 1 a new blend of entirely different wools was introduced. Immediately the acidity dropped. On February 5, after the old surplus stock had been used up, acidity fell below the control limit and continued out of control thereafter until corrective measures were taken on February 8. At this time the amount of acid introduced into the dye liquor was changed. Although after this all points fell within the control limits, the run of points below the central line indicated that the previous average was not restored until February 15.

Thereafter, with the exception of a brief departure from control on February 21, the pH values continued in satisfactory control. The temporary difficulty on February 21 was traced to two batches of improperly neutralized carbonized (baked with concentrated sulfuric acid) stock. Such stock is acid in relation to stock normally used.

TABLE 4.4 pH OF DYE LIQUOR USED FOR DYEING OF BLANKET WOOLS

Date	$\bar{X}$	R	Date	$\bar{X}$	R
Jan. 30	4.17	0.14	Feb. 14*a*	4.25	0.11
31*a*	4.15	0.30	14*b*	4.26	0.26
31*b*	4.08	0.20	15*a*	4.10	0.18
Feb. 1*a*	4.07	0.09	15*b*	4.14	0.23
1*b*	4.13	0.10	16*a*	4.20	0.52
1*c*	4.22	0.24	16*b*	4.24	0.17
2*a*	4.33	0.65	19*a*	4.21	0.46
2*b*	4.33	0.17	19*b*	4.11	0.20
5*a*	4.54	0.58	20*a*	4.07	0.40
5*b*	4.50	0.22	20*b*	4.22	0.12
6*a*	4.54	0.22	21*a*	4.11	1.34
6*b*	4.61	0.18	21*b*	3.72	0.96
7*a*	4.63	0.44	22*a*	4.18	0.35
7*b*	4.61	0.20	22*b*	4.29	0.31
8*a*	4.37	0.23	23*a*	4.17	0.20
8*b*	4.54	0.23	23*b*	4.14	0.13
8*c*	4.29	0.32	26*a*	4.32	0.26
9*a*	4.35	0.62	26*b*	4.26	0.08
9*b*	4.31	0.28	27*a*	4.16	0.51
12*a*	4.32	0.20	27*b*	4.25	0.25
12*b*	4.36	0.40	28*a*	4.28	0.09
13*a*	4.27	0.40	28*b*	4.26	0.15
13*b*	4.28	0.38	Mar. 1	4.14	0.11

Averages and ranges of subgroups composed of samples from five Hussong kettles

Example 4.2
Shift in quality level of steel castings

Facts of the case Certain specially treated steel castings were required to meet rigid requirements as to strength and ductility. As part of the acceptance procedure for these castings, the purchaser's contract required that two tensile specimens from each heat be tested to destruction in the testing laboratory. Specifications regarding the results of these tests covered tensile strength, yield point, percent elongation, and reduction in area. $\bar{X}$ and R charts were maintained on yield point and percent elongation. Whenever these charts showed control for a sufficient number of heats and the tests were otherwise satisfactory, the amount of required testing was reduced to two specimens from every fourth heat.

The tests did in fact show control at a satisfactory level for a number of months. Suddenly, points on both $\bar{X}$ charts went out of control.

Analysis and action The responsible supervisors in the production department were sure no changes had been made in production methods. Nevertheless, all the following heats continued out of control on the $\bar{X}$ charts. An attempt was made to throw the blame on the testing laboratory; however, tests made by other personnel on other machines showed the points continuing to fall out of control. This condition continued for some time despite pressure brought by supervisors on operating personnel. Finally someone remarked that a change in the source of quench water for heat treatment had been made just before the time when the first points fell out of control. Although no one believed that this could be the source of the trouble, as a last resort the original source of quench water was restored. Immediately the results of tensile tests of specimens from subsequent heats gave points that fell within the original control limits. From this time on, the process continued in control at the original satisfactory level.

Comments on Examples 4.1 and 4.2

These examples describe cases in which the average values of the measured quality definitely moved outside control-chart limits. Even though in such cases the actual troubleshooting may encounter many difficulties, from a statistical viewpoint these are the simplest types of examples. In fact, when such cases are described in presenting the advantages of control-chart techniques, the question is likely to be asked, "Would not the need for action have been just as clear from the tests as a matter of common sense even if there had been no control chart? Just how did the control chart itself really help?"

This question implies that the tests would have been carried out and that their results would have been conveniently available for analysis, regardless of whether or not a control chart was used. This assumption is often contrary to the facts. Experience shows that the use of the control chart sometimes leads to a more systematic program of testing and measurement; it nearly always leads to a tabulation of test results in a way that makes them more readily available as a basis for action.

But the control chart's contribution to effective action in cases like Examples 4.1 and 4.2 does not depend on any stimulus that it may have given to more systematic procedures in making and recording measurements of quality. The control chart provides a graphic presentation of quality history that gives a clearer picture than could be obtained from any tabulation of test data. Of course the primary contribution of the control chart to such troubleshooting lies in the information given by the control limits. These provide rules for action that are much more definite and much more reliable than any so-called commonsense judgments. When the control chart shows a long period of control followed by several points out of control, the evidence is conclusive that there is a discoverable cause of variation. Such evidence may be followed with confidence in the face of assertions, such as those in Example 4.2, that no change has really taken place. Moreover, the limits provide a definite basis for judgment as to whether or not the cause of trouble has been corrected.

It will be noted that in Example 4.2 the production personnel first insisted there was no trouble; the cause of trouble was found only because some individual whose job was entirely outside production insisted that the hunt be continued. This example is not intended to suggest that an outsider using the control chart can succeed where those intimately connected with production will fail. What the outsider provided here was *insistence* that the source of trouble could be found; the actual identification of the source was made by individuals connected with production.

Experience shows that those closest to production sometimes have blind spots with respect to certain sources of trouble. A quality control engineer with wide experience states that in production conferences regarding processes that the control chart shows have gone out of control, it is common for someone close to the production operation to state: "It may be cause *A* or cause *B* or cause *C*. One thing I am sure of is that it isn't cause *D*. "In such instances, about half the time the source of trouble turns out to be cause *D*.

Examples 4.1 and 4.2 are alike in that it was the evidence of the control chart as to *when* the process went out of control that was the basis of discovering *why* it went out of control.

Example 4.2 illustrates a case in which a purchaser's control chart gave guidance to a supplier. This is a common occurrence and often leads to the use of the control-chart technique by the supplier.

Example 4.1 represents a common situation characteristic of the food industry, the chemical industry, and the ceramic industry as well as the textile industry. This is the situation in which some variation in raw material quality is inevitable. In spite of this variation (such as that of wool from different sources) it is desired to maintain a certain quality level (as of pH of dye liquor) at some stage of the manufacturing operations. This quality level may in itself not be a quality of the finished product but, as in the case of the dye liquor, may influence various desired qualities of the product.

In the case of Example 4.1, control limits on the $\bar{X}$ chart were set on the basis of a previous $\bar{R}$ of 0.30. No R chart was kept, as previous experience had indicated that troubles were always with process average and not with process dispersion. However, if an R chart is plotted for the data of Table 4.4, it will show lack of control on February 2 and 21. Both cases coincide with times when $\bar{X}$ went out of control and result from an unequal influence on different members of the subgroup by a change in the process average. To use the analogy of drawing from a bowl, it is as if while switching from one bowl to another, a few subgroups were drawn with some chips from the old bowl and the remaining chips from the new bowl.

4.4.3 Frequent Irregular Changes in Universe Average with Constant Universe Standard Deviation

Figure 4.7 is derived from the data of Table 3.7 in a way intended to produce an illustration of the effect of numerous changes in the universe average. Table 3.7 gave drawings from a bowl with a constant average of 30 and standard deviation of 10. Figure 4.7 uses these drawings for the first 50 subgroups. The remaining 50 subgroups show irregular variation in universe average, with the amount and duration of each shift having been determined from a table of random numbers. The

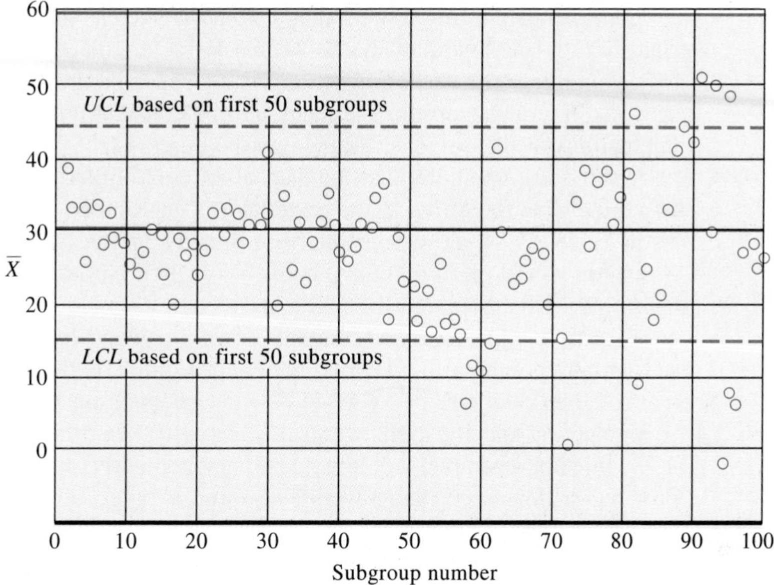

FIGURE 4.7 $\overline{X}$ chart showing the effect of frequent irregular changes in universe average.

universe standard deviation continued at 10 throughout all 100 subgroups. The averages used were as follows:

Subgroups	Universe average μ
1–50	30
51–57	20
58–61	10
62	45
63–69	25
70–72	5
73–81	35
82–85	15
86–93	40
94–96	10
97–100	30

This type of situation is often a result of carelessness in machine setting. As indicated in Fig. 4.7, it results in points falling outside *both* control limits on the $\overline{X}$ chart. If the shifts in universe average occur between subgroups and never happen to fall within a subgroup, the R chart will continue to show control regardless of the variations shown on the $\overline{X}$ chart.

At first glance it might seem that a situation such as that pictured in the control chart of Fig. 4.7 should always be a cause for action on the manufacturing process. However, this is not necessarily true. It all depends on the relationship between the specified tolerances on the one hand and the constant universe dispersion and the fluctuations of the universe average on the other.

Assume, for example, that the data of Table 3.7 referred to the final two digits of a dimension measured in thousandths of a centimeter; let us say that a figure of 27 in Table 3.7 would correspond to a dimension of 21.527 cm. This data transformation would mean that the process σ was approximately 0.010 cm. If the engineering specification were 21.530 ± 0.030 cm, it would be evident that the tolerances were so tight that careful attention would always have to be paid to machine setting; any points falling beyond the control limits on the $\bar{X}$ chart would indicate that some product was being made outside specification limits. On the other hand, if engineering specifications were 21.530 ± 0.100 cm, the situation shown in Fig. 4.7, while not desirable, would not have resulted in the production of unacceptable product; in this case, there would be quite a range of possible carelessness in machine setting without causing any product to fall outside specifications. (Problem 4.21 asks for an explanation of these statements.)

In control of dimensions in manufacturing, the great usefulness of the control charts for $\bar{X}$ and R is in those situations where the specification tolerances are tight compared to the inherent variability of the manufacturing process. Fortunately, in most manufacturing operations this is true of only a small proportion of specified dimensions.

4.5 SHIFT IN UNIVERSE DISPERSION WITH NO CHANGE IN UNIVERSE AVERAGE

The inherent variability of a process may change from time to time even though there is no change in the process average. On any process where the skill and care of the operator is an important factor, a common cause of increase in variability is a change from one operator to another who is less skillful or less careful. In fact, an operator's skill and care may sometimes vary from day to day or from hour to hour.

This type of shift in universe is illustrated in Fig. 4.8, which has been adapted from the data of Table 3.7. For the first 40 subgroups, universe standard deviation is 10; for subgroups 41 to 80, universe standard deviation is 20; for the final 20 subgroups, it is 10 again. The universe average has been held constant at 30 throughout. Figure 4.8 gives both $\bar{X}$ and R charts. Both charts show lack of control, with more points outside control limits on the R chart than on the $\bar{X}$ chart. Extreme runs above the central line on the R chart also give strong evidence of lack of control.

The central lines and control limits on Fig. 4.8 have been set using the data from the first 40 subgroups. If they had been set using the data from the first 80 subgroups, $\bar{R}$ would have been larger and the limits on both the $\bar{X}$ and R charts

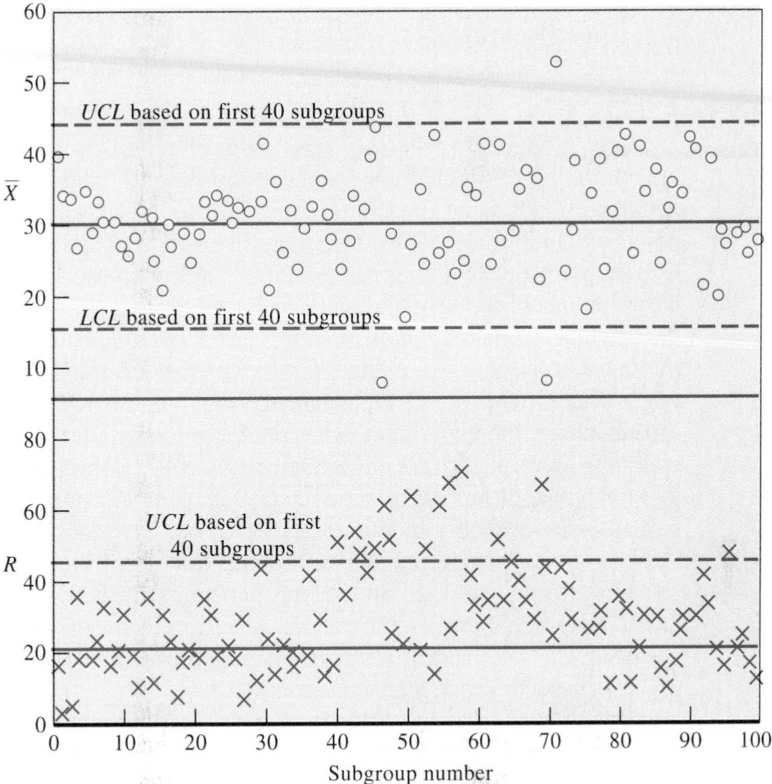

FIGURE 4.8 Control charts for $\overline{X}$ and R showing the effect of a sustained shift in universe dispersion.

would have been wider. With these limits fewer points would have fallen outside the control limits. However, the higher value of the central line on the R chart would have introduced other evidence of lack of control in the form of extreme runs below the central line.

4.5.1 The Importance of the *R* Chart Depends on the Type of Process

The previous section illustrated how a shift in universe dispersion may cause lack of control to be indicated on both the $\overline{X}$ and R charts. Thus, in searching for the causes of an out-of-control condition, most practitioners will view the R chart first. The following quotation from the Western Electric Company Handbook illustrates this philosophy.[†]

[†]"Statistical Quality Control Handbook," p. 11, Western Electric Company, Inc., New York, 1956.

The R pattern is read first, and from this it is possible to identify many causes directly. The $\bar{X}$ pattern is read in the light of the R chart, and this makes it possible to identify other causes. Finally, the $\bar{X}$ pattern and R pattern are read jointly, which gives still further information.

Many production processes tend to have relatively uniform dispersion even though the process centering changes from time to time. Ranges of samples from automatic devices and machines often are found to vary only a little. It is the points on the $\bar{X}$ chart, related to setting, that tend to wander. An automatic machine usually can produce a uniform product for at least a short period of time; it is tool wear and machine setters that need to be controlled, not the inherent variability of the machine. Should the R chart show lack of control, usually it is the result of tool dulling or looseness or severe and otherwise noticeable malfunction necessitating major repair, overhaul, or replacement.

In contrast, there are many other production processes in which it is difficult to maintain uniform process dispersion; in such processes the R chart may be an extremely useful tool for process control. A generalization, subject to exceptions, is that such variability of process dispersion is particularly likely to be found in those processes where the skill and concentration of the operator is vital to consistent performance. Odd and erratic performance likely will show up on the R chart while the $\bar{X}$ chart continues to attest to mean settings or targets. Generally speaking, the first step in improving such processes should be to try to bring the process dispersion into statistical control.

The reader should not conclude from this discussion that plotting of the R chart data may be dismissed out of hand when the product of automatic processes is to be analyzed. It always is the result of process data analysis over a period of time that suggests when sampling frequencies may be reduced or the actual plotting of data points for either $\bar{X}$ or R may be discontinued.

4.5.2 Effect of Gradual Reduction in Process Dispersion on Control Limits on $\bar{X}$ and R Charts

On processes where the skill and care of the operator have a major influence on process dispersion, the mere introduction of a control chart often causes a gradual reduction in the dispersion of the quality characteristic charted. Example 2.2 described a case where this happened. Such a change is a useful contribution of the control chart to improved quality, even though this contribution does not depend on the use of the chart as a statistical tool to diagnose the causes of troubles.

Nevertheless, such cases create certain practical problems in the interpretation of the control chart for R. As long as the subgroup size is six or less, the lower control limit on the R chart is always zero regardless of the value of $\bar{R}$. Thus, with the usual size of subgroups, a decrease in universe dispersion cannot be reflected in points falling below the lower control limit on the R chart. Here reliance must be placed on regular reviews of the value of $\bar{R}$ used to set limits on both the $\bar{X}$ and

R charts. Often the need to review $\bar{R}$ is suggested by extreme runs below the central line on the R chart. It should be remembered that whenever, because of evidence of a decrease in universe dispersion, the value of $\bar{R}$ used in computing control limits is recalculated and reduced, this tightens the control limits on the $\bar{X}$ chart as well as on the R chart.

4.6 CHANGES IN UNIVERSE AVERAGE AND UNIVERSE DISPERSION

When universe dispersion as well as universe average is shifting, it is obvious that lack of control will be indicated in both charts, the R chart as well as the $\bar{X}$ chart. Such a condition was illustrated in Fig. 1.2 of Example 1.1. This state of affairs is common in the first stages of the use of the control chart for variables for analysis of many manufacturing operations.

Where several assignable causes of variation exist, the elimination of some of the causes will decrease the number of out-of-control points but will not eliminate all of them. In such circumstances, the continuance of some points out of control may be discouraging to anyone anxious for quick results. Rather than causing discouragement, the control chart is better viewed as an indication that further improvement is possible and as an incentive to keep hunting for more sources of trouble.

4.6.1 Assignables Causes of Variation May Be Due to Errors of Measurement

In Example 4.2, the production department first tried to lay the blame on faulty inspection as the source of the indication of lack of control. This common production alibi proved to be wrong. Nevertheless, there is always the possibility that it may be right.

It should be emphasized that the control-chart analysis, like any other statistical test, is applied to a set of numbers. Anything that affects these numbers affects the control chart. The numbers are influenced by variations in the measurements, just as they are influenced by variations in the quality characteristic being measured. The universe from which samples are being drawn is the result of a cause system that includes measurement causes as well as production causes.

Therefore, an error in measurement may be an assignable cause of variation in the figures resulting from the measurements. In describing the various types of lack of control, reference was made to production situations that might give rise to each type. Similarly, each might have been associated with some sort of inspection error. An error in setting a measuring device may make a sudden shift in universe average. Frequent errors in setting may make irregular shifts in the average. Some types of wear of measuring devices may increase universe dispersion. Other types of wear may give rise to trends in averages.

Some changes affecting $\overline{X}$ chart (stable variability)	Some causes affecting R chart

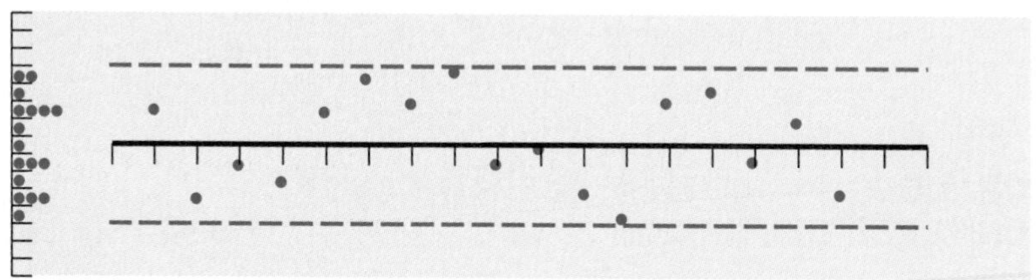

(*a*) *Recurring cycles*

Some changes affecting $\overline{X}$ chart (stable variability)	Some causes affecting R chart
1. Temperature or other recurring changes in physical environment	1. Scheduled preventive maintenance
2. Worker fatigue	2. Worker fatigue
3. Differences in measuring or testing devices that are used in order	3. Worn tools
4. Regular rotation of machines or operators	
5. Merging of subassemblies or other processes	

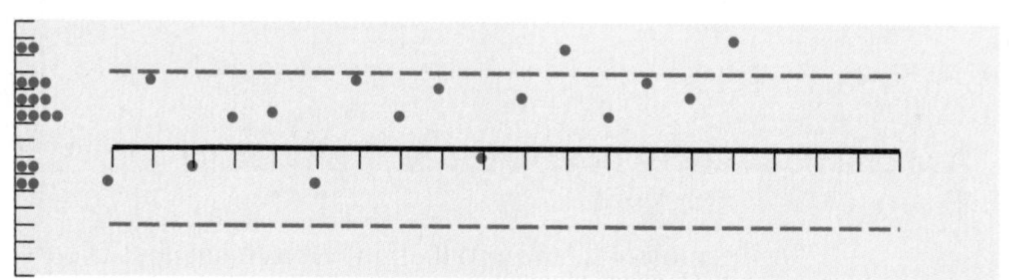

(*b*) *Trends*

Some changes affecting $\overline{X}$ chart (stable variability)	Some causes affecting R chart
1. Gradual deterioration of equipment that can affect all items	1. Improvement or deterioration of operator skill
2. Worker fatigue	2. Worker fatigue
3. Accumulation of waste products	3. Change in proportions of subprocesses feeding an assembly line
4. Deterioration of environmental conditions	4. Gradual change in homogeneity of incoming material quality

FIGURE 4.9 Some interpretations of patterns of $\overline{X}$ and R charts.

4.6.2 Interpreting Patterns of Variation on $\overline{X}$ and R Charts

An important aspect of the use of control charts is the interpretation of the many patterns that they may exhibit. In addition to the standard rule for looking for trouble, we have also discussed secondary or supplementary rules based on the theory of runs.

Some causes affecting $\overline{X}$ chart (stable variability)	Some causes affecting R chart

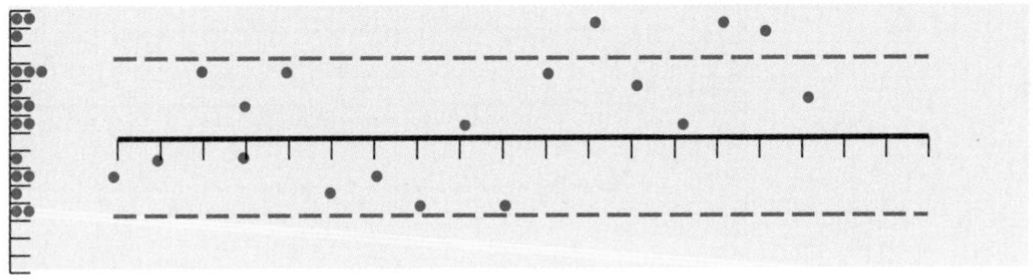

(c) Jumps in process level

1. Change in proportions of materials or subassemblies coming from different sources 2. New worker or machine 3. Modification of production method or process 4. Change in inspection device or method	1. Change in material 2. Change in method 3. Change in worker

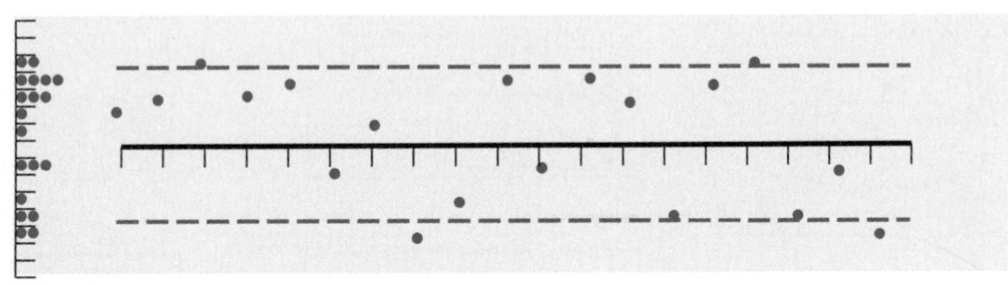

(d) High proportion of points near or outside limits

1. Overcontrol 2. Large systematic differences in material quality 3. Large systematic differences in test method or equipment 4. Control of two or more processes on same chart	1. Mixture of materials of distinctly different quality 2. Different workers using a single R chart 3. Data from processes under different conditions plotted on same chart

FIGURE 4.9 *(continued)*

In the examples and discussions presented, an intimate knowledge of the process being controlled was vital to the effective use of the control chart. The control chart tells *when* to look for trouble but it cannot, by itself, tell *where* to look or *what* cause will be found.

Figure 4.9 illustrates some of the patterns that are frequently seen on control charts and states some possible causes of these patterns. Studies performed at Bell

Some causes affecting $\overline{X}$ chart (stable variability)	Some causes affecting R chart

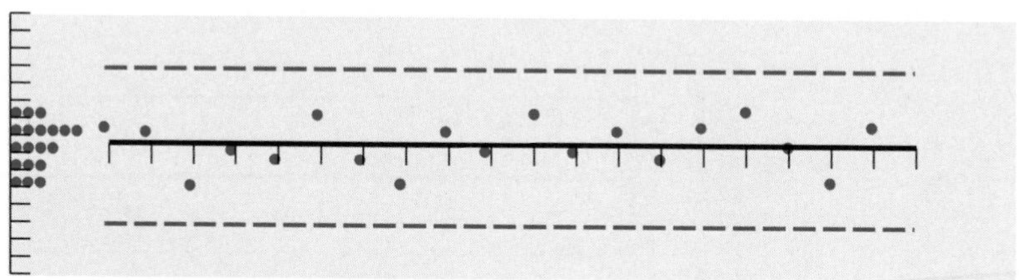

(e) Stratification or lack of variability

1. Incorrect calculation of control limits 2. Incorrect subgrouping; R chart captures more variability than $\overline{X}$ chart	1. Collecting in each sample a number of measurements from widely differing universes

Correlation between charts

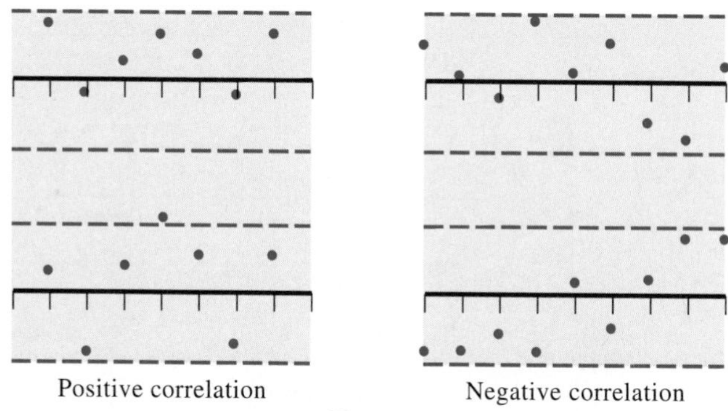

Positive correlation   Negative correlation

(f) Correlation between $\overline{X}$ and R charts for same process

1. Skewness in underlying universe
2. Points generated from the same sample

(g) Correlation between different $\overline{X}$ or R charts

1. Points generated from the same sample
2. Unknown cause-and-effect relationships

FIGURE 4.9 *(continued)*

Telephone Laboratories during the summer of 1964 categorized the most frequent causes of trouble.† Developed as an aid to training young inspectors and engineers,

†The authors are indebted to Dr. Richard L. Patterson, University of Michigan, for supplying the basic data for Fig. 4.9. A more complete discussion of recognition of patterns on control charts is contained in the "Statistical Quality Control Handbook," Western Electric Company, New York, 1956.

the listed causes should be used only as a guide to possible action, not as an authoritative listing of *the* causes of trouble.

4.6.3 The Causing of Product Variability by Too Frequent Adjustment of the Process Centering

In certain types of manufacturing processes, the greatest value of an $\overline{X}$ chart may be that it tells the operator when to leave the process alone. Many processes are similar to the ones illustrated in Fig. 4.4*a, b,* and *c* in having a fairly constant process dispersion under ordinary circumstances combined with a tendency of the process centering to shift. In a number of such processes, the process centering can be changed by one or more simple machine adjustments made by the operator. Without the restraining influence exerted by a control chart, operators frequently have a tendency to adjust the process too often. Example 4.3 describes a case where this occurred.

Example 4.3

Use of $\overline{X}$ chart to prevent too frequent changes in process centering

Facts of the case One of the steps in manufacturing an assembled article was to fill a certain component with a specified weight of a certain dry powder. The specification might be described as $W \pm T$, where W is the desired average weight and T is the allowable tolerance. Several thousand components per hour were filled by an automatic machine. The average filling weight could be increased or decreased by a simple adjustment of the machine. The practice of the operator—working under instructions from a supervisor—was to take one filled component at random every half-hour or so, empty the component, and weigh its powder contents on an accurate balance. Whenever the weight of this sample of one fell below W, the machine was adjusted to give a higher weight; whenever the weight fell above W, the machine was adjusted to give a lower weight. Many of the samples caused some machine adjustment.

Subsequent inspection of finished product showed a substantial fraction of the product falling outside the specified limits of $W \pm T$. One of the persons supervising the operation commented that this failure to meet specifications took place "in spite of the careful attention we are giving to machine adjustment."

Analysis and action When an engineer familiar with control-chart techniques joined the staff of this manufacturing plant, this was the first operation to which $\overline{X}$ and R charts were applied. Samples of five were taken at intervals over a considerable period of time during which no machine adjustment was permitted. Control charts showed that the process was close to being in statistical control. Thereafter the control charts were continued on a regular basis. The operator was instructed to make no machine adjustments as long as points fell within control limits. It turned out that the process was one that ordinarily stayed in statistical control for a fairly long period without requiring adjustment. Not only did it prove possible to hold all product within the original specification limits, but the tolerance limits were later narrowed in a way that improved the performance of the assembled product. It was evident that the "careful attention" to machine adjustment had been the cause of unnecessarily large variation in product quality.

Comment on Example 4.3

Although in fact the example describes one particular case, the reader might interpret it as giving a rough description of many similar cases that have occurred in a wide variety of circumstances. Whenever process adjustments are made based on criteria that do not recognize the inherent variability of a process, there is danger that these adjustments will be too frequent and that they will increase the amount of product variability. One quality control consultant described this type of condition as a "sitting duck."

A simple model based on a sequence of drawings from Shewhart's bowl may help the reader to visualize what was happening in Example 4.3 before the introduction of the control charts. Assume that with no machine adjustments, filling weights follow the distribution shown in Table 3.7, with a centering at 30 and a spread from 0 to 60. (These figures might represent the final two digits of a weight expressed, say, in thousandths of a gram.) Whenever the centering of the distribution is raised or lowered, all other values of the distribution are raised or lowered by an equal amount.

Table 4.5 uses drawings 141 to 156 from Shewhart's bowl to illustrate the adjustments to the process centering that would have been made if an adjustment had been made whenever a single observed value fell above or below the desired value of 30. Each observed value from the original bowl is increased or decreased to correspond to the difference between the current process centering and 30. The arithmetic of the table should be self-explanatory. It will be observed from column *E* that changes are being made at every observation. It will be observed from column *F* that these changes are introducing considerable variability into the centering of the process.

4.7 A POSSIBLE VIEW OF THE QUESTION ANSWERED BY A CONTROL CHART

The following discussion uses the $\overline{X}$ chart as an example, assuming the special case in which the universe dispersion remains constant and the universe average may shift. However, the points that are brought out in the discussion are equally applicable to other types of charts and to cases where changes occur in the dispersion.

The 100 subgroups from Shewhart's bowl listed in Table 3.7 and charted in Fig. 4.2 gave control limits on the $\overline{X}$ chart of approximately 45 and 15. Suppose the 101st subgroup has an $\overline{X}$ of 43. Should we interpret this as telling us that the universe has changed (possibly by the substitution of a set of chips that has a μ of more than 30)? Our control-chart limits answer this question with a "No." However, if the $\overline{X}$ for the 101st subgroup should be 48, the answer would be "Yes."

Either the universe has changed or it has not changed. We are in the position of having to make a judgment on this point based on the evidence of the $\overline{X}$ of a sample of four. Sometimes this judgment will be correct; sometimes it will be incorrect.

TABLE 4.5 CALCULATIONS OF ADJUSTMENTS TO CENTERING OF SHEWHART'S NORMAL BOWL, DRAWINGS 141–156. IT IS ASSUMED THAT SAMPLES OF ONE HAVE BEEN TAKEN AND AN ADJUSTMENT MADE WHENEVER OBSERVED VALUE FALLS ABOVE OR BELOW 30

Bowl drawing number *(A)*	Marking on chip *(B)*	Correction to chip marking, preceding *F* minus 30 *(C)*	Adjusted chip value, *B* + *C* *(D)*	Change in process centering based on adjusted chip value, 30 minus *D* *(E)*	Revised process centering, preceding *F* plus *E* *(F)*
					30
141	21	0	21	+ 9	39
142	31	+ 9	40	−10	29
143	39	− 1	38	− 8	21
144	25	− 9	16	+14	35
145	17	+ 5	22	+ 8	43
146	44	+13	57	−27	16
147	54	−14	40	−10	6
148	13	−24	−11	+41	47
149	36	+17	53	−23	24
150	48	− 6	42	−12	12
151	19	−18	1	+29	41
152	41	+ 11	52	−22	19
153	25	− 11	14	+16	35
154	31	+ 5	36	− 6	29
155	38	− 1	37	− 7	22
156	30	− 8	22	+ 8	30

If we conclude that the universe has changed when it really is unchanged, this conclusion is described by writers on statistics as a *Type I* or α error. If we conclude that the universe has not changed when it really has changed, this conclusion is described as a *Type II* or β error.

With limits on an $\bar{X}$ chart at $\bar{\bar{X}} \pm 3\sigma_{\bar{X}}$, it is evident that Type I errors will be infrequent. On the other hand, if moderate changes in the centering of the process occur, Type II errors may happen fairly often. The existence of Type II errors with 3-sigma limits and a moderate shift in μ may be illustrated by looking back at Fig. 4.5, where we assume that μ shifted from 30 to 40 for subgroups 41 to 80. This was a shift of approximately one standard deviation. Only 9 of the $\bar{X}$ values for subgroups 41 to 80 fell outside the limits that had been established on the basis of subgroups 1 to 40; the first point outside limits did not occur until subgroup 45.

Modern statistical theory provides many different tests of hypotheses that samples come from the same or different universes. Control charts used in manufacturing may be thought of as relatively simple statistical tests; there are many tests of greater complexity and sophistication. A dilemma in all such tests is to strike some sort of rational balance between Type I and Type II errors. There are good reasons why the common practice in manufacturing is to set control-chart limits far enough apart so that Type I errors will be rare.

4.7.1 Some Practical Aspects of the Tightness of Limits on Control Charts Used in Manufacturing

A typical action based on out-of-control points on control charts for variables is to hunt for the source of trouble. Such hunting was illustrated in Examples 4.1 and 4.2. Sometimes a process may be shut down until it is believed that the assignable cause of variation has been found and corrected.

It is rare for a manufacturing process to behave for long periods of time like the drawing of chips from a bowl. At best, small changes in any process may be expected to occur from time to time—in the centering of the process and possibly in its dispersion and the form of the frequency distribution of any quality characteristic. The assignable causes of such small changes may be extremely difficult to find.

Moreover, trouble, in the sense of assignable causes of substantial changes in quality, is a common state of affairs in manufacturing. Even those large changes that cause nearly all points to fall outside 3-sigma limits may require much hunting and hard work before the assignable causes are found. This was illustrated in Example 4.2.

Under such circumstances, it seldom pays to hunt for trouble without a strong basis for confidence that trouble is really there. The real basis for the use of 3-sigma limits on control charts for variables in industrial quality control is experience that when closer limits, such as 2-sigma, are used, the control chart often gives indication of assignable causes of variation that simply cannot be found; whereas when 3-sigma limits are used and points fall out of control, a diligent search will usually disclose the assignable causes of variation.

Although wide limits, such as 3-sigma, are generally appropriate, special circumstances arise where narrower limits may be desirable. In principle, the tightness of control limits is an economic problem involving the balancing of the various economic consequences of different possible limits. If enough information about the relevant costs and other economic matters were available in each case, the tightness of limits could be tailor-made for each control chart. Unfortunately, all the relevant economic information is rarely obtainable.

4.7.2 Reducing the Chance of a Type II Error by Increasing the Subgroup Size

Consider an $\bar{X}$ chart in the type of case illustrated in Fig. 4.5, where σ remains constant and μ changes. A decision to use relatively wide limits, such as 3-sigma, ensures that Type I errors will be rare. This is true regardless of subgroup size.

However, the larger the subgroup size, the narrower the limits on the $\bar{X}$ chart and the greater the sensitivity of the $\bar{X}$ chart to shifts in μ.[†] In other words, an increase in subgroup size reduces the frequency of Type II errors without the penalty of too frequent Type I errors.

[†]Generally speaking, the advantages of the $\bar{X}$ chart cannot be obtained with a subgroup size of one. Nevertheless, some writers have misrepresented the control-chart technique by analyzing the sensitivity of such charts to changes in μ on the assumption (not explicitly stated) that all Shewhart charts use one as the subgroup size. For example, see Fig. 5 in H. M. Truax, "Cumulative Sum Charts and Their Application to the Chemical Industry," *Industrial Quality Control,* vol. 18, no. 6, p. 21, December 1961.

Cases sometimes arise where relatively large subgroup sizes are justified by the need for prompt detection of small shifts in μ. But there are usually good reasons for the common practice by favoring subgroup sizes of four or five.

4.7.3 Basis of Selection of Subgroups

The discussion throughout this chapter has implied *time* of production as the basis for selection of rational subgroups. This is a natural and logical basis, but it is not always a sufficient one. The conclusions to be drawn from any control chart depend on the basis of selection of subgroups; a process may appear to show control with one plan of subgrouping and not show control with some other plan.

Moreover, there are many situations in which the order of production is no longer known but where it is still possible to use the control chart to advantage.

4.8 NONPRODUCT APPLICATIONS OF CONTROL CHARTS FOR VARIABLES

Applications of Shewhart's control-chart procedures to processes not involving manufactured product date back to Deming's work with the U.S. Census Bureau in the late 1930s. (See Chap. 18.) It often happens in nonproduct applications that information is needed that corresponds in a general way to the knowledge of process capability in manufacturing; it is desired to know a value of μ and σ that could be achieved under the operation of a constant system of chance causes. Of course, the issue of the stability of the variable in question is at the heart of the use of most control charts. If there is reason to believe that assignable causes of variation exist and that desirable results can be achieved by identifying and eliminating these assignable causes, there is good reason for initiating $\bar{X}$ and R charts. In this respect, nonproduct applications do not differ greatly from conventional applications to manufactured product.

The heightened interest in continuous improvement supported by applications of statistical process control has lead to such interesting areas of application as hospitals (for example, response time in seconds to patient rings and patient delay times while waiting for x-rays) and sewage treatment facilities (such as control of turbidity of secondary clarifier effluent). In the first area, reducing patient waiting time is highly desirable, to say the least; in the second, sewage treatment in the United States is regulated by the Environmental Protection Agency, and violations of standards may lead to fines, penalties, or surcharges.

Two possible differences should be mentioned between many nonproduct applications and the usual applications in manufacturing. Whereas there are usually definite tolerance limits that apply to quality characteristics of product, there may be no such limits in nonproduct applications. Moreover, considerably more imagination may be required to select the appropriate variable in the nonproduct case.

In one successful nonproduct application, the variable was the difference between estimated performance time and actual performance time for the many component operations in several projects being carried out under a system of critical path scheduling. In a nonproduct application described by J. B. Pringle of the

Bell Telephone Company of Canada,[†] the variable was the deviation of the measured transmission loss in decibels on certain toll telephone trunks from the design loss for the same trunks. Although this latter problem had been analyzed for years by conventional statistical methods and substantial improvements obtained, the use of $\bar{X}$ and R charts was responsible for still further improvements. On this point, Mr. Pringle comments as follows:

> By this procedure, it was possible to bring the standard deviation of the distribution of trunk net loss deviations down from about 1.35 dB to 0.93 dB within the trial period. This was within the bogey of 1.0 dB and hence a matter of some satisfaction. But to me, probably the most satisfaction came about the third quarter of the trial, when reviewing the results with the foreman responsible for carrying out the trial. I asked him to estimate the amount of effort being expended on the trial compared to the amount expended previously when it seemed impossible to improve on $\sigma = 1.35$. After some thought, he said:—"Even including the time it took us to learn about the control chart method, I think we are spending only about 80% of the time we spent previously." This is the meaning of the quotation, "Better Quality at Lower Cost."
>
> The essence of the process is that action is taken only when it is highly probable that assignable causes are present and can be found. This directs effort into constructive channels, and reduces effort expended in pursuing minor causes, difficult to find, and which do not contribute much to total variability. Only after major contributing causes are found and removed, is there much to be gained from going after minor things. But the Control Chart, by itself, only points out where action can be taken with profit. It is the action taken that improves things. If we don't act to discover and remove causes, we won't get improvement.

4.9 CONFLICTING EXPRESSIONS FOR THE STANDARD DEVIATION OF A SET OF NUMBERS

Persons familiar with previous editions of this book, and who have experience using control charts for variables dating back prior to 1978, will recall that certain changes have been made in the symbols used to identify parameters and in the choice of sample statistics used to describe dispersion. Early writings and standards for control charting used a prime (′) notation to denote parameters. Thus the mean μ was denoted as $\bar{X}'$; the standard deviation σ was denoted as σ'. The change was consistent with changes instituted by ASQC and ANSI beginning in 1978. However, the latest revision of the ASTM manual STP 15D (1976)[‡] retains the prime notation. By 1985, most books and articles on quality control published in the United States had changed away from using the prime notation.

[†] J. B. Pringle, "S. Q. C. Methods in Telephone Transmission Maintenance," *Annual Convention Transactions* 1961, pp. 151–158, American Society for Quality Control, Milwaukee, Wis., 1961.

[‡] "ASTM Manual on the Presentation of Data and Control Chart Analysis," STP 15D (1976), American Society for Testing and Materials, Philadelphia, Pa. STP 15D is the fourth revision of the "ASTM Manual on Presentation of Data" first published in 1933. The original manual was the first attempt at standardizing data analysis procedures in the United States. Control-chart procedures were added in 1935, first as a supplementary document and then combined in the manual in 1937. The ASTM symbols, definitions, and nomenclature were adopted in 1941 and 1942 into the American War Standards on Quality Control, Z1.1, Z1.2, and Z1.3.

The second, and more confusing, change is related to the choice of statistics used to describe dispersion. We have named s the *sample standard deviation* and defined it as

$$s = \sqrt{\frac{\Sigma(X_j - \bar{X})^2}{n-1}}$$

This name and definition are consistent with writings in applied statistics where the mean of an unknown distribution must be estimated from a sample average.

In early editions we used the *root-mean-square* (RMS) *deviation* to designate this quantity, where

$$\sigma_{\text{RMS}} = \sqrt{\frac{\Sigma(X_j - \bar{X})^2}{n}}$$

The definition of the standard deviation of a set of numbers as the root-mean-square deviation from the arithmetic mean is in the classical tradition of writings on mathematical statistics. When the techniques and terminology of statistical quality control were being developed in the 1920s, it was natural that the current definition of standard deviation should be used. Moreover, because an estimate of σ is so important in judging process capability, and because σ_{RMS} and s both give biased estimates of σ, neither has any mathematical advantage over the other.

This usage of σ_{RMS} was introduced into the "ASTM Manual on Presentation of Data" (STP 15A), along with tables of factors to be used in control charting, in 1937 and adopted in 1941 in the American War Standards on Quality Control. The original (1951) version of ASQC Standard A1, "Definitions and Symbols for Control Charts," followed suit and adopted the definitions and tables of the ASTM Manual. In the first revision of the ASQC Standard, s was defined as the sample standard deviation and was recommended to replace σ_{RMS}. The 1976 revision of the ASTM Manual (STP 15D) followed the ASQC lead.

An unbiased estimator of σ obtained from s was given as

$$\hat{\sigma} = \frac{\bar{s}}{c_4}$$

$$\text{Since } \sigma_{\text{RMS}} = \sqrt{\frac{(n-1)}{n}}\, s, \quad \bar{\sigma}_{\text{RMS}} = \sqrt{\frac{(n-1)}{n}}\, \bar{s}$$

Thus

$$\hat{\sigma} = \frac{\bar{\sigma}_{\text{RMS}}}{c_4\sqrt{(n-1)/n}}$$

The σ_{RMS} symbol was used in the sixth edition of this book. In earlier editions, and in most quality control literature, σ was the symbol used. The RMS subscript was not used. Thus σ was defined as the root-mean-square statistic and, after correction, was used to estimate the parameter σ'. Tables of factors, like those in App. 3, were used to calculate control limits and estimate dispersion.

There are two important reasons why even a beginning reader in the subject of statistical quality control should be aware of the changes in notation and in the definition of sample dispersion. The first is that, at one time or another, anyone involved with quality control is going to be exposed to standards, articles, and texts that used the earlier symbols and definitions. Those who choose to read the extensive and rich literature of quality control, especially the landmark works of pioneers in the field, should be aware of the differences in how sample dispersion has been calculated. Awareness of the changes is protection against making mistakes and helps allay questions.

The second reason relates to the use of computer software in statistical process control. It sometimes is difficult to learn from reading the materials accompanying software packages or, for that matter, material accompanying pocket calculators which of the definitions of sample dispersion is being used. Usually somewhere in the description or instructions for these programs and routines there is a formula to indicate which definition is used, that is, an answer is given to the question, "Is n in the denominator, or $(n - 1)$?" Some pocket calculators allow for both. Usually the instructions will tell the user to apply the routine with $(n - 1)$ when working with deviations from a sample average and the routine with n when the mean μ parameter is known.

PROBLEMS

4.1. Control charts for $\bar{X}$ and s are maintained on the breaking strength in pounds in a certain destructive test of a particular type of ceramic insulator used in vacuum tubes. The subgroup size is 15. The values of $\bar{X}$ and s are computed for each subgroup. After 12 subgroups, $\Sigma\bar{X} = 1{,}307$ and $\Sigma s = 198.2$. Compute the values of the 3-sigma limits for the $\bar{X}$ and s charts, and estimate the value of σ on the assumption that the process is in statistical control.

4.2. Control charts for $\bar{X}$ and R, based on a subgroup size of 4, are to be used to control a process. The standard deviation of this process is 10. An aimed-at value of the mean ($\bar{X}_0$) is to be 250.

(*a*) Determine control limits for the $\bar{X}$ and R charts.

(*b*) Determine the probability of a point falling within the $\bar{X}$ chart control limits if the actual μ is 0.5σ below the aimed-at value of 250.

(*c*) If it is 1.0σ below 250.

(*d*) If it is 2.0σ below 250.

(*e*) If it is 2.5σ below 250.

(*f*) Sketch these probabilities as a function of the location of the true mean (horizontal scale), and comment on the usefulness of this diagram in relation to the detection of shifts in the mean.

4.3. The high-voltage output of a certain power supply unit for copy machines is specified as 350 ± 5 V dc. Subgroups of four units are drawn approximately every hour and tested. After 30 subgroups, $\Sigma\bar{X} = 10{,}560.0$ and $\Sigma R = 86.5$.

(*a*) Determine control limits for $\overline{X}$ and R charts, and estimate the value of σ.
Answer: 349.89, 354.11; 0, 6.57; 1.40.

(*b*) Assuming the process is in control and that variation in the product follows a normal distribution, what proportion of defective product is being made?
Answer: 0.0162.

4.4. (*a*) Can the process described in Prob. 4.3 meet specifications if it is recentered at 350?

(*b*) Calculate the control limits for $\overline{X}$ and R based on recentering at 350.

(*c*) If these new control limits are used but no adjustment is made in the process, compute the Type II error probability that this fact will not be detected on the first subgroup plotted against the new control limits.
Answer: 0.5557.

4.5. An automatic screw machine turns out round-head bolts with a specified shank diameter of 9.00 ± 0.04 mm. The process has been operating in control at an estimated μ of 9.00 mm and an $\overline{R}$ of 0.0206 mm. The subgroup size is four.

(*a*) Calculate the $\overline{X}$ and R chart control limits.
Answer: 8.985, 9.015; 0, 0.470.

(*b*) If the mean of the process shifts to 9.02 mm, compute the Type II error probability that the shift will not be detected on the first subgroup drawn after it occurs.
Answer: 0.1587.

(*c*) What proportion of defective product is being produced at this new value of μ assuming that the product is normally distributed?
Answer: 0.0227.

4.6. Control charts for $\overline{X}$ and s are maintained on the resistance in ohms of a certain rheostat coil based on a subgroup size of 5. After 30 subgroups, $\Sigma\overline{X} = 58{,}395$ and $\Sigma s = 1{,}216$.

(*a*) Determine the central lines and 3-sigma control limits for this process.

(*b*) Estimate the value of σ, assuming that the process is operating in statistical control.

(*c*) Assuming that the distribution generated by the process is approximately normal, what proportion of the rheostat coils meets specifications of 2,000 ± 150 ohms?

4.7. In order to meet government regulations, the contained weight of a product must at least equal the labeled weight 98% of the time. Control charts for $\overline{X}$ and s are maintained on the weight in ounces of the contents using a subgroup size of 10. After 20 subgroups, $\Sigma\overline{X} = 731.4$ and $\Sigma s = 10.95$. Compute 3-sigma control limits for $\overline{X}$ and s, and estimate the value of σ, assuming that the process is in statistical control. If the label weight is 36 oz, and assuming that the process generates a normal distribution, does it meet federal requirements?

4.8. Control charts for $\overline{X}$, R, and s are to be maintained on drawings from a bowl of chips the distribution of which is approximately normal. The subgroup size is 4. μ is 50.00 and σ is 4.00. Assume that the 3-sigma limits are to be based on μ and σ. Compute the value of the upper control limit, the central line, and the lower control limit for $\overline{X}$, R, and s charts, respectively.

4.9. After the bowl drawing referred to in Prob. 4.8 has continued for some time, a frequency distribution is made of the $\overline{X}$ values. Estimate the average and standard deviation of this distribution. A frequency distribution is also made of the s values. Estimate the average and standard deviation of this distribution.

4.10. The statement is made on page 125 that the "range of a subgroup of 2 is always $\sqrt{2}$ times the sample standard deviation." Explain why this is true.

4.11. Use the factors in Table C to determine the values of A, A_2, and A_3 for a subgroup size of 5.

4.12. Use the factors in Table C to determine the values of A, A_2, and A_3 for a subgroup size of 9.

4.13. Use the factors in Table C and Eq. (4.2) for σ_s to determine the factors B_5, B_6, B_3 and B_4 for a subgroup size of 12.

4.14. Use the factors in Table C and Eq. (4.2) for σ_s to determine the factors B_5, B_6, B_3 and B_4 for a subgroup size of 6.

4.15. Compute 3-sigma control-chart limits for $\bar{X}$, R, and s for drawings 201 through 300 (25 subgroups) from Shewhart's normal bowl in Table 3.7. Estimate the value of σ using both ΣR and Σs. Is a state of statistical control indicated for these 25 values of $\bar{X}$, R, and s? If these control limits are used for charting the last 25 subgroups (drawings 301 through 400), is there any reason to conclude that the two groups of drawings might not come from the same universe? Explain your answer.

4.16. $\bar{X}$ and R charts have been maintained on a certain quality characteristic. All points have fallen within control limits on both charts. A sudden change in the process occurs that increases μ by 1.5σ but does not change σ. In answering the following questions, assume that the quality characteristic is normally distributed both before and after the change and that the control limits are based on observations made before the shift in process centering.

(*a*) If the subgroup size is 2, approximately what percentage of points would you expect to fall outside control limits on the $\bar{X}$ chart because of the change in μ?

(*b*) Answer the same question assuming a subgroup size of 4.

(*c*) Answer the same question assuming a subgroup size of 8.

4.17. Plot control charts for $\bar{X}$ and R for the data in Prob. 3.1. Does this process appear to be in control?

4.18. Calculate 2-sigma control limits for $\bar{X}$ from the Shewhart normal bowl data given in Table 3.7. What proportion of the mean values would you expect to fall outside these limits purely by chance? What proportion actually fall outside the 2-sigma limits? Comment on your results.

Answer: 40.2, 20.0; 4.55%; 9%.

4.19. Calculate the position of the central lines and control limits for $\bar{X}$ and R for the data of Table 4.2, assuming that no information prior to January 30 was available. Plot both $\bar{X}$ and R charts with these limits. Comment on the difference between the results so obtained and the results shown in Fig. 4.6.

4.20. Compute control limits for $\bar{X}$ for Fig. 4.5, basing the limits on the first 80 subgroups rather than on the first 40. For these 80 subgroups $\bar{\bar{X}}$ 34.7 and $\bar{R} = 20.0$. What change does this make in the points falling outside the control limits?

4.21. On page 140 it is stated that a specification of 21.530 ± 0.100 permits a considerable range of carelessness in machine setting but that a specification of 21.530 $\pm$ 0.030 calls for close attention to machine setting. Explain this statement. If necessary, refer to Tables 3.6 and 3.7 and Fig. 4.7 in your explanation.

4.22. Compute control limits for $\bar{X}$ and R for Fig. 4.8, basing the limits on the first 80 subgroups rather than the first 40. For these 80 subgroups $\bar{\bar{X}} = 29.50$ and $\bar{R} = 30.00$. What points fall outside the control limits? Are there any significant extreme runs above or below the central line?

4.23. Why is it that an increase in universe dispersion with no change in universe average throws the $\bar{X}$ chart out of control as well as the R chart?

4.24. The $\bar{X}$ and R charts of Fig. 4.2, Example 4.1, show lack of control. Of the various types of shifts in the universe illustrated in Fig. 4.4, which ones might be expected to result in control charts such as these? Explain.

4.25. You are shown what purports to be a control chart for $\bar{X}$ on a certain quality characteristic of a manufactured product. This control chart contains 50 subgroups. You observe that all the $\bar{X}$ values are close to the central line on the chart and none are near the 3-sigma limits. In fact, when you draw 1-sigma limits (only one-third of the distance from the central line to the control limits shown), all the points fall within these narrow limits. (This is called hugging.)

Would such a chart make you suspicious that something was wrong? Why? What possible explanations occur to you that might account for an $\bar{X}$ chart of this type?

4.26. $\bar{X}$ and R charts have been maintained on a certain quality characteristic. All points have fallen within control limits on both charts. A sudden change in the process occurs that decreases μ by 1.0σ but does not change σ. In answering the following questions, assume that the quality characteristic is normally distributed both before and after the change and that the control limits are based on observations made before the shift in process centering.

(*a*) Approximately what percentage of points would you expect to fall outside control limits on the $\bar{X}$ chart because of the change in μ if the subgroup size is 2?

(*b*) Answer the same question assuming that the subgroup size is 4.

(*c*) Answer the same question assuming that the subgroup size is 8.

(*d*) Relate the answers to the above questions to Type II error and the effect of subgroup size on the detection of shifts in μ on the $\bar{X}$ chart.

4.27. Problem 4.17 involved the plotting of $\bar{X}$ and R charts for the 30 subgroups of Prob. 3.1. All points on the $\bar{X}$ chart fell within the control limits; one point on the R chart fell outside the limits. Compute a revised $\bar{R}$ eliminating this out-of-control value of R. From this revised $\bar{R}$, estimate the value of σ that might be expected if the process dispersion could be held under statistical control.

Answer: 4.07, 1.75.

The specification for the "on" temperature of this thermostatically controlled switch is 54 ± 4. If the process could be centered at 54 with the σ you have just estimated, and if statistical control could be maintained, what percentage of switches would you expect to fall outside these specification limits? Assume that the normal distribution is applicable. Among the 150 switches observed, how many actually fell outside the specification limits?

Answer: 2.2%, 2.7%.

4.28. Automatic machinery is used to fill and seal 10-oz cans of a certain liquid product. The process standard deviation is 0.20 oz. To ensure that every can meets or exceeds this 10-oz minimum, the company has set a target value for the process of 11.0 oz.

(*a*) At this process average of 11 oz, what percent of the cans will have less than 10.5 oz of product? Assume that contained weights are normally distributed.
(*b*) If the quality control section samples these cans in subgroups of four, what will 3-sigma control limits be for the $\overline{X}$ chart?
(*c*) Assuming that $\pm 3\sigma$ natural tolerance limits on the process cover virtually all the filled cans, what is the minimum value to which the process average may be lowered to ensure that virtually no cans are filled with less than a minimum of 10 oz?
Answer: (a) 0.62%; (b) 11.3%, 10.7%; (c) 10.6%.

☆**4.29.** (*a*) Compute the central lines and 3-sigma control limits for $\overline{X}$ and R charts for the data of Prob. 3.10.
Answer: 193.20, 199.63, 206.06; 0, 11.08, 23.38.
(*b*) Plot the control charts and determine if the process is in control.

4.30. This is a continuation of Prob. 4.29.
(*a*) Estimate the value of σ from the range data of Prob. 3.10.
Answer: 4.764
(*b*) How does this estimate of σ compare to the value of σ calculated in either Prob. 3.14 or 3.15? Explain your answer based on the conclusions drawn in Prob. 4.29.

4.31. This is a continuation of Prob. 4.29.
(*a*) Estimate the value of σ from the range data of Prob. 3.10.
(*b*) Assuming that the distribution of timing signal is approximately normal, what proportion of the devices would you expect to meet specifications of 190.5 and 210.5 ms?
Answer: 0.9613.
(*c*) What proportion of the devices tested actually met these specifications?
Answer: 0.96.

4.32. This is a continuation of Prob. 4.29.
(*a*) If the mean of the distribution shifts to 205 ms, compute the probability that the shift will not be detected on the $\overline{X}$ chart on the first subgroup after the shift occurs. Assume no change in σ, and base your analysis on the $\overline{X}$ control limits found in Prob. 4.29.
Answer: 0.6915.
(*b*) Assuming a normal distribution of timing signals, what proportion of nonconforming devices would be produced at this new value of the mean?
Answer: 0.1268.

☆**4.33.** (*a*) Compute the central lines and 3-sigma control limits for $\overline{X}$ and R charts for the data of Prob. 3.19.
(*b*) Plot the control charts, and determine if the process is in control.

4.34. This is a continuation of Prob. 4.33.
(*a*) Estimate the value of σ from the range data of Prob. 3.19.
(*b*) How does this estimate of σ compare to the value of σ calculated in either Prob. 3.24 or 3.25? Explain your answer based on the conclusions drawn in Prob. 4.33.

4.35. This is a continuation of Problem 4.33.
(*a*) Estimate the value of σ from the range data of Prob. 3.19.
(*b*) Assuming that the distribution of voltage output is approximately normal,

what proportion of the units would you expect to meet specifications of 347.25 and 352.75 V?

(*c*) What proportion of the units tested actually met these specifications?

4.36. This is a continuation of Prob. 4.33.

(*a*) If the mean of the distribution shifts to 347 V, compute the probability that the shift will not be detected on the $\bar{X}$ chart on the first subgroup plotted after the shift takes place. Assume no change in σ and base your analysis on the $\bar{X}$ control limits calculated in Prob. 4.33.

(*b*) Assuming a normal distribution of voltage output, what proportion of nonconforming units would be produced at this new value of the mean?

4.37. Control charts for $\bar{X}$ and s are maintained on the shear strength in pounds of test spot welds. The subgroup size is 4 (spot welds). Values of $\bar{X}$ and s are computed and plotted for each subgroup. After 25 subgroups, $\Sigma\bar{X} = 13{,}050.00$ and $\Sigma s = 660.00$.

(*a*) Compute the values of 3-sigma limits for the $\bar{X}$ and s charts, and estimate the value of σ on the assumption that the process is in statistical control.

(*b*) If the process generates a normal distribution of shear strengths, what proportion of the spot welds do not meet a minimum shear strength of 450 lb?

4.38. Control charts for $\bar{X}$ and s have been run on a certain process for a long period of time. The subgroup size is 6. Assume an aimed-at mean $\bar{X}_0$ of 25.750 and a known standard deviation σ of 0.005.

(*a*) Calculate the central lines and control limits for $\bar{X}$ and s control charts.

(*b*) On a particular run of this process, the actual mean is 25.755. Use Table *A* to find the probability that any one subgroup average $\bar{X}$ plotted during the run of product will fall above the *UCL* found in part (*a*).

4.39. $\bar{X}$ and R charts are to be run on the thickness of an oxide base coating on silicon wafers. Specifications call for a thickness between 600 and 1,200 angstroms (Å). After 30 subgroups of 4 wafers each have been inspected, one subgroup for each batch of 40 wafers, $\Sigma\bar{X} = 22{,}350$ Å and $\Sigma R = 4{,}320$ Å.

(*a*) Calculate the central lines and control limits for these charts.

(*b*) Assuming that the process is operating in control and that product output is approximately normally distributed, what proportion of product does not meet specifications?

4.40. Specifications on the process described in Prob. 2.10 are 1,140 ± 10. Shafts with diameters less than 1,130 must be scrapped; shafts with diameters greater than 1,150 can be reworked, but the rework operation must be done by skilled lathe operators. The resulting rework cost is as great as the direct cost of manufacture.

(*a*) Determine the percentage of rework and of scrap product assuming a normal distribution for the process.

(*b*) 100% inspection has been carried on up to this point. What would be your advice regarding the level of inspection, centering, and so forth, for the immediate future?

(*c*) A Material Review Board, made up of representatives from the plant and the customer, has decided that shafts can be used successfully with diameters as large as 1,155. How would this information affect your decisions in (*b*)?

4.41. $\bar{X}$ and R charts are used to control a process by drawing subgroups of five units every 2 h. Specifications on one critical characteristic are 2,119 ± 10. Product over specs may be reworked; if undersized, it must be scrapped. After 50 subgroups, $\Sigma\bar{X} = 106{,}200.0$ and $\Sigma R = 581.5$.

(*a*) Determine 3-sigma control limits for $\bar{X}$ and R charts.

(*b*) Assuming that the process is in control and normally distributed, estimate σ and determine the percent of product that must be reworked and that must be scrapped.

Answer: 5.0; 15.87% reworked, 0% scrapped.

4.42. (*a*) In the situation described in Prob. 4.41 someone suggests that the specifications be changed such that the upper spec limit USL_X equals the upper control limit on the $\bar{X}$ chart, and the lower spec limit LSL_X equals the lower control limit. Assuming that this change would be accepted by the design department, do you think it is a good suggestion? Why or why not?

(*b*) Someone else suggests that the problem can be solved by merely changing from 3-sigma to 2-sigma control limits on the $\bar{X}$ chart. Is this a good suggestion? Why or why not?

(*c*) A supervisor points out that, by changing the subgroup size from 5 to 10, both control limits will easily fall within specifications. Comment on this suggestion.

(*d*) Suggest an approach that you believe would be desirable. What further investigations are indicated?

4.43. $\bar{X}$ and R charts have been maintained on a certain quality characteristic. All points have fallen within limits on both charts. A sudden shift occurs in the process that increases μ by 2.1σ. Assuming that the quality characteristic is normally distributed both before and after the shift, that σ remains constant, and that the control limits were determined prior to the shift:

(*a*) Approximately what percentage of points would you expect to fall outside the upper control limit on the $\bar{X}$ chart if the subgroup size is five?

(*b*) Answer (*a*) for a subgroup size of nine.

(*c*) Comment on the effect that subgroup size has on the ability of the $\bar{X}$ chart to detect shifts in μ.

(*d*) Comment briefly on the effect of this type of shift on a s or R chart.

(*e*) Why is it that a shift in universe dispersion with no change in universe average may cause $\bar{X}$ to show lack of control as well as R?

4.44. A plot of Type II error probability as a function of shift magnitude in multiples of σ is a useful tool in illustrating the sensitivity of various control charting schemes. Assuming a normal distribution, find the probability of Type II error for an $\bar{X}$ chart for shift magnitudes of $0^+\ \sigma$, 0.5σ, 1.0σ, 1.5σ, and 2σ for the following cases:

(*a*) 3-sigma control limits and a subgroup size of 4.

(*b*) 2-sigma control limits and a subgroup size of 4.

(*c*) 3-sigma control limits and a subgroup size of 10.

(*d*) Sketch these three curves on a single graph with Type II error probability on the ordinate scale and shift magnitude on the abscissa. Discuss these curves.

4.45. Control charts for $\bar{X}$ and R are to be run on a certain process known to have a σ of 0.0035 using a subgroup size of 5. Specifications on the item are 6.5000 ± 0.0110. The nominal dimension 6.5000 is to be the target mean $\bar{X}_0$.

(*a*) Calculate trial control limits for $\bar{X}$ and R.

(*b*) After operation in control for some period of time, μ suddenly shifts to 6.5050. What is the probability of detecting a shift of this magnitude on the first subgroup drawn after the shift occurs?

(*c*) What fraction of the product does not meet specifications after the shift described in part (*b*) occurs? Assume that the distribution of product is approximately normal.

4.46. Control-chart data have been recorded and plotted on $\overline{X}$ and R charts for a certain process using a subgroup size of 5. After 30 subgroups, $\Sigma\overline{X} = 1{,}518.0$ and $\Sigma R = 104.7$.

(*a*) Find control limits for the $\overline{X}$ and R charts, and estimate the value of σ, assuming that the process is in control.

(*b*) If the mean of the process should suddenly shift to 48.75, what is the probability that the shift will be detected on the first subgroup plotted after it occurs?

4.47. A resistor manufacturing process has been operating in statistical control for a long period of time. Specification limits are 350 ± 5 ohms, and σ is estimated to be 1.4 ohms.

(*a*) What proportion of nonconforming product is being made if the process is centered at 350? Assume that the distribution is approximately normal.

4.48. Mark 500 chips as follows:

Marking	Frequency	Marking	Frequency	Marking	Frequency
28	1	22	58	16	22
27	2	21	73	15	11
26	5	20	78	14	5
25	11	19	73	13	2
24	22	18	58	12	1
23	39	17	39		

Place the chips in a bowl and stir thoroughly. Draw 5 chips, and record the readings. Replace the chips, stir thoroughly, and draw 5 more. Repeat this until you have a record of 50 subgroups of 5 each. Compute $\overline{X}$ and R for each subgroup as you go along, and plot a control chart. After the first 20 subgroups, compute control limits and put them on the chart. After the 50 subgroups have been drawn, compute $\overline{R}$ for the entire set of 50. From this estimate σ. Compute the true σ from the distribution in the bowl, and compare with the σ estimated from the drawings.

4.49. Mark 500 chips as follows:

Marking	Frequency	Marking	Frequency	Marking	Frequency
32	1	26	58	20	22
31	2	25	73	19	11
30	5	24	78	18	5
29	11	23	73	17	2
28	22	22	58	16	1
27	39	21	39		

Continue the control charts for $\overline{X}$ and R that we started in Prob. 4.48 by drawing 25 subgroups of 5 chips from this new bowl. Replace the chips after each drawing. Continue to use the control limits established in Prob. 4.48. How many points fall out of control on each chart? Do you get any extreme runs on either side of the central line? What has happened to universe average and universe dispersion?

4.50. Mark 500 chips as follows:

Marking	Frequency	Marking	Frequency	Marking	Frequency
36	1	25	24	14	19
35	1	24	29	13	15
34	1	23	33	12	11
33	1	22	36	11	8
32	2	21	39	10	6
31	4	20	40	9	4
30	6	19	39	8	2
29	8	18	36	7	1
28	11	17	33	6	1
27	15	16	29	5	1
26	19	15	24	4	1

Continue the control charts for $\overline{X}$ and R that were started in Prob. 4.48 by drawing 25 subgroups of 5 chips each from this new bowl. Replace the chips after each drawing. Continue to use the control limits established in Prob. 4.48. (For purposes of this problem neglect your drawings in Prob. 4.49.) How many points fall out of control on each chart? Do you get any extreme runs on either side of the central line? What has happened to universe average and universe dispersion?

☆**4.51.** Use your personal computer (or scientific calculator) to generate sets of normally distributed pseudo-random numbers with the same means and standard deviations represented by the sampling bowls described in Probs. 4.48, 4.49, and 4.50. Your instructor may then ask you to "mix and match" some of these values to illustrate the sensitivity of control charts using, perhaps, different subgroup sizes and different measures of dispersion.

5

SOME FUNDAMENTALS OF THE THEORY OF PROBABILITY

There are many difficulties and troubles with which a factory management has to contend—dies which wear; bearings that get loose; stock which is undersize, oversize, or dirty; loose fixtures; careless, tired, or untrained employees. For these reasons it would seem that there is no mathematical method which takes into account all these factors. However there is a kind of mathematics which is applicable in just such conditions, and that is the mathematics of probability.

—L. T. Rader[†]

5.1 PROBABILITY HAS A MATHEMATICAL MEANING[‡]

The statement that tomorrow will probably be a hot day is perfectly clear and understandable, whether you agree with it or not. So also is the statement that Smith is more likely than Jones to receive a promotion. In general, the word *probability* and its derivative and related words such as *probable, probably, likelihood, likely,* and *chance* are used regularly in everyday speech in a qualitative sense, and there is no difficulty in their interpretation.

But consider a statement that the probability is 0.98 that the shear strength of a spot weld will be above 480 lb if two 0.040-gage test strips of duralumin are

[†]L. T. Rader, "Putting Quality into Quantity," *American Machinist,* vol. 87, pp. 92–93, Oct. 28, 1943.

[‡]The explanation of probability developed in the initial sections of this chapter was considerably influenced by a volume of mimeographed notes on the subject prepared by Paul Coggins and R. I. Wilkinson for use in an out-of-hour course given for engineers of the New York Telephone Company.

welded on Sciaky machine No. 18 provided statistical control is maintained on the welding operation. Or consider a statement that if a sample of 5 is taken at random from a lot of 50 pieces that contains exactly 3 nonconforming pieces, the probability is 0.724 that the sample will contain no nonconforming pieces. In such statements, *probability* is used in its quantitative or mathematical sense. It is evident that some special explanation of the meaning of *probability* is necessary before these statements can be understood. A critical consideration will show that the two statements not only call for more explanation but need somewhat different explanations.

Two different traditional definitions of *probability* in its mathematical sense may be given. One may be described as the *frequency definition*; the other, as the *classical definition.*

5.1.1 Traditional Definitions of Probability

From the standpoint of its useful applications in industry, probability may be thought of as relative frequency in the long run. This may be phrased somewhat more precisely as follows:

Assume that if a large number of trials are made under the same essential conditions, the ratio of the number of trials in which a certain event happens to the total number of trials will approach a limit as the total number of trials is indefinitely increased. This limit is called the probability that the event will happen under these conditions.

Mathematically, this a posteriori definition may be expressed as follows: Let E represent the occurrence of some event and $n(E)$ represent the total number of times E occurs in n trials. Then the probability of the event E, $P(E)$, equals $n(E)/n$ as n increases without bound.

It may be noted that this limit is always a fraction (or decimal fraction), which may vary from 0 to 1. A probability of 0 corresponds to an event that never happens under the described conditions; a probability of 1 corresponds to an event that always happens.

It is because *probability* describes relative frequency in the long run that the concept is so useful in practical affairs. But its use would be severely limited if the only way to estimate any probability were by a long series of experiments. Most mathematical manipulations of probabilities are based on another definition, which may be stated as follows:

If an event may happen in a ways and fail to happen in b ways, and all these ways are mutually exclusive and equally likely to occur, the probability of the event happening is $a/(a+b)$*, the ratio of the number of ways favorable to the event to the total number of ways.*

This is called the *classical* or a priori *definition.* It represents the approach to the subject developed by the classical writers on the mathematics of probability, many of whom wrote particularly about probabilities associated with games of chance. Experience shows that where properly used, this definition permits the successful forecasting of relative frequency in the long run without the necessity of a long set of trials prior to each forecast.

The statement in the preceding section about the probability of a given strength for spot welds could only be justified after the fact, or a posteriori, on the basis of a considerable record of measurements of spot-weld strength from Sciaky machine No. 18; it would be impossible to enumerate a number of equally likely ways in which the strength could be above or below 480 lb. On the other hand, the statement about the sample of 5 from the lot of 50 is based on a counting of equally likely ways in which the sample of 5 might contain no nonconforming articles or one or more such articles; even though not based on the evidence of actual trials, a statement of this sort may be made with strong confidence that the stated probability is really the relative frequency to be expected in the long run.

5.1.2 Axiomatic Definition of Probability

It is possible to raise philosophical and practical objections to both the traditional definitions of probability. For example, in the frequency definition, how long a series of trials should one have to estimate relative frequency in the long run? In the classical definition, how can one tell which ways are equally likely? Such types of objections have led many of the modern writers on probability to view it merely as a branch of abstract mathematics developed from certain axioms or assertions. The establishment of any relationship between actual phenomena in the real world and the laws of probability developed from the axioms is viewed as an entirely separate matter from the mathematical manipulations leading to the probability theorems.[†]

In spite of the different definitions that have been used by mathematicians, the same probability theorems are developed from all the definitions. Generally speaking, persons who use control charts, acceptance sampling procedures, and other statistical techniques in industry will find it satisfactory to think of probability as meaning relative frequency in the long run.

5.1.3 A Simple Example Relating the "Equally Likely" and "Relative-Frequency" Concepts

Table 3.7 gave the actual numbers on each of the first 400 chips drawn from Shewhart's normal bowl. In the technical language of statistics (as first explained in Chap. 1), we describe the data of Table 3.7 as having been recorded "by variables."

To illustrate certain aspects of the two traditional definitions of probability (and also to illustrate several other matters to be discussed later in this book), it will be helpful to classify the same data "by attributes." The reader will recall that a record by attributes merely classifies observations as conforming or failing to conform to certain specified requirements. Let us assume that the drawing of chips from Shewhart's bowl corresponds to a manufacturing process and that any chip numbered 20

[†]For a clear discussion of the relationship of the axiomatic definition of probability to the two traditional definitions, see G. P. Wadsworth and J. G. Bryan, *Applications of Probability and Random Variables,* 2d ed., pp. 2–9, McGraw-Hill Book Company, New York, 1974.

or less represents a nonconforming product. (The process capability in relation to an LSL_X of 20.5 is similar to the one shown in distribution *B* of Fig. 2.6.)

An examination of the markings on the chips (shown in Table 3.6) tells us that 170 of the 998 chips are marked 20 or less. If we assume that each of the 998 chips is equally likely to be drawn (a reasonable assumption when the chips are alike and are mixed thoroughly after each drawing), the classical definition gives us $\frac{170}{998} = 0.17034$ as the probability of drawing a nonconforming item.

In Table 5.1 the first 400 drawings from Shewhart's bowl that were shown in Table 3.7 have been divided into successive subgroups of 10, with the number nonconforming recorded for each subgroup. For the purpose of illustrating the relative-frequency concept of probability, the final three columns of the table (designated *E, F,* and *G*) are the ones of interest. These columns record the total number observed, the total number of nonconforming items found, and the ratio of total nonconforming items to total observed. The last figure (from column *G*) is plotted in Fig. 5.1.

Column *G* and Fig. 5.1 illustrate the meaning of the "limit" referred to in the frequency definition of probability. Under the special circumstances of the bowl-drawing experiment, we know the true fraction nonconforming in the bowl to be 0.17034. Obviously a single draw could give us only 0 or 1 as an estimate of this fraction. If the first draw happens to be a conforming item (as it did), the estimate will continue to be 0 until the first nonconforming item is drawn; thereafter it will always be between 0 and 1. Similarly, the only estimates obtainable from a draw of 5 are 0, 0.2, 0.4, 0.6, 0.8, and 1.0. It seems clear that we need a fair number of drawings before we can expect the relative frequency to give us a satisfactory estimate of the fraction nonconforming in the bowl.

However, another matter that may not be so evident is the special meaning attached to the word *limit* in the frequency definition of probability. This is not a limit in the conventional mathematical sense of a function always getting closer and closer to the limit as some variable increases or decreases even though the limit is never quite reached with any finite value of the variable. It is rather what is called a *statistical limit* (or, sometimes, a *stochastic limit*).† It will be noted from column *G* of Table 5.1 and from Fig. 5.1 that the computed figure for relative frequency passes through the true proportion nonconforming of 0.17034 several times.

Of course, the more observations we make, the smaller the influence of the most recent observation on the computed figure for relative frequency. Intuitively, it seems reasonable that the greater the number of observations used in computing a relative frequency, the better our estimate of an unknown true relative frequency in a bowl (or other source of our observations). It seems sensible to have more confidence in an estimate from, say, 400 observations than in one from, say, 100 observations. Nevertheless, this does not mean that 100 observations could not, by chance, give us an excellent estimate of the proportion of nonconforming items in the bowl and 400 observations could not, by chance, give us a poorer estimate.

†For a graph showing a statistical approach to a limit in 1,000 drawings with a new value of relative frequency computed after each drawing, see W. A. Shewhart, *Economic Control of Quality of Manufactured Product,* p. 438, D. Van Nostrand Company, Inc., Princeton, N.J., 1931.

TABLE 5.1 EXAMPLE OF SUCCESSIVE ESTIMATES OF RELATIVE FREQUENCY BASED ON 400 DRAWINGS FROM SHEWHART'S NORMAL BOWL

A chip marked 20 or less is designated as a nonconforming item. Observations are considered in subgroups of 10. Fraction nonconforming in bowl is 0.17034

Subgroup number (*A*)	Size of subgroup (*B*)	Number of noncon-forming items in subgroup (*C*)	Fraction noncon-forming in subgroup (*D*)	Cumulative number observed (*E*)	Cumulative number of noncon-forming items (*F*)	Cumulative fraction noncon-forming (*G*)
1	10	0	0.0	10	0	0.000
2	10	1	0.1	20	1	0.050
3	10	2	0.2	30	3	0.100
4	10	1	0.1	40	4	0.100
5	10	3	0.3	50	7	0.140
6	10	3	0.3	60	10	0.167
7	10	3	0.3	70	13	0.186
8	10	1	0.1	80	14	0.175
9	10	2	0.2	90	16	0.178
10	10	0	0.0	100	16	0.160
11	10	2	0.2	110	18	0.164
12	10	1	0.1	120	19	0.158
13	10	4	0.4	130	23	0.177
14	10	3	0.3	140	26	0.186
15	10	2	0.2	150	28	0.187
16	10	2	0.2	160	30	0.188
17	10	1	0.1	170	31	0.182
18	10	1	0.1	180	32	0.178
19	10	3	0.3	190	35	0.184
20	10	3	0.3	200	38	0.190
21	10	2	0.2	210	40	0.190
22	10	1	0.1	220	41	0.186
23	10	2	0.2	230	43	0.187
24	10	1	0.1	240	44	0.183
25	10	0	0.0	250	44	0.176
26	10	1	0.1	260	45	0.173
27	10	1	0.1	270	46	0.170
28	10	4	0.4	280	50	0.179
29	10	1	0.1	290	51	0.176
30	10	3	0.3	300	54	0.180
31	10	1	0.1	310	55	0.177
32	10	0	0.0	320	55	0.172
33	10	0	0.0	330	55	0.167
34	10	1	0.1	340	56	0.165
35	10	1	0.1	350	57	0.163
36	10	1	0.1	360	58	0.161
37	10	2	0.2	370	60	0.162
38	10	4	0.4	380	64	0.168
39	10	2	0.2	390	66	0.169
40	10	1	0.1	400	67	0.168

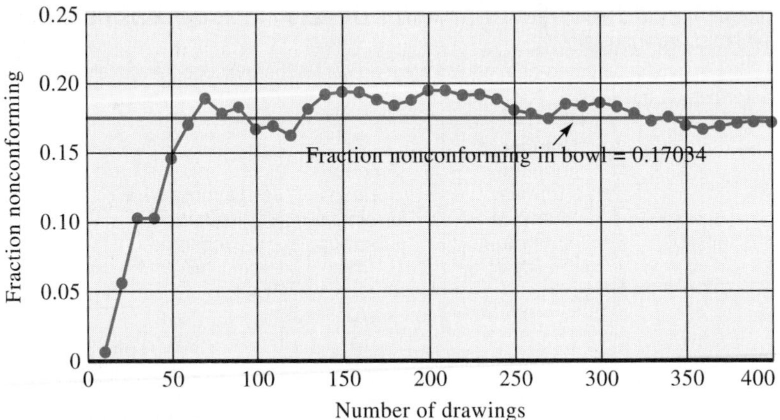

FIGURE 5.1 Cumulative estimates of fraction nonconforming from 400 drawings, considered by tens, from Shewhart's bowl (assuming that a chip numbered 20 or less is a nonconforming item).

Incidentally, it should be mentioned that the reader should not attach any significance to the fact that in Fig. 5.1 the relative frequencies start at zero and gradually work *up* to a value close to the proportion nonconforming in the bowl. The early estimates might as readily have been too high, and the relative frequencies could have worked *down* to somewhere near the true value.

If we do not know the true fraction nonconforming in the bowl (the usual condition), the best we can do is to use the relative frequency up to the most recent observation as our estimate of the probability.

However, in the usual case where we have only the observations and no other information about the "bowl" from which the observations are made, it always is appropriate to consider the question of whether all the observations came from the same bowl (or the same other source). In other words, were all the observations made "under the same essential conditions"? There is no meaningful relative frequency in the long run if a bowl that contains a different distribution has been substituted for the initial bowl at some time during the observations. The reader will recognize that control charts constitute one type of test to judge whether a constant system of chance causes has been operating during a period of observation. The p chart, explained in Chap. 6, is specifically applicable to fraction rejected.

5.2 MODERN CONCEPTS OF PROBABILITY THEORY

Modern writers in the field of probability theory prefer a foundation in set theory as a basis for the development of probability theorems. This foundation rests on the few relatively simple axioms, or definitions, presented in the following paragraphs.

First, we shall define the set of all possible outcomes of an experiment or test as the *probability space* for the experiment and designate the probability space by S. In Fig. 5.2, each of the boxes (a), (b), (c), and (d) represents a probability space.

Since the probability of any event occurring in S is between 0 and 1, the probability of S must equal 1 since S includes all possible outcomes. The first axiom therefore states:

$$P(S) = 1 \tag{5.1}$$

As examples of probability spaces, consider the following:

1. Tossing a coin. The set of all possible outcomes includes getting a head or a tail, assuming that the coin cannot land on end, or

   $$S = \{H, T\}$$

2. Tossing a single die. S, in this case, contains the discrete numerical quantities 1, 2, 3, 4, 5, and 6, or

   $$S = \{1, 2, 3, 4, 5, 6\}$$

FIGURE 5.2 Venn diagrams illustrating probability spaces, events, and elementary outcomes.

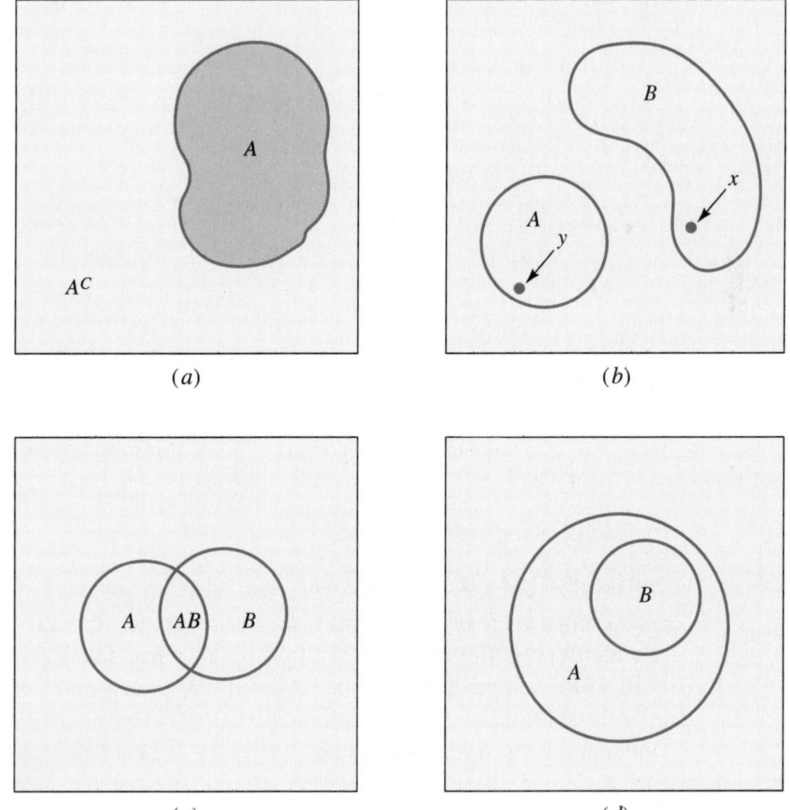

3. Life of a vacuum tube. S includes the continuous time scale from $t = 0$ to $t = \infty$.

Correspondingly, it is impossible for any outcome outside the probability space to occur. In the language of probability, this impossible outcome is termed a *null* outcome and usually designated Φ (the Greek capital letter phi). Thus

$$P(\Phi) = 0 \tag{5.2}$$

An *event,* which will be designated by a capital letter other than the letter S, is a subset of outcomes contained within S. In Fig. 5.2, the areas designated A and B are events.

The points x and y in Fig. 5.2*b* may also be termed events but are of a somewhat different nature and will be discussed in subsequent paragraphs.

Since an event is a subset of the probability space S, the probability of an event is greater than (or equal to) zero and less than (or equal to) one. That is

$$0 \le P(A) \le 1 \tag{5.3}$$

Consider the examples used previously in describing probability spaces.

1. Tossing a coin. If we let the event A equal the obtaining of a head on the tossing of a coin, then

   $$A = \{H\}$$

   which is a subset of $S = \{H, T\}$.
2. Tossing a single die. In this case we may describe several events. For example, the event A may be described as the occurrence of a 1, 2, or 3 on the toss. That is,

   $$A = \{1, 2, 3\}$$

   which is a subset of

   $$S = \{1, 2, 3, 4, 5, 6\}$$

   or A might be

   $$A = \{1, 4, 6\} \qquad \text{or} \qquad A = \{5\}$$

   In this latter case, A is an event and also an *elementary outcome* since the occurrence of a single face of the die is the smallest unit into which we can divide the probability space with probability finite and greater than 0.
3. Life of a vacuum tube. In this case we are concerned with time, which is an event measured on a continuous scale from $t = 0$ to $t = \infty$. We may then describe an event A, for example, where

   $$A = \{0 \le t \le 400 \text{ h}\} \qquad \text{(the life } t \text{ is between 0 and 400 h)}$$

or another event B where

$B = \{100 \text{ h} \le t \le 1000 \text{ h}\}$ (life is between 100 and 1,000 h)

or

$C = \{t \ge 1000 \text{ h}\}$ (life is greater than or equal to 1,000 h)

An *elementary outcome* is the occurrence of an event that is the smallest unit into which the probability space may be divided. Two elementary outcomes x and y are illustrated on the Venn diagram, Fig. 5.2*b*. The probability of an elementary outcome from any given experiment is always greater than or equal to zero (≥ 0). Using our previous examples:

1. Tossing a coin. The smallest units into which we can divide the probability space S are

   $S = \{H, T\}$

   Note that these single-point outcomes are *mutually exclusive*. That is, if either outcome does occur (e.g., a head turns up), then the other cannot occur simultaneously (we cannot also get a tail).
2. Tossing a single die. The probability space, in this case, may be divided into six mutually exclusive single-point outcomes:

   $S = \{1, 2, 3, 4, 5, 6\}$

   Each of these outcomes is an elementary outcome with finite probability of occurrence. Note that each may also be called an event, in which case the six events also are mutually exclusive.
3. Life of a vacuum tube. The smallest unit into which we may divide the life of a vacuum tube, or alternatively divide the continuous time scale, is an infinitely small unit.

As an example, the measured life may be 413.69 h. This measurement is limited by the capability of the measuring instrument, which has, in reality, measured the life t somewhere between

$413.68499 \cdots \le t \le 413.69499 \cdots$

The nature of continuous scales dictates that, for practical purposes, we divide them into reasonable nonoverlapping segments. Several examples of this type of procedure were discussed in Chap. 3 in forming the cells and cell boundaries for calculating frequency distributions. That discussion indicated that as cell boundaries became narrower, the expected frequency became smaller. The same logic applies to mathematical statements about outcomes of experiments measured along a continuous scale. As the range of the outcome becomes narrower, and approaches 0, the probability of the outcome becomes smaller and approaches 0.†

5.2.1 Mutually Exclusive Events

Two events are mutually exclusive if the occurrence of one event precludes the occurrence of the other. The Venn diagrams in Fig. 5.2 illustrate some events that are mutually exclusive and some that are not.

In Fig. 5.2*a* the set S contains two events, the event A and the event A^c (read as "A complement"), which is the event that A does not occur. Mathematically

$$S = A + A^c$$

by definition; then

$$P(S) = P(A) + P(A^c) = 1 \tag{5.4}$$

Thus the probability that any event A, defined in the probability space S, plus the probability of *not* A, A^c, equals the probability of S, which is always 1, or

$$P(A^c) = P(S) - P(A) = 1 - P(A) \tag{5.5}$$

In Fig. 5.2*b*, the events A and B are also mutually exclusive since the occurrence of A precludes B from occurring. As an example, the following events may be defined for the tossing of a single die:

$$S = \{1, 2, 3, 4, 5, 6\}$$
$$A = \{1, 2, 3\}$$
$$B = \{5, 6\}$$

The definitions of A and B also define their complements:

$$A^c = \{4, 5, 6\} = S - A$$
$$B^c = \{1, 2, 3, 4\} = S - B$$

Furthermore, the definitions of A and B may be used to define the *compound events:*

1. Both A and B $\quad AB = \{\Phi\}$
2. Both A^c and B^c $\quad A^cB^c = \{4\}$

†Writers sometimes say that the probability of a precise single-point outcome on a continuous scale *is* 0. This statement is made merely for reasons of simplicity. The null or impossible event is the only event that has probability *equal to* 0. All other events are, by implication at least, defined over some finite range and therefore have some finite probability of occurrence.

3. Both A and B^c $\quad AB^c = \{1, 2, 3\}$
4. Both A^c and B $\quad A^cB = \{5, 6\}$

When the die is rolled, one of the faces, one of the elementary outcomes between one and six, will turn up. Note, however, that in no case will the outcome fall within the event A *and* within B. Thus the *compound event AB* is impossible (a *null* event).

This statement leads directly to the definition of mutually exclusive events. *Given any two events A and B, each of which has some finite probability of occurrence, if the compound event AB is an impossible event* (Φ), *then A and B are mutually exclusive.* Thus if $AB = \Phi$, then

$$P(AB) = P(\Phi) = 0 \tag{5.6}$$

The events A and B in Fig. 5.2*c* and *d* are not mutually exclusive. In Fig. 5.2*d,* in fact, the event B is fully contained within the event A. Thus the event AB equals the event B, and

$$P(AB) = P(B)$$

5.2.2 Dividing a Probability Space into Mutually Exclusive Events

Description of the occurrence of events in Fig. 5.2*c* is not simple; the event A could occur, or the event B, or the compound event AB.

Suppose it were desired to describe an event that is composed of at least one of events A or B. It would be necessary to include the portion of the probability space covered by A and also that covered by B. Since there is an area of overlap, AB, adding area A directly to area B would result in double counting of the area AB. The solution is obvious. In order to avoid double counting, the area AB is subtracted from the sum of areas A and B. This statement is precisely the definition of the *union* of two events,

$$A \cup B = A + B - AB$$

and may be extended by similar logic to the union of three or more events. Stated in terms of the probabilities of the respective events,

$$P(A \cup B) = P(A) + P(B) - P(AB) \tag{5.7}$$

The reader should note that Fig. 5.2*b* illustrates the special case of Eq. (5.7) where $AB = \Phi$. Thus the union of two *mutually exclusive* events A and B may be expressed

$$P(A \cup B) = P(A) + P(B) \text{ if } AB = \Phi \tag{5.8}$$

Usually it is convenient in application to divide a probability space into a set of mutually exclusive events in order to make use of the simplicity of the expression for union of events given in Eq. (5.8).

The probability space in Fig. 5.2*c* may be so divided as follows:

1. Only event A occurs.
2. Only event B occurs.
3. Both A and B occur.
4. Neither A nor B occurs.

Consider first the compound event AB. AB can occur only if the outcome lies within both the events A and B. AB may, therefore, be *conditioned* upon A having occurred, or it may be *conditioned* upon B having occurred. That is,

$$P(AB) = P(B|A)P(A) = P(A|B)P(B) \tag{5.9}$$

The first expression is read as "the probability of AB equals the probability of B, given A has occurred, times the probability of A." The vertical line stands for the word *given,* and the expression $P(B|A)$ is the *conditional probability* of the event B *given* A.

Figure 5.2*c* may now be divided into the four mutually exclusive events previously listed.

1. Only event A occurs. In order to isolate A, it is necessary to form the compound event A and *not* B, or AB^c. Mathematically, this is the *intersection* of A and B^c, or

   $$AB^c = A - AB$$

   and

   $$P(AB^c) = P(A) - P(AB) = P(B|A)P(A)$$

2. Only event B occurs. The same argument may be used in this case, resulting in

   $$A^cB = B - AB$$

   and

   $$P(A^cB) = P(B) - P(AB) = P(B) - P(A|B)P(B)$$

3. Both A and B occur. From the previous discussion, this is the compound event AB, and

   $$P(AB) = P(A|B)P(B) = P(B|A)P(A)$$

4. Neither A nor B occurs. This is the compound event *not A* (A^c) and *not B* (B^c), A^cB^c. By the laws of set theory, it can be shown that the intersection of the complements of two events equals the complement of the union of the events themselves. That is,

$$A^cB^c = (A \cup B)^c$$

thus

$$P(A^cB^c) = P[(A \cup B)^c] \tag{5.10}$$

Using Eqs. (5.5), (5.7), and (5.10),

$$\begin{aligned} P(A^cB^c) &= \mathrm{P}[(A \cup B)^c] = 1 - [P(A) + P(B) - P(AB)] \\ &= 1 - P(A) - P(B) + P(AB) \end{aligned}$$

If the four mutually exclusive and exhaustive events have been properly defined, the sum of their respective probabilities should be 1. That is, if

$$S = AB^c + A^cB + AB + A^cB^c$$

then

$$P(S) = P(AB^c) + P(A^cB) + P(AB) + P(A^cB^c)$$

Making the appropriate substitutions,

$$\begin{aligned} P(S) &= P(A) - P(AB) + P(B) - P(AB) + P(AB) + 1 - P(A) - P(B) + P(AB) \\ &= 1 \end{aligned}$$

Consider the tossing of a single die for which the following events, and their respective probabilities, are defined:

$$\begin{aligned} S &= \{1, 2, 3, 4, 5, 6\} & P(S) &= 1 \\ A &= \{2, 3, 4\} & P(A) &= \tfrac{3}{6} \\ B &= \{4, 5\} & P(B) &= \tfrac{2}{6} \\ AB &= \{4\} & P(AB) &= \tfrac{1}{6} \end{aligned}$$

Either intuitively or by enumeration, the mutually exclusive and exhaustive set may be defined as follows:

1. $AB^c = \{2, 3\}$
2. $A^cB = \{5\}$
3. $AB = \{4\}$
4. $A^cB^c = \{1, 6\}$

Assuming that the die is not loaded and that each face is as likely to turn up as any other, application of the classical definition of probability yields:

1. $P(AB^c) = \frac{2}{6}$
2. $P(A^cB) = \frac{1}{6}$
3. $P(AB) = \frac{1}{6}$
4. $P(A^cB^c) = \frac{2}{6}$

$P(S) = 1$

Alternatively, the same results could have been found using only the defined events and their respective probabilities. In this case, the defined outcomes are:

1. $P(S) = 1$
2. $P(A) = \frac{3}{6}$
3. $P(B) = \frac{2}{6}$
4. $P(AB) = \frac{1}{6}$

Using the previous derivations of the four mutually exclusive and exhaustive subsets:

1. $P(AB^c) = P(A) - P(AB) = \frac{3}{6} - \frac{1}{6} = \frac{2}{6}$
2. $P(A^cB) = P(B) - P(AB) = \frac{2}{6} - \frac{1}{6} = \frac{1}{6}$
3. $P(AB) = \frac{1}{6}$
4. $P(A^cB^c) = 1 - P(A) - P(B) + P(AB) = 1 - \frac{3}{6} - \frac{2}{6} + \frac{1}{6} = \frac{2}{6}$

If the set is formed of mutually exclusive events, then the union of the set may be found as the sum of their respective probabilities. Furthermore, if the set of events exhausts the probability space, the sum of their respective probabilities will be 1. Thus

$$P[(AB^c) \cup (A^cB) \cup (AB) \cup (A^cB^c)] = P(AB^c) + P(A^cB) + P(AB) + P(A^cB^c)$$
$$= \frac{2}{6} + \frac{1}{6} + \frac{1}{6} + \frac{2}{6} = 1$$

The probabilities defined for the events *A*, *B*, and *AB* could also be used to find the conditional probabilities $P(B|A)$ and $P(A|B)$. Rearranging Eq. (5.9):

$$P(B|A) = \frac{P(AB)}{P(A)} = \frac{\frac{1}{6}}{\frac{3}{6}} = \frac{1}{3}$$

$$P(A|B) = \frac{P(AB)}{P(B)} = \frac{\frac{1}{6}}{\frac{2}{6}} = \frac{1}{2}$$

Thus, if A has already occurred, the face on the die must read 2, 3, 4. In order for B to occur also, the face must be 4, which can happen in only one of three ways, $P(B|A)$, therefore, must equal $\frac{1}{3}$. A similar argument may be used to describe $P(A|B)$.

5.3 SOME THEOREMS OF THE THEORY OF PROBABILITY

Three important theorems, derived in texts on mathematical statistics from a set theoretical base, are here stated and illustrated by reference to the drawings from Shewhart's normal bowl discussed in Chap. 3.

Theorem 1. The addition theorem or theorem of total probabilities
The probability of the occurrence of either one or any number of mutually exclusive events is the sum of the probabilities for the separate events.

As an example, consider the drawing of a single chip from Shewhart's normal bowl. The events A and B are defined as follows:

A = a chip marked 50

B = a chip marked 51

Assuming that any one of the 998 chips in the bowl is as likely to be drawn as any other,[†] and since there are 5 chips marked 50 and 4 marked 51,

$$P(A) = \frac{5}{998}$$

$$P(B) = \frac{4}{998}$$

The occurrence on a single draw of a chip marked either 50 or 51 is the union of the events A and B. Thus

$$P(A \cup B) = P(A) + P(B)$$
$$= \frac{5}{998} + \frac{4}{998} = \frac{9}{998}$$

where A and B are mutually exclusive.

[†]This assumption of a *purely random draw* is not a trivial one either when drawing chips from a bowl or when drawing a sample from a batch of product.

Theorem 2. The theorem of compound probabilities or multiplication theorem

If a compound event is made up of a number of separate and independent subevents, and the occurrence of the compound event is the result of each of the subevents happening, the probability of occurrence of the compound event is the product of the probabilities that each of the subevents will happen.

Returning to the Shewhart normal bowl example, consider a case in which two chips are drawn in succession with the second drawn after the first has been returned to the bowl and the bowl stirred. The events *A* and *B* are defined as

A = occurrence of a chip marked 50 on the first draw

B = occurrence of a chip marked 50 on the second draw

In this case, $P(A) = P(B) = \frac{5}{998}$. The two draws, or events, are *independent* since the occurrence of one does not affect the probability of occurrence of the other. The event of drawing a chip marked 50 on both draws is the compound event *AB* and has probability

$$P(AB) = P(A)P(B) = \left(\frac{5}{998}\right)\left(\frac{5}{998}\right) = \frac{25}{996{,}004}$$

Theorem 3. The multiplication theorem for conditional probabilities

If a compound event is made up of two (or more) dependent subevents, the probability of occurrence of the compound event is the product of the probability of the first multiplied by the probability that if the first has occurred the second will also happen.

Consider the successive drawings from Shewhart's normal bowl discussed following Theorem 2. The definitions of events *A* and *B* were based on replacement of the first chip before drawing the second. If the first chip had not been replaced, the bowl would have been partially exhausted, and if event *A* had occurred, there would be only 4 chips marked 50 remaining. The probability of *B*, therefore, will depend on the results of the first drawing and the occurrence of *A*. Thus, if

$$P(A) = \tfrac{5}{998}$$

$$P(B|A) = \tfrac{4}{997}$$

$$P(B|A^c) = \tfrac{5}{997}$$

then

$$\begin{aligned} P(AB) &= P(A)P(B|A) \\ &= \left(\frac{5}{998}\right)\left(\frac{4}{997}\right) = \frac{20}{995{,}006} \end{aligned}$$

Note that this conditional form could have been used to describe the case explained following Theorem 2. The example could have been stated

$$P(A) = \tfrac{5}{998}$$

$$P(B|A) = \tfrac{5}{998}$$

$$\begin{aligned} P(AB) &= P(A)P(B|A) \\ &= \left(\tfrac{5}{998}\right)^2 \end{aligned}$$

Theorem 2, therefore, is a special case of Theorem 3 where the events A and B are independent. The mathematical statement equating the two cases yields the definition of *statistical independence.* Thus

$$P(AB) = P(A)P(B|A) = P(A)P(B)$$

if and only if $P(B|A) = P(B)$; that is, the two events are independent.

5.4 INFINITE AND FINITE UNIVERSES

Early in this book, there was a brief introduction to certain aspects of acceptance sampling by attributes. Example 1.3 dealt with a process in which there was a constant probability 0.04 of producing a nonconforming item. The output from this process was submitted for acceptance inspection in lots of 50 items. From each lot of 50 items, a random sample of 5 items was inspected. If there were no nonconforming items in the sample of 5, the lot was accepted; otherwise, it was rejected.

Example 1.3 analyzed one weakness of this acceptance sampling procedure by showing that the product accepted would be, on the average, little better than the product submitted. Numbers computed by the mathematics of probability were used in developing this conclusion.

Although the point was not brought out in the discussion in Chap. 1, this example illustrated both the concept of an infinite or unlimited universe and the concept of a finite or limited universe. The distinction between these two concepts is important in any explanation of the mathematics of probability.

In effect, column B of Table 1.3 gives the respective probabilities of various numbers of nonconforming items in lots of 50 from a process in which the probability of a nonconforming item is 0.04. (For instance, the statement that in the long run 275 lots out of 1,000 will have exactly 2 nonconforming items is another way of saying that the probability of exactly 2 such items is 0.275.) The probabilities in column B were based on the concept of an *infinite* universe; each lot of 50 was assumed to be a random sample from an unlimited number of items 4% of which are nonconforming.

In contrast, the probabilities in column E of Table 1.3 are based on the concept of samples of 5 from *finite* universes of 50 items. For the purpose of Example 1.3, it was necessary to know the respective probabilities of acceptance when lots of 50 contained various numbers of nonconforming items.

We shall see that in computing desired probabilities for problems in industrial quality control, there will be many cases where we assume an infinite universe for each case where we assume a finite universe. In part, this will be because, as in column B of Table 1.3, the assumption of an infinite universe is correct in principle for the particular application. In part, also, this will be because probability calculations involving large finite universes are often quite burdensome; in many such cases, simpler calculations assuming an infinite universe give results that are close enough for practical purposes.

Nevertheless, some aspects of the subject are easier to understand if an explanation of the mathematics of probability starts with examples that assume relatively small finite universes. We shall therefore examine a number of cases of sampling from finite universes before proceeding to the more important and more general subject of sampling from an infinite universe.

5.4.1 Formula for Combinations

For the solution of many problems in probability it is necessary to know how many different sets of r objects can be chosen from n objects.

For example, consider a lot of 50 pieces from which a sample is to be drawn. How many different ordered sets of 5 can be drawn? (In other words, how many sets are possible that differ either in the pieces included or in the order in which the pieces were drawn?) The first draw might be any one of 50; the next, any one of the remaining 49; the next, any one of 48; the next, any one of 47; and the final draw any one of 46. The total number of possible ordered sets is obviously the product of these numbers; this is called the number of *permutations*. The symbol P_5^{50} is read as "the number of permutations of 50 things taken 5 at a time." This is

$$P_5^{50} = (50)(49)(48)(47)(46) = 254{,}251{,}200$$

In general, the number of permutations of n things taken r at a time is given by the formula

$$P_r^n = (n)(n-1)(n-2)\cdots(n-r+1) = \frac{n!}{(n-r)!}$$

Here the expression $n!$, read as "factorial n" or "n factorial," is used for the product of the first n integers. By definition, $0! = 1$. From this an expression may be written for the number of permutations of n things taken all at a time:

$$P_n^n = n!$$

Sets without regard to the order of drawing are called *combinations*. To find the number of different combinations of 5 that may be drawn from a lot of 50, it may first be noted that any one combination of 5 has 5! = (5)(4)(3)(2)(1) = 120 possible permutations. As each combination includes 120 possible permutations, the total number of combinations can be computed by dividing the total number of permutations by 120:

$$C_5^{50} = \frac{50!}{5!45!} = \frac{(50)(49)(48)(47)(46)}{(5)(4)(3)(2)(1)} = \frac{254{,}251{,}200}{120} = 2{,}118{,}760$$

In general, the number of combinations of n things taken r at a time is given by the formula

$$C_r^n = \binom{n}{r} = \frac{n!}{r!(n-r)!}$$

5.4.2 Application of the Combination Formula to Probability Problems

Example 1.3 gave some probabilities associated with the common sampling acceptance procedure of inspecting 5 articles from each lot of 50. With the use of the combination formula, it is now possible to illustrate how such probabilities are computed.

Consider a lot of 50 articles containing 3 nonconforming articles. A sample of 5 is selected at random from the lot. What are the respective probabilities of 0, 1, 2, and 3 nonconforming articles occurring in the sample of 5?

The total number of different samples of 5 has already been calculated as 2,118,760. If the sample is selected at random, all these may be considered to be equally likely. To compute the respective probabilities we must find how many of these different possible samples contain exactly 0, 1, 2, and 3 nonconforming articles.

Consider a sample containing 0 nonconforming articles. Such a sample must come from the 47 articles that conform to specifications. From these articles the number of different possible samples of 5 is

$$C_5^{47} = \frac{47!}{5!42!} = \frac{(47)(46)(45)(44)(43)}{(5)(4)(3)(2)(1)} = 1{,}533{,}939$$

A sample containing exactly 1 nonconforming article must include 4 articles from the 47 acceptable ones and 1 article from the 3 rejectable ones. Thus there are

$$C_4^{47}C_1^3 = \frac{47!3!}{4!43!1!2!} = \frac{(47)(46)(45)(44)}{(4)(2)} = 535{,}095$$

A sample containing exactly 2 nonconforming articles must include 3 articles from the 47 good ones and 2 articles from the 3 bad ones. Of such samples, there are

$$C_3^{47}C_2^3 = \frac{47!3!}{3!44!2!1!} = \frac{(47)(46)(45)}{2} = 48{,}645$$

A sample containing exactly 3 nonconforming articles must include 2 articles from the 47 good ones and all 3 bad articles. There are

$$C_2^{47}C_3^3 = \frac{47!}{2!45!} = \frac{(47)(46)}{2} = 1{,}081$$

The respective probabilities are

$$P_0 = \frac{1{,}533{,}939}{2{,}118{,}760} = 0.72398$$

$$P_1 = \frac{535{,}095}{2{,}118{,}760} = 0.25255$$

$$P_2 = \frac{48{,}645}{2{,}118{,}760} = 0.02296$$

$$P_3 = \frac{1{,}081}{2{,}118{,}760} = 0.00051$$

$$\text{Total} = 1.00000$$

(Here P_0 means the probability of exactly 0 nonconforming articles, P_1 means the probability of exactly 1 nonconforming article, etc.)

As the probabilities of all the mutually exclusive alternatives have been computed, their sum must be unity. This check should be made for all such calculations.

5.4.3 Handling Factorials

Factorials are very large numbers. For instance 15! = 1,307,674,368,000. Many small pocket calculators will handle up to 69!; others, even programmables, make no provision for the direct calculation of factorials. Most mini- and microcomputers are limited to about 35! before an overflow condition occurs. Needless to say, all computers and many programmable pocket calculators require that special programs be written (or software acquired) to calculate factorials.

When probability calculations involving factorials are made infrequently, it is convenient to use logarithms of factorials such as those given in Table *H,* App. 3.[†]

As an illustration, consider the calculation of P_0 from the preceding discussion.

$$\frac{C_5^{47} C_0^3}{C_5^{50}} = \frac{47!45!}{42!50!}$$

Solving this problem by logarithms of factorials from Table *H,*

log 47!	59.4127	
		115.4905
log 45!	56.0778	
log 42!	51.1477	
log 50!	64.4831	115.6308
log P_0		9.8597 – 10

and

$P_0 = 0.724$

5.5 THE HYPERGEOMETRIC PROBABILITY DISTRIBUTION

The example used in the preceding discussion involved drawing a sample of size $n = 5$ from a lot of size $N = 50$ that contained $D = 3$ nonconforming items. The probabilities associated with drawing r nonconforming items in the sample were calculated where r varied from 0 to 3. Such problems are sufficiently common that the mathematical formula describing them has been named the *hypergeometric probability law.*

This law of probability may be stated as follows:

$$p(r \mid N, D, n) = \frac{C_r^D C_{n-r}^{N-D}}{C_n^N} = \frac{\binom{D}{r}\binom{N-D}{n-r}}{\binom{N}{n}} \tag{5.11}$$

which says: *the probability of r nonconforming items in a sample of size n is equal to the product of the possible combinations of nonconforming items,* (C_r^D), *times*

[†]For factorials of numbers above 1,000, use the first term of Stirling's approximation to the factorial: $n! = n^n e^{-n}\sqrt{2\pi n}$. Here, e is 2.71828. See T. C. Fry, *Probability and Its Engineering Uses,* 2d ed., pp. 121–124, D. Van Nostrand Company, Inc., Princeton, N.J., 1965.

the possible combinations of conforming items, (C_{n-r}^{N-D}), *divided by the possible combinations of samples of size n that can be drawn from lots of size N,* (C_n^N). *N, D,* and *n* are *parameters* of the probability law; they must be known, or hypothesized, ahead of time.

The vertical line in $p(r \mid N, D, n)$ stands for *given that*; and the statement itself reads as "the probability of *r* occurring given that *N, D,* and *n* are specified."

The previous statement of the hypergeometric probability law is usually referred to as the *probability density function* (p.d.f.) or *mass function.* While tables of such functions are published by many sources, *cumulative density functions,* or *cumulative distribution functions* (c.d.f.), are more common and frequently more useful.

As an illustration of the use of the cumulative distribution functions, consider the example $N = 50$, $D = 3$, $n = 5$ used previously. We would like to find the probability that no more than 1 nonconforming item will be found in the sample of 5. To do so, two terms of the hypergeometric probability law will have to be calculated and summed. In general, the mathematical expression is

$$P(r \leq c \mid N, D, n) = \sum_{r=0}^{c} \frac{\binom{D}{r}\binom{N-D}{n-r}}{\binom{N}{n}} \tag{5.12}$$

Substituting the values for the example,

$$\begin{aligned} P(r \leq 1 \mid 50, 3, 5) &= \sum_{r=0}^{1} \frac{\binom{3}{r}\binom{50-3}{5-r}}{\binom{50}{5}} \\ &= \frac{\binom{3}{0}\binom{47}{5}}{\binom{50}{5}} + \frac{\binom{3}{1}\binom{47}{4}}{\binom{50}{5}} \\ &= 0.724 + 0.253 = 0.977 \end{aligned}$$

If the value $c = 1$ were used as an acceptance number for a plan that specified testing samples of 5 drawn from lots of 50 items, then lots containing 3 nonconforming items would be accepted with probability equal to 0.977.

Special computer programs are available through ASQC and others for computing hypergeometric probabilities. Many of these programs take advantage of the fact that any calculation may contain no more than $c + 1$ terms and successively multiply and divide individual terms of factorials in an iterative sequential procedure. Since factorials are not calculated directly, hypergeometric probabilities may be calculated for very large values of N.

5.5.1 Changes in Probabilities due to Partial Exhaustion of a Lot by a Sample

The problems in the three preceding sections that were solved with the help of the combinatorial formula might also have been solved by computing the probabilities on each draw and applying the theorem of conditional probabilities.

This may be illustrated by again considering the calculation of the probability that a sample of 5 will contain 0 nonconforming items if drawn from a lot of 50 containing 3 such items. The probability that the first article drawn will be acceptable is $\frac{47}{50}$. If the first draw is acceptable, the probability that the second will be acceptable is $\frac{46}{49}$. If the first and second are both acceptable, the probability that the third will be acceptable is $\frac{45}{48}$, etc. This gives the probability that all 5 are acceptable as

$$P_0 = \left(\frac{47}{50}\right)\left(\frac{46}{49}\right)\left(\frac{45}{48}\right)\left(\frac{44}{47}\right)\left(\frac{43}{46}\right) = \left(\frac{45}{50}\right)\left(\frac{44}{49}\right)\left(\frac{43}{48}\right) = 0.72398$$

It is evident that, because each new drawing changes the proportion of acceptable and rejectable articles in the remaining portion of the lot, the probability changes from draw to draw. This partial exhaustion of the lot by the sample is recognized by the above type of calculation, or by calculations that use the combinatorial formula in the way previously illustrated. However, if the lot is large enough compared to the sample, the change in probability from one draw to the next is of negligible importance. Consider a sample of 5 taken from a lot of 5,000 containing 300 nonconforming items. The probability that this sample contains 0 nonconforming items is

$$P_0 = \left(\frac{4{,}700}{5{,}000}\right)\left(\frac{4{,}699}{4{,}999}\right)\left(\frac{4{,}698}{4{,}998}\right)\left(\frac{4{,}697}{4{,}997}\right)\left(\frac{4{,}696}{4{,}996}\right) = 0.73381$$

For all practical purposes, this is $(4{,}700/5{,}000)^5$, or $(0.94)^5$, which equals 0.73390.

The foregoing calculation provides a good transition from the case of the finite universe to the case of the infinite universe. It is evident that in taking a small sample from a finite universe of 5,000 items 6% nonconforming, no appreciable error is introduced if calculations are simplified by assuming that the probability of a nonconforming item is 0.06 on every draw. In other words, calculations are made assuming an infinite universe 6% nonconforming. We shall introduce the subject of the infinite universe by examining the very useful binomial distribution.

5.6 THE BINOMIAL AS A PROBABILITY DISTRIBUTION

Probability problems in which the probability of occurrence of an event may be assumed to be constant may be solved by the use of a formula that depends on the familiar binomial theorem. The reader may recall from algebra that

$$(a+b)^n = a^n + \binom{n}{1}a^{n-1}b + \binom{n}{2}a^{n-2}b^2 + \ldots + b^n$$
$$= \sum_{k=0}^{n} \binom{n}{k} a^{n-k}b^k$$

This binomial expansion is the basis of a probability distribution that is of great importance in statistical quality control.

Let p be the symbol for the probability that a particular event will happen. In applications to statistical quality control, this generally is the probability of a rejectable item, that is, an item not conforming to specifications.

Let q be the symbol for the probability that the same event will not happen. In applications to statistical quality control, this generally is the probability of an acceptable article, that is, of an item conforming to specifications. Because $p + q = 1$, it follows that $q = 1 - p$.

By substitution and a slight rearrangement of the terms in the binomial expansion, we have

$$(p+q)^n = \sum_{k=0}^{n} \binom{n}{k} p^k q^{n-k}$$

The p.d.f. for the binomial distribution may be stated as

$$p(r \mid n, p) = \binom{n}{r} p^r q^{n-r} \tag{5.13}$$

and the c.d.f. as

$$P(r \leq c \mid n, p) = \sum_{r=0}^{c} \binom{n}{r} p^r q^{n-r} \tag{5.14}$$

In the literature of statistics, the binomial as a probability distribution frequently is described as the point binomial or as the Bernoulli distribution.

Table 5.2 illustrates the use of the p.d.f. Eq. (5.13) to calculate the probabilities of exactly r (0 through 5) occurrences in n (equals 5) trials when the probability of occurrence p is 0.06. The reader should note that the probability of 0 occurrences in Table 5.2 using the binomial formula is exactly the value found in Sec. 5.5.1 when it is assumed that there is no depletion of the lot by withdrawing the sample of 5.

5.6.1 Average and Standard Deviation of the Binomial

If a product is 6% rejectable ($p = 0.06$) and many samples of 5 are drawn from this product, it seems obvious that the expected average number of rejectable items per sample will be (5)(0.06) = 0.3. This may be verified in a particular case by a calculation using the binomial distribution of Table 5.2.

TABLE 5.2 ILLUSTRATION OF THE BINOMIAL AS A PROBABILITY FORMULA
Probability of exactly *r* occurrences in *n* trials

r	General expression	Value when $p = 0.06$ and $n = 5$
0	$\binom{n}{0} p^0 q^{n-0}$	$(0.94)^5 = 0.7339040224$
1	$\binom{n}{1} p^1 q^{n-1}$	$5(0.06)(0.94)^4 = 0.2342246880$
2	$\binom{n}{2} p^2 q^{n-2}$	$10(0.06)^2(0.94)^3 = 0.0299010240$
3	$\binom{n}{3} p^3 q^{n-3}$	$10(0.06)^3(0.94)^2 = 0.0019085760$
4	$\binom{n}{4} p^4 q^{n-4}$	$5(0.06)^4(0.94) = 0.0000609120$
5	$\binom{n}{5} p^5 q^{n-5}$	$(0.06)^5 = 0.0000007776$
		Total = 1.0000000000

Number of rejectable items†	Relative frequency, that is, probability of occurrence	Rejectable items × relative frequency
0	0.7339040224	0.0000000000
1	0.2342246880	0.2342246880
2	0.0299010240	0.0598020480
3	0.0019085760	0.0057257280
4	0.0000609120	0.0002436480
5	0.0000007776	0.0000038880
Totals	1.0000000000	0.3000000000

†Average number of rejectable items = 0.3/1 = 0.3.

Where many sets of n trials are made of an event with a constant probability of occurrence p, the expected average number of occurrences in the long run is np; that is, np is the average of the binomial. Translated into terms of a problem in statistical quality control, if many random samples of size n are taken from a product having a fraction rejectable p, the expected average number of rejectable items per sample is np.

In mathematical terms, np is the *expected value,* or *mathematical expectation,* of x where $x = 0$ if an item is acceptable and $x = 1$ if it is rejectable. The expected value of x can be developed directly from the binomial formula by recognition of the fact that the mathematical expectation $E(x)$ is a first-degree operator equivalent to center of gravity in mechanics. In general terms, this operator can be expressed as follows:

$$\mu_x = E(x) = \begin{cases} \int_{-\infty}^{+\infty} xf(x)\,dx & \text{where } x \text{ is continuous} \\ \sum_x xp(x) & \text{where } x \text{ is discrete} \end{cases}$$

Since the binomial is a discrete distribution, we shall use the second form. Designating r as equal to the number of rejectable items found in a sample of size n that is p fraction rejectable, the expected value of r can be found from

$$E(r) = \sum_{r=0}^{n} r \binom{n}{r} p^r (1-p)^{n-r}$$

where

$$\sum_{r=0}^{n} \binom{n}{r} p^r (1-p)^{n-r} = 1$$

This summation may be written

$$\sum_{r=1}^{n} np \binom{n-1}{r-1} p^{r-1} (1-p)^{(n-1)-(r-1)}$$

and since n and p are constant parameters of the distribution, they may be moved outside the summation sign, leaving

$$np \sum_{r=1}^{n} \binom{n-1}{r-1} p^{(r-1)} (1-p)^{(n-1)-(r-1)}$$

Returning to the original expansion of the binomial,

$$(a+b)^n = \sum_{k=0}^{n} \binom{n}{k} a^k b^{n-k}$$

$$\sum_{r=1}^{n} \binom{n-1}{r-1} p^{(r-1)} (1-p)^{(n-1)-(r-1)} = (p + (1-p))^{n-1} = 1$$

Therefore

$$\mu_{np} = E(r) = np$$

and since the mathematical expectation is a linear operator,

$$\mu_p = E\left(\frac{r}{n}\right) = p$$

That is, the expected value of the number of rejectable items found divided by the sample size is the parametric value of the fraction rejectable.

The expression for standard deviation of the frequency distribution that results from the binomial is derived in standard works on mathematical statistics. It is $\sqrt{npq} = \sqrt{np(1-p)}$.

In the case where n is 5 and p is 0.06, the standard deviation is

$$\sigma_{np} = \sqrt{5(0.06)(0.94)} = \sqrt{0.282} = 0.531037$$

The standard deviation given by the formula may be checked by the following calculation from the distribution:

Fraction rejectable	Number of rejectable items	Rejectable items squared	Relative frequency	Relative frequency × rejectable items squared
0.00	0	0	0.7339040224	0.0000000000
0.20	1	1	0.2342246880	0.2342246880
0.40	2	4	0.0299010240	0.1196040960
0.60	3	9	0.0019085760	0.0171771840
0.80	4	16	0.0000609120	0.0009745920
1.00	5	25	0.0000007776	0.0000194400
Total	. .	. .	. . . .	0.3720000000

$$\sigma_{np} = \sqrt{0.372 - (0.3)^2} = \sqrt{0.372 - 0.09} = \sqrt{0.282} = 0.531037$$

For purposes of mathematical development, the expression for standard deviation will be developed from the mathematical expectation of the variance, σ^2_{np}.

$$\sigma^2_{np} = \text{Var}\,(r) = E[(r - np)^2]$$

This expression is related to the moment of inertia, or second moment about the center of gravity, from mechanics. Notice, also, that the expression inside the expectation operator, which itself is a linear operator, is a second-degree expression. The variance operator, therefore, is a second-degree mathematical operator. This fact will prove useful later in developing expressions for the standard deviation of μ_{np} and μ_x.

Using the binomial distribution, it is desired to find the expected value of $(r - np)^2$, where r, n, and p were defined previously.

$$E[(r - np)^2] = E[r^2 - 2r(np) + (np)^2] = E(r^2) - 2E[r(np)] + E(np)^2$$

The foregoing paragraph showed that $E(r) = np$, from which we may deduce that $E(np) = np$ and $E(np)^2 = (np)^2$. Thus the expression for variance yields

$$E[(r - np)^2] = E(r^2) - (np)^2$$

leaving $E(r^2)$ yet to be found.

The mathematical expectation equation may again be applied, where, for convenience, r^2 is set equal to $r(r - 1) + r$. Thus

$$\begin{aligned} E(r^2) &= E[r(r-1)] + E(r) \\ &= \sum_{r=0}^{n} r(r-1)\binom{n}{r} p^r (1-p)^{n-r} + \sum_{r=0}^{n} r\binom{n}{r} p^r (1-p)^{n-r} \\ &= n(n-1)p^2 \sum_{r=2}^{n} \binom{n-2}{r-2} p^{(r-2)} (1-p)^{(n-2)-(r-2)} + np \\ &= n(n-1)p^2 + np \end{aligned}$$

Substituting this result into the expression for variance,

$$\begin{aligned} \text{Var}(r) &= n(n-1)p^2 + np - (np)^2 \\ &= np - np^2 = np(1-p) \end{aligned}$$

and the standard deviation is

$$\sigma_{np} = \sqrt{np(1-p)}$$

It is important to distinguish between the average number of occurrences of the event in n trials np and the relative proportion of occurrence or probability of occurrence p. In statistical quality control, this is a distinction between average number of rejectable items in the samples and the average fraction rejectable. In any sample, the fraction rejectable is the number of rejectable items divided by the sample size n. The standard deviation of the fraction rejectable is, of course, the standard deviation of the number of rejectable items, $\sqrt{npq}$, divided by the sample size n. This is

$$\sigma_p = \frac{\sqrt{npq}}{n} = \sqrt{\frac{pq}{n}} = \sqrt{\frac{p(1-p)}{n}} = \frac{\sqrt{p(1-p)}}{\sqrt{n}}$$

This is an important formula in connection with the control chart for fraction rejected. Its mathematical development may be drawn from

$$\sigma_p^2 = \text{Var}\left(\frac{r}{n}\right) = E\left\{\left[\left(\frac{r}{n}\right) - p\right]^2\right\}$$

In this case, however, it is much easier to work with the second-degree variance operator than with the first-degree expectation operator.

Any constant may be drawn outside a second-degree operator merely by squaring the constant; thus

$$\text{Var}\left(\frac{r}{n}\right) = \left(\frac{1}{n}\right)^2 \text{Var}\,(r)$$

and since the Var (r) has previously been derived,

$$\text{Var}\left(\frac{r}{n}\right) = \left(\frac{1}{n}\right)^2 np(1-p)$$

$$= \frac{p(1-p)}{n}$$

yielding

$$\sigma_p = \sqrt{\frac{p(1-p)}{n}}$$

5.6.2 An Experimental Example of the Meaning of the Average and Standard Deviation of the Binomial

The data of Table 5.1 may be used to clarify certain aspects of the foregoing formulas. Consider columns *C* and *D* of that table. Both apply to the results of samples of 10. Because each chip drawn was replaced before the next drawing, the binomial is applicable in principle; there was a constant probability p of drawing a nonconforming item.

Column *C* gives the number of nonconforming items in each sample of 10. The result of the 40 samples may be arranged into a frequency distribution as follows:

Number of nonconforming items	Frequency
4	3
3	7
2	9
1	16
0	5
Total	40

If we calculate the average and standard deviation of this distribution, we find the average to be 1.675 and the standard deviation to be 1.127. (The tabular method illustrated for the binomial is appropriate for these calculations.) If we use the formulas for average and standard deviation that have been given for numbers of occurrences, we can find the expected values of these statistics if we are able to obtain a great many such samples of 10. The expected average

$$\mu_{np} = np = 10(0.17034) = 1.7034$$

The expected standard deviation

$$\sigma_{np} = \sqrt{np(1-p)} = \sqrt{10(0.17034)(0.82966)} = 1.189$$

In this particular experiment, the observed values seem fairly close to the expected values.

Column D of Table 5.1 gives the fraction nonconforming in each sample of 10. The fractions nonconforming may be summarized into the following frequency distribution:

Fraction nonconforming	Frequency
0.4	3
0.3	7
0.2	9
0.1	16
0.0	5
Total	40

The similarity to the previous frequency distribution is apparent. The average and standard deviation of the distribution of fractions nonconforming are 0.1675 and 0.1127, respectively. Obviously each figure is one-tenth the corresponding figure from the distribution of numbers of nonconforming items. In general, in a binomial distribution these statistics of a distribution of fractions of occurrences will be $1/n$ times the corresponding statistics of numbers of occurrences. The expected average fraction of occurrences is, of course, the p of 0.17034. The expected standard deviation is

$$\sqrt{\frac{p(1-p)}{n}} = \sqrt{\frac{(0.17034)(0.82966)}{10}} = 0.1189$$

5.6.3 The Binomial as a Basis for Approximate Estimates of Probabilities in Sampling from Finite Lots

The binomial describes the situation that exists when the probability of a nonconforming item is constant from draw to draw. As already pointed out, this is never quite true in sampling from finite lots. Nevertheless, the binomial often provides a good enough approximation to serve as a practical basis for judgments about various sampling plans.

In the preceding discussion, the binomial example was carried to many decimal places. This was necessary to illustrate how μ_{np} and σ_{np} gave exact results for the average and standard deviation of the binomial distribution. However, practical calculations for industrial quality control never require so many decimal places. For the common calculations to judge the quality protection given by alternative sampling plans, three decimal places, or in some instances only two decimal places, are adequate.

Our illustration of the use of the combination formula gave the probabilities of 0, 1, 2, and 3 nonconforming items in drawing samples of 5 from lots of 50 that were 6% nonconforming. Table 5.2 gave the probabilities of 0, 1, 2, 3, 4, and 5 nonconforming items in drawing samples of 5 from an infinite lot that was 6% nonconforming. Where these probabilities are expressed only to three decimal places, the differences between them do not seem to be serious.

Nonconforming items	Probabilities from lot of 50	Probabilities from infinite lot
0	0.724	0.734
1	0.252	0.234
2	0.023	0.030
3	0.001	0.002

In general, the more the lot tends to be exhausted by the sample, the greater the difference between the correct probabilities and the approximate ones computed by the binomial theorem on the assumption of an infinite lot. From the standpoint of practical action based on these probabilities, it may well be true that it is satisfactory to assume that the probabilities that apply to an infinite lot also apply to samples of 5 from finite lots of 50 or more.

5.7 THE POISSON LAW AS A PROBABILITY DISTRIBUTION

Certain types of frequency distributions occur in nature, both in quality control work and elsewhere, that are closely fitted by a formula known as the Poisson law. The situations to which it has been shown to be applicable are so numerous and so diversified that the Poisson law has sometimes been called the law of small numbers. Some representative examples are shown in Table 5.3.

If c is the count of the occurrences of some event of interest, and μ_c[†] is the parametric value of the rate of occurrence, then the Poisson p.d.f. may be stated as

$$p(c \mid \mu_c) = \frac{\mu_c^c}{c!} e^{-\mu_c} \tag{5.15}$$

and the c.d.f. as

$$P(r \leq c \mid \mu_c) = \sum_{r=0}^{c} \frac{\mu_c^r}{r!} e^{-\mu_c} \tag{5.16}$$

First we shall consider some examples of Poisson distributed data and then use one of the examples to compare the theoretical frequency with the actual.

Table 5.3*a* is based on 33 years of records for 10 rainfall stations widely scattered throughout the midwestern United States. It gives the number of 10-min periods in a year having half an inch or more of rainfall. c is the number of such cloudbursts in a station-year; the frequency is the number of station-years having respectively 0, 1, 2, 3, 4, and 5 such excessive rainstorms. The average number of such storms per station-year $\bar{c}$ is 1.2. Table 5.4 shows the close agreement between the frequencies observed for the 330 station-years and the frequency computed by the Poisson for the same average and the same number of station-years.

Table 5.3*b* is the classic example of the Poisson series that has been quoted in many books on probability and statistics. It was compiled by Bortkewitsch, who wrote on the subject of the Poisson law in 1898. He found from the records of the Prussian army the number of men killed by the kick of a horse in each of 14 cavalry corps in each of 20 successive years, and after discarding the records for 4 corps that were considerably larger than the others, he treated the rest as one series of samples. c is the number of Prussians killed in this way in a corps-year. The frequency is the number of corps-years having exactly c cavalrymen killed. This was the first example of the applicability of the Poisson to accident statistics.

Table 5.3*c* gives the number of articles turned in per day to the lost and found bureau of a large office building (excluding Sundays and holidays, and June, July, and August, when there might be a considerable reduction in the average population of the building). The frequency is the number of days with exactly c lost articles turned in.

[†]Frequently the Poisson rate parameter is referred to as the intensity parameter and is designated by the lowercase Greek letter λ.

TABLE 5.3 EXAMPLES OF DISTRIBUTIONS TO WHICH THE POISSON LAW IS APPLICABLE

a. Excessive rainstorms†

c	Frequency
0	102
1	114
2	74
3	28
4	10
5	2

b. Deaths from kick of a horse

c	Frequency
0	109
1	65
2	22
3	3
4	1

c. Lost articles‡

c	Frequency
0	169
1	134
2	74
3	32
4	11
5	2
6	0
7	1

d. Vacancies in the U.S. Supreme Court§

c	Frequency
0	59
1	27
2	9
3	1

e. Calls from groups of 6 coin-box telephones‡

c	Frequency
0	8
1	13
2	20
3	37
4	24
5	20
6	8
7	5
8	2
9	1

f. Errors in alignment found at aircraft final inspection

c	Frequency	*c*	Frequency
0	0	10	2
1	0	11	5
2	0	12	2
3	1	13	3
4	4	14	1
5	3	15	2
6	6	16	1
7	7	17	0
8	7	18	1
9	5		

†E. L. Grant, Discussion of "Rainfall Intensities and Frequencies," by A. J. Schafmayer and B. E. Grant, *Transactions, American Society of Civil Engineers,* vol. 103, p. 388, 1938.

‡Taken from Frances Thorndike, "Applications of Poisson's Probability Summation," *The Bell System Technical Journal,* vol 5, pp. 604–624, October 1926.

§W. A. Wallis, "The Poisson Distribution and the Supreme Court," *Journal of the American Statistical Association,* vol. 31, p. 376, June 1936.

Table 5.3*d* shows the vacancies in the United States Supreme Court, either by death or resignation of members, from 1837 to 1932. The frequency is the number of years in which there were exactly *c* vacancies.

Table 5.3*e* shows the number of telephone calls per 5-min interval from a group of six coin-box telephones in a large railway terminal. The data were taken for the

TABLE 5.4 COMPARISON OF OBSERVED FREQUENCIES WITH THEORETICAL FREQUENCIES BY THE POISSON
Data of Table 5.3*a* on excessive rainstorms

c	Observed frequency in 330 trials	Total number of rainstorms	Summation terms of Poisson, $\mu_c = 1.2$	Individual terms of Poisson, $\mu_c = 1.2$	Expected frequency in 330 trials
0	102	0	0.301	0.301	99
1	114	114	0.663	0.362	119
2	74	148	0.879	0.216	71
3	28	84	0.966	0.087	29
4	10	40	0.992	0.026	9
5	2	10	0.998	0.006	2
6	0	0	1.000	0.002	1
Total	330	396	.....	1.000	330

$\bar{c} = \frac{396}{330} = 1.20$

period from noon to 2 P.M. for seven days, not including a Saturday or Sunday. The frequency is the number of 5-min intervals in which exactly c calls were originated.

Table 5.3*f* gives the numbers of errors in alignment discovered by inspectors at the time of final inspection of an airplane. The frequency is the number of planes for which exactly c such errors were listed on the inspector's "squawk sheet."

At first impression, these six Poisson illustrations may seem to have very little similarity to one another. However, a more critical examination shows that they have very definite characteristics in common. In each, a count was made of the number of occurrences of an event that had many opportunities to occur but was extremely unlikely to occur at any given opportunity. There were many 10-min periods in a year; it was unlikely that any particular one would bring a cloudburst of half an inch or more of rain. There were many contacts between a cavalryman and a horse during a year of history of a cavalry corps; it was unlikely that the horse would make a fatal kick at any particular contact. There were many people passing through the large office building during a day; it was unlikely that any one person would find a lost article and turn it in to the lost and found bureau, and so forth.

In all these cases, there is the concept of the existence of a large n and a small p, even though it may be impossible to assign definite values to either n or p. Even in cases where it is not impossible to determine definite values of n and p, it may simply be of no advantage to determine n and p and to calculate a binomial based on them; the Poisson will serve as well.

It will be noted that in all instances an effort has been made to keep constant a quantity that might be called *the area of opportunity for occurrence.* Thus the 4 cavalry corps that were much larger than the other 10 were eliminated from the tabulation of deaths of cavalrymen; the days in which the office building was believed to have a less than normal population were eliminated from the tabulation of articles turned in to the lost and found bureau; the record of calls from the group of coin-box phones was limited to an apparently homogeneous 2-h period in the middle of the day and Saturdays and Sundays were eliminated.

By similar reasoning, it would not have been appropriate to combine alignment "squawks" observed on two airplanes of different designs.

5.7.1 The Poisson Distribution as an Approximation to the Binomial

Calculations involving the use of binomials are often burdensome; this is particularly true if many terms are involved and if n is large. Fortunately, a simple approximation may be obtained to any term of the binomial. This approximation is often called *Poisson's exponential binomial limit.* The larger the value of n and the smaller the value of p, the closer the Poisson approximation. It is admirably suited to the solution of many problems that arise in industrial quality control.

The Poisson formula may be derived from the binomial theorem in the following manner. In the binomial formula, let $c = np$, then

$$\frac{n!}{r!(n-r)!}p^r(1-p)^{n-r} = \frac{n!}{r!(n-r)!}\left(\frac{c}{n}\right)^r\left(1-\frac{c}{n}\right)^{n-r}$$

$$= \frac{n!}{(n-r)!n^r}\left(\frac{c^r}{r!}\right)\left(1-\frac{c}{n}\right)^n\left(1-\frac{c}{n}\right)^{-r}$$

At this stage, the *limit* of each term is taken, allowing n to go to infinity and holding $c = np$ constant.

$$\lim_{n\to\infty}\frac{n!}{(n-r)!\,n^r} = 1$$

$$\lim_{n\to\infty}\frac{c^r}{r!} = \frac{c^r}{r!}$$

$$\lim_{n\to\infty}\left(1-\frac{c}{n}\right)^n = e^{-c}$$

where e is 2.71828^+, the base of natural or napierian logarithms; and

$$\lim_{n\to\infty}\left(1-\frac{c}{n}\right)^{-r} = 1$$

Thus, as n increases without bound, the limiting case of the binomial distribution, stated

$$B(r\,|\,n, p) = \binom{n}{r} p^r (1-p)^{n-r}$$

is the Poisson distribution

$$P(r\,|\,c = np) = \frac{c^r}{r!} e^{-c} = \frac{(np)^r}{r!} e^{-np}$$

It will be remembered that $\mu_{np} = np$ is the average value of the expected number of occurrences. For ordinary use in discussion of the Poisson law throughout this book, $\mu_c = c$ is used for this average instead of μ_{np}.

5.7.2 Use of Tables and Computer Programs for Solution of Poisson Problems

Molina's tables[†] give individual terms and summation terms of the Poisson formula to six decimal places for values of μ_c (μ_{np}) up to 100. Table G in App. 3 of this book gives summation terms to three decimal places, following the formula

$$P(r \leq c\,|\,\mu_c) = \sum_{r=0}^{c} \frac{\mu_c^r}{r!} e^{-\mu_c}$$

Individual Poisson terms must be obtained by subtracting adjacent summation terms. Thus

$$p(c\,|\,\mu_c) = P(r \leq c\,|\,\mu_c) - P(r \leq c - 1\,|\,\mu_c)$$

In Table G the values of μ_c go by intervals of 0.02 from 0 to 0.10, by intervals of 0.05 from 0.10 to 1.00, by intervals of 0.1 from 1.0 to 2.0, by intervals of 0.2 from 2.0 to 8.0, by intervals of 0.5 from 8.0 to 15.0, and by intervals of 1 from 15 to 25. The approximate Poisson distribution corresponding to any value of μ_c under 25 that is not given in Table G may be obtained by interpolation between the distributions for the two adjacent values of μ_c that are given in the table. A value for μ_c equals 5.28, for example, is assumed to be 0.4 [that is, (5.28 − 5.20)/(5.40 − 5.20)] of the way from the 5.20 value to the 5.40 value. Such a linear interpolation in Table G will generally give values that are either correct or in error by not more than one unit in the third decimal place. Individual Poisson terms must be obtained by subtracting the adjacent summation terms.

[†]E. C. Molina, *Poisson's Exponential Binomial Limit,* D. Van Nostrand Company, Princeton, N.J., 1942.

5.7.3 Average and Standard Deviation of the Poisson

The average (expected value) and standard deviation of the Poisson distribution are $\mu_c = \mu_{np}$ and $\sqrt{\mu_c} = \sqrt{\mu_{np}}$, respectively. These parameters may be derived by considering the Poisson either as the limiting case of the binomial or as a distribution in its own right.

The mean and standard deviation of the binomial were found to be $\mu_{np} = np$ and $\sigma_{np} = \sqrt{np(1-p)}$, respectively. In the limiting case where $E(c) = E(np)$, a constant,

$$\text{Mean} = E(np) = np = E(c) = c$$

$$\begin{aligned}\sigma_c &= \sqrt{np(1-p)}\\ &= \sqrt{c\left(1-\frac{c}{n}\right)}\\ &= \sqrt{c}\end{aligned}$$

Considered in its own right, the mean of the Poisson may be derived from

$$E(r) = \sum_{r=0}^{\infty} r\,\frac{\mu_c^r}{r!}\,e^{-\mu_c}$$

and the standard deviation from

$$\sigma_c = \sqrt{\text{Var}\,(r)} = \sqrt{\mu_c}$$

where

$$\text{Var}\,(r) = E[(r-\mu_c)^2] = \sum_{r=0}^{\infty} r^2\,\frac{\mu_c^r}{r!}\,e^{-\mu_c} - \mu_c^2$$

The Poisson is therefore a distribution for which the standard deviation $\sqrt{\mu_c}$ is always the square root of the average μ_c.

5.8 THE NORMAL DISTRIBUTION

The basic probability distribution underlying the calculation of Shewhart control-chart limits for variables data is the *normal* or *gaussian* distribution. The p.d.f. for this distribution is

$$f(x) = \frac{1}{\sqrt{2\pi}}\,e^{-x^2/2} \tag{5.17}$$

A plot of this function over its full range $-\infty$ to $+\infty$ produces the familiar symmetrical bell-shaped curve. It is a continuous curve over its full range and has a mean value of zero and a standard deviation of one. Fortunately, almost the entire area under this curve lies on the range of z between -3 and $+3$. Table *A* of App. 3 tabulates the area under this curve from $-\infty$ to the value of z indicated in the left-hand column, that is, from $z = -3.59$ to $z = +3.59$. The c.d.f. for the table values is

$$F(z) = \int_{-\infty}^{z} \frac{1}{\sqrt{2\pi}}\, e^{-x^2/2}\, dx \tag{5.18}$$

In mathematical terms, the table is referred to as a tabulation of the cumulative left-hand standard normal curve, designated $N(0,1)$, that is, $N(\mu = 0, \sigma = 1)$.

For practical usage, however, it is necessary to convert from mean values and standard deviations other than zero and one, respectively. This procedure, called *normalizing,* involves substituting

$$z = \frac{X - \mu}{\sigma}$$

into the general formula, yielding

$$F(z) = \int_{-\infty}^{(X-\mu)/\sigma} \frac{1}{\sqrt{2\pi}}\, e^{-x^2/2}\, dx$$

Thus the values read from the table represent the area under the curve from $-\infty$ to $z = (X - \mu)/\sigma$.

Two important mathematical relationships may be developed from the mean μ and standard deviation σ of the normal distribution. First, the expected value of the average of a sample of size n equals the expected value of an individual measurement μ. This may be demonstrated as follows:

$$\mu_{\overline{X}} = E(\overline{X}) = E\left(\frac{1}{n}\,\Sigma X_j\right) = \frac{1}{n}\,E\left(\Sigma X_j\right)$$

since E is a linear operator. Thus

$$\mu_{\overline{X}} = \frac{1}{n}\,(n\mu_X) = \mu_X$$

The second important relationship is that between the variance of distribution of sample averages and that of the distribution of individual items σ^2. Remembering that Var is a second-degree operator,

$$\text{Var}(\bar{X}) = \text{Var}\left(\frac{1}{n}\,\Sigma\,X_j\right) = \frac{1}{n^2}\,\text{Var}\left(\Sigma\,X_j\right)$$

$$= \frac{1}{n^2}\,\Sigma\,\text{Var}(X_j) = \frac{1}{n^2}\,(n\sigma^2) = \frac{\sigma^2}{n}$$

Thus the standard deviation of the distribution of sample averages is equal to the standard deviation of the distribution of individual measurements divided by the square root of the sample size. This relationship has already been extensively used in the development of $\bar{X}$ control charts.

5.8.1 The Central Limit Theorem

From the standpoint of process control, the *central limit theorem* is one of the most powerful tools of mathematical statistics. It has been and will be used extensively throughout this book. Simply stated, the theorem says:

Irrespective of the shape of the distribution of a universe, the distribution of average values, $\bar{X}$'s, of subgroups of size n, ($\bar{X}_1$, $\bar{X}_2$, $\bar{X}_3$, . . . , $\bar{X}_k$), drawn from that universe will tend toward a normal distribution as the subgroup size n grows without bound.

Thus we may use the normalized statistic $z = \sqrt{n}(\bar{X} - \mu)/\sigma$ and a normal table such as Table *A* of App. 3 to evaluate probabilities related to the distribution of sample averages.

The reader will recognize that the $\bar{X}$ statistic is used for the control chart for means. As discussed in Chap. 3 and illustrated in Figs. 3.7 to 3.9 for distributions of sample averages from normal, rectangular, and triangular universes, the value of n does not have to be very large before the normal distribution may be applied. While this fact is very useful in analyzing probabilities, it does not form the basis for Shewhart control charts with $\pm 3\sigma$ limits.[†]

The mathematical proof of the central limit theorem is beyond the scope of this book. The discussion and illustrations in Chap. 3 clearly demonstrate not only the fact but also the speed at which the limit is approached.

5.8.2 The Normal Curve as an Approximation to the Binomial

One of the several common derivations of the normal curve is as a limit of the binomial distribution as n is increased indefinitely. The greater the value of n, the better the estimate of the binomial that can be made from a normal curve area table such as Table *A*, App. 3. It should be recognized, however, that the normal curve is always a symmetrical distribution. The binomial is symmetrical only in the special case where $\mu_p = p = \frac{1}{2}$. For a given value of n the normal curve gives

[†]For an excellent discussion of the power of Shewhart control charts, see the monograph by Donald J. Wheeler, *Myths, Foundations, and Competitors for Shewhart's Control Charts,* 1991, Statistical Process Controls, Inc., Knoxville, Tenn.

a better approximation when p is close to $\frac{1}{2}$ than when p is close to 0 or 1. However, if n is large enough—say in the hundreds or more—the normal curve may be used for a wide range of values of p and will give an approximation to the binomial that is good enough for practical purposes in many problems in industrial sampling. Where p is close to 0 or 1, the approximation will be considerably less reliable in the extreme tails of the distribution than near the center of the distribution. In previous mathematical expectation calculations the following relationships were found:

Distribution	Expected value	Standard deviation
Normal	μ	σ
Binomial	$\mu_{np} = np$	$\sqrt{np(1-p)}$

The form for z, any deviate in the standard normal table, was

$$z = \frac{X - \mu}{\sigma}$$

which may be approximated to the binomial by substituting the binomial statistics equivalent to μ and σ. This substitution leads to

$$z = \frac{X - np}{\sqrt{np(1-p)}}$$

A problem remains, however, in relating the mathematical form of the normal curve, a continuous function, to that of the binomial, a discrete function. That is, given some discrete value r, the binomial formula

$$p(r) = \binom{n}{r} p^r (1-p)^{n-r}$$

has some finite value. The corresponding mathematical operation with the normal curve function yields nothing; that is,

$$P(z = r) = \int_{(r-np)/\sqrt{np(1-p)}}^{(r-np)/\sqrt{np(1-p)}} \frac{1}{\sqrt{2\pi}} e^{-x^2/2}\, dx = 0$$

An approximation method for circumventing this problem is to make a *continuity correction* and use the area under the normal curve from $r - \frac{1}{2}$ to $r + \frac{1}{2}$. Therefore, we designate

$$z_1 = \frac{r - np - 0.5}{\sqrt{np(1-p)}}$$

$$z_2 = \frac{r - np + 0.5}{\sqrt{np(1-p)}}$$

and

$$P(z = r) = F(z_2) - F(z_1) = \int_{z_1}^{z_2} \frac{1}{\sqrt{2\pi}} e^{-x^2/2}\, dx$$

The method of using Table *A* to approximate the binomial is illustrated in the following example.

Samples of 45 are being taken from a stream of products. This product is, on the average, 25% nonconforming in the sense that one-fourth of the product normally fails to conform to a particularly severe specification that is being applied as part of the acceptance procedure. The following two questions are representative of the two types of probability calculations that might be required:

1. What is the probability that a sample of 45 will contain 13 nonconforming items?
2. What is the probability that a sample of 45 will contain 13 or more nonconforming items?

It is evident that the binomial distribution is applicable and that $n = 45$ and $p = 0.25$. To answer question 1 by calculation from the binomial distribution itself, it must be recognized that $r = 13$.

$$\begin{aligned} P(13) &= \binom{n}{r} p^r (1-p)^{n-r} \\ &= \frac{45!}{13!32!}(0.25)^{13}(0.75)^{32} \\ &= 0.1093 \end{aligned}$$

To approximate this answer from Table *A*, it is first necessary to compute the average and standard deviation of the binomial distribution.

$$np = 45(0.25) = 11.25$$

$$\sqrt{np(1-p)} = \sqrt{45(0.25)(0.75)} = 2.905$$

These values may now be substituted into the formulas previously developed.

$$z_1 = \frac{13 - 11.25 - 0.5}{2.905} = 0.4303$$

$$z_2 = \frac{13 - 11.25 + 0.5}{2.905} = 0.7745$$

The approximation of $P(13)$ may be taken from Table *A*.

$$\begin{aligned} F(z_2) &= F(0.7745) &&= 0.7807 \\ F(z_1) &= F(0.4303) &&= \underline{0.6665} \\ P(13) &= F(z_2) - F(z_1) &&= 0.1142 \end{aligned}$$

This figure, 0.1142, is an approximation to the correct figure of 0.1093, the probability of exactly 13 nonconforming items under the stated conditions.

A fairly long and tedious calculation is required to find the correct answer to question 2 regarding the probability of 13 or more nonconforming items. It is necessary to calculate 33 terms of the binomial corresponding to values of r from 13 to 45 inclusive and to take the sum of these terms. A somewhat shorter calculation, still fairly long, requires the evaluation of the 13 terms corresponding to values of r from 0 to 12 inclusive, the summation of these terms, and the subtraction of the sum from 1.

In the present instance these long calculations can be avoided by the use of "Applied Mathematics Series No. 6, Tables of the Binomial Probability Distribution," published by the Government Printing Office and available from the Superintendent of Documents, Washington, D.C. This volume, containing 387 large pages, gives to seven decimal places both the individual values and the summation values of the terms of the binomial where n is 49 or less. The values of μ_p go from 0.01 to 0.50 by intervals of 0.01. The table for $r = 13$ and $n = 45$ gives the correct answer to question 2 as 0.3251992.†

The approximate answer that the normal curve gives for question 2 requires only a slight modification of the calculations already made to answer question 1. It has already been pointed out that, as applied to a continuous distribution such as the normal curve, the probability of exactly 13 nonconforming items should be interpreted as meaning the probability of from 12.5 to 13.5 nonconforming items. It follows that the probability of 13 or more nonconforming items should be interpreted to mean the probability of 12.5 or more. It has already been calculated that 0.6665 of a normal distribution having the same μ and σ as this particular binomial will be below 12.5. It follows that the approximate probability of 13 or more nonconforming items is $1 - 0.6665 = 0.3335$.

Similarly, the Poisson distribution may be approximated by the normal by applying the same continuity corrections. In using the binomial with parameters μ_p and σ_p, it is necessary to multiply both by n to get them into the standard binomial form prior to applying the continuity corrections and using the normal approximation. To use $(p - \mu_p + \frac{1}{2})/\sigma_p$ is incorrect for obvious reasons.

5.9 DECIDING ON THE METHOD TO BE USED FOR CALCULATING PROBABILITIES IN INDUSTRIAL SAMPLING PROBLEMS

In the attributes inspection of industrial product, each item either conforms to specifications or fails to conform. In sampling for acceptance purposes, the decision on acceptance or rejection of the product sampled will commonly depend on

†For additional individual and summation values of the binomial, see H. G. Romig, *50–100 Binomial Tables*, John Wiley & Sons, Inc., New York, 1953. This publication of the Bell Telephone Laboratories covers values of n from 50 to 100 by steps of 5. Extensive tables of summation values are given in *Cumulative Binomial Probability Distribution*, Harvard University Press, Cambridge, Mass., 1955. This Harvard volume includes values of n from 100 to 200 by steps of 10, from 200 to 500 by steps of 20, and from 500 to 1,000 by steps of 50.

the number of nonconforming items in the sample or samples. In judging the merits of any proposed sampling acceptance scheme, it is appropriate to calculate the probabilities of acceptance assuming a number of different percentages of nonconforming items in the product sampled. Some calculations of this type have already been illustrated; many more are given in the chapters on acceptance sampling by attributes.

In one type of problem it is specified that a random sample of size n is to be drawn from a lot of size N containing a specified percentage of nonconforming items. It is desired to find the probability that the sample will contain exactly r nonconforming items or, perhaps, r or more or r or less. In this type of problem, the theoretically correct answer requires consideration of lot size N as well as sample size n. Methods of making such calculations were illustrated on pages 183–186; probabilities so calculated are described as *hypergeometric* probabilities. As pointed out on page 193, in problems of this type it is often good enough for practical purposes to assume that the sample is drawn from an infinite lot rather than from a finite lot and to use the binomial to give an approximation to the desired hypergeometric probabilities.

In another type of problem the random sample of size n is assumed to be drawn from a stream of product containing a specified percentage of nonconforming items. Here the binomial is applicable in principle rather than as an approximation.

While the use of mini- and microcomputers at the workstation is gaining widespread acceptance, there are still many instances in which lack of access to such equipment may delay important action. The use of approximation methods and readily available tables such as Tables A and G may be quicker and easier for finding a first-cut solution. The hand calculation of hypergeometric probabilities, for example, is extremely tedious and should be avoided.

No simple general rules can be laid down as to when to use approximate methods. It is always a matter of balancing the savings in time and cost of calculations against the error introduced by the approximation. The following suggestions are merely intended to give some general guidance.

1. In principle, hypergeometric probabilities are required whenever a sample is drawn from a finite lot of stated size N. The decision on whether to compute hypergeometric probabilities or to assume an infinite lot should be influenced by the size of N, by the ratio n/N, and by the number of values of r for which calculations are required. The smaller the ratio of sample size to lot size, the less the error introduced by assuming that the sample is drawn from an infinite lot. The amount of time saved by an approximation is relatively small if calculations are required for only one or two values of r and much greater if many values of r are involved. A possible rule of thumb is to view the binomial as giving an acceptable approximation to hypergeometric probabilities whenever n/N does not exceed 0.1.[†]

[†]For some 700 pages of tables giving hypergeometric probabilities, see G. J. Lieberman and D. B. Owen, *Tables of the Hypergeometric Probability Distribution*, Stanford University Press, Stanford, Calif., 1961.

2. Whenever the binomial applies in principle or is considered to give a satisfactory approximation to hypergeometric probabilities, the decision must be made among (*a*) calculation by the binomial, (*b*) use of the Poisson distribution as an approximation to the binomial, and (*c*) use of the normal curve as an approximation to the binomial. For anyone who has the sets of binomial tables (with sufficient values of n and p) or the computer access that we have mentioned, it is simple to use the binomial. Without these tables, or for values of n and p not given, a decision for or against calculation by the binomial depends in part on the number of terms to be evaluated. If many terms are to be evaluated in an "r or more" or "r or less" type of problem, it is usually more practical to use an approximate method.
3. The great timesaver in all such calculations is the Poisson distribution. With a table such as Molina's table or Table *G*, approximate answers can be obtained very rapidly. This statement applies to calculations of the probability of exactly r occurrences as well as to "r or more" or "r or less" occurrences. In many industrial sampling problems, the approximate values given by the Poisson distribution are good enough. The larger the n and the smaller the p ($p < 0.1$), the closer the approximate answer will be to the true probability.
4. The normal curve as an approximation to the binomial gives a more rapid answer than the Poisson approximation only when the μ_{np} of the problem is greater than the maximum μ_{np} in the available Poisson table or diagram. (This maximum is 25 in Table *G* and 100 in Molina's tables.) The normal curve is somewhat better adapted to rapid calculation of cumulative terms than of individual terms. The larger the n and the nearer the p to 0.5, the closer the approximate answer will be to the true probability. A rough working rule of thumb is to avoid using the normal curve as an approximation to the binomial whenever np is less than 5, or as an approximation to the Poisson when np is less than 10.

5.10 RELATIONSHIP BETWEEN CONTROL CHARTS AND CERTAIN OTHER STATISTICAL TECHNIQUES

This book deals chiefly with two types of statistical tools that have been found to be broadly useful in controlling the quality of manufactured product. These two tools, control charts and acceptance sampling procedures, have many different aspects. We have already observed a variety of possible control charts and a considerable diversity of possible objectives for such charts. We shall discover a similar variety of methods and objectives when we examine the subject of acceptance sampling. Nevertheless, these are only two of a large number of types of statistical techniques that are useful in the physical, biological, and social sciences as well as in manufacturing.

Many of the persons who read this book will already have read one or more general treatises on statistical inference. Doubtless most of the other readers will do so.

This section aims to help the reader to bridge the gap between the general literature of probability and statistical inference and the literature of control charts and acceptance sampling. More specifically, the objectives are as follows:

1. To emphasize certain basic differences between the usual purposes of control charts and of some other common statistical techniques.
2. To introduce certain general terminology of the literature of statistical inference.
3. To relate certain symbols and formulas commonly used in connection with control charts to certain corresponding symbols and formulas that are common in the literature of statistical inference.

5.10.1 A Classification of the Problems of the Statistician

Enoch B. Ferrell of Bell Telephone Laboratories, writing in *Industrial Quality Control,* makes an interesting distinction among four classes of statistical problems.† We have paraphrased his classification as follows:

1. *Descriptive statistics* involves the summarizing of a set of numbers by means of relatively few numbers, for example, by a measure of central tendency and a measure of dispersion.
2. *Probability* a priori involves prediction of the characteristics of samples drawn at random from a known universe. It is as if there were a bowl full of known numbers and it is proposed to draw samples from this known bowl.
3. *Statistical inference* involves estimating the parameters of a universe on the basis of a sample drawn from that universe. It is as if there were a bowl full of unknown numbers and it is desired to draw conclusions about these unknown numbers on the basis of a sample or series of samples drawn from the bowl.
4. Although a quality controller often is confronted with the three foregoing types of problems, one of the basic statistical problems is quite different. It is *to determine the presence or absence of statistical control.* The controller has certain observed numbers and needs to know whether it is reasonable to treat them as a sample from one universe or whether they should be treated as coming from two or more universes. In other words, was there just one bowl or were there several bowls from which these numbers were drawn?

The following sections contain explanations of a number of terms and concepts used in the literature of estimation and statistical inference. Many of the concepts associated with the terms have been introduced in earlier chapters even though the following discussion may constitute the first mention in this book of the term itself.

†E. B. Ferrell, "Control Charts Using Midranges and Medians," *Industrial Quality Control,* vol. 9, no. 5, p. 30, March 1953.

5.11 RANDOM VARIABLES

G. P. Wadsworth and J. G. Bryan give the following clear explanation of this important concept.†

> On an abstract level, ordinary mathematics is concerned with independent variables, the values of which may be chosen arbitrarily, and dependent variables, which are determined by the values assigned to the former. In the concrete domain, science aims at the discovery of laws whereby natural phenomena are interrelated, and the value of a particular variable can be determined when pertinent conditions are prescribed. Nevertheless, there exist enormous areas of objective reality characterized by changes which do not seem to follow any definite pattern or have any connection with recognizable antecedents. We do not mean to suggest an absence of causality. However, from the viewpoint of an observer who cannot look behind the scenes, a variable produced by the interplay of a complex system of causes exhibits irregular (though not necessarily discontinuous) variations which are, to all intents and purposes, random. Broadly speaking, a variable which eludes predictability in assuming its different possible values is called a random variable, or synonymously a variate. More precisely, a random variable must have a specific range or set of possible values and a definable probability associated with each value.

The reader will note that the stipulation that a random variable requires a definable probability for each value implies the existence of some single "bowl" as a source of the variable mentioned in Ferrell's cases 2 and 3.

The reader who recalls the discussion in Chap. 3 of sampling from a statistically controlled process will recognize that a great many different random variables may be generated by a single universe, or "bowl." For example, we illustrated not only the behavior of individual drawings from Shewhart's normal bowl but also the behavior of $\overline{X}$ values, s values, and R values from samples of 2, 4, 8, and so forth. All these $\overline{X}$, s, and R values were random variables.

5.12 POINT ESTIMATES AND ESTIMATORS

In Ferrell's case 3, it is desired to use various statistics of a sample (or of a number of samples) to estimate various parameters that help to describe a universe. A single value used as an estimate of a given parameter is called a *point estimate.* The sample statistic used in making the estimate is called an *estimator.*

For instance, it might be desired to estimate the arithmetic mean of an unknown universe on the basis of a sample taken from the universe. Some possible estimators would be the sample arithmetic mean, the sample median, and the sample midrange. In judging the merits of proposed estimators, it is desirable to consider whether they are *unbiased* and whether they are *efficient.*

†G. P. Wadsworth and J. G. Bryan, *Applications of Probability and Random Variables,* p. 40, McGraw-Hill Book Company, New York, 1974.

5.12.1 Unbiased and Biased Estimates

A statistic provides an unbiased estimate of a universe parameter if the arithmetic mean of the sampling distribution of the statistic is equal to the value of the universe parameter. The arithmetic mean of a sample gives an unbiased estimate of the arithmetic mean of the universe.

In contrast to the arithmetic mean, we noted in Chap. 4 that the sample standard deviations from Shewhart's normal bowl tended to give a biased estimate of the standard deviation in the bowl. The amount of the bias evidently decreased as the sample size increased. To correct for this bias when the $\bar{s}$ of a set of samples or the s of one sample is used to estimate the σ of a normal universe, the reader will recall that it is necessary to divide $\bar{s}$ or s by the c_4 factor given in Table *B* of App. 3. This c_4 factor, which is only 0.7979 for a sample size of 2, is 0.9975 for a sample size of 100 and approaches unity as the sample size is increased. It will be recalled that the c_4 factors are based on the assumption that the universe is normal.

5.12.2 Efficiency of an Estimate

Consider the arithmetic mean, the median, and the midrange of a sample as possible estimators for the arithmetic mean of a normal universe. All of these happen to be unbiased. However, if a great many samples of, say, size 5 were taken from a given normal universe, the dispersions of the medians and midranges would be greater than the dispersion of the arithmetic means. (We already know that the standard deviation of the arithmetic mean, $\sigma_{\bar{X}}$, would be $\sigma/\sqrt{5}$, or approximately 0.45σ. It can be shown that the respective standard deviations of medians and midranges are approximately 0.54σ and 0.51σ when samples of 5 are taken from a normal universe.)

Mathematical statisticians define the most *efficient* estimator of a parameter as the estimator that has the minimum variance. (It will be recalled that the variance is the square of the standard deviation.) The *efficiency* of any other unbiased estimator is defined as the ratio of the variance of the sampling distribution of the efficient estimator to the variance of the sampling distribution of the other estimator. When the arithmetic mean of a normal distribution is estimated from samples of 5, the efficiency of the median is 0.697 and the efficiency of the midrange is 0.767. Dixon and Massey give efficiencies of these estimators for sample sizes from 2 to 20.[†]

The efficiency of a particular estimator depends on the form of the universe. For example, the midrange of a sample is more efficient than the arithmetic mean of the sample as an estimator of the arithmetic mean of a rectangular universe.

The reader should note that the variances of the sampling distributions, rather than the standard deviations of these distributions, are used by mathematical statisticians in defining the efficiency of an estimator. Therefore, the ratio of standard

[†]W. J. Dixon and F. J. Massey, Jr., *Introduction to Statistical Analysis*, 3d ed., McGraw-Hill Book Company, New York, 1969.

deviations of the respective sampling distributions is the square root of the efficiency. For instance, the ratio of $\sigma_{\overline{X}}$ to the standard deviation of midranges in samples of 5 from a normal distribution is $\sqrt{0.767} = 0.88$.

5.13 THE PROBLEM OF SELECTING A PARAMETER TO DESCRIBE UNIVERSE DISPERSION AND OF CHOOSING AN ESTIMATOR FOR THAT PARAMETER

A source of possible confusion to readers of books on statistics is that two different parameters—namely, standard deviation and variance—may be used to describe the dispersion of a universe. At first glance, this does not appear to be a source of trouble. The standard deviation of a universe is defined as the root-mean-square deviation from the arithmetic mean of the universe; the variance of the universe is defined as the square of its standard deviation. Because universe variance is readily computed from universe standard deviation, and vice versa, it might appear to be immaterial which parameter is preferred. Nevertheless, because the choice of the parameter is related to the choice of the estimator, it really makes a difference which parameter is to be estimated.

When values of $(X_j - \overline{X})^2$ from samples are used to estimate universe dispersion, a source of difficulty is that the μ of the universe is unknown; the deviations that are squared must be measured from the $\overline{X}$ of each sample. Except in the occasional case where the $\overline{X}$ of a sample happens to be identical with the unknown μ, $\Sigma(X_j - \overline{X})^2$ will be less than $\Sigma(X_j - \mu)^2$. Some method of compensation for this bias is therefore needed in any statistic based on $\Sigma(X_j - \overline{X})^2$ if the statistic is to be used to estimate the universe standard deviation σ or the universe variance σ^2.

When $\Sigma(X_j - \overline{X})^2$ is calculated for a sample of size n, conceivably the values of X_j in the sample could take on any value within the population except one. In order to arrive at the sample average $\overline{X}$, one value is "determined." Thus only $n - 1$ values of X_j are "free." Statisticians refer to this value $n - 1$ as the *degrees of freedom* of the sample and frequently designate it by the Greek letter ν (nu). Thus the calculation of the sample variance

$$s^2 = \frac{\Sigma(X_j - \overline{X})^2}{n - 1}$$

has $n - 1$ degrees of freedom. The sample standard deviation s usually is defined in statistics as the square root of an unbiased estimate of the sample variance. This definition of s was used in Tables 3.4 and 3.5. In those calculations it would have made very little difference whether s had been calculated (with $n - 1$ in the denominator) or σ_{RMS} (with n in the denominator) or whether the correction factors c_4 and c_2, respectively, were applied. Of much greater importance is whether or not the process was in control during accumulation of the data. Any estimates $\hat{\mu}$, $\hat{\sigma}$, or $\hat{\sigma}^2$ assume that there really was one universe—one bowl as in Ferrell's case 3 rather than several bowls as in case 4. Control charts, with subgroups formed on a rational basis, provide the test for a single universe.

5.14 SAMPLING FROM A NORMAL DISTRIBUTION

In the usual case in quality control, samples are drawn from an unknown population or universe. The power of the central limit theorem permits inferences to be made in probabilistic terms about the distribution of $\overline{X}$ from a controlled process. In addition, inferences are made about the standard deviation of the process using $\overline{R}/d_2$ or $\overline{s}/c_4$. These inferences and the resulting correction factors (d_2, c_4, and so on) are based on the assumption of a normal or nearly normal distribution underlying the process.

When samples are drawn from a normal distribution with mean zero and variance one, the random variable $(n - 1)s^2/\sigma^2$ is approximately χ^2 (chi-square) distributed with a mean of ν $(= n - 1)$ and a variance of 2ν. The density function is

$$f(\chi^2) = \frac{1}{2^{\nu/2}\,\Gamma(\nu/2)}\,(\chi^2)^{(\nu/2)-1}\,e^{-\chi^2/2} \qquad \text{for } \chi^2 > 0 \tag{5.19}$$

where $\Gamma(a) = (a - 1)!$, the gamma function of a. Table *B*, App. 3, gives values for $\chi^2_{\gamma,\nu}$ for a variety of values of the cumulative probability γ and degrees of freedom ν according to the formula

$$P(\chi^2 \leq \chi^2_{\gamma,\nu}) = \int_0^{\chi^2_{\gamma,\nu}} f(\chi^2)\,d\chi^2 = \gamma$$

where the test statistic is

$$\chi^2 = \frac{\nu s^2}{\sigma^2}$$

One frequently used test is to determine whether there is a significant difference between s^2 and a previously determined or assumed process variance σ^2. It should immediately be apparent to the reader that, with the appropriate choices of tail probabilities, a control chart for values of s^2 could be developed using s^2 as the central line and estimate of σ^2.

The many aspects of the use of the χ^2 distribution, and its relative the F distribution, are beyond the scope of this book. (The F distribution provides for the comparison of two sample variances using the ratio of two χ^2's.) Some aspects of the use of this distribution will be discussed in Chaps. 9 and 10.

Another distribution of great usefulness in drawing inferences about processes is Student's t distribution.[†] If a set of measurements X_j are drawn from a normal universe with independent variance σ^2, a statistic such as $\overline{X}$ may be tested against a known or assumed mean μ using the test statistic

$$t = \frac{\overline{X} - \mu}{s/\sqrt{n}}$$

[†]Student was the pen name taken by William S. Gosset, brewmaster for the Guinness brewery of Dublin, Ireland, in his early twentieth-century writings in statistics. He used that pen name his entire life.

where $s/\sqrt{n}$ is an estimator of $\sigma_{\bar{X}}$ and σ^2 has $\nu = n - 1$ degrees of freedom.

The probability density function of t is

$$f(t) = \frac{1}{\sqrt{\nu\pi}} \frac{\Gamma[(\nu+1)/2]}{\Gamma(\nu/2)} \left(1 - \frac{t^2}{\nu}\right)^{-(\nu+1)/2} \qquad -\infty < t < +\infty \tag{5.20}$$

One of the many uses of the t distribution is to develop confidence limits about an estimate of μ.

5.14.1 Point Estimates and Interval Estimates

When $\bar{X}$ ($\bar{\bar{X}}$) and $\bar{s}$ ($\bar{s^2}$) are used as estimators of μ and σ (σ^2), respectively, we are making *point estimates* of the parameters. Sometimes, particularly if the total number of measurements in the sample is relatively small, it may be desirable to accompany the point estimate with an *interval estimate* that indicates how close the estimate is to the true parameter. *Confidence intervals* and *confidence limits* may be used for this purpose.

If there is some rational basis for assuming a process to be in control and that the measurements are nearly normal in distribution, Student's t distribution may be used to establish confidence limits around $\bar{X}$ using the probability statement

$$P\left(\bar{X} - t_{\alpha/2,\nu} \frac{s}{\sqrt{n}} \le \mu \le \bar{X} + t_{\alpha/2,\nu} \frac{s}{\sqrt{n}}\right) = 1 - \alpha$$

This statement is read as "the probability is $1 - \alpha$ that the mean μ lies within the interval defined." There is a risk α, then, that the true mean is outside the interval.

Table 5.5, calculated from Student's t distribution, gives values for $t_{\alpha/2,\nu}$ for values of $1 - \alpha$ of 90, 95, and 99%.

To illustrate the application of Table 5.5, consider the last four drawings from Shewhart's normal bowl as given in Table 3.7. For these drawings—namely, 23, 23, 27, and 36—$\bar{X}$ is 27.25 and s is 6.1. If this were all the information we had from the bowl, our point estimate of μ would be 27.25, the average of the sample. The various confidence intervals from Table 5.5 and for $\nu = n - 1 = 3$ degrees of freedom would be

$$1 - \alpha \text{ confidence interval} = \pm\, t_{\alpha/2,\nu} \frac{s}{\sqrt{n}}$$

$$90\% \text{ confidence interval} = \pm\, 2.353 \left(\frac{6.1}{2}\right) = \pm\, 7.18$$

$$95\% \text{ confidence interval} = \pm\, 3.182 \left(\frac{6.1}{2}\right) = \pm\, 9.71$$

$$99\% \text{ confidence interval} = \pm 5.841 \left(\frac{6.1}{2} \right) = \pm 17.82$$

The confidence limits are obtained by applying the computed confidence interval to the point estimate of the parameter $\bar{X}$. In this case the limits are

$$1 - \alpha \text{ confidence limits} = \pm t_{\alpha/2,\nu} \frac{s}{\sqrt{n}}$$

$$90\% \text{ confidence limits} = 27.25 \pm 7.18, \text{ or } 20.07 \text{ and } 34.43$$

$$95\% \text{ confidence limits} = 27.25 \pm 9.71, \text{ or } 17.54 \text{ and } 36.96$$

$$99\% \text{ confidence limits} = 27.25 \pm 17.82, \text{ or } 9.43 \text{ and } 45.07$$

TABLE 5.5 VALUES OF STUDENT'S *t* DISTRIBUTION FOR 90, 95, AND 99% CONFIDENCE LIMITS ($t_{\alpha/2,\nu}$)

ν	90% limits, $t_{0.05,\nu}$	95% limits, $t_{0.025,\nu}$	99% limits, $t_{0.005,\nu}$
1	6.314	12.706	63.657
2	2.920	4.303	9.925
3	2.353	3.182	5.841
4	2.132	2.776	4.604
5	2.015	2.571	4.032
6	1.943	2.447	3.707
7	1.895	2.365	3.499
8	1.860	2.306	3.355
9	1.833	2.262	3.250
10	1.812	2.228	3.169
11	1.796	2.201	3.106
12	1.782	2.179	3.055
13	1.771	2.160	3.012
14	1.761	2.145	2.977
15	1.753	2.131	2.947
16	1.746	2.120	2.921
17	1.740	2.110	2.898
18	1.734	2.101	2.878
19	1.729	2.093	2.861
20	1.725	2.086	2.845
21	1.721	2.080	2.831
22	1.717	2.074	2.819
23	1.714	2.069	2.807
24	1.711	2.064	2.797
25	1.708	2.060	2.787
50	1.676	2.009	2.678
100	1.660	1.985	2.626
200	1.653	1.972	2.601
500	1.648	1.965	2.586
∞	1.645	1.960	2.576

It should be noted that, in this case, the known mean of the distribution ($\mu = 30$) falls within each of the pairs of confidence limits. In any one case, either μ falls within the chosen confidence limits or it does not. The statement "the probability is 0.90 that μ is within the 90% confidence limits in this case" has no operational meaning. In any series of samples the point estimate $\overline{X}$ will vary in position and the range of the confidence interval will vary. Thus the best we can expect is that μ will be between the limits 9 out of 10 times when 90% confidence limits are calculated. Had we chosen drawings 213–216 for the calculations ($\overline{X} = 36.00$ and $s = 2.9$), μ would have been outside the confidence limits at 90 and 95% and within the limits at 99%.

It is important to remember that these calculations assume that a normal (or nearly normal) distribution underlies the process, the sample has been drawn at random, and the process is operating in a state of statistical control, that is, there is only one bowl as in Ferrell's case 3.

5.15 THEORY OF EXTREME RUNS

If, in tossing a coin, the probability of a head on 1 toss is $\frac{1}{2}$, the probability of heads on both of 2 successive independent tosses is $(\frac{1}{2})^2$. Similarly, the probability of heads on all of 7 successive independent tosses is $(\frac{1}{2})^7 = \frac{1}{128}$. As the probability of tails on every toss for 7 tosses is also $\frac{1}{128}$, the probability of a run of either 7 heads or 7 tails in any given set of 7 tosses is $\frac{1}{128} + \frac{1}{128} = \frac{1}{64}$.

The *median* of a set of numbers has been defined (Chap. 3) as the midnumber, the one so located that half the numbers are above it and the other half below it. If numbers were written on a set of chips, the probability is $\frac{1}{2}$ that a single chip drawn at random would fall above the median. Or if measurements are made on a quality characteristic that is statistically controlled, the probability is $\frac{1}{2}$ that any one measurement will be above the universe median. Whenever a frequency distribution is symmetrical, the median and average (arithmetic mean) are the same. In this case, the probability is $\frac{1}{2}$ that any one measurement will fall above the universe average, or that the average of any subgroup will fall above the universe average. Extreme runs of subgroup averages above or below the universe average are as likely as extreme runs of heads or tails in coin tossing. Thus the probability is $\frac{1}{64}$ that 7 successive subgroup averages will fall on the same side of the universe average.

This type of reasoning is the basis for the rules given in Chap. 4 regarding interpretation of extreme runs on a control chart. For example, if the probability is assumed as $\frac{1}{2}$ that any subgroup average will fall above the universe average, the probability that at least 10 out of 11 subgroups would fall on the same side of the universe average can be computed by adding the first two and last two terms of the binomial, $(\frac{1}{2} + \frac{1}{2})^{11}$.

P_{11}, the probability of all 11 above $\quad = \left(\frac{1}{2}\right)^{11} \quad = \frac{1}{2{,}048}$

P_{10}, the probability of exactly 10 above $= 11\left(\frac{1}{2}\right)^{10}\left(\frac{1}{2}\right) = \frac{11}{2{,}048}$

P_1, the probability of exactly 10 below (1 above) $= 11\left(\frac{1}{2}\right)\left(\frac{1}{2}\right)^{10} = \frac{11}{2{,}048}$

P_0, the probability of all 11 below (0 above) $= \left(\frac{1}{2}\right)^{11} = \frac{1}{2{,}048}$

The probability that 10 or more out of 11 will fall on the same side of the universe average $= \frac{24}{2{,}048}$

$$\frac{24}{2{,}048} = 0.0117 \qquad \text{or approximately } \frac{1}{85}$$

As many distributions of industrial quality characteristics are not symmetrical, it is not strictly correct to state that the probability is always $\frac{1}{64}$ that, with no change in the universe, 7 successive subgroup averages will be on the same side of the universe average, or that the probability is always $\frac{1}{85}$ that at least 10 out of 11 will be on the same side. Just as the control-chart limits are best interpreted as rules for action rather than as means of estimating exact probabilities, so also should the extreme runs mentioned in Chap. 4 be interpreted in this way.

As pointed out in Chap. 4, these runs may be expected to occur as a matter of chance with no change in universe average oftener than would a departure from 3-sigma limits. The relative frequencies of the runs and the out-of-control points may be judged by noting that, assuming a normal universe, the probability of a point outside 3-sigma limits is 0.0027 or about $\frac{1}{370}$. If it were desired to set up rules for extreme runs that were as unlikely as departures from 3-sigma limits, such runs could readily be calculated. For example, 9 points in a row on the same side of the average have a probability of $\frac{1}{256}$; 10 points have a probability of $\frac{1}{512}$.

Significant extreme runs are not necessarily with reference to the central line of the control chart. It is suggested in Chap. 11 that two points in succession beyond a 2-sigma limit constitute an extreme run that is a more significant indication of an assignable cause than a single point outside of 3-sigma limits. The statement may be verified by the following calculations. If the probability of one point *above* the *upper* 2-sigma limit is 0.0228, the probability of two in a row is $(0.0228)^2 =$ 0.00052. As the probability of two points in a row *below* the *lower* 2-sigma limit is also 0.00052, the probability of two points in succession outside the *same* 2-sigma limit (either above the upper limit or below the lower limit) is 0.00052 + 0.00052 = 0.00104, or about $\frac{1}{960}$.[†]

[†]Probability tests may also be applied to runs in which each value is greater (or less) than the preceding value. For probability tables and a diagram applicable to this type of run, see P. S. Olmstead, "Distribution of Sample Arrangements for Runs Up and Down," *Annals of Mathematical Statistics,* vol. 17, pp. 24–33, March 1945.

PROBLEMS

Note: In all the following problems, use the method that is correct in principle unless the problem statement specifies the use of an approximate method. Your instructor will specify whether it is permissible to use computer programs in obtaining numerical solutions to these problems. In any case, the problems should be formulated by hand, using the appropriate formulas, to ensure that you understand them because it is important to the understanding of the fundamentals of probability involved.

5.1. What is the probability that a single draw from Shewhart's normal bowl will yield a chip marked 29, 30, or 31? See Table 3.6 for the distribution in this bowl.
Answer: $\frac{120}{998}$.

5.2. If two successive draws from Shewhart's normal bowl are made with the first chip drawn replaced before the second draw, what is the probability that a 30 will be obtained on both draws?
Answer: 1,600/996,004.

5.3. If two successive draws are made with the first chip drawn not replaced before the second draw, what is the probability that a 30 will be obtained on both draws?
Answer: 1,560/995,006.

5.4. A sample of 3 is to be selected from a lot of 20 articles. How many different samples are possible?
Answer: 1,140.

5.5. A sample of 30 is to be selected from a lot of 200 articles. How many different samples are possible?
Answer: 4.096×10^{35}.

5.6. What is the probability of getting exactly 3 sixes in a throw of 6 dice? What is the probability of 3 or more sixes?
Answer: 0.0536; 0.0623.

5.7. What are the respective probabilities of getting 0, 1, 2, 3, 4, 5, 6, and 7 heads in a toss of 7 coins? Assume the probability of heads in one toss to be $\frac{1}{2}$.
Answer: $\frac{1}{128}; \frac{7}{128}; \frac{21}{128}; \frac{35}{128}; \frac{35}{128}; \frac{21}{128}; \frac{7}{128}; \frac{1}{128}$.

5.8. From past records it is estimated that the probability that a flood of 10,000 sec-ft or more will occur in any year on a certain stream is $\frac{1}{5}$. What is the probability that such a flood will occur at least once in the next 5 years?
Answer: 0.672.

5.9. Jill proposes the following dice game to Jack. Jack will throw 5 dice. If 3 or more of them do not turn up the same face (that is, 3 or more sixes, 3 or more fives, and so on), Jill wins and Jack will pay her 10 cents. If 3 or more of the 5 dice do turn up the same face, Jack wins and Jill will pay him 40 cents. What is the probability that Jack will win? As the game is proposed, who has the better of the bet?
Answer: 0.213; odds of 4 to 1 are favorable to Jack, as fair odds would be 3.69 to 1.

5.10. (*a*) How many different hands of 13 cards might you have out of a standard deck of 52 playing cards?
(*b*) What is the probability of a 13-card hand without an ace, king, queen, or jack?
(*c*) What is the probability of a 13-card hand containing all four aces?

(*d*) What is the probability of a 13-card hand containing one or more aces?
Answer: (a) 635,013,560,000; (b) 0.00364; (c) 0.00264; (d) 0.696.

5.11. In the discussion on probability spaces and events, several illustrations were given of the formation of events. For each of the experiments below, define the probability space by listing all possible elementary outcomes. Also list all the events that can be defined on each space. Illustrate the elementary outcomes and the probability of each.
(*a*) The tossing of a single coin
(*b*) The tossing of a single die
(*c*) The simultaneous tossing of two coins
(*d*) The tossing of a pair of dice

5.12. A random sample of 4 is to be selected from a lot of 12 articles, 3 of which are nonconforming. What is the probability that the sample will contain exactly 1 nonconforming article?
Answer: $\frac{28}{55}$, or 0.51.

5.13. A random sample of 20 is to be selected from a lot of 150 articles, 15 of which are nonconforming. What is the probability that the sample will contain 2 or more nonconforming articles?
Answer: 0.627.

5.14. A controlled manufacturing process is generating 0.2% nonconforming. What is the probability of finding 2 or more nonconforming pieces among 100 pieces?
Answer: 0.017.

5.15. Use the Poisson distribution to obtain an approximate answer to Prob. 5.14.
Answer: 0.018.

5.16. A random sample of 10 articles is taken from a stream of product 2% nonconforming. What is the probability that the sample will contain no nonconforming articles?
Answer: 0.817.

5.17. An acceptance plan calls for the inspection of a sample of 75 articles out of a lot of 1,500. If there are no nonconforming articles in the sample, the lot is accepted; otherwise, it is rejected. If a lot 1% nonconforming is submitted, what is the probability that it will be accepted? Solve this using the Poisson distribution as an approximation.
Answer: 0.472.

5.18. An acceptance plan calls for the inspection of a sample of 115 articles out of a lot of 3,000. If there are 6 or less nonconforming articles in the sample, the lot is accepted; with 7 or more, it is rejected. If a lot 5% nonconforming is submitted, what is the probability that it will be rejected? Solve using the Poisson distribution as an approximation.
Answer: 0.354.

5.19. If the probability is 0.033 that a single article will be nonconforming, what are the respective probabilities that a sample of 100 will contain exactly 0, 1, 2, 3, 4, 5, 6, 7, 8, 9, 10, and 11 nonconforming articles? Assume that the Poisson distribution is a satisfactory approximation to the binomial in this case, and solve by interpolation in Table *G*.
Answer: 0.037; 0.122; 0.201; 0.2205; 0.182; 0.1205; 0.0655; 0.0315; 0.013; 0.0045; 0.002; 0.0005.

5.20. If the probability is 0.25 that a single article will not conform to a special severe test specification, what is the probability that a sample of 50 will contain exactly 15 nonconforming articles?
Answer: 0.089.

5.21. Using the normal curve as an approximation to the binomial, answer the question in Prob. 5.20. Also find the approximate probability of 15 or more nonconforming articles.
Answer: 0.093; 0.257.

5.22. Using the Poisson distribution as an approximation to the binomial, answer the question in Prob. 5.20. Also find the approximate probability of 15 or more nonconforming articles.
Answer: 0.081; 0.275.

5.23. (*a*) A sample of 6 is selected from a lot of 30 articles. How many different samples are possible?
(*b*) A sample of 36 is selected from a lot of 180 articles. How many different samples are possible?
(*c*) Discuss the problem of answering the foregoing questions when solving them on a calculator or minicomputer.

5.24. If two successive draws from Shewhart's normal bowl (see Table 3.6) are made with the first chip drawn replaced before the second draw, what is the probability that a 24 will be obtained on both draws? Find the probability of a 24 on both draws if the first chip drawn is not replaced before the second draw.

5.25. Solve Prob. 5.24 using draws of a chip marked 30 rather than 24.

5.26. Solve Prob. 5.24 using draws of a chip marked 6 rather than 24.

5.27. A six-sided die has two sides painted red, two painted black, and two painted yellow. Define the elementary outcomes and the probability space. What probabilities would you assign to each elementary outcome?

5.28. (*a*) If the die described in Prob. 5.27 is rolled three times, what is the probability that a yellow side appears only on the third roll?
(*b*) Only on the sixth roll in six rolls?
(*c*) Within the first three rolls?

5.29. In a hand of 13 cards drawn from a standard 52-card deck, what is the probability of exactly 3 diamonds?

5.30. One urn contains three white and three black balls; a second urn contains four white balls and one black ball.
(*a*) If one ball is chosen from each urn, what is the probability that they will be the same color?
(*b*) If an urn is selected at random, what is the probability that a ball drawn from that urn will be black?
(*c*) If an urn is selected at random and two balls are drawn (without replacement) from it, what is the probability that they will both be the same color?

5.31. Assuming that the ratio of one sex to total children is exactly $\frac{1}{2}$, what is the probability that in a family of five children:
(*a*) All will be of the same sex?
(*b*) The two oldest will be boys and the three youngest will be girls?
(*c*) Exactly three of the children will be boys?

5.32. A lot of 50 items contains 5 nonconforming units. A random sample of 10 is drawn, and 1 nonconforming unit (the expected average number of nonconforming items in the sample) is found. Using the hypergeometric distribution, the probability of this event happening at random is 0.43134.
(*a*) Use the Poisson approximation to estimate this probability.
(*b*) Use the normal approximation to estimate this probability.
(*c*) Which approximation method is better in this case?

5.33. It is proposed that on an $\overline{X}$ chart the process be considered out of control if 2 points in a row fall beyond a special set of limits (called warning limits) at $+2\sigma_{\overline{X}}$ and $-2\sigma_{\overline{X}}$. This secondary rule would be used in addition to the usual rule of 1 point falling beyond a control limit.
(*a*) What is the probability that 2 points in a row will fall between a given $2\sigma_{\overline{X}}$ limit and the corresponding $3\sigma_{\overline{X}}$ control limit when the process is correctly centered?
(*b*) What is the probability that 2 points in a row will fall between a given $2\sigma_{\overline{X}}$ limit and corresponding $3\sigma_{\overline{X}}$ control limit if the process mean shifts by $+ 1\sigma$? Assume a subgroup size of 5.

5.34. (*a*) A lot of 50 articles contains 3 items that do not meet specifications ($\mu_p = \frac{3}{50} = 0.06$). A sample of 10 items is to be drawn at random from the lot. What is the probability that all 10 items will be accepted?
(*b*) A process generating 6.0% nonconforming items is sampled at random intervals in subgroups of 10 items. What is the probability that a given subgroup will contain all acceptable items?
(*c*) Use Table *G* (the Poisson table) to approximate the probability asked for in part *(b)*.

5.35. If the probability is 0.04 that a single article is nonconforming in a continuous manufacturing process, what are the respective probabilities that 0, 1, 2, 3, 4, and 5 articles are nonconforming in a random sample of 5 items taken from the process?

5.36. Use the Poisson approximation to the binomial to solve Prob. 5.35. Comment on the difference in results so obtained.

5.37. Use the normal approximation to the binomial to solve Prob. 5.35. Comment on the difference in results so obtained.

5.38. The probabilities computed in Prob. 5.35 constitute a binomial distribution. (Why?) Compute the average and standard deviation of this binomial following the tabular pattern given in the calculations on page 187. Compare these calculated values with values given by the formulas np for the mean and $\sqrt{np\,(1-p)}$ for the standard deviation.

5.39. Random samples of 100 items are drawn from a continuous process that is known to produce 20% nonconforming items. Determine the probability of finding exactly 15 nonconforming items in a sample:
(*a*) Using the exact binomial distribution.
(*b*) Using the normal approximation to the binomial.
(*c*) Using the Poisson approximation to the binomial.
(*d*) Comment on the relative accuracy of the approximations.

5.40. Change Prob. 5.39 so that the probability sought is that of more than 15 nonconforming items in a sample rather than exactly 15. Do not attempt to solve part *(a)*, but comment on the difficulties presented in making an exact binomial calculation in this case.

5.41. Make a calculation similar to Table 5.4 to compare the observed frequencies of Table 5.3*c* with the theoretical frequencies of the Poisson distribution.

5.42. Derive the mean and standard deviation for the Poisson distribution following the mathematical expectation procedure.
Hint: The pattern of the derivation follows very closely that used to derive the mean and standard deviation for the binomial.

5.43. A lot of 100 items contains 10 nonconforming items. A random sample of 20 items is drawn, and 2 nonconforming items (the expected average number of nonconforming items in a sample of 20) are found.
(*a*) Use the exact hypergeometric distribution to find the probability of this event.
(*b*) Use the binomial approximation to the hypergeometric to estimate this probability.
(*c*) Use the Poisson approximation to estimate this probability.
(*d*) Use the normal approximation to estimate this probability.
(*e*) Comment on the relative accuracy of the various approximation formulas in this case.

5.44. Consider the sampling plan for which the lot size is 40, the sample size is 10, and lot acceptance is based on finding no nonconforming items.
(*a*) On the assumption that the lot is 20% nonconforming, use the correct formula to find the probability that the lot will be accepted.
(*b*) Use a Poisson approximation to find the probability that the lot will be accepted.
(*c*) Comment on the relationship between the answers found in *(a)* and *(b)*.

5.45. A large number of samples of 150 items each are taken from a process generating 5% nonconforming items.
(*a*) What is the expected number of nonconforming items per sample?
(*b*) What is the upper limit of the number of nonconforming items in a sample that you would expect to find exceeded only 5% of the time assuming no shift in the process average? Use the normal approximation to the binomial to find the limit.
(*c*) Solve part *(b)* using the Poisson approximation to the binomial.

5.46. In Chap. 4, several "decision rules" related to the theory of extreme runs are presented. The statement was that "whenever in n successive points on a control chart, at least r are on the same side of the central line," where:

	n	r
(*a*)	7	7
(*b*)	11	10
(*c*)	14	12
(*d*)	17	14
(*e*)	20	16

Assume the probability to be $\frac{1}{2}$ that a subgroup average will fall above the central line. If there has been no shift in the universe average, what probability is associated with the random occurrence of each of these events?

5.47. Assuming that the probability is 0.0559 that a point will randomly fall above a $1.5\sigma_{\overline{X}}$ limit on a control chart for $\overline{X}$ if the universe average has not shifted, develop a decision rule similar to that shown in Prob. 5.46*a* based on the theory of runs that yields a probability of occurrence approximately equivalent to that in Prob. 5.46*a*.

Answer: Two points in row; 0.00625.

5.48. Certain shipments of insulators were subject to inspection by a high-voltage laboratory. One of the important tests was a destructive one. The procedure adopted for inspection was as follows: Select six insulators at random from the large lot, and test them. If all six pass the test, accept the lot. If two or more fail the test, reject the lot. If only one insulator fails, take a second sample of six insulators. If all six insulators in the second sample pass the test, accept the lot; otherwise, reject it.

(*a*) What is the probability of accepting lots in which 2% are nonconforming?

(*b*) What is the probability of acceptance if 20% are nonconforming? Comment on the effectiveness of the procedure.

5.49. A stipulated acceptance procedure calls for examining 25 articles from a lot of 1,000 articles. If none of the 25 articles is classified nonconforming, the lot is accepted; otherwise, it is rejected. Assume that a lot containing 10% nonconforming articles is submitted for acceptance.

(*a*) Using hypergeometric probabilities, compute the probability of acceptance.

(*b*) Using the binomial distribution as an approximation to the method that is correct in principle, compute the approximate probability of acceptance.

(*c*) Using the Poisson distribution as an approximation to the binomial, compute the approximate probability of acceptance.

(*d*) Using the normal distribution as an approximation to the binomial, compute the approximate probability of acceptance.

5.50. Solve Prob. 5.49 changing the lot size from 1,000 to 200 and making no other change in the problem statement.

5.51. Discuss your results in Prob. 5.49 and 5.50 in relation to the question of which approximation, if any, to use in computing such probabilities of acceptance.

5.52. (*a*) If the probability that a single article drawn from a continuous manufacturing process does not meet specifications is 0.10, what is the exact probability that a sample of 50 articles drawn from that process will contain four such items?

(*b*) Using the Poisson as an approximation to the correct distribution, answer the question in part (*a*).

(*c*) Using the normal distribution as an approximation to the correct distribution, answer the question in part (*a*).

5.53. The $\overline{X}$ control chart for a process uses standard $3\sigma_{\overline{X}}$ limits and a subgroup size of 4.

(*a*) What is the probability that 8 points in a row will fall between the central line on the chart and the lower control limit when the process is correctly centered?

(*b*) What is the probability if the mean is actually located 0.6σ below the central line on the chart?

(*c*) What is the probability that no point in a sequence of 8 subgroups will fall below the lower control limit based on the shift described in part (*b*)?

(*d*) Comment on the relative usefulness of simple run tests in detecting process mean shifts using this information as an example.

5.54. Lots of 500 items are formed from a process under statistical control at a value of μ_p of 0.02. Samples of 50 units are drawn at random and tested.

(*a*) Use the binomial formula to find the probability that a sample will contain exactly 2 rejectable items.

(*b*) Use the normal approximation to the binomial to find the probability that a sample will contain exactly 2 rejectable items.

(*c*) Use the Poisson approximation and Table *G* (or the Poisson formula) to find the probability that a sample will contain exactly 2 rejectable items.

5.55. A lot contains 60 items of which 10% do not conform to specifications. A sample of 10 items is to be drawn from the lot, and the lot is accepted if no item is rejected. Find the probability of acceptance by the following methods:

(*a*) The correct hypergeometric formula.

(*b*) The binomial formula as an approximation to the hypergeometric.

(*c*) The Poisson distribution as an approximation (Table *G*).

(*d*) The normal distribution using a mean (μ_{np}) of *np*. The standard deviation of the hypergeometric is $\sqrt{np(1-p)}\,/\sqrt{(N-\text{n})/(N-1)}$.

5.56. A batch of 50 units contains 5 units that do not meet specifications. A sample of 10 units is drawn at random from the batch.

(*a*) Calculate the exact probability that the sample of 10 units will contain 1 unacceptable unit.

(*b*) Use the binomial formula to find the approximate probability requested in *(a)*.

(*c*) Use the Poisson formula (or Table *G*) to find the approximate probability requested in *(a)*.

5.57. A batch of 50 items contains 1 item that does not meet specifications. A random sample of 5 items is drawn from the batch and tested.

(*a*) Use the correct probability distribution to calculate the probability that the 1 nonconforming item will be drawn in the sample.

(*b*) Use the binomial approximation to calculate this probability.

(*c*) Use the Poisson approximation to calculate this probability.

(*d*) By what percentage does each of these approximation methods overestimate (+) or underestimate (–) the correct figure?

5.58. Forty 64K programmable read-only memory chips (PROMs) are produced on a single silicon wafer. After they have been cut apart, mounted, and sealed in their packages, a sample of 10 units is drawn at random and tested. If one or more units in the sample is found to be defective, all units in that batch must be inspected before packing and shipment.

(*a*) Assuming that a batch contains 4 defective units, what is the exact probability that none will be found in the sample and the lot thus shipped without further inspection?

(*b*) If the process described above were a continuous process generating a μ_p of 0.06, what would be the exact probability of rejecting none in a sample of 10 units?

(*c*) Use the formula for the Poisson distribution or Table *G* to find the probability called for in part *(b)* as an approximation to that answer.

6

THE CONTROL CHART FOR FRACTION REJECTED

Some time ago when we first began to get steamed up about this matter of statistical control, we called on a number of large automotive parts manufacturers with the intention of getting production records from which simple control charts could be plotted. The results were quite startling. In the first place, even the well-managed plants which ordinarily have all kinds of records found that they had no record of the percentage defective by lots. Some information was available, of course, but it was woefully inadequate and could not be used to indicate past experience. The tragic thing about it all is that the executives in these several plants had assured the writer that they had specific data on their quality control and knew exactly the range in process average. In several instances, further investigation revealed that the process average was considerably higher than they had estimated, in fact much higher than permissible.

—Joseph Geschelin[†]

6.1 SOME PRACTICAL LIMITATIONS OF CONTROL CHARTS FOR VARIABLES

In spite of the advantages of the $\overline{X}$ and R charts, both as powerful instruments for the diagnosis of quality problems and as a means for routine detection of sources of trouble, it is evident that their use is limited to only a small fraction of the quality characteristics specified for products and services.

[†]Joseph Geschelin, "Statistical Method Points to Process Control by Spotting Variables in Manufacture," *Automotive Industries,* vol. 67, pp. 166–169, Aug. 6, 1932.

One limitation is that they are charts for *variables,* that is, for quality characteristics that can be measured and expressed in numbers. Many quality characteristics can be observed only as *attributes,* that is, by classifying each item inspected into one of two classes, either conforming or nonconforming to the specifications.

Moreover, even for those many quality characteristics that can be measured, the indiscriminate use of $\overline{X}$ and R charts would often be totally impracticable, as well as uneconomical. For example, the inspection department in one manufacturing plant had the responsibility for checking over 500,000 dimensions. Although any one of these dimensions could have been measured as a variable and was therefore a possible candidate for $\overline{X}$ and R charts, it is obvious that there could not be 500,000 such charts. No dimension should be chosen for $\overline{X}$ and R charts unless there is an opportunity to save costs—costs of spoilage and rework, inspection costs, costs of excess material—or otherwise to effect quality improvements that would in some way more than compensate for the costs of taking the measurements, keeping the charts, and analyzing them.

6.2 CONTROL CHARTS FOR ATTRIBUTES

There are several different types of control charts that may be used in either of the cases described in the foregoing paragraphs:

1. The p chart, the chart for fraction rejected as nonconforming to specifications
2. The np chart, the control chart for number of nonconforming items
3. The c chart, the control chart for number of nonconformities
4. The u chart, the control chart for number of nonconformities per unit

The c and u charts are discussed in Chap. 7.

The cost of collecting data for attributes charts is likely to be less than the cost of collecting data for $\overline{X}$ and R charts, because the attributes chart generally uses data already collected for other purposes. $\overline{X}$ and R charts, on the other hand, require special measurements for control-chart purposes. The cost of computing and charting may be less since one attributes chart may apply to any number of quality characteristics observed at an inspection station, whereas separate $\overline{X}$ and R charts are necessary for each measured quality characteristic.

In addition to the cost advantages that may be gained, the use of attributes control charts provides management with useful records of quality history. Many managerial decisions need to be based on a knowledge of the quality level currently maintained and on prompt information about changes that occur in the quality level. Such information may be required in order to demonstrate capability to meet requirements both prior to award of a contract and later during fulfillment.

6.3 THE CONTROL CHART FOR FRACTION REJECTED

The most versatile and widely used attributes control chart is the p chart. This is the chart for the fraction rejected as nonconforming to specifications (the so-called fraction defective). It may be applied to quality characteristics that can be observed only as attributes. It may also be applied to quality characteristics that are considered as attributes—for example, dimensions checked by go and not-go gages—even though they might have been measured as variables. As long as the result of an inspection is the classification of an individual article as accepted or rejected, a single p chart may be applied to one quality characteristic or a dozen or a hundred.

Fraction rejected p may be defined as the ratio of the number of nonconforming articles found in any inspection or series of inspections to the total number of articles actually inspected. Fraction rejected is nearly always expressed as a decimal fraction.

Percent rejected is $100p$, that is, 100 times the fraction rejected. For actual calculation of control limits, it is necessary to use the fraction rejected. For charting, and for general presentation of results to shop personnel and to management, the fraction is generally converted to percent.

6.4 THE BINOMIAL AS A PROBABILITY LAW THAT DETERMINES THE FLUCTUATIONS OF FRACTION REJECTED

Suppose 10,000 beads of the same size and density are placed in a container. Of these, 9,500 are white and 500 red. White may be considered to represent articles that conform to specifications and red to represent nonconforming articles. Let samples of 50 beads be drawn at random from this container. If the 50 beads are replaced after each sample has been drawn and all beads are thoroughly mixed before the next drawing, the theory of probability enables us to calculate the relative frequency in the long run of getting exactly 0, 1, 2, 3, 4, 5, and so on, red beads. As explained in Chap. 5, the binomial gives a very close approximation to these probabilities. If the container held an infinite number of beads, 5% of which were red, the binomial would apply exactly.

Any one sample of beads drawn from the container is a sample from a very large quantity of beads 5% red. As a matter of chance, variations in the number of red beads are inevitable from sample to sample. In a similar way, we may think of a day's production (or other lot) of any manufactured article or part as a sample from a larger quantity with some unknown fraction that do not conform to specifications. This unknown universe nonconforming fraction depends on a complex set of causes influencing the production and inspection operations. As a matter of chance, the nonconforming fraction in the sample may vary considerably. As long as the nonconforming fraction in the universe remains unchanged, the relative frequencies of nonconforming fractions in the samples may be expected to follow the binomial law.

6.5 CONTROL LIMITS FOR THE p CHART

As a general mathematical model, the Shewhart control-chart model with 3-sigma limits may be expressed as

$$UCL_y = E(y) + 3\sigma_y$$
$$\text{Central line}_y = E(y)$$
$$LCL_y = E(y) - 3\sigma_y$$

where y is the random variable, or control statistic, to be plotted on the control chart, for example, $\overline{X}$, R, p, c, and so forth. $E(y)$ is the expected value of the statistical variable, and σ_y is the standard deviation of the variable y.

In the case of the control chart for $\overline{X}$, the basic model becomes

$$UCL_{\overline{X}} = \mu_{\overline{X}} + 3\sigma_{\overline{X}}$$
$$\text{Central line}_{\overline{X}} = \mu_{\overline{X}}$$
$$LCL_{\overline{X}} = \mu_{\overline{X}} - 3\sigma_{\overline{X}}$$

where the statistical variable $y = \overline{X}$, $E(y) = \mu_{\overline{X}}$, and $\sigma_y = \sigma_{\overline{X}} = \sigma/\sqrt{n}$. The reader will remember that the distribution of the statistic $y = \overline{X}$, even with small subgroup sizes, tends toward a normal distribution because of the power of the central limit theorem. Since the chance (probability) that an exactly normally distributed random variable will fall outside either 3-sigma control limit is 0.0027, such an occurrence purely by chance on an $\overline{X}$ chart would be very rare indeed.

The $\overline{X}$ chart, however, is the only instance in the application of the Shewhart control-chart model for which the distribution of the random variable can be shown to tend toward the normal distribution. In all other cases—for example, the R chart, s chart, p chart, and the like—it is appropriate to say simply that the occurrence of a point falling outside 3-sigma limits at random would be very unlikely.[†]

The mean, or expected value, of the binomial fraction is p, and its standard deviation is $\sqrt{p(1-p)/n}$ (see Chap. 5, Sec. 5.6.1). Thus, 3-sigma limits for the p chart are

$$\left.\begin{aligned} UCL_p &= p + 3\sqrt{\frac{p(1-p)}{n_i}} = p + 3\,\frac{\sqrt{p(1-p)}}{\sqrt{n_i}} \\ LCL_p &= p - 3\sqrt{\frac{p(1-p)}{n_i}} = p - 3\,\frac{\sqrt{p(1-p)}}{\sqrt{n_i}} \end{aligned}\right\} \quad \text{for subgroup } i$$

[†]This is not to say that it is impossible to associate probability values with points outside control limits for charts other than the $\overline{X}$ chart. Such calculations are very tedious and time-consuming, especially in p chart applications with varying sample size, and contribute little to the application of the basic decision rule. The general statement may be made, however, that they are of the same order of magnitude as the probabilities of a point falling outside 3-sigma limits in drawing from a normal universe. This is the true power of the Shewhart model.

The value of p in these formulas refers to the parametric, or universe, value of the statistic in question. When combined with the subscript o (that is, p_o), it refers to the standard or aimed-at value chosen for control-chart purposes.

In cases where a standard value is not used, the observed value $\bar{p}$ may be. The observed value is necessarily used in the calculation of trial control limits. In these cases, the control limits are

$$UCL_p = \bar{p} + 3\sqrt{\frac{\bar{p}(1-\bar{p})}{n_i}} = \bar{p} + 3\frac{\sqrt{\bar{p}(1-\bar{p})}}{\sqrt{n_i}}$$

$$LCL_p = \bar{p} - 3\sqrt{\frac{\bar{p}(1-\bar{p})}{n_i}} = \bar{p} - 3\frac{\sqrt{\bar{p}(1-\bar{p})}}{\sqrt{n_i}}$$

where $\bar{p} = \Sigma r_i / \Sigma n_i$. Each point plot is $p_i = r_i / n_i$, the number of units rejected divided by the number of units inspected in the subgroup.

Industrial practice in the use of the p chart generally bases control limits on either 3 sigma or some other multiple of sigma. Except for very small subgroups, the calculation of probability limits (such as described for $\bar{X}$, R, and s charts in Chap. 10, Sec. 10.5) for a p chart is too burdensome a job. However, if probability limits are desired, they may be obtained by direct use of the binomial distribution.

As in the case of the $\bar{X}$ and R charts, the use of 3-sigma limits rather than narrower or wider limits is a matter of experience as to the economic balance between the cost of hunting for assignable causes when they are absent and the cost of not hunting for them when they are present. Although in most cases 3-sigma limits are best, special cases arise in which the use of narrower limits, such as 2-sigma, is desirable. The need for narrower limits arises out of the use of the p chart as an instrument for executive pressure on quality.

6.6 PROBLEMS INTRODUCED BY VARIABLE SUBGROUP SIZE

The larger the subgroup, the more likely that the fraction nonconforming in the subgroup will be close to the fraction nonconforming of the universe. This obvious general principle is expressed in mathematical terms by the statement that the standard deviation of p, like the standard deviation of $\bar{X}$, varies inversely with $\sqrt{n}$.

For both the $\bar{X}$ chart and the p chart, the appropriate 3-sigma control limits depend on the subgroup size (see Fig. 10.1). The practical difference is that because most measurements used for $\bar{X}$ charts are taken for control-chart purposes, it is usually possible to keep subgroup size constant. Many p charts, on the other hand, use data taken for other purposes than the control chart; where subgroups consist of daily or weekly production, the subgroup size is almost certain to vary.

As a practical matter, whenever subgroup size is expected to vary, a decision must be made as to the way in which control limits are to be shown on the p chart. There are three common solutions to this problem, as follows:

1. Compute new control limits for every subgroup, and show these fluctuating limits on the control chart. This is illustrated in the first 2 months of the 4 months shown in Example 6.1.
2. Estimate the average subgroup size for the immediate future. Compute one set of limits for this average, and draw them on the control chart. Whenever the actual subgroup size is substantially different from this estimated average, separate limits may be computed for individual subgroups. This is particularly necessary if a point for an unusually small subgroup falls outside control limits. This scheme is illustrated in the final 2 months of the period shown in Example 6.1. Estimates of future average subgroup size must be revised from time to time; where each day's production is one subgroup, it is customary to revise these estimates monthly.
3. Draw several sets of control limits on the chart corresponding to different subgroup sizes. A good plan is to use three sets of limits, one for expected average subgroup size, one close to the expected minimum, and one close to the expected maximum. In Example 6.1, for instance, limits might be drawn corresponding to subgroup sizes of 1,000, 2,500, and 4,000. This scheme is not satisfactory unless the data are shown on the same sheet with the control chart, with the figures for each subgroup on the same line the plotted point for the subgroup, so that the subgroup size may be seen at a glance. Because with this scheme it is not immediately evident from inspection of the chart which points are out of control, it is desirable to use a special symbol to mark out-of-control points. For instance, if a dot were used for each point on the chart, a circle might be drawn around the dot for each point out of control.

The main objection raised against separate calculation of correct control limits for each subgroup is the time and effort consumed in calculating and drawing the limits. Actually, even inexpensive pocket or desk calculators provide for automatic extraction of a square root; thus such calculations should take very little time. All process control computer software provides for automatic calculation of limits. Thus hand calculation usually is relegated to the status of a teaching tool in most practical applications.

Example 6.1

Illustration of calculations necessary for control chart for fraction rejected

Facts of the case This example applies to a 4-month record of daily 100% inspection of a single critical quality characteristic of a part for an electrical device. It is intended chiefly as an illustration of how control limits are calculated with variable production, and of the setting and revision of standard values for fraction rejected.

When, after a change in design, the production of this part was started early in June, the daily fraction rejected was computed and plotted on a chart. At the end of the month, the average fraction rejected $\overline{p}$ was computed. Trial control limits were computed for each point. A standard value of fraction rejected p_o

was then established to apply to future production. During July new control limits were computed and plotted daily on the basis of the number of parts n inspected during the day. A single set of control limits was established for August, based on the estimated average daily production. At the end of August, a revised p_o was computed to apply to September, and the control chart was continued during September with this revised value.

Calculation of trial control limits Table 6.1 shows the number inspected and the number rejected as nonconforming to specifications each day during June. The fraction rejected each day is the number of parts rejected (r_i) divided by the number inspected that day (n_i). For example, for June 6, $p_i = 31/3{,}350 = 0.0093$. The percent rejected is $100p = 0.93\%$.

TABLE 6.1 COMPUTATION OF TRIAL CONTROL LIMITS FOR CONTROL CHART FOR FRACTION REJECTED

Data on a single quality characteristic of a part of an electrical device

Date	Number inspected n	Number rejected	Fraction rejected p	$3\sigma = \frac{3\sqrt{\bar{p}(1-\bar{p})}}{\sqrt{n}}$	UCL $\bar{p}+3\sigma$	LCL $\bar{p}-3\sigma$
June 6	3,350	31	0.0093	0.0062	0.0207	0.0083
7	3,354	113	0.0337	0.0062	0.0207	0.0083
8	1,509	28	0.0186	0.0092	0.0237	0.0053
9	2,190	20	0.0091	0.0077	0.0222	0.0068
11	2,678	35	0.0131	0.0069	0.0214	0.0076
12	3,252	68	0.0209	0.0063	0.0208	0.0082
13	4,641	339	0.0730	0.0053	0.0198	0.0092
14	3,782	12	0.0032	0.0058	0.0203	0.0087
15	2,993	3	0.0010	0.0066	0.0211	0.0079
16	3,382	17	0.0050	0.0062	0.0207	0.0083
18	3,694	14	0.0038	0.0059	0.0204	0.0086
19	3,052	8	0.0026	0.0065	0.0210	0.0080
20	3,477	27	0.0078	0.0061	0.0206	0.0084
21	4,051	44	0.0109	0.0056	0.0201	0.0089
22	3,042	70	0.0230	0.0065	0.0210	0.0080
23	1,623	12	0.0074	0.0089	0.0234	0.0056
25	915	9	0.0098	0.0119	0.0264	0.0026
26	1,644	1	0.0006	0.0088	0.0233	0.0057
27	1,572	22	0.0140	0.0090	0.0235	0.0055
28	1,961	3	0.0015	0.0081	0.0226	0.0064
29	2,440	3	0.0012	0.0073	0.0218	0.0072
30	2,086	1	0.0005	0.0079	0.0224	0.0066
Totals	60,688	880	......	......	......	......

$$\bar{p} = \frac{\text{total number rejected}}{\text{total number inspected}} = \frac{880}{60{,}688} = 0.0145$$

$$3\sqrt{\bar{p}(1-\bar{p})} = 3\sqrt{(0.0145)(0.9855)} = 0.3586$$

At the end of the month, the average fraction rejected $\bar{p}$ is computed. It should be emphasized that the correct way to calculate $\bar{p}$ is to divide the total number rejected in the period by the total number of parts inspected during the period. Whenever the subgroup size (in this case, the daily number inspected) is not constant, it is incorrect to average the values of p_i.

The standard deviation was calculated on the basis of this observed value of $\bar{p}$, 0.0145. Note that the value of $3\sqrt{\bar{p}(1 - p)}$ was calculated only once to apply to all calculations of control limits. Thus, for June 6, the 3-sigma limits are $\pm 0.3586/\sqrt{3{,}350}$, or ± 0.0062.

The daily values of p_i and the control limits for each day are shown in Fig. 6.1. In this figure, percent rejected (100p) rather than fraction rejected has been plotted. Because percent rejected is more readily understood by both shop and administrative personnel, it is usually desirable for fraction rejected to be converted to percent rejected for all plotting.

FIGURE 6.1 Control chart for percent rejected—4 months' production of an electrical device.

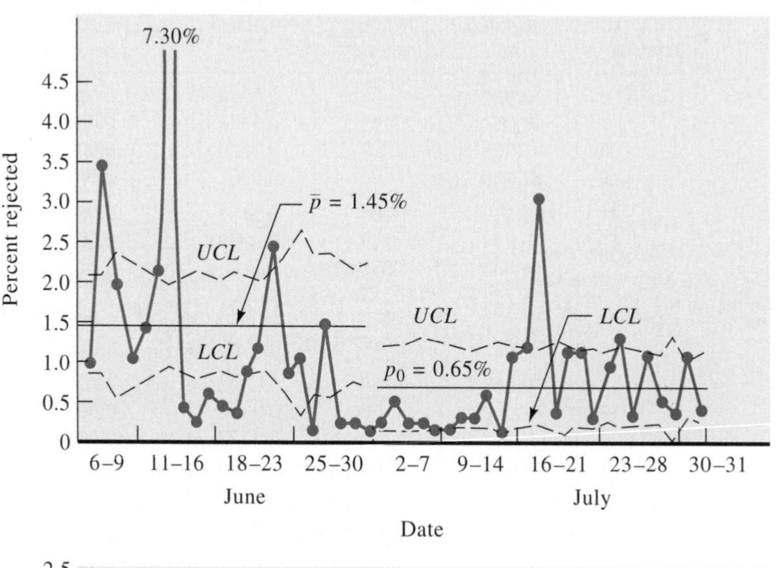

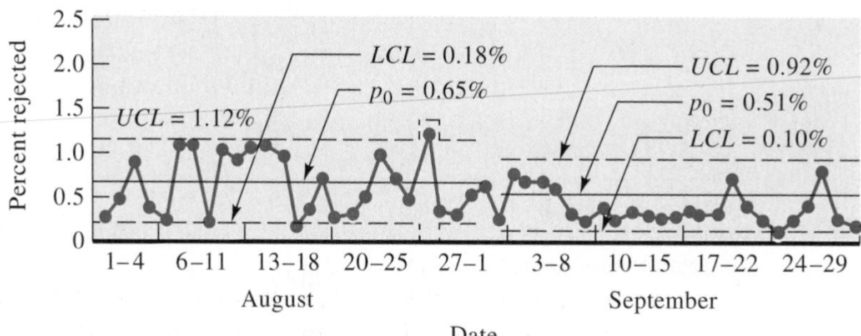

Determination of standard value p_o If all the points fall within the trial control limits, the standard value p_o may be assumed to be equal to $\bar{p}$.

Here many points fell outside the trial control limits. In such cases, the decision as to the value of p_o to be used calls for judgment as to what process average fraction rejected can be maintained in the future, provided the occasional assignable causes of bad quality can be eliminated. An aid to such judgment may be obtained by computing a revised value of $\bar{p}$, eliminating the days on which p_i fell above the upper control limit.

With these days—June 7, 12, 13, and 22—eliminated, the remaining number of rejects is 290 and the remaining number inspected is 46,399. The revised $\bar{p} = 290/46{,}399 = 0.0063$.

After consideration of this and of the previous record on similar parts of slightly different design, it was decided to assume $p_o = 0.0065$.

Calculation of control limits based on standard fraction rejected p_o Table 6.2 gives the daily numbers inspected and rejected during July and shows the calculation of control limits based on the standard fraction rejected. This calculation appears to be almost identical with that shown in Table 6.1. The value of p_o is used in the calculation of limits in Table 6.2 wherever $\bar{p}$ was used in Table 6.1.

The practical difference is that where $\bar{p}$ is used, no control limits can be computed until $\bar{p}$ is known, that is, not until the end of the period. Where a standard value p_o is established in advance, the limits can be computed each day and drawn on the control chart as the day's point is plotted. In this way, the control chart provides a basis for immediate action whenever a point goes outside the control limits.

Establishment of control limits based on expected average subgroup size Although the correct position of 3-sigma control limits on a p chart depends on subgroup size (in this case, a subgroup is the number of parts inspected each day), the calculation of new limits for each new subgroup consumes some time and effort. Where the variation in subgroup size is not too great (for example, where the maximum and minimum subgroups are not more than 25% away from the average), it often may be good enough for practical purposes to establish a single set of control limits based on the expected average subgroup size. In this way, limits may be established at the start of a period (for instance, a month) and projected ahead for the entire period.

At the end of July the situation was reviewed to consider the possibility of doing this. It was decided that daily output was well enough stabilized to justify the use of a single set of control limits during August. Average daily production during July had been 61,701/25 = 2,468. The estimated average daily output during August was 2,600; this was assumed as the value of n for calculation of control limits. As $\bar{p}$ during July had been 393/61,701 = 0.0064, no change was made in the p_o of 0.0065. The calculations for the control limits for August are shown in Table 6.3.

Whenever control limits are set in this way on an expected average value of n, any points on the control chart that are either outside the limits or just inside

TABLE 6.2 COMPUTATION OF DAILY CONTROL LIMITS BASED ON STANDARD VALUE OF FRACTION REJECTED p_o

Data on a single quality characteristic of a part of an electrical device

Date	Number inspected n	Number rejected	Fraction rejected p	$3\sigma = \frac{3\sqrt{p_o(1-p_o)}}{\sqrt{n}}$	UCL $p_o + 3\sigma$	LCL $p_o - 3\sigma$
July 2	2,228	4	0.0018	0.0051	0.0116	0.0014
3	2,087	9	0.0043	0.0053	0.0118	0.0012
5	2,088	3	0.0014	0.0053	0.0118	0.0012
6	1,746	2	0.0011	0.0058	0.0123	0.0007
7	2,076	1	0.0005	0.0053	0.0118	0.0012
9	2,164	1	0.0005	0.0052	0.0117	0.0013
10	2,855	5	0.0018	0.0045	0.0110	0.0020
11	2,560	5	0.0020	0.0048	0.0113	0.0017
12	2,545	14	0.0055	0.0048	0.0113	0.0017
13	1,874	1	0.0005	0.0056	0.0121	0.0009
14	2,329	24	0.0103	0.0050	0.0115	0.0015
16	2,744	30	0.0109	0.0046	0.0111	0.0019
17	2,619	77	0.0294	0.0047	0.0112	0.0018
18	2,211	5	0.0023	0.0051	0.0116	0.0014
19	1,746	19	0.0109	0.0058	0.0123	0.0007
20	2,628	28	0.0107	0.0047	0.0112	0.0018
21	2,366	5	0.0021	0.0050	0.0115	0.0015
23	2,954	23	0.0078	0.0044	0.0109	0.0021
24	2,586	32	0.0124	0.0047	0.0112	0.0018
25	2,790	8	0.0029	0.0046	0.0111	0.0019
26	2,968	30	0.0101	0.0044	0.0109	0.0021
27	3,100	13	0.0042	0.0043	0.0108	0.0022
28	1,359	4	0.0029	0.0065	0.0130	0.0000
30	3,940	39	0.0099	0.0038	0.0103	0.0027
31	3,138	11	0.0035	0.0043	0.0108	0.0022
Totals	61,701	393	......	......	......	......

Standard fraction rejected $p_o = 0.0065$

$3\sqrt{p_o(1-p_o)} = 3\sqrt{(0.0065)(0.9935)} = 0.241$

the limits require more critical examination to see whether the limits as drawn really apply to these points. Whenever the subgroup size is larger than the assumed average value of n, the true limits are inside those drawn. Whenever the subgroup size is smaller, the true limits are outside.

Such a calculation was made for August 27, when p was 0.0116. This is above the upper control limit of 0.0112 that was computed for the assumed daily production of 2,600. A revised upper control limit for this day based on the actual production of 1,205 is 0.0134. The revised limits for this day are indicated on the control chart (Fig. 6.1); they show that the point was actually not out of control.

TABLE 6.3 RECORD OF DAILY FRACTION REJECTED WITH CONTROL LIMITS COMPUTED ON STANDARD DAILY PRODUCTION AND ON STANDARD VALUE OF FRACTION REJECTED p_o
Data on a single quality characteristic of a part of an electrical device

Date	Number inspected n	Number rejected	Fraction rejected p	Date	Number inspected n	Number rejected	Fraction rejected p
Aug. 1†	3,068	6	0.0020	Sept. 1‡	2,539	3	0.0012
2	776	3	0.0039	3	2,425	16	0.0066
3	2,086	16	0.0077	4	1,537	9	0.0059
4	3,652	10	0.0027	5	2,852	17	0.0060
6	2,606	3	0.0012	6	2,953	16	0.0054
7	2,159	21	0.0097	7	2,649	5	0.0019
8	2,745	27	0.0098	8	2,835	4	0.0014
9	2,606	3	0.0012	10	2,752	6	0.0022
10	2,159	21	0.0097	11	892	1	0.0011
11	2,745	22	0.0080	12	3,186	7	0.0022
13	3,114	30	0.0096	13	2,646	5	0.0019
14	1,768	18	0.0102	14	2,714	4	0.0015
15	3,208	29	0.0090	15	2,878	5	0.0017
16	2,629	2	0.0008	17	2,384	6	0.0025
17	3,576	9	0.0025	18	2,639	5	0.0019
18	2,262	15	0.0066	19	3,160	7	0.0022
20	3,294	5	0.0015	20	1,895	11	0.0058
21	3,026	5	0.0017	21	4,287	13	0.0030
22	2,713	10	0.0037	22	2,917	3	0.0010
23	2,687	24	0.0089	24	2,479	1	0.0004
24	3,824	23	0.0060	25	1,991	2	0.0010
25	3,265	12	0.0037	26	3,280	10	0.0030
27	1,205	14	0.0116	27	2,195	15	0.0068
28	3,035	7	0.0023	28	2,570	3	0.0012
29	2,793	6	0.0021	29	3,323	3	0.0009
30	3,295	14	0.0042				
31	3,227	18	0.0056				
Totals	73,523	373	......	Totals	65,978	177	......

†For August:

Estimated average daily production = 2,600

Standard fraction rejected $p_o = 0.0065$

$$3\sigma = \frac{3\sqrt{p_o(1-p_o)}}{\sqrt{n}} = \frac{3\sqrt{(0.0065)(0.9935)}}{\sqrt{2,600}} = 0.0047$$

$UCL = p_o + 3\sigma = 0.0065 + 0.0047 = 0.0112$

$LCL = p_o - 3\sigma = 0.0065 - 0.0047 = 0.0018$

‡For September:

Estimated average daily production = 2,700

Standard fraction rejected $p_o = 0.0051$

$3\sigma = 0.0041$

$UCL = 0.0051 + 0.0041 = 0.0092$

$LCL = 0.0051 - 0.0041 = 0.0010$

Further revision of p_o During August, the average fraction rejected $\bar{p}$ was 373/73,523 = 0.0051. No points fell above the upper control limit. This value, 0.0051, was therefore assumed as p_o to apply to September. Control limits for September were based on an estimated average subgroup size of 2,700. (Daily production during August had been 73,523/27 = 2,723.) Daily values for September with calculated control limits are shown in Table 6.3 on the preceding page and plotted in Fig. 6.1.

The process quality during September improved even more. Although only two points fell below the lower control limit during the month, confirmation of the existence of a new better level of quality was given by an extreme run for 11 points—from September 7 to 19—below the central line. For the month, the process average $\overline{p}$ was 177/65,978 = 0.0027. This justified a further downward revision of p_o to 0.0027 for October. The data and the control chart for October are not shown here.

6.7 CHECKLIST OF NECESSARY STEPS IN CONNECTION WITH CONTROL CHART FOR FRACTION REJECTED

It will be helpful in gaining a full understanding of the procedures followed in Example 6.1 to review the decisions and calculations that must be made and the actions that must be taken in initiating and operating a control chart for attributes data. The major events occur in a sequence somewhat as follows:

1. Decisions preparatory to the control chart
   a. Determination of the purpose of the chart
   b. Selection of inspection station and quality characteristics to be charted
   c. Decisions on the selection of subgroups
   d. Choice of type of chart to be used (p or np)
   e. Decision regarding calculation of control limits
   f. Setting up the forms for recording and charting the data
2. Starting the control chart
   a. Recording the data
   b. Calculation of the subgroup control statistics (p_i)
   c. Calculation of the average value of the control statistic ($\overline{p}$)
   d. Calculation of trial control limits
   e. Plotting the points and trial control limits
3. Continuing the control chart
   a. Selecting a standard value of the control statistic (p_o)
   b. Calculation of control limits
   c. Plotting the points and limits
   d. Interpretation of lack of control
   e. Periodic review and revision of the standard value (p_o)
4. Reports and actions based on the control chart
   a. Actions to bring process into control at a satisfactory level
   b. Review of specifications in relation to the capabilities of a production process
   c. Information to management regarding the quality level

6.7.1 Decisions Preparatory to the Control Chart

A. *Determination of the Purpose of the p Chart*

As applied to 100% inspection, a control chart for fraction rejected may have any or all of the following purposes:

1. To discover the average proportion of nonconforming articles or parts submitted for inspection over a period of time.
2. To bring to the attention of management any changes in this average quality level.
3. To discover those out-of-control high spots that call for action to identify and correct causes of bad quality.
4. To discover those out-of-control low spots that indicate either relaxed inspection standards or erratic causes of quality improvement that might be converted into causes of consistent quality improvement.
5. To suggest places for the use of $\bar{X}$ and R charts to diagnose quality problems.

The p chart as applied to sampling inspection on a lot-by-lot basis may have any or all of the purposes cited for 100% inspection. An additional purpose usually is:

6. To afford a basis for judgment whether successive lots may be considered as representative of a process. This judgment may properly influence the severity of customer acceptance criteria (discussed in Chap. 12) and affect contractual compliance stipulations.

Example 6.2

An illustration of the selection of quality characteristics for charts for fraction rejected

Facts of the case In this example, the product manufactured was an automatic pressure switch used to open and close a valve in a gas line in order to maintain gas pressure within given limits. An inspection station was located in the assembly line just before the cover was bolted onto this device. This was not final inspection in the sense that it involved a check of the overall functioning of the device. However, it was the final chance before the completion of the device to identify a number of nonconformities that would prevent the functioning of a completed pressure switch.

At this inspection station, 31 different possible nonconformities to specifications might be observed. A device for which the inspector finds one or more nonconformities is a *reject.* For each reject, the inspector's record sheet showed the particular nonconformities found, designated by initials or some other appropriate symbol (such as *BL* for broken lead, *DR* for defective receptacle, and so forth).

When the p chart was first applied at this inspection station, the average proportion rejected was around 15%. It was decided to start with 11 p charts. A chart was maintained for each of the following nine nonconformities: (1) points off, (2) broken lead, (3) defective receptacle, (4) defective bar, (5) high tube, (6) cracked base, (7) loose bar, (8) close bar, and (9) defective tube. The other 22 of the 31 possible nonconformities were grouped together in a single chart (10) for miscellaneous. In addition, a chart was kept for total percent rejected. The total number of rejects recorded from any day's inspection might, of course, be less than the total nonconformities observed because a device containing two or more nonconformities would be classified as only one reject. For this reason, the fraction rejected shown on the total chart was generally less than the sum of the fractions rejected shown on the 10 constituent charts.

Prompt attention to out-of-control points on some of the charts resulted in a substantial improvement in the average quality. In a short time, "points off" was reduced from 4.5 to 0.5% and "total rejects" reduced from 15 to 9%. After the charts had been in operation for 3 months, it was decided that separate charts for individual nonconformities could be abandoned except for the three quality characteristics that were giving the most trouble. Several months later the total rejects level was reduced to 3%. After this time only a single control chart for total fraction rejected was used.

In the establishment of these control charts, a problem was created by the fact that all rework as well as all new work was inspected at this station. Experience showed that the chances of rejection of a reworked article were considerably greater than the chances of rejection of a new article. There was a tendency on the part of the production department to accumulate rework for several days; then a large number of reworked devices would come to the inspection station on the same day. If these had been included in the inspection record, the days receiving considerable rework would have shown lack of control. Moreover, the average quality level would have appeared to be worse than it really was, because of the inclusion of two or more rejections of the same article. And the number of items inspected would have appeared to be greater than the total production. For these reasons, the maintenance of the control chart made it necessary to put special tags on all rework items and to keep separate inspection records for inspection of rework. The results of this rework inspection were not included in the control-chart data.

B. Selection of Inspection Stations and Quality Characteristics to Be Charted

It is a common practice for a p chart to be applied at an inspection station where many different quality characteristics are to be checked. In such cases, a decision must always be made on the question faced in Example 6.2, that is, whether to have one control chart or several. A single control chart is the most common solution, with the thought that any investigation of the causes of rejections may look to the supporting data on the inspection record sheet.

Occasionally, however, it will pay to use separate control charts for certain selected nonconformities. Nearly always it is true that troubles with different quality characteristics differ in their influence on costs. Some nonconformities may be corrected by simple inexpensive rework operations; others require costly rework; others involve the scrapping of the article inspected. Separate control charts may help to concentrate attention on those nonconformities that are responsible for the greatest costs. A single control chart that includes all nonconformities observed at an inspection station will have its variations (and its showings of lack of control) influenced more by the most common causes of rejection than by the most costly ones.

In some cases it may pay to have one control chart for spoilage and another for rework. This may be true, for example, in checking dimensions with go and not-go gages, where rejection by the go gage means rework and rejection by the not-go means spoilage (or vice versa, depending on the dimension to be checked).

Another possible breakdown for control-chart purposes may be on the basis of the effect of a nonconformity on the functioning of a part or product. An example is the classification into critical, major, minor, and incidental defects, discussed in Chap. 7 and again in Chap. 12. This breakdown is most common where acceptance sampling plans use acceptance criteria that depend on the seriousness of a nonconformity to specifications.

In all cases, the determination of which inspection stations should employ p charts should be based on a consideration of whether the accomplishment of the various purposes seems likely to have a sufficiently favorable effect on overall costs to justify the expense of maintaining the chart.

C. *Decision on the Selection of Subgroups*

As in the case of the Shewhart control chart for variables, in the control chart for fraction rejected the most natural basis for selecting rational subgroups is the order in which production takes place.

A common basis for subgrouping is the one illustrated in Example 6.1 in which each subgroup consisted of the items inspected in a day. This is a good basis wherever the inspection operation is an integral part of the production process, so that the order of inspection is substantially the same as the order of production. Sometimes a control chart showing daily percent rejected may be supplemented by charts showing weekly and monthly figures. The daily chart may be used as a basis for current action on the manufacturing process by production supervisors, methods analysts, and operators; the weekly chart may be used by manufacturing executives such as department heads; the monthly chart may be used in quality reports to top management.

Where production is not on a continuous basis, a satisfactory alternative basis of subgrouping may be to consider each production order as one subgroup.

In the sampling inspection of a large lot of purchased articles from a single source, it is often desirable to use the p chart as a test for homogeneity in order to judge whether the sample may be considered to be representative of the entire lot.

For example, consider the receipt of a shipment of 100,000 bolts packed in 50 containers each holding 2,000 bolts. Even though the purchaser of the bolts has no way of knowing the order of production, there is a strong likelihood that the bolts have not been thoroughly mixed since production took place. If 50 bolts were taken at random from each box of 2,000 and tested for conformity to specifications, each such sample of 50 would be an appropriate rational subgroup for use in a p chart. If the control chart showed control, the conclusion would be that either (1) the bolts had been well mixed or (2) although not well mixed, they came from a production process that was in statistical control. Either conclusion would be satisfactory from the standpoint of considering the 2,500 bolts inspected as representative of the lot of 100,000.

In all control charts, rational subgroups should be selected in a way that tends to minimize the chance for variation within any subgroup. A possible assignable cause of variation in any inspection operation is difference in inspectors. This is particularly true in inspection by attributes. In visual inspection, where judgment plays an important part, there is a great chance for differences among inspectors. Even with go and not-go gages, inspectors may differ considerably. If each subgroup is taken in a way to reflect the work of only one inspector, the p chart may sometimes be used as a useful check on inspection standards.

D. Choice between Chart for p and Chart for np

The choice among the control-chart techniques for attributes data is based partially on convenience in interpretation of the chart and partially on the choice of the probability distribution that best fits the circumstances. In Example 6.1, the p chart was appropriate because the number of items inspected varied daily and the statistic of interest was the fraction (or percent) rejected. Thus a charting technique based on the binomial distribution was a rational choice.

Whenever subgroup size is variable, the control chart must show the fraction rejected (or proportion rejected) rather than the actual number of rejects. If actual numbers of rejects were plotted, the central line on the chart (as well as the limits) would need to be changed with every change in subgroup size. However, if subgroup size is constant, the chart for actual numbers of rejects may be used. Such a chart is called a chart for np or pn. (The fraction rejected p was obtained by dividing the actual number of rejects by the subgroup size n. The actual number of rejects may therefore be represented by np, the quantity that, divided by n, gives p.)

A chart for np may be used for data such as those shown in Table 6.4. This table gives the results of inspection of a sheet-metal part for an aircraft turbo-supercharger skin. The part was inspected after being shaped by a drop hammer.

As explained in Chap. 5, the standard deviation of the *number* of occurrences in n trials of an event with a constant probability of occurrence p (in other words, the standard deviation of the *number* of rejects) is $\sqrt{np(1-p)}$. The standard deviation of the *proportion* of occurrences (in other words, the *fraction* rejected) is $\sqrt{p(1-p)/n}$. Thus the appropriate model for 3-sigma control limits on an np chart is

$$UCL_{np} = np + 3\sqrt{np(1-p)}$$
$$LCL_{np} = np - 3\sqrt{np(1-p)}$$

Since no standard value for p was established for the data in Table 6.4, the average fraction rejected $\bar{p}$ is used as the best available estimate of p.

Figure 6.2 shows a chart for np for the data of Table 6.4. If a chart for p were drawn for the same data, it would look exactly like the chart for np except for the graduations on the vertical scale. Each unit on the vertical scale would represent $\frac{1}{200}$ (that is, $1/n$) as much as it does on the chart for np. It is evident that there is no fundamental difference in the appearance of the np and the p charts or in the information they give.

Where subgroup size is constant, there might be two possible reasons for preferring an np chart to a p chart. One reason is that the np chart saves one calculation for each subgroup, the division of number of rejects r_i by subgroup size n_i to get p. The other reason is that some people may understand the np chart more readily.

However, it often happens that in a manufacturing plant there are many places where, because of variable subgroup size, only a p chart is applicable, and a few places where, because subgroup size is constant, either type of chart may be used. In such cases, the possibility of confusion from having two types of charts might outweigh the slight advantage of the np chart, and it would be better to use the p charts even for constant subgroup size.

TABLE 6.4 CALCULATIONS FOR *np* CONTROL CHART FOR NUMBER OF REJECTS
Inspection in drop-hammer department of sheet-metal part

Production order number	Lot size n	Number of rejects r
1	200	23
2	200	15
3	200	17
4	200	15
5	200	41
6	200	0
7	200	25
8	200	31
9	200	29
10	200	0
11	200	8
12	200	16
Totals	2,400	220

$$np = \frac{220}{12} = 18.3$$

$$\bar{p} = \frac{220}{2{,}400} = 0.0917$$

$$3\sigma_{np} = 3\sqrt{n\bar{p}(1-\bar{p})}$$
$$= 3\sqrt{(200)(0.0917)(0.9083)}$$
$$= 12.2$$

$$UCL = n\bar{p} + 3\sigma_{np} = 18.3 + 12.2 = 30.5$$

$$LCL = n\bar{p} - 3\sigma_{np} = 18.3 - 12.2 = 6.1$$

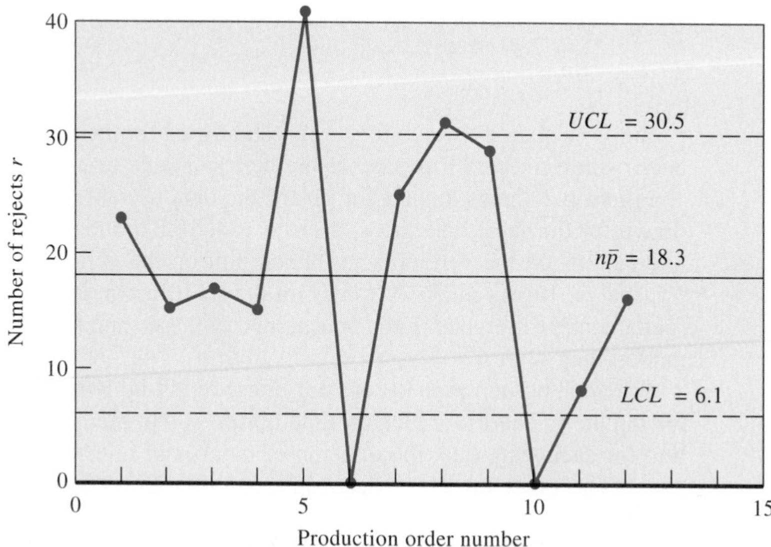

FIGURE 6.2 An *np* chart shows actual numbers of rejects rather than percent rejected.

E. Decision Regarding Calculation of Control Limits

For any p chart with variable subgroup size, a decision must be made whether to compute new control limits for each subgroup or to adopt one of the less accurate methods that have been described.

In making this decision, it should be recognized that only the separately computed limits are really correct. The objections raised in some manufacturing plants to separately computed limits have been twofold, namely, (1) the difficulty (and, therefore, cost) of their calculation and (2) the difficulty in explaining to many of the people who see the control charts the reasons why control limits vary from day to day.

The difficulties and costs of separate calculations of control limits for each subgroup tend to be overestimated. As already explained, there are a number of ways in which the job of making these computations can be simplified. With proper organization of the calculations, the extra costs of computing separate limits may be negligible. The use of computer programs to plot control charts, of course, eliminates this objection.

On the other hand, where the p charts are posted in places available to shop personnel, there are real difficulties in explaining varying control limits. (Sometimes it may be even harder to explain them to the top executives than to the machine operators.) This may be a good reason for establishing one set of limits based on expected average subgroup size, particularly if such limits do not have to be changed frequently and if variations in subgroup size are not so great as to call for numerous calculations of separate limits for individual subgroups (such as the one shown in Fig. 6.1 for August 27). However, where subgroup sizes vary enough so that numerous variations in limits are inevitable, it may be better practical psychology to have everyone get used to the idea that control limits *always* vary with subgroup size, rather than to have control limits seem to vary in some cases and not in others.

F. Setting Up the Forms for Recording and Charting the Data

In those instances where data are entered directly into a computer at the workstation or inspection station, it may not be necessary to manually record the results of inspection. In many others, however, even when charting will be accomplished at a computer workstation, a certain amount of recording is necessary. It always is preferable to have some standard form rather than loose scraps of paper that tend to get lost or confused as time passes. Figure 6.3 illustrates a form used for recording data for a control chart for fraction rejected. This form contains space for all the information that is essential for preparing the control chart itself.

Often it is desirable that information regarding the particular nonconformities observed be included in a record sheet showing items inspected and numbers of items rejected. Figure 6.4 illustrates a sample record form containing columns giving this type of information. Manually recording this information may be an intermediate step in using computer software to prepare control charts and Pareto diagrams.

A compact presentation is provided by a combination of data sheet and control chart in one form such as is illustrated in Fig. 6.5. This particular form shows weekly figures as well as daily figures.

6.7.2 Essential Steps in Starting the Control Chart

The steps in starting the control chart were illustrated in Example 6.1. Briefly stated, they are:

A. *Record the data* for each subgroup on number inspected and number of rejects. Any occurrences that might be clues to an explanation of points out of control

FIGURE 6.3 A simple form for *p* chart data.

RECORD SHEET FOR *p* CHART

Name of product or part __________ Part No. ______

Characteristics measured __________

Inspection station __________ Recorded by ______

Lot No.	Date	Number inspected	No. of rejects	Percent rejected	Control limits		Remarks
					Upper	Lower	

or to changes in the quality level should be noted on the data sheet as supplementary remarks.

B. *Compute p for each subgroup:*

$$p_i = \frac{\text{number of rejects in subgroup}}{\text{number inspected in subgroup}} = \frac{r_i}{n_i}$$

C. *Compute $\bar{p}$, the average fraction rejected:*

FIGURE 6.4 Record of attributes inspection including columns showing different types of nonconformities observed. *(Reproduced from Carl L. Gartner, "Quality Control in Television Receiver Manufacturing," Industrial Quality Control, November 1951.)*

LOT BY LOT
SAMPLING RECORD

DEFECTS IN FIRST SAMPLE

SAMP. PLAN No. DR – 7

SUMMARY PERIOD 3 mos.

PART No. 03001570

DESCRIPTION Cap, Pa, .005mf 25% 600V

VENDOR

	1949 YEAR DATE	REC. REPORT No.	LOT SIZE	FIRST SAMPLE SS₁	FIRST SAMPLE DEF.	TOTAL SAMPLE SS₁ + SS₂ + +	TOTAL SAMPLE TOTAL DEF.	Out of tol. on high side	Out of tol. on low side	Wrong lead length	Pin holes in wax coating	INSPECTOR'S INITIALS	DISPOSITION OF LOT PASS. REJECT, ETC.	REMARK No.
1	8/15	0014	1000	35	0							TH	P	
2	8/25	0090	700	35	0							BK	P	
3	9/5	0170	1500	50	1	150	3	1				BK	Ⓡ	1
4	9/20	0220	1800	50	1	150	2		1			CD	P	
5	10/5	0278	1250	35	1	105	1				1	TH	P	
6	10/22	0315	1155	35	2			1		1		BK	Ⓡ	2
7	10/31		7405	240	5			2	1	1	1			3
8														
9	11/1	0407	1200	35	0							TH	P	
10	11/13	0438	1500	50	0							TH	P	
11	11/29	0500	2100	50	0							BK	P	
12	12/5	0539	2900	50	4			2		2		BK	Ⓡ	4
13	12/10	0581	3500	75	0							TH	P	
14	12/17	0644	2700	50	0							TH	P	
15	1/8	0773	2005	50	1				1			TH	P	
16	2/1		15905	360	5									3
17														
18	2/3	0838	1550	50	0							CD	P	
19	2/21	0907	1275	35	0							CD	P	
20	3/5	0982	1475	50	0							TH	P	
21														
22														
23														
24														
25														

Remarks:
1. Returned to vendor.
2. Rush - Balance of lot inspected.
3. Entered into summary report.
4. Rejection waived on lead lengths. 100% electrical inspection of lot.

$$\overline{p} = \frac{\text{total number of rejects during period}}{\text{total number inspected during period}} = \frac{\Sigma r_i}{\Sigma n_i}$$

Wherever practicable, it is desirable to have data for at least 25 subgroups before computing $\overline{p}$ and establishing trial control limits.

D. *Compute trial control limits* for each subgroup on the basis of observed average fraction rejected $\overline{p}$.

E. *Plot each point as obtained.* Plot trial control limits as soon as calculated, and note whether the process appears to be in control.

It often happens that when the decision is made to use a p chart for any manufacturing operation, data are available for the period immediately past. If so, the steps just outlined should be applied to this past record. This permits putting the p chart to work at once as an effective instrument for process control and avoids a period for which no control limits are currently available.

FIGURE 6.5 This form combines p chart data with a control chart. *(Courtesy General Electric Company.)*

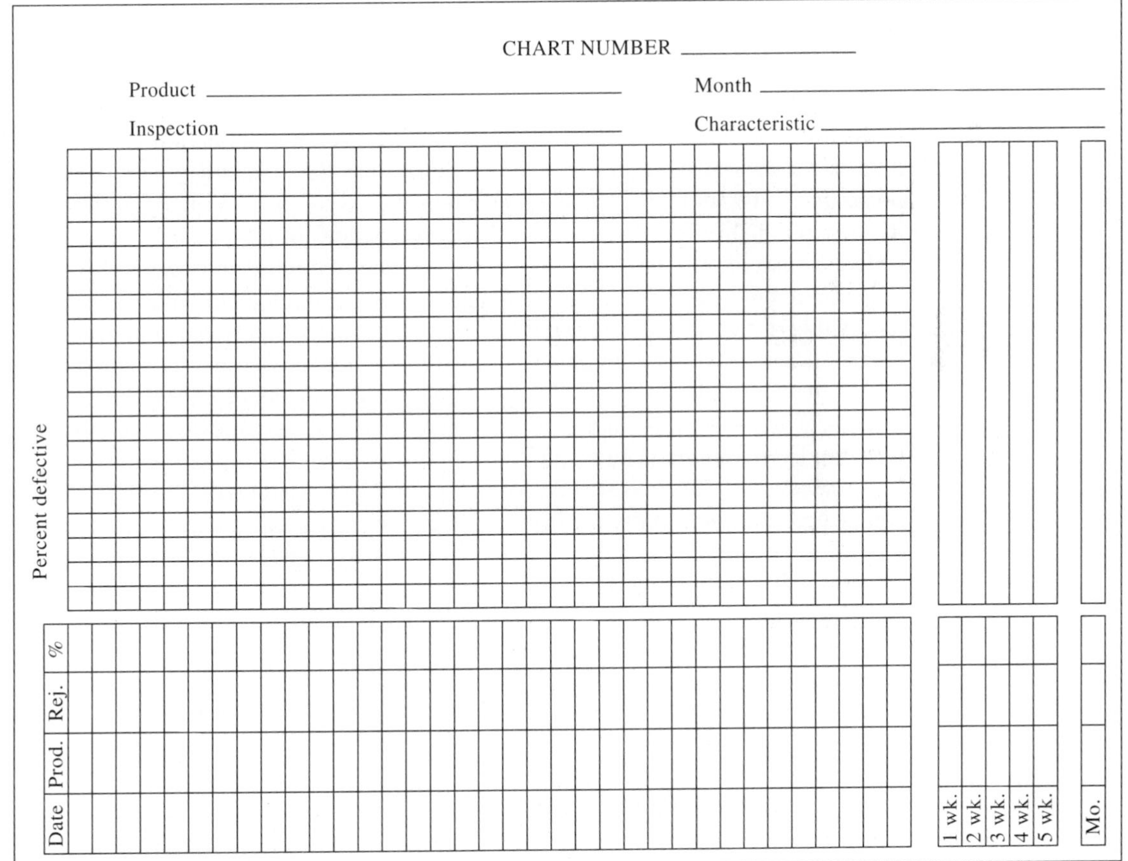
CHART NUMBER ______

Product ______ Month ______

Inspection ______ Characteristic ______

Percent defective

Date | Prod. | Rej. | %

1 wk. | 2 wk. | 3 wk. | 4 wk. | 5 wk. | Mo.

6.7.3 Continuing the Control Chart

A. *Selecting a Standard Fraction Rejected*

The p chart is not, as generally used, merely a test for the presence or absence of assignable causes of variation. It is also a basis for judging whether the quality level is at some desired objective.

In the setting of a standard fraction rejected p_o, the two purposes sometimes seem to be in conflict. For example, if p_o should be set at 0.02 and if the process actually is in statistical control (that is, there are no assignable causes of variation from subgroup to subgroup) at a substantially higher fraction rejected, such as 0.05, the majority of the points on the control chart may fall above the upper control limit. With the p chart used to establish a standard quality level, points may be expected to fall outside control limits for either of two reasons: (1) the existence of assignable causes of variation or (2) the existence of a quality level that is different from the assumed standard p_o. This interpretation of the p chart should be kept in mind in the establishment and revision of the standard fraction rejected p_o.

When a preliminary period has been completed and trial limits have been computed on the basis of $\bar{p}$, the control chart may show any condition from an excellent state of control with all points falling within control limits to an apparently hopeless absence of control with very few points within limits.

If the chart shows control, p_o should be assumed equal to $\bar{p}$. This is generally desirable even though $\bar{p}$ is considered too high a fraction rejected to be satisfactory in the long run. For any standard really to be accepted by production personnel as a basis for action, there needs to be evidence that the standard is attainable. As illustrated in Example 6.1, p_o may later be reduced as soon as efforts to improve the general quality level have resulted in lower values of $\bar{p}$.

If the chart shows apparently hopeless absence of control, it is generally better to continue the p chart for a time without any control limits (and without any standard value of fraction rejected) until the situation can be improved somewhat. For control limits to be respected, there needs to be evidence that it is possible to stay within the control limits most of the time. Until such evidence exists, the drawing of control limits on a p chart can be of little help and may hurt the control-chart program by creating a state of mind antagonistic to control limits.

In most cases, the control chart for the preliminary period will show a condition somewhere between the two extremes of perfect statistical control and complete absence of control. There will be a few points outside control limits, even though the majority fall within limits. This common situation was illustrated in Example 6.1. In such cases, the best procedure is to eliminate the points above the upper control limit and then to recompute $\bar{p}$. (In Example 6.1, this revised $\bar{p}$ was 0.0065 in contrast to the original $\bar{p}$ of 0.0145.) Judgment may then be applied to the revised $\bar{p}$ when establishing the standard fraction rejected p_o to be used in the immediate future.

B. *Calculation of Control Limits*

Once p_o is established, 3-sigma control limits are computed on the basis of this standard value.

Where subgroup size is variable, limits should be computed separately for each subgroup unless some plan for approximate limits is adopted such as one of those described earlier in this chapter.

In some instances management may elect to use tighter limits than 3 sigma. If 2-sigma limits are to be used, the figure 2 is substituted for 3 in the appropriate formulas.

C. *Plotting the Points and Limits*

As soon as the data are obtained, points and limits should be plotted promptly on the control chart. Promptness is particularly important when the charts are posted in the shop, where they may be seen by operating personnel and supervisors.

In charts exhibited in the shop, it is often desirable to omit the lower control limit. It is almost impossible to make clear to some operating people just why a point should be classed as out of control when it refers to quality that is better than the standard.

In the discussion of the $\overline{X}$ and R charts, it was stated that it is generally desirable not to draw lines connecting the points that represent the successive subgroups. The contrary is true in the case of the p chart; a line connecting the points is usually helpful in interpretation of the chart. Such a line assists in the interpretation of trends; this may be almost as important on the p chart as the interpretation of control limits.

D. *Interpretation of Lack of Control*

There may be erratic changes in the quality level for an occasional subgroup even though quality is otherwise maintained at the standard fraction rejected p_o. Such changes are shown by points outside control limits and are evidence of assignable causes of variation.

In most p charts that extend over any considerable length of time, there are also definite sustained shifts of average fraction rejected to a new level either better or worse than standard. A shift to a better level was illustrated in Example 6.1. Such departures from the standard fraction rejected p_o are often evident merely from inspection of the control chart without application of any formal statistical tests. Extreme runs above or below the central line, as well as points outside control limits, may be used to provide tests that supplement observation of the chart.

For purposes of a statistical test, any consecutive set of subgroups may be combined into a single subgroup. In this way the average fraction rejected of a set of subgroups may be tested to see whether it varies by more than 3 sigma from the standard fraction rejected.

For instance, the September data in Example 6.1 might be tested to see whether the observed fraction rejected during September, 0.0027, may be explained as a chance variation from a process that has its average at the assumed p_o of 0.0051. With all September combined into a single subgroup, the subgroup size n is 65,978. The lower control limit corresponding to this subgroup size and to a p_o of 0.0051 is computed as follows:

$$LCL = p_o - \frac{3\sqrt{p_o(1-p_o)}}{\sqrt{n}} = 0.0051 - \frac{3\sqrt{(0.0051)(0.9949)}}{\sqrt{65{,}978}}$$

$$= 0.0051 - 0.0008 = 0.0043$$

It is evident that the September figure of 0.0027 is a long way outside the limits. (The difference of 0.0024 between the p_o of 0.0051 and the actual p of 0.0027 is really a 9-sigma difference.) As this makes it clear that a new improved quality level has been established, it is possible to proceed with confidence to set a new value of p_o.

E. *Periodic Review and Revision of p_o*

The standard fraction rejected should be reviewed from time to time. This might be done at irregular intervals whenever there seems to be enough evidence to justify a change. Where many p charts are being used, it is usually better practice to ensure periodic review by establishing a regular review period. In Example 6.1, this review period was once a month. Where a subgroup consists of one day's production, it may be satisfactory to have a 2 months' review period, with half the charts reviewed each month. Where subgroups consist of production orders or of lots submitted for inspection, the frequency of subgroups will influence the proper length of the review period. This period might then be stated as once every 20 subgroups, once every 40 subgroups, and so on.

Whenever there is sustained evidence of a decrease in average percent rejected and it is clear that this decrease reflects real quality improvement rather than relaxed inspection, it is a good idea to revise p_o downward. This helps to supply an incentive to hold this new and better level. On the other hand, where there is sustained evidence of a poorer quality level, the quality control engineer should be reluctant to revise p_o upward. An upward revision should not be made without evidence that changes have taken place that seem to make it inevitable that, with the same attention to quality as before, the percent rejected will increase. Some possible changes of this type are tighter specification limits, more adequate enforcement of existing limits, and poorer incoming materials. The value of p_o should not be increased merely on the basis of a poorer quality level that seems to have resulted from reduced attention to quality on the part of production personnel.

6.7.4 Reports and Actions Based on the Control Chart

A. *Action to Bring a Process into Control at a Satisfactory Level*

Experience shows that the mere introduction of a p chart often causes some quality improvement. This improvement may result from the influence of the chart in focusing the attention of production personnel on the quality level and may have no relation to the actual use of the control limits. This influence is most likely to be effective when the chart is new.

In the long run, much of the quality improvement attributable to the use of the p chart will come from concentration of attention on assignable causes of trouble indicated whenever a point on the chart falls above the upper control limit. Such out-of-control points are known as *high spots.* Often they are reported to production supervisors and to management in regular forms known as *high spot reports.*

Frequently the discovery and correction of assignable causes of poor quality are really technical jobs. In such cases, it may do no good merely to bring pressure on the production supervisor by means of a high spot report. The supervisor may already know of the trouble; what is needed is technical help in discovering its causes.

For this reason, any p chart program may need to be reinforced by methods engineers or other technical specialists who are available to give immediate attention to the most urgent high spots. In this connection, it should be emphasized that the only clue given by the p chart as to the *cause* of lack of control is the *time* at which lack of control was observed. This is in contrast to the $\bar{X}$ and R charts, which, as pointed out in Chap. 4, are often very effective instruments for the diagnosis of the causes why product fails to meet specifications. The p chart, therefore, may point to the place for effective use of $\bar{X}$ and R charts.

Low spots on the control chart (that is, points below the lower control limit) call for a different kind of attention from that given to high spots. They sometimes point to faulty inspection and may indicate the necessity of providing better inspection standards or providing better inspector training. In other cases they may be worth examining to find the reasons why quality for one subgroup was so much better than the standard; a knowledge of these reasons may help to bring about more permanent quality improvement.

B. Review of Design and Specifications in Relation to the Capabilities of a Production Process

The control chart for p may exhibit fairly good control over a period of time, but this control may be at an average fraction rejected that is too high to be satisfactory. This suggests that the situation can be improved only by fundamental changes of some sort.

Such a fundamental change might be in the design of the product. For example, a p chart used in the manufacture of an oxygen pressure gage gave an average fraction rejected that was too high. It was suggested that this might be corrected by the use of a somewhat heavier Bourdon tube in the gage. When this change in design was made, the average fraction rejected was immediately reduced to half its previous figure.

Or it might be a change in specifications. For instance, a review of the needs of the product might indicate that tolerances were tighter than necessary on certain dimensions.

Or it might be a change in the production process through the substitution of new tooling or new machinery.

These matters may call for joint study of the problem by representatives of design, production, and inspection. In such cases, many excellent results have been reported from the use of Quality Circles and continuous improvement teams.

C. *Information to Management Regarding Quality Level*

The facts obtained for purposes of a p chart constitute information that should always be available to management. But—as suggested in the quotation at the start of this chapter—frequently this information is not available in any form.

The usual difficulty is that although numbers of rejects may have been recorded regularly, no record shows the numbers inspected for each group of rejects. To compare quality levels at different times, it is necessary to know the fractions rejected at the various times. Every fraction has a numerator and a denominator. The numerator, number of rejects, is commonly recorded. The denominator, number inspected, is often omitted.[†] One great advantage of the p chart is that it requires the denominator as well as the numerator to be recorded and thus supplies information to management regarding the current quality level and the changes in that level.

In a plant where there are a number of departments, and many control charts for fraction rejected are maintained in each department, it may be desirable to prepare charts that in some way summarize all the p charts in each department. Such summary charts may be useful to top executives who would not have time to examine each individual p chart.

6.8 SENSITIVITY OF THE *p* CHART

In order to make effective use of the control chart for fraction rejected as a help in process control, there must be some rejects in the sample observed. It is obvious that the better the quality, the larger the sample in order to find some rejects in the majority of the samples. If only 0.1% of the product is rejected, the sample must be at least 1,000 before there will be an average of one reject per sample. In contrast, a sample of 5 will give an average of 1 reject per sample if 20% of the product is rejectable. It is evident that with very good quality, the p chart is useful in detecting lack of control only if the sample is large; with poor quality, the p chart may be useful with small samples.

In addition, the larger the sample, the more closely it may be expected to reflect the universe. Consider, for example, samples of various sizes drawn from product that is 2% rejectable. A sample of 5 will generally be either 0 or 20% nonconforming (that is, it will contain either 0 or 1 rejects). A sample of 20 will generally be from 0 to 10% nonconforming. A sample of 100 will generally be from 0 to 6% nonconforming. A sample of 2,000 will generally be from 1.06 to 2.94% nonconforming. A sample of 50,000 will generally be from 1.81 to 2.19% nonconforming. In the preceding statements, "generally" applies to 3-sigma limits. For this reason, the smaller the subgroup size, the less sensitive the p chart to changes in the quality level and the less satisfactory it is as an indicator of assignable causes of variation.

Where it is desirable to use a control chart for a single measurable quality characteristic and a choice must be made between the p chart and the $\bar{X}$ chart, it is

[†]Occasionally attempts are made to use a total production figure as the denominator. Unless 100% inspection is in force, such attempts give a biased and misleading view of quality.

important to recognize that the $\overline{X}$ chart will give useful results with a much smaller sample. From a statistical point of view, variables are much superior to attributes; actual measurements on a few parts are as good as gagings on many parts with go and not-go gages.

6.9 NONPRODUCT APPLICATIONS OF *p* AND *np* CHARTS

Charts for percent rejected or for numbers of nonconformities sometimes are applied with advantage to the control of clerical and service processes. The following discussion is not based on one specific case but is a composite of observations based on a number of reported cases.

In many clerical operations, a certain number of errors seem to be inevitable. Just as in manufacturing, the skill, training, and fatigue of the individual will play an important part in the proportion of errors made. However, differences among individuals tend to be particularly important when one is dealing with clerical errors. The variability in materials and in machines that often is so important in manufacturing tends to be absent or of minor importance in many clerical operations. It follows that rational subgrouping for control charts for clerical operations ought to separate the work of different individuals whenever practicable.

In the discussion of p charts it was explained that these charts often are applied to the results of 100% inspection. Conceivably, control charts might also be applied to 100% verification of clerical work. Nevertheless, the usual condition is that 100% verification of clerical work is deemed to be unduly costly. Many types of clerical errors tend to be discovered, if at all, as a result of subsequent operations; the discovery of such errors is comparable to defective manufactured parts being discovered in the assembly department or defective final product being discovered by customers.

If it is desired to reduce the frequency of clerical errors in work not subject to 100% verification, a first step is to use sample verification to estimate how many such errors are being made on the average. A second step is to secure a basis for judgment on whether the differences in proportion of errors from individual to individual, from day to day, and so on (that is, the variations among rational subgroups), are so great that they cannot be attributed to chance causes. A control chart is useful in both steps; just as in manufacturing, the chart provides an estimate of the process average and offers evidence about the presence or absence of statistical control.

In the health services area, np charts have been used to control and improve operating room delays, and p charts have served to assist in the improvement (reduction) of delays in obtaining x-rays for emergency room patients. In this latter case, an excessive delay was arbitrarily defined as a wait time in excess of 15 minutes, and p was the proportion of emergency room patients who waited excessively plotted on a per shift per day and on a per day basis. A team was assembled to study the causes and propose improvements to the system.

6.10 *p* CHARTS ARE NOT SUITABLE FOR ALL DATA ON FRACTION REJECTED

Control limits on the p chart are based on the use of the binomial as a probability distribution. As explained in Chap. 5, the binomial assumes a constant probability of occurrence of whatever event is under consideration—in the case of the p chart, a constant probability of occurrence of a rejected article or part. If the probability is to be constant from one article to another, each article must be a separate unit independent of the preceding and succeeding articles. With this probability constant, rejections will tend to occur at random rather than in bunches. As long as successive articles continue to be independent of one another and the quality level does not change, practically all the points on the p chart will fall within 3-sigma limits.

In some instances, however, it is obvious from the way in which manufacturing or inspection operations are carried on that successive units measured are not independent of one another. In such cases, most of the points on a conventional p chart may fall outside control limits. Here the control-chart limits tell—quite correctly—a fact that is already evident without the control chart, namely, that the probability of one article being rejected is influenced by whether or not the immediately preceding articles were rejectable. If this fact is already known and it is evident that nothing can be done about it, the information given by the p chart limits is of no practical help. The use of the p chart with conventional limits may even do harm, as it may tend to discredit the control chart in the minds of operating personnel and thus handicap the use of the control chart in cases where it would really be helpful.

Four representative cases where conventional p charts were unsuitable because of the foregoing type of difficulty are described in the following paragraphs:

1. In the manufacture of a certain type of rubber belt, a large mold was used that produced 2,300 belts at one time. Conditions of curing varied throughout the mold; if one belt was rejectable, it was likely that many of the adjacent belts also would be rejectable.
2. In the manufacture of a certain type of pile floor covering, daily figures were recorded for total yards produced and for the total yardage of output that was classified as substandard. In this case an imperfection that caused the classification of product as substandard might persist through many successive yards; rejectable yards were not independent of one another. Moreover, there was no natural unit of counting the product (no n for the binomial); for instance, feet or meters might have been used as readily as yards.
3. In the sheet-metal inspection booth of a large aircraft factory, records were kept showing daily numbers of parts inspected and numbers rejected. An average of 160,000 parts per day were inspected. Some 1,250 different parts—including practically everything of sheet metal that went into the airplane—were subject to this inspection. Over a period of time the rejections averaged about 3%. Because of the many different parts that passed through this booth, with differ-

ent cause systems affecting the different parts, it was evident that the probability of a rejectable item could not be constant from part to part. Control limits based on a p of 0.03 and an n of 160,000 would have been 0.0313 and 0.0287. Actual daily fractions rejected varied from about 0.05 to about 0.01.

4. In the 100% inspection of detonators in an ammunition plant, inspection was performed in lots of 10,000. The average fraction rejected was 0.0223. This gave control limits of 0.0268 and 0.0178. Actual lot fractions rejected varied from 0.055 to 0.008 with approximately two-thirds of the points falling outside the control limits. A study of the production process made it evident that quite different cause systems were influencing the different lots.

Even though the conventional p chart cannot be used to advantage in situations like these, it may still be desirable to keep a quality record in the form of a run chart that will be available to production supervision and to management. Generally speaking, this may be done to best advantage simply by plotting percent rejected without control limits.

PROBLEMS

Note: Problems suggested for computer solution are marked with a star (☆).

☆ **6.1.** An electronics company manufactures several types of cathode ray tubes on a mass production basis. During the past month, tube Type A has caused considerable difficulty. The following table contains data from 21 days of this troublesome period. Compute the central line and 3-sigma control limits for a p chart for this tube process. 100 units are inspected each day.

Day	Tube type *A* fraction rejected	Day	Tube type *A* fraction rejected
1	0.22	12	0.46
2	0.33	13	0.31
3	0.24	14	0.24
4	0.20	15	0.22
5	0.18	16	0.22
6	0.24	17	0.29
7	0.24	18	0.31
8	0.29	19	0.21
9	0.18	20	0.26
10	0.27	21	0.24
11	0.31		

Answer: 0.392, 0.260, 0.128.

☆ **6.2.** A certain product is given 100% inspection as it is manufactured, and the resultant data are summarized by the hour. In the following table, 16 h of data is recorded. Calculate the central line and variable control limits of a p chart using 3-sigma control limits, and indicate the values that are out of control.

Hour	Number of units inspected	Number of units outside specs
1	48	5
2	36	5
3	50	0
4	47	5
5	48	0
6	54	3
7	50	0
8	42	1
9	32	5
10	40	2
11	47	2
12	47	4
13	46	1
14	46	0
15	48	3
16	39	0

Answer: $\bar{p} = 0.050$; $3\sigma_p = 0.654/\sqrt{n}$.

☆ **6.3.** Example 6.1 states that "because the daily production is not constant, it would be incorrect to average the values of p." Compute the correct value of $\bar{p}$ from the following data:

Lot	Number inspected	Number of units outside specs	p
1	1,200	18	0.015
2	750	40	0.053
3	150	26	0.173
4	75	15	0.200
5	225	23	0.102
Total	2,400	122	

Compare your correct $\bar{p}$ with the unweighted average value of $\bar{p}$. Why do the two figures differ? Why is the unweighted average value of $\bar{p}$ an unsatisfactory measure of the process average fraction nonconforming?

6.4. A manufacturer purchases small bolts in cartons that usually contain several thousand bolts. Each shipment consists of a number of cartons. As part of the acceptance procedure for these bolts, 400 bolts are selected at random from each carton and are subjected to visual inspection for certain nonconformities. In a shipment of 10 cartons, the respective percentages of rejected bolts in the samples from each carton are 0, 0, 0.5, 0.75, 0, 2.0, 0.25, 0, 0.25, and 1.25. Does this shipment of bolts appear to exhibit statistical control with respect to the quality characteristics examined in this inspection?

Answer: $UCL_p = 0.016$; $LCL_p = 0$.

6.5. An item is made in lots of 200 each. The lots are given 100% inspection. The record sheet for the first 25 lots inspected showed that a total of 75 items did not conform to specifications.

(*a*) Determine the trial control limits for an *np* chart.

(*b*) Assume that all points fall within the control limits. What is your estimate of the process average fraction nonconforming μ_p?

(*c*) If this μ_p remains unchanged, what is the probability that the twenty-sixth lot will contain exactly 7 nonconforming units? That it will contain 7 or more nonconforming units?

Answer: (*a*) $UCL_{np} = 8.157$, $LCL_{np} = 0$; (*b*) 0.015; (*c*) 0.022, 0.034.

6.6. Daily inspection records are maintained on production of a special-design electronic device. 100 items have been inspected each day for the past 21 days. A total of 546 items failed during a particularly severe heat stress test. The four highest and lowest values of *p* are:

Highest	Lowest
0.46	0.18
0.33	0.18
0.31	0.20
0.31	0.21

(*a*) Compute the central line and 3-sigma trial control limits for a *p* chart. Is the process operating in control?

(*b*) Recommend an aimed-at value of p_o and 3-sigma control limits for continued use of the *p* chart.

Answer: (*a*) 0.392, 0.260, 0.128; (*b*) 0.380, 0.250, 0.120.

6.7. The test results described in Prob. 6.6 are from a special severe heat stress chamber that is designed in such a way that 25% of the product will fail when, in fact, it is satisfactory for its intended use. If the minimum stress specification is 750 units and the process standard deviation is known to be 8 units, what should be the lower limit of the test? Assume this stress characteristic to be normally distributed.

6.8. A large number of samples of 300 items each are taken from a process that has a percentage nonconforming of 10%.

(*a*) What is the expected average number of nonconforming units per sample?

(*b*) Find the 3-sigma control limits for an *np* chart to control this process.

(*c*) What is the upper limit of the number of nonconforming items in a sample that, in the long run, you would expect to find exceeded only 5% of the time? Use a Poisson approximation.

6.9. The Selectator Corporation produces synthetic and natural gut casings for a process meat packer. Natural gut materials are visually inspected upon receipt, graded, and sent to processing. After processing, all finished casings are tested under pressure on a special device to ensure a specified strength before shipping to the meat packer. During the past month 25 lots of 300 casings each have been subjected to 100% inspection. A total of 1,000 casings burst during test.

(*a*) Find 3-sigma limits for a control chart for p.

(*b*) Assuming that all points fall within these limits, what is your estimate of the process average fraction nonconforming μ_p?

6.10. In the manufacture of certain special-duty transformers, units are required to meet a number of specifications related to temperature rise, output voltages, voltage and current ripple, on-off recovery times, and the like. Approximately 200 units are produced and subjected to a final inspection daily. At the end of 20 working days, 190 units have been rejected out of 4,150 units produced and inspected.

(*a*) Determine 3-sigma trial control limits for a p chart based on the estimated average daily production of 200 units.

(*b*) Only one point on the control chart falls outside limits. On that day, 30 nonconforming units were found in 200 units inspected. Investigation uncovered the fact that a voltage pot setting was being incorrectly adjusted. What aimed-at values of p_o and control limits would you recommend for the following period based on an average daily production of 200 units?

6.11. The new item startup procedure of a certain electronics plant calls for 100% inspection for at least the first 4 months or until process control is established at an economically acceptable level of nonconforming product. A total of 960 units were found to not meet specifications during the first 20 working days. The number of units produced during this time period was 31,985.

Determine the central line and trial control limits for a p chart based on the average number of units produced per day.

Answer: 0.043, 0.030, 0.017.

6.12. After the control limits found in Prob. 6.11 were plotted on the p chart, three points were found to be above the *UCL*. On these days, a total of 4,950 items were produced and 265 nonconforming units found.

(*a*) What value of an aimed-at p_o would you recommend for the next production period?

(*b*) Give a general formula for calculating control limits during the next production period. Reduce this formula to the point that only n needs to be found in order to plot the limits.

6.13. In Prob. 6.8*b*, find the probability that a shift to 20% nonconforming product will be detected on the first sample drawn after the shift occurs.

6.14. Assume that the process described in Prob. 6.11 was found to be in control but that the level of nonconforming product shifts from 3 to 4.5%.

(*a*) Use Table *A* to estimate the probability that this shift will be detected on the first subgroup drawn after the shift occurs. Base this calculation on the average value of n found in Prob. 6.11.

(*b*) What is the probability that this shift will be detected within the first three subgroups drawn?

Answer: (*a*) 0.6628; (*b*) 0.9617.

6.15. A manufacturer wishes to maintain a process average of 0.5% nonconforming product or less. 1,500 units are produced per day, and 2 days' runs are combined to form a shipping lot. It is decided to sample 250 units each day and use an *np* chart to control production.

(*a*) Find the 3-sigma control limits for this process.

(*b*) Assume that the process shifts from 0.5 to 4% nonconforming product. Use Table *G* to find the probability that the shift will be detected as the result of the first day's sampling after the shift occurs.

(*c*) What is the probability that the shift described in (*b*) will be caught within the first 3 days after it occurs?

6.16. A parts manufacturer requires that 100% final inspection be performed during the first 3 months of production on any new or modified part. A total of 750 items were found to be nonconforming to specifications in the first 50 lots of a new item produced in lots of 800 units.

(*a*) Compute 3-sigma trial control limits for an *np* chart for this process.

(*b*) If the process average remains unchanged, what is the probability that the fifty-first lot will contain more than 20 nonconforming units?

(*c*) If the process average value of μ_{np} increases by 8, what is the probability that the fifty-first lot inspected will exceed the upper control limit found in (*a*)? Use Table *G*.

6.17. Receiving inspection is performed on a certain high-volume part using a *p* chart based on a standard value (p_o) of 0.02, 3-sigma limits, and a standard sample size of 80.

(*a*) Compute control limits for the chart.

(*b*) A group of lots is received from a process that was generating 4% nonconforming product. What is the probability that this higher value of μ_p will not be detected within the first five lots inspected? Use Table *G*.

☆**6.18.** The table on the next page gives the results of daily inspection of a vacuum tube. The standard value of fraction defective p_o established at the start of the month was 0.04. The estimated daily average production was 1,600 tubes. Establish a single set of control limits based on these figures, and plot a control chart. Compute separate control limits for any points that seem to you to require them. On the basis of this month's data, what would you recommend as the value of p_o to use for the following month?

If solved on a computer, calculate and plot the true control limits. Then find the limits based on the estimated average daily production, and draw them on your chart.

Date	Number inspected	Number of defectives	Fraction defective
Nov. 2	531	25	0.0471
3	1,393	62	0.0445
4	1,422	61	0.0428
5	1,500	73	0.0487
6	1,250	46	0.0368
7	2,000	58	0.0290
9	685	28	0.0408
10	2,385	89	0.0373
11	2,150	89	0.0414
12	2,150	58	0.0270
13	2,417	115	0.0476
14	2,549	115	0.0451
16	2,331	75	0.0322
17	2,009	81	0.0403
18	2,198	86	0.0392
19	2,271	67	0.0295
20	1,948	41	0.0210
21	2,150	77	0.0358
23	1,700	49	0.0288
24	2,214	68	0.0307
25	2,394	82	0.0343
26	1,197	56	0.0468
27	850	27	0.0318
28	848	30	0.0353
30	850	33	0.0388
Total	43,392	1,591	....

6.19. A p chart is run on the results of 100% inspection of a small low-voltage transformer using a target value of p_o of 0.012 (1.2%). Average daily production, which forms one subgroup, is 2,500 units.

(*a*) Calculate 3-sigma control limits based on the average subgroup size.

(*b*) On the first day back at work after a 3-day holiday weekend, 2,000 units are produced. Of this number, 32 are rejected. Does the process appear to be out of control? Why or why not?

(*c*) If the value of p found in part (*b*) is the actual process average, what is the probability that this shift would be detected on the day it occurred? Assume that the shift was in effect the entire day.

6.20. In a certain manufacturing operation, a p chart with 2-sigma rather than 3-sigma control limits is to be used based on a standard p_o of 0.015. 100% inspection is to be used during the initial period with 1 day's inspection results constituting a subgroup. Average daily production is expected to be 600 units.

(*a*) Compute control limits for the p chart on the basis of the average daily production.

(*b*) Use Table G to find the approximate Type I error probability under the assumption that the process is operating at a mean of 0.015.

(c) Assume that the process is actually operating at a μ_p of 0.025. What is the probability that a point will fall within the control limits found in part (a)? That 5 points in a row will all fall within these control limits (Type II error)?

(d) Under the conditions described in part (c), does a point out of the control limits necessarily mean that the process is out of control? Why or why not?

☆**6.21.** In the production of bottles, one standard covers the clarity of the glass once the bottles have cooled. A sample of 200 bottles is inspected from each glass melt production batch. The results of inspection of 20 batches are given below.

Batch	Rejects	Batch	Rejects	Batch	Rejects	Batch	Rejects
1	21	6	26	11	18	16	23
2	17	7	11	12	25	17	19
3	28	8	16	13	27	18	35
4	12	9	24	14	12	19	23
5	19	10	14	15	21	20	15

(a) Compute the central line and 3-sigma control limits for a p chart, and test the process for control.

(b) Recommend a standard value for p_o and 3-sigma control limits for the next period of time assuming that the process can be maintained in statistical control.

☆**6.22.** It is proposed to use an np chart for the situation described in Prob. 6.21. Calculate the central line and 3-sigma control limits for the chart, and test for control. Can you see any advantage to using an np chart in this case rather than a p chart? Recommend a standard value for np_o and 3-sigma control limits for the next period of time assuming that statistical control can be maintained.

☆**6.23.** Compute the central line and 3-sigma control limits for an np chart for the data of Prob. 6.1.

6.24. A p chart is used to control brake pad assemblies used in automotive production. The average number of units inspected each shift is 500, and the chart uses a target value p_o of 0.010.

(a) What are the control limits on the p chart based on the average number of units inspected on a shift?

(b) Assume that the process actually is operating at a μ_p of 0.018. Use Table G to estimate the probability that any one point plot would fall within the control limits when 500 units are inspected (Type II error).

(c) What is the probability that five consecutive points would fall within the control limits?

6.25. An np chart is used to control a surgical-sponge production process. Sponges are produced in lots of 500 and are subject to 100% visual inspection before packaging and sterilization. The latest 30 lots produced yielded 280 rejected sponges.

(a) Calculate the central line and control limits to monitor this process.

(b) Use Table G to find the approximate probability of a Type I error.

(c) What is the approximate probability of a Type II error if the mean should shift to 16.0?

6.26. 100% inspection is maintained on a certain assembly used in an automotive braking system. A p chart is used to plot the results of inspection of five essential characteristics. An aimed-at value of p_o of 0.015 is used.
(*a*) Compute control limits for the results of 1 day's inspection when 650 units were produced.
(*b*) Use Table *G* to find the probability of making a Type II error on a single point plot if the process is actually producing 4% nonconforming units.

6.27. Plastic cable connectors are checked automatically for contact point location and quantity of sealant by an optical scanning device. All items manufactured are passed through the automatic inspection system. Rejected items are segregated as to the type of fault. The results of 20 days' inspection show 800 units rejected out of 100,000 produced. Average daily production was 5,000 units.
(*a*) Calculate control limits for a p chart based on the average daily production.
(*b*) Assume that this process has been operating in control for some time with the control limits found in part (*a*). Suddenly the process mean shifts to 2.0%. What is the probability of a point out of control on the first subgroup after the shift occurs? Use a normal approximation and Table *A*.
(*c*) What is the probability that the point-out rule will detect the shift described in part (*b*) within 5 subgroups?

6.28. A control chart for np is run on batches of a certain item purchased regularly from an outside source. Each batch of 500 units is subjected to 100% inspection as it is received. An estimate of μ_p from the most recent 30 batches received is 0.010.
(*a*) Calculate control limits for the np chart.
(*b*) Use Table *G* to find the approximate probability that, if batches should come into receiving at an average μ_p of 0.03, this fact would be detected on the first such batch inspected.

6.29. A p chart has been used to monitor the quality of output of a certain mechanical device. At present, the process is operating at a constant μ_p of 0.02. An average of 350 units are produced each day and subjected to 100% inspection.
(*a*) Calculate the control limits based on the average daily output.
(*b*) On one particular day 450 units were inspected of which 22 were rejected. Was the process operating in control that day?
(*c*) On a day when 350 units are inspected, the process shifts to a μ_p of 0.04. Use Table *G* to find the probability that this shift will be detected on this subgroup. Assume that the shift occurred at the beginning of the day.

6.30. An automated inspection procedure checks four characteristics on all units of a certain item. Production averages 2,000 units per day. The results of each day's inspection are used to plot a p chart with an aimed-at value of μ_p of 0.005.
(*a*) What control limits would be used for an average day when 2,000 units are produced?
(*b*) A sudden change in the process increases μ_p to 0.015. What is the probability that the point plot for that day will fall outside the UCL_p? Assume that the change occurred before the start of production that day, and use Table *A* (normal approximation).

6.31. In the manufacture of semiconductor devices a circuitry pattern must be transferred from a mask to a silicon wafer. These patterns are "built up" in distinct levels requiring very careful alignment of the mask with previous patterns. Each day's output is inspected, and the results are used to plot a p chart.

After 20 days, a total of 800 wafers were rejected out of 6,000 produced.

(*a*) Calculate the central line and control limits for the p chart based on average daily production.

(*b*) A sudden malfunction causes the mean μ_p to shift to 0.25. Use the normal approximation to estimate the probability that this shift will be detected on the first subgroup. Use the value of n found in part (*a*).

(*c*) It comes to your attention that four operators perform the alignment function. What information other than just a simple count of rejected and inspected units might you want to show on the inspection report?

7

THE CONTROL CHART FOR NONCONFORMITIES

Some processes in nature exhibit statistical control. Radioactive disintegration is an example. The distribution of time to failure of vacuum tubes and of many other pieces of complex apparatus furnish further examples. But a state of statistical control is not a natural state for a manufacturing process. It is instead an achievement, arrived at by elimination one by one, by determined effort, of special causes of excessive variation.

—W. Edwards Deming†

7.1 THE PLACE OF THE *c* CHART IN STATISTICAL PROCESS CONTROL

The $\overline{X}$ and R control charts may be applied to any quality characteristic that is measurable. The control chart for p may be applied to the results of any inspection that accepts or rejects individual items of product. Thus both these types of charts are broadly useful in any statistical quality control program.

The control chart for nonconformities, generally called the *c chart,* has a much more restricted field of usefulness. In many manufacturing plants there may be no opportunities for its economic use, even though there are dozens of places where X and R charts and p charts can be used advantageously.

Nevertheless, there are certain manufacturing and many business situations in which the c chart is definitely needed. To decide whether or not to use a c chart in

†W. Edwards Deming, "On Some Statistical Aids toward Economic Production," *Interfaces,* vol. 5, no. 4, p. 5, August 1975, The Institute of Management Sciences, Providence, R.I.

any individual case, it is first necessary to determine whether its use is appropriate from the viewpoint of statistical theory. If so, then it is necessary to judge whether the c chart is really the best technique to use for the purpose at hand.

7.2 DISTINCTION BETWEEN A NONCONFORMING ARTICLE AND A NONCONFORMITY

As already explained, a *nonconforming article* (a *defective* in the restricted technical sense of the word) is an article that in some way fails to conform to one or more given specifications. Each instance of the article's lack of conformity to specifications is a *nonconformity* (a *defect* in the restricted technical sense of the word). Every nonconforming article contains one or more nonconformities. Where it is appropriate to make a total count of the number of nonconformities (or errors or mistakes) in each article, or in each group of an equal number of similar articles, it may be reasonable to use a control-chart technique based on the Poisson distribution. This means using either a c chart or a u chart.

The np chart, which was explained in Chap. 6, applies to the number of rejected items in subgroups of constant size. In contrast, the c chart applies to the number of nonconformities in subgroups of constant size. Each subgroup for the c chart usually is a single article; the variable c is the number of nonconformities observed in one article. But a c chart subgroup may be two or more articles. It is essential only that the subgroup size be constant in the sense that the different subgroups have *substantially equal opportunity for the occurrence of nonconformities.* When the opportunity for occurrence of nonconformities changes from subgroup to subgroup, the u chart for nonconformities per unit is available.

7.3 LIMITS FOR THE *c* CHART ARE BASED ON THE POISSON DISTRIBUTION

Chapter 5 explained that Poisson's exponential bionomial limit was useful not only as a limit of the binomial but also as a probability distribution in its own right. Table 5.4 illustrated a number of examples of this. In all these examples, a count was made of the number of occurrences of some event that had many opportunities to occur but that was extremely unlikely to occur at any given opportunity.

In many different kinds of manufactured articles, the opportunities for nonconformities are numerous, even though the chances of a nonconformity occurring in any one spot are small. Whenever this is true, it is correct as a matter of statistical theory to base control limits on the assumption that the Poisson distribution is applicable. The limits on the control chart for c are based on this assumption. Some representative types of nonconformities to which the c chart may be applied are as follows:

1. c is the number of nonconforming rivets in an aircraft wing or fuselage.
2. c is the number of breakdowns at weak spots in insulation in a given length of insulated wire subjected to a specified test voltage.

3. c is the number of surface imperfections observed in a galvanized sheet or a painted, plated, or enameled surface of a given area.
4. c is the number of "seeds" (small air pockets) observed in a glass bottle.
5. c is the number of imperfections observed in a bolt of cloth.
6. c is the number of surface imperfections observed in a roll of coated paper or sheet of photographic film.
7. c is the number of errors made in completing a form.

7.3.1 Calculating Limits on Control Charts for *c*

The standard deviation of the Poisson distribution is $\sqrt{\mu_c}$ (see Chap. 5). Thus 3-sigma limits on a c chart are as follows:

$$UCL = \mu_c + 3\sqrt{\mu_c}$$
$$LCL = \mu_c - 3\sqrt{\mu_c}$$

When a standard value of average number of nonconformities per unit c_o is not used, μ_c may be estimated as equal to the observed average $\bar{c}$. This is always done in the calculation of trial control limits. In this case the control limits are

$$UCL = \bar{c} + 3\sqrt{\bar{c}}$$
$$LCL = \bar{c} - 3\sqrt{\bar{c}}$$

As the Poisson distribution is not symmetrical, the upper and lower 3-sigma limits do not correspond to equal probabilities of a point on the control chart falling outside limits even though there has been no change in the universe. This fact has sometimes been advanced as a reason for the use of probability limits on c charts. The use of 0.995 and 0.005 probability limits has been favored.[†]

The position of limits corresponding either to these probabilities or to any other desired probabilities may readily be determined from Table G in App. 3. The use of this table is explained in Chap. 5.

The theoretical conditions for the applicability of the Poisson distribution call for the count of the number of occurrences of an event that has an infinite number of opportunities to occur and a very small constant probability of occurrence at each opportunity. (For practical purposes, "infinite" may be interpreted as meaning very large.) As already emphasized, the area of opportunity for occurrence at each count of occurrences must remain constant. However, as we have noted, the count may be of the occurrence of all of several different events, each with its own

[†]A diagram on logarithmic paper giving these limits is shown in "Control Chart Method of Controlling Quality during Production, American Standard Z1.3—1958," p. 16, American National Standards Institute, New York, 1958. This standard was reaffirmed by ANSI in 1975. Another diagram that shows these limits on rectangular coordinate paper is given in L. E. Simon, *An Engineers' Manual of Statistical Methods,* p. 73, John Wiley & Sons, Inc., New York, 1941.

very large number of opportunities to occur and each with a different small probability of occurrence at every opportunity.

In a large proportion of the applications of the Poisson distribution to statistical quality control (or, for that matter, to all other practical affairs), it is possible to pick minor flaws in the theoretical applicability of the Poisson law to the actual situation. It may be evident that the number of opportunities for the occurrence of a nonconformity (or other event being counted) falls far short of being infinite. Or it may be clear that the unknown probability of occurrence of a nonconformity is not quite constant. Or it may not be possible to keep the area of opportunity exactly constant. As long as these are only minor failures to meet the exact conditions of applicability, the results obtained by assuming that the Poisson distribution is applicable are likely to be good enough for practical purposes.

Slight departures of the actual distribution from the true Poisson law usually will cause the standard deviation to be slightly greater than $\sqrt{\bar{c}}$. Limits based on $3\sqrt{\bar{c}}$ may really be at a little less than 3 sigma. This fact in itself generally does not justify discarding $3\sqrt{\bar{c}}$ or $3\sqrt{c_0}$ as a basis for calculating limits. In some situations to which the c chart is applied, such as records of numbers of nonconformities observed in inspections of complex assemblies, this use of limits a little tighter than 3 sigma may actually be desirable. As pointed out in Chap. 4, the economic basis of any control-chart limits is experience that the procedures used for their computation strike a satisfactory balance between the costs of two kinds of errors, namely, looking for assignable causes when they are really absent and not looking for them when they are really present.

In general, the c chart limits used in this book are based on $3\sqrt{\bar{c}}$ or $3\sqrt{c_0}$.

Example 7.1
Control chart for nonconformities

Facts of the case Table 7.1 gives the numbers of errors of alignment observed at final inspection of a certain model of airplane. Figure 7.1 gives the control chart for these 50 observations. The alignment errors observed on each airplane constitute one subgroup for this chart.

The total number of alignment errors in the first 25 planes was 200. The average $\bar{c}$ is $\frac{200}{25} = 8.0$. Trial control limits computed from this average are as follows:

$$UCL = \bar{c} + 3\sqrt{\bar{c}} = 8 + 3\sqrt{8} \simeq 16.5$$

$$LCL = \bar{c} - 3\sqrt{\bar{c}} = 8 - 3\sqrt{8} = \text{negative, therefore no } LCL$$

Whenever calculations give a negative value of the lower control limit of a control chart for attributes, no lower control limit is used. In effect, the chart will exhibit only an upper control limit.

As none of the first 25 points on this chart is outside the trial control limits based on these points, the standard number of defects c_o may be taken as equal

TABLE 7.1 AIRCRAFT ALIGNMENT ERRORS OBSERVED AT FINAL INSPECTION

Airplane number	Number of alignment errors	Airplane number	Number of alignment errors
201	7	226	7
202	6	227	13
203	6	228	4
204	7	229	5
205	4	230	9
206	7	231	3
207	8	232	4
208	12	233	6
209	9	234	7
210	9	235	14
211	8	236	18
212	5	237	11
213	5	238	11
214	9	239	11
215	8	240	8
216	15	241	10
217	6	242	8
218	4	243	7
219	13	244	16
220	7	245	13
221	8	246	12
222	15	247	9
223	6	248	11
224	6	249	11
225	10	250	8
Total	200	Total	236

FIGURE 7.1 Control chart for nonconformities *c*. Data on aircraft alignment errors observed at final inspection.

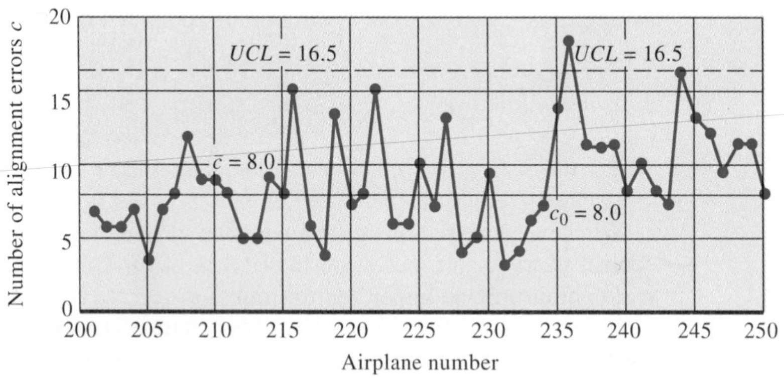

to $\bar{c}$ and the control chart continued for the following period with a central line of 8.0 and an upper control limit of 16.5.

One point (airplane No. 236) out of the next 25 is above the upper control limit. The average during this period was $\frac{236}{25} = 9.44$. (Even omitting the out-of-control value, the average is 9.08.) Of the final 16 points corresponding to airplanes 235 to 250, 12 are above the standard c_o, 3 are exactly at the standard, and only 1 is below. It seems evident that there has been a slight but definite deterioration in quality (or increase in the strictness of inspection) during this period.

Comment on Example 7.1

In applying the formula for the upper control limit, the actual calculation yields 16.4853. Since actual counts of nonconformities must be integer-valued, calculations yielding 16.0015 or 16.9985, for example, mean that a count of 16 should yield a point below the *UCL* and a count of 17 a point above the *UCL*. A convenient way to ensure that these points nearest to a control limit will show clearly as points in or out of control is to place the limit midway in the range. Thus a control limit at 16.5 will indicate clearly that a count of 16 is in control and one of 17 is out. In a case such as this, the previously established standard value of c_o should nearly always be continued despite the evidence of quality deterioration. The principle stated in the preceding chapter regarding the p chart also applies to the c chart, namely, that the standard value should not be revised in the direction of poorer quality merely because production personnel seem to be giving less attention to quality. On the other hand, if a definite tightening up of inspection standards had the effect of making quality *seem* poorer, even though it was really no worse than before, an upward revision of c_o might be justified.

The checklist of steps and comments and suggestions relative to the p chart on pages 234 to 248 obviously apply also to the c chart and do not need to be repeated here. It should be emphasized that extreme runs should be looked for in the c chart, just as in the other control charts. Other significant patterns in the data also should be noted.

7.4 THE COMBINATION OF POISSON DISTRIBUTIONS

In discussing the setting of control limits, it was stated that small departures of the actual distribution from the true Poisson law usually will cause the standard deviation to be slightly greater than $\sqrt{\bar{c}}$. This departure from theoretical justification exists in all applications in which mixtures of inspection units each with different areas of opportunity for nonconformities are included. The effect of this departure is discussed in the following paragraphs.

Assume that the average number of surface imperfections observed in a piece of enameled ware of a certain size is 0.5 and that the frequency of these imperfections follows the Poisson law. If 1,000 such pieces are examined, the expected frequencies of the various numbers of imperfections may be determined from Table *G*, App. 3, to be as follows:

Number of imperfections	Frequency in 1,000 observations
0	607
1	303
2	76
3	12
4	2
Total	1,000

Now assume that the average number of surface imperfections is 1.5 in another piece of enameled ware having three times the area of the first. The expected frequencies of various numbers of imperfections in 1,000 pieces are as follows:

Number of imperfections	Frequency in 1,000 observations
0	223
1	335
2	251
3	125
4	47
5	15
6	3
7	1
Total	1,000

Suppose that these two pieces of enameled ware pass the same inspection station and that the results of the inspection are recorded with no identification of whether the imperfections were observed on the small or the large pieces. If 1,000 pieces of each size pass the station, the expected frequencies in the total of 2,000 will be the sum of the two frequency distributions just given. The average number of surface imperfections per observed piece will of course be 1.0, the average of 0.5 and 1.5.

However, this combined distribution will not follow the Poisson law. This may be demonstrated by comparing it with the expected frequency distribution of 2,000 observations on a quality characteristic that does follow the Poisson law and that has an average value $\bar{c}$ of 1.0. Clearly, if the standard deviation of the mixed distribution is calculated from the data in the second column, the result will be numerically greater than the same calculation made from the data of the third column:

Number of imperfections	Expected frequencies in 2,000 observations: 1,000 observations with $\bar{c} = 0.5$ and 1,000 observations with $\bar{c} = 1.5$	2,000 observations with $\bar{c} = 1.0$
0	830	736
1	638	736
2	327	368
3	137	122
4	49	30
5	15	6
6	3	2
7	1	0
Totals	2,000	2,000

On the other hand, assume that one small and one large piece of enameled ware are fastened together in an assembly. If many such assemblies are made and inspected for surface imperfections, the average number of imperfections per assembly will, of course, be 0.5 + 1.5 = 2.0. If the pairing of the small and large pieces is done at random, the frequency distribution of the number of imperfections in the assemblies will follow the Poisson law. (This statement can be proved mathematically. However, no simple demonstration of it is given here, similar to the demonstration just given that the sum of two Poisson distributions having different averages is not a Poisson distribution.)

These two illustrations show one kind of combination of Poisson distributions that *does not* yield a Poisson distribution, and another combination of Poisson distributions that *does.* In the first case, two products, for which the area of opportunity for a nonconformity was different, passed the same inspection station. This situation is analogous to merging the output of two production lines prior to inspection. The average that may be expected to result from the combination is the *weighted average* of the two processes. That is,

$$\mu_{\text{comb}} = \frac{n_1\mu_{c1} + n_2\,\mu_{c2}}{n_1 + n_2} = \frac{1{,}000(0.05) + 1{,}000(1.5)}{1{,}000 + 1{,}000} = 1.0$$

The result is a distribution that *does not* follow the Poisson law and *does not* have a standard deviation equal to that for the Poisson distribution where $\mu_c = 1.0$, that is, $\sigma_c = \sqrt{1.0}$.

In the second case, where assembly is assumed to take place at random, the resulting combination forms what is called a *linear combination* of random variables each of which follows the same law. This situation may be explained by first considering the drawing of a unit from Part 1, one of the smaller dishes, for example. This dish has d_1 imperfections. We can *expect* that upon drawing a dish at random from Part 2, the larger dish, and assembling it with the dish selected from Part 1, the combination will have $d_1 + \mu_{c2}$ total imperfections. Expanding this concept to cover all random assemblies of Parts 1 and 2, any assembly at random may be *expected* to have $\mu_{c1} + \mu_{c2}$ total imperfections. By definition, this type of combination is called a *linear combination* where

$$\mu_{\text{comb}} = \mu_{c1} + \mu_{c2} = 0.5 + 1.5 = 2.0$$

and will be Poisson-distributed with a standard deviation equal to $\sqrt{\mu_{\text{comb}}}$ if the parts that make up the assembly are Poisson-distributed. Problem 7.2 at the end of this chapter demonstrates the theory just discussed. The comparison of a calculated standard deviation, following the procedures described in Chap. 3, Sec. 3.4, with a theoretical one often is useful in deciding whether or not a constant system may be assumed.

Because the control chart for c uses limits based on the Poisson law, the preceding remarks on various ways of combining Poisson distributions have a bearing on quality control. They indicate that if Poisson limits are to be used, care should be taken to keep approximately constant the area of opportunity for the occurrence of a nonconformity. However, the c chart need not be restricted to a single type of nonconformity but may be used for the total of many different kinds of nonconformities observed on any unit. This adds another field of usefulness to those listed already, namely, a count of the total number of nonconformities of all types in complex products such as tractor subassemblies, television sets, computer chips, computer software, and the like.

7.5 CONDITIONS FAVORABLE TO THE ECONOMIC USE OF THE CONTROL CHART FOR NONCONFORMITIES

The c chart has been used to advantage in four different types of situation, as follows:

1. It has been applied to a count of nonconformities all of which must be eliminated following 100% inspection. In this use, the c chart is primarily an instrument for reducing cost of rework incident to correcting the nonconformities and, to a lesser extent, for reducing costs of inspection incident to identifying the nonconformities. The chart serves to keep management and production supervisors informed about the current quality level, indicates whether or not the process appears to be in control, and serves as a basis for executive pressure to improve the general quality level and to eliminate out-of-control points. The c chart applied in this way sometimes calls attention to the lack of definite inspection standards or to irregularities in the application of inspection standards. A typical example of this kind of use of the c chart is its application to nonconformities of all types observed in inspection of subassemblies and final assemblies of many complex products, systems, and processes.
2. Where a certain number of nonconformities per unit are tolerable, even though it is desired to hold their number to a minimum, the c chart may be applied to periodic samples of production. Here the chief objective is the improvement of

the quality of outgoing product, leading possibly to fewer rejections by customers' inspection and to a generally better consumer acceptance of the product. Like the application to 100% inspection, this use of the c chart gives management up-to-date information on the quality level and helps to increase uniformity of product by putting pressure on out-of-control points. A typical example of this kind of use is found in a paper mill in which a sample consisting of one roll of coated paper per shift is carefully examined for surface defects.

3. It has been applied for special short studies of the variation of quality of a particular product, operation, or procedure.
4. It has been applied to sampling acceptance procedures based on defects (that is, nonconformities) per unit. This application is referred to in Chap. 13.

7.6 ADAPTATIONS OF THE *c* CHART TO VARIATIONS IN THE AREA OF OPPORTUNITY FOR A NONCONFORMITY

The quantity c is the number of nonconformities observed in some specified inspection. Often this inspection is of a single unit of product, such as an airplane, a television set, a coil of wire, or a roll of coated paper. In this common case where the subgroup size is unity, c is both the number of nonconformities and the number of nonconformities per unit. As already explained, the units should be alike in size and in the apparent likelihood of the existence of a nonconformity, in order that the area of opportunity for a nonconformity be constant from unit to unit.

However, it is not really necessary that the subgroup be a single unit of product. The unit for control-chart purposes (that is, the subgroup) may be 10 product units, or 100, or any other convenient number. Total nonconformities for each subgroup may be plotted just as if the subgroup were a single unit of product. Control charts for c using a fixed multiple of units are frequently used when the probability of a nonconformity is so small that a single unit of product is likely to have no nonconformities. As long as the number of product units does not change from subgroup to subgroup and each unit is essentially identical with all others, no special problem is created.

Whenever there is an evident change in the area of opportunity for occurrence of a nonconformity from subgroup to subgroup, the conventional c chart showing only total number of nonconformities is not applicable. It is necessary to create some standard measure of the area of opportunity. For example, if a number of units constitute a subgroup of size n, where n varies from subgroup to subgroup, nonconformities per unit (c/n) may be an appropriate control statistic. In such cases, if total nonconformities observed in each subgroup were plotted, the central line on the chart as well as the control limits would have to change from one subgroup to another. This would make the chart confusing and hard to interpret.

The symbol u is used to represent nonconformities per unit c/n where c is the count of nonconformities found and n may be the number of items, the number of square centimeters, or whatever standard unit of measure is used to establish the constant area of opportunity for the occurrence of a nonconformity. The central line on the u chart will be μ_u with standard 3-sigma limits of

$$UCL = \mu_u + \frac{3\sqrt{\mu_u}}{\sqrt{n_i}}$$

$$LCL = \mu_u - \frac{3\sqrt{\mu_u}}{\sqrt{n_i}}$$

Control limit lines on such a chart will vary with subgroup size, just as they do on the $\bar{X}$ chart and the p chart. When a standard value of u is to be used, u_o is substituted for μ_u into the previous equations. When the average value $\bar{u}$ from a series of subgroups is to be used for trial control limits to test for a constant chance cause system and estimate μ_u, $\bar{u}$ is found from

$$\bar{u} = \frac{\Sigma c_i}{\Sigma n_i} = \frac{\text{total nonconformities found}}{\text{total measure units inspected}}$$

and the trial control limits are

$$UCL = \bar{u} + \frac{3\sqrt{\bar{u}}}{\sqrt{n_i}}$$

$$LCL = \bar{u} - \frac{3\sqrt{\bar{u}}}{\sqrt{n_i}}$$

It should be noted that the statistic u does not follow the Poisson distribution.[†] However, the statistic nu does. Thus probabilities may be associated with specific points falling within or outside control limits using Table G in App. 3 and using the values of n for the individual subgroups.

Example 7.2 illustrates the calculations required for this adaptation of the chart for nonconformities.

Example 7.2

A control chart for nonconformities per unit with variable subgroup size

Facts of the case This is an application to nonconformities observed in aircraft subassembly. The problem of the variable subgroup size was created by the difference in the number of employees on the three shifts. These were roughly in the ratio of 3 on the day shift to 2 on the swing and 1 on the graveyard shift. The num-

[†]In Chap. 5 it was shown that the statistic c was Poisson-distributed with mean equal to μ_c and standard deviation equal to the square root of the mean, $\sqrt{\mu_c}$. It therefore follows that since the standard deviation of u is equal to $\sqrt{\mu_c}/\sqrt{n}$, the square root of its mean divided by $\sqrt{n}$, u cannot be Poisson-distributed.

ber of units assembled in each production center varied from shift to shift in something like this proportion and also showed some variation from day to day.

Table 7.2 gives the record of the number of nonconformities (which, in this aircraft plant, were picturesquely described as *squawks*) recorded by the inspection department for each shift in one production center for a period of 8 days. The number of units produced is also given. To measure this production on each shift, the establishment of a system of weighting different assembly operations was required. For example, if one production center carried out operations 251, 252, and 253, it might happen that on one day, because of the irregular flow of parts to the department, the day shift would work chiefly on operations 251 and 252, while the swing shift might concentrate on 253. On another day, this might be reversed. By giving each operation an appropriate weighting factor, the actual production on the different operations may be converted to an equivalent number of standard production units as follows:

Assembly operation number	Weighting factor	Production on assembly operation	Equivalent units produced on shift
251	0.45	6	2.7
252	0.30	8	2.4
253	0.25	4	1.0
Totals	1.00	. . .	6.1

That is, these operations actually carried out on the shift are judged to be equivalent to carrying out all three of the required assembly operations on 6.1 airplanes.

In Table 7.2*c,* the total nonconformities observed on each shift are divided by n, the number of equivalent units produced, to get nonconformities per unit u. In plotting the value of u on the control chart (Fig. 7.2), a different symbol is used for each shift. The standard value of nonconformities per unit u_0 of 3.2 was established by past performance; this is used as the central line on the control chart. Table 7.2 shows the calculation of the upper and lower control limits for each subgroup.

Comment on Example 7.2

The comparison in one chart of the performance of the three shifts may improve the general quality level by stimulating competition among the shifts. Moreover, if one shift is out of line with the others, such a chart will make this fact quickly evident to management.

If no consideration had been given to the advantages of combining the shifts on one chart and separate charts had been kept for each shift, the slight variations in subgroup size within each shift from day to day might have been neglected. Each chart might then have been a conventional c chart, showing number of nonconformities and having constant limits. Or if it were desired to express quality in terms

TABLE 7.2 COMPUTATION OF CONTROL-CHART LIMITS FOR DATA ON NONCONFORMITIES OBSERVED ON AIRCRAFT SUBASSEMBLIES
Standard value of nonconformities per unit u_0 is 3.2

Date	Shift	Nonconformities observed on shift c	Equiv. units produced n	Nonconformities per unit u	$3\sigma = \frac{3\sqrt{u_0}}{\sqrt{n}}$	$UCL = u_0 + 3\sigma$	$LCL = u_0 - 3\sigma$
June 9	*D*	13	6.0	2.2	2.2	5.4	1.0
	S	12	4.3	2.8	2.6	5.8	0.6
	G	7	2.9	2.4	3.2	6.4	0.0
11	*D*	19	5.5	3.5	2.3	5.5	0.9
	S	14	4.4	3.2	2.6	5.8	0.6
	G	9	2.0	4.5	3.8	7.0	0.0
12	*D*	18	5.5	3.3	2.3	5.5	0.9
	S	13	4.0	3.2	2.7	5.9	0.5
	G	6	2.0	3.0	3.8	7.0	0.0
13	*D*	24	6.1	3.9	2.2	5.4	1.0
	S	15	4.9	3.1	2.4	5.6	0.8
	G	6	2.9	2.1	3.2	6.4	0.0
14	*D*	16	6.6	2.4	2.1	5.3	1.1
	S	11	4.1	2.7	2.7	5.9	0.5
	G	20	2.5	8.0	3.4	6.6	0.0
15	*D*	16	4.3	3.7	2.6	5.8	0.6
	S	29	4.2	6.9	2.6	5.8	0.6
	G	3	2.2	1.4	3.6	6.8	0.0
16	*D*	21	6.1	3.4	2.2	5.4	1.0
	S	20	4.2	4.8	2.6	5.8	0.6
	G	2	1.8	1.1	4.0	7.2	0.0
17	*D*	14	2.9	4.8	3.2	6.4	0.0
	S	10	1.9	5.3	3.9	7.1	0.0
	G	3	1.0	3.0	5.4	8.6	0.0
Totals		321	92.3	. . .	. . .	. . .	. . .

of nonconformities per unit, constant limits might have been plotted on the basis of average subgroup size.

The points in Fig. 7.2 have not been connected. If they were to be connected, it would be better to draw separate lines for each shift rather than a single line connecting all the points. If the symbols used for the three shifts differ enough from one another, the chart is more easily read if the points are not connected. A good plan is to use a different color for each shift.

This is one of the many cases where, even though the strict applicability of the Poisson distribution might be questioned by a statistical theorist, the limits based

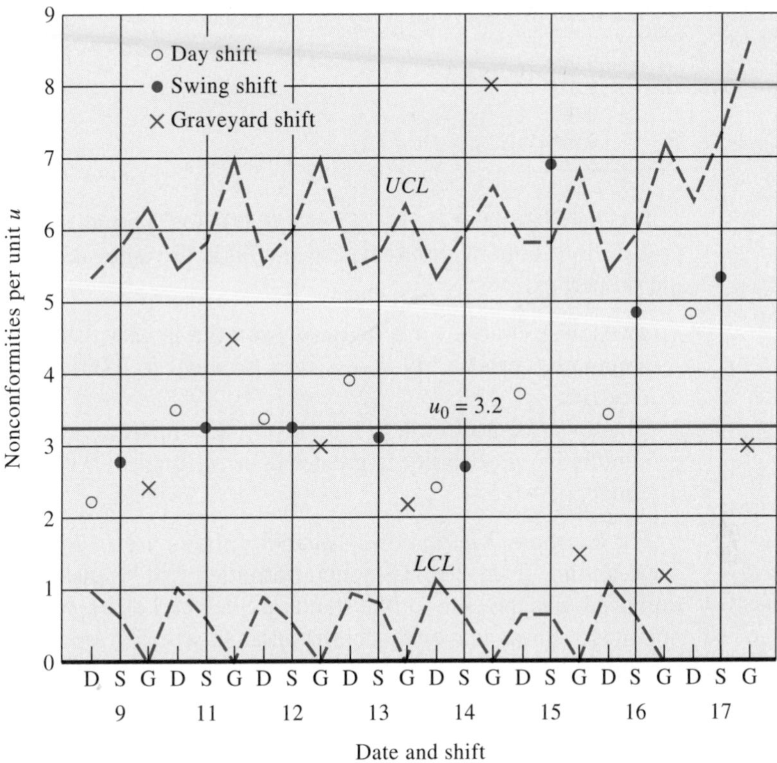

FIGURE 7.2 Adaptation of c chart to variable subgroup size—data of Table 7.2.

on the Poisson law are good enough for practical purposes. The Poisson distribution might be questioned because the equivalent production units depend on the weighting factors adopted for the various assembly operations. Ideally, the weighting factors should be chosen in a way to be proportional to the relative frequency of nonconformities in each of the assembly operations being weighted. Actual weighting, based on some such factor as standard direct labor hours for each operation, may measure this well enough in most instances unless certain of the assembly operations are much more difficult than others.

7.7 PROBABILITY LIMITS FOR *c* AND *u* CHARTS

Table G in App. 3 may be used to find control-chart limits for c and u charts based on probability limits. For example, the standard value used in Example 7.1 was $c_o = 8.0$. Entering Table G at a μ_c of 8.0, the following cumulative probabilities are obtained for pertinent values of c:

c	Probability
1	0.003
2	0.014
15	0.992
16	0.996

If symmetrical 0.995 ($1 - \alpha/2$) and 0.005 ($\alpha/2$) limits are desired, so-called 99% probability limits, we must seek from Table *G* values of *c* that satisfy the following properties:

1. The lower control limit is found from the largest value of *c*, c^-, for which the cumulative probability is less than or equal to 0.005 ($\alpha/2$). Then LCL_c equals $c^- + 0.5$.
2. The upper control limit is found from the smallest value of *c*, c^+, for which the cumulative probability is greater than or equal to 0.995 ($1 - \alpha/2$). Then UCL_c equals $c^+ + 0.5$.

For Example 7.1 the 99% probability limits are $LCL_c = 1.5$ and $UCL_c = 16.5$. Thus finding from 2 to 16 nonconformities will result in a point within control limits. As was the case for the standard Shewhart chart with 3-sigma limits, setting noninteger limits leaves no confusion as to whether a point lies within or outside the limits.

Two observations should be made about the calculation of these probability limits. First, they are not 0.995 and 0.005 symmetrical probability limits. They are 0.996 and 0.003 probability limits. Interpolation between $c = 15$ and $c = 16$ and again between $c = 1$ and $c = 2$ will not make them symmetrical because the count of nonconformities is integer-valued. If μ_c is relatively small, it is possible to come close to the desired probabilities, but it is not possible to obtain them exactly.

The second observation is that these limits are different from those found in Example 7.1; in this case, narrower. Thus use of these limits will not always lead to the same conclusions about statistical control or the same estimates of μ_c.

Table *G* may also be used to find probability limits for the *u* chart in Example 7.2. Suppose approximate 0.995 and 0.005 probability limits are desired. As previously explained, μ_u is not Poisson-distributed but $n\mu_u$ is. Thus, for the *G* shift on June 11, $n\mu_u = (2.0 \times 3.2) = 6.4$, and from Table *G* we can read:

c	Probability
0	0.002
1	0.012
13	0.994
14	0.997

Appropriate 99% control limits are

$$UCL_u = \frac{14.5}{2.0} = 7.25$$

$$LCL_u = \frac{0.5}{2.0} = 0.25$$

The same general comments may be made about the results of this application of probability limits as were made about the c chart application. The probabilities associated with points out of control are not exactly those desired, and they are not exactly symmetrical.

7.8 THE *u* CHART FOR NONCONFORMITIES PER MULTIPLE UNITS

An adaptation of the u chart useful in clerical as well as industrial application is the ku chart for nonconformities per multiple units. An example of this type of chart involves counts of mistakes made per 100 worker-hours. Where nonconformities (or mistakes) are very infrequent and the count of units produced is very large, it is necessary to scale the data if useful charts are to be prepared. The control statistic may be defined:

$$ku = \frac{c}{n/k} = \frac{\text{no. of nonconformities}}{\text{no. of measure units/scaling factor}}$$

where k is the scaling factor; for example, 100 worker-hours and n is the total worker-hours in the subgroup.

The formulas for control limits are

$$UCL_{ku} = k\mu_u + 3\sqrt{\frac{k\mu_u}{n/k}}$$

$$LCL_{ku} = k\mu_u - 3\sqrt{\frac{k\mu_u}{n/k}}$$

A standard value u_o or an average value $\bar{u}$ developed from sampling may be substituted as appropriate for μ_u in the above formulas.

For a process with a standard value of 4 mistakes per 100 worker-hours, a subgroup of 250 worker-hours would have control limits of

$$UCL_{ku} = 4 + 3\sqrt{\frac{4}{\frac{250}{100}}} = 7.79$$

$$LCL_{ku} = 4 - 3\sqrt{\frac{4}{\frac{250}{100}}} = 0.21$$

7.9 LISTING INDIVIDUAL NONCONFORMITIES ON THE FORM CONTAINING A *c* OR *u* CHART

If not too many different types of nonconformities are possible, it sometimes is a good idea to combine a *c* or *u* chart with a record of the number of nonconformities observed of each type. Figure 7.3 is such a chart, applicable to sewing machine cabinet covers. It will be noted that, although this is a *u* chart, the sample size, 10, is constant.

7.10 THE INTRODUCTION OF A CONTROL CHART MAY MOTIVATE QUALITY IMPROVEMENT

Incidentally, Fig. 7.3 illustrates a common occurrence when a chart for *p, np, c,* or *u* is first applied to a product. This occurrence is a substantial improvement in the average quality level, with this improvement unrelated to any actions taken on points outside the control limits. Such an improvement usually is caused by factory personnel taking an increased interest in the quality characteristics being charted. Sometimes this type of improvement, unrelated to statistical techniques, is the most useful consequence of the introduction of a control chart.

It will be observed that, although neither section of Fig. 7.3 shows any points outside control limits, the average number of defects per unit in the second section of the chart is less than one-third the average number shown in the first section some 15 months earlier. The author states that this improvement was secured "through better handling, a greater desire to produce quality, and a superior polishing operation."

Example 7.3

_c_ chart used to control errors on forms

A process improvement team decided to undertake a study to reduce errors made on certain complex forms generated in their department. There were many places on the form where errors could be made. Some errors were of greater consequence than others and were thus classified and recorded on a data sheet similar to that shown in Fig. 7.3. Although one person was responsible for completing each particular form, a number of people were involved in filling out the type of form. A sample of four forms was selected each day from each person's output, using random numbers to determine the particular forms to be selected. Each form selected was given a careful verification by a well-qualified person, and the number of errors found and type were recorded.

A *c* chart similar to Fig. 7.1 was plotted where *c* was the total number of errors found on the four forms inspected for each individual's daily output. Ini-

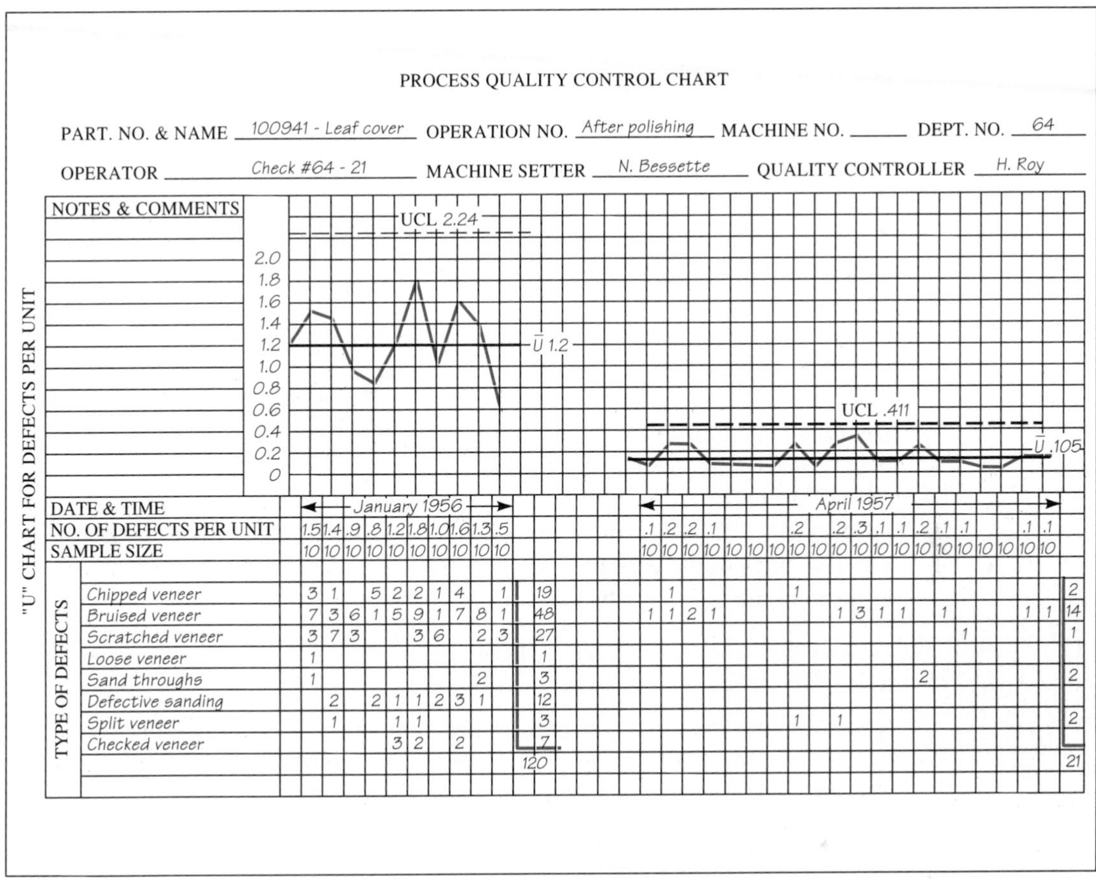

FIGURE 7.3 Example of *u* chart showing numbers of blemishes of various types on sewing machine cabinet covers. (*Reproduced from Robert Chateauneuf, "Modern QC Pays Off in Woodwork," Industrial Quality Control, vol. 17, no. 3, pp. 19–25, September 1960.*)

tially, individual results were not identified on the chart but could be reviewed, in retrospect, from the original data sheets. (Such an identification of different shifts was illustrated in Fig. 7.2.) Early indications were that some retraining was needed. Later, more detailed studies led to certain changes in the form, which proved beneficial in eliminating many errors.

Comment on Example 7.3

Where the chart is to be available only to management, each individual may be identified by a different symbol. Before long, evidence will accumulate on the quality level expressed in average number of errors per form. The control chart also will show whether certain persons or certain days yield more errors than would be expected because of chance fluctuations from the process average.

The evidence about the quality level may turn out to be favorable or unfavorable. Perhaps it will be found that the error rate is so low that no action is required; if so, the sample verification and the related control chart can be discontinued. On the other hand, the error rate may turn out to be much higher than management had expected—possibly much higher than considered to be tolerable.

If so, and if there are a number of out-of-control high points on the chart, a first action should be to try to bring the process into statistical control. If certain individuals are responsible for most of the high points, these persons may need to be retrained or perhaps shifted to other duties. If eliminating such high points does not improve the quality level enough, action may be called for comparable to a fundamental change in a manufacturing process.

We already have noted that the mere introduction of a control chart in manufacturing sometimes is accompanied by an appreciable process improvement that is unrelated to any managerial use of the data charted. Such an improvement also can occur in nonproduct applications, particularly where each posted chart applies to the work of a group of persons and where pride of workmanship causes the group to want to improve its showing.

7.11 CLASSIFICATION OF NONCONFORMITIES AND THEIR WEIGHTING

Some defects (that is, nonconformities) are more serious than others. In an overall picture of quality by departments to be presented to management, there may be an advantage in weighting defects according to some scale that measures their seriousness. Such a plan has been in use for many years in the Bell Telephone System. The classes of defects used are described by H. F. Dodge, as follows:[†]

Class "A" Defects—Very serious.
Will render unit totally unfit for service.
Will surely cause operating failure of the unit in service which cannot be readily corrected on the job, e.g., open induction coil, transmitter without carbon, etc.
Liable to cause personal injury or property damage.

Class "B" Defects—Serious.
Will probably, but not surely, cause Class *A* operating failure of the unit in service.

[†]H. F. Dodge, "A Method of Rating Manufactured Product," *The Bell System Technical Journal,* vol. 7, pp. 350–368, April 1928. This article, which has been reprinted as Bell Telephone Laboratories Reprint B-315, gives clear detailed directions for the use of such a plan based on demerits per unit.

See also the following:

H. F. Dodge and M. N. Torrey, "A Check Inspection and Demerit Weighting Plan," *Industrial Quality Control,* vol. 13, no. 1, pp. 5–12, July 1956; and reprinted in *Journal of Quality Technology,* vol. 9, no. 3, July 1977. This article states the demerit weights, used since the early 1930s, as Class *A*—100, Class *B*—50, Class *C*—10, and Class *D*—1.

D. A. Hill, "Control of Complicated Products," *Industrial Quality Control,* vol. 8, no. 4, pp. 18–22, January 1952.

Will surely cause trouble of a nature less serious than Class *A* operating failure, e.g., adjustment failure, operation below standard, etc.
Will surely cause increased maintenance or decreased life.

Class "C" Defects—Moderately serious.
Will possibly cause operating failure of the unit in service.
Likely to cause trouble of a nature less serious than operating failure.
Likely to cause increased maintenance or decreased life.
Major defects of appearance, finish, or workmanship.

Class "D" Defects—Not serious.
Will not cause operating failure of the unit in service.
Minor defects of appearance, finish, or workmanship.

Modern procedures often divide the various possible defects of a product (in the sense of nonconformity to specifications) into three or four classes, depending on the seriousness of the different defects. A. V. Feigenbaum describes a fourfold classification as follows.[†]

A critical characteristic is one which threatens loss of life or property or makes the product nonfunctional if it was outside prescribed limits.

A major characteristic is one which makes the product fail to accomplish its intended function if outside prescribed limits.

A minor characteristic is one which makes the product fall short of its intended function if outside prescribed limits.

An incidental characteristic is one that will have no unsatisfactory effect on customer quality.

Another formal classification of defects is discussed in detail in Chap. 12, Sec. 12.8.

Once such a classification of all defects is established, demerits may then be assigned to each class of defect. Control charts may be plotted for demerits per unit rather than for defects per unit. Quality rating schemes based on demerits per unit may be an important feature of both supplier rating and quality audit procedures.

7.12 *Q* CHARTS FOR QUALITY SCORES AND *D* CHARTS FOR DEMERIT CLASSIFICATIONS

Two modifications of the *c* chart (or *u* chart) may prove useful where multiple characteristics of varying importance are recorded. Most frequently such charts are used as management tools to indicate where attention should be concentrated and to indicate relative levels of quality over a period of time. They may be applied to incoming materials and products as well as to in-process control.

The first of these charts is the *Q* chart, or quality score chart. Under this scheme, weighting factors w_j are assigned to each defect class *j*. Defect weights may be based simply on the type of nonconformity or may differ by the extent of the noncon-

[†]A. V. Feigenbaum, *Total Quality Control,* 3d ed., p. 253, McGraw-Hill Book Company, New York, 1983.

formity as well as by class. For example, one defect class may be a surface scratch. The weighting applied to surface scratches would increase with the length of the scratch as well as its depth or visibility to the customer. Very specific definitions of the weights must be prepared in such cases to achieve uniformity in inspection.

The control statistic for the Q chart is

$$Q_i = \sum_j w_j c_{ij} \qquad \text{for subgroup } i$$

The control limits and central line are

$$UCL_Q = \Sigma w_j \bar{c}_j + 3\sqrt{\Sigma w_j^2 \bar{c}_j}$$
$$\bar{Q} = \Sigma w_j \bar{c}_j$$
$$LCL_Q = \Sigma w_j \bar{c}_j - 3\sqrt{\Sigma\, w_j^2 \bar{c}_j}$$

where $\bar{c} = \Sigma_i\, c_{ij}/k$, and $0 < w_j < 1$, and i is the sample or subgroup number (i = 1, 2, . . . , k). Thus Q is a weighted average of the count of nonconformities. Aimed-at or standard values c_{oj} may be substituted for $\bar{c}_j$ in these calculations.

The demerit control chart, or D chart, formulates identically with the Q chart except in the weighting factors applied. Frequently it is used in conjunction with a formal classification of characteristics such as those described by Dodge and Feigenbaum. Whereas the Q chart applies weightings to nonconformities (perhaps by intensity of the nonconformity), the D chart assigns demerits that usually range between 1 and 100. A possible scheme for these weightings is

Critical—100
Major— 40
Minor— 20
Incidental—1

The control statistic is

$$D_i = \sum_j d_j c_{ij} \qquad \text{for subgroup } i$$

The control limits and central line are

$$UCL_D = \Sigma d_j \bar{c}_j + 3\sqrt{\Sigma d_j^2 \bar{c}_j}$$
$$\bar{D} = \Sigma d_j \bar{c}_j$$
$$LCL_D = \Sigma d_j \bar{c}_j - 3\sqrt{\Sigma d_j^2 \bar{c}_j}$$

where $\bar{c}_j = \Sigma_i\, c_{ij}/k$, and i is the sample or subgroup number. Aimed-at or standard values c_{oj} may be substituted for $\bar{c}_j$ in these calculations. Using $\bar{c}_j$ implies an intent to illustrate how a process is performing. Using c_{oj} implies an intent to apply pressure to bring a process to no worse than the stipulated level.

Points out of control on Q and D charts are harder to interpret than on the standard c chart and do not necessarily direct attention to the causes of a problem. It may prove more desirable to use several c charts for process analysis purposes, one for each classification, reserving the Q and D chart for upper-level management reporting.

7.12.1 Nonproduct Use of the Demerits-per-Unit Concept

Manufacturing applications of charts for demerits per unit include quality rating of manufacturing departments, quality rating of suppliers, and manufacturers' quality audit of their own finished products. The techniques of computing and charting demerits per unit may also be applied in nonproduct situations.

The usefulness of this technique can be illustrated by describing an application that was made to a fairly complex servicing operation being carried out by teams of workers in widely separated locations. There were many ways in which the actual carrying out of the operation could differ from what was considered to be ideal. Some of the differences were much more serious than others.

Management suspected that the quality of these servicing operations was much better at some locations than at others but had not taken action because of the lack of a suitable way to measure this quality. To provide a basis for such measurement, a demerit rating scheme was drawn up, listing various possible deviations from ideal performance and assigning demerits to each deviation. This rating scheme was given a trial run in one location and modified in a way that made it possible for it to be applied by a pair of trained observers at any location. When applied to a random sample of servicing operations at each location, the resulting control chart, as was anticipated, showed great lack of statistical control. It was then possible to continue control charts for demerits per unit for a period of time at each of the poorer locations until an adequate improvement had been made in the quality of servicing.

7.13 USE OF $3\sqrt{c}$ FOR APPROXIMATE CALCULATION OF CONTROL LIMITS IN SITUATIONS INVOLVING THE BINOMIAL DISTRIBUTION

It often happens that a quick calculation of control limits is useful in some situation where no actual control chart has been plotted. If this is needed in any case to which the binomial is applicable as a probability distribution, it is handy to remember that the standard deviation of the Poisson $\sqrt{c}$ is an approximation to $\sqrt{np(1-p)}$. Even though the binomial itself is only an approximation to the correct probability obtainable by the use of combinatorial formulas, $3\sqrt{c}$ often provides a basis for rapid mental arithmetic about control limits that may be good enough for practical purposes.

For example, assume that product submitted under a purchase contract has been 0.4% nonconforming. The question is raised as to what variation in number of nonconforming items may be expected in lots of 1,000. The expected average number of such items in such lots is obviously 4. As $3\sqrt{4} = 6$, the answer is that nonconforming items might generally be expected to fall between 0 and 10 in lots of 1,000. The occurrence of 11 or more nonconforming items in such a lot would be evidence of an assignable cause of variation.

7.14 APPLICABILITY OF *c* CHART TECHNIQUE IN FIELDS OTHER THAN STATISTICAL PROCESS CONTROL

Like all control charts, the c chart gives evidence regarding the quality level, its variability, and the presence or absence of assignable causes of variation. This is useful information as a basis for action in many other fields besides industrial quality control. The universal nature of the Poisson distribution as the law of small numbers makes the c chart technique broadly useful.

For example, it may be applied to such diverse phenomena as pollution abatement, industrial accidents, and highway accidents. In all these fields, action should often be based on evidence of assignable causes of variation, and it is important to note changes in the average value of the variable being studied.

PROBLEMS

7.1. The following table gives the numbers of missing rivets noted at an aircraft final inspection:

Airplane number	Number of missing rivets	Airplane number	Number of missing rivets	Airplane number	Number of missing rivets
201	8	210	12	218	14
202	16	211	23	219	11
203	14	212	16	220	9
204	19	213	9	221	10
205	11	214	25	222	22
206	15	215	15	223	7
207	8	216	9	224	28
208	11	217	9	225	9
209	21				

Find $\bar{c}$, compute trial control limits, and plot a control chart for c. What value of c_o would you suggest for the subsequent period?

Answer: $UCL_c = 25.28$ (25.5); $LCL_c = 2.80$ (2.5); $c_o = 12.96$

7.2. Page 266 gives the expected frequencies in 2,000 observations, 1,000 of which came from a Poisson distribution with $\bar{c} = 0.5$ and 1,000 from a Poisson distribution with $\bar{c} = 1.5$. Find the standard deviation of this distribution.

The $\bar{c}$ of this combined distribution is, of course, 1.0. Assume that a c chart is established with the upper control limit at $3\sqrt{\bar{c}}$ above the central line on the chart. Consult the table on page 267 to see how many of the 2,000 observations would fall above this upper limit. What objections can you see to the use of a single control chart in this type of situation?

7.3. Use Table *G*, App. 3, to find 0.995 and 0.005 probability limits for a *c* chart when $\mu_c = 5.8$. Also when $\mu_c = 12.0$.
Answer: 13.5, 0.5; 22.5; 3.5.

7.4. Use Table *G*, App. 3, to find 0.95 and 0.05 probability limits for a *c* chart when $\mu_c = 4.2$. Also when $\mu_c = 9.5$.
Answer: 8.5, 0.5; 15.5, 4.5.

7.5. Using the short method described in Chap. 3, Sec. 3.4.4, calculate the average and standard deviation of the combined Poisson distributions (2,000 observations) shown in Sec. 7.4, p. 267. Use these results to calculate the central line and 3-sigma control limits for a *c* chart. How do these values compare with limits calculated in the prescribed manner, that is, $c \pm 3\sqrt{c}$ where $c = 1.0$?

7.6. Prepare a control chart for the data of Table 7.1, Example 7.1, using the results of all 50 inspections. How does this chart compare with Fig. 7.1, and how does it affect the conclusions about process control? The argument could be made that waiting for the results of the 50 inspections before calculating control limits resulted in waiting too long. Comment on the question of how much data is too much when processes have yet to be brought under statistical control.

7.7. A control chart for nonconformities per unit *u* uses probability limits corresponding to probabilities of 0.975 and 0.025. The central line on the control chart is at $\mu_u = 2.0$. The limits vary with the value of *n*. Determine the correct position of these upper and lower control limits when $n = 5$.
Answer: 3.5, 0.7.

7.8. A *c* chart is used to monitor the number of surface imperfections on sheets of photographic film. The chart presently is set up based on $\bar{c}$ of 2.6.
(*a*) Find 3-sigma control limits for this process.
(*b*) Use Table *G* to determine the probability that a point will fall outside these control limits while the process is actually operating at a μ_c of 2.6.
(*c*) If the process average shifts to 4.8, what is the probability of not detecting the shift on the first sample taken after the shift occurs?
Answer: *(a)* 7.44 (7.5), none; *(b)* 0.005; *(c)* 0.887.

7.9. Solve Prob. 7.8 using 0.005 and 0.995 probability limits rather than 3-sigma limits. What is the actual probability of a Type I error in this case as opposed to the planned probability 0.01?

7.10. Explain the difference in the results obtained in Prob. 7.8 and 7.9.

7.11. A *c* chart is used to monitor surface imperfections on porcelain enameled water heater cabinets. Each cabinet is checked for nonconformities of a certain classification and the count entered on the *c* chart. Two limits are used on the chart: a control limit at $+ 3\sigma$ and a warning limit at $+ 2\sigma$. If a point falls above the control limit or if two points in a row fall between the warning limit and the control limit, the process is stopped until the problem is identified and corrected. The central line is set at an aimed-at value c_o of 1.5.
(*a*) Find the values of the warning limit and the control limit.
(*b*) If the process suddenly shifts to a mean value of 4, what is the probability that a point will fall above the control limit?

(*c*) Under the circumstances described in part (*b*), what is the probability that two points in a row will fall between the warning limit and the control limit?
(*d*) What is the combined probability of detection of this shift within the first two units inspected after the shift occurs?

7.12. A textile manufacturer initiates use of a c chart to monitor the number of imperfections found in bolts of cloth. Each bolt is the same length, width, weave, and fiber composition. A total of 160 imperfections were found in the last 25 bolts inspected. The four highest and lowest counts were:

Count of imperfections	
Highest	Lowest
20	4
16	4
10	5
9	5

(*a*) Calculate 3-sigma control limits for this process.
(*b*) Is this process in control? If not, what aimed-at values of c_o and control limits would you recommend for the next period?

7.13. (*a*) Solve Prob. 7.12 using 0.025 and 0.975 probability limits rather than 3-sigma limits (Table *G*).
(*b*) Explain the difference in the conclusions arrived at in part (*a*) and those arrived at in Prob. 7.12.

7.14. Both 3-sigma control limits and 2-sigma warning limits are used on a c chart. The decision rules state that an out-of-control condition will be declared if (1) one point falls outside a 3-sigma limit or (2) two points in succession fall between a given set of warning and control limits (for example, between $+ 2\sigma$ and $+ 3\sigma$). The aimed-at value c_o is 14.
(*a*) Calculate the warning and control limits.
(*b*) If the process mean suddenly shifts downward to a μ_c of 6.0, what is the probability that the next value of c plotted will fall below the lower control limit?
(*c*) In the circumstances described in part (*b*), what is the probability that the next two values of c will fall between the lower warning and control limits?
(*d*) What is the combined probability that the shift described in part (*b*) will be detected within the next two samples?

7.15. A u chart is to be used to control a corrugated paper product line. End product is produced in rolls of varying length 48 in wide. Nonconformities include surface imperfections, improper gluing, improper tension setting on the corrugated inner core, and the like. The control statistic is nonconformities per 100 ft with one roll constituting a sample. After 20 rolls have been inspected, the total count of nonconformities is 290 in a total of 9,300 ft inspected.
(*a*) Find the value of $\bar{u}$ in nonconformities per 100 ft.

(*b*) Set up the formulas for 3-sigma control limits, and reduce them to the most convenient form for calculating specific limits.

(*c*) Find the control limits for the following three representative samples, and determine if the points are in control.

Length of roll, ft	Count of nonconformities
250	7
500	34
150	11

7.16. The hydraulic shop of a large aircraft maintenance facility maintains control charts on maintenance workers based on maintenance errors per standard worker-hour required to refurbish hydraulic parts. Since assemblies of many sizes and degrees of complexity flow through the shop, no other measure of quality performance seems feasible. A chart is maintained on each worker based on a random sampling of 5 items daily. The inspector records the item description code, the worker code, number and type code for errors found, and the standard worker-hours required to refurbish the assembly. Each day the statistic c/n is plotted on the worker's control chart where c is the count of errors found in 5 assemblies and n is the total worker-hours required for the 5 assemblies.

(*a*) After the first 4 weeks of operation, the record for one worker is $\Sigma c = 22$ and $\Sigma n = 54$. Determine the central line and 3-sigma control limits for this worker's chart. Reduce the calculation to the simplest form for direct calculation of the limits.

(*b*) On a certain day during the 4-week period, $c = 2$ errors and $n = 4.3$ standard worker-hours. Make the necessary calculations to determine if the point for this day falls within control limits.

7.17. A manufacturer wishes to operate a control chart on nonconformities found per 100 units produced where each point on the chart represents the output of a shop for a day. Production in a certain shop varies from 800 to 1,500 units per day. The average number of nonconformities per 100 units for the past 3 months is 3.8.

(*a*) Set up the general formula for 3-sigma control limits for this process, and reduce it to its simplest form.

(*b*) Calculate the control limits for a day during which 900 units are produced.

(*c*) Someone suggests that a simpler control-chart model could be used. The suggestion is that c be defined as the number of nonconformities found during a given day divided by the number of hundreds of units produced during a day. The average value $\bar{c}$ could then be obtained by averaging the values of c and the model $\bar{c} \pm 3\sqrt{\bar{c}}$ used to calculate control limits. Discuss why this suggestion is inappropriate. Calculate control limits for this model based on an average $\bar{c}$ of 3.8 and compare these results with those found in part (*b*).

7.18. The manager of a dye plant for 65-35 dacron polyester fabric for the garment industry believes that the plant will lose a most important contract if the rate of dye imperfections exceeds 4 per 100 yd of material more than 2% of the time.

(*a*) Use Table *G* to find an aimed-at value u_o to be used as a central line for a control chart to monitor this process.
(*b*) Calculate 3-sigma control limits for a bolt of material that is 500 yd long.
Answer: (a) 1.52; *(b)* 3.17, none.

7.19. Assume that the centering of the process described in Prob. 7.18 shifts from 4.6 nonconformities per 100 units to 7.6.
(*a*) What is the probability that this shift will be detected on the first subgroup plotted after the shift occurs? Use Table *A*, and assume 900 units were produced on that day.
(*b*) What would be the probability of detection if 1,500 units had been produced on that day?
(*c*) Discuss the difference in the results you obtained in parts *(a)* and *(b)*.
Answer: (a) 0.8315; *(b)* 0.9726.

7.20. Solve Prob. 7.1 using 0.025 and 0.975 (95%) probability limits rather than 3-sigma limits. Use Table *G* to find these control limits. Explain why the conclusion reached in determining a prospective aimed-at value c_o differs from that reached in Prob. 7.1.

7.21. Listed in the following table are the number of bolts of cloth produced on a daily basis in a small textile mill and the corresponding number of imperfections found in these bolts. Use these data to estimate μ_u, and then compute trial control limits for a *u* chart.

Day	Bolts of cloth produced	Number of imperfections	Day	Bolts of cloth produced	Number of imperfections
1	20	37	6	22	31
2	20	23	7	23	37
3	20	30	8	33	24
4	21	28	9	23	36
5	22	34	10	21	27

7.22. Solve Prob. 7.21 using 0.025 and 0.975 (95%) probability limits rather than 3-sigma limits. Derive these limits by the use of Table *G*.

7.23. A *c* chart is used to control imperfections in the glass face of television picture tubes. A target value of c_o of 1.0 is used with five identical-size tubes forming a subgroup. A *c* chart may be used in this case since the subgroup size is constant.
(*a*) Find 3-sigma control limits for this process.
(*b*) Find the probability of detecting a shift in the mean value of this process to a μ_c of 5.0 on the first subgroup drawn after the shift takes place.
(*c*) What is the probability of detecting the shift within the first four subgroups drawn?

7.24. A *u* chart is used to monitor the quality of bolts of cloth of varying length at receiving inspection in a garment factory. Nonconformities include flaws in the fabric weave as well as dye imperfections. The control statistic is nonconformities per 100 linear yd of 48-in-wide material.
(*a*) After inspection of 8,000 yd of material, the count of nonconformities is 150. Calculate standard 3-sigma control limits for a particular bolt of cloth 850 yd in length.

(*b*) The particular bolt described in part (*a*) contained 35 nonconformities. Is there reason to believe that it is significantly worse than the other inspected bolts? Why or why not?

(*c*) Assume that the process had shifted to a μ_u of 3.5 when the 850-yd bolt was made. Use Table *A* to find the approximate probability that this shift would have gone undetected (Type II error).

7.25. A *c* chart for nonconformities is to be used to control an automobile windshield-forming operation. It is to have 98% probability control limits and 90% probability warning limits. The μ_c of the process has been holding steady at 1.6.

(*a*) Calculate the control limits and warning limits for this process using Table *G*.

(*b*) If the process is operating at the μ_c stated, 1.6, what are the probabilities of a point falling at random: above the *UCL*; between the *UCL* and the *UWL*?

(*c*) Answer the questions in part (*b*) assuming that the process is operating at a μ_c of 5.0.

7.26. A *c* chart is used to monitor surface imperfections on a certain electrical insulator. Past history indicates a mean of 2.5.

(*a*) Calculate standard 3-sigma limits for this process.

(*b*) Use Table *G* to find the Type I error probability if the process is operating at a μ_c of 2.5.

(*c*) What is the Type II error probability if the process is operating at a μ_c of 6.8?

7.27. A clothing manufacturer inspects incoming material using a well-established demerits scheme. Since both the width and length of rolls of material may vary, the control-chart procedure used to evaluate each roll uses a count of demerits per 100 yd^2 as the control statistic. The maximum acceptable process level of 15 demerits per 100 yd^2 is used as the central line on the chart.

The results of inspecting four rolls of material are:

Roll	Width, in	Length, yd	Demerits
1	36	200	38
2	48	140	45
3	48	360	136
4	36	200	46

For each roll, compute the values of the control statistic, the control limits, and determine whether or not it complies with the company's standard of acceptability.

7.28. A *c* chart is used to monitor surface imperfections on china plates after firing in a continuous-flow kiln. A subgroup of 10 plates is drawn every half-hour and inspected. When operating in control, the process is able to maintain a mean of 2.8. 95% probability limits are used on the *c* chart.

(*a*) Determine control limits for a *c* chart to control this process.

(*b*) What is the actual Type I error probability for this process as opposed to the design error probability of 0.05?

(*c*) What is the Type II error probability if the process mean suddenly shifts to 9.5?

7.29. A u chart is used to control imperfections in the preparation of mats for advertising copy to be used in print media. The control statistic is number of flaws per 100 cm^2 of mat area. A target value of μ_o of 1.0 flaws per 100 cm^2 is used.

(*a*) A particular mat subjected to inspection is 18 by 26 cm and was found to have 16 flaws. Calculate control limits for this sample, and test for conformance to the standard.

(*b*) At the time that this inspection was performed, the process actually was operating at a μ_u of 3.0 flaws per 100 cm^2. What is the probability of not detecting this fact from the sample? Use Table *G* (where $\mu_c = nu$).

7.30. A c chart is to be used to control soldering imperfections on a certain mass-produced circuit board. After 30 circuit boards have been inspected, a total of 42 bad solder joints was found.

(*a*) Calculate 3-sigma control limits and the central line for the c chart.

(*b*) Find the probability of a point out of control on this chart if the process should suddenly shift to a μ_c of 4.0.

(*c*) How would you detect a shift in the process average to a μ_c of 0.6?

7.31. A c chart is used to monitor the number of surface imperfections at final inspection on a class of television receiver cabinets. Inspection is on a sampling basis with six consecutive cabinets forming a subgroup. About 30% of product output is inspected. The most recent 15 subgroups contained 112 imperfections.

(*a*) Calculate the central line and control limits for this process.

(*b*) Assume there is a sudden shift in the process to a μ_c of 14. What is the Type II error probability of not detecting this shift on the first subgroup after the shift occurs?

(*c*) Assuming that the shift described in part (*a*) is sustained until detected, the probability of detecting it within k subgroups is $1 - P(\text{Type II error})^k$. At what value of k do we have about a 50/50 chance of detection? [This figure is the approximate average run length (ARL) for this c chart when only the single-point-out rule is used.]

7.32. A producer of metal medallions and commemorative coins uses c charts to control imperfections on large orders for single items. All imperfections are recorded but not all cause rejection and ultimate destruction of the item. Thirty items constitute an inspection unit. After 20 inspection units have been inspected and the data recorded, the total count of imperfections is 30.

(*a*) Calculate control limits for this process.

(*b*) What is the probability of a Type I error for this chart? Use Table *G*.

(*c*) Find the probability of a Type II error should the process shift to a μ_c of 4.0.

8

RATIONAL SUBGROUPING

The ultimate object is not only to detect trouble but also to find it, and such discovery naturally involves classification. The engineer who is successful in dividing his data initially into rational subgroups based upon rational hypotheses is therefore inherently better off in the long run than the one who is not thus successful.

—W. A. Shewhart[†]

8.1 THE INFORMATION GIVEN BY THE CONTROL CHART DEPENDS ON THE BASIS USED FOR SELECTION OF SUBGROUPS

One possible view of a control chart is that it provides a statistical test to determine whether the variation from subgroup to subgroup is consistent with the average variation within the subgroups. If it is desired to determine whether or not a group of measurements is statistically homogeneous (that is, whether they appear to come from a constant system of chance causes), subgroups should be chosen in a way that appears likely to give the maximum chance for the measurements in each subgroup to be alike and the maximum chance for the subgroups to differ one from the other.

This may be demonstrated in a striking way. Take a set of measurements that have been subgrouped according to order of production and that show definite lack of control as based on the evidence of $\bar{X}$ and R charts. Write each measurement on

[†]W. A. Shewhart, *Economic Control of Quality of Manufactured Product,* p. 299, 1931, by Litton Educational Publications, Inc., quoted here by permission of Van Nostrand Reinhold Company.

a chip, put the chips in a bowl, mix them thoroughly, and draw the chips out one by one without replacement. Record the values written on the chips in the order drawn. Plot $\overline{X}$ and R charts from these recorded values. If you have done a good job of mixing the chips, these new charts will show control. By this mixing, you have substituted chance causes for the original assignable causes as a basis for the differences among subgroups.

The basis of subgrouping calls for careful study, with a view to obtaining the maximum amount of useful information from any control chart. As already pointed out, the most obvious rational basis for subgrouping is the order of production.

8.2 TWO SCHEMES INVOLVING ORDER OF PRODUCTION AS A BASIS FOR SUBGROUPING

Where order of production is used as a basis for subgrouping, two fundamentally different approaches are possible:

1. The first subgroup consists of product all produced as nearly as possible at one time; the next subgroup consists of product all produced as nearly as possible at a later time; and so forth. For example, if the subgroup size is five, whoever makes measurements at hourly intervals may measure the last five items that were produced just before each hourly interval. This is possible on machine parts, for example, if the parts are placed in trays in the order of production. Otherwise the same result may be obtained by waiting for five items to come off the machine and measuring them as they come.
2. One subgroup consists of product intended to be representative of all the production over a given period of time; the next subgroup consists of product intended to be representative of all the production of approximately the same quantity of product in a later period; and so forth. Where product accumulates at the point of production, someone may choose a random sample from all the product made since the last visit. If this is not practicable, there might be five visits (if $n = 5$) approximately equally spaced over a given production quantity or time, with one measurement made at each visit; these five measurements constitute one subgroup.

The first method follows the rule for selection of rational subgroups of permitting a minimum chance for variation within a subgroup and a maximum chance for variation from subgroup to subgroup. It can be expected to give the best estimate of a value of σ that represents the ideal capabilities of a process obtainable if assignable causes of variation, the special causes, from one subgroup to another can be eliminated. Moreover, it provides a more sensitive measurement of shifts in the process average; it makes the control chart a better guide to machine setting or to other actions intended to maintain a given process average. Thus the first method is more ideally suited to analysis of a process and to process control.

However, if subgrouping is by the first method and a change in process average takes place after one subgroup is taken and is corrected before the next subgroup, the change will not be reflected in the control chart. For this reason, the second method is sometimes preferred where one of the purposes of the control chart is to influence decisions on acceptance of product. In such cases a question to be asked before choosing between the two schemes is whether or not two compensating shifts in process average are really likely to occur between subgroups. If so, and if there is no other way to detect bad product that might be produced between subgroups (for instance, failure of a part to fit into an assembly), the second method of subgrouping is desirable in spite of the other advantages of the first method. In machining operations, such compensating shifts between subgroups are more likely to occur in hand-operated machines than on automatic machines.

Where the second method of subgrouping is used, the interpretation of points out of control on the R chart is somewhat different from that in the first method. With the second method a shift in the process average during the period covered by a subgroup may cause out-of-control points on the R chart even though there has been no real change in the process dispersion.

The choice of subgroup size should be influenced, in part, by the desirability of permitting a minimum chance for variation within a subgroup. In most cases, more useful information will be obtained from, say, five subgroups of 5 than from one subgroup of 25. In large subgroups, such as 25, there is likely to be too much opportunity for a process change within a subgroup. Nevertheless, if process changes do not occur within subgroups, large subgroup sizes have the advantage mentioned in Chap. 4 of reducing the risk of so-called Type II errors in interpreting control charts. Whenever relatively large subgroup sizes are used for this purpose, it is particularly desirable to use the first method of subgrouping.

In many cases where the first method of subgrouping is really better, it may be necessary to use the second method because of practical reasons associated with the taking of the measurements. This was illustrated in Example 2.3.

8.3 QUESTION ADDRESSED BY THE SHEWHART CONTROL CHART

The question asked by the Shewhart control chart may be stated as follows:

> Is the pattern of variation *among* the subgroups consistent with the averaged pattern of variation *within* the subgroups?

If all points fall within the control limits, the answer is yes; that is, we should act as if there are no special or assignable causes of variation present. If one or more points fall beyond a control limit, the answer is no; that is, special or assignable causes are present, and they should be discovered and corrective action taken. However, in order for these decisions to yield meaningful results, the user must know what it is that the *within* subgroup variation is measuring and how this differs from what the *among* subgroup variation is measuring. The chart then must be designed such that useful decision-making information results from interpreting the chart results.

Consider an R or s chart where the subgroups have been drawn in order of production according to method (1) as described in Sec. 8.2. The measure of dispersion for each subgroup becomes a point plot on the chart, and after a suitable number of subgroup statistics have been collected, an average is calculated. This figure is the *averaged pattern of variation within the subgroups.* It becomes the central line on the chart and is used to determine the positioning of the control limits. The test for control thus asks whether the variation observed in going from one point plot to another, the *among* or *between subgroup variation,* is consistent with the pattern established by $\bar{R}$ or $\bar{s}$. If the answer is in the affirmative, then it is reasonable to estimate σ from $\bar{R}/d_2$ or $\bar{s}/c_4$.

The $\bar{X}$ chart for these data is analyzed in the same way. But since the variability of the $\bar{X}$'s from a controlled process $\sigma_{\bar{x}}$ is determined by the dispersion of the process σ, the control limits about the grand average $\bar{\bar{X}}$ are placed at $\pm 3\sigma_{\bar{x}}$ about the central line. Thus any decision based on the results of the $\bar{X}$ chart must be interpreted in light of what was observed on the chart for dispersion.

For any dynamic process, subject to change and not previously submitted to statistical control, the averaged within subgroup variation is the correct method for estimating the standard deviation of the process. Estimating σ from an s statistic calculated from the entire set of measurements will yield the same result, or nearly so, only if the process was in statistical control throughout the period of sampling. Otherwise s will be larger and will include any changes in μ that occurred during the period.

The formation of rational subgroups of data depends on the purpose intended for the charts. It is based on four elements, each of which helps determine where, when, and how subgroups are to be drawn, data recorded, and charts drawn and interpreted. They are *(a)* the circumstances surrounding the determination of need, *(b)* the sources of variation present in the process, *(c)* the use to be made of the knowledge acquired from the charts, and *(d)* the questions to be addressed by the charts. This chapter and Chap. 9 address a number of examples of different formations of subgroups, each of which is rational for the intended purpose. First, it is worthwhile to consider some possible sources of variability in process data.

8.4 SOURCES OF VARIABILITY†

Much of the discussion of process capability will concentrate on the identification and analysis of the sources of common causes of variation, using control-charting techniques. It is worthwhile, therefore, to consider the possible sources of variation in manufactured product. In doing so, it is helpful to view these sources in a manner similar to that shown in Fig. 8.1.

†This breakdown of process variability was suggested by the procedures described in Leonard A. Seder and David Cowan, "The Span Plan Method, Process Capability Analysis," Pub. No. 3, American Society for Quality Control, Milwaukee, Wis., 1956.

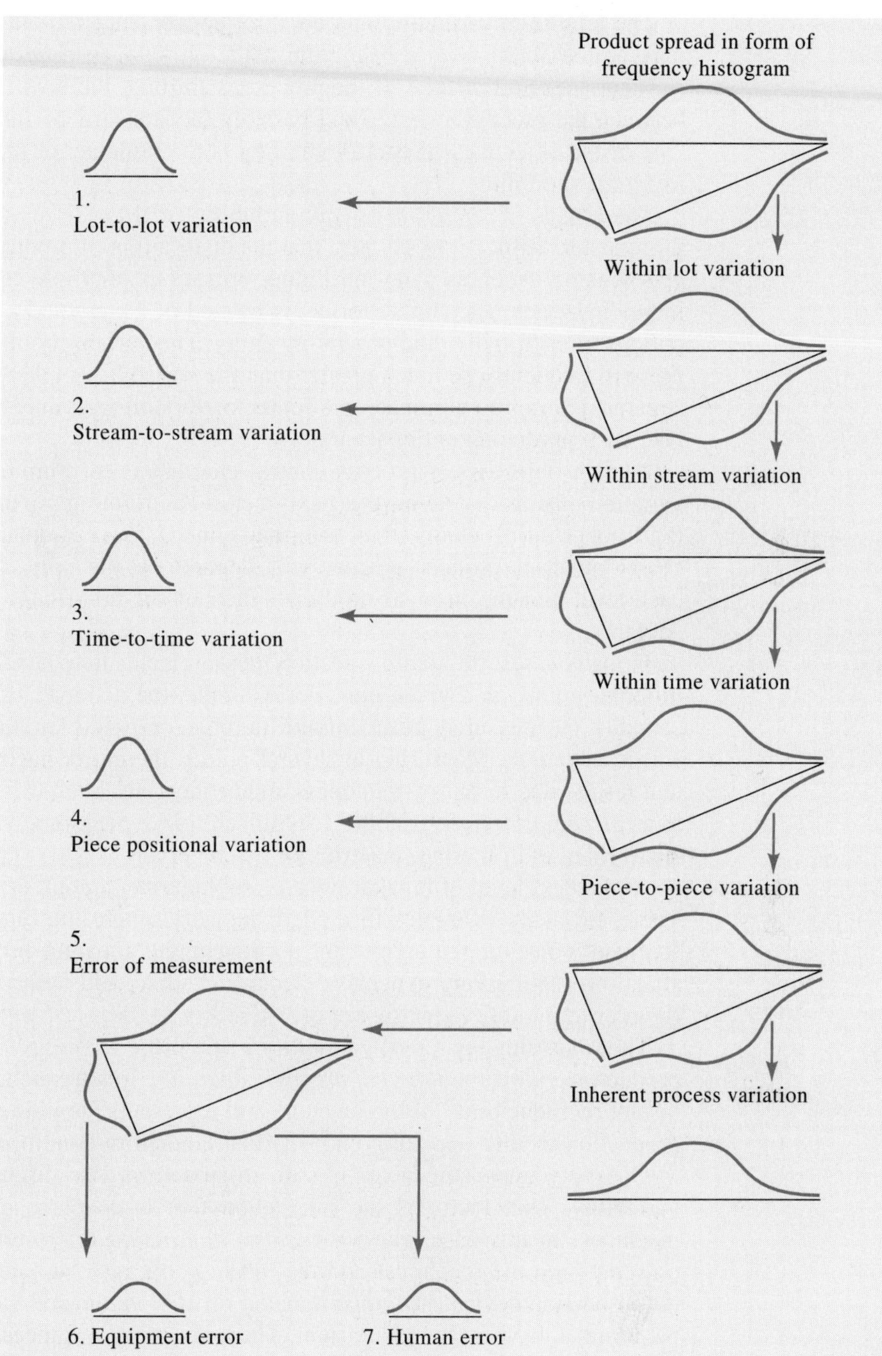

FIGURE 8.1 Diagram of breakdown of sources of variability in manufactured product. *(This figure is suggested by and similar to fig. 6 in L. A. Seder and D. Cowan, "The Span Plan Method, Process Capability Analysis," Pub. No. 3, American Society for Quality Control, Milwaukee, Wis., 1956.)*

The long-term variation in product, for convenience termed the *product spread,* may be measured from a histogram made up from inspection data taken over a substantial period of time. If the process is shifting, there will be some difference between the process average, and possibly the standard deviation, from lot to lot. One of the objectives of control charting is to eliminate, or markedly reduce, this *lot-to-lot* variability.

The effects of *stream-to-stream* variability will be discussed at length later in this chapter. Suffice it to say here that the distribution of product flowing from several streams—such as gang-machining operations, multiple-cavity molds, several individual machines, and so forth—is formed of a weighted average of the distributions of each individual producing entity. The variability of this weighted average will frequently be much greater than the variabilities inherent in the individual streams. In order to eliminate this source of variability, it may be necessary to analyze each producing entity separately.

A main objective of process control charting is to minimize the *time-to-time* variation that is listed as the next factor contributing to product spread. The types of control charts that are maintained, for example separate $\bar{X}$ and R charts on each product stream, will depend on the spread of the process at each level and the need to minimize each of the contributing factors discussed so far.

In many cases, physical inspection measurements may be taken at a great many different points on a given unit. For example, the diameter of a shaft might conceivably be measured at an infinite number of points around its diameter and across its length. Measuring at several points, therefore, need not lead to consistent results due to out-of-roundness and/or tapering. Such differences are referred to as *piece positional* variability. Significant piece positional variation may necessitate changes in tooling, material, or machinery.

At the next level of breakdown are problems of *inherent error of measurement.* There are many examples in modern industry where the inherent error of measurement constitutes a significant portion of the apparent product spread. Many situations require very expensive electronic, sonic, and optical measuring devices in order to minimize this source of variability.

The remaining source of variability is the piece-to-piece variability of a single production entity, the *inherent process capability.* It is frequently referred to as the instant reproducibility of the machine and represents the ultimate capability of the production facility operating under virtual laboratory conditions.

One very important factor is still missing from this discussion of sources of variability. That factor is the interaction that takes place between people and machines. In this case, *person-machine interaction* refers to the interaction not only between the machinist and the machine but also between the inspector and the inspection device. In semiautomated or fully automated cases, this effect may be minimal or virtually nonexistent. There are many instances, however, where to ignore the person in the system would be to ignore the largest single source of variability.

8.5 ORDER OF PRODUCTION IS NOT ALWAYS A SUFFICIENT BASIS FOR SUBGROUPING

The reason why order in time is a good basis of subgrouping is that its use tends to disclose assignable causes of variation that come and go. However, there may be other assignable causes of variation that are not disclosed merely by taking subgroups in the order of production.

Two or more apparently identical machines may have different process averages, different process dispersions, or other differences in their patterns of variation. If these machines contribute to a stream of product in a way that subgroups taken from that stream contain approximately constant numbers from each machine, the differences among the machines will not be disclosed by the control charts. The principle here is that assignable causes, if they are to be indicated by the charts, must influence some but not all of the subgroups.

For this reason, consideration often needs to be given to the question of different subgroups for different machines each doing the same operation, or for different spindles on the same machine, or for different cavities in a mold, or for different operators or different inspectors or different shifts. In some cases separate control charts may be needed rather than merely separate subgroups.

The extent to which it pays to make this type of breakdown is a matter for judgment in each individual case. The decision depends on whether it is normally difficult or easy to meet specified tolerances, on the costs of keeping and analyzing the control charts, on whether it is practicable or economical to correct certain known assignable causes of variation, and on other matters that vary from case to case. One case in which it did pay to make this breakdown is described in Example 8.1.

Example 8.1
Setting of thermostatic controls. Compensation for the differences between operators

Facts of the case Thermostatic controls for an electrical device were all adjusted on two banks of units that soak the thermostats at a given temperature level. After soaking for a few minutes, each thermostat was adjusted by an operator until a light flashed on the adjusting unit; a locknut was subsequently put on the control to hold this adjustment. Each bank of units required its own operator. These controls were produced on two shifts. Thus four operators used two banks of units.

The specifications stated the temperatures at which the thermostatically controlled switches should turn the electrical device on and off and gave tolerances for these temperatures. Each finished device was checked by a testing set operating on the go and not-go principle to determine whether these tolerances were met. Whenever an out-of-tolerance thermostat was found on this final inspection, it had to be removed from the completed device, reset, and reassembled into the device.

Several weeks might elapse between the original adjustment of a thermostatic control and its final assembly into a completed device. Hence it was not economical to depend only on the 100% inspection at final assembly as a check on the thermostat setting; any continued systematic error in setting could be responsible for many defective thermostats before it was detected at final assembly and hence could cause much costly rework.

For this reason, samples of the thermostatic controls were taken immediately after setting and checked on a test panel that permitted the measurement of actual on and off temperatures. Control charts for $\bar{X}$ and R were plotted. At first the scheme of subgrouping was to take five thermostats that had just been set; a subgroup would generally contain some thermostats adjusted by each of the two operators on the current shift.

This proved successful in detecting trouble from time to time and in obtaining prompt correction of the trouble. Usually the assignable causes were of a type that could be corrected by maintenance work on one of the adjusting banks.

Analysis and action However, even when the process stayed in control, some of the thermostats were outside specified tolerances. With σ estimated from the $\bar{R}$ of the control charts, the tolerance spread ($USL_x - LSL_{\bar{x}}$) appeared to be about 5σ.

It was decided to use a plan of subgrouping by operators and by adjusting banks; all the thermostats in any subgroup came from one operator and one bank. This disclosed the fact that on each shift the thermostats from one operator showed a consistently higher average on-and-off temperature than those from the other operator. By shifting operators from one bank to the other it was determined that this was a personal difference between operators and not a difference between banks. It was evidently a difference in reaction time to the flashing of the light in the adjusting unit.

The temperature levels in the adjusting banks were then established in a way that allowed the difference in temperature level to compensate for the difference between the "hot" and "cold" operators. The scheme of subgrouping by operators reduced $\bar{R}$ and σ. The change in temperature level of the banks to compensate for differences between operators tended to keep the same process average for all operators and thus kept the process in control with the narrower control limits. As the new σ was about five-sixths the previous σ, the tolerance spread ($USL_x - LSL_x$) was now six times the new σ. With careful attention to routine use of the control charts to prevent shifts in the process average, it was now possible to make all the thermostats within tolerances and to avoid rework costs at final assembly.

Comment on Example 8.1

In this example, order in time was a necessary but not a sufficient basis for subgrouping. After it proved possible to compensate for differences in operators, a single chart was maintained but the data were stratified with each subgroup representing a single bank and one operator. Separate control charts were not maintained on each machine-operator combination. If the tolerances had not been so tight compared to the dispersion of the process, it might not have been necessary to go to the extra trouble of keeping subgroups by operators.

8.6 NEED FOR DISCRIMINATION IN THE SELECTION OF SUBGROUPS

Where there is trouble in meeting tolerances and control is shown by a control chart based merely on the order of production, it may still be possible to diagnose and correct trouble by changing the basis of subgrouping in an effort to stratify the data or by keeping separate charts for different sources of measurements, such as different machines, different spindles on the same machine, or different operators. But because such breakdowns usually increase the costs of taking and analyzing data, it may pay to avoid them in cases where tolerances are easily held most of the time.

Moreover, there is no point in continuing to take subgroups in a way that has the effect of disclosing assignable causes of variation that it is impracticable or uneconomical to remove. The need for discrimination in selection of subgroups is illustrated in Example 8.2.

Example 8.2

Thickness of pads on half-ring engine mount. An example in which subgroups were taken in a way that made evidence of lack of control have no practical value

Facts of the case Figure 8.2 shows a rough sketch of a half-ring that was part of an assembly used in connection with the mounting of an airplane engine. This half-ring contained four contact pads. It was desired to control the thick-

FIGURE 8.2 Half-ring engine mount.

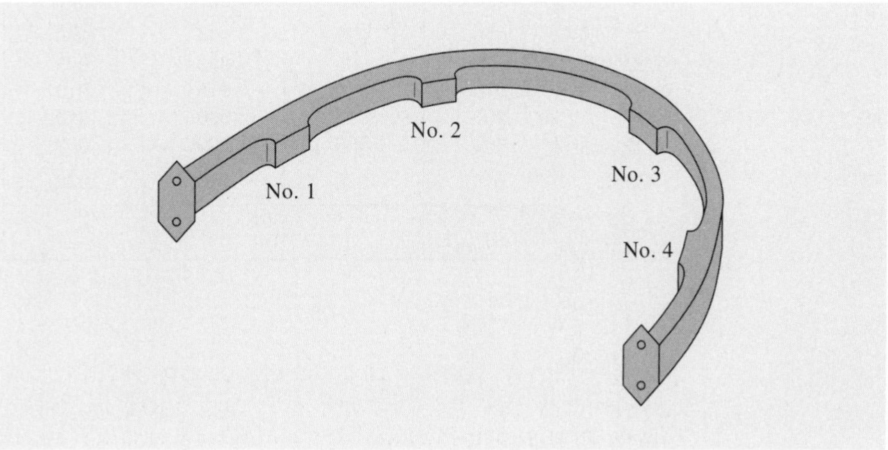

TABLE 8.1 THICKNESS OF EACH PAD ON HALF-RING ENGINE MOUNT
Measurements in units of 0.0001 in above 1.5000 in

Half-ring number	Pad 1	Pad 2	Pad 3	Pad 4	$\overline{X}$	R
1	933	937	938	935	936	5
2	897	898	915	913	906	18
3	840	900	900	930	892	90
4	900	905	902	900	902	5
5	879	852	873	871	869	27
6	903	890	892	908	898	18
7	930	940	930	920	930	20
8	890	895	897	895	894	7
9	890	900	850	900	885	50
10	900	915	900	905	905	15
11	901	916	901	900	904	16
12	920	890	905	895	902	30
13	920	890	910	880	900	40
14	929	921	924	928	925	8
15	927	914	925	931	924	17
16	907	896	895	908	902	13
17	902	900	903	905	902	5
18	903	900	914	900	904	14
19	870	930	920	920	910	60
20	925	930	920	930	926	10
21	880	895	910	885	892	30
22	890	900	895	895	895	10
23	940	935	930	940	936	10
24	930	935	938	930	933	8
25	915	921	918	927	920	12
26	895	930	925	925	919	35
27	910	907	905	913	909	8
28	905	916	902	928	913	26
29	925	930	910	925	922	20
30	924	928	882	927	915	46
31	925	931	924	930	928	7
32	900	905	925	925	914	25
33	910	910	915	910	911	5
34	900	905	900	910	904	10
35	900	950	920	900	918	50
36	940	938	940	938	939	2
Total	. . .	. . .	. . .	. . .	32,781	774

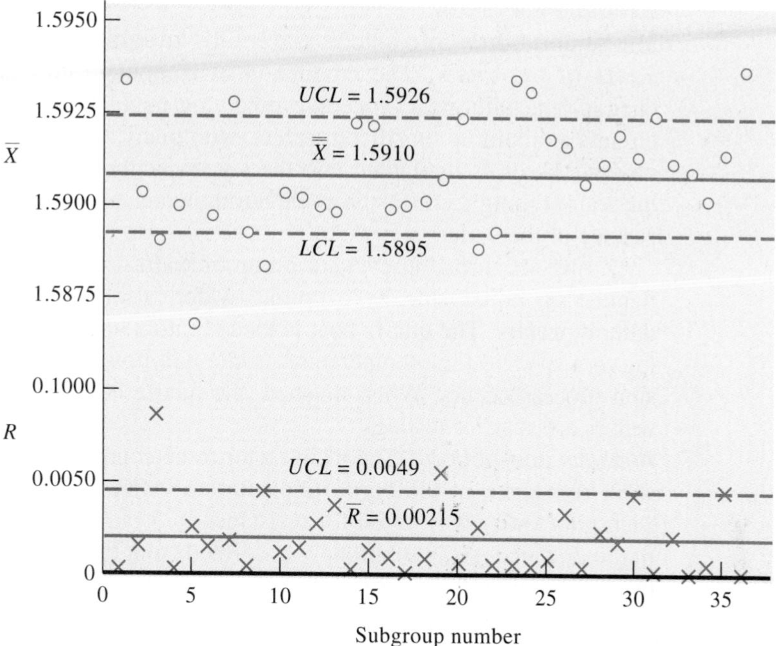

FIGURE 8.3 $\overline{X}$ and R charts for thickness of pads on half-ring engine mount—data of Table 8.1.

ness of these pads. All four pads were machined at the same time and were supposed to have the same thickness.

The thickness of each pad was measured with a micrometer to the nearest ten-thousandth of an inch. The four pads on one half-ring were considered to be one subgroup. The measurements for 36 half-rings are shown in Table 8.1. The $\overline{X}$ and R charts plotted from these measurements are shown in Fig. 8.3.

These charts appeared to show the process badly out of control. However, a more critical consideration of the matter showed that it was practically certain this type of subgrouping would show lack of control, and that this showing of lack of control did not provide a useful guide to action.

Control charts answer the question, "Is the variation among the subgroups consistent with the variation within the subgroups?" With subgroups taken as they were, this question might have been phrased, "Is the variation among the half-rings consistent with the variation from pad to pad within each half-ring?" The answer, "No," given by the control chart might have been guessed without help from any control charts. That is, it was to be expected from the fact that the four pads on any half-ring were machined together; it was reasonable to expect more variation among the half-rings than among the pads. Moreover, it was no help to know that this was true, as nothing could be done about it by attention to out-of-control points.

Example 8.3
Implanting boron into silicon wafers for integrated-circuit chips

Facts of the case The production of computer chips is a highly complex chemical-metallurgical procedure involving many steps and processes. Microcircuits are built up on silicon wafers, with many circuits on each wafer. In one process, boron is implanted into the silicon wafer and then driven deeper into the wafer through a diffusion cycle during which an oxide layer is grown on the surface of the wafer.

During the diffusion cycle, a group of wafers are loaded into a quartz boat flanked on either side by a "pilot" wafer, also implanted with boron, and dummy wafers. The quartz boat is loaded into a tubular furnace where, according to a specific time-temperature profile and flow of reactant gases, the diffusion process occurs. When finished, the quartz boat is removed and the pilot wafers are sent for testing.

Analysis and action The main factor to be controlled in this process was the sheet resistance of the boron implantation. After stripping the oxide coating, each pilot wafer was tested for resistance in 5 places using a four-point probe. In the first analysis, a subgroup was formed from the 10 measurements from the two pilots taken from a batch run. $\bar{X}$ and R charts indicated a definite lack of control, with many points out on the $\bar{X}$ chart and some out on the R chart.

Since the data were organized in such a way as to allow them to be separated by location of each pilot in the quartz boat, separate $\bar{X}$ and R charts using a subgroup size of 5 were prepared for each pilot. These charts still showed a substantial lack of control on the $\bar{X}$ charts, but the situation improved on the R charts. Also there appeared to be some difference between the performance of the two pilot locations.

Finally, the average of the 5 measurements taken from each pilot was used as a single data point, and two pilots from successive runs were used to form a subgroup. Moving average and moving range charts, discussed in Chap. 10, were used. The $\bar{X}$ and R charts for both pilots indicated control, and since the measured resistances fell comfortably within specifications, this method of subgrouping was continued. So long as control could be maintained at the nominal resistance value, satisfactory product would result. Occasionally a point would go out on either the $\bar{X}$ or the R chart. In each case, an assignable cause was found and some adjustment made either in the implantation process or in the diffusion process. Since the chart results for both pilots tended to give the same indications, it was decided that control charts could be run on only one pilot with occasional checks of the data of the other.

Comment on Examples 8.2 and 8.3

In both of these cases, what might have appeared to be a natural plan of subgrouping was, in fact, an unsatisfactory plan. The initial plans called for a subgroup to comprise the four pads on a single half-ring in Example 8.2 and the five locations on each of two pilot wafers in Example 8.3.

Pad thicknesses at the four locations on a half-ring tended to vary in unison, with an increase or decrease at one location accompanied by equal increases or decreases at the other locations. Thus a single chart for one location, with a subgroup formed from three to five successive half-rings, would have served to control the general thickness.

In the case of the pilot silicon wafers, differences in resistance across the surface of a wafer were minimal, and while there were positional differences between the locations of the two pilot wafers, they were too small to result in significant numbers of product units beyond specifications. Major differences appeared between batch runs of the wafers. Thus a single chart for one pilot wafer position in a quartz boat, and two to five quartz boats (wafer batch runs) forming a subgroup, might well have been used to control resistance.

Electrical resistance could not be checked after the initial implantation process and before the diffusion process. Since each of these processes involves a number of steps, and therefore possibilities for introducing variation, this provides an excellent opportunity for experimentation to determine the optimal combination of factors to yield the correct level of performance with minimum variability.

In the situation described in Example 8.2, a slight change in tooling resulted in a reduction in the variation of pad thickness from pad to pad within a half-ring. It then proved satisfactory to use a single control chart on which the variable was thickness of pad 2, and a subgroup consisted of this measured thickness from four successive half-rings. Had the pad thickness not varied in unison within a half-ring, four control charts, one for each pad, might well have been used.

Many situations exist like Examples 8.2 and 8.3. In such situations the decision regarding the number of control charts to be used cannot be made without studying the behavior of the variables and having a clear idea of the objectives of the charts. There was no feasible way that the variation in resistance from batch run to batch run could be made consistent with the variation across the surface (the five measurements) of a single pilot. Thus, as one engineer stated, "The $\bar{X}$ chart always shows lack of control. It's to be expected in this business." The choice of subgrouping made the control charts meaningless and tended to discredit the entire procedure. Once subgrouping was changed to include the results of two successive batch runs, a state of control could be established and maintained with virtually all product within limits. When a point went out of control, a basis for action was established that led to meaningful corrections.

If the variations from one location to another on either a half-ring or a silicon wafer had been completely unrelated, it is clear that one chart could not have done the job. Actual situations nearly always fall between these two extremes. Whether one chart or several are required depends on the degree of relationship (in statistical language, the correlation) between the fluctuations at the several points and the tightness of the specifications.

It may be remarked that it is always an advantage if one control chart can be made to do the work of several. Not only is there a saving in the cost of clerical labor in computing and charting, but the practical difficulties in having charts analyzed and in securing action based on that analysis also are reduced. It is easier to get people to study one chart and take action on it than to get them to study four related charts and take action on them.

Sometimes in this type of situation a practical answer may be to take enough data to permit the several control charts to be made, but actually to keep only one chart as a routine matter. Then, if at any time the situation calls for critical study, the data will be available for the construction of the other charts for purposes of analysis.

8.7 IDENTIFICATION ON A CONTROL CHART OF DIFFERENT SOURCES OF SUBGROUPS

In Example 8.1, subgrouping was finally by operators and by banks of adjusting units. Four sources of subgroups were possible, namely, operator *A* and bank 1, operator *B* and bank 2, operator *C* and bank 1, and operator *D* and bank 2. However, subgroups for all four combinations were plotted on the same control chart. It is common for different subgroups to come from different shifts, and it is not unusual for them to come from different machines or different operators.

In all such cases it is desirable to differentiate the various subgroup sources on the control charts, so that any consistent differences may be readily observed by someone looking at the charts. This may be done by the use of different types of symbols (such as circles, dots, and crosses) or different colors to represent each source of subgroups.

8.8 PRECISION, REPRODUCIBILITY, AND ACCURACY OF METHODS OF MEASUREMENT†

The only way in which the value of any quality characteristic—dimensions, hardness, tensile strength, weight, percentage of chemical impurity, and so on—may be determined is by some form of measurement. If measuring devices and those who used them were perfect, it would be possible to make a direct determination of the variability of the true values of the measured quality characteristic. Actually, the measured values reflect errors of measurement as well as variation in the quantity measured.

The phrase *method of measurement* may be thought of as including not only the measuring devices and the procedures specified for their use but also their manipulation by the particular user or users. Any method of measurement of an industrial quality characteristic may be expected to have some pattern of variability. Under favorable conditions this pattern may be observed by repeating a measurement many times on a quality characteristic that remains unchanged. For instance, Fig. 8.4 shows a frequency distribution obtained from 100 micrometer measurements all on the thickness of the same steel strip. Different methods of measurement will have different patterns of variation.

†The meanings of these three terms are not standardized among writers on errors of measurement. Regardless of the terminology used, it is important to distinguish among these three concepts.

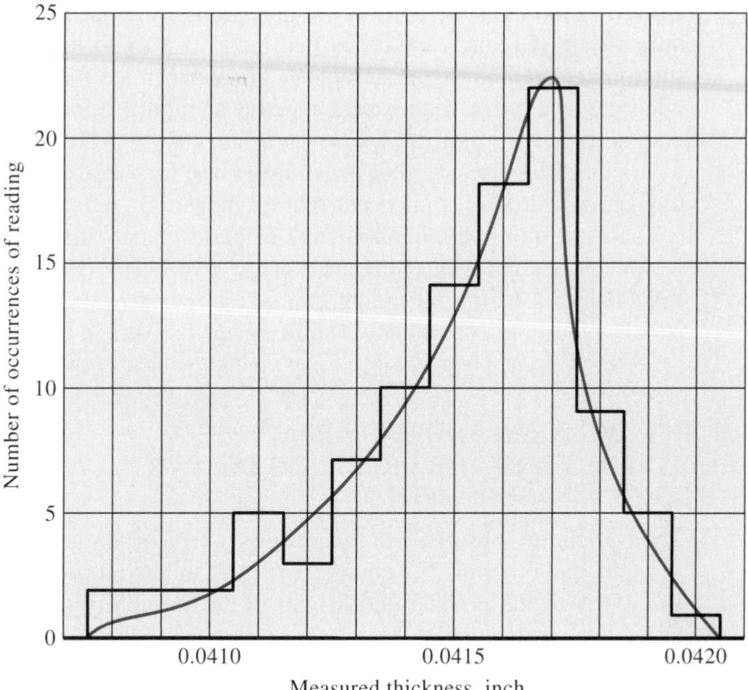

FIGURE 8.4 Frequency distribution of 100 micrometer measurements of the same dimension made by the same experienced inspector. *(Reproduced from "Dimensional Control," published by Sheffield Corporation, Dayton, Ohio, 1942.)*

The *precision* of a method of measurement refers to its variability when used to make repeated measurements under carefully controlled conditions. Where practicable, as in the case referred to in Fig. 8.4, these repeated measurements should be made on a single quality characteristic of one particular article. Otherwise, as in chemical tests or other destructive tests, they should be made on a homogeneous sample. A numerical measure of precision is the standard deviation of the frequency distribution that would be obtained from such repeated measurements. This may be referred to as $\sigma_{\text{error of measurement}}$.

The *reproducibility* of a method of measurement refers to the consistency of its pattern of variation. This may be judged to best advantage by control charts for $\overline{X}$ and R or $\overline{X}$ and s. Where repeated measurements show erratic patterns of variation, the method of measurement used is not reproducible. Any statement regarding the precision of a method of measurement implies that the method of measurement is reproducible.

The *accuracy* of a method of measurement refers to its absence of bias—to the conformity of its results to the true value of the quality characteristic being measured. In an accurate method, the average value obtained from a set of measurements should differ from the true value by not more than would be expected as a chance variation in the light of the precision of the particular method. A practical difficulty in judging accuracy is that the only way to find the "true" value is by some other method of measurement. Presumably the method employed to determine the true value should be a method of high precision that is believed to be without bias.

In our initial discussion of the control chart for variables as applied to an industrial process, it was pointed out that an industrial quality characteristic will have (1) basic variability; (2) consistency of performance, that is, presence or absence of statistical control; and (3) an average level. This is also true of a method of measurement. In the foregoing discussion, *precision* refers to basic variability, *reproducibility* to consistency, and *accuracy* to average level.

8.9 RELATIONSHIP BETWEEN THE VARIABILITY OF MEASURED VALUES AND THE PRECISION OF THE METHOD OF MEASUREMENT

The variability observed in measured values of dimensions, hardness, tensile strength, and other quality characteristics of industrial product is due in part to the variability of the product and in part to the variability inherent in the method of measurement. It is helpful to think of a measured value as the sum of two variables, the quantity measured and the error of measurement.

As these two variables are likely to be independent of one another as indicated in Fig. 8.1, the formula for the standard deviation of the sum of two independent variables may be used to advantage. For this purpose it might be written

$$\sigma_{\text{measured value}} = \sqrt{\sigma^2_{\text{true value}} + \sigma^2_{\text{error of measurement}}}$$

This formula always supplies a useful point of view toward any problem involving errors of measurement. Its numerical use in solving practical problems depends on obtaining a reliable estimate of the magnitude of the standard deviation of chance errors of measurement.

One possible numerical application of the formula for standard deviation of the sum of two variables is in the determination of tolerance limits that are intended to allow for errors of measurement as well as for fluctuations in the true value of a quality characteristic. Another is in judging the ratio of the variability of true values to the variability of measured values. This latter application is illustrated in Example 8.4.

Example 8.4
Influence of chance errors of measurement on variability of measured values

Facts of the case In a chemical plant, the method employed for laboratory analysis of the nonvolatile content of a certain product was not the most precise one possible. It was a rapid and economical method that was considered satisfactory by the laboratory director. Nevertheless, whenever a laboratory analysis showed the product to be outside specification limits, the production department was inclined to place the blame on the lack of precision of the laboratory methods rather than on any fault in the manufacturing operations. For this reason, a change to a more precise method was under consideration.

A determination of the precision of the analytical procedure indicated that the standard deviation of the error of measurement was 3.3 units. A control chart on the measured value for the manufactured product supplied an estimate that when the process remained in control, σ was 9.5 units. These values were substituted in the formula given above to estimate the σ of the true value, as follows:

$$9.5 = \sqrt{\sigma^2_{\text{true value}} + (3.3)^2}$$

$$\sigma_{\text{true value}} = \sqrt{79.36} = 8.9$$

This indicated that the chance errors inherent in this analytical procedure caused the measured values to be about 7% more variable than they would have been if the analytical procedures had always given perfect results. The adoption of a more precise analytical procedure with a standard deviation of two-thirds the existing one would have reduced the variability of the measured values by about 3%, as shown by the following calculation:

$$\sigma_{\text{measured value}} = \sqrt{(8.9)^2 + (2.2)^2} = 9.2$$

On the basis of this analysis, it was decided to continue with the less accurate and more economical analytical procedure.

8.9.1 Gage *R* and *R* Studies

Those whose work is devoted to metrology and gage and instrument calibration have devised special procedures for dealing with error of measurement. One primary concern is where more than one person is using a measurement system. In Example 8.4, it was assumed that the inspection system operated consistently. Thus the analysis was focused on determining if it was economical to change the measurement system. In many cases, inspections are performed by a number of individuals, and differences among them are an issue of importance. *Gage R and R studies* may be designed to illuminate these differences.

While this division is not accepted universally in quality assurance literature, the study approach divides what we have defined as measurement reproducibility into two portions. The *reproducibility* portion of error of measurement derives from differences among users of the gage or system. The other portion, called *repeatability,* derives from the characteristics of the gage itself averaged over the users. We can formulate this division as follows:

$$\sigma^2_{\text{error of measurement}} = \sigma^2_{\text{repeatability}} + \sigma^2_{\text{reproducibility}}$$

TABLE 8.2 ANALYSIS MODEL FOR A GAGE *R* AND *R* STUDY, THREE INSPECTORS, FIVE SELECTED PARTS, TWO READINGS

	Inspector A			Inspector B			Inspector C		
Part	Readings	$\bar{X}$	R	Readings	$\bar{X}$	R	Readings	$\bar{X}$	R
1	0.20 0.10	0.15	0.10	0.10 0.10	0.10	0.00	0.15 0.15	0.15	0.00
2	0.20 0.30	0.25	0.10	0.15 0.25	0.20	0.10	0.25 0.25	0.25	0.00
3	0.15 0.10	0.12	0.05	0.20 0.10	0.15	0.10	0.20 0.15	0.18	0.05
4	0.30 0.30	0.30	0.00	0.25 0.30	0.28	0.05	0.20 0.25	0.22	0.05
5	0.20 0.15	0.18	0.05	0.15 0.05	0.10	0.10	0.15 0.05	0.10	0.10

$\bar{\bar{X}}_a = 0.200 \qquad \bar{\bar{X}}_b = 0.166 \qquad \bar{\bar{X}}_c = 0.180$

$\bar{R}_a = 0.060 \qquad \bar{R}_b = 0.070 \qquad \bar{R}_c = 0.040$

$$\sigma_{\text{repeat}} = 1/d_2[1/3\ (\bar{R}_a + \bar{R}_b + \bar{R}_c)]$$
$$= 1/1.128[1/3(0.060 + 0.070 + 0.040)] = 0.050$$

$$\sigma_{\text{reproduce}} = 1/d_2[\bar{\bar{X}}_{\max} - \bar{\bar{X}}_{\min}]$$
$$= 1/1.693[0.200 - 0.166] = 0.020^{\dagger}$$

The implication is that multiple measurements are to be taken by each user on a common set of parts, perhaps by recycling them two or three times for inspection by each. The analyst evaluates the results following a specific pattern that is illustrated in Table 8.2 for three inspectors/operators and five parts.

The data of Table 8.2 are not taken from a real process and are intended only to illustrate the analytical procedure and the steps that might be taken in setting up the measurement system testing procedure and carrying out an analysis. Needless to say, the reproducibility portion of the measurement variability can be substantial when persons of differing training and experience are using the system. Example 8.5 discusses a real situation involving substantial inspector differences in the framework of an analysis using attributes data rather than variables.

The combined measure of error for this example is

$$\sigma_{\text{error of measurement}} = \sqrt{(0.050^2 + 0.020^2)} = \sqrt{0.0029} = 0.054$$

While many studies of the type illustrated are made using only a limited number of measurements, it is preferable that sufficient data be obtained so that there is clear evidence of differences between measurement equipment and operator/inspectors. As the pattern of analysis indicates, there is considerable averaging going on, which can either hide real differences or suggest that there are differences when only random process variation is operating.

† The value of d_2 used in this formula, 1.693, applies to a subgroup size of three, the three grand averages, and an infinite universe. Since there is only one subgroup, a larger value of d_2, 1.91, is correct. The matter of adjustment for small numbers of subgroups was discussed in Chap. 4, Sec. 4.2.7, and illustrated for a subgroup size of five in Table 4.3.

Example 8.5

Influence of inspector inconsistency on in-process lot acceptance

Facts of the case At one stage in the production of integrated-circuit chips, a sample of silicon wafers is drawn from each lot for visual inspection. Seventeen people divided among three shifts perform this inspection. Accepted lots proceed to the next operation; rejected lots are returned to the previous operation for rework. While each lot consists of one type of product, there are many products and a large variation among the lots in inspection difficulty and rework rate.

To obtain a feel for consistency among inspectors, each was assigned at random to lots to be inspected. In this way inspectors were not always looking at the same level of difficulty or product type. This method of assignment was designed to minimize the effect of differences among products. Thus any substantial differences among inspectors would become apparent.

Analysis and action Data for a p chart were recorded for each inspector on the number of silicon wafers inspected and the number rejected. The resulting chart showed wide variation with 12 of the 17 inspector points outside 3-sigma control limits. Thus it was obvious that either the 17 inspectors were producing extremely inconsistent results or there was great variation in the quality of the lots.

A second test was designed in which each inspector checked the same set of 20 silicon wafers. Reject rates varied from 0 to 18 with an average of 7.06. One point was out on the np chart, leading to the conclusion that substantial inspector differences were in fact present.

As a result of this study, operational definitions were developed for the various defect types and a notebook of examples detailing marginal cases was put together by a group made up of design and manufacturing engineers. These materials were used for inspector training and as a reference book to achieve the desired consistency.

Example 8.6

Estimating equipment error of measurement

Facts of the case A critical dimension on a hemispherical part is its polar height. The measurement system involves an experienced inspector setting each part on a surface plate and using a preset dial indicator gage to measure polar height. The gage is preset at 0.5 in and reads in thousandths of an inch above 0.5 in.

Analysis Thirty parts produced in succession were taken from the process and measured twice by the inspector. The results of these measurements are given in Table 8.3. Only the range chart was prepared for these data pairs. The resulting R chart showed control with an $\overline{R}$ of 0.29 and a UCL of 0.95. Thus the error of measurement σ estimated from $\overline{R}/d_2$ was 0.26.

Next, the first readings for each part were divided into subgroups of three parts, and $\overline{X}$ and R charts were prepared. The data for the 10 subgroups are shown in Table 8.4.

TABLE 8.3 POLAR HEIGHT READINGS OF HEMISPHERICAL PARTS
Measurements in thousandths of an inch above 0.500 in

Part no.	Reading 1	Reading 2	$\bar{X}$	R	Part no.	Reading 1	Reading 2	$\bar{X}$	R
1	5.1	5.6	5.35	0.5	16	6.5	7.1	6.80	0.6
2	4.0	3.9	3.95	0.1	17	6.2	5.9	6.05	0.3
3	4.6	4.5	4.55	0.1	18	3.8	4.5	4.15	0.7
4	2.8	3.3	3.05	0.5	19	3.4	3.6	3.50	0.2
5	7.6	7.8	7.70	0.2	20	4.6	4.7	4.65	0.1
6	5.8	5.5	5.65	0.3	21	6.2	5.6	5.90	0.6
7	4.4	4.9	4.65	0.5	22	5.8	5.6	5.70	0.2
8	5.3	5.1	5.20	0.2	23	3.0	2.6	2.80	0.4
9	2.7	2.7	2.70	0.0	24	7.4	7.3	7.35	0.1
10	5.8	5.4	5.60	0.4	25	3.3	3.6	3.45	0.3
11	4.3	4.4	4.35	0.1	26	3.3	3.1	3.20	0.2
12	5.2	5.1	5.15	0.1	27	4.1	3.8	3.95	0.3
13	5.0	5.1	5.05	0.1	28	3.1	3.1	3.10	0.0
14	5.1	4.8	4.95	0.3	29	7.4	7.1	7.25	0.3
15	5.5	5.6	5.55	0.1	30	3.8	2.9	3.35	0.9
					Totals	. . .	. . .	144.65	8.7

TABLE 8.4 $\bar{X}$ AND R CHART DATA FOR FIRST READINGS FROM DATA OF EXAMPLE 8.6
Subgroup size 3

Subgroup	$\bar{X}$	R	Subgroup	$\bar{X}$	R
1	4.57	1.1	6	5.50	2.7
2	5.40	4.8	7	4.73	2.8
3	4.13	2.6	8	5.40	4.4
4	5.10	1.5	9	3.57	0.8
5	5.20	0.5	10	4.77	4.3
			Totals	48.37	25.5

The grand average $\bar{\bar{X}}$ is 4.84, and 3-sigma limits for the $\bar{X}$ chart are $UCL = 7.44$ and $LCL = 2.24$. $\bar{R}$ is 2.55, and control limits for the R chart are $UCL = 5.75$ and $LCL = 0$. Since there is no evidence of lack of control, the estimate of σ from the measured values is $\bar{R}/d_2 = 2.55/1.693 = 1.51$.

Given this information, the σ of the true value may be found from

$$\sigma_{\text{true}} = \sqrt{(1.51)^2 - (0.26)^2} = \sqrt{2.213} = 1.49$$

and the error of measurement is $100(0.26/1.51) = 17.3\%$ of the variability of the measured value.

Comments on Examples 8.4, 8.5, and 8.6

The three examples change the focus and context of the analysis from that aimed at control of a process to analyzing the results of tests in the light of limitations on measurement processes. As Example 8.4 illustrates, questions arise frequently about the accuracy of measurement processes and procedures. Precision always is demanded by production personnel in process control situations. The relative precision of different measurement systems and their differing costs are important considerations when recommendations about new equipment must be made to management.

Example 8.5 focused attention on potential variation among people. Statistical process control charts, p and np charts, were used to analyze differences among inspectors. Whenever an inspection process is heavily dependent on the skill and training of people, as was the case of the visual inspection of the silicon wafers, excellent training and continual reevaluation of performance and retraining are important actions. Many situations in service industries and other nonmanufacturing processes are very much like that described in Example 8.5. It should be noted that, when it was decided to limit the influence of causes of variation, the rationality of subgrouping changed completely. The second method of subgrouping was designed for the single purpose of assessing inspection consistency in preparation for developing new inspection aids and training practices.

Example 8.6 stresses the differences in control-chart procedures when the purpose of data collection and analysis change. No $\bar{X}$ chart was plotted for the data of Table 8.2, only the R chart. The reason should be obvious; the R chart was measuring the variation in the measurement system, the combination of person and instrument. Had an $\bar{X}$ chart been plotted, it would have shown total lack of control because it included part-to-part variation as well. It would have said that "the dimension being measured is completely inconsistent with the variability of the measurement system." Had that not been the case, the measurement system would have been useless, and the inspection process would have been a waste of time and effort.

The results in Example 8.6 indicate that apparent total variability was increased by nearly 20% because of lack of precision of the measuring system. The experiment was not designed to isolate human error from equipment error or to compare inspectors who might be using the instrument, thus only the combination or system error could be evaluated. On the assumption that most of the measurement error was due to poor equipment precision and that the dimension being checked was truly critical to the functioning of the part, it would seem worthwhile to investigate alternative methods of testing.

PROBLEMS

8.1. In Example 8.1, why should $\bar{R}$ have been decreased by the change in the method of subgrouping? With the subgrouping by operators but with no change in the temperature levels of the adjusting banks to compensate for the differences among operators, what would have been the appearance of the $\bar{X}$ and R charts?

With the original plan of subgrouping, how was it that the differences between operators did not cause points to go out of control on the $\overline{X}$ chart?

8.2. In Example 8.2 someone made the suggestion that the variable ($\overline{X}$) used for the control charts should be the average of the four pad thicknesses on a half-ring, and that a subgroup should consist of four such averages. What objection can you see to this suggestion?

8.3. Prepare $\overline{X}$ and R control charts for the thickness of pad 2 in Table 8.1, Example 8.2. Use a subgroup size of 3. This will give you 12 subgroups. Discuss the difference between the results you obtain and those indicated in Fig. 8.3.
Answer: 880.7, 912.64, 944.6; 0, 31.3, 80.5.

8.4. Prepare $\overline{X}$ and R control charts for the thickness of pad 3 in Table 8.1, Example 8.2. Use a subgroup size of 3. This will give you 12 subgroups. Discuss the difference between the results you obtain and those indicated in Fig. 8.3.

8.5. Prepare $\overline{X}$ and R control charts for the thickness of pad 4 in Table 8.1. Example 8.2. Use a subgroup size of 3. This will give you 12 subgroups. Discuss the difference between the results you obtain and those indicated in Fig. 8.3.

8.6. One convenient method for plotting data such as that of Table 8.1, Example 8.2, is on what has been termed a group control chart. The technique is valid only when there are no significant differences among the σ's of the producing units (measured in pad thicknesses, in this case) that cannot be corrected. $\overline{X}$ and R for each subgroup for each source are calculated, but it is only the highest and lowest values of $\overline{X}$ and R that are plotted. Usually, the high values will be connected by straight lines as will the low values, and the producing source that yielded the high and low values will be identified on the chart. Plot group control charts for $\overline{X}$ and R for the data of Table 8.1, using a subgroup size of 4. This will give you 9 subgroups for each pad.

8.7. In Probs. 2.10 and 4.40 you were asked to analyze the capability of a process to meet specifications. Assume that, after the initial analysis called for, you observe the process during the time that the next 10 sample subgroups are drawn. The parts are being made on three different semiautomatic machines. Finished parts move on a single conveyor to the inspection station where each unit is measured as it arrives. Five successive measurements compose a subgroup. At the suggestion of the supervisor, operators occasionally take a part and check it themselves, adjusting their machine as seems appropriate.

(*a*) What is your opinion of the estimates of μ and σ found in Prob. 2.10, assuming that these procedures were followed when the first 30 subgroups were formed?

(*b*) What suggestions would you make relative to the formation of subgroups and analysis of process capability?

(*c*) What suggestions would you make relative to manufacturing procedures?

8.8. The standard deviation of the measured values of a quality characteristic is 40.0 units. However, the standard deviation of the error of measurement of this characteristic has been determined to be 12.0 units.

(*a*) Estimate the value of the true σ of this quality characteristic.

(*b*) How much improvement in the measuring technique would be required to reduce the overall standard deviation to within 2% of the true standard deviation?

8.9. In the manufacture of glass bottles for carbonated beverages, glass thickness of the walls is a critical characteristic for protecting the process during filling and capping operations as well as for handling and consumer protection. Thickness is measured at four positions around the wall, and the average is calculated. Five such averages constitute a subgroup. A subgroup is drawn every half-hour. The data are used to calculate limits for $\bar{X}$ and R charts.

(*a*) After 25 subgroups, $\Sigma\bar{X} = 15.50$ cm and $\Sigma R = 2.28$ cm. Calculate control limits and central lines for control charts for $\bar{X}$ and R, and estimate σ.

(*b*) Results of plotting the limits found in part (*a*) on the charts indicate that the process is operating in control. Use the data to test whether or not the value of $\bar{\bar{X}}$ found in (*a*) would occur at random when the aimed-at value (target value) is 0.60 cm. Use a 2-sigma confidence region.

(*c*) Each data point was an average of four wall-thickness measurements. What weakness might this procedure have in light of the discussions of rational subgrouping?

8.10. Control charts for $\bar{X}$ and R have been used to monitor a process for a long time. The process is sampled in subgroups of 4 items at intervals of about 2 h. Control limits have been based on an aimed-at mean of 3.750 and an $\bar{R}$ of 0.1647.

(*a*) Estimate σ from this control chart data.

(*b*) If the standard deviation of error of measurement is known to be 0.015, what is the actual standard deviation of the process?

(*c*) Assuming the process output to be normally distributed, what actual percentage of product will not meet specifications of 3.810 ± 0.240?

(*d*) To minimize scrap loss, it is decided to adjust the process centering to exactly 3σ above the lower specification limit. Where should the process be centered?

8.11. As an alternative to the dial gage inspection system discussed in Example 8.6, a more sophisticated new computer gage may be used. The same 30 hemisphere parts were measured on this gage in an experimental procedure yielding the results given below. Readings are in thousandths of an inch above 0.500 in.

Part no.	Reading 1	Reading 2	Part no.	Reading 1	Reading 2	Part no.	Reading 1	Reading 2
1	4.2	4.1	11	4.0	4.0	21	5.7	5.6
2	5.1	5.1	12	5.0	5.0	22	6.2	6.1
3	4.8	4.8	13	4.1	4.2	23	4.1	3.9
4	3.3	3.3	14	5.7	5.5	24	6.8	6.9
5	6.4	6.4	15	4.3	4.2	25	3.1	3.0
6	5.0	4.8	16	6.0	5.8	26	2.5	2.5
7	4.0	4.0	17	5.7	5.7	27	5.0	4.9
8	6.2	6.1	18	4.2	4.3	28	3.5	3.3
9	3.5	3.5	19	2.9	3.0	29	8.3	8.1
10	4.9	4.9	20	5.0	5.2	30	5.0	4.9

Analyze these data following the pattern of the study discussed in Example 8.6. That is, prepare tables similar to Tables 8.3 and 8.4, and plot the indicated control charts.

8.12. The discussion of Example 8.6 stated that, when the two readings on each part are treated as a subgroup, an X chart for the paired readings is not appropriate. Prepare an $\overline{X}$ chart for the paired readings in Prob. 8.11, and discuss the interpretation of this chart. What does a lack-of-control condition mean in this circumstance, and how would this information be used by management?

8.13. Compare the results obtained in Example 8.6 to those obtained in Prob. 8.11. Discuss the results in terms of the tightness of specifications on polar height of the part. Under what circumstances would the less expensive and less precise measurement system be preferable?

8.14. For a certain sensitive inspection operation, a gage R and R study has been performed as part of a retraining exercise for three operators. Twenty parts, selected by the manufacturing engineers, were used. Each operator used the same measurement device to check each part. After one set of readings was taken, the parts were rearranged and submitted to the operators for reinspection. The 20 parts were coded so that the engineers could arrange the inspection data in paired readings for each operator. Results of the analysis gave the following measures for each operator.

Operator	1	2	3
$\overline{\overline{X}}_j$	223.0	222.8	226.0
$\overline{R}_j$	10.0	12.5	12.0

Calculate the standard deviations of repeatability, reproducibility, and total error of measurement.

8.15. Visit a manufacturing plant in which multiple machine or multiple spindle operations are performed. Discuss the operation with production and inspection personnel. Prepare a short report describing the operation, the critical measured characteristics, potential causes of nonconformance, and how you would proceed with a process capability study. Prepare to discuss your report in class.

9

STATISTICAL ANALYSIS OF PROCESS CAPABILITY AND FOR PROCESS IMPROVEMENT

The saving in inspection costs and personnel which often results from the introduction of quality control has been so extensively publicised that there has been a tendency to lose sight of the real object of the system, which is to improve quality, increase uniformity, reduce or prevent the production of scrap, and to provide a running commentary on the performance of machines and operators, invaluable for shop floor investigations and factory planning.[†]

9.1 PROCESS CAPABILITY AS A STEP TOWARD PROCESS IMPROVEMENT

How do we differentiate among statistical process control, process capability analysis, and process improvement activities? In reality, it cannot be done. Most of Part Two of this book has emphasized the use of statistical techniques to control ongoing production or service activities. Nevertheless, we have regularly compared 6σ of a controlled process with its specification range ($USL_X - LSL_X$) to determine whether or not a process was *capable* of meeting its specifications, when correctly centered. Furthermore, nearly all examples and problems have led to *improvement* in process performance even if the effort was not intended to produce that result.

When most analysts speak of process improvement, they are referring to reducing the dispersion of the process and/or making it more *robust,* that is, making the output less sensitive to variation in the values of input parameters, some of which

[†]"Quality Control on Hand-Operated Machines," *Production and Engineering Bulletin,* vol. 3, pp. 25–31, January 1944 (British Ministry of Labour and National Service and the Ministry of Production).

may be difficult to control. This chapter focuses on tools and procedures used to evaluate and report process capability and some of the procedures used to improve processes. Statistical design of experiments (DOE)—a principal tool, or kit of tools, for process improvement—will only be mentioned and described by a short example. Complete coverage of this complex subject is beyond the scope of this book.[†]

9.1.1 Objectives of an Analysis of Process Capability

The basic statistical problem in process quality control is that of establishing a state of control over the manufacturing process, that is, eliminating special causes of variation and then maintaining that state of control through time. This type of action is illustrated in Fig. 9.1*a*. Of no less importance is the problem of adjusting the process to the point where virtually all the product output meets specifications. This second problem, illustrated in Fig. 9.1*b*, is the realm of capability improvement. That is, once a state of control has been established, attention turns to the question, "Is the output meeting specifications, and if not, can the process be adjusted to a level where it will?"

Actions that result in a change or adjustment in a process, directed at eliminating common causes, are frequently the result of some form of capability study. The comparison of natural tolerance limits with specification limits and the natural tolerance range with the specification range may lead to any of the following possible courses of action:

1. *No action.* If the natural tolerance limits fall well within the specification limits, usually no action will be required. Frequently in such cases, control may be relaxed; a modified control chart, discussed in Chap. 10, or a p chart, the subject of Chap. 6, may be substituted for the conventional $\bar{X}$ and R charts.
2. *Action to adjust centering.* When the natural tolerance range is about the same as the specification range, a relatively simple adjustment to the centering of the process may be all that is necessary to bring virtually all product within specifications.
3. *Action to reduce variability.* This is usually the most complex action. In those cases where several product streams merge into one line prior to inspection, similar to examples discussed in this chapter, action may involve the relatively uncomplicated task of bringing the several streams under control separately at some standard $\bar{X}_0$. In other cases, a complex analysis of the sources of variation may be required resulting in changes of methods, tooling, materials, and/or equipment.
4. *Actions to change specifications.* This is a design decision but one that should not be ignored by quality control personnel. Simply because specifications are stipulated in writing does not necessarily mean they are inviolate. On the other

[†]The book *Statistics for Experimenters: An Introduction to Design, Data Analysis, and Model Building*, by G. E. P. Box, W. G. Hunter, and J. S. Hunter (John Wiley & Sons, Inc., New York, 1978) is an excellent treatise on the subject of experimental design and is quite readable for engineers and business analysts.

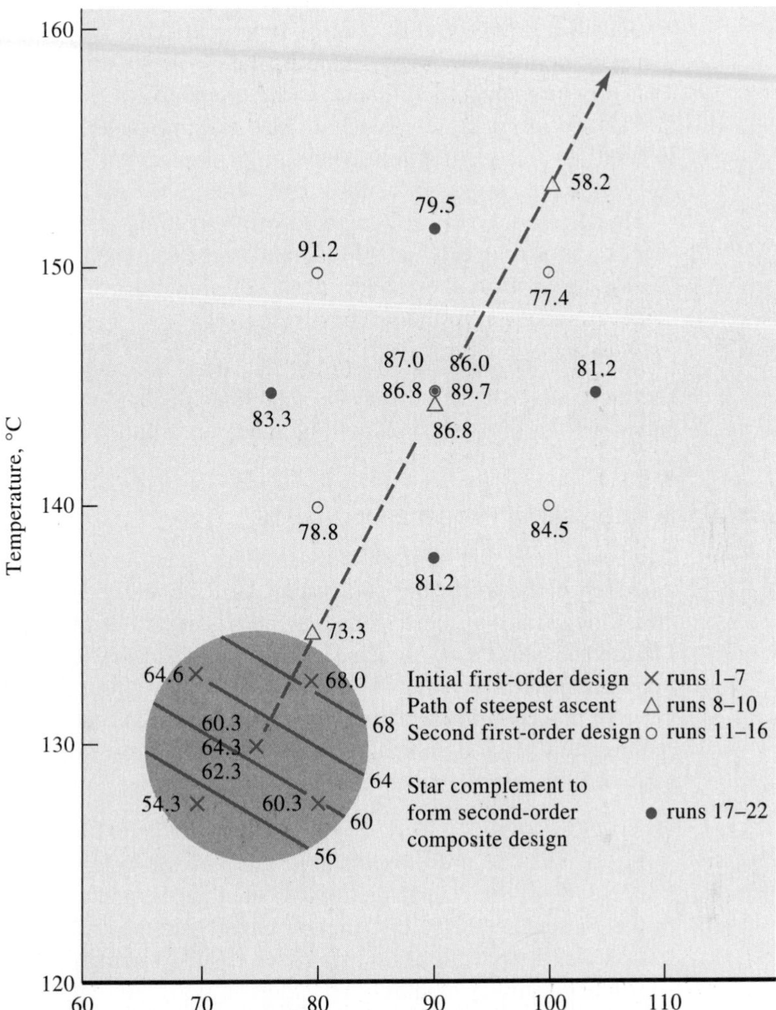

FIGURE 9.1 Illustration of the effect of elimination of (*a*) special and (*b*) common causes of variation. *("Continuing Process Control and Process Capability Improvement," July 1983. Statistical Methods Office, Operations Support Staffs, Ford Motor Company.)*

hand, quality control and manufacturing personnel cannot callously ignore them without running the risk of causing real trouble. There is a limit, however, to the amount of time and investment that should be put into the types of analyses and adjustments discussed in action 3 before design groups should be made aware of the problem. In some cases, specifications may be set tighter than necessary. In others, slight design changes may be less costly, or more feasible, than changes in machinery or tooling. It may be feasible on occasion to totally redesign the product unit within the manufacturer's capability to produce. Needless to say, there are important economic trade-offs between action items

3 and 4 that require close coordination and communication among design, manufacturing, and quality control functions. This was one of the main driving forces behind the development of the Quality Circles concept in Japan, which is now implemented in many U.S. companies.

5. *Resignation to losses.* When all else fails, management must be content with a high loss rate. Attention at this point focuses on scrap and rework costs, the economical level of control that should be exercised over the production process, and the costs associated with screening all product, realizing that some nonconforming product still is likely to be accepted. It is most important in this case to have good estimates of the form and shape of the distribution of product as an aid to proper process centering.

In nearly all cases, the resulting decisions are management decisions. One or another course of action may be recommended by quality control or production personnel, but the decisions will be made and funded by higher management.

9.1.2 Steps in the Analysis of Process Capability

The basic steps that appropriately may be taken in performing a capability study are given in the following paragraphs. Unfortunately, no two studies are identical; therefore, certain of these steps may be eliminated or the order may be changed to suit the needs of the particular situation. A good knowledge of both the techniques to be used and the operation of the manufacturing process is required. Each study must be planned ahead of time. Decisions must be made regarding sampling procedures, timing, and collection of data; and data must be recorded in such a manner that there will be no confusion as to how, where, or when they were taken.

1. *Establish control over the process.* The statistical tests discussed in this book begin with the presumption that samples are being drawn from a single universe. Until the control charts indicate that the presumption is true, estimates of the parameters of the distribution underlying the process are virtually meaningless. All the pitfalls regarding rational subgrouping apply, with special emphasis, to capability studies.
2. *Analyze process data.* Estimates are made of the process average and dispersion, and frequency histograms may be plotted to get a feel for the form of the distribution of product. This step may be carried out whether or not control is indicated. Even though control is not indicated, a frequency histogram gives an indication of the product spread. When combined with control-chart information, prime candidates for reduction of this spread may become apparent.
3. *Analyze sources of variation.* Study of the component sources of variation and their magnitudes may range from relatively simple tests to extremely complicated experimental designs carried on over long periods of time. This is the point at which knowledge of the characteristics of the manufacturing process becomes most important. Utilizing the general outline illustrated in Fig. 9.1, the analyst should determine those factors and operations that are most likely to be contributing to the process spread. Data collection is then planned in such a

way as to isolate these potential sources and evaluate them first. Since the results of any part of the capability study may suggest further investigation in other directions, it is important to identify carefully the timing and source of all measurements. This may make it possible to go back and recombine data based on insights gained from the planned study.

Example 9.1
Process capability study of glass stresses in glass-to-wire seal

Facts of the case This study is concerned with the preservation of integrity of glass-to-wire seals on certain miniature electrical resistance components. Finished bare components are sealed in a glass envelope in a semivacuum on infrared heat source assembly machines. During the sealing process, an inert forming gas flows through the forming head to ensure an inert atmosphere within each unit of product when the seal is completed. Each assembly machine has four head positions and thus can handle four product units at a time.

Integrity of the seals is predicted by measuring stress in the glass seals once the test units have cooled. These stresses are either tension or compression. Tension occurs at the inside of the seal next to the wire, while compression occurs at the outer surface of the seal. Both stresses are measured by a polariscope, which measures the angle of retardation in degrees, produced by the stress lines, of a light source passed through a seal. Previous study had shown that angles of retardation between 7.5 and 25.0 degrees and 0 and 30.0 degrees under tension and compression, respectively, would adequately protect the integrity of the seals.

Analysis Initially, samples were collected from each machine twice daily. The tension and compression in both top and bottom seals were measured and recorded. The data collected were plotted on $\bar{X}$ and R charts by assembly head position for each machine. A subgroup size of two, consisting of the two samples taken during one day, was used. These charts showed wide variation from machine head to machine head as well as from day to day on any one head position.

The next step in the analysis was to isolate a single machine for experimental purposes. Control charts for the next twelve subgroups of two taken from each of the four head positions are shown in Fig. 9.2. In order to simplify discussion of the analysis, only the test results of the top seals are presented. Similar results, analyzed concurrently, were obtained for the bottom seals. It can be seen from the very apparent pattern on the $\bar{X}$ charts that something causes temporary shifts in level and that the cause appears to affect all heads at the same time. However, the R chart for each requirement remains essentially stable. This indicates that the changes in level are likely to be controllable shifts. That is, whatever affects one seal tends to affect all seals in the same way.

Of the substantial number of variables that could affect the process, observation indicated that the least controlled was the flow rate of the inert forming gas. Thus, the next step was to evaluate the effect of this one variable on stress

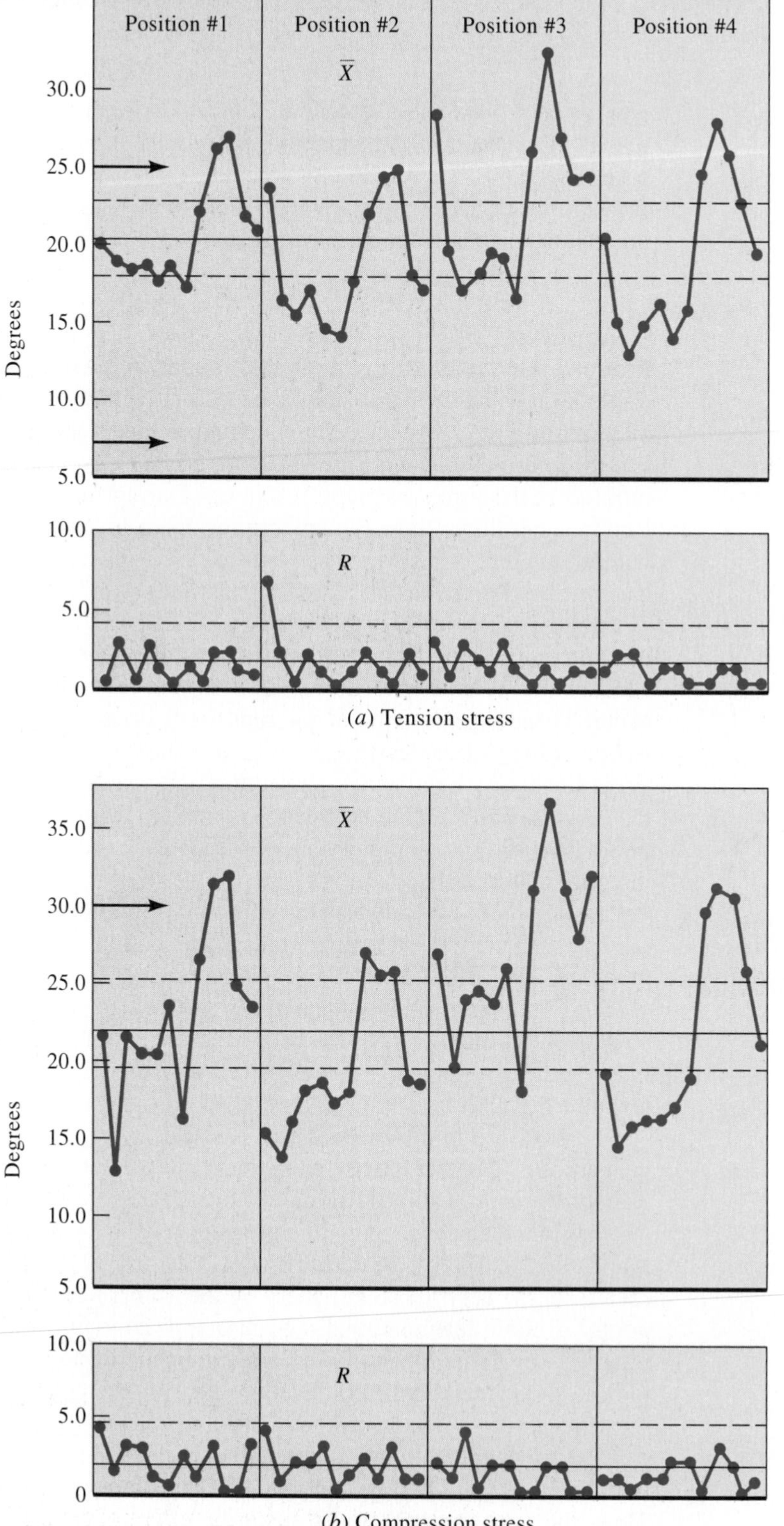

FIGURE 9.2 Stress in glass-to-wire seal by polariscope analysis of angle of retardation, machine no. 1, top seal.

characteristics. Three series of samples were run at carefully controlled flow rates using a single head on the test machine, a single rack of glass from the same melt, and product units from the same lot. The results of these tests are shown in Fig. 9.3 for the three selected flow rates of 2.6, 4.2, and 5.8 liters per minute.

From the control-chart patterns it readily can be seen that variations in the forming-gas flow rate are a prime factor having a significant effect on the level of stresses in the end seals. Also, the charts show a positive correlation between forming-gas flow rate and the level of stress; that is, when forming-gas flow rate is increased, both tension and compression are increased.

Action Further samples were taken from the production floor in much the same manner as in the first test. However, as each sample was taken, the flow rate of the forming gas was measured. These data were plotted as a function of flow rate indicating a nearly linear relationship between forming-gas flow rate

FIGURE 9.3 Effect of varied flow rate of forming gas on glass-to-wire seal. Analysis of angle of retardation by polariscope, machine no. 1, head position no. 1.

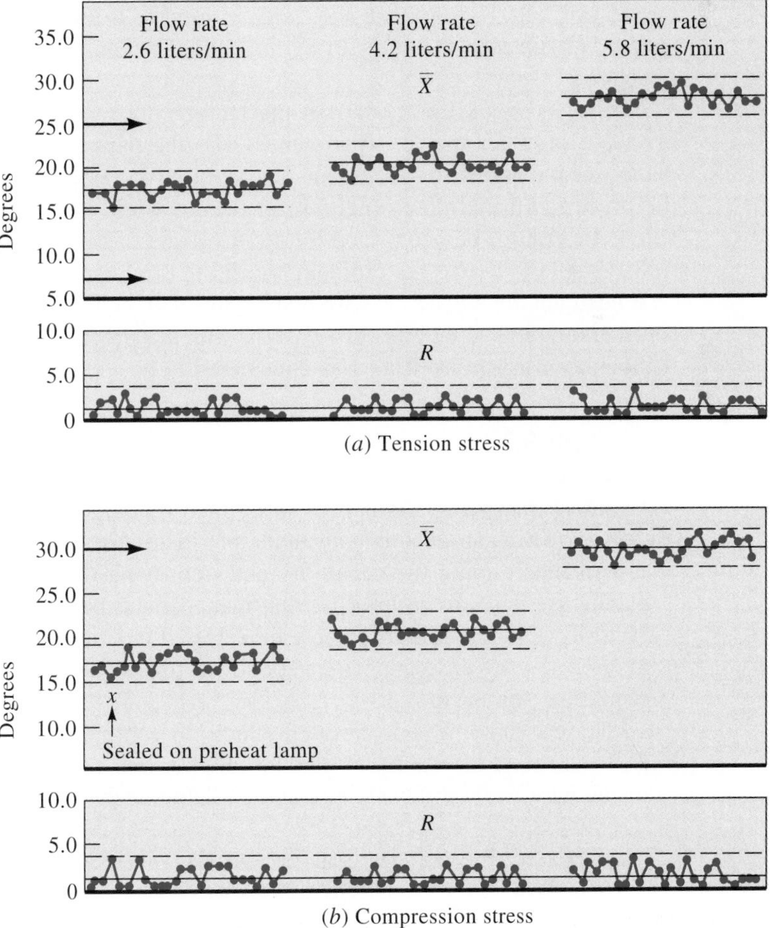

and stress in the glass seals. This enabled the engineers to establish tolerances on forming-gas flow rate necessary to maintain stress requirements within engineering limits. These limitations introduced no problems in the ability to make satisfactory seals in all other respects.

A means for limiting forming-gas flow rate was designed and installed on all the machines. Control-chart data since the installation of these controls show that the stress requirements have been maintained at a satisfactory level.

Comment on Example 9.1

It is the usual case that the need for a process capability study is evidenced by a high rate of defective production. In this case, the failure of glass-to-wire seals was evidenced by a foreshortened life of the resistance devices. The ultimate result of the study was a much more uniform product in general and the virtual elimination of the problem.

In Example 9.1, careful observation of the process during early stages of study allowed the engineers to bypass many of the factors contributing to variation discussed in Chap. 8. Attention was quickly directed to the main contributing factor. This was facilitated in large part by the lack of uncontrolled variation in subgroup ranges.

The point should be made that the factory floor can seldom operate with the precision of a laboratory experiment. Once the parameters of the problem were determined, engineering judgment was required to establish feasible tolerances on the process. Furthermore, engineering design work was required to develop the automatic control device to control the flow rate of the forming gas. Quality control was able to define the problem but did not, by itself, solve it.

It need not always be the case that savings can be made only at obvious trouble spots. Example 9.2 reports on a case in which a process capability and improvement study was run on what was considered to be a satisfactory process.

Example 9.2

Using $\bar{X}$ and R control charts to make a process capability study†

Facts of the case As part of an in-plant training course in statistical process control (SPC), an outside consultant and instructor was asked to take on a project of determining the capability and improving the performance of a bridge cable wire galvanizing process that management believed to be working satisfactorily. Usually, initial projects are selected in areas where success is virtually assured. However, management believed that selecting a known trouble spot would be "too easy."

The galvanizing bath, through which 30 continuous wire lines passed, was 18 ft wide and 60 ft long. While the supplier and customer each tested samples of the wire for both tensile strength and zinc coating, this case study considers

†Paul C. Clifford, "A Process Capability Study Using Control Charts," *Journal of Quality Technology,* vol. 3, no. 3, July 1971. Used by permission of the American Society for Quality Control.

only the results of the standard ASTM Preece dip test for zinc coating thickness. Specification required a minimum 4-min acid immersion of the test specimens while maintaining the integrity of the coating. Investigation revealed that rejections by the customers' tests were rare. If a sample did not pass, a second sample was drawn and retested. Failure on the second test resulted in rejection of a roll of wire or possibly an entire shipment.

Initial consideration involved brainstorming with both production and inspection personnel and an opinion-driven Pareto analysis. The initial study was set up to test variation through time and across the bath. The 30 lines of the bath were divided into 5 sections, 6 wire lines per section, with subgroups of 5 samples formed from one line from each section. Since the process operated 24 h per day, a subgroup was drawn each hour. This gave 4 drawings from each line during a 24-h period.

Five bowls, each containing numbers for each line in a bath section, were used to determine which lines would be sampled each hour. Thus, by drawing, without replacement, one number from each bowl, all lines would be sampled over a 6-h period. The numbers were replaced and drawn again for each of the remaining three 6-h periods. In this way, the pattern of drawing was never repeated, and each line was represented four times.

Table 9.1 shows the results of the first day's sampling that were used for the first phase of the study. The numbers in parentheses tell which line was present in the sample. Samples were left in the test bath until the zinc coating was penetrated and the time recorded to the nearest half minute. A histogram of the results and control charts for $\bar{X}$ and R for the time samples are shown in Figs. 9.4 and 9.5.

Analysis and action The day's worth of data contained in Fig. 9.5 does not indicate control of either dispersion or centering. These data were collected on a Monday, after a weekend shutdown, which may have accounted for the erratic gyrations for the first few hours. The only point out on the R chart occurred at the second hour. With this point eliminated, the $\bar{R}$ reduced to 2.0. This value was used as a target for the rearranged charts of Fig. 9.6. The histogram of Fig. 9.4 has a mean of 6.8 min and a sample standard deviation of 1.2 min. Using the $\bar{R}$ of 2.0, the estimate of σ from within subgroup variation is 0.9. This combination of influences suggests that the data are not normally distributed and that there likely are differences due to time and to location.

The $\bar{X}$ chart of Fig. 9.5 exhibits what might be a surge-depletion pattern over time. In such a pattern, a common occurrence in plating operations, the bath is recharged periodically, producing a sudden jump in results followed by a tapering off as time progresses. This is apparent at subgroups 7, 13, and 19.

As Table 9.1 indicates, wire line 16 was responsible for the four lowest test readings in each 6-h period. The average of these readings is 4.8 min, which plots well below the LCL in Fig. 9.6. Line 21 also appears low. Further investigation by engineering determined that those lines had become contaminated with oil as a result of poor housekeeping at the tank entry end of the line. With an average range of 2.0 for within-line variability, there is further evi-

TABLE 9.1 SAMPLING SEQUENCE AND RESULTS OF PREECE DIP TEST FOR GALVANIZED WIRE LINES
Figures given are in minutes that samples remained in acid bath†

Sample no. (time)	Bath section number					$\bar{X}$	R
	I	II	III	IV	V		
1	6.5 (1)	8.5 (8)	6.5 (15)	8.0 (23)	8.0 (28)	7.5	2.0
2	7.5 (5)	10.0 (12)	‡4.5 (16)	8.5 (19)	9.0 (30)	7.9	5.5
3	6.5 (2)	5.5 (10)	6.0 (17)	5.0 (21)	6.0 (27)	5.8	1.5
4	7.0 (6)	9.5 (11)	10.0 (18)	9.5 (22)	7.5 (25)	8.7	3.0
5	7.0 (3)	7.5 (9)	7.5 (13)	9.0 (20)	8.0 (29)	7.8	2.0
6	5.5 (4)	5.5 (7)	5.5 (14)	6.0 (24)	6.5 (26)	5.8	1.0
7	7.5 (1)	8.5 (12)	7.0 (18)	8.0 (23)	6.0 (30)	7.4	2.5
8	6.5 (3)	7.5 (8)	6.0 (13)	8.0 (19)	7.5 (28)	7.1	2.0
9	7.0 (6)	7.0 (11)	5.5 (17)	7.0 (20)	7.0 (26)	6.7	1.5
10	6.0 (5)	6.0 (9)	7.0 (14)	6.5 (21)	5.5 (25)	6.2	1.5
11	5.5 (4)	5.5 (10)	8.0 (15)	6.5 (22)	8.5 (27)	6.8	3.0
12	7.5 (2)	6.5 (7)	‡5.0 (16)	6.0 (24)	7.5 (29)	6.5	2.5
13	8.0 (3)	7.0 (8)	7.5 (13)	8.0 (23)	8.0 (25)	7.7	1.0
14	7.5 (1)	7.0 (11)	6.5 (14)	8.0 (22)	7.5 (26)	7.3	1.5
15	7.0 (6)	7.0 (10)	7.5 (18)	6.5 (20)	6.5 (29)	6.9	1.0
16	5.5 (4)	6.0 (7)	‡5.0 (16)	5.5 (19)	9.0 (30)	6.2	4.0
17	7.5 (2)	6.5 (9)	6.0 (17)	4.5 (21)	5.5 (28)	6.0	3.0
18	5.5 (5)	6.5 (12)	5.5 (15)	6.5 (24)	6.0 (27)	6.0	1.0
19	9.0 (1)	8.0 (11)	7.5 (18)	6.5 (22)	6.5 (30)	7.5	2.5
20	7.0 (3)	8.0 (8)	8.0 (13)	6.0 (19)	6.5 (29)	7.1	2.0
21	7.5 (6)	6.0 (12)	8.0 (14)	5.5 (24)	7.5 (28)	6.9	2.5
22	6.0 (2)	6.0 (9)	7.0 (15)	7.0 (20)	7.5 (27)	6.7	1.5
23	7.0 (5)	5.5 (7)	5.0 (17)	6.0 (23)	7.0 (25)	6.1	2.0
24	6.5 (4)	5.5 (10)	‡4.5 (16)	4.5 (21)	6.0 (26)	5.4	2.0

†The number in parentheses is the number of the wire line on which the test was performed.
‡Low readings.

dence that σ might be held to 0.9 min. With that information, and the knowledge that any reading of 3.75 min or greater will pass the 4.0 specification, the distance in multiples of σ from the mean to the lower specification is 3.39. Thus it appeared possible to reduce the value of the mean, which would allow for an increased throughput, if statistical control could be established and maintained.

Comments on Example 9.2

In a period of 6 months, the team was able to increase productivity by about 5%, reduce the amount of zinc consumed by 15%, and reduce variability by 10%, thus producing a more uniform product.

It remains a poor idea to begin implementation of SPC in a new area and with new people with a project where no trouble appears to exist. Nevertheless, this

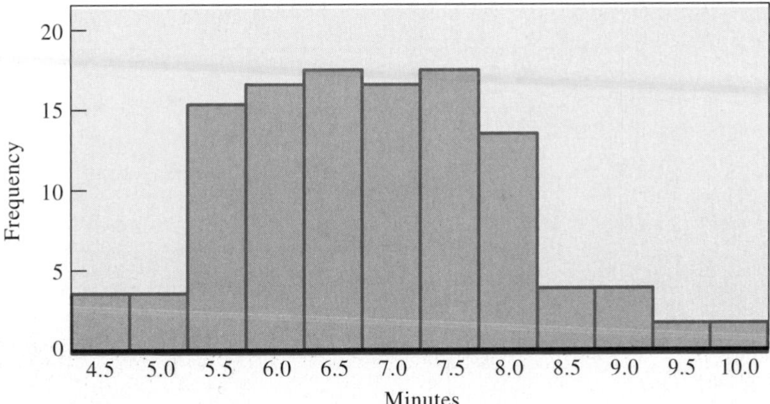

FIGURE 9.4 Histogram of results of Preece dip test times from Table 9.1. *(P. C. Clifford, "A Process Capability Study Using Control Charts," Journal of Quality Technology, vol. 3, no. 3, July 1971. Used by permission of the American Society for Quality Control.)*

example points up the fact that any process can be improved if it has not been the subject of continuous improvement using statistical tools. In this case, it appeared better to defer to management judgment as to where to begin than to object.

It is important to note that the sampling procedure was designed before data were taken. Existing data were limited to accept/reject decisions with few rejections. After one day's sampling, and the analysis that followed, a decision had to be made about whether to continue with the investigation. The sampling results were used, as they would be in any process capability study situation, to estimate what *might* be accomplished if the initial indications were near correct. This violates one rule which says, "Do not conduct capability studies until the process is in statistical control." That rule is based on the fact that, until control is achieved, no *reliable* estimate of σ can be obtained. Frequently, however, time and money can be saved with a little speculation about what might be achieved followed by much hard work to establish control and verify the conjectures.

9.1.3 Stages in the Analytical Procedure

Normally, a capability study proceeds in stages. As implied by the discussion of the steps in the procedure, these stages usually proceed from those involving simpler tests and adjustments to those involving more complex and, therefore, more expensive changes.

Usually the study will terminate as soon as the adjusted production spread is comfortably less than the specification range, in short, as soon as the immediate problem is solved. In the long run, this may be an unfortunate result. A basic knowledge of the capability of manufacturing and testing equipment, verified and

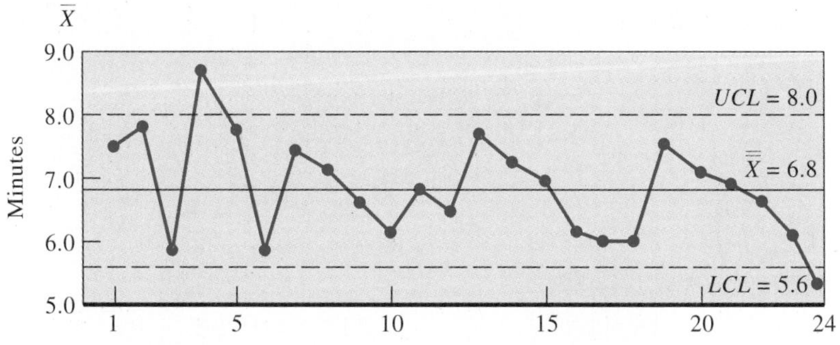

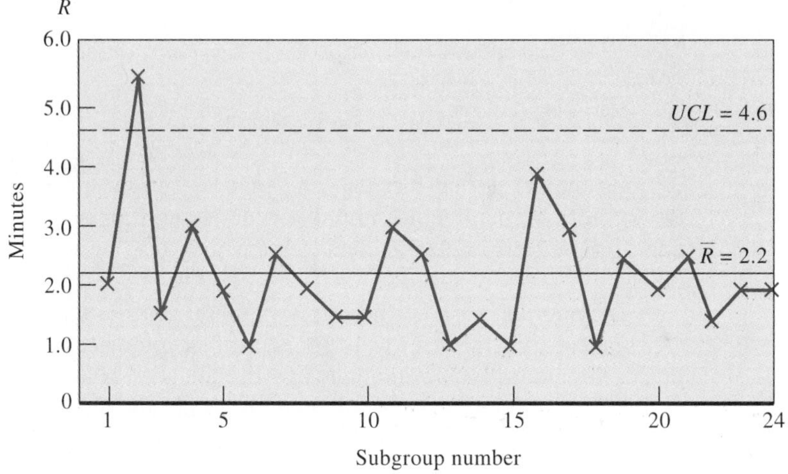

FIGURE 9.5 $\bar{X}$ and R charts for Preece dip test times for hourly samples of Table 9.1, op. cit.

adjusted from time to time as more data become available, can be of great help to the design, manufacturing, and quality control functions in setting specifications, costing out contracts and new products, production planning, and decisions regarding the purchase of new facilities. This is in addition to its usual value in setting standards for control charts on new production runs.

9.2 PROCESS CAPABILITY AND PERFORMANCE INDEXES

In many of the examples presented thus far, the importance of comparing 6σ from a process in statistical control, the so-called natural tolerances of the process, with the specification range $(USL_X - LSL_X)$ has been stressed. The smaller 6σ is in relation to $(USL_X - LSL_X)$, the more capable the process so long as it is properly centered at the nominal or target dimension. Operating under the Deming philosophy of never-ending quality improvement, an objective of manufacturing should be to

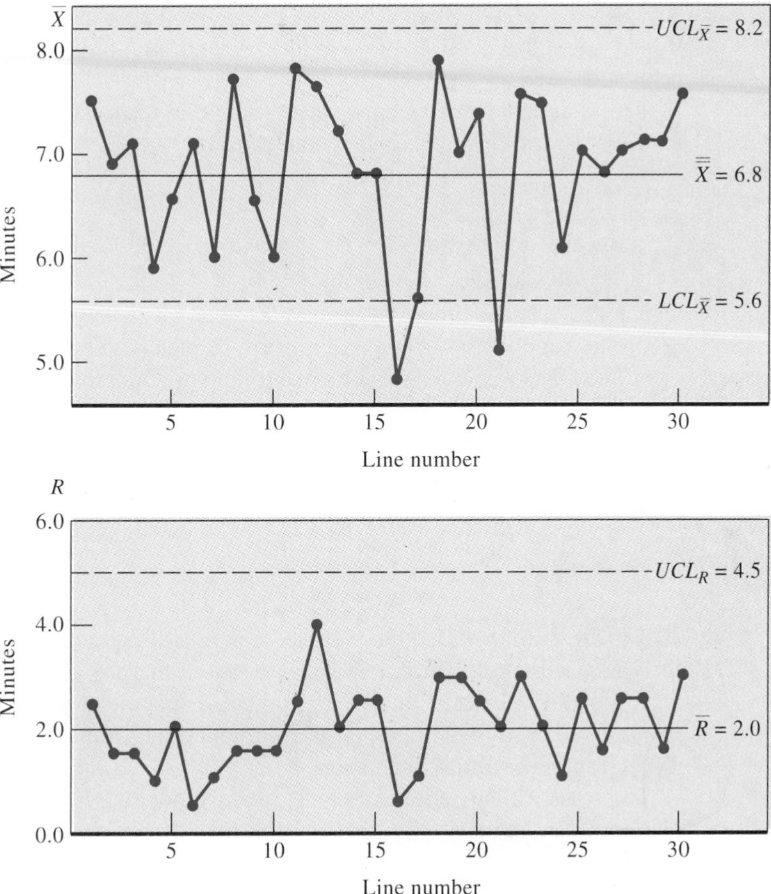

FIGURE 9.6 $\overline{X}$ and R charts for Preece dip test data arranged by wire line in subgroups of four measurements, op. cit.

seek methods to continually reduce 6σ even when 6σ is well within $(USL_X - LSL_X)$. In the discussion of indexes that follows, we have used the alternative symbols U for USL_X and L for LSL_X in order to simplify the expressions.

The Japanese have formalized this type of comparison into a series of process indexes that are rapidly being adopted in the United States.† The first of these, designated C_p, is most closely related to process capability *potential* as discussed thus far. It is found from

$$C_p = \frac{U - L}{6\hat{\sigma}}$$

†See Victor E. Kane, "Process Capability Indices," *Journal of Quality Technology,* vol. 18, no. 1, January 1986.

At any value greater than 1, a process has the potential for meeting specifications if held in control at an $\overline{X}_0$ of $(U + L)/2$. A minimum target value for C_p of 1.33 has been suggested.

The relationship between μ, the estimate of process centering, and the target value $\overline{X}_0$ is designated as k. It is found from

$$k = \frac{|\overline{X}_0 - \hat{\mu}|}{\min(\overline{X}_0 - L,\ U - \overline{X}_0)}$$

where the vertical lines (|) around $\overline{X}_0 - \hat{\mu}$ mean the "absolute value," that is, ignore the sign of the value, and min (a, b) means select the minimum value of a or b. The index k always will be greater than or equal to zero and has a target value of zero.

Where specification tolerances are symmetrical about the nominal value, the formula for k reduces to

$$k = \frac{|\overline{X}_0 - \hat{\mu}|}{(U - L)/2}$$

This is the form in which the index is seen most frequently. However it is not valid for one-sided specification tolerances or when there is a single specification limit. Also it should be noted that when one-sided specification tolerances are used, the target value of $\overline{X}_0$ will not equal the nominal dimension unless there is no possible dispersion beyond that dimension.

Three additional indexes, which relate more to process current performance than to potential capability, are as follows:

$$C_{pL} = \frac{\hat{\mu} - L}{3\hat{\sigma}}$$

$$C_{pU} = \frac{U - \hat{\mu}}{3\hat{\sigma}}$$

$$C_{pk} = \min(C_{pL}, C_{pU})$$

As with the factor k, each of these indexes takes into account the most recent estimate of process centering. C_{pL} and C_{pU} relate the difference between the current estimate of process mean and the respective specification limits, L and U, to half the natural process spread. The minimum target value for each is one. C_{pk} provides the minimum of these two values. Companies that have adopted process capability indexes within their own plants frequently specify their usage in supplier purchase contracts, especially C_{pk}.[†]

[†]One additional capability index, C_{pm}, which is closely related to the Taguchi economic loss function, is discussed in Chap. 17.

It should be noted that each index tends to imply that the distribution of product units from a controlled process may be represented by a normal reference distribution; that is, virtually all product will fall between the mean μ and $\pm 3\sigma$. (A check of Table A will show that all but 0.27% of the normal curve falls between $\pm 3\sigma$.) A histogram of a substantial number of measurements from a controlled process is required to determine whether virtually all product falls within ±3-sigma limits. These symmetrical limits may not fit processes that tend to produce skewed distributions. Also, their usage and target value interpretation must be modified in situations where only a single specification limit, either U or L, is specified.

In testing for steel hardness, for example, there is a metallurgical limit on hardness for various grades of steel alloy. Usually the grade of steel selected for a product is based on this hardness limit (as well as other physical and economic properties). A lower specification limit is indicated with the stated intent to achieve a target as close as possible to the physical limit. In such cases, C_{pL} is the only applicable index and a highly skewed distribution is likely to result (see Fig. 3.9*c*).[†]

When Just-In-Time inventory management is employed, it is the supplier's responsibility to show that shipped product meets specifications. Purchase contracts with JIT suppliers generally will require documented evidence of compliance with each shipment made. The requirement of a process capability index calculation often is part of this documentation. More meaningful evidence of compliance would include a histogram from samples from a lot or series of consecutively produced lots. A procedure such as the Shainin lot plot method carried out by the supplier would provide valuable information to the purchaser in addition to any required process capability index. The lot plot method is discussed in Chap. 15.

9.2.1 Process Capability Studies Using Attributes Data

The discussion so far has focused on using variables control charts and indexes to analyze process capability. In many cases in industry these studies arise out of the initial use of control charts and acceptance procedures based on attributes data. When attributes data are collected and categorized by type of nonconformity as shown in Fig. 9.3, data are available on the frequency of the various nonconformities. These data may be used to direct attention to the most common problems.

Two important tools in this type of capability analysis are the Pareto diagram and the cause and effect diagram. They, and the procedures used in their development, frequently are employed to identify those problems that should be isolated and controlled separately with either attributes or variables control charts.

[†]See Allan Brooks, "Heat Treating Shows Why SPC Is No Cure-All for Manufacturers," *Production Engineering*, September 1985.

9.2.2 Pareto Analysis

Pareto analysis is a technique to assist in determining where to begin an attack on a problem or to display the relative importance of problems or conditions. The resulting chart is used to choose a starting point for problem solving, monitor success, or identify basic causes of problems. It is the result of data collection in which possible anomalies have been identified and count (attributes) data have been collected using check sheets, computer process control results, or some other data collection and analysis method.

An example of a Pareto diagram, involving the production of a small shaft, is shown in Fig. 9.7. Nonconformance codes are identified by the letters "A" to "G" and "O." The letter "O" applies to several types of nonconformity that occur too infrequently to identify separately. The steps used in performing the analysis are:

1. Identify types of nonconformances. If past attribute control-chart data have been categorized as in Fig. 7.3, developing the list may be quite easy. Otherwise, new procedures of data collection may have to be instituted and data collected over some period of time before the analysis can be performed.
2. Determine the frequencies for the various categories.
3. List the nonconformities in descending order of frequency. Each designated nonconformity is listed separately, reserving the letter "O" for a grouping of those of trivial occurrence frequency.
4. Calculate the frequency percentage for each category and the cumulative frequency.
5. Set up scales for the Pareto diagram. In Fig. 3.5, the scale on the left side gives the actual frequency of occurrence in the sample; the right-side scale applies to the cumulative percent frequency.
6. Plot the Pareto frequency bars and the cumulative frequency percentage.

If the Pareto diagram is set up following the steps indicated, it will direct attention to the most frequent nonconformities but not necessarily to the most important. When the list contains some that might be considered extremely serious and others that might be considered trivial, a weighting scheme should be used to modify the frequency count and ordering resulting from steps 2 and 3. Weighting schemes and classifications of nonconformities (defects) were discussed in Chap. 7, Sec. 7.11. Costs resulting from errors and nonconformances also must be taken into account. Thus more than one Pareto diagram covering the same process may be needed.

"Before" and "after" improvement Pareto diagrams form a good basis for determining the success of an improvement effort. If shown superimposed or side by side, the number of units inspected in each case should be the same and the left-hand scale of counts of defects, nonconformances, and the like, should be the same. In this way, the overall improvement is clearly illustrated.

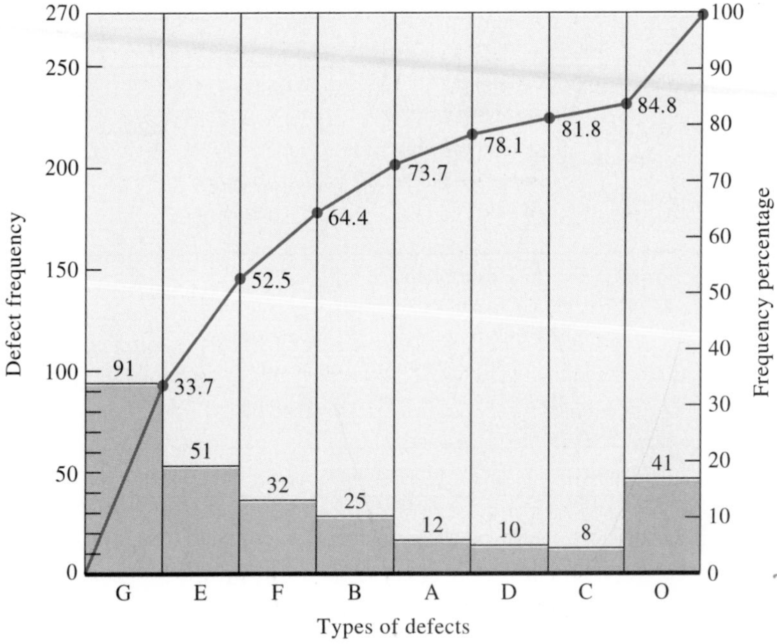

FIGURE 9.7 Example of a Pareto diagram. See Fig. 9.8.

9.2.3 The Cause and Effect (CE) Diagram†

Once a nonconformity has been isolated for further study, attention shifts to analyzing how the deviation from specifications was caused. Sometimes the reason is obvious; sometimes a considerable amount of investigation is required to uncover the cause or causes. The Japanese have developed and used cause and effect analysis, and the resulting diagram, as a formal structure for uncovering problem areas. Usually these were manufacturing-related problems of too much variation in a process. However, their use has spread to all aspects of business activity when it is necessary to sort out causes and organize them into mutual relationships. An example CE diagram, pertaining to the shaft in the Pareto analysis, in shown in Fig. 9.8. This diagram applies to one of the categories of nonconformity found in the Pareto analysis, "wrong shaft diameter."

The steps in a cause and effect analysis are:

1. Define the problem. The problem may be obvious; but uncovering its root cause may require the use of other tools such as histograms of data results, control-chart results, Pareto diagrams, and so on.

†Cause and effect analysis was developed by Kaoru Ishikawa initially for use by Quality Circles in discovering potential causes of a problem. The diagram also is known as an Ishikawa diagram or fishbone diagram because of its resemblance to the skeleton of a fish. See Chap. 18 for further mention of Dr. Ishikawa.

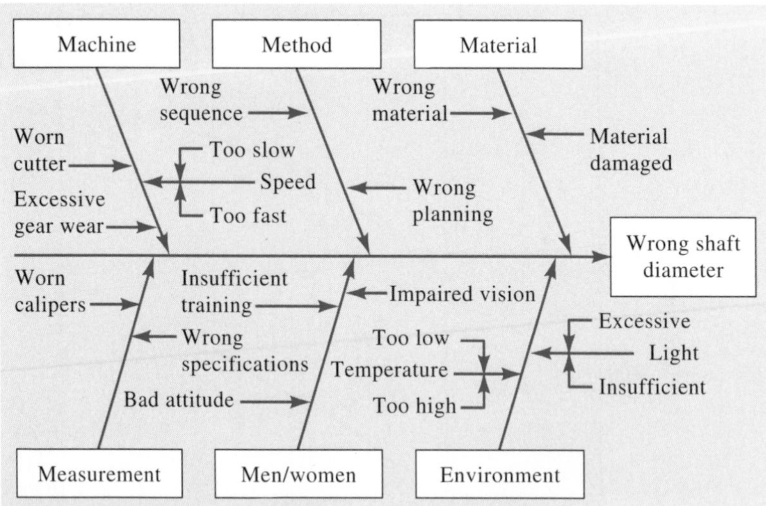

FIGURE 9.8 Example of a cause and effect diagram. Reprinted from *The Two-Day Statistician,* 1986, by Jack B. ReVelle. Used by permission of Hughes Aircraft Company.

2. Select the method of analysis. Often the method of analysis involves brainstorming with a team of representatives from production, engineering, inspection, and any other business unit potentially involved with the item in question.
3. Draw the problem box and prime (center) arrow.
4. Specify the major categories of possible sources contributing to the problem.
5. Identify the possible causes of the problem.
6. Analyze the causes, and suggest corrective action.

Whenever some form of brainstorming is to be used to uncover problem causes, it is helpful to structure the discussion by first laying out the major categories of problem cause. In a manufacturing environment, these may include the "5 M's and an E" grouping: machines, methods, materials, men/women, measurement, and environment. Environment refers not only to the environment in which the item is produced but also to the environment in which it may be expected to function. For service processes, the "3 P's and an E" grouping may be helpful: procedures, policies, people, and equipment. Frequently a flowchart of the process is used to generate understanding and a common basis for discussion. Then the participants in the session are invited to contribute ideas in each of the categories. Through discussion and a process of elimination, the number of causes may be reduced to a consensus set of "most likely" causes.

The *action and effect diagram* is a logical extension of the cause and effect diagram. Once corrective actions have been formulated, they may be entered onto the action and effect diagram. This diagram appears much like the CE diagram except that the action arrows point away from the prime arrow box and the actions taken and resultant effects are indicated for each contributing factor. The diagram acts as a double check to ensure that all causes have been accounted for and countermeasures implemented.

9.3 QUALITY BY DESIGN: DESIGN AND INSPECTION SPECIFICATIONS

The three functions in manufacturing of specification, production, and inspection were mentioned in Chap. 1, Sec. 1.1. In any discussion of the use of specifications, it is helpful to recognize the distinction between a design specification and an inspection specification. This distinction was clearly pointed out by Shewhart.[†] The design specification deals with what is desired in a manufactured article; in other words, it deals with the so-called specification function. In contrast, the inspection specification deals with the means of judging whether what is desired is actually obtained; in other words, it deals with the inspection function. Parallels exist in service activities as well. In the case of a service activity, the word *provision* may be substituted for *production* and the word *verification* for *inspection.* Thus each service activity has a specification function, a provision function, and a verification function. Design specification deals with the processes and features of the service to be provided; inspection specification deals with the process of verifying that the service has been provided in the manner intended.

9.3.1 Some Problems in the Enforcement of Design Specifications

Although the following quotation from William B. Rice[‡] relates to the specification of dimensional tolerances and their enforcement some years ago, it describes a condition that may occur at any time with respect to many different types of quality characteristics.

> In many machine shops and metal working plants there have grown up over a period of years certain practices with regard to tolerances which can bear re-examination. The blueprints may call for one tolerance; inspection gages may allow another usually wider tolerance, and the foreman may be even more liberal. Each of the three parties views the operation from a different standpoint. The engineer sees the problem as one of design; the inspector tries to maintain an acceptable quality; the foreman is under pressure to produce in quantity. Much of the spoilage and reworking which cost American industry so much in time, money and manpower arise from a failure to coordinate the viewpoints of the designing, inspection and production departments.

[†]W. A. Shewhart, "Some Aspects of Quality Control," *Mechanical Engineering,* vol. 56, pp. 725–730, December 1934.

[‡]W. B. Rice, "Setting Tolerances Scientifically," *Mechanical Engineering,* vol. 66, pp. 801–803, December 1944.

One reason why engineers, foremen and inspectors have different ideas about what tolerances should be permitted is that there are several ways in which inspection can be done. Take as an illustration a half-inch turning on a four-spindle automatic screw machine. If four successive pieces are taken off the same spindle, the diameters of the four turnings will probably vary by less than a thousandth of an inch. If, however, one turning is taken off each spindle, the measurements may vary by as much as two thousandths. Again, suppose that a run of 50,000 turnings has been completed; if a large sample (say 1,000) is taken at random from the entire lot, the difference between the largest and smallest diameters measured may be six or seven thousandths.

Another reason for disagreement is that the dimensions of the work which can actually be produced depend upon many factors. A new machine will hold more closely than an old one; skilled workers can do more accurate work than green hands can; experienced foremen are able to get better results than untried men in supervisory jobs; material, too, has a strong bearing on quality, as have the kind and quantity of tooling, personnel relationships, morale, pay, and a thousand other factors which come and go, fluctuating throughout the process in usually unpredictable fashion.

In general, product or design engineers often seem to do their planning and drafting with reference to somewhat ideal conditions, assuming good machines, well-trained workers and skilled supervision, or else they use reference tables which may tacitly assume such factors. Therefore tolerances on blueprints tend to be conservative and are sometimes actually impossible to meet with any degree of economy in the manufacturing process. Nearly ideal conditions may be attained during some small part of the process, but almost never for any extended period of time.

Production men, on the other hand, knowing from experience that a careless operator, a soft spot in the steel, a slight mis-adjustment of the tools, a loose collet or any one of a multitude of other troubles will cause an automatic to do out-of-tolerance work, are inclined to ask for more liberal tolerances; if these are not forthcoming they just "do the best they can." Unfortunately they do not always stop to figure how they can make that best better. Being human, and under pressure to turn out large volume, they are prone to accept what they have as the best they can get.

The process inspector, caught in the crossfire of these opposing interests, usually has to compromise. The workmen on the floor, with whom the inspector is in daily contact, argue that too strict adherence to blueprint standards will slow down or even stop production, and that another thousandth won't make any difference anyway. The engineering department, more remote but exerting its influence through the blueprint which the inspector sees and uses constantly, calls for certain dimensions to which all pieces must adhere. Who can blame the inspector for letting an occasional out-of-tolerance piece go by under such conditions?

Two dangers arise, however, under such circumstances. First, neither the inspector nor the foreman knows every step in the manufacturing process, hence their judgment as to what can and what cannot be allowed to pass is usually poor. Second, laxity is encouraged which may destroy the validity and nullify the value of inspection. The idea of concrete, objective standards which result in usable articles economically produced should be basic to the inspection function. Violating this idea undermines the fundamental purpose of inspection.

9.3.2 Establishing Manufacturing Tolerances in the Consumer Goods Industries

A situation that is common in many industries is clearly described by O. P. Beckwith.[†] He classifies the possible conditions with regard to customer inspection of final product into two groups, namely:

> (1) Where the product is not subject to immediate, systematic, and quantitative evaluation of quality and, therefore, is not subject to systematic acceptance and rejection. (2) Where the product is subject to immediate, systematic, and quantitative evaluation of quality and, therefore, is subject to systematic acceptance or rejection.

His comments on the first condition are as follows:

> Condition No. 1 has been quite general in the textile industry in the past; that is, goods are produced, but the consumer, in general, does not make immediate systematic, and quantitative determination of their conformity to quality standards. Instead, the manufacturer is forced to maintain quality by the customer's long-time, haphazard, qualitative experience with the manufacturer's output. If, through good manufacturing methods, quality (for example, wear life of a fabric) is maintained at a standard level ±10%, it is not likely that increasing uniformity of product to ±5% will be readily discerned by the individual consumer. Therefore, if a manufacturer is enjoying a profitable business with a quality variation of ±10%, the arguments toward adoption of methods and processes designed to give a more uniform production having ±5% variation are somewhat abstract, even though it might prove over a period of years to show savings because of lower claims for defective merchandise. In such situations the need for close control or statistical control of quality is not readily apparent, unless it happens that competitors adopt quality-control techniques which give better products.
>
> Although many such situations exist in textile mills today, there are still cogent arguments for adoption of statistical quality control in these mills. The relationship of each department in a mill to the next one in the manufacturing operation is that of producer and consumer. The picking department produces prepared wool for its consumer, the card room; the card room, for the spinning room; the spinning room for the dyehouse. Even though there may be no testing at these points or systematic attempts to evaluate quality, nevertheless the production of each department is subject to immediate, systematic, and somewhat quantitative evaluation of the product in the next department. If stock is not well picked, the card room will complain and point to higher costs necessary because card speeds must be reduced to process the insufficiently opened and blended stock. If the roving is not well carded, the spinning department complains of allowances that must be made to spinners for their lowered production. Thus, it is seen that even in mills where the need for closely controlling final product quality is not sufficient to induce action by management, closely controlling quality within departments of the mill can be appreciated and the economies effected demonstrated relatively quickly.

[†] O. P. Beckwith, "A Fresh Approach to Quality Control," *Textile World,* vol. 94, pp. 79–82, January 1944.

When a designer makes a conscious decision to incorporate a margin of safety into one or more tolerance limits, in effect, that decision is an economic decision. The extent of the margin of safety that is appropriate depends on certain cost factors and requires consideration of the acceptance procedures that are to be used. From a broad viewpoint, the design specification and the inspection specification may be thought of as two different aspects of a problem in engineering economy. A brief discussion of the economic analysis of this type of problem is given in Chap. 17, following the explanation of various schemes for acceptance sampling.

9.3.3 Some Bad Consequences of Careless Setting of Specification Limits

Specification limits are often set by designers with little or no critical consideration of the various problems involved. In some instances, this may be because designers are too absorbed in other matters to have time to give much attention to tolerances. Lacking time and information, designers feel they are on the safe side by not being too liberal in their tolerances. They often feel that they can count on complaints from production and inspection to tell them of any cases where tolerances turn out to be tighter than can be held by production.

In other instances, the use of unnecessarily tight tolerances may be a deliberate policy on the part of designers. They may be conscious of the difference between the blueprint tolerances and those that actually are enforced. Therefore, in order actually to get what they think they need, these designers tend to specify closer tolerances than they believe are necessary.

As William Rice suggests, an unfortunate result of this policy may be the creation of lack of respect for specifications on the part of production and inspection personnel. Sometimes it is critically important that specifications be met. Production and inspection personnel, tending to ease up a bit on any tolerances that are difficult to meet, and not understanding the reasons behind the designer's specifications, may ease up as readily on the critical ones as on those less important. On the other hand, enforcing tolerances that are really too tight tends to increase costs.

9.3.4 Various Meanings of Quality of Products and Services in Relation to Design and Inspection Specifications

In the popular sense of the word, particularly as applied to consumer products, the word *quality* means general excellence. More particularly, it may mean excellence in relation to certain things that a consumer wants. There are a number of different types of reasons why products and services may have unsatisfactory quality in this popular sense. There may be certain characteristics desired by a consumer (for example, strength, durability, appearance, performance of some sort) that were not designed into the product because the producer did not intend to do so. Or the characteristics may not have been achieved by the designer even though they were intended to be so. Or they may have been designed into the product but the design specifications may not have been carried out. Because this popular concept of quality applies to design specifications as well as to inspection specifications, it may be difficult to use numbers to express quality in this sense.

In contrast, if one is concerned solely with inspection specifications, it is common practice to attach numbers to the concept of quality. In the sense of quality of conformance, all products and services are either good or bad. Nondefective (good) product conforms to specifications; defective (bad) product does not conform. The percent rejected is a numerical measure of quality of conformance.

9.3.5 The Common Condition of a Twilight Zone near the Specification Limits

Part Three of this book deals with a variety of plans that involve the use of sampling in acceptance inspection. In general, such plans are based on the concept of quality of conformance; each unit of product or service is classed as either good or defective.

Nevertheless, matters are seldom quite so straightforward as this simple classification seems to imply. It is fairly common for design specifications to contain some margin of safety. That is, they are more severe than the designer really thinks is necessary for the functioning of the product or service under ordinary circumstances. Where this is the case, product or service slightly outside specification limits falls in a twilight zone that will actually be good enough for most purposes. The formal allowance for some percentage of rejects that is a necessary part of any acceptance sampling plan may be based in part on the assumption that such a twilight zone exists.

9.3.6 Why Designers Should Indicate the Relative Importance of Various Specifications

The quotation from W. B. Rice described a condition in which process inspectors, under pressure from production personnel, tended to ease up a bit on the enforcement of specifications. Such easing up is likely to be defended by production and inspection personnel on the grounds that product a bit outside of specifications is within the twilight zone of product that is good enough for practical purposes. As Rice points out, one of several objections to this practice is that the easing up on enforcement may occur in the wrong places because supervisors and process inspectors may have poor judgment as to which specifications are the critical ones.

In certain industries the practice has grown up of having a "material review board" or "plant salvage committee" that decides whether parts that do not conform to specifications may be approved for use in an assembled product. Here again, the approval of nonconforming parts may imply that it is believed that there is an out-of-tolerance twilight zone from which some parts are acceptable. Such a board or committee usually includes a representative of the design department who has a veto power in any decisions. A representative of design can therefore make an ex post facto judgment as to which specifications are the important ones.

In general, the basis for decisions on acceptance will be improved if designers stipulate *in advance* which of the numerous specifications are the most important from the viewpoint of the satisfactory functioning of completed product. Thus some companies have adopted the practice of placing a special mark or sign next to those specifications considered to be critical. This topic is discussed further after three short examples that illustrate certain points made in the last few pages.

Example 9.3
Failure of a successful control-chart application

Facts of the case A certain small plant not only manufactured its own main product but also made precision parts for several larger companies. The chief inspector of the small plant attended a statistical quality control course. On returning to the plant, a control chart for $\bar{X}$ and R was started on a dimension (which we may call dimension A) on which a substantial percentage of the parts were being produced outside specified close tolerances. Dimension A proved to be badly out of control. By working a few hours on each shift of the several shifts for some weeks, the sources of trouble were finally diagnosed, and after much effort, corrective action was secured. Dimension A was brought into control and brought within specified tolerances.

However, the purchaser who had rejected some lots of this part in the past continued to reject as many lots as ever. Investigation disclosed that dimension A was one that, even though specified by the purchaser to very close tolerances, was really of minor importance. The purchaser could actually accept parts far outside specified tolerances on dimension A without causing any trouble; for this reason the dimension was subject only to a very perfunctory check by the receiving inspection of the purchasing company and actually had not been the cause for past rejection of lots.

The real trouble had been dimension B on which *no* departures from specified tolerances could be permitted. Although the supplier's quality level on dimension B was good, it was not perfect; a single out-of-tolerance part in a sample was cause for rejection of a lot. As the chief inspector had not realized this fact, control charts had not been applied to dimension B.

The chief inspector was very discouraged. From the viewpoint of the owner of the plant and of the supervisors and machine operators who had cooperated in the troubleshooting activity, the lack of this tangible success tended to discredit the use of statistical quality control. As a result no further use of the control chart was made; the technique was not applied to dimension *B* or to other quality characteristics on which it might have been used to advantage.

Example 9.4

A case of a designer's error†

Facts of the case A homemaker purchased a high-priced electric stove made by the *X* Company, a well-known large manufacturer. This stove had two ovens. Normally, only the smaller oven was used. However, some months after the purchase of the stove, the large oven was needed to cook a turkey. The pan containing the turkey rested on a grating supported by projections on the two walls of the oven. The purchaser was extremely unhappy when the grating, turkey, and pan fell to the floor of the oven because the inner wall of the oven was not stiff enough to support its side of the loaded grating.

When a complaint was made to the retailer who had sold the oven, it was explained that the *X* Company had discovered this weakness in this particular model of stove and had provided a kit to be installed without charge. The retailer's service technician installed this kit. However, several months later when another attempt was made to cook a turkey, it was found that the kit had not been effective; the turkey, pan, and grating again fell to the bottom of the oven.

This time the retailer explained that the guarantee period had expired and that an alteration of the stove could not be made without charge, but that the company had now discovered the change that was needed to increase the stiffness of the inner wall of the large oven. When the service technician came to make the estimate of the cost of this alteration (about $50), the owner of the stove spoke quite unfavorably of the *X* Company. "Well," said the technician, "The *X* Company is usually pretty good but it sure pulled a boo-boo when it designed the oven for this stove!"

Example 9.5

A case of an imperfect design specification and a satisfactory product

Facts of the case In a fairly complex electrical product for household use, there were many company-designed parts that the *Y* Company purchased from outside suppliers. For one such part purchased from supplier *A*, the *Y* Company's designer specified an aimed-at value of a certain mechanical quality characteristic with tolerances of plus or minus 10%. Perhaps because special apparatus was needed to test this quality characteristic, no tests of this characteristic were made by the *Y* Company's receiving department. The parts purchased from supplier *A* were installed in the complex product and functioned satisfactorily.

†Examples 9.4 and 9.5 are adapted from E. L. Grant and L. F. Bell, "Some Economic Aspects of Quality Standards," *ASQC Convention Transactions 1960,* pp. 231–236, American Society for Quality Control, Milwaukee, Wis., 1960.

After this state of affairs had continued for several years, a few of these parts were tested as one of the operations in a quality audit. When none of these parts fell within specification limits, the parts in stock and the incoming lots were tested. The group making the audit finally reached the opinion that it was likely not one of the many components purchased from supplier *A* over a period of years had ever fallen within the specification limits. Moreover, the quality audit indicated that the finished product would have functioned less satisfactorily if supplier *A* actually had made this part in accordance with the designer's specifications.

Some Comments on Examples 9.3, 9.4, and 9.5

Comment on certain interesting cost matters and organizational matters illustrated by these examples will be put off until Chap. 17. The following comments deal only with topics covered in this chapter.

Example 9.3 illustrates how a particular specification that was recognized as being too tight was not enforced by the receiving department of the purchaser. It also illustrates the kind of adverse consequences that may occur if a purchaser gives no information to suppliers about the relative importance of different design specifications.

Even if all the stoves of the particular model discussed in Example 9.4 had shown perfect quality of conformance to specifications, the stoves would not have satisfied their purchasers. The "boo-boo" made by the designer caused their quality of design to be unsatisfactory as judged by any reasonable standard. (Incidentally, the purchaser did not pay the \$50 demanded by the retailer to correct the designer's error. In effect, the support for the grating in the large oven was redesigned by using four empty cans to hold it up at its four corners.)

In contrast to Example 9.4, the components of the complex product made by supplier *A* in Example 9.5 were 100% defective in the sense of showing a complete lack of conformance to specifications. Moreover, the quality of design was somewhat poorer than might reasonably have been expected. Nevertheless, in the sense in which *quality* ordinarily is used in everyday speech, the quality of the finished product was satisfactory. Surprisingly, the failure to check on quality of conformance offset the inadequacies in quality of design.

These three examples have been inserted at this point to ensure that any readers who are unfamiliar with business will not conclude from this discussion of specifications and tolerances that the whole story about quality is necessarily told by a division of product into two categories, namely, conforming and nonconforming to design specifications.

9.4 STATISTICAL METHODS MAY HELP IN SETTING BETTER SPECIFICATION LIMITS

The designer's considerations in establishing any specification limits may be classified into three groups, namely, (1) those related to the service needs of the article or part for which specifications are being written, (2) those related to the capabilities of the production process to produce to any given specification limits, and (3) those related to the means to be used for determining whether the specifications are actually met by the product.

The fundamental basis of all specification limits is, of course, the service need of the part or article. This is not primarily a statistical matter. However, it often happens that the service need can be judged more accurately with the aid of statistical methods. The viewpoint that every quality characteristic is a frequency distribution is always helpful. This is particularly true in matters involving the interrelationships of specification limits. The usefulness of statistics in this respect is illustrated later in this chapter in Examples 9.7 and 9.8.

There is no use specifying desired tolerances on any quality characteristic without some prospect that these tolerances can be met. Whenever production methods will not meet the proposed tolerances, this fact needs to be known and considered before such tolerances are adopted. The inability of any process to meet its quality specifications is sure to be responsible for extra costs. These may be costs of spoilage and rework, or cost of changing to another more expensive production process, or—in the case of dimensional tolerances—costs incident to giving up the idea of interchangeable manufacture and adopting selective fitting of parts. If such costs are to be undertaken, this should be done deliberately after weighing the facts rather than unconsciously because the capabilities of a production process are unknown. One of the major contributions of statistical quality control to design can be in the information the control chart gives about the capacity of a production process to meet any given tolerances. This point was a major topic of Chap. 8 and is expanded later in this chapter.

The third matter that was stated as relevant in setting specification limits is the means to be used for determining whether the specification limits are actually met. The relevance of the inspection specification in determining the design specification is not always understood. Both design engineers and inspection executives sometimes say, "It is the designer's job to specify what is needed; it is the inspector's job to devise tests and acceptance procedures to find out whether the designer's specifications have been met; let each stick to his or her own job." This fails to take into account the fact that, although the designer may not devise tests and acceptance procedures, he or she should not be indifferent to the tests and acceptance procedures that are to be applied. These constitute useful information for the designer for the same reason that he or she should know something about the capabilities of the production process; both influence the likelihood of obtaining what is specified. Teamwork is an absolute requirement in today's highly competitive environment.

9.5 SOME COMMON METHODS OF INTERPRETATION OF A PILOT RUN AS A BASIS FOR SETTING TOLERANCES

In the development of any new design of a manufactured product, it is often possible to improve the design and avoid production difficulties by means of a pilot run. If production methods used on this pilot run are similar to those that are to be used later in actual quantity production, it is good sense to use the pilot run to review the proposed specification limits for the various quality characteristics. Thus it may be possible to anticipate and avoid situations in which the design tolerances are closer than can be met by the production departments.

The number of units produced in such a pilot run will depend on the costs involved, and might vary from two or three to several thousand. Suppose 100 units are produced, and actual measurements are made on each unit for every quality characteristic for which tolerances have been proposed. What tabulations and calculations are required relative to the measurements on each quality characteristic? How should the measurements be interpreted with regard to the setting of tolerances? Different individuals might answer these questions in different ways.

Without benefit of any statistical analysis, a common answer might be to take the highest and lowest measured values on the pilot run as indicating the upper and lower limits of the production process.

Those who have been introduced to conventional elementary statistics, including frequency distributions and the normal curve, might reject this simple method as unsatisfactory. Instead they might group the measurements into a frequency distribution and compute the average and standard deviation. They might then compute 3-sigma limits on either side of the average and state that such limits would include practically all the items produced. Or they might even go so far as to make the precise statement that they will include 99.73% of the items produced.

Individuals who have been introduced to the Shewhart control chart will be inclined to insist that the order of production of the units in the pilot run of 100 not be lost sight of. The measurements might then be divided into rational subgroups—perhaps 25 subgroups of 4—and might then be tested for control by means of charts for $\bar{X}$ and R or $\bar{X}$ and s. If no points fall outside control limits, σ might be estimated, either as $\bar{R}/d_2$ or as $\bar{s}/c_4$. Then, $\bar{\bar{X}} \pm 3\sigma$ will give an estimate of upper and lower limits for individual values in a controlled process—the so-called natural tolerances of the process. Those who have made these calculations might state that if control is maintained, practically all the items produced will fall within these estimated natural tolerance limits. Or, as in the case of the frequency distribution analysis, they might go farther and state that 99.73% of the cases will fall within these limits.

9.5.1 Errors in Common Interpretations of a Pilot Run

All these methods of interpreting a pilot run miss the point to some extent. The frequency distribution method is likely to be better than taking as the limits merely the upper and lower measured values in the pilot run. The control-chart method is much more realistic than the frequency distribution method, in that it recognizes that there is no basis for an inference that a frequency distribution from an uncontrolled process will repeat itself. But no method of analysis can be found that justifies the positive statements that the process will hold within certain limits.

The justifiable statement is a negative one to the effect that without a fundamental change in the process, it will be impossible to hold the process within limits closer than certain specified values. The control-chart method provides the basis for such a statement if the process is in control. If the proposed tolerances are closer than such limits, it is clear that either the tolerance range must be widened or the process must be changed; or if 100% inspection is possible, the decision must be made to accept the inevitability of producing some nonconforming product and doing a 100% sorting job of separating good from bad.

Even though no positive statement can be justified that a production process will hold within given limits, a positive decision must be made by the designer regarding the tolerances to be specified. In using the evidence of the pilot run to help in this decision, the designer should recognize that even though the natural tolerances as computed from $\overline{\overline{X}} \pm 3\sigma$ happen to be within the proposed specification limits, this fact does not *ensure* that these proposed limits can always be met. The estimate of what tolerances actually can be held is partly a matter of statistics and partly a matter of engineering judgment.

The statistical questions involved deal with the reliability of estimates from a limited sample and with the form of the frequency distribution from a controlled process. Not a great deal can be told about the exact shape of a frequency curve from the evidence of a short pilot run, such as one providing only 100 measured values. It would take tens of thousands of values all in control to define the extreme portions of a frequency curve well enough to justify a statement such as one that in the long run 99.73% of the values will fall within certain specified limits. Skewed distributions of industrial quality characteristics often have several times 0.27% of the distribution outside one 3-sigma limit even though none of the distribution is outside the other 3-sigma limit.

The questions of engineering judgment involved in estimating what tolerances can be held deal with the difficulties of maintaining statistical control in the light of such matters as expected tool wear, operator variability, variability of incoming materials, and so forth. Any differences between the conditions of the pilot run and those of actual production should also be considered.

Example 9.6

Establishment of tolerance limits by pilot runs

Facts of the case Nearly all the product of a company manufacturing electronic devices was sold to one customer. New designs of devices were continually being developed. In each new design, it was necessary to establish tolerances on the many different measurable quality characteristics. With regard to many of the characteristics, the usual practice had been for the customer to set specification limits at the design value ±10%. It nearly always happened that when the device got into production, there would be difficulty in meeting some of the tolerance limits that had been specified in this way.

After this manufacturer had started to use statistical quality control in connection with production operations, the control chart was applied to the analysis of the pilot runs used in development work on new devices. This analysis disclosed those characteristics for which the ±10% tolerances seemed likely to cause trouble. It also disclosed other characteristics for which it seemed probable that tolerances could be much closer than ±10%, so that some specified tolerances might be tightened where it was advantageous to do so. On this basis, definite suggestions were made to the customer regarding tolerances for all quality characteristics. These suggestions were usually accepted. As a result, specifications on new designs were better adapted to the capabilities of the production process, the percentage nonconforming was reduced, and increased uniformity was obtained on certain important quality characteristics.

9.6 TWO STATISTICAL THEOREMS OF GREAT IMPORTANCE IN THE INTERRELATIONSHIP OF TOLERANCES

A dimension on an assembled product may be the sum of the dimensions of several parts. Or an electrical resistance may be the sum of several electrical resistances of parts. Or a weight may be the sum of a number of weights of parts. In this common situation, what should be the relationship of the tolerances of the parts to the tolerances of the sum? This question may be answered with the aid of a theorem of mathematical statistics that is extremely useful in many problems of quality control. This theorem states that the standard deviation of the sum of any number of independent variables is the square root of the sum of the squares of the standard deviations of the independent variables:

$$\sigma_{\text{sum}} = \sqrt{\sigma_1^2 + \sigma_2^2 + \sigma_3^2 + \cdots + \sigma_n^2}$$

For example, consider an assembly containing a dimension *AD* that is the sum of the dimensions of parts *AB*, *BC*, and *CD* (see Fig. 9.9). Assume that the dimension of each part is statistically controlled with averages and universe standard deviations as follows:

	μ	σ
AB	1.450	0.0010
BC	0.865	0.0008
CD	1.170	0.0007

Assume that upper and lower specification limits are at 4-sigma distance from the ave Specifications are then as follows:

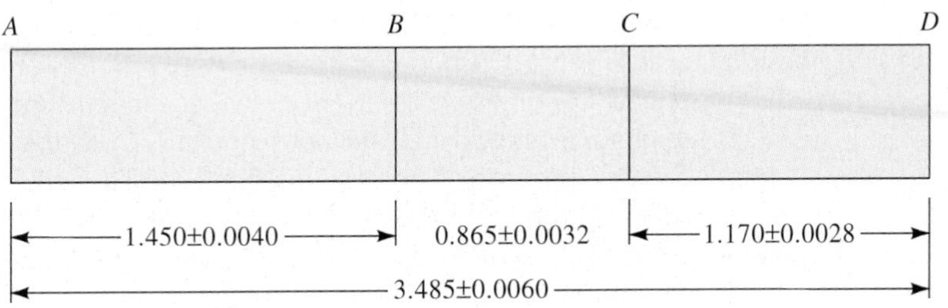

FIGURE 9.9 Dimension *AD* is built up from the assembly of parts having dimensions *AB, BC,* and *CD.*

	Specification	U	L	Tolerance range $U - L$
AB	1.450 ± 0.0040	1.4540	1.4460	0.0080
BC	0.865 ± 0.0032	0.8682	0.8618	0.0064
CD	1.170 ± 0.0028	1.1728	1.1672	0.0056
Total		3.4950	3.4750	0.0200

Someone considering merely the specified tolerances and not the distribution pattern of the dimensions might conclude that the appropriate specification for the overall dimension *AD* was 3.485 ± 0.010. This conclusion is incorrect because it fails to recognize the probability aspects of the situation. If the assembly of the parts is at random, large *AB* parts are just as likely to be assembled with small *BC* parts as with large ones. If the probability is small of the occurrence of an extreme maximum value of one part, the probability that three parts all having maximum values will be assembled together is very much smaller. The probabilities of all kinds of combinations of parts are reflected in a simple way by the formula for the standard deviation of the sum of independent variables:

$$\sigma_{AD} = \sqrt{\sigma_{AB}^2 + \sigma_{BC}^2 + \sigma_{CD}^2}$$

$$= \sqrt{(0.0010)^2 + (0.0008)^2 + (0.0007)^2} = 0.0015$$

If the specification limits for the sum *AD,* like the limits for the parts, are placed at 4-sigma distance from the average, the specifications for *AD* will be 3.485 ± 0.006. This ± 0.006 is in sharp contrast with the ± 0.010 obtained by simple addition of the allowable maximum and minimum values of the parts.

Some individuals in a manufacturing plant may find it difficult to believe that the laws of chance make the appropriate tolerance range for an assembly less than the sum of the tolerance ranges for the component parts. To convince them, it may be helpful to make an experimental verification of the theorem for the standard deviation of the sum of independent variables. Such an experiment at Westinghouse Electric Corporation is reported by B. Epstein.† Fifty of each of three components of an assembly were drawn at random from storage lots. These were gaged, and averages and standard deviations of the dimensions of each component were obtained. The 3-sigma limits on assembled dimensions were then predicted by formula. The components were then selected at random and assembled, and the assemblies were gaged. The 3-sigma limits on the dimensions of the assemblies checked closely with the limits predicted by the use of statistical theory.

Whenever it is reasonable to assume that the tolerance ranges of the parts are proportional to their respective σ values, such tolerance ranges may be combined by taking the square root of the sum of the squares. Thus

$$\begin{aligned}(U-L)_{AD} &= \sqrt{(U-L)^2_{AB} + (U-L)^2_{BC} + (U-L)^2_{CD}} \\ &= \sqrt{(0.0080)^2 + (0.0064)^2 + (0.0056)^2} = 0.012\end{aligned}$$

The actual situation in setting tolerance limits is often the reverse of the one just described. Frequently the desired tolerance range is known for the overall quality characteristic (such as a dimension, electrical resistance, weight), and the problem is to set tolerance ranges on the component parts that can be expected to produce the desired range on the whole. This may be determined without much difficulty by trial-and-error solution of the formula. However, a direct solution is possible if the standard deviations of the various parts are assumed to be equal. In this case the formula for the standard deviation of the sum of n independent parts becomes

$$\sigma_{\text{sum}} = \sqrt{n\sigma^2_{\text{part}}}$$

From this, it is evident that

$$\sigma_{\text{part}} = \sqrt{\frac{\sigma^2_{\text{sum}}}{n}}$$

If tolerance ranges are assumed proportional to standard deviations,

$$(U-L)_{\text{part}} = \sqrt{\frac{(U-L)^2_{\text{sum}}}{n}}$$

†B. Epstein, "Tolerances in Assemblies," *American Machinist,* vol. 90, no. 1, pp. 119–121, Jan. 3, 1946.

Another useful theorem† of mathematical statistics deals with the standard deviation of the difference between independent variables, which is as follows:

$$\sigma_{1-2} = \sqrt{\sigma_1^2 + \sigma_2^2}$$

It will be noted that the standard deviation of the difference is the same as the standard deviation of the sum. Example 9.8 gives an illustration of the use of this difference theorem in a problem involving the interrelationship between nominal dimensions.

A word of caution is called for on the application of these useful theorems to tolerance problems. The theorems assume independence of the variables being added and subtracted. In some assembly operations this independence does not exist. For example, suppose a dimension on an assembled product is the combined thickness of four metal washers. If these washers all come from the same production process and are assembled in the order of production, the thickness of the four washers in any assembly will not be independent of one another unless the process of producing them is in strict statistical control. Otherwise there will be a tendency for thick washers to be assembled together in some assemblies and thin washers assembled together in others. If this tendency exists, the theorem for the standard deviation of the sum of independent variables is not applicable to the problem of setting tolerances on the overall dimension.

Example 9.7
Tolerances at intermediate stages in manufacturing electrical cable

Facts of the case The manufacture of armored electrical cable requires a number of operations, such as insulation of wire, spinning strands of insulated wire together, applying various types of insulating and water-resisting coatings, and applying the armored covering. Customers' specifications on dimensions apply only to the finished diameter. However, to meet specified tolerances on the finished diameter, it is desirable to set tolerance limits on the diameters after each of the manufacturing operations. Such limits, properly established, should be helpful to the various manufacturing departments. In order that the limits be respected, they should represent a practical goal at each stage of manufacturing, and they should be set in such a way that departure from the intermediate limits really causes difficulty in meeting specifications on the completed product.

In the case where an arbitrary guess is used as the basis for such tolerance limits on intermediate diameters, it is likely that the limits so established will be inconsistent with one another. At one stage in manufacturing, they may be tighter than it is practicable to hold. At another, the manufacturing departments may believe that even though the intermediate tolerances are not met, it will

†For a simple proof of these two theorems, see W. Mendenhall and R. L. Schaeffer, *Mathematical Statistics with Applications,* pp. 185–188, Duxbury Press, North Scituate, Mass., 1973.

still be possible to make enough corrections in the remaining operations to meet the overall specification. If the intermediate tolerances have these faults, they may not be respected by the manufacturing departments. If a customer's specified tolerances on final diameter are tight and are enforced, this state of mind on the part of the manufacturing departments may result in a substantial percentage of rejected product.

A rational approach to a consistent set of tolerance limits would require an estimate of the natural tolerances of each operation. For example, if operation 3 increases the diameter from 0.300 to 0.500 in, an $\bar{X}$ control chart might be maintained on which the variable was the increase of diameter produced by operation 3. This would give a basis for estimating 6σ for operation 3. The 6σ values for the increase in diameter from each of the other operations might be similarly estimated. An analysis of the tolerances should then start with the specification limits on the final diameter and work backward operation by operation.

This is illustrated in Table 9.2. The figures used in this table have been modified from an actual case in a way that serves to conceal any confidential information. The calculations in this table start with the assumption that it is desired that the specified tolerance range of 0.050 on the final diameter should represent 6σ. The tolerance range at the start of the last operation (operation 8) is computed to be consistent with this specification and with the natural tolerance of the final operation (0.016). A similar calculation is made for each preceding operation. The calculations depend on the use of the theorem for the standard deviation of the sum of two variables:

$$\sigma_{\text{sum}} = \sqrt{\sigma_1^2 + \sigma_2^2}$$

In this case, σ_{sum} (the universe standard deviation of the final diameter) and σ_2 (the universe standard deviation of the last operation) are known or assumed, and it is desired to find a consistent σ for the diameter just before the final operation. Hence

$$\sigma_1 = \sqrt{\sigma_{\text{sum}}^2 - \sigma_2^2}$$

Table 9.2 is representative of the first of a series of trial calculations for setting the tolerances on the intermediate operations. It uses (column *C*) an allowance of 6σ for each operation from operation 8 to operation 2. This results in a computed tolerance range for operation 1 of 0.015, whereas the 6σ tolerance range of operation 1 is 0.008. Thus, if the computed specifications of column *I* were to be adopted, this would amount to allowing a little over 11σ range for operation 1 and only a 6σ range for each of the other seven operations. It is evident that a second trial calculation is necessary.

This new calculation might be based on an arbitrary increase of all the column *C* figures by some selected percentage. In this way each of the eight operations would be treated alike with regard to the increased latitude allowed in the

TABLE 9.2 EXAMPLE OF TRIAL COMPUTATION OF TOLERANCE LIMITS AT INTERMEDIATE STAGES IN MANUFACTURE OF ELECTRICAL CABLE

Final diameter specified as 0.750 in ± 0.025

Operation number *(A)*	Nominal diameter at completion of operation *(B)*	Estimated natural tolerance range (6σ) of increase in diameter due to operation *(C)*	Specified tolerance range of diameter at completion of operation $X_{max} - X_{min}$ *(D)*	Computed squared tolerance range *(E)*	C^2 *(F)*	$E - F$ *(G)*	Computed tolerance range before start of operation $\sqrt{G}$ *(H)*	Computed specification at start of operation *(I)*
8	0.750	0.016	0.050	0.002500	0.000256	0.002244	0.047	0.710 ± 0.023
7	0.710	0.021	0.047	0.002244	0.000441	0.001803	0.042	0.640 ± 0.021
6	0.640	0.018	0.042	0.001803	0.000324	0.001479	0.038	0.585 ± 0.019
5	0.585	0.007	0.038	0.001479	0.000049	0.001430	0.038	0.550 ± 0.019
4	0.550	0.013	0.038	0.001430	0.000169	0.001261	0.036	0.500 ± 0.018
3	0.500	0.028	0.036	0.001261	0.000784	0.000477	0.022	0.300 ± 0.011
2	0.300	0.016	0.022	0.000477	0.000256	0.000221	0.015	0.100 ± 0.007
1	0.100	0.008	0.015	0.000221	0.000064	0.000157		

tolerances. Or, if experience indicated that it was difficult to maintain control on two of the operations and easy to maintain control on the other six, the entire increase in latitude of tolerances might be thrown into an increase in the column *C* figures for the two operations on which control was difficult. Several trial calculations might be required to obtain a satisfactory set of computed specifications. The test of a consistent set is agreement between the computed tolerance range for operation 1 and the figure shown in column *C* for the estimated natural tolerance range of this operation. Or, stated a little differently, the final computed figure in column *G* (shown as 0.000157 in Table 9.2) should be zero.

Example 9.8

An illustration of the statistical relationship among specified nominal dimensions, tolerances, and fits and clearances

Facts of the case Many hand tools used by mechanics involve sockets and attachments. W. B. Rice describes a study[†] made to determine the difference between the specified nominal dimensions of the socket and the corresponding attachment, and to establish the tolerances for each dimension. He explains, as follows:

> The socket has a square hole into which the end of the attachment fits. There is a ball and spring assembled on the male square (attachment) which catches in a ball check hole in the female square (socket) so that the two tools become a single driving unit. Important dimensions are: On sockets, the inside dimension of the female square; on attachments, the width across the flats and the height of the ball. . . . In actually working out this problem, there were several variables involved, but for the sake of clarity in presentation only the distance across male and female squares will be considered. The type of attachment studied was an extension, a straight metal bar with a male square at one end.
>
> The first step was to make several hundred random assemblies of finished sockets and extensions to determine what maximum and minimum clearances between male and female squares were necessary for practical use in a mechanic's work. This was done by a field survey. When the effect of ball height and other variables were removed, it was found on the $\frac{1}{2}$-in drive series that a minimum of 0.004 in and a maximum of 0.015 in clearance between male and female squares were required.

Rice explains that this study of consumer requirements was followed by a study of the variability of the dimensions of the sockets and extensions as they were then being produced. Shewhart control charts for variables were maintained on each dimension until a state of control had been reached and the natural tolerances determined.

For the extensions (male squares) the average dimension $\overline{\overline{X}}_M$ was 0.5005 in. The standard deviation σ_M of this dimension (as estimated from $\overline{R}/d_2$) was 0.0015 in. For the sockets (female square) the average dimension $\overline{\overline{X}}_F$ was

[†]W. B. Rice, "Setting Tolerances Scientifically," *Mechanical Engineering*, vol. 66, pp. 801–803, December 1944.

0.5120 in. The standard deviation σ_F was 0.0010 in. The distribution curves for these two dimensions were approximately normal. This is illustrated graphically in Fig. 9.2*a*.

Establishing the nominal dimensions for sockets and extensions As shown in Fig. 9.2*a*, the 3σ limits for the individual extensions are 0.496 and 0.505. The 3σ limits for sockets are 0.509 and 0.515. If it is assumed that, because the distributions are approximately normal, practically all the values will fall within limits of $\pm 3\sigma$, it is evident that there will practically never be found an extension and socket with less than the desired clearance of 0.004 in. The maximum extension dimension is 0.505 in, and the minimum socket dimension is 0.509 in.

On the other hand, it is clear that occasionally an extension will be paired with a socket with more than the maximum desired clearance of 0.015 in. The minimum extension dimension is 0.496 in, whereas the maximum socket dimension is 0.515 in.

It therefore appears that the existing setting of the nominal dimensions of extensions and sockets at 0.5005 and 0.5120 in, respectively, tends toward fits that are too loose. Generally speaking, the tighter the fit, as long as there is enough clearance for extensions actually to go into sockets, the better satisfied the mechanic will be and the better the reputation of the line of tools. Hence consideration should be given to a possible decrease in the difference between the nominal dimensions.

Rice suggests that the principle governing the decision of how far to go in the change of nominal dimensions may be found in the answer to the question, "What risk does the manufacturer want to take that, somewhere, sometime, a mechanic will pick up a socket and extension with less than 0.004 in . . . clearance between them?"

The theorem for the standard deviation of the difference of two independent variables may be used advantageously to estimate this risk associated with any given difference between the nominal dimensions.[†] This standard deviation is

$$\sigma_{F-M} = \sqrt{\sigma_F^2 + \sigma_M^2} = \sqrt{(0.0010)^2 + (0.0015)^2} = 0.0018$$

The existing average clearance between sockets and extensions is

$$\overline{\overline{X}}_F - \overline{\overline{X}}_M = 0.5120 - 0.5005 = 0.0115$$

If, as in this case, both variables are distributed normally, the difference between the variables (that is, the clearance between sockets and extensions) will also be distributed normally. From this fact the probabilities may be calculated of a smaller clearance than the minimum desired value of 0.004 and of a larger clearance than the maximum desired value of 0.015.

[†]From this point on, the analysis presented here differs slightly from that given by Rice.

A clearance of 0.004 is a departure of 0.0075 (that is, 0.0115 – 0.004) from the average clearance. This is 0.0075/0.0018, or 4.17 times σ_{F-M}. The probability of as great a departure in one direction from the average is so small that it is outside the limits of Table *A,* App. 3, but may be determined from a six-place table of normal curve areas to be 0.000015. Thus too tight a fit might be expected about one time in 67,000.

A clearance of 0.015 is a departure of 0.0035 from the average clearance. This is 1.94 times σ_{F-M}. The probability of such a departure is given by Table *A* as 0.0262, or about one time in 38.

The distribution of this clearance between sockets and extensions is illustrated in Fig. 9.10*c*. This figure shows the dimension centered on the nominal dimension 0.0115, prior to revision of specifications, and with upper and lower natural tolerances of 0.0169 and 0.0061, respectively. Probabilities associated with clearance less than 0.004 and greater than 0.015 may be changed simply by shifting the distribution of differences to the left, that is, by narrowing the difference between the nominal dimensions of the distributions of the individual parts as shown in Fig. 9.10*b*. Calculations for other assumed clearances between average dimensions give the following probabilities of fits that are too tight or too loose:

$\bar{\bar{X}}_F - \bar{\bar{X}}_M$, in	Probability of smaller clearance than 0.004 in	Probability of larger clearance than 0.015 in
0.0115	0.0000+	0.0262
0.011	0.0001–	0.0132
0.0105	0.0002–	0.0062
0.010	0.0004	0.0027
0.0095	0.0011	0.0011
0.009	0.0027	0.0004
0.0085	0.0062	0.0002–
0.008	0.0132	0.0001–
0.0075	0.0262	0.0000+

This type of probability analysis supplies a rational basis for a decision as to the difference between the nominal dimensions. It is clear that the existing average clearance of 0.0115 should be somewhat reduced. A reduction of 0.002 to 0.0095 in gives a probability of 0.0011 (about 1 chance in 900) that a socket and extension selected at random will have a smaller clearance than the desired minimum of 0.004. In deciding whether or not to adopt this average clearance of 0.0095, it is also helpful to observe that if it is adopted, the probability of a clearance of 0.003 in is 0.0002; and the probability of a clearance of 0.002 in is negligible. All these probabilities are subject to the limitation that they assume statistical control of the production process and a normal distribution of the quality characteristics involved.

If 0.0095 is adopted as the average clearance, it remains to establish the average value of the nominal dimension. If this were to be 0.511 for the sock-

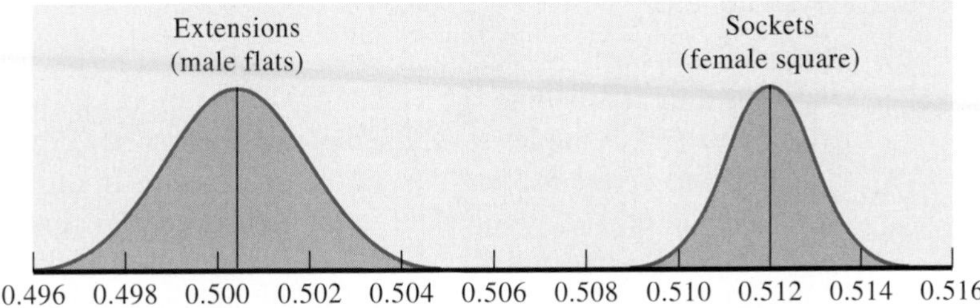

(*a*) Distribution of dimensions of extentions and sockets before revision of specifications

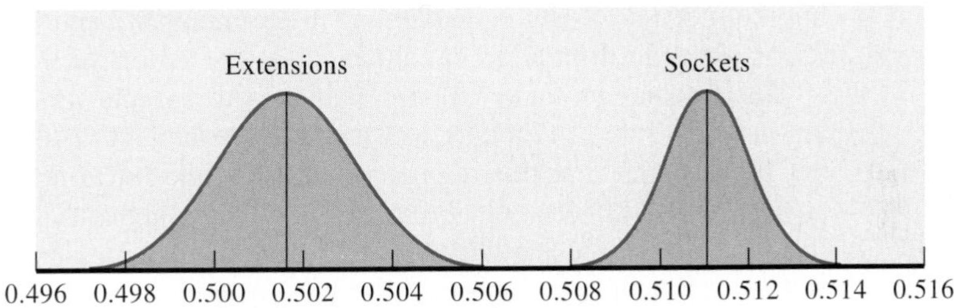

(*b*) Statistical analysis indicates that a closer fit between sockets and extensions is advantageous

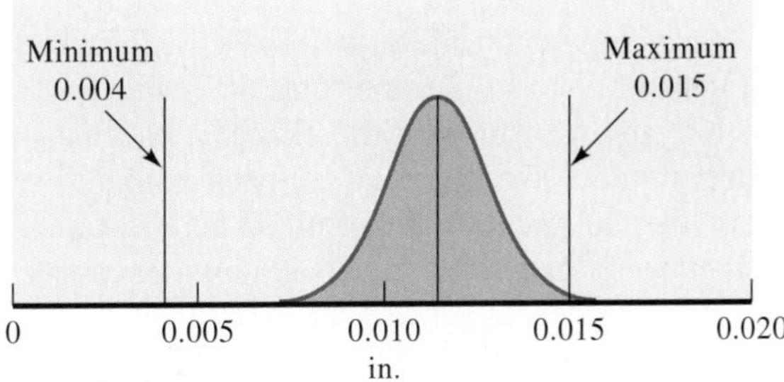

(*c*) Distribution of the clearance between extensions and sockets before revision of specifications

FIGURE 9.10 Frequency curves representing distribution of dimensions of sockets and extensions. Example 9.8.

ets, the nominal dimension of extensions would be 0.5015. Figure 9.10*b* shows the relative positions of the frequency curves if these values are adopted.

If 100% inspection is to be used, the tolerances on extensions should be specified as ±0.0045 and on sockets as ±0.0030. For reasons explained in Chap. 8, if sampling inspection is to be employed, somewhat tighter tolerance limits may be desirable.

9.7 EXPERIMENTATION

The main emphasis in statistical process control is on securing and maintaining control and understanding the variability of the process. However, we have studied a number of examples in which there were other objectives. In some instances, the capability of the process was of major concern. In others, it was the error of measurement or the judgmental error introduced by individuals when visual inspections were performed.

In each of the examples discussed in this and the previous chapter, some degree of planned experimentation was involved, and control charts were used to evaluate the results and make decisions. Sometimes this involved shifting people from one machine or duty to another, as in Example 8.1. Sometimes it involved changing and controlling process input variables, an in Example 8.3, or simply redesigning the method of data collection, as in Examples 8.2 and 9.1, so that the control charts yielded information on which corrective action decisions could be based.

The adjustment of and experimentation with a single variable, such as the forming-gas flow rate in Example 9.1, may yield excellent results. However, it should not be assumed that this approach, called a one-factor-at-a-time approach, will always work well or at all. Processes involving chemical, metallurgical, and biological reactions in particular may be extremely complex and, like the silicon wafers in Example 9.1 and the wire lines in Example 9.2, may involve too many adjustable variables. When it is known that a number of variables, or factors, influence the results of interest, many one-factor experiments are necessary to study the effects of all. Furthermore, the optimal combination of variables may never be revealed. This is true particularly if there are interactions among the input variables.

When many controllable variables are involved and/or it is suspected that varying their levels causes some form of interaction, experiments need to be designed carefully prior to making the changes. Normally, operating and engineering personnel do not have the technical background to set up these designs. The services of an internal or external consultant and/or statistician thoroughly grounded in the techniques of experimental design are a necessity.

Any extended exposition of experimental design is well beyond the scope of this book. However, we do wish to provide the reader with some background on a few major contributors to the science and offer some feel for the importance of this subject in improving production processes. To accomplish the latter, we present an example condensed from G. E. P. Box, W. G. Hunter, and J. S. Hunter.†

9.7.1 Two Major Contributors to Industrial Experimentation

Perhaps the most celebrated example of the application of experimental design is the often reported ceramic tile design improvement experiment performed by Dr. Genichi Taguchi in Japan in 1953. The problem resulted from the inability to

†G. E. P. Box, W. G. Hunter, and J. S. Hunter, *Statistics for Experimenters,* op. cit., pp. 510–525.

maintain a uniform temperature throughout a kiln during the firing of clay tiles. Completed batches exhibited too much variation in the dimensions of the end product. This excessive variation resulted from an uneven temperature distribution throughout the kiln. The objective of the experimental design was to reduce variability among the tiles by changing the input materials in order to make the end product less variable. While the root cause had been identified, redesigning the oven was not economically feasible. The experiments resulted in identifying an optimal mix of materials that made the dimension of end product tiles less sensitive to variation in firing temperature.

Two important contributions to product and service design optimization that may be credited to Taguchi are his concept of a *robust product design* and his philosophy of presenting experimentation in a pragmatic way. A robust product design is one for which output values of the characteristics of interest are insensitive to moderate variation in input factors or variables. Thus the designer is driven to seek combinations of inputs for which variation of output is relatively flat. In the clay tile example, the problem was variability in a dimension, and the solution was to improve the "recipe."

Taguchi's second contribution may be his most important because it led him to develop models and present design techniques in such a way that they could be taught easily to engineers and managers in industry in Japan. Thus, by the early 1980s Americans visiting Japan began to see the "new" statistical methodology being used to optimize productive output. Many U.S. companies eagerly sought his help in implementing the new tools.

A second major contributor to experimental design philosophy and application is the British statistician Dr. George E. P. Box, who, for many years, has been a member of the faculty of the University of Wisconsin in Madison. He is credited with having developed *evolutionary operation* methodology and the techniques of *response surface analysis*. Evolutionary operation involves experimenting with operating processes. It is assumed that the operating process is performing in a satisfactory manner at the start of experimentation. The controllable variables are changed marginally in a scientifically designed pattern to gain insight into the effect on one or more performance characteristics. Thus the objective is to improve output while changing input variables marginally in such a way that no serious disruption in output can occur. This concept meshes perfectly with a continuous improvement management philosophy.

Response surface methodology is intended to assist in determining optimal combinations of the controllable variables to maximize or minimize some one performance characteristic. If there are several performance characteristics, those remaining are treated as constraints on the problem.

Example 9.9
Improving the yield of a chemical reaction

Facts of the case In a certain chemical reaction, the two controllable variables were time in the reaction vessel and temperature during processing. Past experience indicated that a yield of about 62 g could be obtained at 130°C for 75 min. Suspecting that this yield might be considerably improved, a series of experiments was set up.

Analysis Initially an experiment was set up using times of 70 and 80 min and 127.5 and 132.5°C. These four runs plus three at the standard levels of 75 min and 130°C, a total of seven runs made in random order, gave the yields indicated by the x's on Fig. 9.11.

FIGURE 9.11 A diagram illustrating an experimental design intended to maximize yield from a chemical process. (*Taken by permission from G. E. P. Box, W. G. Hunter, and J. S. Hunter, "Statistics for Experimenters," p. 515, John Wiley & Sons, Inc., New York, 1978.*)

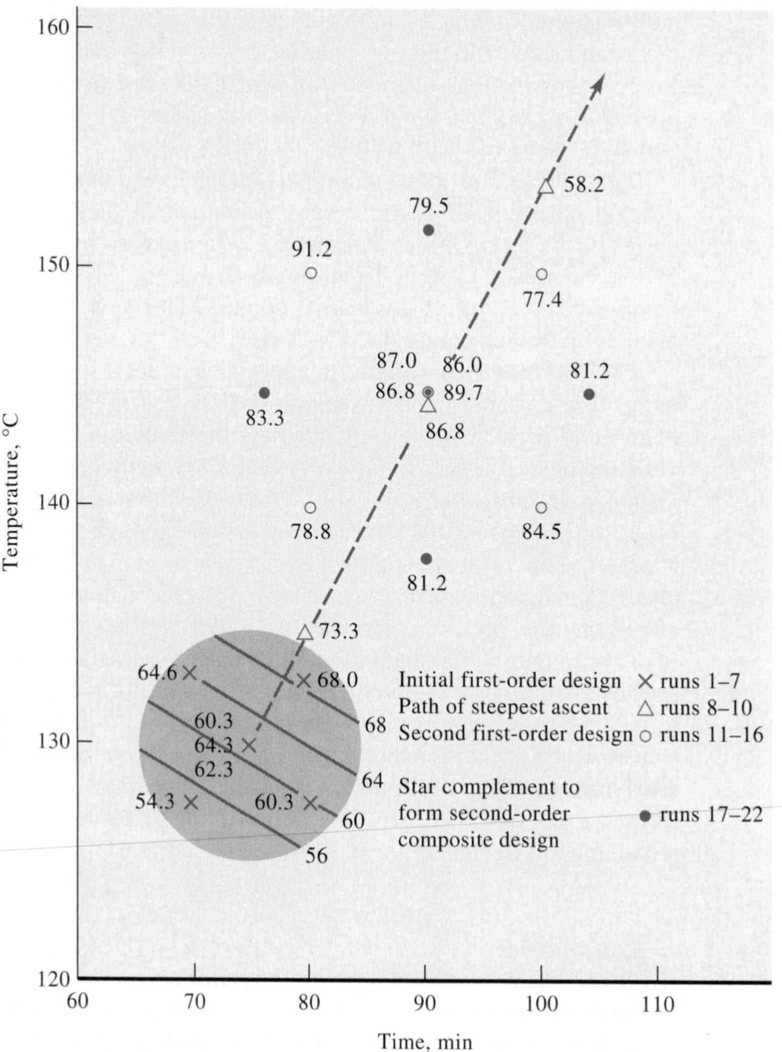

From these data, a line could be developed indicating the direction of movement needed to improve yield. This is the dashed line in Fig. 9.11. In statistical language this line is called a *linear regression line.*

In the second stage of the experiment, runs were made along this line. Run 8, made at 134.8°C for 80 min, yielded 73.3 g. Run 9 was made at 153.9°C for 100 min and yielded 58.2 g. Apparently the maximum point had been overshot. Run 10, made at 144.4°C for 90 min, yielded 86.8 g. Thus it appeared that the maximum yield following this *line of steepest ascent* would be in the vicinity of the run 10 levels. These run levels are indicated in Fig. 9.11 by diamonds.

A new experiment was set up to explore the region around the run 10 level. Six additional runs were made, two at 145°C for 90 min and four of the combinations of 140 and 150°C and 80 and 100 min. These runs and their respective yields are indicated by the open circles on Fig. 9.11. The results provided data to make a linear mathematical model of process yield response and also to test for nonlinearity. From this information, six more runs, indicated by the filled circles, provided further information for the development of a nonlinear model. Thus contour lines of equal yield, much like the isobar lines on a meteorologist's map or the elevation lines on a topographical map, could be plotted on the figure.

Comment on Example 9.9

The focus of attention in Example 9.9 was on the yield of the process. Suspecting that it could be improved, a scientifically designed statistical experiment was set up and performed in stages. The details of how the levels of time and temperature were determined have been left out. Suffice it to say that their choice was based on sound statistical principles covered in any good book on experimental design.

Two additional characteristics were measured and recorded for each run: color and viscosity. In the final decision on target levels of time and temperature, these additional performance characteristics were treated as constraints. That is, limits were set for color and viscosity, and the levels of time and temperature were chosen to maximize yield subject to those constraints. Naturally, the validity of the experiments and of the resulting mathematical models rests on operating the process in control at the new levels for a period of time of adequate length.

The study involved an analysis of means, the levels of yield, and an analysis of variances s^2. From these analyses, initial linear mathematical models and prediction equations were derived. Ultimately a quadratic model was developed. Very simple tests for the analysis of means and analysis of variances are discussed in Chap. 5.

PROBLEMS

9.1. Ceramic items are "baked" in a gas-fired tubular-shaped oven in lots of 100 units. The items are lined up in a single row on a special high-temperature pallet, they are slid into the oven, the oven is closed, and the temperature is increased manually according to a time-temperature scale. After cooling, a sample of 5 units is drawn from a lot and strength-tested by breaking them in a special testing device. To be effective in use, breaking strength must exceed 5 psi. If any item breaks at less than 5 psi, the lot from which it came is scrapped. About 10% of the lots have been scrapped in recent times.

(*a*) A decision is made to run a process capability study. Data on the next 20 lots are tested using the same procedure, and a sample size of 5, and $\overline{X}$ and R charts are drawn. At this point, $\Sigma\overline{X} = 126.00$ psi and $\Sigma R = 28.44$ psi. Calculate control limits for $\overline{X}$ and R for this process, and estimate σ.

(*b*) Assuming a normal distribution of product, what proportion would you estimate does not meet specifications?

(*c*) Calculate the pertinent process capability indexes for this process, and interpret their meaning relative to your answer to part *(b)*.

(*d*) Assuming that the R chart shows good control, where should the process be centered to ensure that no more than 0.1% of the items do not meet specifications (assume a normal distribution)? Calculate the same process capability indexes at this target mean setting.

(*e*) Assume now that the $\overline{X}$ chart has a number of points out of control on both sides. What further investigation would you recommend?

9.2. In the mounting of microelectronic chips in their package, a critical characteristic is the attachment of lead wires from the chip to the contact pins prior to sealing the package. If the contact is not sufficiently solid when sealed, the leads will separate, rendering the chip inoperable. Study has indicated that, if the leads can withstand at least 6 g tension pull, they will hold during the sealing operation. Samples of 5 units are drawn from the process and the leads tested to destruction. After 20 subgroups have been drawn, $\Sigma\overline{X} = 138.40$ g and $\Sigma R = 11.64$ g.

(*a*) Calculate the central lines and control limits for $\overline{X}$ and R for this process, and estimate the standard deviation σ.

(*b*) Estimate the proportion of leads that do not meet the specification of a minimum 6-g attachment strength. Assume that a normal distribution underlies the process.

(*c*) Calculate the appropriate process capability indexes for this process. Which indexes discussed in the text do not apply and why?

(*d*) The actual proportion of rejected product exceeds the estimate found in part (*b*) by a considerable amount. Further investigation reveals that several operators and machines generate product over two shifts of operation and that inspection was performed on receipt of batches of product at the package-sealing operation. Suggest improved procedures for sampling designed to uncover the true capabilities of the attachment process.

9.3. A critical dimension in the production of a certain part specifies a length of 60.00 ± 0.05 mm. It is proposed to use $\overline{X}$ and R charts to control this dimension. After 30 subgroups of 4 items each have been taken, $\Sigma\overline{X} = 1799.40$ mm and $\Sigma R = 1.86$ mm.

(*a*) Calculate trial control limits for this process.

(*b*) Calculate σ, and estimate the proportion that does not meet specifications on the assumption that unit measurements follow a normal distribution.

(*c*) Calculate the appropriate process capability indexes, and comment on the relationship between these indexes and the results found in part *(b)*.

(*d*) In actuality about 15% do not meet specifications, mostly below the lower limit of 59.95 mm, and the process shows a definite lack of control with respect to $\overline{X}$ with a run of seven points above $\overline{\overline{X}}$ and points out below the lower control limit. Further analysis of the data yields

$$n = \Sigma f = 120 \qquad \Sigma fd = -240 \qquad \text{and} \qquad \Sigma fd^2 = 2{,}360$$

(using the short method described in Chap. 3, p. 89) with an origin at 60.00 mm and a cell width of 0.01 mm. Estimate μ and σ from these data.

(*e*) Can you offer a tentative explanation of the difference between the σ estimated in part (*b*) and that estimated in part (*d*)? How might a histogram of the data appear? How do these results affect the interpretation of the process capability indexes calculated in part (*c*)?

9.4. A process has demonstrated that, when held in control, it can maintain a σ of 0.15 cm. A certain part has specifications of 15.00 ± 0.50 cm.

(*a*) Using a target mean of 15.00 cm, find control limits for $\overline{X}$ and R charts based on a subgroup size of 4 units.

In answering the following questions, assume that the actual mean setting μ is 14.90 cm.

(*b*) What proportion of units would you expect to not meet specifications, assuming that the process generates an approximately normal distribution of measurements?

(*c*) What is the probability that any value of $\overline{X}$ will fall within the control limits (the probability of Type II error)?

(*d*) Test the natural process spread (6σ) against the specification range. Is this process capable? Also calculate the process capability indexes C_p and C_{pk}.

9.5. A spherical plastic ball is to be produced against a specification of 1.500 ± 0.010 cm. Spherical diameter is measured by a dial indicator set to exactly 1.500 cm and measures deviations from its zero setting in plus and minus 0.001-cm increments. $\overline{X}$ and R charts are used to control this process. After 20 subgroups of 8 units each have been inspected, $\Sigma\overline{X} = -0.040$ and $\Sigma R = 0.114$.

(*a*) Calculate the central lines and control limits for the results of these dial gage readings.

(*b*) Assume that the results of charting the data indicate that the process is in control and that the normal distribution may be used to make inferences about its capability. Estimate the standard deviation of the process, and calculate the process capability indexes C_p and C_{pk}. What is your estimate of percent nonconforming product?

9.6. A certain manufacturing process has been operating in control at a mean μ of 65.00 mm with upper and lower control limits on the $\overline{X}$ chart of 65.225 and 64.775, respectively. The process standard deviation is known to be 0.15 mm, and specifications on the dimension are 65.00 ± 0.50 mm.

(*a*) What is the probability of not detecting a shift in the mean to 64.75 mm on the first subgroup sampled after the shift occurs? The subgroup size is four.

(*b*) What proportion of nonconforming product results from the shift described in part (*a*)? Assume a normal distribution of this dimension.

(*c*) Calculate the process capability indexes C_p and C_{pk} for this process, and comment on their meaning relative to parts (*a*) and (*b*).

9.7. Final specifications applied to the process discussed in Prob. 2.10 were L = 1,130 and U = 1,155. Calculate the process capability indexes C_p, C_{pU}, C_{pL}, k, and C_{pk} for this process. Interpret the indexes in the light of the solution to Prob. 4.40 and 8.7.

9.8. Specifications on the dimension of a certain part are 101.550 ± 0.200. Parts produced outside specifications must be scrapped. Two automatic screw machines produce these parts at a rate of 100 units per hour each. Items from both machines are discharged into a single tote box from which the inspector selects a subgroup of 5 parts every half-hour. Adjustments to both machines are made only on the approval of the inspector.

(*a*) After 50 subgroups have been drawn, $\Sigma\overline{X} = 72.25$ and $\Sigma R = 6.90$. To simplify the arithmetic, 100 has been subtracted from each value of X. Determine the control limits for $\overline{X}$ and R.

(*b*) Assuming that no points are outside the control limits on either chart and that no significant runs are apparent, on the basis of a normal distribution of this characteristic, what would you estimate the fraction nonconforming to be?

(*c*) Recentering the process would reduce this estimate of fraction nonconforming to what level?

(*d*) How would you approach the task of improving this process capability study?

9.9. Suppose it is desired to establish the relationship between nominal dimensions of extensions and sockets on hand tools in a situation similar to Example 9.8 but involving another size. Assume that a study of consumer requirements indicated a desirable minimum clearance of 0.005 in and a maximum clearance of 0.0018 in. Assume that a study of past production showed an average dimension of extensions of 0.7510 with a standard deviation of 0.0015 and an average dimension of sockets of 0.7650 with a standard deviation of 0.0010. Both distributions are statistically controlled and approximately normal.

Compute the respective probabilities of a smaller clearance than the desired minimum and a larger clearance than the desired maximum for average clearances of 0.015, 0.014, 0.013, 0.012, 0.011, and 0.010. Would you recommend any change in the existing average clearance? Explain your reasoning.

9.10. Two mating parts, A and B, have dimensions specified as 2.610 and 2.615 in, respectively. Control-chart analysis indicates the standard deviations of A and B to be 0.0012 and 0.0015 in, respectively. If the distributions of A and B are normal and centered about the specified dimensions, and if parts are assembled at random, find the probability of interference between the two distributions.

Answer: 0.0047.

9.11. Control-chart analysis indicates that the standard deviations of the distributions of dimensions of two mating parts, C and D, are 0.0008 and 0.0020 in, respectively. It is desired that the probability of a smaller clearance than 0.002 in should be 0.005. What distance between the average dimensions of C and D should be specified by the designer? Assume normal distributions and random assembly. With this distance specified, what is the probability that two parts assembled at random will have a greater clearance than 0.012 in?

Answer: 0.0076 in; 0.02.

9.12. Table 3.6 gives the distribution of markings of chips in Shewhart's bowl. Assume that these represent the excess over 0.0800 in of the thickness of metal washers measured to the nearest 0.0001 in; for example, 30 corresponds to a thickness of 0.0830.

The standard deviation σ of this distribution is approximately 0.0010. Table 3.7, in which chips (that is, washers) are drawn in groups of 4, illustrates the distribution that might be expected if washers were assembled at random in groups of 4. The sum of the four figures in each group, added to 0.3200, gives the thickness of one assembly.

Table 3.8 gives a frequency distribution that can be adapted to find the frequency distribution of the thickness of the 100 assemblies if appropriate changes are made in the cell boundaries. Find the standard deviation of the frequency distribution of assembly thicknesses as adapted from Table 3.8. How does this standard deviation check with the one that would be anticipated by the use of the theorem for the sum of 4 independent variables?

9.13. A certain mass-produced electrical product contains a circuit in which the resistance is the sum of the resistances of three components *F, G,* and *H*. The average resistances of *F, G,* and *H* are 90, 180, and 300 ohms, respectively; the estimated σ values are 4.5, 9, and 15 ohms, respectively.

(*a*) If the three components are assembled at random, what would you expect to be the average resistance of the circuit? How would your answer change if the assembly were not done at random?

(*b*) If the three components are assembled at random, what would you expect to be the σ of the resistance of the circuit? How would your answer change if the assembly were not at random?

(*c*) It is apparent that components *F, G,* and *H* are alike in having a σ value that is 5% of the average resistance. If assemblies are at random, will the σ value of the resistance of the circuit also be 5% of the average resistance of the circuit? Explain.

9.14. Two parts *A* and *B* are received in an assembly operation where each part is permanently attached to the other. When the combined width of the parts does not meet the required specification of 10.000 ± 0.020 in, the assembled product must be scrapped. The width of part *A* is normally distributed with μ equal to 3.000 in and a σ of 0.004 in. The width of part *B* is also normally distributed with μ equal to 7.000 in and a σ of 0.006 in. Assembly is at random. Determine the percentage of the assembled product that will have to be scrapped.

Answer: 0.0056.

9.15. The average clearance specified between two mating parts *A* and *B* is 0.0075 in. The distributions of *A* and *B* are considered to be normal, and from control charting of the two parts, their respective standard deviations have been determined to be 0.0014 and 0.0026. Assembly is at random. Perform the necessary calculations to determine the probability of interference between the two parts.

9.16. Manufacturer *A* produces a metal piece whose dimension is normally distributed with a μ of 8.500 in and an $\bar{R}$ of 0.004 in based on a subgroup size of four. Manufacturer *B* produces a second metal piece with a dimension that is also normally distributed with μ equal to 6.500 in and an $\bar{R}$ value of 0.005 in based on a subgroup size of nine. Company *C* purchases these two parts and assembles them together to obtain a combined dimension of 15.00 in. What percent of the combined assemblies would you expect to have a dimension in excess of 15.006 in?

Answer: 0.0099.

9.17. It has been suggested that, when extremely high performance of certain missile components is required, a boundary on the stress requirement be set at 6 standard deviations of the stress requirement above the average stress. The average strength requirement of the article would then be set at 5 standard deviations of the strength of the component above this boundary.

A certain critical electronic component must operate in a salt air environment at an average temperature stress of 30°C. The standard deviation of this operating temperature is believed to be 5°C.

(*a*) What must be the minimum acceptable average strength, in terms of average failing temperature, of this component? Assume that the standard deviation of strength in this case is 3°C.

(*b*) How much of a safety margin does the requirement provide in multiples of the standard deviation of the combined strength-stress characteristic?

9.18. A certain automatic filling operation fills metal containers with a sticky plastic compound. Each container is supposed to contain 6 lb of compound (96 oz). Once a container is filled, it is impracticable to empty it to find the exact weight of its contents because some of the compound adheres to the sides and bottom of the container. For this reason, indirect methods of analyzing filling weights were used in the following study. The purpose of the study was to find out whether more overfill than necessary was being used to satisfy the weight specification or there was not enough overfill.

One part of the study dealt with the weights of the containers themselves. A random sample of 100 was taken from the most recent shipment of containers (purchased from an outside supplier). Each container was weighed, and the following frequency distribution was obtained:

Weight, oz	Frequency	Weight, oz	Frequency
22.0	1	20.25	16
21.75	1	20.0	15
21.5	2	19.75	12
21.25	5	19.5	8
21.0	7	19.25	3
20.75	13	19.0	2
20.5	15		

In the second part of the study, a random sample of 8 filled containers was taken from each day's production for a 12-day period. The filled containers were weighed, and the following control-chart data were obtained. (The figures given are the excess in ounces above 110 oz. Thus, the first figure, 6.2, means 116.2 oz. The 110 figure has been restored in showing the values of $\bar{X}$.) No changes in the setting of the filling machine were made during the 12-day period.

Day	Weights								$\bar{X}$	R	s
1	6.2	8.9	9.9	5.8	8.1	9.3	6.4	8.6	117.90	4.1	1.56
2	7.0	7.6	8.3	7.5	8.0	6.6	6.5	7.9	117.42	1.8	0.66
3	6.6	6.7	8.9	6.4	8.4	6.7	6.9	7.4	117.25	2.5	0.92
4	8.5	8.9	6.9	6.3	6.8	7.0	9.1	7.4	117.61	2.8	1.03
5	6.8	8.2	8.9	7.9	8.1	9.7	7.5	7.3	118.05	2.9	0.92
6	8.0	6.8	5.6	5.0	7.8	6.0	8.5	7.4	116.89	3.5	1.25
7	6.3	7.5	6.7	7.0	6.8	7.5	6.5	6.4	116.84	1.2	0.47
8	5.2	7.7	8.1	8.3	7.4	7.5	8.4	7.6	117.52	3.2	1.02
9	8.1	5.7	7.9	7.0	8.1	8.2	8.4	7.7	117.64	2.7	0.89
10	8.3	5.4	7.0	8.4	5.6	8.9	7.4	7.4	117.30	3.5	1.27
11	7.6	6.3	7.6	7.0	6.7	9.2	5.7	7.2	117.16	3.5	1.05
12	7.4	6.9	7.7	8.9	8.7	7.1	7.7	7.2	117.70	2.0	0.74

$\bar{\bar{X}} = 117.44$ oz

$\bar{R} = 2.81$ oz

$\bar{s} = 0.982$ oz

(*a*) Compute the average and standard deviation of the weights of the containers shown in the frequency distribution.

(*b*) Make the necessary calculations to judge whether or not the weights of the filled containers exhibit statistical control. It is not necessary to plot control charts.

(*c*) What can you say about the inherent variability of the filling operation itself? For example, can you estimate the standard deviation of the weights of the *contents* of the containers (as distinct from the weights of the filled containers)? If so, what is your estimate of this figure?

(*d*) Overfill is defined as the excess of the contents of the containers above 96 oz. What is your estimate of the average overfill?

(*e*) If the distribution of the weights of contents of containers is normal and can be maintained in statistical control, what is your estimate of the percentage of containers having contents less than 96 oz?

(*f*) If the manufacturer is willing for 5% of the containers to have slightly less than 96 oz of contents, on the basis of the limited information available to you, what would you recommend as the aimed-at value for average overfill?

10

SOME SPECIAL PROCESS CONTROL PROCEDURES

You need not be a mathematical statistician to do good statistical work, but you will need the guidance of a first class mathematical statistician. A good engineer, or a good economist, or a good chemist, already has a good start, because the statistical method is only good science brought up to date by the recognition that all laws are subject to the variations which occur in nature. Your study of statistical methods will not displace any other knowledge that you have; rather, it will extend your knowledge of engineering, chemistry, or economics, and make it more useful.

—W. E. Deming[†]

10.1 SOME MISCELLANEOUS TOPICS

This chapter deals with a number of more or less unrelated topics that should be covered in any presentation of statistical methods for process control. Some of these are forms of the Shewhart technique for process control that vary slightly from the simple $\overline{X}$, R, and p charts described in the preceding chapters. Other topics deal with schemes for plotting and analysis that are based on other statistical procedures or were devised with principal objectives other than control of a process. The material is divided into 10 subtopics to emphasize its relationship to other topics presented in this book.

[†]W. E. Deming, "Some Principles of the Shewhart Methods of Quality Control," *Mechanical Engineering,* vol. 66, pp. 173–177, March 1944.

10.2 SOME SPECIAL TOPICS ON SHEWHART CONTROL CHARTS FOR VARIABLES

This section deals with a group of variations on the basic Shewhart charts for variables. They include corrections needed when subgroup size is not constant, places where $\bar{X}$ charts are not appropriate, the use of warning limits and zone tests, and plotting subgroup totals rather than averages.

10.2.1 Control Charts with Variable Subgroup Size

Wherever possible it is desirable to have a constant subgroup size. If this cannot be done, the limits on both $\bar{X}$ and R charts (or $\bar{X}$ and s charts) should be variable limits. Such variable limits are illustrated in the $\bar{X}$ chart of Fig. 10.1, which shows limits corresponding to subgroup sizes of 10, 15, and 20 for ultimate strength of suspension insulators.

Once σ has been estimated, these limits for various sample sizes may be obtained by using the factors and formulas of Table *F*, App. 3. Where the data used to estimate σ include subgroups of various sizes, a satisfactory working rule (although not precisely correct as a matter of statistical theory) is to calculate R/d_2 for each subgroup, using the appropriate d_2 factor from Table *C*, App. 3, for the size of the subgroup in question. The estimate of σ is the average of these values of R/d_2. If sample standard deviation, rather than range, is the measure used for

FIGURE 10.1 $\bar{X}$ chart illustrating variation of control limits with sample size. The variable X is the ultimate strength of suspension insulators. (*Reproduced from J. J. Taylor, "Statistical Methods Applied to Insulator Development and Manufacture," Transactions, American Institute of Electrical Engineers, vol. 64, pp. 495–499, July 1945.*)

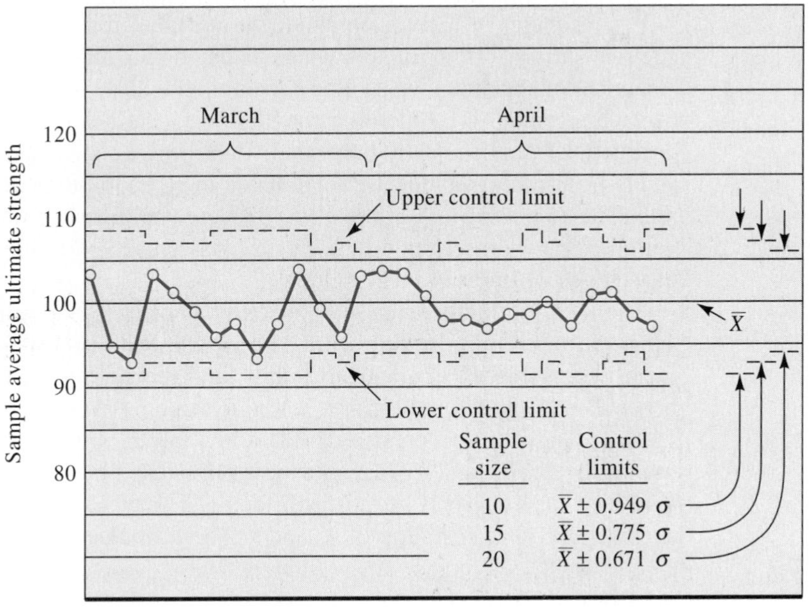

subgroup dispersion, s/c_4 may be computed for each subgroup; the estimate of σ is the average value of s/c_4.

10.2.2 *R* Charts or *s* Charts Where $\overline{X}$ Charts Are Not Appropriate

In some cases subgroups may be comparable in their dispersion even though not comparable in their averages. This is true, for example, of many standard chemical analyses that are made in duplicate, triplicate, or quadruplicate. Each two, three, or four analyses of a given sample may be thought of as a subgroup. If samples having somewhat different chemical content are analyzed, the averages of the subgroups are not comparable. The dispersion of the subgroups, however, reflects the ability of an analyst and an analytical procedure to reproduce results by several similar determinations. The control chart for R or s provides a basis for judging whether this dispersion seems to be influenced by a constant-cause system. Calculations for the central lines and control limits for R or s are no different from such calculations for any other control charts. Section 8.9 and Example 8.4 further illustrate this situation.

10.2.3 $\overline{X}$ and *s* Charts with Large Subgroups

It sometimes happens that data are at hand on averages and standard deviations of some measured variable from a number of different sources. It may be desired to apply a test for homogeneity to these figures to see if there is clear evidence that the different sources seem to represent different cause systems. The control charts for $\overline{X}$ and s constitute a simple test for this purpose.

For example, in a certain manufacturing plant it was desired to compare the strength and uniformity of spot welds made on nine apparently identical machines. A number of specimen welds of sheet aluminum alloy of a certain gage were made on each machine. The ultimate strength of each weld was determined by testing to destruction in a shear testing machine. The averages and standard deviations of the welds from each machine were tabulated in a report summarizing this investigation. They are shown in Table 10.1. Because the investigator was not acquainted with the Shewhart techniques, no record was preserved of the order of production of welds or of order of measurement.

The shear strengths from each machine constitute one rational subgroup. The subgroups are large, ranging from 111 to 128. With such large subgroups, the c_4 factor (see Table *C*) is practically unity, and the best estimate of σ (that is, $\bar{s}/c_4$) becomes $\bar{s}$. A_3 therefore becomes $3/\sqrt{n}$, B_4 becomes $1 + (3/\sqrt{2n})$, and B_3 becomes $1 - (3/\sqrt{2n})$. These expressions, stated in Table *E*, are evident from the explanation of the A_3, B_4, and B_3 factors as given in Chap. 4.

Where the subgroup sizes are different, it may be advisable to use weighted averages for the calculation of $\overline{\overline{X}}$ and $\bar{s}$. The formulas for these are as follows:

$$\overline{\overline{X}} = \frac{n_1\overline{X}_1 + n_2\overline{X}_2 + n_3\overline{X}_3 + \cdots + n_k\overline{X}_k}{n_1 + n_2 + n_3 + \cdots + n_k}$$

$$\bar{s} = \sqrt{\frac{(n_1 - 1)s_1^2 + (n_2 - 1)s_2^2 + (n_3 - 1)s_3^2 + \cdots + (n_k - 1)s_k^2}{n_1 + n_2 + n_3 + \cdots + n_k - k}}$$

However, unless the differences in subgroup size are large, the calculations are simpler and the results are nearly the same if $\bar{\bar{X}}$ is estimated as the simple unweighted average of the $\bar{X}$ values and $\bar{s}$ is estimated as the simple unweighted average of the s values. No absolute rule may be given as to when this simplification is satisfactory; a rough rule is to figure unweighted averages unless the largest subgroup is at least twice the smallest. It is clearly satisfactory to compute unweighted averages for the data of Table 10.1. Thus

$$\bar{\bar{X}} = \frac{\bar{X}_1 + \bar{X}_2 + \bar{X}_3 \cdots + \bar{X}_k}{k} = \frac{6{,}435}{9} = 715$$

$$\bar{s} = \frac{s_1 + s_2 + s_3 \cdots + s_k}{k} = \frac{572}{9} = 64$$

The question now arises whether, because of the different subgroup sizes, different limits should be computed for each subgroup. Again the computations are much simpler if one set of limits is computed based on average subgroup size. This simplification is usually satisfactory for a start; separate limits for individual subgroups may be calculated later for any doubtful cases. The average subgroup size $\bar{n}$ must be computed:

$$\bar{n} = \frac{1{,}103}{9} = 123$$

TABLE 10.1 SHEAR STRENGTHS OF SPOT WELDS MADE BY NINE DIFFERENT MACHINES

Machine	Number of tests n	Average shear strength, lb $\bar{X}$	Standard deviation s
A	128	743	63
B	127	695	47
C	126	711	67
D	114	668	51
E	126	736	80
F	126	791	58
G	126	686	50
H	111	801	92
J	119	604	64
Totals	1,103	6,435	572

The factors A_3, B_4, and B_3 may now be computed:

$$A_3 = \frac{3}{\sqrt{n}} = \frac{3}{\sqrt{123}} = 0.27$$

$$B_4 = 1 + \frac{3}{\sqrt{2n}} = 1 + \frac{3}{\sqrt{246}} = 1.19$$

$$B_3 = 1 - \frac{3}{\sqrt{2n}} = 1 - \frac{3}{\sqrt{246}} = 0.81$$

From these factors, the control limits may be computed:

$$UCL_{\bar{x}} = \bar{\bar{X}} + A_3\bar{s} = 715 + 0.27(64) = 732$$

$$LCL_{\bar{x}} = \bar{\bar{X}} - A_3\bar{s} = 715 - 0.27(64) = 698$$

$$UCL_{\sigma} = B_4\bar{s} = 1.19(64) = 76$$

$$LCL_{\sigma} = B_3\bar{s} = 0.81(64) = 52$$

These limits are plotted on the control charts of Fig. 10.2. Both charts definitely show lack of control; in fact, only one point falls within the control limits on the $\bar{X}$ chart. It is quite evident that even though these spot-welding machines are identical in their design, they perform differently with regard to both average strength and uniformity of strength of welds.

It should be remarked in passing that a tabulation of averages and sample standard deviations of large numbers of measurements such as that in Table 10.2 is of little value for prediction without the knowledge that each source of measurement is itself in control. For example, without any evidence of control on machine *A*, it is not safe to assume that the strength of future spot welds on machine *A* will fall within limits suggested by the average of 743 lb and the standard deviation of 63 lb. It would have been much better if the order of production had not been lost and the results on each machine had been subjected to the control-chart analysis.

10.2.4 Warning Limits on Control Charts

Some writers on statistical quality control have advocated the use of two sets of limits on $\bar{X}$ charts. The outer limits, sometimes called *action limits,* are the conventional limits, usually at 3 sigma. The inner limits are recommended as *warning limits* and are usually at 2 sigma.

On the conventional $\bar{X}$ chart with only one set of limits, the chart seems to give only two kinds of advice. Either it says, "Look for trouble," or it says, "Leave the process alone." This has the virtue of definiteness. However, any such definite advice is sure to be wrong part of the time. Limits placed at 3 sigma are seldom wrong when they say, "Look for trouble," but are much more often wrong when they say, "Leave the process alone."

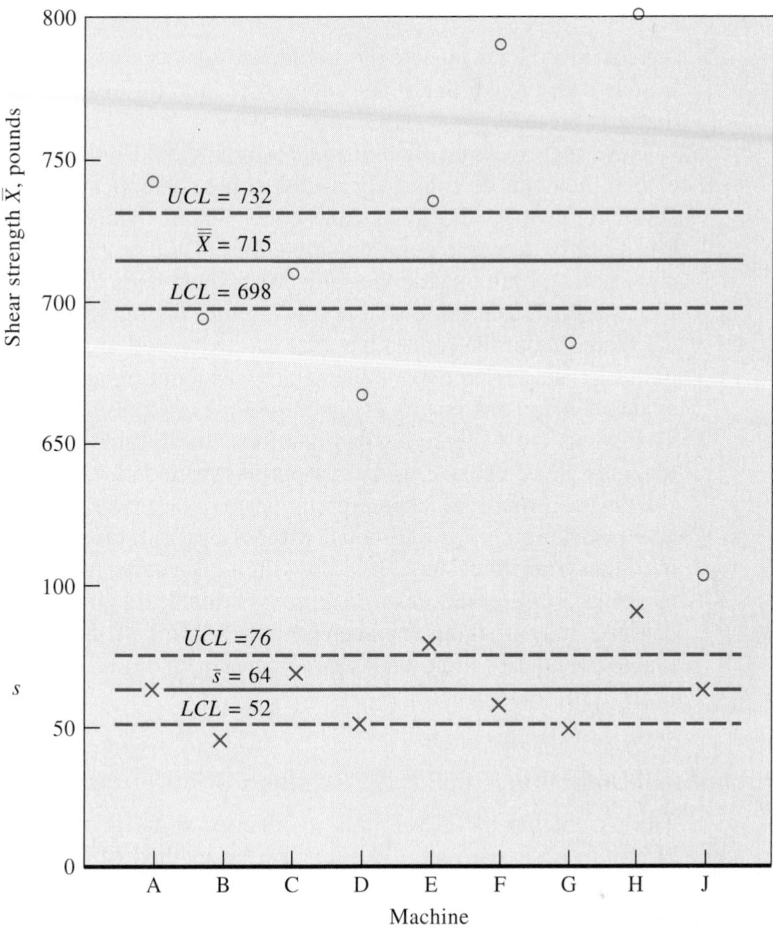

FIGURE 10.2 $\bar{X}$ and s control charts for shear strength of spot welds—data of Table 10.1.

The inner limits or warning limits seem to add a third kind of advice. This might be phrased, "Start being suspicious that trouble is brewing."

At first thought, the idea of having warning limits on $\bar{X}$ charts may seem attractive. Nevertheless, there is a sound reason for the common practice of having only one set of limits and having these limits at or near 3 sigma. This reason is the greater definiteness of a single set of limits. Two sets of limits tend to be confusing with regard to the exact action to be taken when a point falls between the inner and the outer limits. This is particularly true if many of the people in a facility who are using the $\bar{X}$ and R control charts as a basis for action are not fully clear as to the principles underlying these charts. Inner limits will be exceeded at least 5% of the time as a matter of chance. If a single point just outside the inner limits is to be used as a basis for hunting for trouble, there is bound to be unproductive hunting that may tend to destroy confidence in the control charts. Usually so much trouble really exists that it does not pay to hunt for trouble without strong evidence that it is present.

Nevertheless, even though inner limits should not be drawn on most control charts, they can be extremely useful in the sophisticated interpretation of control charts by people who understand control-chart theory. Here the clue to action is given not by a *single* point outside either of the inner limits but rather by two or more points, both outside the same inner limit. This is really a matter of sizing up extreme runs; it is somewhat comparable to the interpretation of extreme runs on the same side of the central line that was explained in Sec. 4.4.1. For example, two points in succession outside the same inner limit on an $\overline{X}$ chart give even stronger evidence of a shift in process average than a single point outside the outer limit. Two points out of three beyond one inner limit, three out of seven, or four out of ten, may all be considered as appropriate grounds for action.

However, this type of interpretation may be made by the quality control engineer or other qualified individual without confusing matters for colleagues by having inner limit lines actually drawn on all $\overline{X}$ charts. When a suspicious sequence of points is observed close to a conventional 3-sigma limit, the quality control engineer may imagine an inner limit two-thirds of the distance from the central line to the control limit. Or, such a line may be drawn lightly on the portion of the chart to be studied.

10.2.5 Problems of Detecting Small but Sustained Shifts

The use of the theory of runs to identify shifts in the universe was discussed briefly in Chap. 4. Later, in Chap. 5, the method of calculating the approximate probabilities associated with runs of points on one side of the central line of an $\overline{X}$ chart was presented. Such tests may be useful because an $\overline{X}$ chart with 3-sigma limits and a small subgroup size (e.g., two to five) is not sensitive to small shifts in the process centering.

Consider an example in which a controlled and correctly centered process shifts to a new centering $+\frac{1}{2}\sigma$ greater than the old centering. σ remains constant through the shift, and the centering holds constant at the new level after the shift. A combination rule is used that says: Look for an assignable cause of a shift in a given direction if either of these conditions occurs:

1. A point falls outside a control limit.
2. Eight points in a row fall between the central line and a single control limit.

Assuming that subgroups of four are drawn from a normally distributed universe, the probability that a single point will fall outside a given control limit is 0.00135, and the probability that a point will fall between the central line and one control limit is 0.49865 when the process is correctly centered. After the $+\frac{1}{2}\sigma$ shift, the respective probabilities increase to 0.0228 and 0.8186.

Under the rules as stated, it is conceivable that either rule could signal the shift on the first sample after the shift. That is, a single point could fall outside the upper 3-sigma control limit (with probability 0.0228), or the eighth point in a row could be between the central line and the upper control limit (with probability 0.8186) if the previous seven points before the shift had fallen in the same region (with probability 0.49865^7). Thus the approximate combined probability of detection on the first sample after the shift is

$$0.0228 + (0.49865)^7(0.8186) = 0.0291$$

The probability of detection on the second sample is the probability of not detecting the shift on the first sample multiplied by the sum of the probabilities of a point outside the upper 3-sigma limit and eight points in a row between the central line and the upper limit, or

$$0.9709[0.0228 + (0.49865)^6(0.8186)^2] = 0.0321$$

The probability of detection by the second sample is the sum of the respective individual probabilities, or

$$0.0291 + 0.0321 = 0.0612$$

The following tabulation shows the respective probabilities of detection on the first eight sample subgroups and the cumulative probabilities to and including the subgroup listed. It should be noted that these probabilities are accurate only if the shift is sustained at the new level until detected.

Subgroup number	Probability of detection: On that subgroup	Probability of detection: By that subgroup
1	0.0291	0.0291
2	0.0321	0.0612
3	0.0373	0.0985
4	0.0456	0.1441
5	0.0585	0.2026
6	0.0778	0.2804
7	0.1048	0.3852
8	0.1380	0.5232

Some companies have formalized the use of the theory of runs into standard operating procedures. The following paragraphs describe the procedure used by Western Electric Company as published in the company's "Statistical Quality Control Handbook."†

† "Statistical Quality Control Handbook," Western Electric Company, Inc., 1956, sec. I, part B, and sec. II, part F. Copies may be purchased from ASQC Quality Press, P.O. Box 3066, Milwaukee, Wis.

In addition to the usual 3-sigma control limits, the distance between the central line and each limit is divided into three equal zones: *A, B,* and *C*. Each zone has a specific decision rule for points falling within or beyond that zone. Considering the distance between μ and the *UCL* for $\overline{X}$, zone *A* lies between +2 sigma and +3 sigma above the central line. An increase in process average is signaled when two out of three points fall within or beyond zone *A*. Zone *B* lies between +1 sigma and +2 sigma above the central line. An increase in the process average is signaled if four out of five points fall within or beyond zone *B*. Zone *C* lies between the central line and +1 sigma above the central line. An increase in the process average is signaled if eight points in a row fall within or beyond zone *C*. The reader will note that the zone *C* rule is equivalent to the rule discussed earlier in this section, that is, eight points in a row on one side of the central line.

These rules may be condensed as follows:

Zone	Rule	Proportion in zone assuming normal distribution	Probability of Type I error when using rule
	1 point beyond limit	0.00135	0.00135
$UCL_{\overline{X}}$			
A	2 of 3 in zone *A* or beyond	0.02135	0.0015
B	4 of 5 in zone *B* or beyond	0.1360	0.0027
C	8 in a row in zone *C* or beyond	0.3413	0.0039
μ			

Source: Adapted by permission from "Statistical Quality Control Handbook," Western Electric Company, Inc., 1956.

The figures in the two right-hand columns are based on an $\overline{X}$ chart with 3-sigma limits and drawings from a normal universe. It should be noted that there is a degree of redundancy in the rules. For example, the rule for zone *A* includes the possibility that at least one of the two points in or beyond the zone will fall beyond the 3-sigma limit. Such an occurrence would automatically signal a shift, thus nullifying the effect of the zone rule.

The probabilities shown apply to Type I error; that is, when the process is operating in control and correctly centered. The sum of the probabilities associated with the use of each rule, 0.0094, is an approximation of the Type I error associated with their joint application, but somewhat overstates the true probability because of the aforementioned interaction between decision rules.

As described, the rules apply to the occurrence of positive shifts in μ. Negative shifts in μ are accounted for simply by changing all "+" signs to "–" signs, references to the *UCL* to *LCL*, and the word *increases* to *decreases*. In other words, the rules to detect decreases in μ are the mirror image of those shown.

We have made specific mention of several combinations of decision rules that may be used in control charting to detect shifts in this and preceding articles in this book. Examples of the types of calculations that may be useful in approximating the effectiveness of these combination rules have been illustrated. It should be noted, however, that the real justification of the use of the rules based on runs such as those adopted by Western Electric is similar to the justification of the use of

3-sigma limits on $\bar{X}$ and R charts, namely, that experience indicates that, in general, when the rules say "hunt for trouble," a source of trouble can be found. When the rules say "leave the process alone," it often happens that no source of trouble (that is, no assignable cause of variation) can be discovered.

10.2.6 The Problem of Misinterpretation of the Relationship between Control-Chart Limits and Specification Limits

One source of confusion appears almost universally wherever the control chart for $\bar{X}$ is introduced on production operations. Whenever specifications apply to individual values (as is always true of dimensions and usually true of other quality characteristics), the specification limits tend to be confused with the control-chart limits. This confusion often exists in the minds of shop personnel, inspectors, engineers, and even managers. It leads to a diversity of troubles.

It has already been pointed out in Chap. 2 that where the specification tolerances are tighter than the natural tolerances (that is, than the 3-sigma limits on individual values), some nonconforming product is sure to be made even by a controlled process. A controlled process may also make nonconforming product if the process average is not properly centered with respect to upper and lower specification limits or properly located with respect to a single specification limit. Thus there may be many cases where the control chart shows the process in control and some of the product is bad. If this is not understood, the control chart may give a false assurance that all is well.

If specification limits are drawn on the $\bar{X}$ control chart, there is a natural tendency to compare the subgroup *averages,* plotted on the chart, with the specification limits. This sometimes leads to the false conclusion that whenever an average plots within specification limits, all the product is within specifications.

A misinterpretation opposite to this is also made occasionally. This is to compare measured individual values with the control limits that apply to averages, and to conclude that trouble exists whenever an individual value is outside the limit for averages.

The spread of the control limits for averages is less than the spread of individual values. It is therefore often less than the spread of the specification limits. This condition sometimes leads to the incorrect conclusion that the use of the $\bar{X}$ chart amounts to the use of working tolerances that are closer than the specification tolerances.

The ideal preventive for these various errors is for the people who use the control chart to understand clearly that averages are different from individual values and that control limits mean something entirely different from specification limits. In this book the first illustrations of the $\bar{X}$ chart were deliberately chosen in a way to bring this out. The reader may recall the contrast between Examples 1.1 and 1.2. In Example 1.1 practically all the product was within specification limits even though there were many points out of control. In Example 1.2 much of the product did not conform to specifications even though all points were within control limits. Some such illustrations are desirable in any introduction of personnel to the

$\overline{X}$ control chart, not only in the short in-plant courses frequently given to production and inspection supervisors but also in presentation of the technique to individuals.

Unless all the people exposed to $\overline{X}$ charts are familiar with control-chart principles, any specification limits for individual values drawn on such charts may constitute a troublesome source of misunderstanding.

10.2.7 Plotting Subgroup Totals

One scheme that has been used in many plants is to plot on the control chart the sum of the n measurements in each subgroup rather than the average of these measurements. Where totals are plotted, the values on the chart do not appear to shop personnel as if they were comparable with specification limits; hence there is little chance for confusion on this point.

This type of chart is merely a conventional $\overline{X}$ chart with the scale magnified n times. The values for the central line and limits are the $\overline{X}$ chart values multiplied by n. Any conclusions to be drawn from the $\overline{X}$ chart may also be drawn from the chart for totals.

This variation of the $\overline{X}$ chart is particularly useful where a machine operator is using the control chart under definite instructions to leave the machine settings alone as long as a process shows control and to stop production and get help from some definite source (for example, the machine setter or the maintenance department) whenever a point goes out of control.

A minor advantage of the chart for totals is a saving of the arithmetical operation of dividing the total of each subgroup by the subgroup size. A minor limitation is that the method should not be used where the subgroup size is variable.

10.2.8 Precontrol—A Poor Answer to the Wrong Question

The precontrol technique attempts to relate the output of a process to the specifications on individual parts and to use the results of the inspection of individual units to control the process. The distance from the lower to the upper specification $(USL_X - LSL_X)$ is divided into three zones. With the process targeted at $\mu_o = (USL_X + LSL_X)/2$, half way between specifications, precontrol (PC) lines are set at $(USL_X - \mu_o)/2$ and at $(\mu_o - LSL_X)/2$. The distance between the PC lines is called the "green" zone. So long as the dimensions of inspected items fall within this zone, the process is left alone. The two zones between the PC lines and their respective specification limits, USL_X and LSL_X, are called the "orange" zones. If the dimension of one item falls within an "orange" zone, the next successive item is inspected. If that dimension falls within an "orange" zone, not necessarily the *same* "orange" zone, the process is reset. The zones beyond USL_X and LSL_X are called "red" zones. If the dimension of an item falls beyond either USL_X or LSL_X, and thus in a "red" zone, the process is reset. Various rules, much like those employed in acceptance sampling from continuous production, are used to determine when it is feasible to switch from 100% inspection to the sampling of some fraction of production. Sampling for acceptance purposes from continuous production is the subject of Chap. 14. In that chapter we will discuss the use of continuous sampling schemes to determine sampling frequency rates for process control purposes.

Justification for the procedure is claimed by arguing that, if $(USL_X - LSL_X)$ is equal to 6σ, if the process generates a normal distribution of unit dimensions, and if it is correctly centered midway between specification limits, then about 86% of the time individual measurements will fall in the "green" zone. This may be verified from Table *A* by recognizing that the PC lines will be at $\mu_o \pm 1.5\sigma$ and that contained with this range lies 0.8664, or roughly 86%, of the normal distribution. Falling beyond each PC line is 0.0668, or roughly 7%, of the normal distribution. The probability that two points in a row will fall within any "orange" zone is 0.0045, or about 1 chance in 200. Only 0.00135, or 0.135%, of the normal distribution will fall beyond one specification limit under the conditions described.

Six major objections to the application of this technique can be made. They are as follows:

1. Unfortunately, not many manufacturing processes generate a normal distribution of measurements even when held in excellent control. Thus a major premise for recommending the procedure is suspect immediately.
2. If there is enough experience in controlling the process to be able to establish σ a priori—that is, process capability has been establishing along with the requirement that $6\sigma \leq (USL_X - LSL_X)$—then standard control charts and control limits can be established and there is no need for the PC lines.
3. There is no necessary relationship between natural tolerances of a process and specification limits on dimensions. Actually, the two may be quite different. We have stressed throughout this book the *desirability* of having the natural tolerance spread, 6σ, factored into design decisions about specification tolerances, but we have always tried to keep clear the difference between the two entities.
4. A number of users of the technique have assumed that it can be applied *without* knowledge of σ or of whether or not the process *can* be held in control. They wish to use it with new processes and for short production runs without the intent to ever establish statistical control and process capability.
5. It may have the dangerous effect of convincing shop personnel and engineers that there is no need to apply even the simplest statistics in "controlling" processes, that somehow the same results can be obtained but more simply.
6. Application of the technique will lead to overcontrol problems—if $(USL_X - LSL_X)$ actually is less than the true, but unknown, 6σ natural tolerance range—or to undercontrol problems—if $(USL_X - LSL_X)$ is greater than 6σ. The true facts of the case will be unknown to anyone. In the first instance, managers will believe that production is really working at the job of controlling the process. In the second instance, since practically no rejects or process resets are occurring, they will think that the process never goes out of control.

The best advice we can give is to recommend almost any technique discussed in this book in preference to precontrol. The pattern of variation is a characteristic of the *process* and should never become confused with design tolerances. To automatically assume that 6σ equals whatever specification tolerances the engineers, or the customers, place on production drawings is to err seriously.

10.3 SOME RELATED SPECIAL PROCEDURES

In this section, some procedures closely related to Shewhart control charts for variables are discussed. They include charts for individual measurements, using the median statistic as the measure of central tendencies, and using box plots to show within-unit or subgroup variation when the sample size is large and, perhaps, highly correlated.

10.3.1 Charts for Individual Measurements

Where charts for averages are misunderstood by shop personnel, one possible way to avoid this misunderstanding is not to plot averages at all but rather to plot individual measurements. Figures 1.1*a* and 1.3*a* are charts of this type. Specification limits applying to individual measurements may be properly shown on such charts. If sampling is by subgroups, as illustrated in Figs. 1.1*a* and 1.3*a*, the universe standard deviation σ may be estimated from $\bar{R}$ or $\bar{s}$. Control limits for individual measurements may then be drawn at $\mu \pm 3\sigma$. These also are the natural tolerance limits of the process. A point outside such limits may be considered as evidence of an assignable cause of variation. It would be rare, indeed, for a measurement to fall outside these limits if the process is in control and generating approximately a normal distribution of measured values.

Such a chart may be better than nothing, but it is much less satisfactory than a conventional $\bar{X}$ chart based on a subgroup size of four or five. If the limits on a chart for individual values are set at $\mu \pm 3\sigma$, the chart is relatively insensitive to substantial shifts in process average. Although greater sensitivity to such shifts may be gained by the use of narrower limits and/or zone tests, such sensitivity is gained only by increasing the chance of false indications of lack of control (so-called Type I errors). Unless a chart for individual measurements is accompanied by a range chart, say a two-point moving range (*mR*) chart, it is difficult to discover whether there have been changes in process dispersion. Control charts for moving averages and moving ranges are discussed in Sec. 10.5. It will be remembered that the accepted valid means for estimating σ is through averaging within-subgroup variation. In general, charts for individual measurements are inferior to conventional control charts because they fail to give as clear a picture of changes in a process or quick evidence of assignable causes of variation.

A further problem in using control charts for individual measurements results when the underlying distribution may not be assumed to be normal or near normal. Examples of such distributions are tests for flatness and hardness, which are definitely nonnormal and, in fact, are truncated. A piece of metal may not be flatter than dead flat; an upper limit on hardness of a grade of steel is dictated by the metallurgy of the alloy. In such cases, $\pm\ 3\sigma$ limits on individual measurements make no sense at all and actually represent a misapplication of the control-chart technique. Furthermore, the use of process capability indexes, discussed in Chap. 9, may lead to strange results and even stranger decisions by management.[†]

10.3.2 Combination of Chart for Individual Measurements and Chart for Medians

If it is desired to plot individual measurements, it is suggested that a control chart also be plotted that reflects the central tendency of subgroups ($\bar{X}$, median, or midrange). Such a chart might be separate. However, the control chart for central tendency might also be superimposed on the chart for individual measurements.

A convenient plan for such superimposed charts is to combine a chart for individual measurements with a chart for medians. The convenience of the median chart for this purpose is greatest where the subgroup contains an odd number of measurements such as 3, 5, or 7. If so, the median may be quickly identified and circled as shown in Fig. 10.3. The data charted are from Example 1.1.

[†]See, for example, Allan Brooks, "Heat Treating Shows Why SPC Is No Cure-All for Manufacturers," *Production Engineering,* September 1985.

FIGURE 10.3 Control chart for medians superimposed on chart showing individual measurements and specification limits—data of Example 1.1. *(Adapted, with some changes, from a similar chart in P. C. Clifford, "Control Charts without Calculations: Some Modifications and Some Extensions," Industrial Quality Control, vol. 15, no. 11, pp. 40–44, May 1959.)*

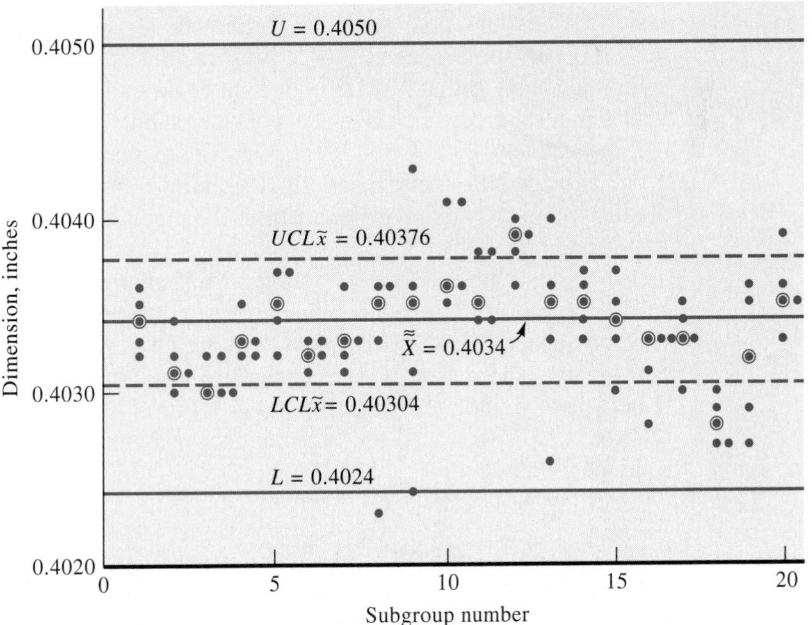

On such a combined chart, the individual values may be examined with reference to the tolerance limits shown on the chart and the circled medians may be examined with reference to the control limits for medians.

10.3.3 Control Charts for Medians within Subgroups Using Medians of Statistics of Sets of Subgroups

Enoch B. Ferrell of Bell Telephone Laboratories has proposed the use of the midrange as the measure of central tendency of each subgroup. He has also proposed that the estimate of the central tendency of a universe be based on the median of the midranges of a set of subgroups and that the estimate of universe dispersion be based on the median of the ranges of the same set of subgroups.[†]

The midrange of a set of numbers is the average (arithmetic mean) of the maximum and minimum numbers in the set. The reader will recall that the median was defined in Chap. 3 as the magnitude of the middle case—the value that has half the observations above it and half below it.

Paul C. Clifford has suggested modifying Ferrell's procedure.[‡] Medians of subgroups are used rather than midranges. The median of the subgroup medians (or, alternatively, the median of the individual observations) is used as the estimate of the central tendency of the universe. Just as in Ferrell's procedure, the median of the ranges is used as an estimate of universe dispersion.

Table 10.2 gives factors for 3-sigma limits for charts for subgroup medians. These factors assume that sampling has been from a normal universe. One wavy line over an X (that is, $\widetilde{X}$) is used to indicate a subgroup median. Two wavy lines over an X (that is, $\widetilde{\widetilde{X}}$) indicate the median of a set of subgroup medians.

An advantage of the median as a measure of central tendency of a subgroup is that it can be found quickly with no arithmetic for subgroups containing odd numbers of observations such as 3 or 5. A characteristic of the median as a measure of central tendency is that it gives no weight to the extent of the extremity of extreme values in a subgroup. For instance, consider subgroup 8 in Table 1.1; the X values are, respectively, 23, 33, 36, 35, and 36. The extreme value 23 influences the subgroup average (arithmetic mean) of 32.6 and the midrange of 29.5. In contrast, the median value of 35 is no different from the median that would have existed if the first X had been, say, 33.

When the central tendency of subgroups is charted, the foregoing characteristic of the median usually is a disadvantage as compared with charting $\bar{X}$. Nevertheless, this characteristic of the median might conceivably be viewed as an advantage if for some reason it were desired to give little weight to occasional "wild shot" values in individual measurements.

Figure 10.3 illustrated the combination of a chart for medians and a chart for individual values.

[†]E. B. Ferrell, "Control Charts Using Midranges and Medians," *Industrial Quality Control,* vol. 9, pp. 30–32, March 1953. Equations and factors for midrange charts are given in this article.

[‡]P. C. Clifford, "Control Charts without Calculations: Some Modifications and Some Extensions," *Industrial Quality Control,* vol. 15, no. 11, pp. 40–44, May 1959.

TABLE 10.2 FACTORS AND FORMULAS FOR 3-SIGMA LIMITS FOR CONTROL CHARTS FOR MEDIANS OF SUBGROUPS USING MEDIAN RANGE AS A MEASURE OF SUBGROUP DISPERSION[†]

n	A_5	n	A_5
		6	0.562
2	2.224	7	0.520
3	1.265	8	0.441
4	0.829	9	0.419
5	0.712	10	0.369

$$UCL_{\tilde{X}} = \tilde{\tilde{X}} + A_5 m(R)$$
$$LCL_{\tilde{X}} = \tilde{\tilde{X}} - A_5 m(R)$$

[†]Reproduced from P. C. Clifford, "Control Charts without Calculations: Some Modifications and Some Extensions," *Industrial Quality Control,* vol. 15, no. 11, pp. 40–44, May 1959.

10.3.4 Box Plots as a Means of Displaying Data Variability from Large Subgroups

It sometimes happens that data are accumulated into samples with a large number of data points, often highly correlated, where the variability within the sample is not useful for controlling a process through time but is important in judging uniformity within units of output or processed batches. Examples include resistance measurements on batches of oxidized silicon wafers, as in Example 8.2, the resistance of individual wires bound into cables, the variation in samples drawn from different locations in a tank or a batch, and the variation of, say, 12 diameter measurements taken at 15° intervals on a unit that is supposed to be a perfect circle. The average of such samples may be used as a single data point for controlling the process through time, but the within-sample variation may not be used to establish control limits.

Sometimes called a box and whisker diagram, a box plot[†] uses the median of a set of data, the interquartile range, and the extreme values to describe pictorially its pattern of variation. Thus the information from even a large set of data may be summarized into just five numbers. Figure 10.4 shows a simple form of box plot for the data of Table 1.1. Four grand samples have been formed consisting of five subgroups each of five measurements. This gives four samples. Upper and lower specification limits have been plotted on the chart. The horizontal line through each box represents the median value of each set of twenty-five measurements. The ends of the box are at the upper and lower interquartile marks. Thus the measure height of the box includes half the measured values. The dots at the end of each whisker indicate the largest and smallest measurements in each grand sample.

A median value that plots toward the end of a box indicates that the data are skewed in that direction, either high or low; particularly long whiskers, such as in

[†]Development of this procedure is credited to John W. Tukey. See his 1977 book, *Exploratory Data Analysis,* Addison-Wesley Publishing Company, Reading, Mass.

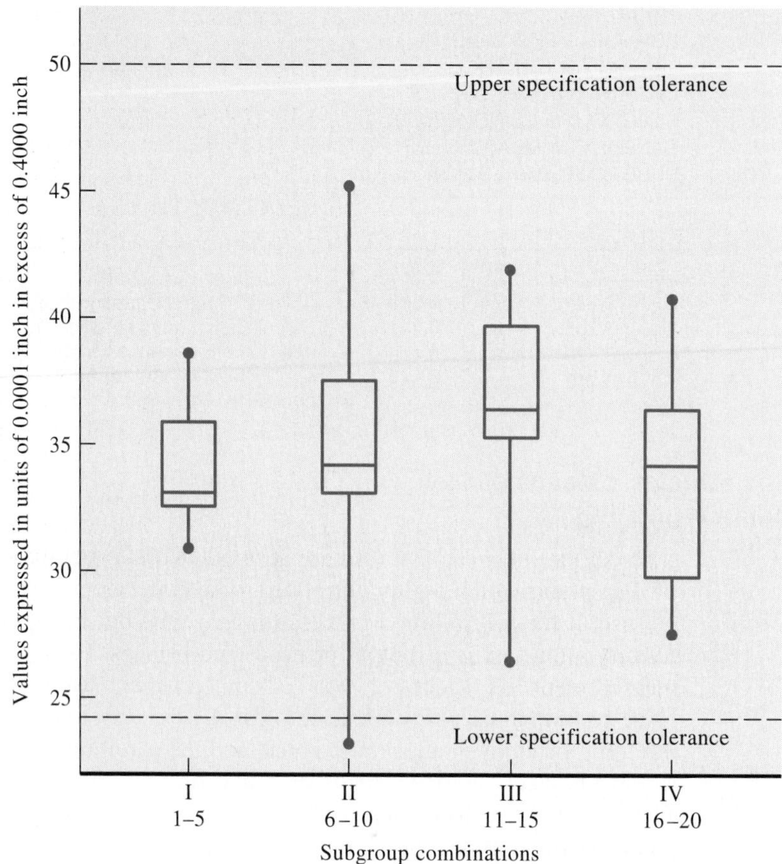

FIGURE 10.4 Illustration of box plots of measurements of pitch diameter of threads on aircraft fitting—data of Table 1.1. Grand samples of 25 have been formed from the data.

grand samples II and III, suggest that there are outliers. This would lead to an interpretation of Fig. 10.4 not unlike that of Figs. 1.1 and 1.2—that is, there are likely to be differences in the way the process is operating through time—but Fig. 10.4 lacks the definiteness provided by the control charts in Fig. 1.2.

10.4 A GENERAL TEST FOR HOMOGENEITY

10.4.1 The Control Chart as a General Test for Homogeneity

It is impossible to give too much emphasis to the importance of keeping track of the order of production whenever measurements are made of any quality of manufactured product. Ideally, measurements should be planned with this in view. Practically this may not be possible. This is particularly true when a purchaser wishes to apply the control-chart analysis to an incoming shipment of product of which there is no knowledge of the order of production.

Suppose, for instance, that an aircraft manufacturer receives a shipment of 100,000 bolts packed in 50 boxes each containing 2,000 bolts. Wishing to test 200 of these for some quality, such as Rockwell hardness number or tensile strength, the manufacturer will naturally pick four bolts from each of the 50 boxes.

Suppose the results of these tests are plotted on $\bar{X}$ and R charts, with the 4 bolts from each box constituting a subgroup, and these control charts show a state of control. The interpretation of this showing of control is not the same as if the basis of subgrouping had been the order of production. Here the control charts are simply a test for homogeneity. This homogeneity may have been obtained by a constant-cause system during production with the bolts packed in boxes in order of production, or it may have been obtained by a thorough mixing of the bolts before they were packed even though they came from several different cause systems.

The control-chart analysis may also be applied to data already at hand that were taken with no thought of the control chart, provided there is some rational basis for subgrouping. Here also the control charts are a general test for homogeneity. For example, they have been applied by Shewhart to published data regarding the determination of fundamental physical constants such as the velocity of light.[†] This type of application in a manufacturing setting is illustrated in Fig. 10.2 in an example dealing with the shear strength of spot welds. Applications to inspection error were discussed in Chap. 9.

10.5 PROBABILITY LIMITS ON CONTROL CHARTS FOR VARIABLES

10.5.1 Two Questions for Decision Regarding Control Limits on Charts for $\bar{X}$, R, and s

Two questions that often have been debated among statisticians, as well as among persons engaged in quality control work in industry, are as follows:

1. Shall control-chart limits be described as multiples of the standard deviation of the statistic charted or in terms of the estimated probabilities associated with the limits?
2. How far apart should the limits be spaced, all things considered?

The first question is primarily one of semantics. We have already suggested that, in principle, the second question is primarily an economic one.

10.5.2 Probability Limits on $\bar{X}$ Charts

If we assume that $\bar{X}$ values follow the normal distribution when all samples have been taken from the same universe, we can use Table A to find the multiple of $\sigma_{\bar{X}}$

[†]W. A. Shewhart (edited by W. E. Deming), "Statistical Method from the Viewpoint of Quality Control," p. 68, The Graduate School, Department of Agriculture, Washington, 1939.

associated with any stipulated probabilities. The reader will recall that in Chap. 3 it was pointed out that $\overline{X}$ values are normally distributed when the universe is normal; they are approximately normal even from nonnormal universes when the sample size is four or more.

For instance, if it is desired that the probability be 0.001 that without a change in the universe a point will fall above the upper control limit, we look in Table *A* for the multiplier of σ corresponding to 0.001. This is 3.09. Because the factors A_2, A_3, and A in Tables *D, E,* and *F* are based on 3-sigma limits, it is merely necessary to multiply them by 3.09/3 = 1.03 to obtain corresponding factors based on 0.001 probability limits. It should be remembered that to the probability of 0.001 that a point will fall above the upper limit must be added the equal probability of 0.001 that a point will fall below the lower limit. This means a probability of 0.002 of a point outside limits, or 0.998 within limits. The British particularly have used limits based on $3.09\sigma_{\overline{X}}$.

Or if it is desired that the probability be 0.025 that a point fall above the upper limit (corresponding to 0.05 that it fall outside limits), Table *A* gives 1.96 as the appropriate multiple of $\sigma_{\overline{X}}$. The multiplier to be applied to the A_2, A_3, and A factors is then 1.96/3 = 0.653. Factors based on $1.96\sigma_{\overline{X}}$ have been used by the British as inner control limits or warning limits.

10.5.3 Probability Limits on Charts for *R* and *s*

The distribution of $\overline{X}$ values, which is normal or approximately so, is symmetrical. Therefore probability limits on an $\overline{X}$ chart are like 3-sigma limits in being equidistant from the central line on the chart. In contrast, because the distributions of R and s are not symmetrical even when the universe is normal, it is necessary to have separate factors for the upper and lower control limits if the probabilities of extreme variations are to be made equal. Table 10.3 gives such factors for probability limits for s charts; Table 10.4 gives them for R charts. Three sets of symmetrical probabilities (giving unsymmetrical limits) are 0.001 and 0.999, corresponding to a 0.002 probability that with no change in the universe a point will fall outside of the limits; 0.005 and 0.995, corresponding to a 0.01 probability of a point falling outside; and 0.025 and 0.975, corresponding to an 0.05 probability of a point falling outside.

Table 10.3 gives multipliers for the assumed or estimated value of σ to obtain limits on a chart for s. σ may be estimated as $\bar{s}/c_4$.

Table 10.4 gives multipliers for the assumed or estimated value of σ to obtain limits on a chart for R.[†] σ may be estimated as $\overline{R}/d_2$.

It should be emphasized that the probabilities given for these limits are strictly accurate only when sampling from a normal universe. They are also based on the

[†]For extensive tables of the percentage points of the distribution of values of R when sampling from a normal universe, see H. L. Harter, "Tables of Range and Studentized Range," *The Annals of Mathematical Statistics,* vol. 31, no. 4, pp. 1122–1147, December 1960.

TABLE 10.3 FACTORS FOR PROBABILITY LIMITS TO BE USED IN CONTROL CHARTS FOR SAMPLE STANDARD DEVIATION

To obtain limits, multiply the estimated value of σ by the B factor with subscript corresponding to the desired probability

Size of subgroup n	Lower limits			Upper limits		
	$B_{0.001}$	$B_{0.005}$	$B_{0.025}$	$B_{0.975}$	$B_{0.995}$	$B_{0.999}$
2	0.00	0.00	0.03	2.25	2.81	3.30
3	0.04	0.07	0.16	1.92	2.30	2.63
4	0.09	0.15	0.27	1.77	2.07	2.33
5	0.15	0.22	0.35	1.67	1.92	2.15
6	0.21	0.28	0.41	1.60	1.83	2.03
7	0.25	0.33	0.45	1.56	1.76	1.93
8	0.29	0.37	0.49	1.52	1.70	1.86
9	0.33	0.41	0.52	1.48	1.65	1.80
10	0.36	0.44	0.55	1.45	1.62	1.76

TABLE 10.4 FACTORS FOR PROBABILITY LIMITS TO BE USED IN CONTROL CHARTS FOR RANGE

To obtain limits, multiply the estimated value of σ by the D factor with the subscript corresponding to the desired probability

Size of subgroup n	Lower limits			Upper limits		
	$D_{0.001}$	$D_{0.005}$	$D_{0.025}$	$D_{0.975}$	$D_{0.995}$	$D_{0.999}$
2	0.00	0.01	0.04	3.17	3.97	4.65
3	0.06	0.13	0.30	3.68	4.42	5.06
4	0.20	0.34	0.59	3.98	4.69	5.31
5	0.37	0.55	0.85	4.20	4.89	5.48
6	0.53	0.75	1.07	4.36	5.03	5.62
7	0.69	0.92	1.25	4.49	5.15	5.73
8	0.83	1.08	1.41	4.60	5.25	5.82
9	0.97	1.21	1.55	4.70	5.34	5.90
10	1.08	1.33	1.67	4.78	5.42	5.97

assumption that a σ estimated from the data is the true universe standard deviation. For practical control-chart work in industry, where the exact form of the universe is hardly ever known and control limits are often based on the evidence of short series of observations, it must be recognized that these probabilities are approximate rather than exact and may often be substantially in error.

10.5.4 Differing Viewpoints on How to Describe the Limits on Control Charts for Variables

Two points of view are found regarding the best way to describe limits on control charts for $\overline{X}$, R, and s.

One viewpoint is that the appropriate decision to make in setting limits is the numerical value of a probability. This should be the probability that, with no change in the universe, a point will fall within the control limits (or, alternatively, beyond the limits). The advocates of this point of view have usually adopted 0.998 as the desired probability of falling within limits of $\overline{X}$ charts. This has led to limits of $\overline{\overline{X}}$ (or $\overline{X}_0$) $\pm 3.09\sigma_{\overline{X}}$.

An alternative point of view is that, even if the probability associated with any limits could be known accurately, the exact numerical value of the probability is only of incidental interest. The important matter is that there be a definite basis for the establishment of limits and that this basis form a suitable guide to the actions that are to be based on the control charts.

In comparing these viewpoints, the reader should recognize that it really is not the exact value of the probability that constitutes a basis for action. For instance, it would be practically impossible to have any sense of the difference between the effect of probabilities of 0.0010 and 0.00135 that a point would fall above the upper control limit by chance. The real basis for acceptance of control limits, whether called 3-sigma limits or something else, is experience that the operational procedure involved in their computation gives limits that provide a satisfactory basis for action. As far as the practical use of $\overline{X}$ charts in manufacturing is concerned, it makes very little difference whether limits are at 3 sigma or at 3.09 sigma, and it makes even less difference whether they are called 3-sigma limits or probability limits. Both types of limits are used in the same manner and give very nearly the same results.

10.5.5 Some Special Aspects of Probability Limits on Control Charts for *R* and *s*

Although probability limits on $\overline{X}$ charts are like 3-sigma limits in being equidistant from the central line, probability limits on charts for R are not equally spaced with reference to the central line. This statement also applies to the distribution of s values from small subgroups. Both distributions are decidedly skewed.

A point emphasized by some advocates of probability limits is that for the common subgroup sizes of five or less, the lower 3-sigma limit is zero on these charts. In contrast, for the probabilities given in Tables 10.3 and 10.4, a subgroup size of three or more will give a lower control limit greater than zero.

10.6 CONTROL CHARTS FOR MOVING AVERAGES

10.6.1 The Use of Control Limits for Moving Averages

In manufacturing plants many schemes other than the Shewhart control chart have been used for plotting data on quality characteristics. For example, a chemical plant may maintain charts on which are plotted the results of daily analyses made to determine the percentages of certain chemical constituents in its incoming materials, product in process, and finished product. A common variation of this is to plot moving averages rather than daily values. The moving average is particularly appropriate in continuous process chemical manufacture when applied to quality characteristics of raw materials and product in process. The smoothing effect of the moving average often has an effect on the figures similar to the effect on the product of the blending and mixing that take place in the remainder of the production process.

In the introduction of Shewhart techniques into chemical plants, it may be desirable not to disturb the custom of plotting moving averages. However it is appropriate to apply control limits to such moving average charts and to add charts for moving ranges. The calculations for these limits and the interpretation of these charts are similar to the conventional $\bar{X}$ and R charts but differ in certain respects.

Table 10.5 illustrates the calculation of moving averages ($m\bar{X}$) and moving ranges (mR). The figures given are the daily analyses of percentages of unreacted lime (CaO) at an intermediate stage in a continuous manufacturing process. The average given for Sept. 3 is the average of the percentages on the first, second, and third; that for Sept. 4 is the average of the values on the second, third, and fourth; and so forth.[†] The calculation of a 3-day moving average is simplified by carrying a moving total to which is added each day the algebraic difference between the value today and the value 3 days ago.

The daily values are plotted in Fig. 10.5*a*; the moving averages, in Fig. 10.5*b*. A comparison of these two graphs shows the effect of the moving average in smoothing the curve. The more successive points averaged, the greater this smoothing effect and the more the curve emphasizes trends rather than point-to-point fluctuations. The moving range for the same data is plotted in Fig. 10.5*c*.

The control limits in Fig. 10.5*b* and *c* are computed from an $\bar{X}_0$ of 0.17 and an R_0 of 0.065. These figures were established on the basis of the record of the two preceding months. The calculation of limits uses the factors and formulas of Table *D,* App. 3.

The interpretation of a point outside control limits on moving average and moving range charts is the same as a point outside control limits on conventional $\bar{X}$ and R charts. However, because successive points on moving average and moving range charts are not independent of one another, the interpretation of several points in a row outside control limits is obviously not the same. Similarly, runs above or below the central line do not have the same significance on moving average and moving range charts as on conventional $\bar{X}$ and R charts.

As pointed out in previous chapters, whenever a shift in universe average occurs *within* a subgroup rather than *between* subgroups, the R chart tends to show

[†]From a technical statistical viewpoint, the average should always be plotted at the *midpoint* of the period; for instance, the average of the first, second, and third should be assigned to the second, not to the third. The reason for the common practice in industry of assigning the moving average to the final date rather than to the middle date is to have the average always seem up to date rather than behind time. In this case, practical psychology takes precedence over statistical correctness.

TABLE 10.5 CALCULATION OF MOVING AVERAGE AND MOVING RANGE
Data on percent of unreacted CaO at an intermediate stage in a chemical manufacturing process

Date	Daily value	3-day moving total	3-day moving average	3-day moving range	Date	Daily value	3-day moving total	3-day moving average	3-day moving range
Sept. 1	0.24				26	0.17	0.49	0.163	0.10
2	0.13				27	0.18	0.46	0.153	0.07
3	0.11	0.48	0.160	0.13	28	0.13	0.48	0.160	0.05
4	0.19	0.43	0.143	0.08	29	0.28	0.59	0.197	0.15
5	0.16	0.46	0.153	0.08	30	0.16	0.57	0.190	0.15
6	0.17	0.52	0.173	0.03	Oct. 1	0.14	0.58	0.193	0.14
7	0.13	0.46	0.153	0.04	2	0.16	0.46	0.153	0.02
8	0.17	0.47	0.157	0.04	3	0.14	0.44	0.147	0.02
9	0.10	0.40	0.133	0.07	4	0.10	0.40	0.133	0.06
10	0.14	0.41	0.137	0.07	5	0.13	0.37	0.123	0.04
11	0.16	0.40	0.133	0.06	6	0.20	0.43	0.143	0.10
12	0.14	0.44	0.147	0.02	7	0.14	0.47	0.157	0.07
13	0.17	0.47	0.157	0.03	8	0.10	0.44	0.147	0.10
14	0.15	0.46	0.153	0.03	9	0.18	0.42	0.140	0.08
15	0.20	0.52	0.173	0.05	10	0.11	0.39	0.130	0.08
16	0.26	0.61	0.203	0.11	11	0.08	0.37	0.123	0.10
17	0.16	0.62	0.207	0.10	12	0.12	0.31	0.103	0.04
18	0.00	0.42	0.140	0.26	13	0.13	0.33	0.110	0.05
19	0.18	0.34	0.113	0.18	14	0.12	0.37	0.123	0.01
20	0.18	0.36	0.120	0.18	15	0.17	0.42	0.140	0.05
21	0.20	0.56	0.187	0.02	16	0.10	0.39	0.130	0.07
22	0.11	0.49	0.163	0.09	17	0.09	0.36	0.120	0.08
23	0.30	0.61	0.203	0.19					
24	0.21	0.62	0.207	0.19					
25	0.11	0.62	0.207	0.19					

lack of control. Thus a sudden shift in the mean may show up first on the *R* chart. Changes of short duration that do not persist through one entire subgroup may be shown only on the *R* chart and not on the $\overline{X}$ chart. In Fig. 10.5, the indications of lack of control are on the *R* chart.

10.6.2 Combination of Chart for Individual Measurements and Moving Range Chart†

In some processes, it is common to have single measurements spaced some time apart. A case in point was the chemical process with one measurement per day that is used in Table 10.5 and Fig. 10.5 to illustrate charts for moving averages and moving ranges.

†A good discussion of this topic, with several examples, is given in "ASTM Manual on Presentation of Data and Control Chart Analysis," STP 15D, pp. 99, 130–133, American Society for Testing and Materials, Philadelphia, Pa., 1976.

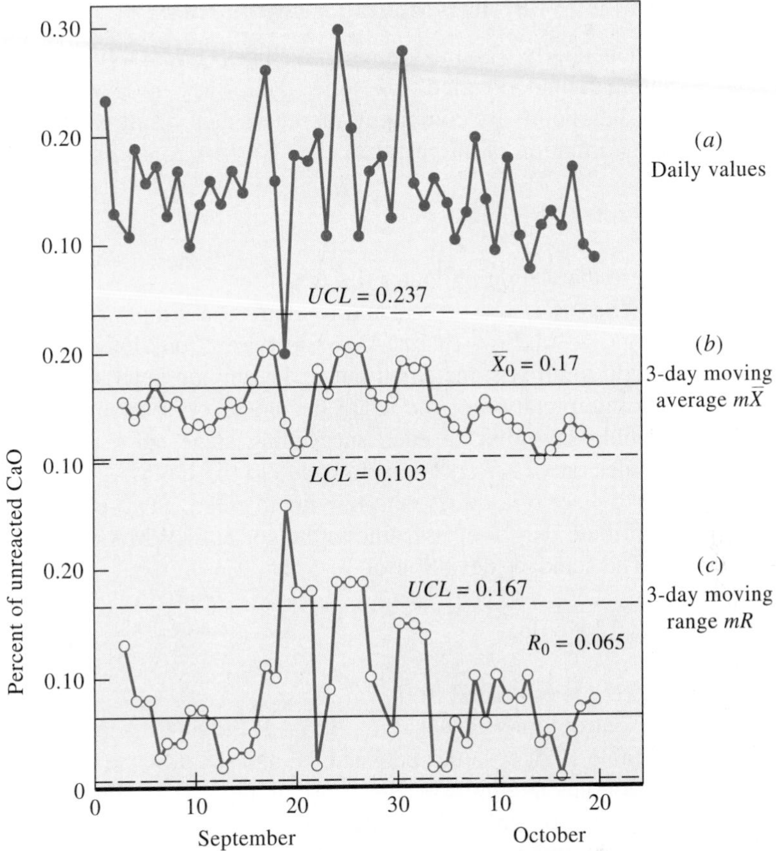

FIGURE 10.5 Illustration of percentage of unreacted CaO: (*a*) individual daily values, (*b*) 3-day moving average $m\bar{X}$ control chart, and (*c*) 3-day moving range *mR* control chart. Data of Table 10.5.

Under such circumstances, it may be desired to plot *X* values rather than to smooth the day-to-day fluctuations by plotting moving averages. The individual *X* values from Table 10.5 were shown without control limits in Fig. 10.5*a*. To add 3-sigma control limits for individuals to such a chart, it is necessary to compute $\bar{X}_o \pm 3\sigma$, with σ estimated from the moving range as $\bar{R}/d_2$. With the standard $\bar{X}_o$ of 0.17 and the standard R_o of 0.065 for subgroups of three used in Fig. 10.5, the upper and lower control limits for individual *X* values are 0.285 and 0.055, respectively. If these limits had been added to Fig. 10.5*a*, this figure would have been a control chart for individual values.

Even though the moving $\bar{X}$ is not used in such a case, it is necessary to use the moving range as a measure of process dispersion. The "ASTM Manual on Quality Control of Materials" recommends a subgroup size of two for the moving range in connection with such a chart for individual values.

10.6.3 An Exponentially Weighted Moving Average Chart[†]

Exponentially weighted moving averages (EWMAs) represent another method for summarizing so-called time series data. They involve a transformation of individual data points by combining the latest data result with a geometrically weighted transformation of all past data. The EWMA control statistic Y_t is found from

$$Y_t = \alpha X_t + (1 - \alpha)Y_{t-1}$$

where the weighting factor $0 \leq \alpha \leq 1$.

When α is one, the resulting chart is a simple chart for individual measurements, as in Fig. 10.5*a*. As α decreases, previous history is given increasing weight until, at some small but undetermined level, the averaging is so smoothed that interpretation of the result becomes very obscure. Simple observation of the formula says that, at each succeeding stage, an r period older data point contributes only $(1 - \alpha)^r$ times its value to the EWMA. Thus, if α is $\frac{1}{3}$, as one might use in place of a three-point moving average, a two-period-old data reading would contribute only $\frac{1}{9}$ of its value to the current EWMA.

The standard deviation of Y_t, σ_y is

$$\sqrt{[\alpha/(2 - \alpha)]}\sigma$$

where σ is estimated from $\overline{R}/d_2$ or $\overline{s}/c_4$.

Control limits are set at $\overline{X}_O \pm 3\ \sigma_y$. If the EWMA technique is applied to the data of Table 10.5, α values between 0.6 and 0.8 produce charts that look most like the three-point moving average chart of Fig. 10.5*b*. Frequently two-point or three-point moving range charts are used to control dispersion and estimate σ.

Some authorities suggest that much insight is lost from the transformation of data that results from use of the EWMA technique. However, it has considerable appeal for those familiar with exponential smoothing.

10.7 $\overline{X}$ CHART WITH A LINEAR TREND

10.7.1 $\overline{X}$ Charts for Trended Universe Average with Constant Standard Deviation

In some machining operations, tool wear occurs at a uniform rate over the period of use of the tool. This tool wear may be one of the factors influencing the average value of some dimension of the product manufactured and may be responsible for a trend in this average. Where subgroups are selected in a way that spaces them uniformly with respect to this wear, control charts for $\overline{X}$ may look something like Fig. 10.6. The R chart, however, is likely to remain in control yielding a constant estimate of σ.

[†]See A. W. Wortham, and L. J. Ringer, "Control via Exponential Smoothing," *The Logistics Review*, vol. 7, no. 3, 1971, pp. 33–40.

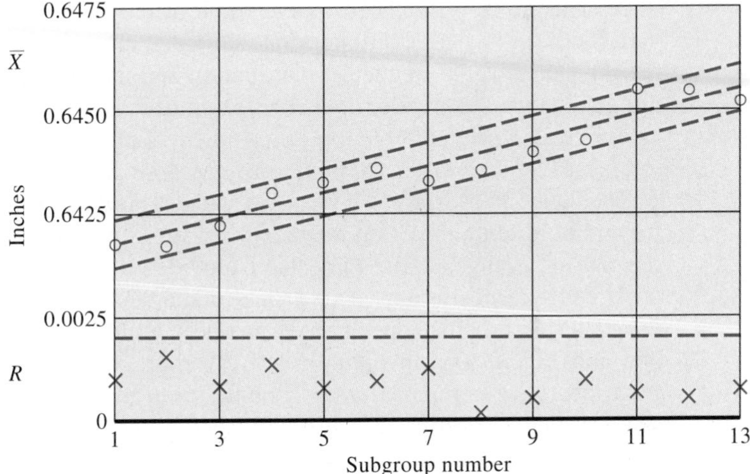

FIGURE 10.6 Illustration of trend line on $\overline{X}$ chart—Example 10.1.

The central line and control limits for the $\overline{X}$ chart, in such a case, should be sloping rather than horizontal. The slope of the central line, or universe average, and σ, estimated from $\overline{R}$, are determined from the measurements themselves. Once known, it becomes possible to determine the initial setting and length of run that together will give the maximum period between machine settings consistent with the specified tolerances. The methods of making the required calculations are illustrated in Example 10.1.

Example 10.1
Determining the initial machine setting when a trend is expected in the value of a dimension

Facts of the case On certain operations in one machine shop, it was a common experience for a definite steady trend in the average value of dimensions of machined parts to be caused by rapid tool wear. On many specified dimensions the spread of the specification limits, $USL_X - LSL_X$, was substantially greater than the natural tolerance range of the process, 6σ. This provided an ample margin of safety against the production of nonconforming product as long as the machine setting held the average value of the dimension somewhere close to a point midway between the specification limits. However, the tendency of the average value of a dimension to shift rapidly as a result of tool wear called for frequent new setups to restore the average value to its desired position. Each new setup involved appreciable costs, both for setup expense and for idle machine time.

The introduction of the control-chart point of view provided a basis for reducing the frequency of setups. The situation here was essentially the one illustrated in Fig. 4.4*c*. The average value of the frequency distribution of the variable X (in this case, a dimension) shifted at an approximately uniform rate, but the shape of the frequency distribution and its dispersion (measured by σ) did not change. From control-chart data properly taken, it was possible to estimate σ. It was also possible to estimate the rate of change of the average $\overline{X}$ and to express this rate of change in terms of an equation.

The estimate of σ made it possible to determine the aimed-at average value for an initial setup to permit the maximum run between setups. The estimate of the rate of change of the dimension provided a basis for decision as to the required frequency of new setups. Once σ and the rate of change of X were estimated for a particular operation, it proved possible to use this information on new jobs. Even jobs involving relatively short runs were benefited, as many such jobs were completed with a single setup in contrast to the two or three setups usually required for similar jobs before the use of the control chart.

Fitting a trend line to an $\overline{X}$ chart Consider the following values of $\overline{X}$ and R for a dimension specified as 0.644 ± 0.004 in. Subgroups of five components were measured every half-hour.

Subgroup number	$\overline{X}$	R	h
1	0.6417	0.0011	−6
2	0.6418	0.0016	−5
3	0.6424	0.0010	−4
4	0.6431	0.0015	−3
5	0.6433	0.0009	−2
6	0.6437	0.0010	−1
7	0.6433	0.0014	0
8	0.6436	0.0004	1
9	0.6441	0.0006	2
10	0.6444	0.0011	3
11	0.6456	0.0009	4
12	0.6457	0.0007	5
13	0.6454	0.0009	6
Totals	8.3681	0.0131	

$$\overline{\overline{X}} = \frac{8.3681}{13} = 0.6437$$

$$\overline{R} = \frac{0.0131}{13} = 0.0010$$

Certain precautions should be taken in the collection of data to be used for the calculation of a trend line and control limits. The production between successive subgroups should be approximately constant.[†] The components included in a subgroup should be produced in succession so that the trend will have little effect on the range of a subgroup.

[†]If the total production up to each subgroup were recorded, it would be possible to plot $\overline{X}$ as a function of production even though the production were not uniform from subgroup to subgroup. In such a case, however, the fitting of a least-squares trend line is somewhat more complicated than the simple method illustrated in Table 10.6. Methods of fitting least-squares lines in such cases are explained in standard texts on statistical methods.

On the conventional $\overline{X}$ chart, the central line is horizontal. With an upward trend of $\overline{X}$ values, the central line must be a sloping line. The position of this central line may be described by an equation of the form $\overline{X} = a + bh$, with the symbol h used to represent the subgroup number (preferably with a revised subgroup numbering using the middle subgroup as the origin). a is the value of $\overline{X}$ when $h = 0$, and b is the slope of the line. The method of least squares provides a satisfactory way of finding the values of a and b corresponding to any given set of measurements.†

Under certain special circumstances, computing a least-squares trend line is a very simple operation. These circumstances are that the observed values of the variable plotted on the vertical axis of the chart (in this case, values of $\overline{X}$) are uniformly spaced along the horizontal axis of the chart (in this case, on the scale of subgroup numbers h); that there are an odd number of subgroups; and that the origin on the horizontal axis is taken as the midpoint of that axis (that is, the 0 value of h is assumed at the middle subgroup). Under these circumstances $a = \overline{\overline{X}}$, and $b = \Sigma h\overline{X}/\Sigma h^2$. The method of calculation is illustrated in Table 10.6.

†Although fair results may often be obtained by drawing a trend line by eye on the chart, the personal equation enters into such fits to the extent that no two people are likely to agree exactly. As the least-squares line is so easy to compute in this special case, it will generally pay to take the slight additional time required to obtain the least-squares fit.

TABLE 10.6 CALCULATION OF EQUATION OF LEAST-SQUARES TREND LINE FOR $\overline{X}$

X expressed in units of 0.0001 in excess of 0.6400

Subgroup number	Revised subgroup number h	Subgroup average $\overline{X}$	$h\overline{X}$	h^2
1	−6	17	−102	36
2	−5	18	−90	25
3	−4	24	−96	16
4	−3	31	−93	9
5	−2	33	−66	4
6	−1	37	−37	1
7	0	33	0	0
8	1	36	36	1
9	2	41	82	4
10	3	44	132	9
11	4	56	224	16
12	5	57	285	25
13	6	54	324	36
Totals	0	481	599	182

$$a = \overline{\overline{X}} = \frac{481}{13} = 37.0$$

$$b = \frac{\Sigma h\overline{X}}{\Sigma h^2} = \frac{599}{182} = 3.29$$

$$\overline{X} = a + bh = 37.0 + 3.29h$$

For simplicity in calculations, the $\overline{X}$ values in Table 10.6 are expressed in units of 0.0001 in excess of 0.6400. The equation $\overline{X} = 37.0 + 3.29h$ is, of course, expressed in these units. In terms of the actual dimension in inches, the equation becomes $\overline{X} = 0.6437 + 0.000329h$.

To plot this line on the control chart (Fig. 10.6), it is necessary to locate two points on the line and connect them. For example,

$$\overline{X} = \begin{cases} 0.6437 + (0.000329)(-6) = 0.6417 & \text{for } h = -6 \\ 0.6437 + (0.000329)(6) = 0.6457 & \text{for } h = +6 \end{cases}$$

Interpreting the control chart as a basis for action The control limits are sloping lines parallel to the central trend line. The upper control limit is $A_2\overline{R}$ above the trend line, and the lower control limit is $A_2\overline{R}$ below it. In this case, $A_2\overline{R} = 0.58(0.0010) = 0.0006$. Figure 10.6 shows sloping limits plotted at this distance from the central line.

To decide on the value of the average dimension to be aimed at in the initial machine setting, it is first necessary to estimate σ. This is $\overline{R}/d_2 = 0.0010/2.326 = 0.00043$. The tolerance spread $USL_X - LSL_X$ may then be compared with σ. In this case,

$$USL_X - LSL_X = 0.6480 - 0.6400 = 0.0080$$

This is equal to 18.6σ.

A decision must then be made as to how many multiples of σ the initial setting should be from the lower specification limit. In some instances, this might be 3σ; in others, it might be a greater multiple of σ such as 4σ. This decision is a function of the economics involved. Where the tolerance spread is great enough to allow a considerable run between settings, as in this case, the additional margin of safety involved in the use of 4σ is likely to be justified. The same multiple of σ should be used to determine the time of a new setup. Using 4σ in this case, we obtain the following directions: On initial setup, aim at

$$\overline{X} = LSL_X + 4\sigma = 0.6400 + 0.0017 = 0.6417$$

Make a new setup when trend line $= USL_X - 4\sigma = 0.6480 - 0.0017 = 0.6463$.

The slope of the trend line, $b = 0.000329$, is the expected change in average dimension from one subgroup to the next. The interval between setups may be estimated as $(0.6463 - 0.6417)/0.000329 = 14$ subgroup intervals. With subgroups spaced half an hour apart, this indicates that a new setup is required every 7 h.

Comment on Example 10.1

When a point falls outside control limits on an $\overline{X}$ chart with a sloping central line, there are two possible interpretations. One is the conventional interpretation that an assignable cause of variation is responsible for the point falling outside the limits. The other is that the limits are based on a trend line that is wrong. For example, a control chart might be started for a new job using a slope of a trend line determined from a previous similar job; it might turn out that the jobs were really sufficiently different for the slopes to differ. As another example, tool wear is a factor that does not necessarily operate at a uniform rate; on jobs where the rate of change of the average value of a dimension is variable, no straight line fitted to $\overline{X}$ values will provide a satisfactory central line for a control chart.

10.8 NARROW LIMIT GAGING

10.8.1 Special Adaptation of the *np* Chart for Patrol Inspection

The reader will recall that the fraction rejected control chart requires a substantially larger sample size than the $\overline{X}$ chart in order to maintain the same level of sensitivity to shifts in the process. Two aspects of sample size determination require a special adaptation of the p chart if it is to be an effective instrument for process control where a high level of quality must be maintained. First, some rejects must be *expected* to show up in the sample. If, for example, the process operates at 0.5% rejected, a sample size of 200 is required before there will be an average of one reject per sample. This factor would suggest the use of a rather large sample size.

Second, the larger the sample size, the more likely it is that a shift will occur during the drawing of the sample. The result, of course, would be an increased chance that shifts occurring between samples would go undetected.

In cases where it is necessary to use inspection by attributes and still detect small changes in the process, circumstances may require the use of small samples. Here the only way in which a p chart or np chart may be used to advantage is by artificially creating a pseudo bad quality level by applying acceptance standards for control-chart purposes that are much more severe than those really imposed by the specifications. That is, there may be established a special severe definition of a "defective" that is used only for purposes of process control by means of the control chart. In this way, the small sample size is not such an obstacle to the use of a control chart based on inspection by attributes.

This scheme has been applied advantageously in certain types of electrical testing and in ballistic testing. In Great Britain it has been applied to dimensional control. Example 10.2 describes an application taken from British literature.[†]

[†]"Quality Control by Limit Gauging," *Production and Engineering Bulletin,* vol. 3, pp. 433–437, October 1944. See also A. E. Mace, "The Use of Limit Gages in Process Control," *Industrial Quality Control,* vol. 8, no. 4, pp. 24–31, January 1952; and E. R. Ott and E. G. Shilling, *Process Quality Control,* 2d ed., McGraw-Hill Book Company, New York, 1990.

Example 10.2

Process control by *np* chart based on gage tolerances that are tighter than specification tolerances

Facts of the case This example deals with an internal diameter specified as 1.008 ± 0.0035 in. Actual measurements were made on 100 components. These showed control with an estimated σ of 0.0011. The following calculations show the relationship between the natural tolerances of the process and the specification tolerances:

$$USL_X - LSL_X = 1.0115 - 1.0045 = 0.0070 \text{ in}$$

$$6\sigma = 6(0.0011) = 0.0066 \text{ in}$$

As the tolerance spread was barely greater than 6σ, it was evident that the centering of the dimension must be maintained accurately at the nominal dimension of 1.008.

A special quality control gage was made with not-go and go dimensions as 1.00925 and 1.00675, respectively. These limits were at $\overline{X}_o \pm 1.15\sigma$. If the distribution were normal and perfect statistical control were maintained with the dimension centered at its nominal value, approximately 25% of the product would be rejected by this quality control gage, even though substantially all the product would be within specification limits.

Samples of 10 were used with this quality control gage. A variation of the *np* chart was used. The upper half of this chart showed the number rejected by the not-go gage, and the lower half showed the number rejected by the go gage. In accordance with the British practice, warning limits as well as action limits were used on this chart.

Figure 10.7 shows a comparison of this chart for number of defectives with an $\overline{X}$ chart based on actual measurements of the same components that were gaged with the special go and not-go gage. This applies to a period when the process was out of control. It will be observed that the lack of control was readily apparent from this *np* chart, with even more points showing out of control than on the chart for averages.

Comment on Example 10.2

In some manufacturing plants, patrol inspection is carried out with go and not-go gages having tighter tolerances than the gages used for acceptance or rejection of finished product. This may provide a good opportunity for the use of the type of *np* chart shown in Fig. 10.7.

The appropriate value of p for use in setting control limits on this *np* chart is not necessarily the average value $\overline{p}$ as determined from the work gages. For best results, it is necessary to estimate σ, the universe standard deviation. (In Example 10.2 this estimate was made by actual measurements on 100 components.) This permits differentiation of circumstances where the centering of a dimension must

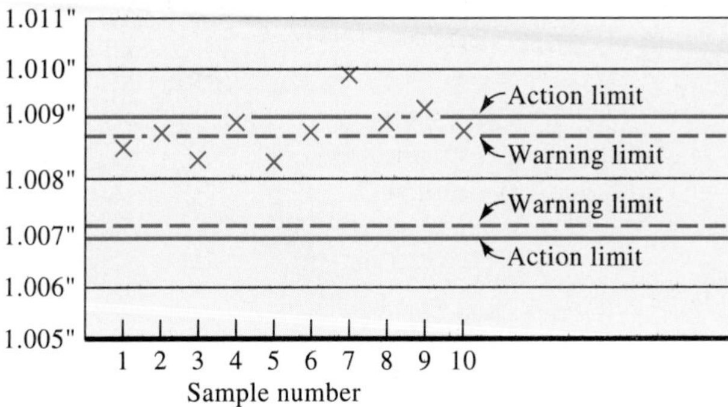

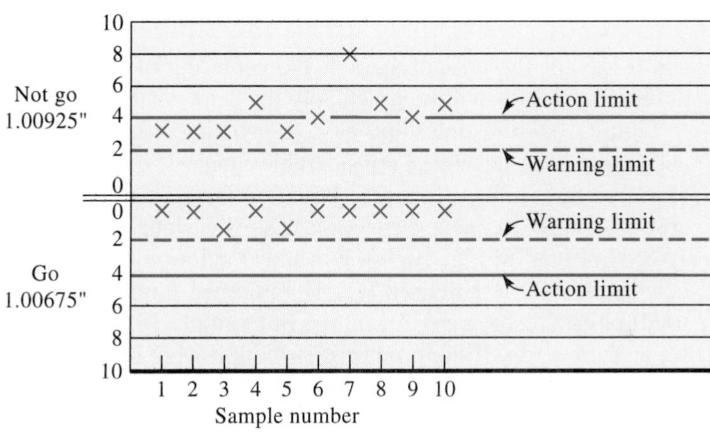

FIGURE 10.7 Comparison of charts for $\overline{X}$ and *np* for out-of-control period—Example 10.2. *(Reproduced from "Quality Control by Limit Gauging," Production and Engineering Bulletin, October 1944.)*

be closely controlled from those circumstances where the average value may be permitted to shift.

Figure 10.8 illustrates the case in which the centering of a process may not shift without producing defective product. If samples of 10 are to be used, the gage limits should be set so that each shaded portion of the area under the frequency curve (representing satisfactory product rejected by the work gage) includes 10 to 15% of the total area under the curve. In this case, if the centering is maintained, the units rejected by the go portion of the work gage will be approximately equal to those rejected by the not-go portion of the gage. If the sum of the shaded areas in Fig. 10.8 is 25% of the area under the frequency curve, p is 0.125 for the go chart

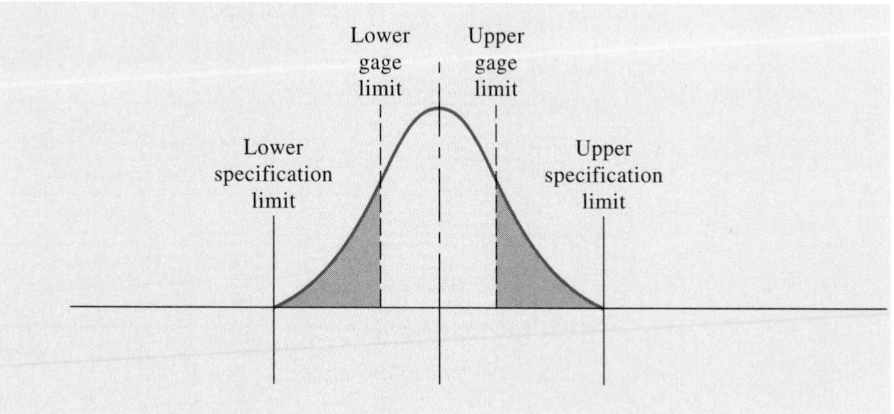

FIGURE 10.8 Relationship between special severe gage limits and specification limits when manufacturing processes can just work to specification limits.

and 0.125 for the not-go chart. If the centering of the process is maintained, the observed $\bar{p}$ values will be practically the same as these p values.

Figure 10.9 illustrates the case in which the centering of a process may be allowed to shift through a considerable range. If control-chart limits are to be set to permit this shift, p should be based on the ratio of each shaded area to the total area under the frequency curve, just as was done in the case illustrated by Fig. 10.8. However, in Fig. 10.9, the actual fraction rejected by the work gages might be considerably less than in Fig. 10.8; that is, $\bar{p}$ might be less than the value of p used for establishing control limits. In Example 10.2, the narrow gage limits were set at $\bar{X}_o \pm 1.15\sigma$. If the process distribution were related to specifications as indicated in Fig. 10.9, these gage limits might have been set as follows:

$$LGL = LSL_X + 3\sigma - 1.15\sigma = LSL_X + 1.85\sigma \qquad UGL = USL_X - 1.85\sigma$$

It may be desirable to increase the multiplier of σ to 2.0 or slightly more to be on the conservative side. Use of gaging as illustrated in Fig. 10.9 assumes that the economic cost function is extremely flat so long as items are within specifications; in other words, there is no economic justification in striving to hold dispersion about the target value to a minimum.

Some writers offer discussions of narrow limit gaging such as that presented as justification for precontrol, the subject of Sec. 10.2.8. Except for certain measurement mechanics, there is no relationship between the two techniques. Patrol inspection is valid only when a process is relatively easy to control, when it has been in statistical control using $\bar{X}$ and R or s charts for some time, and when the user is comfortable about the assumption that the process generates a normal distribution when in control. If anything, it might be more properly referred to as "postcontrol." Patrol inspection is not a good tool when little is known about the process. When much is known, the justification for using the paired np charts is economic, not technical.

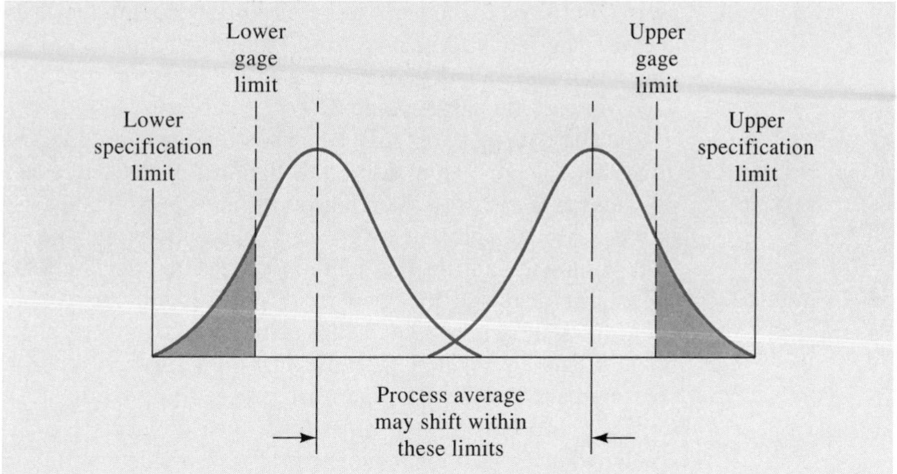

FIGURE 10.9 Relationship between special severe gage limits and specification limits when a manufacturing process has a margin to spare.

Generally speaking, more useful information can be secured from $\overline{X}$ and R charts than can be obtained by the use of np charts using small samples with a special severe definition of a rejectable unit for control-chart purposes. The chief practical advantage of this type of np chart over the $\overline{X}$ and R charts is that the np chart may occasionally be applied to the results of patrol inspection with little or no change in the way this inspection is being carried on. In such circumstances, the np chart may involve lower costs for securing data and a minimum amount of training of personnel and changing of inspection procedures. Unless these conditions favorable to the np chart are present, the $\overline{X}$ and R charts are usually preferable.

10.9 WORKING WITH SHORT PRODUCTION RUNS

Sometimes it is argued that SPC techniques are useful for mass production but may not be applied when production runs are small or where there is a varying mix of products in the product stream. A reason for this concern is that people tend to focus on the individual units of product rather than on the process. Within the limits of its physical capacity, most machines are intended to turn out dimensions over a certain range; and so long as the tooling, materials, and operator remain the same, they will exhibit a nearly constant underlying dispersion as measured by the standard deviation σ.

If σ is constant, or relatively so, over a range of dimensions, it is possible to utilize the measurement results of every unit produced to establish its value. That value becomes a standard σ_0 to be used in establishing control limits for the machine-person-tooling-material combination. Each piece part emerging from the operation may then be controlled against its target dimension. That is, if T is the nominal dimension for a part and the actual dimension produced is X, the deviation $D = (X - T)$ is the statistic used for developing the $\overline{X}$ and R or s charts. Values of D are likely to be both positive and negative and to be distributed approximately normally about zero much as if the deviations had been measured by a dial gage preset at the target value T.

The charts themselves may be standard Shewhart charts, with independent subgroups of four or five measured deviations, or they may be moving average and moving range charts as described in Sec. 10.6. If data are generated very slowly, moving average and range (or s) charts may be preferable. They have the advantage of showing any sudden jump in a measured deviation at the time it is experienced. The results of the chart for dispersion will verify whether the assumptions about the standard value of σ_o are justified.

As discussed in Chap. 8, there is no circumstance under which taking multiple measurements on the same part can be used to provide an estimate of σ for controlling a process through time. Such measurements provide an estimate of the combined piece positional and error of measurement variation and, therefore, cannot be used rationally for process control.

10.10 A SUCCESS CHART FOR PRODUCTION RUNS OF EXTREMELY HIGH QUALITY

The use of standard Shewhart control charts for attributes data—the *p, np, c,* or *u* chart—can yield results that are quite unsatisfactory if the rate at which rejects, or nonconformities, are being generated is very small. As was stated in the discussions of control charts for attributes data, the sample or subgroup size should be large enough that some rejects or nonconformities are expected to show up in every sample. When quality is extremely high, it is not practical to use standard attributes charts because days may go by before inspection results yield more than zero counts of rejects or nonconformities. When rejects are measured in just a few parts per million (ppm), the time, or number of units accepted between each reject, may be very long. For example, a rejection rate of 200 ppm, a value of p of 0.0002, on the average will yield only one reject in 5,000 units produced; that of 3.4 ppm, a stated objective of some major producers, will yield only one reject in 294,118. Operators tend to lose confidence in procedures that require them to record and plot zero's repeatedly on control charts.

An alternative procedure, one that really is not intended for process control purposes but has been shown to be useful in verifying high levels of quality production, is the "success chart," or chart for counts of units between rejects (CBR).[†] Figure 10.10, based on data from the first 15 sets of observations from Table 16.2, shows an example of one form the chart takes. Each column displays the number of units produced, and passed on inspection, before a unit is rejected. Thus each

[†]The first mention of a chart of this type appeared in T. N. Goh, "A Control Chart for Very High Yield Processes," *Quality Assurance,* March 1987. The procedure was developed independently, tested, and published in *Proceedings of: 1993 CMI Tools Symposium,* June 1993, Hughes Aircraft Company, Fullerton, Calif. The paper is titled "Using a 'Success Chart,' Tallies of Consecutive Conforming Items, to Promote Process Capability," and is coauthored by Vera Abatayo, David Fink, Cheryl Jurbala, Robert Saber, and Ronald Schroeder.

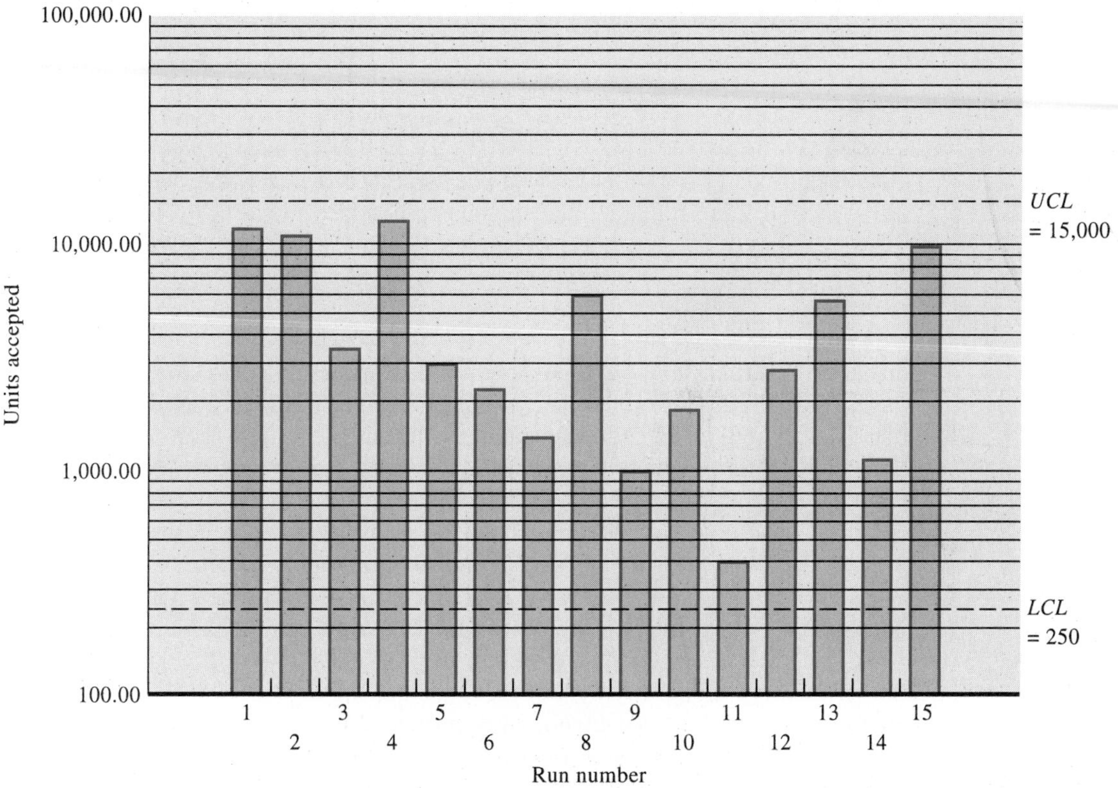

FIGURE 10.10 An example success chart, counts of units produced before a reject, based on a modification of the first 15 sets of observations from Table 16.2.

column shows the run length between rejects expressed in number of units produced. The operator accumulates the data and plots the bars each day. Each run may last for several days or even weeks. When a reject is found, a record is made of the length of the run. This is plotted on the chart and, perhaps, dated; and the count begins all over again. The left-hand scale of Fig. 10.10 is logarithmic. An alternative form, recommended in the Hughes Aircraft paper, is to use a linear scale to simplify operator understanding and avoid plotting errors.

Control limits on the chart are positioned at the 5% and 95% bounds on a process yielding 200 ppm rejects ($p = 0.0002$) and assumed to be constant. Where p is the best estimator of proportion nonconforming, or possibly the target value, the average run length between rejects is

$$\bar{n} = \frac{1}{p}$$

$$= \frac{1}{0.0002} = 5{,}000 \text{ units for our example}$$

If α is the allowable risk level for Type I error, then the control limits are

$$UCL = -\bar{n} \ln \frac{\alpha}{2} \qquad LCL = \bar{n} \frac{\alpha}{2}$$

and, if desired, a central line may be placed at the median, $0.7\,\bar{n}$. When α is 0.10, these control limits reduce to the simple expressions

$$UCL = 3.0\,\bar{n} \qquad LCL = 0.05\,\bar{n}$$

which, for our example, are 15,000 and 250, respectively. The theory behind the constant reject, or failure, rate assumption is discussed in Chap. 16.

The 15 runs plotted on the chart each terminate between the control limits. Thus there is no significant indication that the reject rate differs from the 200 ppm assumed and upon which the samples were based. Nine of the fifteen runs terminate at less than the median, 3,500; and three of the first four terminate above 10,000. Since there is about a 5% chance each that a run will terminate before 250 or after 15,000, a "single point out" rule, as is used with standard Shewhart charts, would not be useful. A *tendency* to go out low, or to cluster near the *LCL,* would signal that the process is not producing at the target rate, 200 ppm in this case. Likewise, a tendency for runs to cluster near or above the *UCL* would signal that the process is significantly better than the target.

Actual control of such high-quality processes must be exercised at the input end. Attempting to control at final inspection, the output end, is not likely to be very useful. Circumstances where the success or CBR chart may prove most useful are:

1. To establish a value of p in ppm when the quality level is extremely high
2. To audit production quality level through time to determine if the high level is being maintained
3. To determine stages where a process has either improved or deteriorated significantly

10.11 COMBINING PROCESS CONTROL AND PRODUCT ACCEPTANCE

10.11.1 Reject Limits for Averages on $\bar{X}$ Charts

One possible method of showing the relationship between $\bar{X}$ values and the specification limits that apply to individual items is through the use of *reject limits* for averages.† If the assumption is made that σ is known and will not change and that

†A. J. Winterhalter, "Development of Reject Limits for Measurements," *Industrial Quality Control,* vol. 1, no. 4, pp. 12–15, January 1945; vol. 1, no. 5, pp. 12–13, March 1945. See also *Engineering Data Book,* sec. 29, "Statistical Methods in Quality Control," issued Sept. 22, 1943, Hunter Pressed Steel Co., Lansdale, Pa.

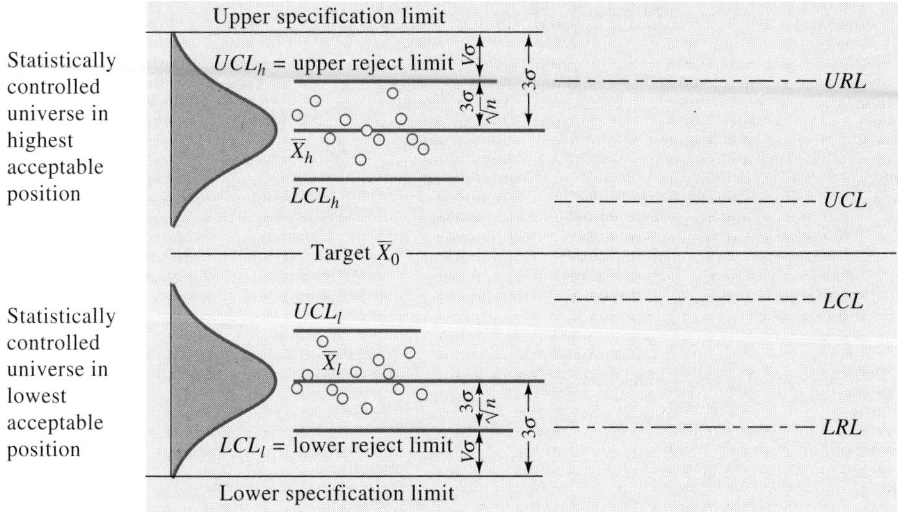

FIGURE 10.11 Relationship among specification limits, control limits, and reject limits.

practically all the product will fall within limits of $\mu \pm 3\sigma$ (or, for that matter, $\mu \pm$ any other desired multiple of σ), it is easy to calculate the highest and lowest values of μ that will permit practically all product to fall within specification limits. The reject limits for averages are certain control-chart limits that would be appropriate if μ should be at each of these computed values. The derivation of these reject limits is explained here. Their application to acceptance inspection is discussed in Chap. 13.

Figure 10.11 illustrates the development of such limits. Assume the universe in its highest acceptable position with universe average *exactly* 3σ below the upper specification limit (μ_h). On an $\overline{X}$ chart with subgroup size n, the upper control limit $UCL_{\overline{X}}$ will be $3\sigma/\sqrt{n}$ above the universe average. This will evidently be the highest possible satisfactory value of the upper control limit and is designated as the upper reject limit for averages, abbreviated as $URL_{\overline{X}}$.

The distance of $URL_{\overline{X}}$ below the upper specification limit for individual values is obviously $3\sigma - (3\sigma/\sqrt{n})$. This may be expressed as $[3 - (3/\sqrt{n})]\sigma$. This factor $[3 - (3/\sqrt{n})]$ is designated as V. Table 10.7 gives values of V corresponding to values of n from 2 to 25. By a similar process of reasoning that assumes the universe in its lowest acceptable position, it may be shown that the lower reject limit for averages, $LRL_{\overline{X}}$, is $V\sigma$ above the lower specification limit.

If an $\overline{X}$ chart shows a state of statistical control and both control limits fall within the two reject limits, this means that practically all the product will fall within specification limits. If the $UCL_{\overline{X}}$ falls above the $URL_{\overline{X}}$, or if the $LCL_{\overline{X}}$ falls below the $LRL_{\overline{X}}$, the conclusion is that even though control is maintained, some product will fall outside one of the specifications. In this way, limits telling the conformance of individual values with specifications may be placed on the chart for averages.

TABLE 10.7 VALUES OF *V*, FACTOR FOR REJECT LIMITS, FOR DIFFERENT VALUES OF *n* SUBGROUP SIZE

n	*V*	*n*	*V*	*n*	*V*
2	0.88	10	2.05	18	2.29
3	1.27	11	2.09	19	2.31
4	1.50	12	2.13	20	2.33
5	1.66	13	2.17	21	2.35
6	1.78	14	2.20	22	2.36
7	1.87	15	2.23	23	2.37
8	1.94	16	2.25	24	2.39
9	2.00	17	2.27	25	2.40

$$URL_{\bar{X}} = USL_{\bar{X}} - V\sigma$$

$$LRL_{\bar{X}} = LSL_{\bar{X}} + V\sigma$$

$$V = 3 - \frac{3}{\sqrt{n}}$$

The right-hand side of Fig. 10.11 shows the positioning of the reject limits in relation to control limits targeted at the midpoint between USL_X and LSL_X. This positioning would be consistent with the drive toward never-ending quality improvement where symmetrical tolerances are specified. Target values could be developed for one-sided specifications and in cases where a linear trend is unavoidable. Whenever reject limits are employed, it must be possible to maintain the *R* (or *s*) chart in statistical control. When process dispersion behaves erratically, reject limits are not appropriate.

10.11.2 Modified Control Limits

In situations where merely meeting specifications is considered economically sufficient, the use of reject limits *in place of* control limits may be considered. Reject limits used in this way have been called *modified control limits.*[†] They have been applied particularly to control of dimensions. Their use is practical only where the spread of the process (frequently estimated as 6σ) is appreciably less than the difference between the two specification limits ($USL_X - LSL_X$).

The idea behind the modified control limits is to permit limited shifts in the process average in cases where the difference between the two specification limits is substantially greater than the spread of a controlled process. This is intended to avoid the cost of stopping production to hunt for trouble whenever the shifts in process average are not sufficient to cause the production of nonconforming product. It is as though in Fig. 10.11 the process were to be allowed to vary from the position marked "highest acceptable position of universe" to that marked "lowest acceptable position."

[†]See M. J. Moroney, *Facts from Figures,* 2d ed., pp. 165–171, Penguin Books, Ltd., Harmondsworth, 1953; and D. Hill, "Modified Control Limits," *Applied Statistics,* vol. 5, no. 1, pp. 12–19, 1956.

However, in the use of modified control limits a larger margin of safety is sometimes introduced than that given by the assumption that the spread of the process is $\pm 3\sigma$ from the process average. One British practice is to set such limits at a distance from the specification limits equal to $3.09\sigma - (1.96\sigma/\sqrt{n})$. This gives multipliers of σ of 1.70, 1.97, 2.10, and 2.21 for $n = 2$, 3, 4, and 5, respectively; this contrasts with the more common practice in the United States of using the respective V factors of 0.88, 1.27, 1.50, and 1.66 given in Table 10.7 for the same subgroup sizes.

Modified control limits seem to have proved particularly useful as applied to intermittent short production runs in machining operations where process dispersion (6σ) has been determined from previous runs. The more $USL_X - LSL_X$ exceeds 6σ, the greater the permissible latitude in machine setting. The use of modified control limits may simplify the problem of maintaining machine settings that are good enough for practical purposes or where a trend exists.

Nevertheless, where the only limits shown on $\bar{X}$ charts are modified control limits, users of these charts should recognize that the charts fail to disclose the presence or absence of statistical control in the manufacturing process. Moreover, the protection given by the reject limits depends on a good estimate of σ; after this estimate has been made, the process dispersion must remain in statistical control.

10.11.3 Acceptance Control Charts[†]

One step beyond the use of modified control limits is the use of limits and sample sizes derived from specified risks associated with various levels of quality. R. A. Freund has termed these charts *acceptance control charts (ACCs)*. If a process is found to be normally distributed, or nearly so, Table *A* may be used to estimate the proportion of product that will not meet specifications when a process is centered at a known mean μ. Correspondingly, this procedure may be inverted to determine the appropriate location for a mean given some stipulated nonconforming proportion.

Before proceeding to the formulas for ACC control limits and sample sizes, it is helpful to look at the implications of the formula for the V factors given in Table 10.7. Under the normality assumption, the location of μ_h in Fig. 10.11 at 3σ below USL_X implies that 0.00135 nonconforming product is acceptable. Thus it may be claimed the μ_h represents an *acceptable process level (APL)* yielding a proportion nonconforming of $p_1 = 0.00135$. Other values of the multiplier of σ, designated z_{p1}, may be found from Table *A* for any stipulated value of p_1.

When 3 is the multiplier of $\sigma/\sqrt{n}$ in the formula for V, the implication is that a risk level α equal to 0.00135 that a point will fall at random above the *URL* is acceptable. In general, then, where α and p_1 are predetermined, a *URL* may be found from

[†]See R. A. Freund, "Acceptance Control Charts," *Industrial Quality Control,* October 1957.

$$URL = USL_X - \left(|z_{p_1}| - \frac{|z_\alpha|}{\sqrt{n}} \right)\sigma$$

The vertical lines on either side of the z factor indicate absolute value of z. (Since α and p are small numbers, the z values from Table A will be negative numbers.)

A similar development for the LRL will yield

$$LRL = LSL_X + \left(|z_{p_1}| - \frac{|z_\alpha|}{\sqrt{n}} \right)\sigma$$

σ is estimated from either $\overline{R}/d_2$ or $\overline{s}/c_4$.

Acceptance control charts combine this generalization with the stipulation of a *rejectable process level (RPL)* and associated risk to develop formulas for *acceptance control limits (ACLs)* and for finding the sample size n that most nearly meets the criteria. Upper and lower specification limits are treated separately. For each, the following quantities must be chosen:

p_1 the proportion nonconforming considered acceptable.

α the risk level that a point will fall at random outside the ACL.

p_2 the proportion nonconforming considered to be unacceptable.

β the risk level that a point will fall within the ACL.

If both USL_X and LSL_X are specified, the values of these inputs should be the same for each. Otherwise different values of n may be found for each specification. Also $USL_X - LSL_X$ must exceed 6σ.

The sample or subgroup size is found from

$$n = \left(\frac{|z_\alpha| + |z_\beta|}{|z_{p_1}| - |z_{p_2}|} \right)^2$$

To be conservative, the value of n chosen should be the next integer larger than the calculation unless the fractional part is small.

The control limits, designated $UACL$ for the upper ACL and $LACL$ for the lower ACL, are found from

$$UACL = USL_X - \left(|z_{p_1}| - \frac{|z_\alpha|}{\sqrt{n}} \right)\sigma \quad \text{or} \quad UACL = USL_X - \left(|z_{p_2}| + \frac{|z_\beta|}{\sqrt{n}} \right)\sigma$$

$$LACL = LSL_X + \left(|z_{p_1}| - \frac{|z_\alpha|}{\sqrt{n}} \right)\sigma \quad \text{or} \quad LACL = LSL_X + \left(|z_{p_2}| + \frac{|z_\beta|}{\sqrt{n}} \right)\sigma$$

The reason that two formulas are required for each $UACL$ and $LACL$ is that the solution for n almost never will yield an integer value. Thus the two formulas for each ACL will give slightly different answers. The conservative action would be to select the lower value of the *UACL* and the higher value of the *LACL*. Again the user should be cautioned that the results of these calculations are heavily dependent on the assumption of a normal distribution underlying the process and of knowledge of σ. Either an R or an s chart should always be run in conjunction with ACCs.

As an example of *ACL* calculations, assume the following information:

$$USL_X = 2.520 \text{ cm} \qquad LSL_X = 2.480 \text{ cm} \qquad \sigma = 0.005 \text{ cm}$$

thus

$$\frac{USL_X - LSL_X}{\sigma} = 8$$

$$p_1 = 0.00135 \quad |z_{p_1}| = 3.00$$
$$\alpha = 0.01 \quad |z_\alpha| = 2.33$$
$$p_2 = 0.02 \quad |z_{p_2}| = 2.05$$
$$\beta = 0.05 \quad |z_\beta| = 1.65$$

From these data inputs,

$$n = \left(\frac{2.33 + 1.65}{3.00 - 2.05}\right)^2 = (4.19)^2 = 17.55$$

Therefore, $n = 18$, so

$$UACL = 2.520 - \left(3.00 - \frac{2.33}{\sqrt{18}}\right)(0.005)$$
$$= 2.520 - 0.01225 = 2.5077$$

or

$$UACL = 2.520 - \left(2.05 + \frac{1.65}{\sqrt{18}}\right)(0.005)$$
$$= 2.520 - 0.01219 = 2.5078$$

and

$$LACL = 2.480 + 0.01225 = 2.4923 \qquad \text{or} \qquad LACL = 2.480 + 0.01219 = 2.4922$$

A conservative choice of control limits would be $UACL = 2.5077$ and $LACL = 2.4923$ with a subgroup size of 18.

10.12 CUMULATIVE SUM CONTROL CHART FOR AVERAGES

The primary Shewhart control-chart rule, look for trouble if a point falls outside the control limits, is a statistically independent decision rule. That is, the decision is based on a point plot of information taken from one and only one sample or subgroup. When a secondary rule is introduced into the decision process, such as look for trouble if seven points in a row are on the same side of the central line, the decision process becomes, in part, statistically *dependent.* This statistical dependence results from the fact that one of the decision rules utilizes the results of the six immediately preceding samples as well as the last. The decision process becomes more sensitive to small but sustained shifts in the statistical measure being controlled.

The *cumulative sum control chart,* frequently called the *CuSum* chart and sometimes abbreviated *CSCC,* was designed to identify slight but sustained shifts in a universe. Originally developed in Great Britain in 1954 by E. S. Page,† the technique incorporates information taken from the latest samples into that taken from a sequence of prior samples.

This chart has an entirely different appearance from any Shewhart control chart. In place of the centerline and symmetrical control limits, a mask is constructed that incorporates a location pointer and two decision lines that angle away from it. Figure 10.12 illustrates the mask and chart for a CSCC for sample averages. The horizontal scale represents the counter i on subgroups; the vertical scale represents the cumulative sum of subgroup average deviations from an aimed-at or standard value $\overline{X}_o$. The two-sided control limits for $\overline{X}$ are the edges AA' and BB', which are the *lower* and *upper* control limits, respectively.

The mask is placed over the control-chart plot in such a way that the point P on the mask lies on the last point plotted and the line OP is parallel to the horizontal axis. As long as none of the previous points lie above AA' or below BB', the process is assumed to be in control. The basis for the CSCC control limits is described in the following paragraph.

Suppose that, after a sustained period of operation in control at the level $\mu_o = \overline{X}_o$, the mean suddenly shifts to a new level of $\mu_1 = \overline{X}_1$, which is above $\overline{X}_o$. Each new point that successively is added into the cumulative sum of the $(\overline{X}_i - \overline{X}_o)$ will cause the sum to increase and will result in a general upward trend on the chart. If this trend is great enough, sooner or later a point will fall below BB', the upper control limit, signaling an upward shift in the process average. Similarly, if the average shifts downward, the cumulative sum will exhibit a downward trend, which will be signaled by one or more points falling above AA'.

†E. S. Page, "Continuous Inspection Schemes," *Biometrika,* vol. 41, pp. 100–115, 1954. Significant contributions to both the theory and the practice of CuSum control charting have been made by E. W. Kemp, P. L. Goldsmith, and H. Whitfield, all of Great Britain. Probably the most expository discussion of the technique, from both a theoretical and a practical standpoint, is the series of three articles by N. L. Johnson and F. C. Leone, "Cumulative Sum Control Charts—Mathematical Principles Applied to Their Construction and Use," *Industrial Quality Control,* pp. 15–21, June 1962; pp. 29–36, July 1962; and pp. 22–28, August 1962.

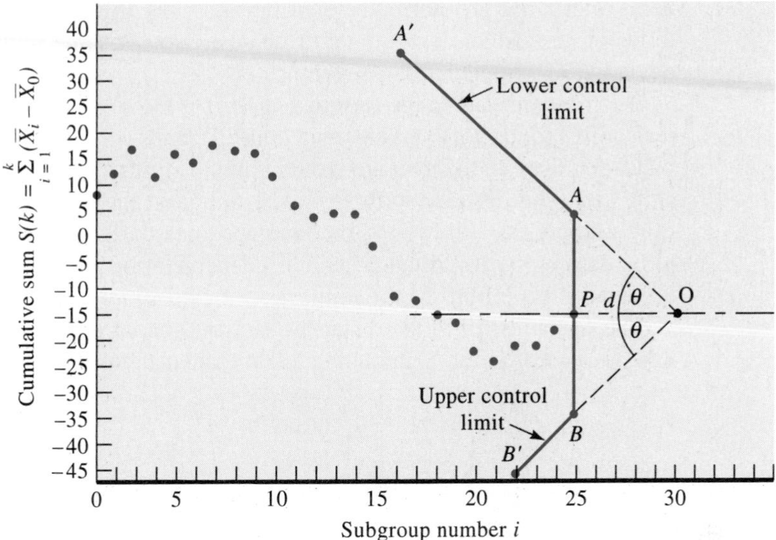

FIGURE 10.12 Cumulative sum control chart for sample averages ($\overline{X}$) data from the first 25 subgroups of Table 3.7.

Two elements are needed to construct the mask. They are θ, the angle between the horizontal and each control limit line, and the distance OP from the vertex of these angles to the locator point P. This section presents the calculations required to prepare CSCCs for sample averages only.[†]

10.12.1 Chart and Mask Construction—CSCC for Sample Average

The reader will remember that the geometry of the Shewhart control chart requires scaling accuracy only with respect to the vertical axis. The horizontal spacing between successive points has no effect on either the accuracy or the interpretation of the chart, although it is customary to employ rectangular coordinate paper that provides equal spacing horizontally between subgroup ordinate lines. The CSCC, on the other hand, requires careful geometric scaling in both directions because of the use of physical measurements along both axes and measured angular relationships between them.

In the chart and mask construction that follows, it is assumed that rectangular coordinate paper is used on which equal spacing between lines in both directions is provided.

A feature of the CSCC is the necessity of specifying the amount of the shift that must be detected with virtual certainty if the values of the quality characteristic

[†]The articles by Johnson and Leone, ibid., also present CSCCs for ranges, variances, and attributes data for nonconforming units and counts of nonconformities.

being measured are normally distributed.[†] This shift may be specified either in terms of an absolute magnitude of change D or as some multiple of the standard deviation of sample averages, δ times $\sigma/\sqrt{n}$.

If the magnitude of the designated shift is expressed as $\delta\sigma/\sqrt{n}$, the value of δ that corresponds to the Shewhart control chart with 3-sigma limits is approximately 6. That is, the average would have to shift by about $\pm 6\sigma/\sqrt{n}$ before such a shift would be detected with virtual certainty on the next subgroup drawn. In setting up the CSCC, however, the user specifies the magnitude of the shift desired to be detected with virtual certainty independent of the choice of sample size and acceptable Type I error probability.

The formulas for d, the distance OP from the vertex of the angles between OP and the locator point P, and θ, the angle that each control limit makes with OP, are

$$d = OP = E(\alpha)\left(\frac{\sigma}{\sqrt{n}D}\right)^2 = \frac{E(\alpha)}{\delta^2}$$

$$\tan\theta = \frac{D}{2y} = \frac{\delta\sigma/\sqrt{n}}{2y}$$

where $E(\alpha)$ = a factor that is a function of the acceptable Type I error probability. Values of this factor for various levels of α are given below. $E(0.0027)$ corresponds to the Type I (α) error probability associated with standard Shewhart 3-sigma limits.

$\frac{1}{2}\alpha$ (one-sided test)	0.00135	0.005	0.01	0.025	0.05
α (two-sided test)	0.0027	0.01	0.02	0.05	0.10
$E(\alpha)$	13.215	10.597	9.210	7.378	5.991

$\sigma/\sqrt{n}$ = the standard deviation of sample averages. It may be estimated from $\bar{R}$ or $\bar{s}$.

$D = \delta\sigma/\sqrt{n}$ = the actual value of the shift magnitude, either plus or minus, that must be detected with virtual certainty.

$\tan\theta$ = the tangent of the angle θ. An abbreviated table of angles and their tangents is given in Table 10.8. A linear interpolation may be made between the listed values without markedly affecting the accuracy of the chart.

y = a scaling factor related to the geometry of the control chart and the mask dimensions. Usually it will be of the same order of magnitude as $\sigma/\sqrt{n}$. A value of about $\sigma/\sqrt{n}$ to $2.5\sigma/\sqrt{n}$ is reasonable.

[†]As previously discussed, the probability of detection of any specific magnitude of shift on a Shewhart control chart may be increased either by increasing the sample size or by reducing the value of the multiplier of the standard deviation of the statistical measure being subjected to control charting (as is done when probability limits rather than 3-sigma limits are used). In the case of the CSCC, this shift magnitude is specified directly.

TABLE 10.8 ABRIDGED TABLE OF ANGLES AND THEIR TANGENTS

θ, degrees	tan θ	θ, degrees	tan θ	θ, degrees	tan θ
26	0.488	41	0.869	56	1.483
27	0.509	42	0.900	57	1.540
28	0.532	43	0.933	58	1.600
29	0.554	44	0.966	59	1.664
30	0.577	45	1.000	60	1.732
31	0.601	46	1.036	61	1.804
32	0.625	47	1.072	62	1.881
33	0.649	48	1.111	63	1.963
34	0.675	49	1.150	64	2.050
35	0.700	50	1.192	65	2.145
36	0.727	51	1.235	66	2.246
37	0.754	52	1.280	67	2.356
38	0.781	53	1.327	68	2.475
39	0.810	54	1.376	69	2.605
40	0.839	55	1.428	70	2.748

The scaling factor y serves two purposes. First, it establishes the geometric relationship between the vertical and horizontal axes of the chart and between $\sigma/\sqrt{n}$ and the chart and mask. Second, it allows the adjustment of θ to some reasonable value. Maintaining θ in the range 30 to 60 degrees, an efficient range for calculating values of tan θ, will yield good results and suppress some of the problems of arithmetic errors and misplotted points that may result when very small or very large values of the tangent are used.

The measures that are plotted on the chart are

k = subgroup number, plotted on the horizontal axis

$$S(k) = \sum_{i=1}^{k} (\overline{X}_i - \overline{X}_o)$$

the cumulative sum of the deviations of the sample averages from the standard value $\overline{X}_o$, plotted on the vertical axis.

In setting up the scales of the chart, there will be y units on the vertical axis for each unit on the horizontal axis. For example, if $y = 0.004$ in and $\frac{1}{4}$-in rectangular coordinate chart paper is to be used, then each $\frac{1}{4}$-in on the vertical axis has the value 0.004 in when each $\frac{1}{4}$-in on the horizontal axis is assigned a subgroup number. Needless to say, proper scaling of the chart requires as much care as proper dimensioning of the mask.

If it were not for the multiple use of the scaling factor y, it would be possible to let it equal $\sigma/\sqrt{n}$, thus simplifying some of the calculations. Unfortunately, y affects θ as well as the vertical scale on the chart. The best that can be said is that it should be of the same order of magnitude as $\sigma/\sqrt{n}$, should be consistent with reasonable chart-scaling procedures, and should yield a value of θ between 30 and 60 degrees, if possible. The use of standard computer software eliminates all of these problems.

10.12.2 Example of Sample Averages

The CuSum technique may be illustrated in its most favorable light by an example in which samples are taken for a period of time from a stable universe that is known to be normally distributed. This period is followed by one in which the centering of the universe has changed to a different level and there has been no change in the universe dispersion.

Let us assume a quality characteristic of a manufactured product that is distributed according to the values of the chips in Shewhart's normal bowl (Table 3.6). Assume that the aimed-at central value $\overline{X}_o$ is 30. Assume that samples of 4 are taken at regular intervals and that the measured values correspond to the first 100 drawings from Shewhart's normal bowl (Table 3.7). Table 10.9 shows the calculations needed to determine the points on a CuSum chart.

The sum of the R values (shown in Table 3.7) for the 25 subgroups is 487. Thus $\overline{R} = \frac{487}{25} = 19.48$, and the estimate of $\sigma/\sqrt{n}$ is $\overline{R}/d_2\sqrt{4} = 4.73$.

We can establish control limits equivalent to 3-sigma Shewhart limits by stipulating that a shift of magnitude D in either direction should be detected with probability 0.0027. Thus a shift in one direction should be detected with probability 0.00135. For purposes of example, let the magnitude of the shift that is to be detected with virtual certainty be 7.5, which is approximately three-fourths standard deviation of the process and $1.5\sigma/\sqrt{n}$. $\overline{X}_o$, the aimed-at value, is set at 30, the true mean of the bowl, and the scaling factor y will be 4. Then

$$d = E(0.0027)\left(\frac{\sigma/\sqrt{n}}{D}\right)^2 = 13.216\left(\frac{4.73}{7.5}\right)^2$$

$$= 5.26 \text{ units on horizontal axis}$$

$$\tan\theta = \frac{D}{2y} = \frac{7.5}{8} = 0.9375$$

and therefore,

$$\theta = 43°9', \text{ or } 43°$$

With these dimensions, the mask illustrated in Fig. 10.12 may be constructed. The chart was originally drawn on $\frac{1}{4}$-in grid chart paper with a horizontal axis of $\frac{1}{4}$-in per subgroup i, and $\frac{1}{4}y = \frac{1}{16}$ in per unit deviation on the vertical scale. The value of $S(k)$, the cumulative deviation from the baseline at each subgroup point, is then plotted. Figure 10.12 shows the mask positioned at the twenty-fifth subgroup point, where $S(k) = -15.50$. All points fall well within the control limits.

TABLE 10.9 ILLUSTRATION OF CALCULATION OF POINTS FOR CUSUM CHART FOR SAMPLE AVERAGES

Subgroup number	Numbers of drawings	Average $\bar{X}$	$\bar{X}_i - \bar{X}_o$	CuSum $\Sigma(\bar{X}_i - \bar{X}_o)$
1	1–4	39.50	9.50	9.50
2	5–8	33.50	3.50	13.00
3	9–12	33.25	3.25	16.25
4	13–16	26.00	–4.00	12.25
5	17–20	34.00	4.00	16.25
6	21–24	28.50	–1.50	14.75
7	25–28	32.75	2.75	17.50
8	29–32	29.25	–0.75	16.75
9	33–36	29.25	–0.75	16.00
10	37–40	26.00	–4.00	12.00
11	41–44	24.75	–5.25	6.75
12	45–48	27.25	–2.75	4.00
13	49–52	30.25	0.25	4.25
14	53–56	30.00	0.00	4.25
15	57–60	24.00	–6.00	–1.75
16	61–64	20.00	–10.00	–11.75
17	65–68	29.50	–0.50	–12.25
18	69–72	27.00	–3.00	–15.25
19	73–76	28.75	–1.25	–16.50
20	77–80	24.25	–5.75	–22.25
21	81–84	27.75	–2.25	–24.50
22	85–88	32.75	2.75	–21.75
23	89–92	30.00	0.00	–21.75
24	93–96	33.25	3.25	–18.50
25	97–100	33.00	3.00	–15.50

Now assume that, after the twenty-fifth subgroup, the process average μ shifts abruptly downward from the previous value of 30.0 to a new value of 22.5. All subsequent measurements will be 7.5 less than the figures in Table 3.7, and each subgroup average will, of course, be 7.5 less than the corresponding average in Table 3.7. Starting with subgroup 26 (drawings 101–104 in Table 3.7), averages for subgroups 26 through 33 will be 21.75, 23.50, 23.50, 24.50, 33.50, 12.50, 27.75, and 17.50. The corresponding CuSum values for these subgroups are –23.75, –30.25, –36.75, –42.25, –38.75, –56.25, –58.50, and –71.00.

Figure 10.13 shows the corresponding CuSum chart for averages from subgroup 15 to subgroup 33. Several of the lower control limit lines are shown for subgroups 26 to 33. The shift in process centering is detected at subgroup 33.

Once an out-of-control condition has been signaled—that is, a point falls below the *UCL* or above the *LCL*—the accumulation of *S(k)* is terminated. After corrective action on the process is taken, the charting procedure begins anew at the baseline.

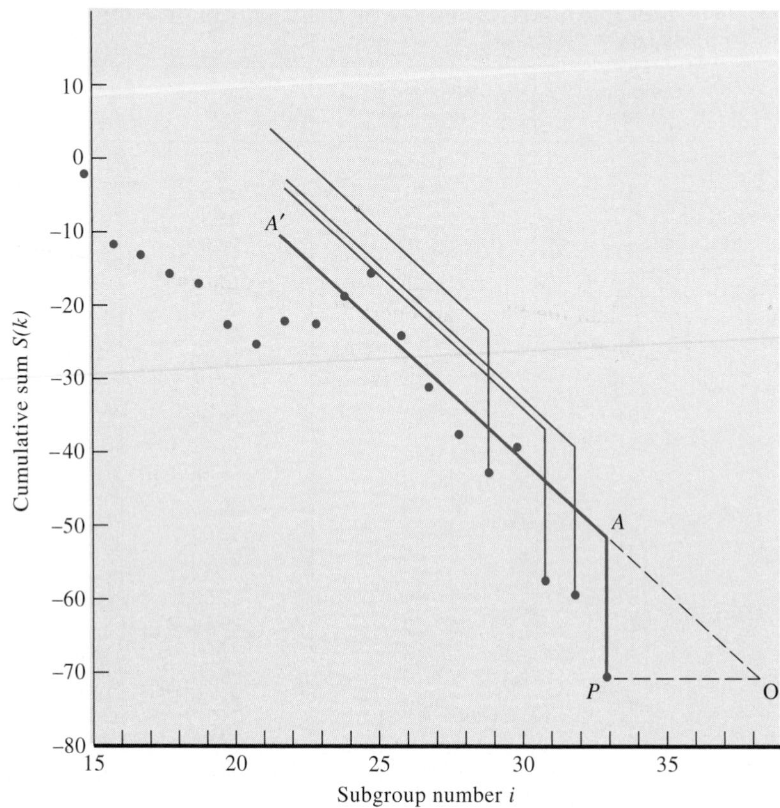

FIGURE 10.13 Continuation of CSCC for $\overline{X}$ to illustrate shift detection (lower control limit only).

10.12.3 Average Run Length as a Measure of Effectiveness

In making comparisons between various analytical schemes for process control, some measure of effectiveness must be developed upon which to base the choice of the best scheme. One such method is to measure the probability of detecting shifts of a given magnitude. Once the sample size and control limits have been determined, it is a relatively uncomplicated matter to calculate the probability that shifts of a specific magnitude will be detected. A number of exercises throughout this book are concerned with such calculations. Usually these exercises have been stated in terms of the probability of detection on the first subgroup drawn after the shift occurs or on any one subgroup drawn after the shift occurs. In the latter case it must be assumed that the shift remains at the new level. This assumption of a sudden and sustained shift, in effect a jump to a new but "in control" level, also was necessary when we discussed run tests on control charts.

If the degree of shift is expressed in terms of multiples of the standard deviation of sample averages $\delta\sigma/\sqrt{n}$, and the spread of the limits is specified—say, 3-sigma limits—then the calculation of the probability of detecting a shift of $\delta\sigma/\sqrt{n}$

applies to all $\bar{X}$ charts with 3-sigma limits. By varying δ and plotting the probability of detection as a function of δ, a curve results that shows the power of the sampling procedure to detect shifts of any magnitude.

Another measure of effectiveness of detecting sustained shifts answers the question, "Assuming that a shift of some magnitude takes place, how long will it be before we can *expect* to detect it?" It is called the *average run length,* abbreviated *ARL,* and is the expected number of subgroups inspected before a shift signal is given. Run length comparisons of Shewhart and CuSum procedures indicate that the CuSum procedure is more efficient in detecting small but sustained shifts in $\bar{X}$ than the Shewhart chart. Such comparisons, however, may be quite misleading on two counts. First, they ignore the almost universal use of run tests on Shewhart charts in combination with the point-out rule. *ARL* comparisons consider the point-out rule only. Second, of necessity they assume a shift from an in-control state at some $\bar{X}_o$ to another in-control state at a new level. This state of affairs rarely, if ever, exists in industry.

PROBLEMS

Some Special Topics on Shewhart Control Charts for Variables

10.1. A certain company manufactures electronic components for television sets. One particular component is made to a critical length of 0.450 in. On the basis of past production experiences, the standard deviation of this dimension is 0.010 in. Because of the critical nature of the dimension, the quality control group maintains warning limits on $\bar{X}$ control chart as well as the normal 3-sigma control limits. The $\bar{X}$ chart is based on subgroups of four samples, and warning limits are maintained at two standard deviations from the mean. Compute the warning limits and the control limits for the $\bar{X}$ chart.
Answer: 0.460, 0.440; 0.465, 0.435.

10.2. Prob. 10.1 asked for the calculation of both warning limits and control limits for an $\bar{X}$ chart. Using those data:

(*a*) What is the probability that a subgroup average will exceed the upper warning limit but not exceed the upper control limit when the process is correctly centered?

(*b*) Policy states that the quality control supervisor should be notified immediately if two successive subgroup averages exceed one of the warning limits or if one subgroup average exceeds either of the control limits. What is the probability that, when there has been no change in the process, the quality control supervisor will have to be notified because of:

(1) Two successive subgroup averages falling between the upper warning limit and upper control limit

(2) A subgroup average exceeding the upper control limit

(*c*) If the process suddenly shifts to 0.460 with no change in the standard deviation, what is the probability that two successive points will exceed the upper warning limit but not exceed the upper control limit?

Answer: (a) 0.0214; (b) 0.0005, 0.0014; (c) 0.1165.

10.3. Calculate control limits and plot an $\overline{X}$ chart for the data of Table 10.1. Base your calculations on the exact sample size n for each subgroup rather than the average value 123. Does using the exact values of n lead to any different conclusions from interpreting your chart than those reached from interpretating the $\overline{X}$ chart in Fig. 10.2?

10.4. In the manufacture of the balance wheel for a speed control device, finished wheels are tested for dynamic balance on a test stand at 6,000 r/min. If a vibration force in excess of 80 g·cm/s^2 results, the balance must be remachined.

Because of the critical nature of this test, control charts for $\overline{X}$ and R are maintained using a subgroup size of five. Warning limits at $\pm$ 2-sigma are used as well as standard 3-sigma control limits. The process has been operating in control at an estimated μ of 65 g·cm/s^2 and a σ of 4.2.

(*a*) Calculate control limits for both the $\overline{X}$ and R charts and warning limits for the $\overline{X}$ chart.

(*b*) With the process centered as indicated, what is the probability that two successive points will fall between the *UCL* and the *UWL* on the $\overline{X}$ chart?

10.5. In the situation described in Prob. 10.4, the mean of the process suddenly shifts to 72 g·cm/s^2.

(*a*) At this new mean level, what proportion of the balance wheels will exceed the 80 g·cm/s^2 limit assuming a normal distribution of this characteristic?

(*b*) What is the probability that this shift in μ will be detected on the first subgroup after the shift occurs?

(*c*) What is the probability that the shift will be detected on the first two subgroups based only on two successive points falling between the *UCL* and the *UWL*?

(*d*) What is the combined probability of detecting this shift within the first two subgroups plotted?

10.6. For a certain process, the natural tolerance range exceeds the specification range. Slight changes in centering will therefore result in substantial product being produced outside specifications. Someone has suggested the use of 1.5-sigma warning limits in addition to 3-sigma control limits on an $\overline{X}$ chart. A shift in process is assumed to have occurred if (1) a point falls outside one of the 3-sigma control limit or (2) three points in a row fall outside one of the warning limits.

(*a*) What is the probability that the application of rule (2) will lead to hunting for trouble when none exists (Type I error)?

(*b*) Assuming an upward shift in μ of 0.4σ and a subgroup size of five, what is the probability that the application of rule (2) will lead to *not* detecting the shift on the first three subgroups drawn after the shift occurs (Type II error)?

10.7. Both 3-sigma control limits and 2-sigma warning limits are maintained on an $\overline{X}$ chart for the internal diameter of a shock tube assembly. The aimed-at value is 35.50 mm, and σ is 0.25 mm. The subgroup size is four.

(*a*) Calculate the control and warning limits for an $\overline{X}$ chart for this process.

(*b*) Calculate the control limits for an R chart to control dispersion.

(*c*) If the actual mean of the process is 35.80 mm, find the probability that any given point would fall above the *UCL* and the probability that two points in succession would fall between the *UWL* and the *UCL*.

10.8. In a certain manufacturing process, $\overline{\overline{X}} = 178.46$ and $\overline{R} = 9.8$. The subgroup size is four, and standard 3-sigma control limits are used. If seven consecutive points all fall above (or below) $\overline{\overline{X}}$, the process average is assumed to have shifted.

(*a*) If μ is 178.46, what is the probability of seven consecutive points falling on one side of $\overline{\overline{X}}$ (Type I error)?

(*b*) If μ shifts to 180.00, what is the probability of seven consecutive points falling on one side of $\overline{\overline{X}}$ (that is, not making a Type II error)?

10.9. Prepare a control chart for subgroup totals for the data of Prob. 3.10.

10.10. Prepare a control chart for subgroup totals for the data of Prob. 3.19.

10.11. An $\overline{X}$ chart used to monitor a particular quality characteristic has an $\overline{\overline{X}}$ value of 32.0 and $3\sigma_{\overline{x}}$ limits at 30.5 and 33.5. The product is sampled in subgroups of four. The analyst considers that a quality characteristic is out of control if one subgroup falls outside either of the control limits or if seven successive subgroup averages fall on one side of the central line.

(*a*) Estimate the μ and σ of the process.

(*b*) With the process in control and assuming a normal distribution, what is the probability that seven successive subgroup averages will fall on the upper side of the central line?

(*c*) If the process average shifts from 32.0 to 32.9 with no change in σ, what is the probability that the next seven successive values of $\overline{X}$ will fall on the upper side of the central line? Assume that the $\overline{X}$ of the subgroup immediately preceding the shift fell below the central line.

(*d*) In the circumstances described in (*c*), what is the probability that at least one $\overline{X}$ value in the next seven subgroups will fall above the upper control limit on the $\overline{X}$ chart?

Answer: (a) 32.0, 1.0; (b) 0.0078; (c) 0.3179; (d) 0.5750.

Some Related Special Procedures

10.12. Plot individual measurements and a superimposed chart for medians using the data of Prob. 3.1. The chart should be similar to Fig. 10.3 except that no specification limits are required. Base 3-sigma control limits for the medians on the median of the ranges.

10.13. Plot individual measurements and a superimposed chart for medians using the data of Prob. 3.10. The chart should be similar to Fig. 10.3 except that no specification limits are required. Base 3-sigma control limits for the medians on the median of the ranges.

A General Test for Homogeneity

10.14. Tests of tensile strengths of malleable iron castings from four foundries gave the following results:

Foundry	Number of tests	Average tensile strength, lb/in^2	Standard deviation s
A	54	58,400	1,600
B	60	57,000	1,550
C	71	57,700	1,190
D	49	56,900	2,080

Use the methods explained in the discussion of Table 10.1 to plot $\bar{X}$ and s charts to judge whether there is clear evidence that the different foundries represent different cause systems. Use simple unweighted averages to determine $\bar{\bar{X}}$ and $\bar{s}$, and base your limits on average subgroup size.

10.15. Solve Prob. 10.14 using the correct weighted averages for $\bar{\bar{X}}$ and $\bar{s}$, and base your limits on the correct values of n. How does your solution in this case compare with that arrived at from interpreting the results in Prob. 10.14?

10.16. A textile manufacturer has six lines producing fiber thread for which one important characteristic is tensile strength. Data obtained over the past 10 days have been summarized in the following table.

Machine	Number of tests	Average tensile strength	Standard deviation
1	56	2.34	0.11
2	70	2.39	0.16
3	75	2.25	0.18
4	102	2.34	0.11
5	86	2.38	0.09
6	90	2.37	0.08

Using the methods explained in the discussion of Table 10.1, plot $\bar{X}$ and s charts for these test data and judge whether there is clear evidence that the different machines represent different chance cause systems. Use simple unweighted averages to determine $\bar{\bar{X}}$ and $\bar{s}$, and base the control limits on the average sample size.

Probability Limits on Control Charts for Variables

10.17. Assume that probability limits rather than 3-sigma limits are to be used for the data of Example 1.2. Where would these limits be on the $\bar{X}$ and R chart:

(*a*) If the probability of a point falling outside the limits in sampling from a normal universe were to be 0.002?

(*b*) If this probability were to be 0.01?

(*c*) If this probability were to be 0.05?

Answer: (a) $UCL_{\bar{x}} = 145.7$; $LCL_{\bar{x}} = 135.5$; $UCL_R = 20.3$; $LCL_R = 1.4$.

10.18. Find 95% probability limits for $\bar{X}$ and R control charts for the data of Prob. 3.10. Does the process appear to be operating in control?

10.19. Calculate 95 and 99% probability limits for $\overline{X}$ and R control charts for the data of Prob. 3.19. Is there any difference in the interpretation of the results based on the use of the two different sets of control limits?

10.20. Control charts for $\overline{X}$ and s are maintained on the weight in ounces of the contents of a cereal container. The subgroup size is 8. After 18 subgroups, $\Sigma\overline{X} = 610.4$ and $\Sigma s = 7.96$.

(*a*) Compute 3-sigma control limits for the $\overline{X}$ chart, and estimate the value of σ.

(*b*) A decision is made to use 95% probability limits on the $\overline{X}$ chart rather than 3-sigma limits. What would be the locations of these limits?

(*c*) If the mean shifts to 33.6, what is the probability of detecting this shift on the first subgroup after the shift occurs if the 3-sigma limits are used?

(*d*) What is the probability of detecting the shift described in (*c*) if 95% probability limits are used?

10.21. A critical dimension in the production of a certain part specifies a length of 60.00 ± 0.07 mm. It is proposed to use $\overline{X}$ and R charts employing 99% probability limits to control this dimension. After 30 subgroups of 4 measurements each have been taken, $\Sigma\overline{X} = 1799.40$ mm and $\Sigma R = 3.24$ mm.

(*a*) Calculate trial control limits for this process.

(*b*) Calculate σ, and estimate the proportion of product that does not meet specifications. Assume that the distribution of unit dimensions is approximately normal.

10.22. $\overline{X}$ and R charts are run on a certain process that, when in control, is known to have a σ of 0.030 mm. The target value of the mean is 6.500 mm. 99.8% probability control limits and 95.0% probability warning limits are to be used on each chart. The subgroup size is five.

(*a*) Calculate the limits for the $\overline{X}$ and R charts.

(*b*) Assume that the process actually is operating at a μ value of 6.525 mm. What is the probability that two points in a row will fall between the upper control limit and the upper warning limit on the $\overline{X}$ chart?

Control Limits for Moving Averages

10.23. For purposes of computing control limits in the moving average illustration of Table 10.5 and Fig. 10.5, $\overline{R}$ was assumed as 0.065. From this $\overline{R}$ what would be your estimate of σ? Assuming μ as 0.17, what are 3-sigma limits on individual daily values? If these limits were drawn on Fig. 10.5*a*, on what dates would points fall outside control limits? How does this compare with the dates on which points fell outside control limits on the moving range chart, Fig. 10.5*c*? How do you explain this relationship?

10.24. Summarized below are daily analyses of CO_2 as CaO at an intermediate stage in a chemical manufacturing process. Compute 3-day moving averages and moving ranges for these data, and establish $\overline{X}$ and R control limits for monitoring this process basing the control limits on a standard process average of 0.660 and an $\overline{R}$ of 0.075.

Date	Percent CO_2 as CaO	Date	Percent CO_2 as CaO	Date	Percent CO_2 as CaO
May 1	0.53	14	0.65	27	0.71
2	0.62	15	0.59	28	0.68
3	0.63	16	0.60	29	0.74
4	0.54	17	0.69	30	0.66
5	0.50	18	0.65	31	0.67
6	0.50	19	0.65	June 1	0.67
7	0.51	20	0.67	2	0.68
8	0.53	21	0.71	3	0.72
9	0.56	22	0.78	4	0.70
10	0.64	23	0.82	5	0.67
11	0.57	24	0.82	6	0.69
12	0.56	25	0.88	7	0.68
13	0.55	26	0.82		

$\overline{X}$ Chart with a Linear Trend

10.25. A certain manufacturing process has exhibited a linear increasing trend. Sample averages and ranges for the past 15 subgroups, taken every 15 min in subgroups of 5 items, are given in the following table:

Subgroup number	Average	Range	Subgroup number	Average	Range
1	198.8	7	9	209.2	7
2	197.6	2	10	207.8	16
3	204.6	10	11	210.0	9
4	203.8	12	12	214.4	8
5	205.6	17	13	211.8	16
6	204.8	9	14	211.8	6
7	205.4	10	15	213.8	8
8	210.6	9			

Fit the linear trend line to these data, and plot a trended $\overline{X}$ control chart with 3-sigma limits.

10.26. Specifications on the process in Prob. 10.25 are 200 ± 30. The process may be stopped at any time and readjusted. If on readjustment the mean is to be set exactly 4σ above the lower specification and the process is to be stopped for readjustment when the mean reaches a level exactly 4σ below the upper specification:

(*a*) Calculate the aimed-at starting and stopping values of $\overline{X}_o$.

(*b*) Estimate the duration of a run between adjustments.

10.27. The following data are from a machining operation that exhibits a steady trend in the average dimension of the items processed. The subgroup size is four, the four samples being taken in succession at 30-min intervals and measured in millimeters.

Subgroup number	$\overline{X}$	R	Subgroup number	$\overline{X}$	R
1	31.42	0.10	8	32.13	0.14
2	31.49	0.13	9	32.19	0.07
3	31.57	0.04	10	32.30	0.10
4	31.71	0.08	11	32.39	0.11
5	31.83	0.11	12	32.51	0.13
6	31.89	0.05	13	32.64	0.12
7	32.00	0.09			

(*a*) Determine the equation of a least-squares trend line fitted to the $\overline{X}$ values.
(*b*) Determine the 3-sigma control limits for an $\overline{X}$ chart.
(*c*) If the dimensions of the part are specified as 32.00 ± 0.50, estimate σ and express the tolerance spread ($USL_X - LSL_X$) as a multiple of the estimated σ.
(*d*) What would you recommend as the aimed-at value of $\overline{X}_o$ in the initial setting if this is to be 3σ above the lower specification limit?
(*c*) If a new setup is to be made when the value of the central line on the $\overline{X}$ chart is 3σ below the upper specification limit, what would you estimate as the length of a production run before a new setup is required?

10.28. A certain dimension was specified as 0.8250 ± 0.0050 in. Experience indicated that tool wear caused a fairly steady trend in the average dimension of parts made on the required type of machining operation. The following values of $\overline{X}$ and R were obtained from subgroups of 4 parts taken from the machine at half-hour intervals:

Subgroup number	$\overline{X}$	R	Subgroup number	$\overline{X}$	R
1	0.8220	0.0004	9	0.8251	0.0005
2	0.8228	0.0007	10	0.8253	0.0013
3	0.8233	0.0005	11	0.8257	0.0003
4	0.8232	0.0010	12	0.8257	0.0008
5	0.8235	0.0011	13	0.8265	0.0011
6	0.8241	0.0012	14	0.8260	0.0010
7	0.8244	0.0014	15	0.8270	0.0007
8	0.8248	0.0004			

Determine the equation of a least-squares trend line fitted to the $\overline{X}$ values. Plot an $\overline{X}$ chart similar to Fig. 10.6 using sloping 3-sigma limits parallel to the trend line.

Estimate σ, and express the tolerance spread ($USL_X - LSL_X$) as a multiple of the estimated σ. What would you recommend as the aimed-at value of $\overline{X}_0$ in the initial setting if this is to be at 3σ above the lower specification limit? If a new setup is to be made when the value of the central line on an $\overline{X}$ chart is 3σ below the upper specification limit, what would you estimate as the length of a production run before a new setup is required?

Narrow Limit Gaging

10.29. It is decided to prepare a set of special patrol inspection go and not-go gages for process control of a critical machining operation. Specifications call for a dimension of 9.650 ± 0.150 cm. Previous application of $\bar{X}$ and R charts indicates an estimate of σ of 0.045 cm and an approximate normal distribution of product when the process is in control. The patrol inspection gages are to be set at $+1\sigma$ for the go gage and -1σ for the not-go gage.

(*a*) Calculate the gage limits.

(*b*) What proportion of the units should fail each of these tests individually when the process is correctly centered?

(*c*) Calculate the control limits for the *np* charts for each gage based on a subgroup size of 12.

Answer: (a) 9.695, 9.605; (b) 0.1587; (c) 5.70, 1.90.

10.30. In the testing of a certain ballistic characteristic of ammunition, it is desired that all values of this quality characteristic fall between 2,720 and 2,900 units. The quality characteristic has been tested on a variables basis with a sample of 4 taken from each lot. $\bar{X}$ and R charts have been plotted from past samples. The R chart has shown excellent statistical control with an $\bar{R}$ of 51 units. The $\bar{X}$ chart has shown a decided lack of statistical control. This lack of control of $\bar{X}$ is to be expected because past experience with other ammunition has indicated that it is very difficult to control the centering of the production process with respect to this particular quality characteristic.

It has been determined that it will be considerably more economical to test 16 items on an attributes basis than 4 items on a variables basis. In this connection it is decided to maintain an *np* chart similar to the lower chart in Fig. 10.7. For purposes of this chart it is desired to conduct the go and not-go testing using special severe test limits similar to the upper and lower gage limits shown in Fig. 10.9. These severe limits are to be established so that the appropriate μ_p is 0.15 for the "above limits" chart (corresponding to the not-go chart of Fig. 10.7) and is also 0.15 for the "below limits" chart (corresponding to the go chart of Fig. 10.7).

(*a*) With specification limits of 2,720 and 2,900 as stated, what should be the values of these special severe test limits?

(*b*) With 16 items tested in each sample, where should the action limits (3-sigma limits) be placed on the *np* chart?

Answer: 2,851.4, 2,768.6; 6.68.

10.31. It is decided to change from variables charts for process control inspection to using specially severe go and not-go gages and control charts for *np*. The process is known to be approximately normally distributed with a nominal dimension of 45.000 mm and a σ of 0.025 mm. It is an external diameter.

(*a*) What should the gage limits be if, when correctly centered, each gage, and its corresponding chart, is to reject 15% of the product?

(*b*) What will be the central lines and control limits on each *np* chart if the subgroup size is 12?

10.32. Specially severe go and not-go gages are to be used for process control replacing the use of direct measurements and $\bar{X}$ and R charts. The standard deviation of the process is known to be 0.020 cm. This is an outside dimension with specifications of 6.500 ± 0.150 cm.

(*a*) Determine the gage dimension assuming that, if the process is centered exactly 3σ from a specification limit, exactly 15% of the product will fail to pass the gage.
(*b*) Calculate the control limits and central line for an *np* chart for each gage based on the average designed 15% failure rate and a sample subgroup size of 15 units.

10.33. Special go and not-go gages are to be used to control a certain machining operation in place of direct measurements and $\overline{X}$ and R charts. Process measurements have been approximately normally distributed when the mean of the process is held in strict control. Its standard deviation is estimated to be 0.010 mm. Specification limits on the operation are 25.200 ± 0.100 mm. Thus the patrol inspection gages will have to be set as shown in Fig. 10.9. With the process mean set at either the high or the low value, the proportion of units that should fail the appropriate gage test is 0.1587. This is an external diameter.
(*a*) Calculate the proper setting for each gage.
(*b*) Where should the central line and upper control limit for an *np* chart be placed for the chart for each gage if the sample size is 12?

Combining Process Control and Product Acceptance

10.34. A shop decides to use reject limits on a modified control chart for $\overline{X}$ for both product acceptance and process control. A random sample of 6 items is drawn from a storage cradle and inspected. If the value of $\overline{X}$ falls outside the reject limits, the tote box is screened for nonconforming units. Previous analysis indicates that this dimension is approximately normally distributed when held in control with an estimated σ of 3.62 mm. The specification range $(USL_X - LSL_X)$ is 50.0 mm.
(*a*) Is this process a reasonable candidate for using reject limits?
(*b*) Calculate the reject limits when $USL_X = 1{,}040$ mm and $LSL_X = 990$ mm.
(*c*) If the process is operating at an actual mean value of 1,035 mm, what fraction of nonconforming units is being produced?
(*d*) What is the probability of detection on any given subgroup when μ is 1,035 mm?
Answer: (a) Yes; (b) 1,033.56, 996.44; (c) 0.0836; (d) 0.8351.

10.35. The specifications for a certain quality characteristic are 220 ± 20. It is decided to initiate $\overline{X}$ and R control charts with the $\overline{X}$ chart based on reject limits. Past evidence indicates that, although the centering of the process shifts from time to time, the dispersion tends to remain constant with a σ of about 9 units.

With a subgroup size of 4, $URL_{\overline{X}} = 240 - 1.5(9) = 226.5$. $LRL_{\overline{X}} = 200 + 1.5(9) = 213.5$. (See Table 10.7.) $UCL_R = 4.70(9) = 42.3$. $LCL_R = 0$. (See Table *F*.)

The control chart is initiated. In the first 50 subgroups no points on the R chart are outside the control limits. On the $\overline{X}$ chart, 4 points are above the upper reject limit and 3 points are below the lower reject limit. The analyst who proposed the charts concludes that the process is "badly out of statistical control."

Discuss the foregoing case with reference to control-chart principles. Is this the type of case in which you would recommend the use of reject limits? Why or why not?

10.36. A certain dimension is specified in inches as 3.5100 ± 0.0050. Control charts for $\overline{X}$ and R indicate that the $\overline{X}$ chart shows lack of statistical control but the R chart always shows control. From the R chart the estimate of σ is 0.0010. If the aimed-at process average $\overline{X}_o$ is to be 3.5100, what should be the upper control limit for $\overline{X}$ with a subgroup size of four? What should be the upper reject limit on the $\overline{X}$ chart assuming the use of 3-sigma reject limits?
Answer: $UCL_{\overline{X}} = 3.5115$; $URL_{\overline{X}} = 3.5135$.

10.37. A certain item must meet a specification of 300 ± 12 units. The process is known to have a standard deviation of 1.0 units. It is decided to sample 8 items each half-hour and to base acceptance of the product on an $\overline{X}$ chart with reject limits while maintaining control of dispersion on a standard R chart.
(*a*) Calculate the reject limits for the $\overline{X}$ chart and control limits for the R chart.
(*b*) Assuming no change in σ and that product output is normally distributed, what proportion of product does not meet specifications if the process is centered at 309 units?
(*c*) What is the probability that this level of centering will not be detected on any given sample?

10.38. Modified control limits for $\overline{X}$ are to be used in a shop that operates intermittently to produce small metal parts. The length of the part is critical in its eventual assembly. Specifications are 19.00 ± 0.03 mm. From past experience with the automatic machine producing the parts, σ is known to be 0.006 mm. To avoid allowing the process mean to drift too much, it is decided to set up the 3-sigma limits such that the mean can get no closer than 3.5σ to either specification limit.
(*a*) Calculate the reject limits and R chart control limits for a subgroup size of five.
(*b*) Assume that the process is operating at a mean level such that 0.5% of the parts exceed the upper specification limit and that product output is approximately normally distributed. What is the probability that an $\overline{X}$ plotted on the chart will fall above the *URL*?

10.39. It is suggested that acceptance control limits rather than modified control limits be used for the process described in Prob. 10.38. Each limit is to be based on a risk α of 0.02 of rejection when the quality level p_1 is 0.005 or better and a risk β of 0.05 of acceptance when the quality level p_2 is 0.015 or worse. Calculate the appropriate sample size n and control limits under these circumstances. How do these results compare with those found in Prob. 10.38?

Cumulative Sum Control Chart for Averages

10.40. Prepare a two-sided CSCC for $\overline{X}$ for the data of Prob. 3.10. Assume that specifications on this item are 190.5 and 210.5 ms (see Prob. 4.31). Use the nominal value, 200.5, as the standard $\overline{X}_o$, and an α risk level of 0.0027 (corresponding to 3-sigma limits on a Shewhart control chart). The magnitude of the shift D, which should be detected with virtual certainty, is 7.8 ms. Use a scaling factor y equal to 5. Plot the chart on a sheet of $\frac{1}{4}$-in grid paper, and draw the mask on a sheet of clear plastic. Is this process in control? Compare the results obtained for the CSCC for $\overline{X}$ with those obtained in Prob. 4.29, 4.30, and 4.31.

10.41. Prepare a two-sided CSCC for $\overline{X}$ for the data of Table 1.1. For this chart, use the nominal dimension 0.4037 in (37 in the units shown in the table) as the standard $\overline{X}_o$ and an α risk level of 0.0027. A shift in the mean of 2.0 (table units) should be detected with virtual certainty. Use a scaling factor y of 1. Draw the chart on $\frac{1}{4}$-in grid paper and the mask on a clear plastic sheet. Does this process show control? Compare your results with those given in Fig. 1.2, and comment on the relative effectiveness of the two procedures.

10.42. Prepare a two-sided CSCC for $\overline{X}$ for the data of Table 3.1. Use 21.5 oz as the standard value $\overline{X}_o$ and an α risk level of 0.0027. A shift in the mean of 0.5 oz should be detected with virtual certainty. (While a contained weight such as this is subject to meeting only a lower specification, too great an overfill means that the company is giving away too much product.) Using a scaling factor y of 1, draw the chart on $\frac{1}{4}$-in grid paper and the mask on a clear plastic sheet. Is control indicated? Compare your results with those given in Fig. 4.1, and comment on the relative effectiveness of the two procedures.

PART THREE

SCIENTIFIC SAMPLING

11

SOME FUNDAMENTAL CONCEPTS IN SCIENTIFIC SAMPLING

The fundamental difference between engineering with and without statistics boils down to the difference between the use of a scientific method based upon the concept of laws of nature that do not allow for chance or uncertainty and a scientific method based upon the concept of laws of probability as an attribute of nature.

—W. A. Shewhart[†]

11.1 THE IMPORTANCE OF SAMPLING

Sampling is used everywhere in quality assurance. Even if all items produced or services performed are inspected for conformance to specifications, we are still sampling from the universe of all like items or services that have been or will be produced. In statistical process control (SPC), the objective is to ensure that a single universe exists, that is, to maintain a constant μ and σ at acceptable levels. It is often the case that not all items are inspected and their conformance verified. Many of the previous examples have indicated that samples were drawn from the process flow; often these samples constituted much less than 100% of the output. The rules and guidelines for forming rational subgroups were presented and discussed without consideration of the proportion of output being inspected.

[†]W. A. Shewhart, "Contribution of Statistics to the Science of Engineering," included in *University of Pennsylvania Bicentennial Conference. Volume on Fluid Mechanics and Statistical Methods,* pp. 97–124, University of Pennsylvania Press, Philadelphia, 1941.

11.1.1 Importance of Sampling for Acceptance Purposes

Inspection for acceptance purposes is carried out at many stages in manufacturing. There may be inspection of incoming materials and parts, process inspection at various points in the manufacturing operations, final inspection by a manufacturer of its own product, and—ultimately—inspection of the finished product by one or more purchasers.

Much of this acceptance inspection is necessarily on a sampling basis. All acceptance tests that are destructive of the item tested must inevitably be done by sampling. In many other instances sampling inspection is used because the cost of 100% inspection is prohibitive. As pointed out in Chap. 1, where there are a great many similar items of product to be inspected, sampling inspection is likely to be better done than 100% inspection because of the influence of inspection fatigue in 100% inspection. An important advantage of modern acceptance sampling systems such as those discussed in Chaps. 12 to 14 is that they exert more effective pressure for quality improvement than is possible with 100% inspection.

11.1.2 Pressure for Quality Improvement Exerted by Rejection of Entire Lots

Inspection, in the sense of sorting product that conforms to specifications from nonconforming product, cannot be relied on to ensure that all accepted product really conforms. Inspection fatigue on repetitive inspection operations often will limit the effectiveness of 100% inspection. Obviously, no *sampling* procedure can eliminate all nonconforming product. It follows that the best way to be sure that accepted product conforms to specifications is to have the product made right in the first place.

Where a producer does not make product right in the first place and, in effect, relies on the consumer to do screening inspection, it often happens that striking quality improvements can be caused by the outright rejection of entire lots of product on the basis of the numbers of nonconforming items found in samples. The rejection of entire lots brings much stronger pressure for quality improvement than the rejection of individual articles. Example 11.1 describes a case where such pressure was exerted in an effective manner.

Example 11.1
Substitution of occasional lot rejections for acceptance of an unsatisfactory process average

Facts of the case A public utility company purchased many measuring devices of a certain type from three competing manufacturers. The specified tolerances for a certain quality characteristic of these devices were fairly tight. (These specifications had been established by the state regulatory commission with the concurrence of the utility company.) However, for many years both the manufacturers of the devices and the utility buying them had acted as if it were believed to be impossible for the manufacturers to make substantially all the devices within the specified tolerances. It was not uncommon for shipments to be accepted containing 5 to 10% of the devices that were defective in the sense of failing to meet these specifications. The invariable procedure was for the utility company to test all the purchased devices in its laboratory. Out-of-tolerance devices usually were adjusted or rebuilt by the utility rather than returned to the manufacturers.

Finally, the management of the utility concluded that it was doing work that it had already paid the manufacturers to do. It seemed unlikely that any improvement in submitted quality would take place either as a result of complaints to the manufacturers or as a result of returning to the manufacturers only the individual out-of-tolerance devices.

The utility therefore notified the three manufacturers that in the future acceptance inspection would be conducted under Military Standard 105B (an earlier version of the standard discussed in Chap. 14).[†] At the time of this example, this military standard and its predecessors had the prestige of many years of use by government and industry. The utility chose a sampling scheme based on a stated "acceptable quality level" that seemed reasonable in relation to the manufacturers' claims for their product. For the time being, lots accepted under this scheme continued to receive 100% testing with adjustment or rebuilding of out-of-tolerance devices.

The long record of quality history from its past testing program made it possible for the management of the utility to predict that at the start a moderate percentage of lots would be rejected by the new sampling scheme. These lot rejections occurred very much as predicted. As anticipated, the manufacturers initially took the view that they were being treated unjustly by the rejection of the good articles contained in the rejected lots. Somewhat illogically (because the actions of the utility were reasonable without reference to any published sampling procedures), the prestige of the military standard was successfully used to justify the actions of the utility. As time went on, the manufacturers were able to diagnose the reasons for their difficulties with this quality characteristic, and the quality of the product submitted to the utility was greatly improved.

11.2 SOME WEAKNESSES OF CERTAIN TRADITIONAL PRACTICES IN ACCEPTANCE SAMPLING

Before the widespread use in industry of modern acceptance sampling systems, inspectors often used a working rule that is still recognized as based on a correct principle. This rule was to permit an inspector's current decisions on acceptance to be influenced by knowledge of the past quality history of the product being sampled. For instance, where the same part or article was purchased from two or more sources, an inspector might check only one or two items in a lot from a source

[†]Military Standard 105B involved acceptance or rejection of entire lots on the basis of the results of tests on a random sample selected from each lot.

considered reliable but might give critical examination to a lot from a source considered to be unreliable.

Although this rule is sound as far as it goes, such an informal system for determining size and frequency of sample and basis of acceptance has obvious limitations. Inspectors' memories of past quality history sometimes may be short and inaccurate. The inspector who tries to remember the quality history of a number of products may die or resign or be transferred to another job. Or the inspector's confidence in past quality may lead to the neglect of current inspection and therefore to failure to discover when quality has changed for the worse.

These limitations suggest the need of definite working rules regarding size and frequency of sample and basis for acceptance or rejection. But it is not sufficient that such rules merely be *definite*. Many of the definite rules that at one time were common in industry were bad because they seemed to give a promise of quality protection that they could not fulfill. Formal schemes such as the once popular one analyzed in Example 1.3 often gave less quality protection than the informal scheme of letting the inspector use personal judgment. For companies with substantial leverage with their suppliers, modern practices often require that suppliers provide quality-related evidence, such as charts and quality indexes, with each shipment.

11.3 PURPOSE OF THIS CHAPTER

The real problem in most acceptance sampling is to design a satisfactory acceptance sampling *system* or, more commonly, to select such a system from a number of possible systems already developed by someone else. Although Example 11.1 necessarily gave the reader an oversimplified view of such systems, it made the essential point that an important aspect of any such system is its influence on the quality of submitted product. To judge the suitability of any proposed acceptance sampling system in a particular case, it is desirable to have an understanding of the strategy and tactics built into the various available types of systems.

This chapter is intended to lay the groundwork for more detailed discussion in Chaps. 12 through 15 of a number of widely used modern acceptance sampling systems. As a prerequisite for an understanding of such systems, the reader needs to know certain terminology, concepts, and probability calculations involved in acceptance sampling. This chapter uses lot-by-lot acceptance sampling by attributes to develop the terminology and concepts and to illustrate some types of probability calculations. Discussion of acceptance sampling from a continuous stream of product is deferred until Chap. 14. Discussion of acceptance sampling by variables is deferred until Chap. 15.

11.3.1 Use of the Words *Defective* and *Defect* in Part Three of This Book

It was pointed out in Chap. 1 that when these words are used in their technical senses dealing with lack of conformity to specifications, they do not necessarily mean *defective* and *defect* in the popular sense. It is common for specifications to contain a margin of safety; therefore, some product that does not meet specifications can be satisfactory for its intended use. The difference between the technical and popular meanings has been a source of confusion and misunderstanding in lawsuits involving product liability.

In the interest of clarity in writing and speaking about topics related to quality of manufactured product, it would be desirable if everyone concerned with quality matters would abandon the use of *defective* and *defect* in the restricted technical sense. In previous chapters, with a few exceptions that generally involve quotations, the authors have been able to substitute appropriate words or phrases. For example, we have referred to *nonconforming* product, percent *rejected,* and numbers of *nonconformities.*

Unfortunately, it does not seem reasonable to continue such substitutions throughout Part Three. The acceptance sampling systems that we examine in the following chapters all make considerable use of the words *defective* and *defect.* Therefore the words should be used in explaining the systems. Nevertheless, the reader should understand that in the coming pages these words are used in their restricted technical senses. A defective item is one that does not conform to specifications in some respect; a defect is a nonconformity to some specification.

11.3.2 Some Symbols and Terms Used in Relation to Acceptance Sampling Plans

The probability principles presented in Chap. 5, and the explanation of control charts for fraction rejected and for numbers of nonconformities given in Chaps. 6 and 7, provide a background for a discussion of the evaluation of acceptance plans involving sampling by attributes. In discussing such plans, the following symbols are used:[†]

N	number of pieces in a given lot or batch.
n	number of pieces in a sample.
D	number of defective pieces (that is, pieces not conforming to specifications) in a given lot of size N.
r	number of defective pieces (that is, pieces not conforming to specifications) in a given sample of size n.
c	acceptance number, the maximum allowable number of defective pieces in a sample of size n[‡] (also denoted by A_c in some cases).
p	fraction defective. In a given submitted lot, this is D/N; in a given sample, it is r/n.
μ_p	true process average fraction defective of a product submitted for inspection.
$\bar{p}$	average fraction defective in observed samples.

[†]Where applicable, the notation used in this text follows that recommended by the American Society for Quality Control in "ASQC Standard A2. Definitions and Symbols for Acceptance Sampling by Attributes," 1987 (ANSI Standard Z1.6-1987).

[‡]See footnote on p. 104 in Chap. 3 for the justification for the use of the symbol c in a different meaning from those previously employed.

P_a	$L(p)$ = probability of acceptance.
β	consumer's risk, the probability of accepting product of some stated undesirable quality. It is the value of P_a at that stated quality.
α	producer's risk, the probability of rejecting product of some stated desirable quality. $\alpha = 1 - P_a$ at that stated quality.
$p_{0.95}$, $p_{0.50}$, $p_{0.10}$, etc.	fraction defective having a probability of acceptance of 0.95, 0.50, 0.10, and so on, under any given acceptance criteria.

11.4 LOT-BY-LOT ACCEPTANCE USING SINGLE SAMPLING BY ATTRIBUTES

In acceptance inspection a defective article is defined as one that fails to conform to specifications in one or more quality characteristics. A common procedure in acceptance sampling is to consider each submitted lot of product separately and to base the decision on acceptance or rejection of the lot on the evidence of one or more samples chosen at random from the lot. When the decision is always made on the evidence of only one sample, the acceptance plan is described as a *single sampling* plan.

Any systematic plan for single sampling requires that three numbers be specified. One is the number of articles N in the lot from which the sample is to be drawn. The second is the number of articles n in the random sample drawn from the lot. The third is the acceptance number c.

This acceptance number is the maximum allowable number of defective articles in the sample. More than c defectives will cause the rejection of the lot. In sampling plans developed without benefit of statistical analysis, c often is specified as zero under the illusion that if the sample is perfect, the lot will be perfect.

In the discussion that follows, sampling acceptance plans of this type are described by these three numbers. For instance, the sampling plan of Example 1.3 is specified in this way as $\begin{cases} N = 50 \\ n = 5 \\ c = 0 \end{cases}$. These three numbers may be interpreted as saying, "Take a random sample of 5 from a lot of 50. If the sample contains more than 0 defectives, reject the lot; otherwise, accept the lot."

Example 1.3 examined this plan critically under a particular assumption, namely, that the plan was used for acceptance of product that on the average was 4% defective. The distribution of defectives among the lots was assumed to follow the laws of chance. (This amounted to an assumption either that the production process was statistically controlled or that the product was well mixed before being divided into lots.) Under this assumption, the lots accepted by the plan proved to be 3.6% defective; this modest improvement in product quality was accomplished at the cost of rejecting 18.5% of the submitted lots. After the defective articles found in samples from the rejected lots were eliminated, the average quality of the remainder of the rejected lots was not appreciably worse than the average of the accepted lots. It was evident that this sampling acceptance plan was not a satisfactory one under the assumed conditions.

11.4.1 The Operating Characteristic (OC) Curve of an Acceptance Sampling Plan Shows the Ability of the Plan to Distinguish between Good and Bad Lots

In judging various acceptance sampling plans it is desirable to compare their performance over a range of possible quality levels of submitted product. An excellent picture of this performance is given by the *operating characteristic curve,* first mentioned in Chap. 1.[†] Such curves are commonly referred to as OC curves.

For any given fraction defective p in a submitted lot, the OC curve shows the probability P_a that such a lot will be accepted by the given sampling plan. Or, stated a little differently, the OC curve shows the long-run percentage of submitted lots that would be accepted if a great many lots of any stated quality were submitted for inspection. Figures 11.1 to 11.3 give the OC curves of a number of single sampling plans.

As explained later in this chapter, in most cases OC curves may also be thought of as showing the probability of accepting lots from a stream of product having a fraction defective p.

11.4.2 Sampling Acceptance Plans with Same Percent Samples Give Very Different Quality Protection

A common practice in industry has been to specify that the sample inspected should be some fixed percentage of the lot, such as 5, 10, or 20%. This specification was generally based on the mistaken idea that the protection given by sampling schemes is constant if the ratio of sample size to lot size is constant. Such specifications often have been associated with an acceptance number of zero.

Figure 11.1 illustrates just how wrong this idea really is. This figure compares the OC curves of four sampling acceptance plans, all of which involve a 10% sample and an acceptance number of zero. The differences in the quality protections provided by these plans are obvious and impressive. They may be emphasized by statements of fact that can be read from the OC curves.

[†]Most of the terminology of acceptance sampling originated in the Bell Telephone Laboratories in the 1920s, where such curves were called "probability of acceptance curves." The phrase "operating characteristic curve," however, originated in the Ballistic Research Laboratories at Aberdeen Proving Ground, Maryland, just before World War II. It was first suggested by a nonstatistician, Col. H. H. Zornig, when he was director of that laboratory, and was used by Gen. (then Major) Leslie E. Simon in his writings on quality control at that time. A number of years later, the phrase was incorporated into the general language of statistical inference and is now used in connection with many kinds of statistical tests of hypotheses.

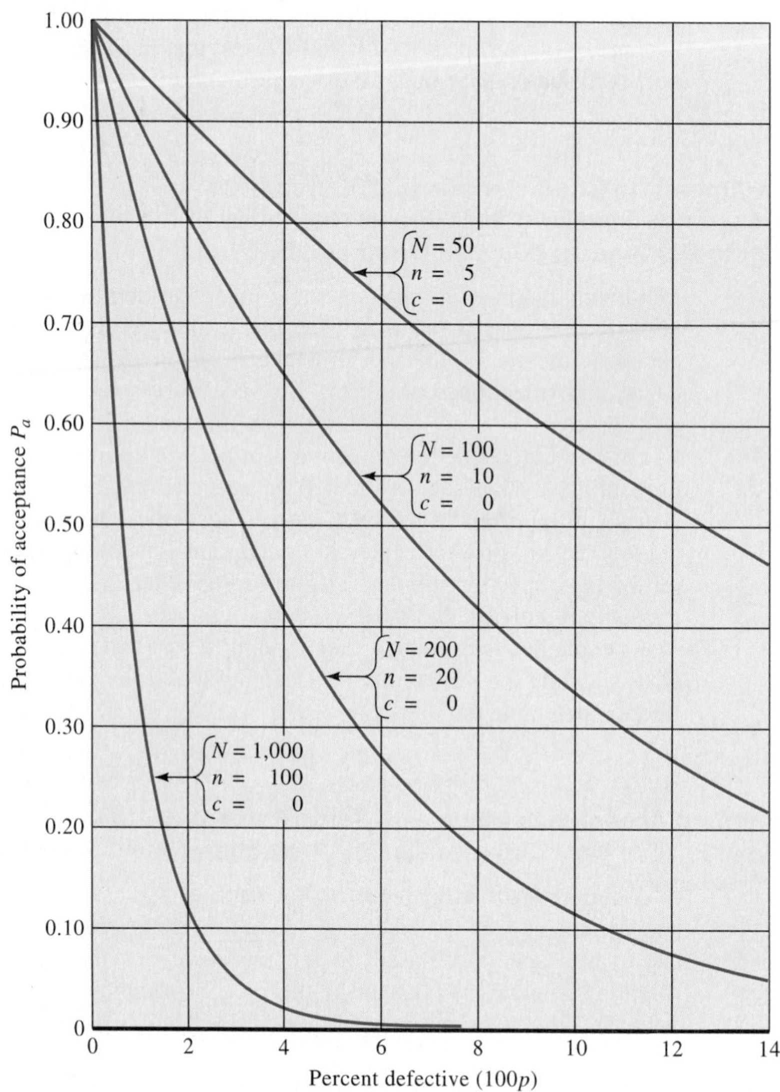

FIGURE 11.1 Comparison of operating characteristic curves for four sampling plans involving 10% samples.

For example, the curves show that lots that are 4% defective will be accepted 81% of the time using a 10% sample from a lot of 50, 65% of the time using a 10% sample from a lot of 100, 42% of the time using a 10% sample from a lot of 200, and less than 2% of the time (actually 1.35%) by a 10% sample from a lot of 1,000, assuming an acceptance number of zero in all cases. Obviously, a producer making product 4% defective would have a strong motive for trying to have its product inspected in lots of 50 rather than in lots of 1,000.

Or, considered in a slightly different way, the curves show the quality of lot that will be passed 50% of the time by each plan. $\left\{\begin{array}{l} N = 50 \\ n = 5 \\ c = 0 \end{array}\right.$ will pass a 12% defective lot half the time, $\left\{\begin{array}{l} N = 100 \\ n = 10 \\ c = 0 \end{array}\right.$ a 6% defective lot, $\left\{\begin{array}{l} N = 200 \\ n = 20 \\ c = 0 \end{array}\right.$ a 3% defective lot, and $\left\{\begin{array}{l} N = 1{,}000 \\ n = 100 \\ c = 0 \end{array}\right.$ a 0.65% defective lot.

These curves are based on computations such as those illustrated in Chap. 5.

11.4.3 Fixed Sample Size Tends toward Constant Quality Protection

From the standpoint of quality protection, the absolute size of a random sample is much more important than its relative size compared to the size of the lot. This fact is illustrated by Fig. 11.2. This figure shows the OC curves of four different sampling plans all having the same sample size 20 but having lot sizes of 50, 100, 200, and 1,000, respectively.

The three upper curves, in which the sample size varies from 20 to 2% of the lot, show close agreement. This agreement is in sharp contrast to the great difference among the curves in Fig. 11.1. These two figures together emphasize the point that it is the absolute size of the sample rather than its relative size that determines the quality protection given by an acceptance sampling plan. The story told by these two figures is particularly striking because it contradicts many preconceived notions on the subject of sampling.

The upper curve corresponding to a lot size of 1,000 is practically identical with the OC curve that would be obtained for an infinite lot size. For example, the probability of acceptance of a 5% defective lot when $N = 1{,}000$ is 0.355. When $N = \infty$, the corresponding probability is 0.358. For a 10% defective lot, the respective probabilities of acceptance are 0.119 and 0.122. (These figures are obtained using hypergeometric probabilities for the lot of 1,000 and using the binomial for the infinite lot.) For a given lot fraction defective and a zero acceptance number, the probability of acceptance computed for a finite lot is always less than that computed for an infinite lot because of the recognition of the partial exhaustion of the lot by the sample.

It is evident that unless a sample is a large proportion of the lot, such as the 10, 20, and 40% samples in the three lower OC curves of Fig. 11.2, it will usually be good enough for practical purposes to compute OC curves as if lot sizes were infinite. Moreover, if the OC curve is viewed as giving probabilities of acceptance of lots from a statistically controlled product having a fraction defective p, the OC curve computed in this way is correct in principle.

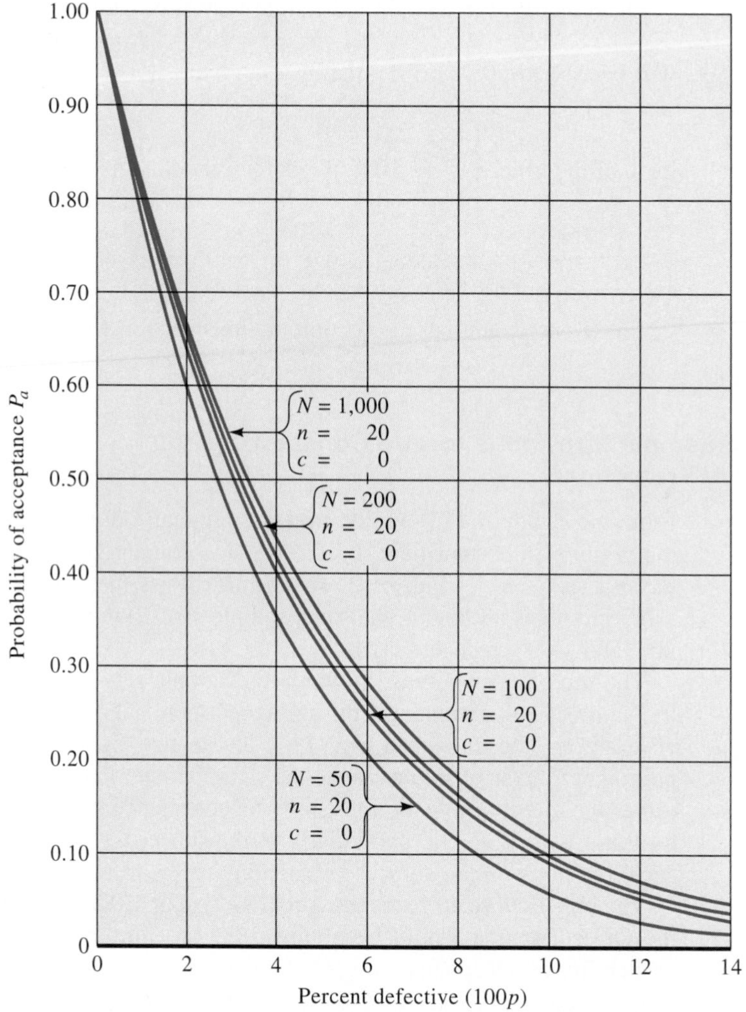

FIGURE 11.2 Comparison of OC curves for four sampling plans involving samples of 20, each with an acceptance number of zero.

11.4.4 No Sampling Plan Can Give Complete Protection against the Acceptance of Defective Product

A practical difficulty in devising an ideal sampling plan is that it is not possible to change the laws of chance.

All lot-by-lot sampling plans are certain to pass some of the lots containing defective product if such product exists in many of the lots submitted for acceptance. This fact needs to be faced by all who specify and use acceptance sampling.

It follows that the selection of an acceptance sampling plan requires a decision on the risks that the user of the plan is willing to face, all things considered. This is an economic decision that depends on a number of matters discussed in Chap. 17. In lot-by-lot sampling by attributes, the risks of acceptance of submitted lots containing any stated percentage of defectives are given by the OC curve of the particular sampling plan. It is therefore appropriate to give careful consideration to the OC curve in the selection of a sampling plan. As pointed out in the following discussion of acceptance sampling, the users of sampling plans often find it convenient to concentrate attention on one or two points on the OC curve rather than on the whole curve.

11.5 OC CURVE OF AN IDEAL SAMPLING PLAN

Suppose it is decided that 2.2% is the maximum tolerable proportion defective in any submitted lots. Once this decision has been made, it might seem that an ideal lot-by-lot sampling scheme would be one that rejected all lots that were worse than 2.2% defective and accepted all lots 2.2% defective or better. The OC curve of such an ideal scheme would be a vertical at $p = 0.022$. Of course, no such scheme can exist short of 100% inspection perfectly executed.

It is, however, possible to select a sampling plan in which a 2.2% defective lot has some desired probability of acceptance. If it should be decided that the consumer can take 1 chance in 10 that if a 2.2% defective lot is submitted the lot will be accepted, the OC curve of the selected plan should pass through the point $P_a = 0.10$, $p = 0.022$ (or $100p = 2.2\%$).

Figure 11.1 shows that the OC curve of the plan $\begin{cases} N = 1{,}000 \\ n = \quad 100 \\ c = \qquad 0 \end{cases}$ passes through this point. If the product is to be submitted for acceptance in lots of 1,000 articles, this plan is a possible one to give the consumer the desired quality protection against accepting 2.2% defective lots.

11.5.1 Conflicting Interests of Consumer and Producer in the Selection of Sampling Plans

There are always two parties to an acceptance procedure, the party submitting the product for acceptance and the party for whom the decision is made regarding acceptance or rejection. In all discussions of acceptance procedures in this book, these parties are concisely referred to as the *producer* and the *consumer.* It should be recognized that for many inspection operations the producer and consumer may be part of the same organization; for instance, the producer might be the machine shop and the consumer the assembly department. In a manufacturer's final inspection of its own product, the sales department may be thought of as the consumer. In acceptance inspection of some purchased product, the producer—in the sense used here—may actually be a dealer who had no connection with the original production.

At first impression, it might seem that the producer and consumer should have completely opposite viewpoints toward the selection of sampling plans. The consumer requires protection against the acceptance of too much defective product. The producer, on the other hand, needs to be protected against the rejection of too much product that conforms to specifications.

As an example, assume that a consumer selects the plan $\begin{cases} N = 1{,}000 \\ n = 100 \\ c = 0 \end{cases}$ in order to be fairly well protected against accepting lots more than 2.2% defective. As already pointed out, if a lot exactly 2.2% defective should be submitted, there is only 1 chance in 10 of its acceptance. The consumer will have even greater protection if lots of poorer quality are submitted.

This plan, however, gives unsatisfactory protection to the producer who submits product considerably better than 2.2% defective. For instance, the OC curve of Fig. 11.3 shows that if the producer submits lots 1% defective, 65% of them will be rejected. (The curve says 35% will be accepted; this is subtracted from 100% to get the percentage rejected.) If lots 0.5% defective are submitted, 41% will be rejected. If lots 0.2% defective are submitted, 19% will be rejected. It is evident that the consumer's protection against accepting product 2.2% defective is obtained only by rejecting a large proportion of any submitted lots that are of much better quality.

A more critical consideration will show that, where there is a continuing relationship between producer and consumer, such substantial rejections of acceptable product in the effort to exclude product not conforming to specifications are not necessarily in the consumer's interest. The consumer is interested in quality, but also in cost. In the long run the costs incident to the rejection of acceptable product tend to be passed on by the producer to the consumer. The consumer may also be interested in having the product now. Any acceptable product rejected by the consumer is not available for immediate use.

It seems clear that, even from the consumer's viewpoint, it is not sufficient to ask that a sampling plan protect the consumer against the acceptance of lots having a higher percentage of defectives than some maximum tolerable figure.

11.5.2 The Acceptance Number Need Not Be Zero

Figures 11.1 and 11.2 emphasized the fact that a perfect sample does not ensure a perfect lot. Once this fact is recognized, the objections sometimes raised to permitting lot acceptance with some defectives in a sample may be shown to have no logical foundation.

The users of modern acceptance sampling procedures recognize certain psychological advantages of allowing at least one defective in a sample. Moreover, the operating characteristics of plans with acceptance numbers greater than zero are superior to those of comparable plans with an acceptance number of zero. For a desired protection against accepting lots containing some stated percentage of defectives, larger acceptance numbers involve larger sample sizes. Plans having

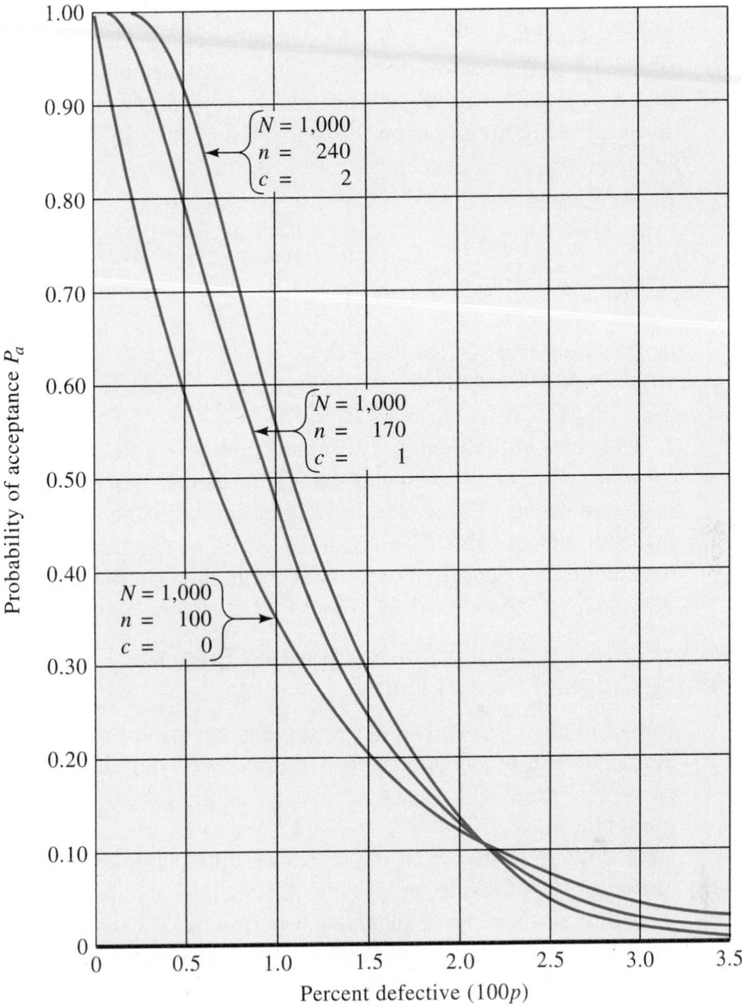

FIGURE 11.3 OC curves for three sampling plans having a 0.10 probability of acceptance of a 2.2% nonconforming lot.

larger sample sizes have greater ability to discriminate between satisfactory and unsatisfactory lots.

For example, consider our consumer who wants protection against accepting lots of 2.2% defective or worse and insists that any 2.2% defective lots submitted shall have only a 0.10 probability of acceptance. Assume that product is submitted in lots of 1,000. This consumer's requirement will be met by using $n = 100$ and $c = 0$, or by using $n = 170$ and $c = 1$, or by using $n = 240$ and $c = 2$, or by various other possible plans.

The OC curves for these three plans are shown in Fig. 11.3. It will be noted that although all three plans give the consumer equal protection against the acceptance of a 2.2% defective lot, the plans with acceptance numbers of 1 and 2 give the producer much better protection against the rejection of lots that are satisfactory in the sense of being greatly superior to 2.2% defective lots. A few figures read from the three curves serve to emphasize this point.

	$\begin{cases} N = 1{,}000 \\ n = 100 \\ c = 0 \end{cases}$	$\begin{cases} N = 1{,}000 \\ n = 170 \\ c = 1 \end{cases}$	$\begin{cases} N = 1{,}000 \\ n = 240 \\ c = 2 \end{cases}$
Lots 1% defective rejected	65%	51%	44%
Lots 0.5% defective rejected	41%	20%	9%
Lots 0.2% defective rejected	19%	3%	0%

The plans with the higher acceptance numbers and better OC curves have higher sample sizes for a given lot size. This means more inspection and hence more inspection cost. If, however, as is often the case, the rejection of a lot by sampling inspection means that the lot is to be 100% inspected, the total amount of resulting inspection may actually be less with the higher sample size and acceptance number.

11.5.3 Generally Speaking, the Larger the Sample Size, the Steeper the Slope of the OC Curve

It is of interest to compare several plans having different sample sizes and having the same ratio of acceptance number to sample size. Consider the plans $\begin{cases} n = 75 \\ c = 1 \end{cases}$ $\begin{cases} n = 150 \\ c = 2 \end{cases}$ and $\begin{cases} n = 750 \\ c = 10 \end{cases}$. Assume a lot size N of 10,000. This lot size is large enough so that even the largest of these samples is a relatively small fraction of the lot. The OC curves of these three plans are shown in Fig. 11.4.

Although these three plans each permit 1.33% (that is, $\frac{1}{75}$) of the sample to be defective, it is evident that they have quite different OC curves. The larger the sample size, the greater the ability of a sampling plan to discriminate between lots of different qualities. For example, a 3% defective lot, if submitted, has more than 1 chance in 3 of acceptance if $n = 75$ and $c = 1$ but is practically certain to be rejected if $n = 750$ and $c = 10$. The larger sample, which protects the consumer against the acceptance of relatively bad lots, also gives the producer better protection against the rejection of relatively good ones. Thus a 0.7% defective lot has 1 chance in 10 of rejection if $n = 75$ and only 2 chances in 100 of rejection if $n = 750$.

11.5.4 Type *A* and Type *B* OC Curves

H. F. Dodge and H. G. Romig make the helpful distinction between two types of OC curves.[†] Type *A* curves give the probabilities of acceptance for various fractions

[†]H. F. Dodge and H. G. Romig, *Sampling Inspection Tables—Single and Double Sampling,* 2d ed., pp. 56–59, John Wiley & Sons, Inc., New York, 1959.

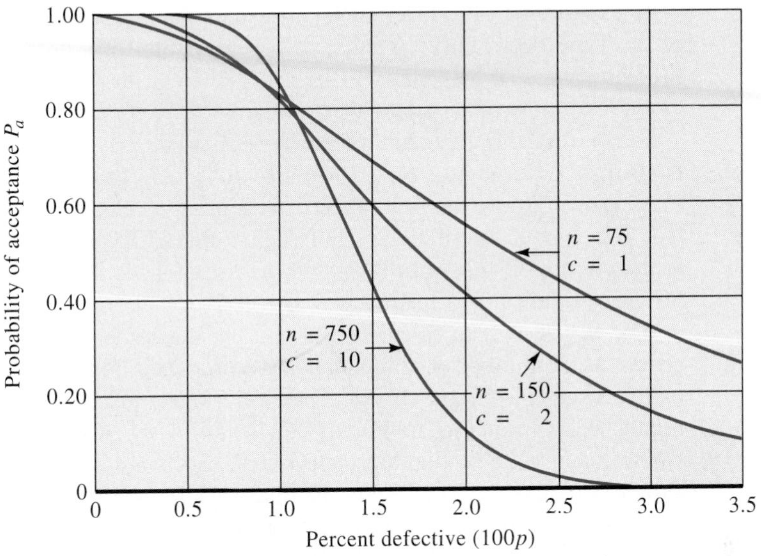

FIGURE 11.4 Comparison of OC curves with different sample sizes, all permitting the same fraction of the sample to be nonconforming.

defective as a function of the *lot* quality of finite lots. In principle, such curves should be computed by hypergeometric probabilities; the binomial or Poisson distributions often give satisfactory approximations. In principle, also, such curves are discontinuous. For example, a lot of 200 items may be 0.5 or 1.0% defective but not 0.8%. In practice, it is common to draw Type *A* curves as continuous.

Type *B* curves give the probabilities of acceptance of a lot as a function of *product* quality. Such curves are calculated as if the lot size were infinite. For Type *B* curves the binomial is exact and the Poisson distribution often gives a satisfactory approximation. Such curves are correctly viewed as continuous.

In the discussion of Fig. 11.2—which gave Type *A* OC curves for several plans that had an n of 20, a c of 0, and various lot sizes—it was pointed out that there was no practical difference between the OC curve for N = 1,000 and $N = \infty$. In general, where the sample size n is not more than one-tenth the lot size N, Type *A* and Type *B* curves may be considered as identical for most practical purposes.

In judging the appropriateness of using Type *B* curves where n is more than one-tenth N, certain assumptions underlying such uses should be recognized. Assume that a statistically controlled process exists that has some fraction defective μ_p. The lots of size N that come from this process are, in a statistical sense, random samples from the process. Each random sample of size n selected from some lot of N items may therefore be viewed as a random sample from the process that has the fraction defective μ_p. It follows that the Type *B* curve gives the expected results of a given sampling plan over the entire range of possible values of process fraction defective.

In examining a Producer's Risk, it usually is appropriate to adopt the viewpoint of the Type *B* OC curve. That is, it is reasonable for a producer to want to know the percentage of lots that will be rejected by a proposed sampling plan for any given level of process quality.

In contrast, the viewpoint of the Type *A* curve may be desirable in evaluating a Consumer's Risk with respect to individual lots. That is, a consumer reasonably may want to know the risk taken of accepting a relatively bad *lot* if it is submitted. The Type *A* curve always falls *below* that of Type *B* in the region of interest to the consumer; thus it follows that the use of Type *B* curves tends to give a conservative figure for Consumer's Risk that is somewhat too high.

The easiest way to calculate and plot OC curves is through the use of computer programs. A number of companies have produced computer software designed to compute and plot these curves. Their offerings and capabilities are reported on a regular basis in quality trade magazines and by the American Society for Quality Control annually in the March (QA/QC Software Directory) issue of *Quality Progress* magazine. In addition, its quarterly *Journal of Quality Technology* regularly includes a Computer Programs department containing a variety of programs useful in quality assurance activities.

11.5.5 Use of Table *G* for Approximate Calculation of Type *B* OC Curves of Sampling Plans

Although the binomial distribution is correct in principle for computing Type *B* OC curves, it is usually satisfactory and convenient to compute approximate probabilities of acceptance by use of a Poisson distribution table such as Table *G*, App. 3.

Table 11.1 illustrates the use of Table *G* in computing the three OC curves shown in Fig. 11.4. The probabilities of acceptance are read directly from Table *G* (or interpolated where necessary) as the figures corresponding to *c* (that is, "*c* or less") occurrences of the event with an average μ_{np}. Table 11.1 gives all probabilities to three decimal places to permit readers to check their use of Table *G*. Nevertheless, it should be recognized that because of the approximate method used, three decimal places are not really justified.[†]

11.6 THE INDEXING OF ACCEPTANCE PLANS BY A SINGLE POINT ON THE OC CURVE

Chapters 12 through 15 include descriptions of a number of different tables of acceptance sampling plans. Certain tables classify acceptance plans in accordance with a single point on the OC curve. Three points on the OC curve have been given particular importance in the design of systems of sampling plans, namely:

[†]The approximation of the Poisson distribution to the binomial is, of course, best for the smallest values of p. Hence the OC curves computed using Table *G* (or other Poisson tables or diagrams) are more accurate at the upper end of the curve than at the lower end. As an example, consider the plan $N = \infty$, $n = 75$, $c = 1$ as applied to product 1% and 5% defective. The correct probabilities of acceptance computed by the binomial are 0.827 and 0.106, respectively. Table 11.1, computed by the Poisson law, shows the approximate values to be 0.827 and 0.112, respectively.

TABLE 11.1 CALCULATION OF APPROXIMATE TYPE *B* OC CURVES FOR THREE SAMPLING PLANS

Fraction defective in lot μ_p	Expected average number of defectives μ_{np} in sample			Probability of acceptance P_a		
	$n = 75$	$n = 150$	$n = 750$	$\begin{cases} n = 75 \\ c = 1 \end{cases}$	$\begin{cases} n = 150 \\ c = 2 \end{cases}$	$\begin{cases} n = 750 \\ c = 10 \end{cases}$
0.002	0.15	0.30	1.5	0.990	0.996	1.000
0.004	0.30	0.60	3.0	0.963	0.977	1.000
0.006	0.45	0.90	4.5	0.925	0.937	0.993
0.008	0.60	1.20	6.0	0.878	0.879	0.957
0.010	0.75	1.50	7.5	0.827	0.809	0.862
0.012	0.90	1.80	9.0	0.772	0.731	0.706
0.014	1.05	2.10	10.5	0.718	0.650	0.521
0.016	1.20	2.40	12.0	0.663	0.570	0.347
0.018	1.35	2.70	13.5	0.610	0.494	0.211
0.020	1.50	3.00	15.0	0.558	0.423	0.118
0.025	1.875	3.75	18.75	0.441	0.278	0.021
0.030	2.25	4.50	22.5	0.343	0.174	0.003
0.035	2.625	5.25	26.25	0.262	0.106	0.000
0.040	3.00	6.00	. . . .	0.199	0.062	
0.050	3.75	7.50	. . . .	0.112	0.020	
0.060	4.50	9.00	. . . .	0.061	0.006	
0.070	5.25	10.50	. . . .	0.033	0.002	
0.080	6.00	12.00	. . . .	0.017	0.001	
0.090	6.75	13.50	. . . .	0.009	0.000	
0.100	7.50	. . . .	. . . .	0.004		

1. The lot (or process) quality for which $P_a = 0.95$. In this book, this quality is referred to as $p_{0.95}$.[†]
2. The lot (or process) quality for which $P_a = 0.50$. In this book, this is referred to as $p_{0.50}$.[‡]
3. The lot (or process) quality for which $P_a = 0.10$. To be consistent with the preceding symbols, this is referred to as $p_{0.10}$. In the literature of acceptance sampling, this quality is most frequently referred to as the lot tolerance fraction defective: $100p_{0.10}$ is described as the *Lot Tolerance Percent Defective* (LTPD).[§]

[†]In the tables developed by the Columbia Statistical Research Group, discussed in Chap. 12, this quality is referred to as the AQL (acceptable quality level).

[‡]This has been referred to by some writers as the "point of control" and by others as the "indifference quality."

[§]More properly, in its original usage by Dodge and Romig, the notion of a tolerance percent defective may be associated with any Consumer's Risk. For example, the lot quality having a P_a of 0.05 could be described as the lot tolerance percent defective associated with a Consumer's Risk of 0.05. However, because of the widely used Dodge-Romig tables (described in Chap. 13) that all assume a Consumer's Risk of 0.10, the term LTPD sometimes is incorrectly used without qualification to mean $100p_{0.10}$.

The same acceptance plan may therefore be referred to by several different product quality values, depending on the point of indexing. There are legitimate reasons for this variability in indexing sampling plans. Nevertheless, this has naturally proved to be a source of confusion to some users of acceptance sampling tables. Further comment on this topic is made in the next three chapters.

11.7 AVERAGE OUTGOING QUALITY AND THE AOQL

The decision to accept a lot, based on the outcome of one or more samples drawn from the lot, leads to rather obvious results. The decision to reject, however, leads to another series of actions and decisions, usually well stipulated in contractual agreements or standard operating procedures, but nevertheless more complex than the actions required by a decision to accept.

Frequently the result of a decision to reject a lot is 100% inspection of that particular lot. Such action is referred to as *screening inspection,* or *detailing.* This is sometimes described as an *acceptance/rectification* scheme. Responsibility for screening inspection may rest with either the producer or the consumer.

The accepted lots will contain approximately the percent defective submitted, although they will be slightly improved by the elimination of any defectives found in the samples whenever c is 1 or more. The rejected lots, after screening, presumably will contain no defectives. Thus, at any given level of incoming quality p greater than 0, there is some lower level of *Average Outgoing Quality* (AOQ). The P_a begins at 1.0, when p equals 0.0, and, as described by the OC curve, descends continuously to 0.0, when p equals 1.0. Thus the value of the AOQ must begin at 0.0, rise continuously to some maximum value, and then descend continuously to 0.0. That maximum value is referred to as the *Average Outgoing Quality Limit* (AOQL).

A common type of approximate calculation to determine the AOQL is illustrated in Table 11.2. This table refers to the plan $\begin{cases} n = 75 \\ c = \ \ 1 \end{cases}$, when N is large compared to n. The probabilities of acceptance were computed in Table 11.1. The right-hand column gives average outgoing quality (AOQ) for each assumed percent defective in submitted lots. The maximum value of the AOQ is 1.12% occurring when submitted lots are 2.2% defective. This maximum value is the AOQL. Figure 11.5 illustrates the variation of AOQ with incoming quality.

The assumptions underlying Table 11.2 may be explained by examining one particular calculation. Consider that incoming lots come from a process that is 0.6% defective. The probability (Type B) that such lots will be accepted on the basis of the sample is 0.925, or $\frac{37}{40}$. In the long run, therefore, only 3 lots out of 40 will be screened. For 40 such lots, 37 will be passed containing, on the average, 0.6% of defectives and 3 will contain no defectives after screening. The approximate AOQ expressed in percent defective will therefore be

$$(P_a)(100p) = (0.925)(0.6\%) = 0.555\%$$

Several simplifying assumptions in the foregoing calculation are as follows:

1. The lot size N is constant. Thus the 3 screened lots may be assumed to be the same size as the 37 unscreened ones.

TABLE 11.2 AVERAGE OUTGOING QUALITY FROM $\begin{cases} N = 75 \\ c = \ \ 1 \end{cases}$ WHEN USED AS AN ACCEPTANCE/RECTIFICATION PLAN

Percent defective in submitted lots, 100*p*	Probability of acceptance, P_a	Average percent defective in accepted product, *AOQ*
0.2	0.990	0.198
0.4	0.963	0.385
0.6	0.925	0.555
0.8	0.878	0.702
1.0	0.827	0.827
1.2	0.772	0.926
1.4	0.718	1.005
1.6	0.663	1.061
1.8	0.610	1.098
2.0	0.558	1.116
2.1	0.533	1.119
2.2	0.509	1.120
2.3	0.486	1.118
2.4	0.463	1.111
2.5	0.441	1.102
3.0	0.343	1.029
3.5	0.262	0.917
4.0	0.199	0.796
4.5	0.150	0.675
5.0	0.112	0.560

2. The screening inspection finds all the defectives in the screened lots, and these defectives are removed.
3. The defective articles removed from the screened lots are replaced with good articles. Thus each screened lot contributes N good articles to the final stream of product.

FIGURE 11.5 Average outgoing quality for acceptance/rectification plan $N = 75$ and $c = 1$.

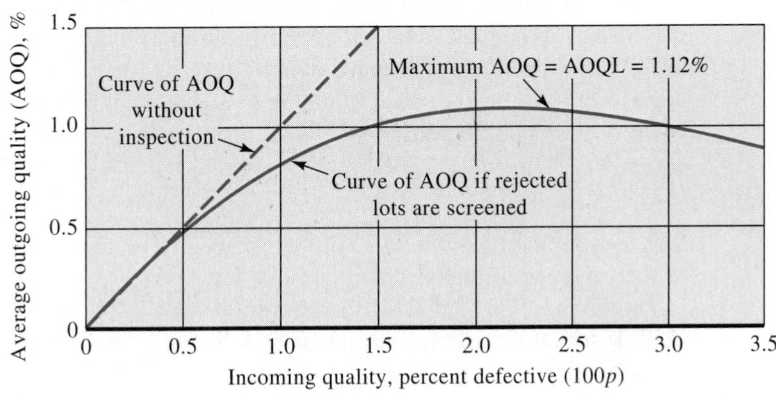

It should be emphasized that any calculation of average outgoing quality gives the expected quality *in the long run.* As its name implies, the AOQ is an *average.* Over a short period, the outgoing quality may be better or worse than the long-run average. Subject to the limitations of the three simplifying assumptions just stated, any acceptance/rectification plan guarantees that, regardless of the incoming quality submitted, the outgoing quality in the long run will not be worse than the plan's AOQL. This guarantee, however, does not apply to short periods.

11.7.1 Some Comments on the Significance of the AOQL

Sampling schemes based on stated AOQL values have gained widespread use in industry. They are used particularly in a manufacturer's inspection of its own product, both in process inspection and in final inspection. They have also been used advantageously in the inspection of lots of purchased product. If the purchaser's sampling inspection takes place in the supplier's plant, the supplier may do the screening of rejected lots and submit a certification of the screening work done. If the purchaser's sampling inspection takes place elsewhere, the purchaser may carry out the screening where necessary, and by agreement charge the cost thereof to the supplier.

AOQL plans have also proved well-adapted to many circumstances outside manufacturing, particularly to the checking of clerical work. Such "nonproduct" applications are discussed briefly in Chap. 13.

At various points in our discussion of lot-by-lot acceptance sampling, we shall have occasion to note that the continued use of a sampling scheme tends to be confined to circumstances where most of the time the quality is good enough for nearly all lots to be accepted on the basis of a sample. AOQL plans are no exception to this rule.

Where nearly all the lots are accepted on the basis of the sample, the AOQ will be only slightly better than the average quality in these accepted lots. If submitted lots are badly out of statistical control, so that the screened lots are generally much worse than the unscreened lots, the screening may effect an important improvement in quality. If the submitted product is in good statistical control and few lots are screened, the outgoing quality will not differ greatly from incoming quality. In any event, the AOQ will nearly always be considerably better than the AOQL.

Even though the simplifying assumptions used in the calculation of AOQLs may depart somewhat from actual facts, it follows that this departure is not a matter of great practical importance. Only rarely will the users of schemes based on the AOQL have need to cash in on the guarantee that outgoing quality will not be worse than the computed AOQL.

11.8 DOUBLE SAMPLING

Single sampling calls for decision on acceptance or rejection of a lot on the basis of the evidence of *one* sample from that lot.

Double sampling involves the possibility of putting off the decision on the lot until a second sample has been taken. A lot may be accepted at once if the first sample is good enough or rejected at once if the first sample is bad enough. If the first sample is neither good enough nor bad enough, the decision is based on the evidence of the first and second samples combined. In general, double sampling schemes will involve less total inspection than single sampling for any given quality protection. They also have certain psychological advantages based on the idea of giving a second chance to doubtful lots.

The additional symbols used in connection with double sampling are:

n_1	number of pieces in the first sample.
c_1	acceptance number for first sample, the maximum number of defectives that will permit the acceptance of the lot on the basis of the first sample.
n_2	number of pieces in the second sample.
$n_1 + n_2$	number of pieces in the two samples combined.
c_2	acceptance number for the two samples combined, the maximum number of defectives that will permit the acceptance of the lot on the basis of the two samples.

An example of the use of these symbols to describe a double sampling plan is

$$\begin{cases} N & 1{,}000 \\ n_1 & 36 \\ c_1 & 0 \\ n_2 & 59 \\ c_2 & 3 \end{cases}$$

This may be interpreted as follows:

1. Inspect a first sample of 36 from a lot of 1,000.
2. Accept the lot on the basis of the first sample if the sample contains 0 defectives.
3. Reject the lot on the basis of the first sample if the sample contains more than 3 defectives.
4. Inspect a second sample of 59 if the first sample contains 1, 2, or 3 defectives.
5. Accept the lot on the basis of the combined sample of 95 if the combined sample contains 3 or less defectives.
6. Reject the lot on the basis of the combined sample if the combined sample contains more than 3 defectives.

11.8.1 Analysis of a Double Sampling Plan

Figure 11.6 shows three Type *A* OC curves involved in the analysis of this double sampling plan.

There are four possibilities for acceptance or rejection of a lot submitted for double sampling, namely:

1. Acceptance after the first sample
2. Rejection after the first sample
3. Acceptance after the second sample
4. Rejection after the second sample

The lowest of the three OC curves in Fig. 11.6 shows the probability of (1) acceptance after the first sample. This is simply the curve for $\begin{cases} N = 1{,}000 \\ n = 36. \\ c = 0 \end{cases}$

FIGURE 11.6 Characteristics of a double sampling plan.

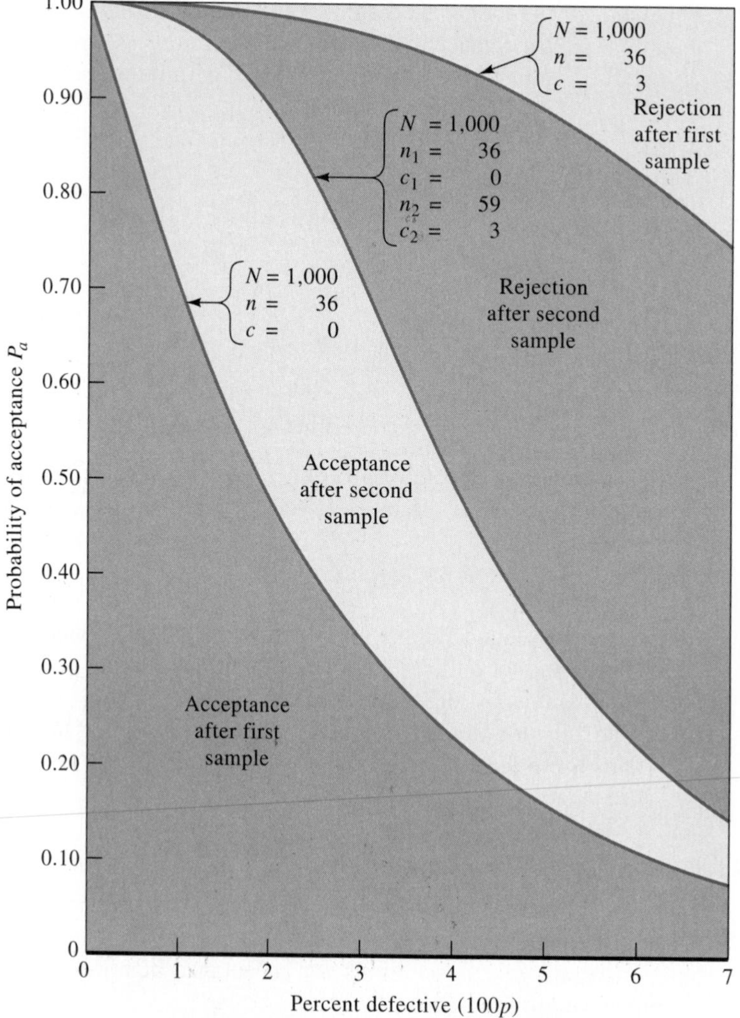

The highest of the three OC curves shows the probability that the lot will not be rejected after the first sample. It is the curve for $\begin{cases} N = 1{,}000 \\ n = 36. \\ c = 3 \end{cases}$

Both these limiting curves may be calculated in the manner illustrated in Chap. 5 for the calculation of curves for all single sampling plans. For any given value of percent defective, the ordinate distance between these two curves corresponds to the probability that for a lot of that percent defective a second sample will be required.

The middle curve of Fig. 11.6 is the actual OC curve of the double sampling plan. To compute points for this curve it is necessary to find the probability that if a second sample is taken, the lot will be accepted.

The lot may be accepted in the following ways:

0 defectives in first sample

1 defective in first sample followed by 0, 1, or 2 defectives in second sample

2 defectives in first sample followed by 0 or 1 defectives in second sample

3 defectives in first sample followed by 0 defectives in second sample

The probability of accepting the lot is the sum of the probabilities of these different ways in which it may be accepted. To compute these, it is first necessary to find the probabilities of 0, 1, 2, and 3 defectives in the first sample of 36.

If not set up and solved using a computer program, the hypergeometric distribution calculations required to obtain Type *A* OC curve probabilities are very time consuming. The method for hand calculation, which involves the use of factorials and Table *H,* was illustrated in Chap. 5. If it is assumed that the binomial is applicable to this sampling problem (in other words, that samples are drawn from an infinite lot or that a Type *B* curve is desired) and that the Poisson distribution is a satisfactory approximation to the binomial, Table *G* may be used. The necessary calculations are illustrated for a single point on the OC curve corresponding to $\mu_p = 0.010$.

In using Table *G* to compute the respective probabilities of results of the first sample, $\mu_{np} = 36(0.010) = 0.36$. Interpolating in Table *G* between the values for μ_{np} of 0.35 and 0.40, we find

$$P_0 = 0.698$$

$$P_{1 \text{ or less}} = 0.948$$

$$P_1 = 0.948 - 0.698 = 0.250$$

$$P_{2 \text{ or less}} = 0.994$$

$$P_2 = 0.994 - 0.948 = 0.046$$

$$P_{3 \text{ or less}} = 1.000$$

$$P_3 = 1.000 - 0.994 = 0.006$$

If there is one defective on the first sample, calculations regarding the second sample should be based on $\mu_{np} = (59)(\frac{9}{964}) = 0.55$. Table *G* gives directly

$$P_{2\text{ or less}} = 0.982$$

If there are two defectives on the first sample, $\mu_{np} = (59)(\frac{8}{964}) = 0.49$. Table *G* gives

$$P_{1\text{ or less}} = 0.913$$

and if there are three defectives on the first sample,

$$P_0 = 0.652$$

The probability of acceptance may now be computed. For comparison, the second set of figures shows the results of Type *A* OC curve calculations. Using the theorem of conditional probabilities:

		Type *B*	Type *A*
0 defectives in first		= 0.698	0.692
1 defective in first, with 0, 1, or 2 in second	= (0.250)(0.982)	= 0.246	0.257
2 defectives in first, with 0 or 1 in second	= (0.046)(0.913)	= 0.042	0.039
3 defectives in first, with 0 in second	= (0.006)(0.652)	= 0.004	0.003
Probability of acceptance of a lot 1.0% defective		= 0.990	0.991

In this case, the difference was negligible between the result of theoretically correct calculations (Type *A*) and the result obtained by the approximate method using the Poisson distribution and Table *G*. Although the check between exact and approximate calculations will not usually be as good as this, the OC curve obtained by the use of the Poisson distribution will ordinarily be close enough for practical purposes.

11.9 CHOOSING A SAMPLING PLAN TO MINIMIZE AVERAGE TOTAL INSPECTION

The question of minimum total inspection depends on the number of rejected lots that must be detailed (that is, 100% inspected). This, in turn, depends on the quality level of the product submitted. In analyzing and evaluating various sampling plans it is convenient to state the problem in terms of the *ATI, average total inspection,* and the *AFI, average fraction inspected.* For single sampling plans the ATI and AFI may be found from

$$\begin{aligned}\text{ATI} &= nP_a + N(1 - P_a)\\ &= n + (N - n)(1 - P_a)\end{aligned}$$

and

$$\text{AFI} = \frac{\text{ATI}}{N}$$

Use of these formulas to choose among sampling plans may be illustrated by using the three plans of Fig. 11.3 under different assumptions as to the quality level of incoming lots. Assume that a large number of lots that are 0.5% defective are submitted for acceptance by each plan. The required values are

N	1,000	1,000	1,000
n	100	170	240
c	0	1	2
P_a	0.59	0.80	0.91
nP_a	59.0	136.0	218.4
$N(1 - P_a)$	410.0	200.0	90.0
ATI	469.0	336.0	308.4
AFI	0.469	0.336	0.308

With the quality level submitted for inspection 0.5% defective, it is evident that the plan $\begin{cases} N = 1{,}000 \\ n = \ \ 240 \\ c = \ \ \ \ 2 \end{cases}$ requires the least total inspection. However, if the quality level submitted should be 0.2% defective rather than 0.5% defective, the minimum total inspection will be required using $\begin{cases} N = 1{,}000 \\ n = \ \ 170 \\ c = \ \ \ \ 1 \end{cases}$. This is shown in the following calculation:

N	1,000	1,000	1,000
n	100	170	240
c	0	1	2
P_a	0.81	0.97	1.00
nP_a	81.0	164.9	240.0
$N(1 - P_a)$	190.0	30.0	0.0
ATI	271.0	194.9	240.0
AFI	0.271	0.195	0.240

It should be noted that the amount of sampling inspection always increases with increasing n. The minimization of AFI results from a trade-off between increased sampling inspection and decreased risk of detailing. In those cases where the purchaser charges detailing costs back to the supplier, there may be strong pressure to minimize the sample size rather than the AFI. In the long run such action will prove to be uneconomical.

For a double sampling plan, the formula for average total inspection is

$$\text{ATI} = n_1 P_a(n_1) + (n_1 + n_2)P_a(n_2) + N(1 - P_a)$$
$$= n_1 P_a + n_2 P_a(n_2) + N(1 - P_a)$$

where $P_a(n_1)$ = probability of acceptance on the first sample
$P_a(n_2)$ = probability of acceptance on the second sample
$P_a = P_a(n_1) + P_a(n_2)$

Referring to the double sampling plan of Fig. 11.6 (N = 1,000, n_1 = 36, c_1 = 0, n_2 = 59, c_2 = 3), if incoming lots are 1.0% defective, the AFI is found from (see p. 448 for Type *B* probabilities of acceptance)

$$\text{ATI} = 36(0.698) + 95(0.292) + 1{,}000(1 - 0.990) = 62.87$$

$$\text{AFI} = \frac{62.87}{1{,}000} = 0.063$$

By varying the value of p for lots entering inspection from 0 to 1, the value of P_a will range from 1 to 0 in accordance with the OC curve for the individual plan. Thus the AFI will range from a minimum of n/N (or n_1/N) to a maximum of 1. Analysis of AFI values of alternative sampling plans at specific values of p permits the user to choose that plan with minimum AFI. Such calculations were used as the basis for specifying the plans contained in the Dodge-Romig tables.

The mathematical relationship between the AFI and the AOQ is

$$\text{AOQ} = p(1 - \text{AFI})$$

The approximate formula for AOQ given previously is correct for an acceptance number of 0. However, when c is greater than 0, some defective units will be found in samples from accepted lots. Naturally these defectives are removed and usually replaced with good items prior to returning the sample to the lot. This reduces the AOQ by some small amount not accounted for in the formula on p. 442. A popular way to account for this modest reduction for single sampling plans is to introduce the correction factor $(1 - n/N)$. Thus

$$\text{AOQ (in percent)} = \left(1 - \frac{n}{N}\right) P_a(100p)$$

Whereas the formula without the correction factor tends to overstate the AOQ by a small amount (except when $c = 0$), that with the correction factor tends to understate it by a small amount.

The AOQ value found from the AFI is correct in all cases for all rectifying inspection plans. When complex sampling plans are to be analyzed, this formula is used rather than direct solution for AOQ. For the previous double sampling plan, the simplest approach is to solve for the AOQ as follows:

$$AOQ = 0.01(1 - 0.063) = 0.00937$$

11.10 MULTIPLE AND SEQUENTIAL SAMPLING

Just as double sampling plans may defer the decision on acceptance or rejection until a second sample has been taken, other plans may permit any number of samples before a decision is reached. The phrase *multiple sampling* is generally used when three or more samples of a stated size are permitted and when the decision on acceptance or rejection must be reached after a stated number of samples. The phrase *sequential sampling* is generally used when a decision is possible after each item has been inspected and when there is no specified limit on the total number of units to be inspected. However, some writers use the two phrases interchangeably. For this reason, plans involving the possibility of decision after each item are referred to in this book as item-by-item sequential plans. The design of an item-by-item plan is illustrated in Chap. 13.

Usually multiple or item-by-item sequential plans can be designed to give OC curves closely similar to the OC curve of any given single or double sampling plan. The following single, double, and multiple plans illustrate a set of matched plans having nearly identical OC curves.

Type of plan	Sample number	Individual sample size	Combined sample size	Acceptance number	Rejection number
Single	1	75	75	2	3
Double	1	50	50	1	4
	2	100	150	3	4
Multiple	1	20	20	†	2
	2	20	40	0	3
	3	20	60	1	3
	4	20	80	2	4
	5	20	100	2	4
	6	20	120	2	4
	7	20	140	3	4

†Acceptance not permitted on first multiple sample.

The foregoing method of describing acceptance sampling plans differs slightly from the one used earlier in this chapter for single and double sampling. It is generally not convenient to use n and c numbers to describe a multiple plan.

Figure 11.7 shows the OC curves for these three attributes plans together with the OC curve for a matched plan using variables criteria. The calculation of the OC curve for the multiple plan follows the same pattern already explained for double sampling. The following calculations apply to a single point on the curve, namely, the probability of acceptance of a 2% defective lot. To simplify the calculation, it is assumed that the lot size is large enough so that the unsampled portion of the lot will still be substantially 2% defective regardless of the results of past samples. Under this assumption, it is not necessary to recalculate μ_{np} based on the number of defectives found in the first sample as was done in Sec. 11.8.1. Each multiple sample consists of 20 articles. To use Table *G*, it should be noted that $\mu_{np} = (20)(0.02) = 0.4$. Table *G* then gives the following useful figures applicable to any sample:

$$P_0 = 0.670$$

$$P_1 = 0.938 - 0.670 = 0.268$$

$$P_2 = 0.992 - 0.938 = 0.054$$

$$P_{3 \text{ or more}} = 1.000 - 0.992 = 0.008$$

FIGURE 11.7 OC curve for the single sampling attributes plan $n = 75$, $c = 2$, and curves for three matching sampling plans.

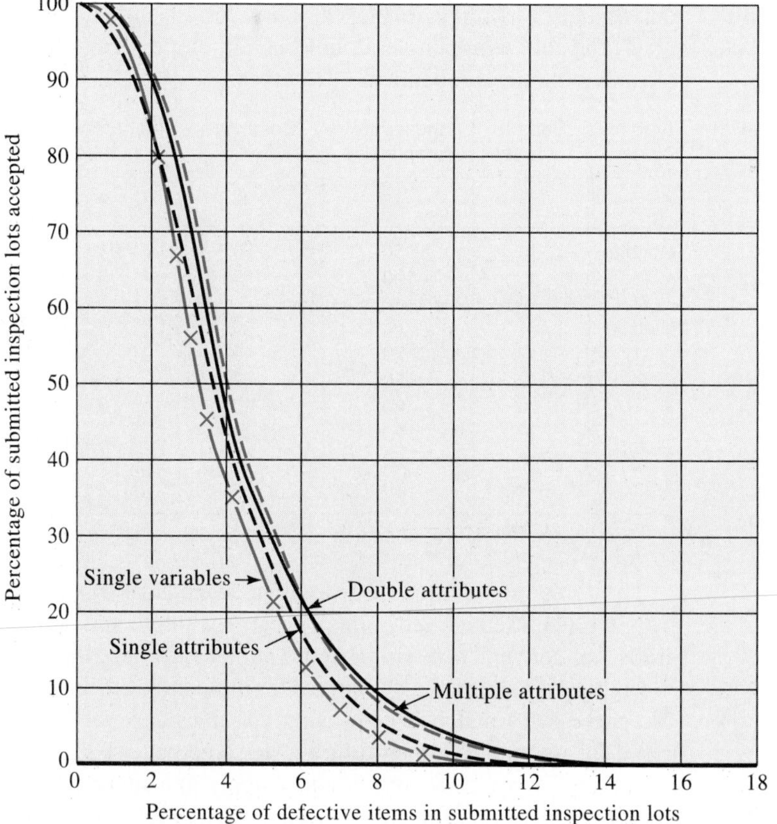

These may be combined by the theorem of conditional probabilities to obtain the following probabilities of acceptance or rejection on each sample:

	Probability of	
Sample number	Acceptance	Rejection
1	0.000	0.062
2	0.449	0.022
3	0.241	0.058
4	0.113	0.010
5	0.000	0.015
6	0.000	0.010
7	0.013	0.007
	0.816	0.184

It will be noted that for 2% defective lots, this particular multiple sampling plan will arrive at a decision with one or two samples (not more than 40 items inspected) more than half the time and will nearly always reach a decision within four samples (with not more than 80 items inspected).

Calculations are along the same general lines as those illustrated in connection with finding Type *B* OC curves in double sampling. For the first three samples, the actual calculations are as follows:

Sample 1

$P_0 = 0.670$ (Continue sampling, entering sample 2 with 0 defectives.)

$P_1 = 0.268$ (Continue sampling, entering sample 2 with 1 defective.)

$P_{2\text{ or more}} = 0.054 + 0.008 = 0.062$ (Reject.)

Note: The probability of continuing sampling is 0.670 + 0.268 = 0.938. The sum of the computed probabilities of the different possible results in sample 2 must be equal to this figure.

Sample 2

$P_{0\text{–}0} = (0.670)(0.670) = 0.449$ (Accept.)

$P_{0\text{–}1} = (0.670)(0.268) = 0.1795$

$P_{1\text{–}0} = (0.268)(0.670) = 0.1795$

0.359 (Continue sampling, entering sample 3 with 1 defective.)

$P_{0\text{–}2} = (0.670)(0.054) = 0.036$

$P_{1\text{–}1} = (0.268)(0.268) = 0.072$

0.108 (Continue sampling, entering sample 3 with 2 defectives.)

$P_{0\text{–}3 \text{ or more}} = (0.670)(0.008) = 0.005$
$P_{1\text{–}2 \text{ or more}} = (0.268)(0.062) = \underline{0.017}$
0.022 (Reject.)
Check: $0.449 + 0.359 + 0.108 + 0.022 = 0.938$

Note: The probability of continuing sampling is 0.359 + 0.108 = 0.467. The sum of the computed probabilities of the different possible results in sample 3 must be equal to this figure.

Sample 3
$P_{1\text{–}0} = (0.359)(0.670) = 0.241$ (Accept.)

$P_{1\text{–}1} = (0.359)(0.268) = 0.096$
$P_{2\text{–}0} = (0.108)(0.670) = \underline{0.072}$
0.168 (Continue sampling, entering sample 4 with 2 defectives.)

$P_{1\text{–}2 \text{ or more}} = (0.359)(0.062) = 0.022$
$P_{2\text{–}1 \text{ or more}} = (0.108)(0.330) = \underline{0.036}$
0.058 (Reject.)
Check: $0.241 + 0.168 + 0.058 = 0.467$

It should be clear to the reader that the only reasonable approach to the calculation of probabilities of acceptance and OC curves for multiple sampling plans is through the use of computer programs designed for the purpose.

The topic of the difference in average amount of sampling inspection under single, double, and multiple sampling is discussed in Chap. 12, which also considers other relative advantages and disadvantages of these competing types of sampling.

An explanation of the calculation of the OC curve of the matching variables plan shown in Fig. 11.7 is deferred until Chap. 15.

11.10.1 Rejection Number for First Sample in Double Sampling

The reader will observe that in the double sampling plans discussed in this chapter, the rejection number is the same for the first sample as for the combined samples. This is in contrast to the multiple sampling plan just explained, which had rejection numbers of 2, 3, or 4, depending on the sample number. We shall see that in nearly all multiple sampling plans, two or more rejection numbers are used.

Administrative simplicity is gained by the policy of having only one rejection number for any given double sampling plan. Before 1963, this policy was adopted for the various tables of double sampling plans that were in popular use.

At some sacrifice of administrative simplicity, the average amount of sampling inspection in double sampling can often be reduced for a given quality protection by making the rejection number for the first sample less than the rejection number for the combined samples. Two such rejection numbers are used in most of the double sampling plans in ABC-STD-105, an international standard for acceptance sampling by attributes adopted in 1963. This ABC standard is discussed in Chap. 12.

11.11 RANDOMNESS IN ACCEPTANCE SAMPLING

The probability calculations used to compute OC curves assume that samples are drawn at random. That is, each item in the lot is assumed to have an equal chance to be selected in the sample. If the items in a lot have been thoroughly mixed, a sample chosen anywhere in the lot meets the requirement of randomness. However, a common condition is that there is no reason to believe that the items have had a thorough mixing. Moreover, it may be impracticable to carry out a thorough mixing because of physical difficulties or excessive costs incident to mixing or for other reasons. Sometimes the best that can be done in drawing a sample is to avoid any obvious type of bias. For instance, if items are packed in layers, it clearly is not sensible to draw the entire sample from the top layers.

If it is practicable to assign a different number to each item in a lot and to draw an item from any place in the lot, a formal scheme for drawing a random sample may be adopted. Such a scheme may use a table of random numbers or some mechanical device for generating random numbers as needed. Computer-generated random numbers may be used so long as a different "seed" number is used each time. Otherwise the same set of pseudo-random numbers will be generated each time the program is run.

11.11.1 Use of Random Numbers to Select a Sample

Random digits can be generated in any way that gives each digit from 0 to 9 an equal chance to be selected. Extensive tables of random digits can be generated rapidly by modern electronic computers. Table *X* in App. 3 contains 2,500 random digits reproduced by permission from an extensive table prepared by the RAND Corporation.

Let us illustrate the use of a table of random digits by a numerical example based on Table *X*. Assume that it is desired to select a sample of 15 from a lot of 750 items. Each item in the lot is identified by a number from 1 to 750. Therefore it is necessary to select 15 random three-digit numbers from 001 to 750.

First it is necessary to determine a starting point in the table. Table *X* contains 50 rows and 50 columns of digits listed in pairs. Assume that a pencil point is placed at random in the table, and the first two-digit number to the right from 1 to 50 determines the row to be selected. The procedure is repeated to determine the starting column. Assume that the 8th row and the 39th column are chosen. Assume that the decision was made in advance that the starting digit chosen will be the first digit of a three-digit number read to the right and that succeeding numbers will be read down the table. The following numbers are obtained. (Numbers that are not between 001 and 750 must be discarded; these are shown in parentheses. A number that has already occurred must also be discarded; there are no such numbers in this example.)

471, 098, 443, 335, 015, 106, (932), 682, (864), 531, 379, (909), 225, 233, 404, (812), 392, (820), (934), 183, (929), 592

The selection of the sample is simplified if the 15 numbers between 001 and 750 are rearranged in order of increasing size as follows:

015, 098, 106, 183, 225, 233, 335, 379, 392, 404, 443, 471, 531, 592, 682

If samplers wish to generate their own random numbers, a convenient method is to use a set of Japanese random dice. Each die is a regular icosahedron. Each digit from 0 to 9 is engraved on two faces so that the 10 digits occur with equal probability. The dice come in sets of three colors. When the dice are used to generate three-digit random numbers, one color may be assigned to each digit.

In cases where it is deemed imperative to use a formal scheme to ensure a random sample, there may be a psychological advantage in having the person who selects the sample use random dice or some other mechanical device. The use of a device that obviously involves the laws of chance may add interest to the job of selecting the sample. Because of this interest, there may be less tendency for the sampler to disregard a stipulated formal set of rules for randomization.

11.11.2 Tables of Random Permutations

In the use of conventional tables of random numbers such as Table *X* or of Japanese random dice, an additional operation is caused in identifying and discarding numbers that occur more than once. This operation is eliminated by the use of a table of random permutations. (Such tables are also useful in design of experiments and in other matters unrelated to acceptance sampling.) An extensive set of such tables has been prepared by L. E. Moses and R. V. Oakford.[†]

11.11.3 Stratified Sampling

In large lots, the difficulties of random selection may be so great that it is advisable to adopt stratified (proportional) sampling. In the volume *Sampling Inspection,* the Statistical Research Group of Columbia University made the following suggestions for this type of sampling:[‡]

1. Draw proportional samples. According to this rule, inspection lots should, wherever possible, be divided into sub-lots on the basis of factors that are likely to lead to variation in the quality of the product. . . . From each sub-lot into which the inspection lot is divided a subsample should be selected. The size of the subsample from each sub-lot should be proportional to the size of the sub-lot.

[†]L. E. Moses and R. V. Oakford, *Tables of Random Permutations,* Stanford University Press, Stanford, Calif., 1963. This 233-page volume contains many random permutations of 9, 16, 20, 30, 50, 100, 200, 500, and 1,000 integers.

[‡]H. A. Freeman, Milton Friedman, Frederick Mosteller, and W. A. Wallis (eds.), *Sampling Inspection,* pp. 48–52, McGraw-Hill Book Company, New York, 1948.

2. Draw sample items from all parts of each sub-lot of the inspection lot.
3. Draw sample items blind.

Such a stratified sample is, of course, different from a random sample. In taking many such stratified samples from a given lot, the average number of defectives will be the same as if samples were drawn at random, but the variation in number of defectives from sample to sample will be less. The result is that the OC curve for any given acceptance criteria will be steeper under stratified sampling than under random sampling. Lots substantially better than those at $p_{0.50}$ will have a somewhat greater chance of acceptance than indicated by the OC curve for random sampling; lots substantially worse than those at $p_{0.50}$ will have a somewhat smaller chance of acceptance. OC curves under stratified sampling cannot be calculated without making an assumption as to the variation of quality from sublot to sublot. Investigations by the Statistical Research Group indicate that for practical situations likely to arise in industrial sampling, the use of stratified samples will seldom make much change from the OC curve computed for random sampling.

PROBLEMS

Many of the following problems request calculation based on the Poisson approximation and the use of Table G for their solution. Others request the use of a specified distribution that is correct in principle. If you have been using computer programs in solving for probabilities based on using the hypergeometric, binomial, or Poisson approximation, you may continue to do so. Where feasible, you might wish to make calculations by both approximate and correct methods even if not called for in the problem. Answers given in the text were found by using the distribution stated in the problem.

11.1. A single sampling plan uses a sample size of 15, and an acceptance number of 1. Using hypergeometric probabilities, compute the respective probabilities of acceptance of lots of 50 articles 2, 6, 10, and 20% defective.
Answer: 1.000; 0.789; 0.524; 0.121.

11.2. The single sampling plan of Prob. 11.1 is used with a relatively large lot. Use Table G to compute the approximate probabilities of acceptance of lots 2, 6, 10, and 20% defective.
Answer: 0.963; 0.772; 0.558; 0.199.

11.3. A single sampling plan has $n = 110$ and $c = 3$. The lot size is large in comparison with sample size. Use Table G to compute the approximate probabilities of acceptance lots 0.5, 1, 2, 3, 4, 5, 6, and 8% defective.
Answer: 0.998; 0.974; 0.819; 0.580; 0.359; 0.202; 0.105; 0.025.

11.4. Plot the OC curve for the sampling plan of Prob. 11.3. What are the approximate values of lot percent defective for which probabilities of acceptance are 0.95, 0.50, and 0.10, respectively?
Answer: 1.2%; 3.3%; 6.1%.

11.5. A double sampling plan is as follows:
(*a*) Select a sample of 2 from a lot of 20. If both articles inspected are good, accept the lot. If both are defective, reject the lot. If 1 is good and 1 defective, take a second sample of one article.

(*b*) If the article in the second sample is good, accept the lot. If it is defective, reject the lot. If a lot 25% defective is submitted, what is the probability of acceptance? Compute this by the method that is theoretically correct rather than by an approximate method.
Answer: 0.860.

11.6. A multiple sampling plan is as follows:

Sample number	Individual sample size	Combined sample size	Acceptance number	Rejection number
1	5	5	†	2
2	5	10	0	2
3	5	15	0	3
4	5	20	1	3
5	5	25	2	3

†Acceptance not permitted on first sample.

Assuming that lot size is large enough for Table *G* to be applicable, compute the probability of acceptance of a 10% defective lot.
Answer: 0.586.

11.7. A double sampling plan is $n_1 = 25$, $c_1 = 1$, $n_2 = 50$, $c_2 = 3$. Compute the probability of acceptance of a 4.0% defective lot. Assume lot size is large in comparison with sample size.
Answer: 0.819.

11.8. A double sampling plan is $n_1 = 150$, $c_1 = 2$, $n_2 = 300$, $c_2 = 4$. Compute the probability of acceptance of a 1.5% defective lot. Assume lot size is large in comparison with sample size.
Answer: 0.623.

11.9. The following multiple sampling plan involves a maximum of 7 samples of 40 each:

Sample number	Combined sample size	Acceptance number	Rejection number
1	40	†	2
2	80	†	2
3	120	0	2
4	160	0	3
5	200	1	3
6	240	2	4
7	280	4	5

†Acceptance not permitted on first or second sample.

Compute the probability of acceptance of a 1.0% defective lot. Assume lot size is larger in comparison with sample size.
Answer: 0.582.

11.10. A single sampling plan uses a sample size of 8 and an acceptance number of zero. Using the correct hypergeometric probabilities, compute the probabilities of acceptance of lots of 50 articles 2, 6, 10, and 20% defective, respectively.
Answer: 0.840, 0.586, 0.401, 0.143.

11.11. The single sampling plan of Prob. 11.10 is used with relatively large lots. Use Table *G* to compute the approximate probabilities of acceptance with lots 2, 6, 10, and 20% defective, respectively.
Answer: 0.853, 0.619, 0.449, 0.202.

11.12. A single sampling plan has $n = 50$ and $c = 1$. The lot size is 500 items. Using hypergeometric probabilities, compute the respective probabilities of acceptance of lots 0.4, 2, 4, and 6% defective, respectively.

11.13. The single sampling plan of Prob. 11.12 is used with relatively large lots. Use Table *G* to compute the approximate probabilities of acceptance of lots 0.4, 2, 4, and 6% defective, respectively.

11.14. For the single sampling plan $n = 150$, $c = 2$, what are the values of $p_{0.95}$, $p_{0.50}$, and $p_{0.10}$? See Table 11.1 for probabilities of acceptance, and interpolate to obtain the values of p.

11.15. For the single sampling plan $n = 75$, $c = 1$, what are the values of $p_{0.95}$, $p_{0.50}$, and $p_{0.10}$? See Table 11.1 for probabilities of acceptance, and interpolate to obtain the values of p.

11.16. Consider the single sampling plan $n = 75$, $c = 1$. Table 11.1 gives the probability of acceptance of a 2% defective lot as 0.558. This computation assumes a very large lot. Using the correct hypergeometric formula, compute the probability of acceptance of a 2% defective lot when $N = 400$.

11.17. Assuming that rejected lots will be screened, compute the AOQ values and AFIs for the results in Prob. 11.1. Sketch the AOQ and AFI functions, and estimate the value of the AOQL. How would these results have been affected if the probability calculations were made using the Poisson approximation, as in Prob. 11.2, rather than the correct hypergeometric formula?

11.18. Assuming that rejected lots will be screened, compute the AOQ values and AFIs for the results in Prob. 11.10. Sketch the AOQ and AFI functions, and estimate the value of the AOQL. How would these results have been affected if the probability calculations were made using the Poisson approximation, as in Prob. 11.11, rather than the correct hypergeometric formula?

11.19. Assuming that rejected lots will be screened, compute the AOQ and AFI values for the results in Prob. 11.8. The lot size is 2,000.

11.20. Assuming that rejected lots will be screened, compute the AOQ values and AFIs for the results in Prob. 11.12. Sketch the AOQ and AFI functions, and estimate the value of the AOQL. How would these results have been affected if the probability calculations were made using the Poisson approximation, as in Prob. 11.13, rather than the correct hypergeometric formula?

11.21. In a double sampling plan, $N = 40$, $n_1 = 5$, $c_1 = 0$, $n_2 = 5$, $c_2 = 1$. Using the correct hypergeometric formula, compute the probability that a lot exactly 10% defective will be accepted by this plan.

11.22. The double sampling plan in Prob. 11.21 is used for relatively large lots. Use Table *G* to compute the approximate probability of acceptance of lots exactly 10% defective.

11.23. Consider the double sampling plan $n_1 = 100$, $c_1 = 0$, $n_2 = 200$, $c_2 = 2$. Use Table *G* to compute the probability of acceptance of 1.5% defective lots assuming the lot size is large in comparison to the sample size.

11.24. A random sample of 10 is to be selected from a lot of 500 articles. Each article in the lot has been assigned a number. Table *X* of App. 3 is to be used to select 10 three-digit numbers between 001 and 500; these numbers will determine the articles to be chosen as the sample. A random choice from Table *X* results in a first number of 288 taken from the 13th, 14th, and 15th digits on line 13. The rule established is to read down the table directly below these digits. Select the required 10 three-digit numbers (of course eliminating all numbers above 500), and arrange them in order of increasing magnitude.

11.25. An electrical manufacturer receives components from three different suppliers. At present, the user accepts lots based on its own sampling plan. The policy is to draw a random sample equal to 10% of the lot size with a lot being rejected if one or more defective components are found. Suppliers *A, B,* and *C* submit components in lots of 100, 250, and 1,000, respectively.

(*a*) Use Table *G* to determine for each supplier the lot fractions defective for which 95, 50, and 10% of the lots will be accepted.

(*b*) Do you consider the user's sampling plan fair? Why or why not?

11.26. A double sampling plan calls for a first sample of 25 items to be inspected. If no defectives are found, the lot is accepted; if two or more defectives are found, it is rejected. Otherwise, a second sample of 50 items is drawn, and the lot is accepted if the combined number of defectives found does not exceed three. Assuming the lot size is large in comparison to the sample size, use Table *G* to find the following:

(*a*) the Producer's Risk (α) at a fraction defective of 0.015.

(*b*) the Consumer's Risk (β) at a fraction defective of 0.10.

11.27. A multiple sampling plan is as follows:

Sample number	Individual sample size	Combined sample size	Acceptance number	Rejection number
1	50	50	†	3
2	50	100	1	3
3	50	150	2	4
4	50	200	3	5
5	50	250	5	6

†Acceptance not permitted on first sample.

Use Table *G* to calculate the approximate probability of acceptance of 1.5% defective lots. Assume the lot size is large in comparison to the combined sample size.

11.28. In the sampling plan described in Prob. 11.27, compute the probabilities of acceptance and rejection on the first two samples and the probability of taking the third sample.

11.29. In the sampling plan described in Prob. 11.27, compute the probabilities of acceptance and rejection on the first three samples and the probability of taking the fourth sample.

11.30. In the sampling plan described in Prob. 11.27, compute the probability of acceptance on the first four samples and the probability of taking the fifth sample.

11.31. Assuming that rejected lots will be screened, compute the AOQ and AFI values for the results in Prob. 11.27. The lot size is 5,000.

11.32. A producer of electronic components for the automobile industry uses the following double sampling plan to inspect batches of incoming integrated circuits:

$$n_1 = 20 \qquad c_1 = 0; \qquad n_2 = 40 \qquad c_2 = 3$$

Use Table *G* to compute the approximate probabilities requested in the following questions. Batches contain 3% rejectable units. Assume the lot size is large in relation to the sample size.

(*a*) What is the probability of acceptance on the first sample?
(*b*) What are the probabilities of acceptance on the second sample and overall?
(*c*) What is the probability of taking the second sample?

11.33. A double sampling acceptance plan specifies $n_1 = 100$, $n_2 = 200$; $c_1 = 1$, $c_2 = 3$. In answering the following questions, assume that incoming lots contain 1.5% nonconforming units and that the lot size is large in relation to the sample sizes so that Table *G* may be used to estimate the probabilities.

(*a*) What is the probability of acceptance on the first sample?
(*b*) What is the probability of having to draw the second sample?
(*c*) What is the probability of acceptance on the second sample and the total P_a?

11.34. Assuming that rejected lots will be screened, compute the AOQ and AFI values for the results in Prob. 11.32. The lot size is 400.

11.35. Assuming that rejected lots will be screened, compute the AOQ and AFI values for the results in Prob. 11.33. The lot size is 3,000.

11.36. For a double sampling plan, $n_1 = 100$, $c_1 = 0$, $n_2 = 100$, and $c_2 = 2$. Assuming that the lot size is large in relation to the sample size and that μ_p is 0.010, use Table *G* to compute the approximate probabilities that:

(*a*) The lot will be accepted on the first sample.
(*b*) The second sample will be taken.
(*c*) The lot will be accepted.

11.37. For a double sampling plan, $n_1 = 40$, $c_1 = 1$, $r_1 = 4$; $n_2 = 60$, $c_2 = 5$, $r_2 = 6$. Use Table *G*, and assume that the lot size is large in relation to the combined sample sizes to find the following approximate probabilities. The value of μ_p is assumed to be 0.05.

(*a*) The probability of acceptance on the first sample.
(*b*) The probability of acceptance on the second sample.
(*c*) The probability that the second sample will have to be drawn.
(*d*) The total probability of acceptance P_a.

11.38. A company specifies that, at receiving inspection, a random sample equal to 10% of the lot size will be drawn and inspected. If one or more units are rejected, the remainder of the lot must be screened (100% inspected) of unacceptable product. The cost of screening is charged to the supplier. Supplier *A* ships in lots of 500 units; supplier *B*, in lots of 1,000 units.

(*a*) What is the probability of acceptance for each supplier if the lots contain 0.3% nonconforming product? (Use Table *G* to find the approximate probabilities.)

(*b*) Do you think the company's sampling plan is fair to its suppliers? Why or why not?

11.39. To illustrate the difference in results obtained from the inspection of isolated lots if the binomial approximation is used to calculate probabilities of acceptance rather than the correct hypergeometric, evaluate the single sampling plan $n = 20$, $c = 0$, for a lot size of 100 units.

(*a*) What is the correct hypergeometric probability of acceptance if $\mu_p = 0.01$ (that is, $D = 1$)?

(*b*) Use the binomial to approximate this probability of acceptance.

(*c*) Solve parts (*a*) and (*b*) for a lot size of 500, and comment on the difference in results for the two lot sizes.

11.40. Power-supply units are purchased under contract from a supplier for use in a special-purpose computer. Units are received in lots of 100 on a weekly basis and are subject to receiving inspection using the single sampling attributes plan $n = 20$, $c = 0$. The contract with the supplier provides for statistical process control by the supplier at a "guaranteed" quality level of 0.5% nonconforming units or less.

(*a*) What is the probability of acceptance of these lots if the incoming μ_p is 0.5%?

(*b*) Results of sampling inspection are plotted on a control chart for *np* using a target value of p_o of 0.005. Calculate the central line and control limits for this chart.

(*c*) Of the last 30 lots received and inspected, 6 were rejected. In 4 of these lots, 1 defective unit was found in the sample. In the other 2 lots, 2 were found in 1 and 3 in the other. What was the average rate nonconforming in the samples ($\overline{p}$)? Comment on whether or not you believe the supplier is meeting contractual requirements.

12

AN AQL SYSTEM FOR LOT-BY-LOT ACCEPTANCE SAMPLING BY ATTRIBUTES

The question is sometimes asked as to why it is necessary to have so many tables. . . . In industry, acceptable quality levels vary and the necessary risks of wrong decisions vary. For each combination of acceptable quality level and risk, a different plan is required; hence the need for many tables. Series of tables that cover nearly all ordinary requirements have been computed and published.

—A. C. Richmond[†]

12.1 SELECTING AN ACCEPTANCE INSPECTION PROCEDURE

The choice among various possible types of acceptance inspection procedures is essentially an economic one. In making a decision regarding acceptance inspection for any particular purpose, it may be desirable to consider not only various possible systems or procedures of acceptance sampling by attributes but also the alternatives of (1) no inspection at all, but imposition of a requirement that statistical evidence of satisfactory quality be provided with each lot; (2) 100% inspection; and (3) possibilities of acceptance sampling by variables. It also is true that a satisfactory evaluation of all the pertinent economic factors is often quite difficult. For this reason, the choice of an acceptance procedure is commonly made on an intuitive basis.

[†]A. C. Richmond, "Acceptance Inspection," Paper No. 8, Fourth National Convention of American Society for Quality Control, June 1950.

An important element of the selection of an acceptance inspection procedure should be the probable contribution of the procedure to quality improvement. The acceptance sampling systems and procedures described in this and the following chapters have often been strikingly successful in leading to such improvements.

12.1.1 Some Further Comments on Indexing of Sampling Systems

It was pointed out in Chap. 11 that three points on the OC curve have been used as methods of indexing sets of sampling plans. These, and other statistical measures of performance, provide the logical basis for combining sets of plans into sampling systems. For example, certain Dodge-Romig tables, described in Chap. 13, employ the $p_{0.10}$ point as the index to two sets of tables. The point $100p_{0.10}$ is termed the *Lot Tolerance Percent Defective* (LTPD), the percent defective that has a probability of acceptance of 0.10.[†] Thus the two sets of tables are indexed on Consumer's Risk β, where β equals 0.10. Other procedures employ the *point-of-control* or *indifference quality* $p_{0.50}$ as the index base.

The system of sampling plans most frequently used in the United States is that of the ABC standard, also known as MIL-STD-105 and ANSI/ASQC Standard Z1.4. That system is indexed on the AQL (acceptable quality level). While not an exact point on the OC curve in terms of Producer Risk, lots submitted at the AQL have a risk of rejection α in the range 0.01 to 0.10.

Selected points on the OC curve need not be the only indexes of sampling plans. Another such index is the *Average Outgoing Quality Limit* (AOQL). Two of the four sets of Dodge-Romig tables described in Chap. 13 are indexed on the AOQL. All sampling plans in the Dodge-Romig tables aim at minimizing the *average total inspection* (ATI) considering both sampling inspection and screening inspection of rejected lots.

12.2 A HISTORICAL NOTE REGARDING ACCEPTANCE SAMPLING SYSTEMS BASED ON THE AQL CONCEPT

The AQL concept was first devised in connection with the development of statistical acceptance sampling for the Ordnance Department of the U.S. Army. The Ordnance tables and procedures were developed in 1942 by a group under the direction of distinguished engineers from the Bell Telephone Laboratories. With some changes and extensions, these became the Army Service Forces tables developed by the same group.[‡] These tables permitted single and double sampling, with

[†]The Dodge-Romig tables refer to this point as p_t rather than $100p_{0.10}$ as used in this text. In order to maintain consistency in this text, all references to p_t have been changed to $100p_{0.10}$ in the tables and discussions.

[‡]G. D. Edwards (then director of quality assurance at Bell Telephone Laboratories), H. F. Dodge, H. G. Romig, and G. R. Gause (then of Army Ordnance) all played an important part in developing these tables.

double sampling preferred wherever practicable. Several of the Army Service Forces tables were included in the first edition of this book.

Statistical sampling tables and procedures developed for the Navy by the Statistical Research Group of Columbia University were first issued in 1945. The general pattern of these tables and procedures was similar to that used by the Army Service Forces. However, multiple sampling schemes were made available, and there were other important points of difference. After the unification of the armed services, these Navy tables were adopted by the Department of Defense early in 1949 as JAN (Joint Army Navy) Standard 105. The tables were made available for public use through the publication of the SRG's volume "Sampling Inspection" mentioned in Chaps. 13 and 15.

MIL-STD-105A superseded JAN-STD-105 in 1950. Although the underlying pattern was similar to the preceding standards, there were again many important changes in detail.[†] The second edition of this book included the master tables from MIL-STD-105A. Only minor changes from 105A were involved in MIL-STD-105B, adopted by the U.S. Department of Defense in 1958, and in MIL-STD-105C, adopted in 1961.

During the years 1960 to 1962, a committee known as the ABC[‡] Working Group—from the military agencies of the United States of America, Great Britain, and Canada—was engaged in developing a common standard for acceptance sampling by attributes to be used by the three countries.[§] The committee worked with the assistance and cooperation of the American and European organizations for quality control. Because this standard may be viewed as somewhat of an evolution from the earlier systems based on the AQL concept, it is used in this chapter to illustrate various aspects of such systems. Tables *K* to *W* in App. 3 are taken from this standard. In the United States, this standard was adopted in 1963 and was designated as MIL-STD-105D. The international designation was ABC-STD-105 until the International Standards Organization changed it to ISO 2859 in 1974. It was adopted for commercial purposes in the United States by the American National Standards Institute in 1971 and was designated ANSI/ASQC Z1.4. Certain additions and changes were incorporated into Z1.4 in 1981. These changes are discussed at various points in this chapter.

Revision E of MIL-STD-105 was issued by the U.S. Department of Defense in 1989. While there were many changes in the ordering of the procedures, only a few substantive changes were made, and no changes were made to the tables developed for the ABC standard and carried over into 105D and Z1.4. The substantive changes are described at appropriate points in the discussion. In order to

[†]For a statement of the main points of difference among the earlier military acceptance procedures, see S. J. Lorber and E. L. Grant, "A Comparison of Military Standard 105A with the 1944 Army Service Forces Sampling Procedures and Tables and with JAN Standard 105," *Industrial Quality Control,* vol. 8, no. 1, pp. 27–29, July 1951.

[‡]For America, Britain, Canada.

[§]The members of the ABC Working Group were G. J. Keefe (chairman), Omberto Cocca, I. D. Hill, Paul Martel, W. G. Milne, and William Pabst, Jr.

stress the importance of this standard and its international character, it is referred to throughout this chapter as the ABC standard (from the name of the working group that prepared it).

It is not merely the use of AQL systems in purchases by governmental organizations that makes such systems important. Systems based on the AQL have been widely adopted by private industry for acceptance sampling of all kinds of products. In most cases, these systems were adapted from one of the military systems.

12.3 SOME DECISIONS MADE IN THE ORIGINAL ESTABLISHMENT OF THE AQL AS A QUALITY STANDARD

The persons who developed the original Army Ordnance procedures made a number of decisions that have remained practically unchanged in the many later systems based on the AQL concept. Some of these decisions were as follows:

1. In order to establish acceptance criteria for any particular quality characteristic of a product, it is first necessary to prescribe a maximum percent defective that, *for acceptance sampling purposes only,* is considered acceptable as a process average. This acceptable quality level usually has been abbreviated AQL.[†]
2. In the absence of unsatisfactory quality history or other reasons for misgivings about the quality of submitted product, the acceptance criteria should be selected with the objective of protecting the producer against the rejection of submitted lots from a process that is at the AQL value or better.
3. Such acceptance criteria generally give the consumer unsatisfactory protection against accepting lots that are moderately worse (sometimes considerably worse) than the AQL. For this reason, more severe acceptance criteria designed to protect the consumer must be used whenever the quality history is unsatisfactory or when there are other good reasons for being suspicious about quality. This concept of *tightened inspection* as an alternative to *normal inspection* is at the heart of all acceptance sampling systems based on the AQL. It is an essential part of any acceptance/rejection procedures where the acceptance criteria are chosen to protect the producer under "normal" conditions.
4. The acceptance criteria for serious defects should be more severe than for trivial defects. In other words, relatively low AQL values should be used for those types of defects that would have serious consequences and relatively high AQL values for those defects that are of little importance. The provision for a *classification of defects* is an essential feature of systems based on the AQL.

[†]The similarity between the initials AQL and AOQL sometimes is a source of confusion. Only the *Q* in these sets of initials stands for the same word, namely, *quality.* In AQL, *A* means *acceptable* and *L* means *level.* In AOQL, *A* means *average* and *L* means *limit.*

5. Economies for the consumer can be realized by permitting *reduced inspection* when the quality history is good enough. This permits the concentration of inspection effort on those products where attention seems to be needed most.
6. In establishing the relationship between lot size and sample size, weight should be given to the greater difficulty of obtaining random samples from large lots and the more serious consequences of a wrong decision on acceptance or rejection of a large lot. For this reason, the relationship between lot size and sample size is based more on empirical grounds than on considerations arising from the mathematics of probability.

12.3.1 Definitions of AQL in the Various Military Standards

AQL is defined as follows in the ABC standard: "The AQL is the maximum percent defective (or the maximum number of defects per hundred units) that, for purposes of sampling inspection, can be considered satisfactory as a process average."

With the addition of the reference to defects per hundred units, this is consistent with the definition given in the original Army Ordnance tables in 1942. It also is consistent with the definition used in the standards of the American Society for Quality Control.[†]

However, other definitions of AQL have also been used. JAN-STD-105 defined AQL as: "Percentage of defective items in an inspection lot such that the sampling plan will result in the acceptance of 95% of submitted inspection lots containing that percentage of defective items." Substantially the same definition was used in the SRG volume "Sampling Inspection."

MIL-STD-105A and 105B contained the following definition: "The acceptable quality level (AQL) is a nominal value expressed in terms of percent defective or defects per hundred units, whichever is applicable, specified for a given group of defects of a product." A similar but slightly different definition appeared in MIL-STD-105C.

The original definition, readopted for the ABC standard, has carried through to the commercial counterpart, ANSI/ASQC Z1.4, and the 1989 version of 105 (105E) with only slight modification.[‡] It is superior because it does the best job of making clear what is implied when an AQL is selected in any AQL system. Much of the criticism leveled against the AQL concept might have been avoided if this definition had been adequately understood. The key words are "for purposes of sampling inspection." They are intended to emphasize that the choice of an AQL is a necessary evil that exists in any sampling inspection; there really is no such thing as an acceptable level of production of defective material. All versions of the

[†]"ANSI/ASQC Standard A2-1987: Terms, Symbols, and Definitions for Acceptance Sampling," American Society for Quality Control, Milwaukee, Wis.

[‡]Actually, the MIL-STD-105E definition stresses two ideas. One is that the system is to be applied to a continuous series of lots, hopefully coming from a process held in statistical control; the other is that the AQL represents an upper limit on what could be considered a satisfactory process average for sampling purposes.

standard contain a "limitation" statement acknowledging that the choice of an AQL does not imply the right to supply defective material.

12.4 SOME ASPECTS OF THE MASTER TABLES REPRODUCED FROM THE ABC STANDARD

Tables *L* to *T* in App. 3 give sample sizes and acceptance and rejection numbers. To enter any one of these tables, it is necessary to know the AQL and sample size code letter. To determine which table should be used, it is necessary to know whether single, double, or multiple sampling is to be used and to know whether normal, tightened, or reduced inspection is to be used.

The AQL values in the standard may be interpreted either as percent defective or as defects per hundred units depending on whether acceptance criteria are to be based on the number of *defectives* observed in a sample or on the number of *defects.* However, AQL values above 10.0 are interpreted as applying to defects per hundred units. Most of our discussion of the standard will assume the more common acceptance criteria based on numbers of defectives with AQL specified as a percent defective.

All the AQL values are multiples of the numbers 1, 1.5, 2.5, 4.0, and 6.5. These numbers are roughly in a geometrical progression ($10^{0.2}$) and correspond to systems of "preferred numbers" in common use for other industrial purposes.

12.5 DETERMINING THE SAMPLE SIZE CODE LETTER

Table *K,* reproduced from the ABC standard, gives the relationship between lot or batch size and the code letter that determines the sample size. The "general inspection levels" on the right-hand side of the table are the ones to use in most cases. The standard states: "Unless otherwise specified, Inspection Level II will be used. However, Inspection Level I may be specified when less discrimination is needed, or Level III may be specified for greater discrimination."

The four special levels, S-1 to S-4 at the left-hand side of the table, are included for the special case where relatively small sample sizes are necessary and large sampling risks can or must be tolerated.

12.5.1 Selecting a Sampling Plan for Normal Inspection

Assume that an AQL of 1.5% has been specified for a certain class of defects. Assume normal inspection with a lot size of 1,000 and inspection level II. Table *K* indicates that the sample size code letter is J.

Table *L* gives acceptance criteria with normal inspection, single sampling. It tells us that for code letter J and a 1.5% AQL, sample size is 80 and the acceptance number is 3. The rejection number is stated as 4. In all these single sampling plans, the rejection number is one more than the acceptance number.

Table *O* gives acceptance criteria with normal inspection, double sampling. Table *R* gives criteria with normal inspection, multiple sampling. For code letter J and an AQL of 1.5%, these are:

Sample number	Sample size	Cumulative sample size	Acceptance number	Rejection number
Double:				
First	50	50	1	4
Second	50	100	4	5
Multiple:				
First	20	20	†	3
Second	20	40	0	3
Third	20	60	1	4
Fourth	20	80	2	5
Fifth	20	100	3	6
Sixth	20	120	4	6
Seventh	20	140	6	7

†Acceptance not permitted at this sample size.

The standard gives OC curves for all single sampling plans for normal and tightened inspection. It is stated that "curves for double and multiple sampling are matched as closely as practicable." The various issues involved in choosing among single, double, and multiple sampling are discussed later in this chapter. For our discussion of a number of aspects of the standard, we shall concentrate our attention on single sampling.

12.5.2 Probabilities of Acceptance of Lots Having AQL Percent Defective

In all AQL systems, the acceptance criteria under normal inspection have been chosen to protect the producer against rejection of lots meeting the quality standard. However, in most AQL systems the Producer's Risk that such lots will be rejected is not the same for all sampling plans.

Figure 12.1 shows a portion of the OC curves for four single sampling plans from the ABC standard all having an AQL of 1%. It is evident that the smaller the sample size, the greater the risk the producer takes that a lot will be rejected when it is exactly 1% defective.

Of course, the Consumer's Risk of accepting a lot much worse than the AQL is also much greater with a small sample size. For example, Fig. 12.1 shows that a 5% defective lot has more than a 0.5 probability of acceptance with $n = 13$ and $c = 0$. In contrast, such a lot is practically certain to be rejected with $n = 500$ and $c = 10$. We have already noted that, when the acceptance number is zero, the OC curve has no point of inflection; it is concave upward through its entire length. If a plan with $c = 0$ were selected to reduce the Producer's Risk of rejection of a 1% defective lot to, say, 0.05, the consumer would have even less protection against accepting bad product than is present with $n = 13$ and $c = 0$. (A P_a of 0.95 for 1% defective product with $c = 0$ would require $n = 5$. With this n, even a 12% defective lot would have a better than even chance of being accepted.)

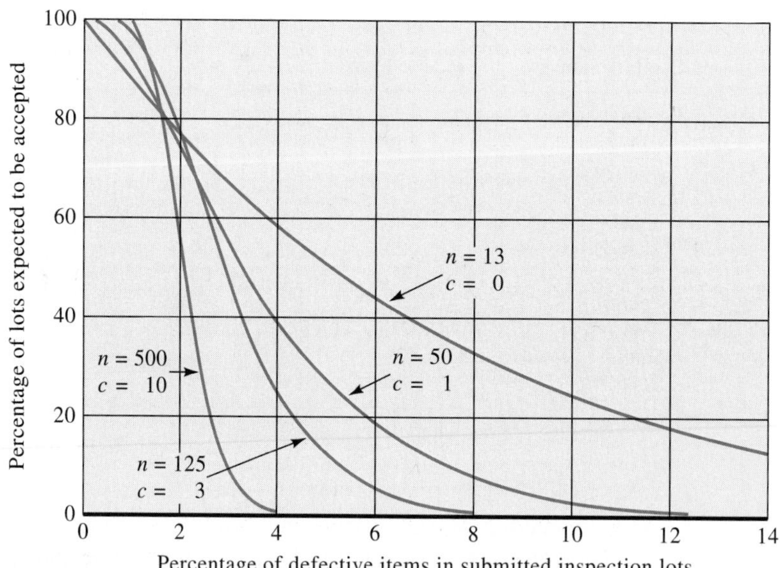

FIGURE 12.1 OC curves for four single sampling plans from the ABC standard, all with an AQL of 1.0%.

In designing an AQL system, it is reasonable to give some weight to the entire OC curve of each of the sampling plans in the system rather than merely to a point near one end of the OC curve, namely, the P_a at the AQL value. In the ABC system, the probability of acceptance at the AQL value in normal inspection varies from about 0.88 for plans with the smaller sample sizes where $c = 0$ to about 0.99 for the large sample sizes and acceptance numbers.

H. F. Dodge proposed an AQL system, discussed briefly in Chap. 13, that eliminates the use of $c = 0$ in single sampling under normal inspection (although $c = 0$ is used in tightened inspection). This elimination makes practicable a standardized aimed-at P_a of 0.95 at the AQL value under normal inspection.

12.5.3 The Relationship between Lot Size and Sample Size in the Military AQL Systems

The discussion in Sec. 12.5 illustrated the use of Table *K*, App. 3, to determine the sample size code letter from the lot size and inspection level in the ABC standard. The code letter then determines the sample size as shown in Tables *L* to *T*.

All the military AQL systems have used a somewhat empirical relationship between lot size and sample size. It is true, of course, that unless a sample is a substantial fraction of its lot, the OC curve of the sampling plan is practically independent of lot size, depending almost entirely on sample size and acceptance number. Nevertheless, computed OC curves always assume random sampling. The use of larger samples for larger lots recognizes that it is relatively difficult to get a small random sample from a very large lot. Moreover, large samples with their steeper OC curves give better discrimination between good and bad lots; the larger the lot size, the more important this discrimination is likely to be.[†]

It should be emphasized that, although the absolute sample size increases with lot size in the military AQL systems, the relative sample size decreases. Consider the following lot sizes under inspection level II applicable to the four plans shown in Fig. 12.1 from the ABC standard:

N	*n*	*c*
51–90	13	0
281–500	50	1
1,201–3,200	125	3
35,001–150,000	500	10

The sample of 13 is 20% of a lot of 65; the sample of 50 is 10% of a lot of 500; the sample of 500 is only 1% of a lot of 50,000.

In Table *K*, the sample size code letter depends only on the lot size and inspection level and is not influenced by the AQL. The military standards have not provided an automatic decrease in sample size with a relaxation of the quality standard.

Nevertheless, an inspection of Tables *L* to *T* indicates that the AQL as well as the code letter may determine the sample size. For example, consider Table *L*, applicable to normal inspection and single sampling. For code letter E, the stated sample size is 13. However, because of the vertical arrows in the table, all the small AQLs require larger sample sizes than 13 when code letter E is specified. Thus n must be 1,250 when the AQL is 0.010%; n must be 200 when the AQL is 0.065%. When the AQL is less than 0.010% (100 parts per million defective), inspection under the standard is not recommended.

12.5.4 Criteria for Shifting to Tightened Inspection and Requalification for Normal Inspection

In acceptance/rejection systems that make use of normal and tightened inspection, it is customary for normal inspection to be used for the first lots submitted.[‡] In effect, the producer is given the benefit of the doubt; if its product meets the stated quality standard for process average, there is relatively little chance that its lots will be rejected.

Because normal inspection will accept nearly all lots submitted that are at the AQL value or better, it will also accept a high percentage of any submitted lots that are moderately worse than the AQL. (The exact protection against such lots is,

[†]For an extended comment on the relationship between lot size and sample size, see H. A. Freeman, Frederick Mosteller, and W. A. Wallis (eds.), *Sampling Inspection*, pp. 171–177, McGraw-Hill Book Company, New York, 1948.

[‡]The ABC standard states: "Normal inspection will be used at the start of inspection unless otherwise directed by the responsible authority."

of course, given by the OC curve, the steepness of which is greatly influenced by sample size.) The consumer's protection against continuing to accept such lots if they are submitted depends on having rules requiring a shift to more severe acceptance criteria whenever there is good evidence that the process quality is worse than the AQL.

The ABC standard states: "When normal inspection is in effect, tightened inspection shall be instituted when 2 out of 5 consecutive lots or batches have been rejected on original inspection (i.e., ignoring resubmitted lots or batches for this procedure)."

The statement regarding requalification for normal inspection is as follows: "When tightened inspection is in effect, normal inspection shall be instituted when 5 consecutive lots or batches have been considered acceptable on original inspection." After tightened inspection has been in effect for 10 consecutive lots, sampling inspection under the standard is terminated until action is taken to improve the process.

MIL-STD-105E (1989) called for discontinuance of sampling inspection after only 5 consecutive lots had been inspected on tightened inspection. Thus, if a return to normal inspection did not occur, inspection under the standard was to be discontinued until corrective action on the process had been taken. Then tightened inspection was to be reinstituted as if the shift to tightened inspection had just occurred.

12.5.5 Acceptance Criteria under Tightened Inspection

In all the military AQL systems using sample size code letters, the code letter in tightened inspection is determined just as in normal inspection. The acceptance criteria under tightened inspection in the ABC standard are shown in Tables *M, P,* and *S.* The relationship between criteria under normal and tightened inspection may be illustrated with reference to the normal plans for a 1% AQL that were shown in Fig. 12.1.

Code letter	Normal		Tightened	
	n	c	n	c
E	13	0	20	0
H	50	1	80	1
K	125	3	125	2
N	500	10	500	8

In the first three cases the criteria under tightened inspection for a 1% AQL are the same as the criteria under normal inspection for the stated code letter for the next lower AQL, namely, 0.65%. In most cases in the ABC standard, the tightened criteria are identical with the normal criteria for the next lower AQL class, although there are a number of exceptions.

Later in this chapter there is an illustration of the difference between the OC curves in normal and tightened inspection. The various AQL systems have differed considerably in this respect.

12.5.6 Criteria for Qualification and Loss of Qualification for Reduced Inspection

Generally speaking, eligibility for reduced inspection should be based on recent quality history indicating average quality considerably better than the AQL. Moreover, it should seem likely that the product to be inspected under reduced inspection will be produced under the same conditions that gave rise to the recent good quality history.

The ABC standard states the following conditions for a shift from normal to reduced inspection:

a. The preceding 10 lots or batches (or more, as indicated by the note to Table VIII)† have been on normal inspection and none has been rejected on original inspection; and
b. The total number of defectives (or defects) in the samples from the preceding 10 lots or batches (or such other number as was used for condition *a* above) is equal to or less than the applicable number given in Table VIII. If double or multiple sampling is in use, all samples inspected should be included, not "first" samples only; and
c. Production is at a steady rate; and
d. Reduced inspection is considered desirable by the responsible authority.

Normal inspection must be reinstated whenever "a lot or batch is rejected" or "production becomes irregular or delayed, or other conditions warrant that normal inspection shall be instituted." A further condition requiring that normal inspection be reinstated is discussed in the next section.

12.5.7 Acceptance Criteria under Reduced Inspection

In all the military AQL systems using sample size code letters, the code letter in reduced inspection has been determined just as in normal inspection. The acceptance criteria under reduced inspection in the ABC standard are shown in Tables *N, Q,* and *T* in App. 3. The relationship between the criteria under normal and reduced inspection may be illustrated with reference to the single sampling normal plans for a 1% AQL that were shown in Fig. 12.1.

†Table VIII of the standard is reproduced in this book as Table *W* of App. 3.

Code letter	Normal			Reduced		
	n	Accept	Reject	*n*	Accept	Reject
E	13	0	1	5	0	1
H	50	1	2	20	0	2
K	125	3	4	50	1	4
N	500	10	11	200	5	8

The reader will observe that the acceptance criteria shown for code letters H, K, and N under reduced inspection all have an area of indecision in which the lot is neither accepted nor rejected. The standard states that, whenever the number of defectives falls in this indecision region (for example, if there should be exactly 1 defective in the sample of 20 with code letter H), the lot in question shall be accepted but reduced inspection shall be discontinued and normal inspection reinstated.

12.5.8 General Comments on Reduced Inspection in Acceptance/Rejection Plans

A provision for reduced inspection is not a necessary part of an acceptance/rejection plan. Nevertheless, such a provision is based on a principle that is economically sound. This principle is to concentrate inspection attention on those products and quality characteristics where the quality history is doubtful and to give less attention where the quality history is very good.

The consumer's savings in inspection costs under reduced inspection are apparent. The producer's advantages are not quite so obvious. However, because the acceptance criteria in reduced inspection are not so stringent, the producer receives added protection against lot rejection. The producer may also have a real sense of accomplishment in having qualified for reduced inspection. Hence, from the consumer's viewpoint, the provisions for reduced inspection in any acceptance program may provide a useful nonfinancial incentive to the producer to improve quality. This point is illustrated in Example 14.1 with reference to reduced inspection in multilevel continuous sampling. Figure 12.2 shows an abbreviated flow diagram of the operation of the switching rules in the ABC standard.

12.6 OC CURVES UNDER NORMAL, TIGHTENED, AND REDUCED INSPECTION

Figure 12.3 compares the OC curves for the various single sampling plans for code letter K, 1% AQL, in the ABC standard. The reader will note that two OC curves are necessary for reduced inspection. The left-hand curve ($n = 50$, $c = 1$) gives the probabilities of lot acceptance accompanied by continuation of reduced inspection for the next lot. The right-hand curve ($n = 50$, $c = 3$) gives the probabilities of failure to reject; in effect, the difference between the two P_a values for any μ_p tells us the probability that a lot at that μ_p will be accepted but normal inspection will be reinstated.

It is evident that the left-hand reduced inspection curve shows almost as good consumer protection as the OC curve for normal inspection with code letter K. (In fact, this reduced plan for code letter K happens to be the normal plan for code letter H.) Therefore, reduced inspection will not be continued for long unless the process quality is good enough for its lots to have passed normal inspection if normal inspection had been in effect. On the other hand, the right-hand curve shows that any particular lot under reduced inspection has a much greater chance of not being rejected than the same lot would have had under normal inspection.

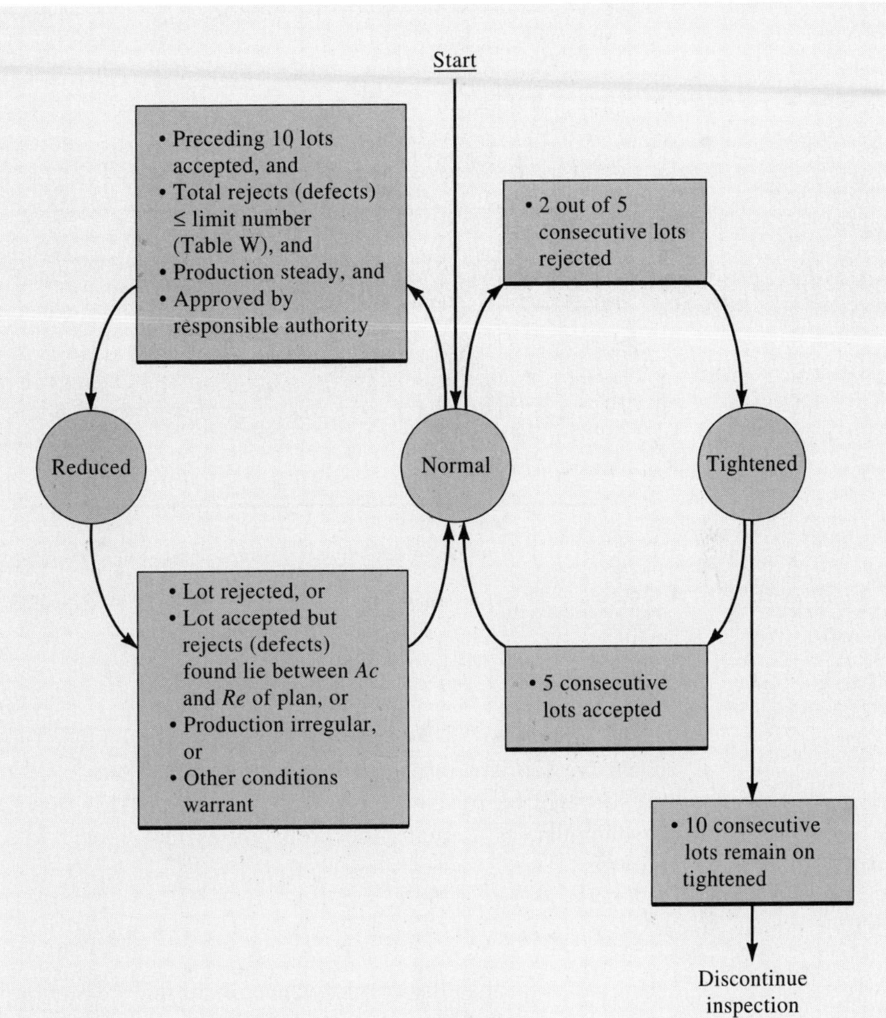

FIGURE 12.2 Schematic diagram of switching rules for ABC standard.

An examination of the OC curve under tightened inspection makes it clear that lots appreciably worse than the AQL still have a fair chance of acceptance. For example, the indifference quality $p_{0.50}$ for this 1% AQL tightened plan is about 2.2% defective. It may be desirable to supplement the tightened inspection criteria by less formal pressures based on a current estimate of the process average.

12.6.1 Calculating Probabilities of Switching, N–T–N

Let us assume single sampling under code letter L and an AQL of 1.0%. The sample size under normal and tightened inspection is 200. The acceptance numbers are 5 under normal inspection and 3 under tightened.

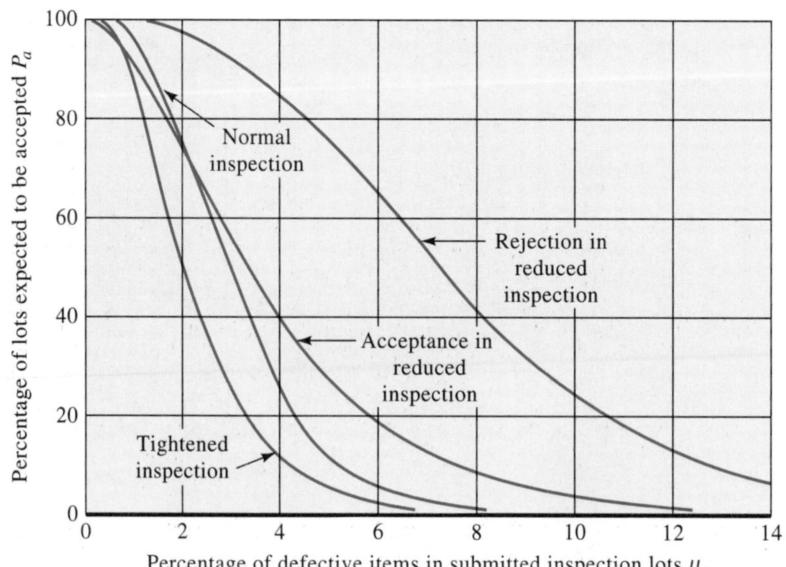

FIGURE 12.3 OC curves under normal, tightened, and reduced inspection; single sampling code letter K, 1% AQL, ABC standard.

The probability of switching from normal to tightened inspection after any 5 successive lots are inspected is the probability that 2 (or more) of 5 consecutive lots will be rejected on normal inspection. This may be formulated, using the binomial, as 1 minus the probabilities that 0 or 1 lots will be rejected as follows:

$$P(\text{N}\rightarrow\text{T}) = 1 - P^5_{a,\text{N}} - 5P_{r,\text{N}}P^4_{a,\text{N}}$$

where $P_{a,\text{N}}$ = probability of acceptance on normal inspection
$P_{r,\text{N}}$ = probability of rejection on normal inspection

If μ_p = AQL = 0.01, then $P_{a,\text{N}} = 0.983$. Hence, with lots drawn from a process operating at the AQL,

$$P(\text{N}\rightarrow\text{T}) = 1 - (0.983)^5 - 5(0.017)(0.983)^4 = 0.0028.$$

This risk is very small.

The probability of switching from tightened to normal inspection after the first 5 lots have been inspected on tightened inspection (that is, 5 consecutive lots accepted) is simply

$$P(\text{T}\rightarrow\text{N}) = P^5_{a,\text{T}}$$

where $P_{a,\mathrm{T}}$ = probability of acceptance on tightened inspection. With the example problem given and no improvement in process quality, the probability of acceptance under tightened inspection is 0.857. Thus the probability of returning to normal inspection after 5 lots is

$$P(\mathrm{T} \rightarrow \mathrm{N}) = (0.857)^5 = 0.4623$$

This is a relatively low value considering that the process is operating at the AQL.†

In order to switch within the first 10 lots inspected, and thus avoid termination of sampling inspection under the standard, the last lot prior to the sequence of 5 lots accepted must be rejected. Prior to that rejected lot in the sequence, it is irrelevant whether the lots are accepted or rejected. This yields a probability of about 0.79 that the switch will be made when μ_p = AQL before inspection is terminated.

A tabulation of switching probabilities at various quality levels for these code letter L, 1.0% AQL, normal and tightened plans gives some insight into the operation of the rules:

Incoming μ_p	$\frac{1}{2}$AQL 0.005	AQL 0.01	2AQL 0.02
$n\mu_p$	1.0	2.0	4.0
$P_{a,\mathrm{N}}$	0.999	0.983	0.785
$P_{a,\mathrm{T}}$	0.981	0.857	0.433
$P(\mathrm{N} \rightarrow \mathrm{T})$	≈0	0.0028	0.2937
$P(\mathrm{T} \rightarrow \mathrm{N})$	0.9085	0.4623	0.0152
P(termination)	0.005	0.207	0.942

These figures demonstrate that, when the quality level is at the AQL or better, a switch from normal to tightened inspection is quite unlikely. Even at a quality level twice that of the AQL the likelihood of switching within the first 5 lots inspected is not great. However, once the switch is made, the chances are not great that a return to normal inspection will be made unless product quality improves to a level better than the AQL.

Similar calculations for switching to and from reduced inspection are more complicated because of the limit number criterion and thus are not illustrated here.

12.6.2 Normal, Tightened, and Reduced Inspection as a Sampling System

When two or more sampling plans are linked together by a set of rules for switching from one plan to another, as with the ABC standard, the linked group may be considered a "sampling system." Figure 12.2 is typical of system flow diagrams in that it shows the various possible *states* in the inspection system—normal, tightened, and reduced inspection—and the *transition* (switching) rules.

†This feature of the switching rule, from tightened to normal inspection, has been criticized by the Japanese, who prefer to use the rules of MIL-STD-105C.

K. S. Stephens and K. E. Larson have analyzed the systems of plans in the ABC standard.[†] While the calculation of various characteristics of the plans are somewhat similar to those illustrated in the previous section, the necessary mathematical models are quite complex.

Figure 12.4 shows the system OC curve for the single sampling code letter K, 1.0% AQL, plans illustrated in Fig. 12.3. To reduce the amount of confusion, only the normal- and tightened-plan OC curves have been drawn in. In general, the system curve is much steeper than that for any individual plan. At quality levels better than the AQL, the system probability of acceptance is better than that indicated for normal inspection. This illustrates the effect of reduced inspection if it is permitted. If reduced inspection is not permitted, the system curve traces that for normal inspection at good quality levels.

As quality deteriorates, the system curve moves downward toward that for tightened inspection and becomes asymptotic to it beyond the indifference quality level $p_{0.50}$. System OC curves are not included in the ABC standard or in its U.S. military equivalent MIL-STD-105E. They are, however, in the 1981 and 1993 versions of ANSI/ASQC standard Z1.4 under the name "Scheme Performance."

A note of caution in using system curves should be introduced at this point. Of necessity, such curves assume that lots are formed from a controlled process at a

[†]K. S. Stephens and K. E. Larson, "Evaluation of the MIL-STD-105D System of Sampling Plans," *Industrial Quality Control,* January 1967, American Society for Quality Control.

FIGURE 12.4 Comparison of normal and tightened OC curves with system (R, N, T) OC curve, single sampling code letter K, 1% AQL, ABC standard.

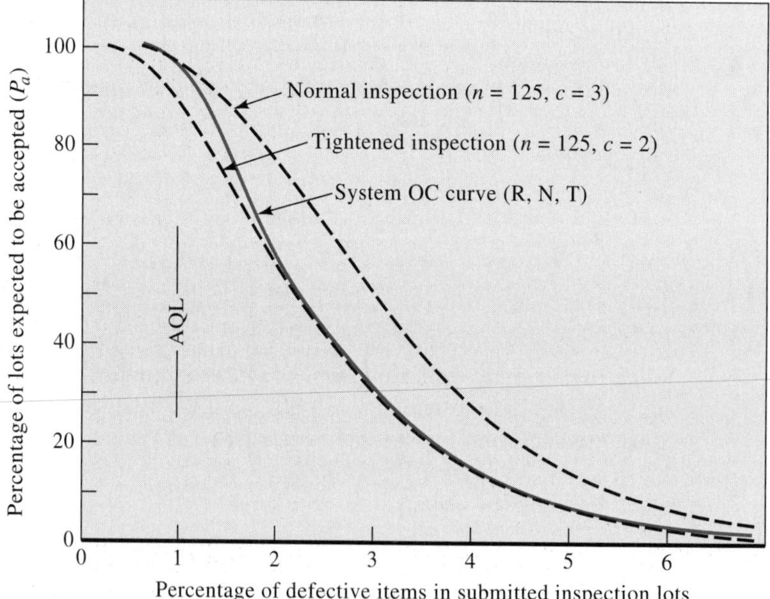

level μ_p and that this level remains constant indefinitely. In calculating the probabilities of shifts to and from tightened inspection in the previous section, the assumption of a constant μ_p over some specified number of lots was used. Thus some compromise with reality was made. This compromise was minimal, however, in comparison to that made in developing system OC curves.

As was illustrated, once a shift to tightened inspection is made, the P_a is reduced considerably. Five consecutive lots must be accepted in order to return to normal inspection, and this shift must take place within ten consecutive lots inspected on tightened inspection. (This termination rule is not reflected in the system OC curves.) This places heavy pressure on the producer to improve quality. As a consequence, the assumption of a constant μ_p is quite unrealistic. Nevertheless, these system curves, when overlaid on curves for the individual plans, may still offer some further insight into the behavior of systems of sampling plans.

12.6.3 Calculation of Estimated Process Average

Most AQL systems have required a formal estimate of the most recent process average (often from the samples from the past 10 lots) to guide the decisions about shifts back and forth among tightened, normal, and reduced inspection. The ABC standard has simplified the administrative rules regarding such shifts and does not make it imperative that an estimated process average be calculated.

Nevertheless, it is a good idea to require calculation of an estimated process average at regular intervals. It is desirable for both producer and consumer to know whether quality is, on the average, better or worse than the AQL and to know whether quality seems to be improving or deteriorating.

The calculated process average from any series of samples is merely the total number of defective units found divided by the total number of sample units inspected. If single sampling is used, it is customary to inspect the entire sample in all cases even though sometimes enough defectives may be found to cause rejection of a lot before all the sample units have been examined. Otherwise, samples from rejected lots will not be given sufficient weight in computing the process average. For the same reason, it is customary to inspect the entire first sample in double and multiple sampling.

In double and multiple sampling, it has been customary to use only the results from first samples in estimating process average. Otherwise, lots requiring more than one sample tend to be given undue weight in the calculation.

The lot-by-lot data used to calculate the process average may also be used to plot a *p* chart (or an *np* chart where appropriate) to supply a basis for judgment as to whether or not the process seems to be in statistical control and successive lots are homogeneous. This is an important aspect of all AQL systems.

12.7 SINGLE, DOUBLE, AND MULTIPLE SAMPLING PLANS IN AQL SYSTEMS

In the AQL systems making use of code letters, the attempt has been made to match OC curves as closely as practicable among the single, double, and multiple sampling plans for any stated code letter and AQL. It follows that the choice of one of these three types of sampling usually should be made on other grounds than the difference in OC curves.

We have already illustrated the use of Tables *O* and *R* of App. 3 to select double and multiple plans applicable to normal inspection in the ABC standard. Tables *P* and *S* may be used in a similar way for tightened inspection, and Tables *Q* and *T* for reduced inspection.

In the double sampling plans in the ABC standard, the second sample is the same size as the first sample. This is a point of difference from all the earlier military AQL systems, in which the second sample in double sampling was invariably twice as large as the first sample. It should be noted that the purpose of a fixed ratio between first and second sample size in double sampling is to achieve administrative simplicity. The reader will recall that in the Dodge-Romig tables, designed to minimize total inspection, there was no fixed ratio of sizes of the two samples in double sampling.

Another aspect of the double sampling plans in the ABC standard is that the rejection number for the first sample is usually less than the rejection number for both samples combined. In all the double sampling plans discussed in Chaps. 11 and 13 and in all the earlier military AQL standards, the rejection number for the first sample was the same as for the two samples combined. In this respect, the ABC standard reduces the amount of inspection under double sampling at some sacrifice of administrative simplicity.

The multiple sampling plans in the ABC standard all provide for seven samples. In all the military AQL standards, all the individual samples in any multiple sampling plan have been the same size.

For any given code letter in the ABC standard, a sample in double sampling is approximately five-eighths the corresponding single sample size; each sample in multiple sampling is one-fourth the corresponding single sample size. There are a few slight exceptions to the foregoing ratios that are caused by the need to have integral values of sample sizes.

12.7.1 Curtailment of Sampling Inspection in Double and Multiple Sampling

For reasons already mentioned in our discussion of the calculation of the process average, it is customary to inspect the entire first sample in double and multiple sampling.

When the rejection number is reached in the second sample in double sampling, it is customary to discontinue inspection. For example, assume the following plan applicable to normal inspection, 1.5% AQL, code letter L, in the ABC standard:

Sample size	Cumulative sample size	Acceptance number	Rejection number
125	125	3	7
125	250	8	9

On the first sample, 6 defectives are found. This requires a second sample. Suppose that 3 more defectives are found by the time 15 items from the second sample have been inspected. Since the rejection number of 9 has been reached, no further inspection is needed. In this instance, curtailment saves inspecting 110 items that would have been examined if the full second sample of 125 had been inspected.

In a similar way it is common to curtail multiple inspection whenever a rejection number is reached on any sample after the first. The advantage of curtailment in double and multiple sampling occurs particularly when product is bad enough to cause a fair percentage of lot rejections.

12.7.2 Comparing the Amount of Sampling Inspection in Single, Double, and Multiple Sampling in the ABC Standard

Figure 12.5, taken from the ABC standard, shows the approximate values of *Average Sample Number* (ASN) for double and multiple sampling relative to the sample size in single sampling.

Each of the 15 sets of curves applies to a set of plans characterized by a particular value of c in single sampling. The horizontal scale in each diagram is the product of single sample size and fraction defective. The ABC standard contains the following statement about these curves: "The curves assume no curtailment of inspection and are approximate to the extent that they are based upon the Poisson distribution, and that the sample sizes for double and multiple sampling are assumed to be $0.631n$ and $0.25n$ respectively, where n is the equivalent single sample size."

A reader who inspects the 15 diagrams in Fig. 12.5 might gain the impression that the average amount of inspection in multiple sampling in the ABC standard is much less than in double sampling. As applied to the portion of the standard that ordinarily will have the greatest use, this is an incorrect impression for the following reasons:

1. The plans from the standard that are likely to be used the most are those corresponding to the relatively low c numbers in single sampling, particularly numbers from 0 to 5. Upon examining Table *L,* the reader will observe that the higher values of c apply to the relatively high AQL values or to the relatively large sample sizes (or both).
2. It is the left-hand portion of the ASN curve—particularly to the left of the AQL value for normal inspection (shown by vertical arrows in Fig. 12.5)—that is of the most interest from a practical point of view. No AQL acceptance sampling plan is likely to be continued for long if the quality of submitted product is consistently much worse than the AQL.

Consider the portions of the ASN curves to the left of the AQL arrows where c is 1, 2, 3, and 5 in single sampling. (Figure 12.5 does not contain a diagram for $c = 0$ because the standard does not have matching double and multiple plans for

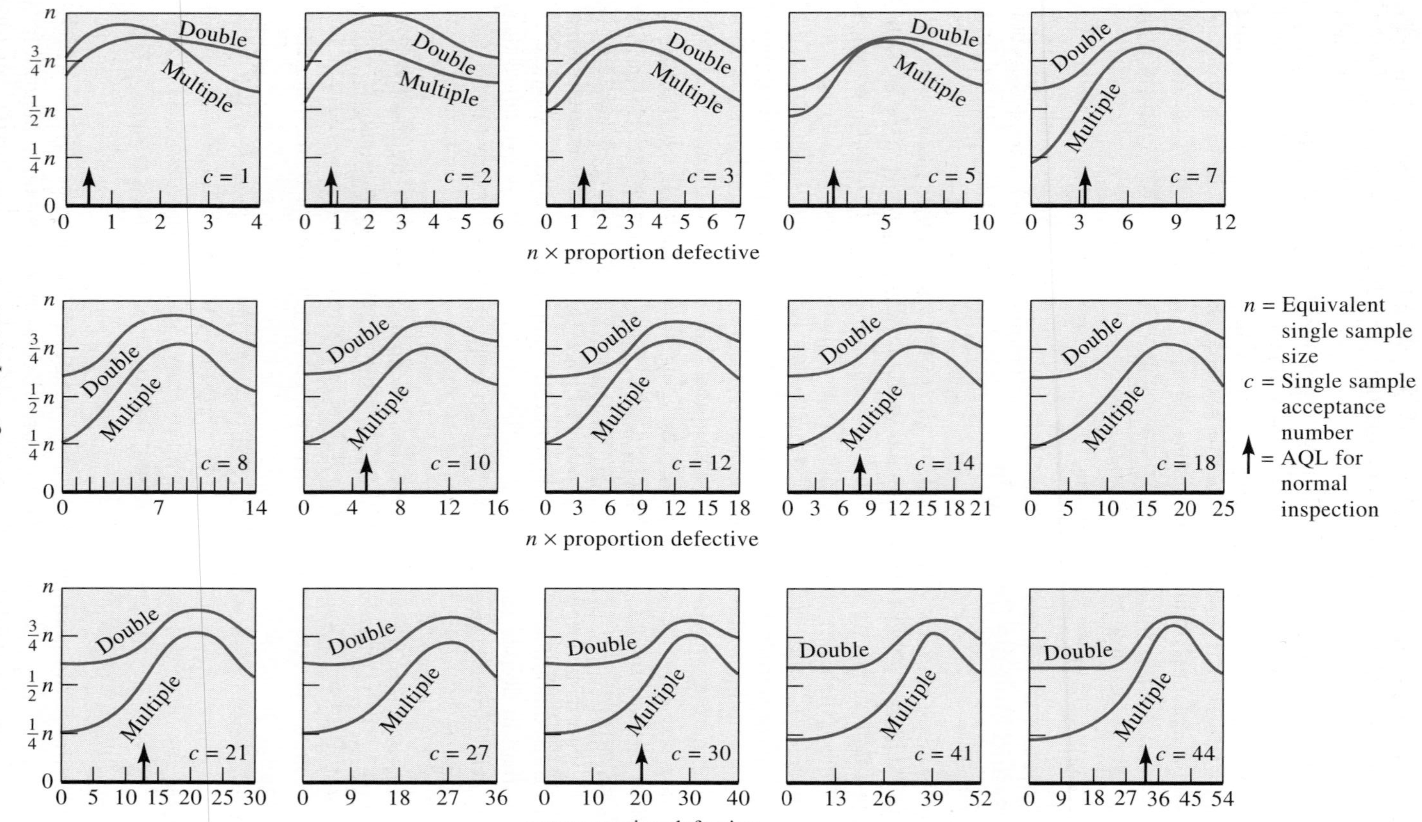

FIGURE 12.5 Table IX of MIL-STD-105E (and ABC standard) showing ratio of average sample size in double and multiple sampling to single sample size (normal and tightened inspection) assuming no curtailment.

this value of *c*.) Where *c* is 1, double sampling involves less inspection than multiple sampling because all the multiple sampling plans corresponding to this *c* value will not accept until the third sample. Where *c* is 2, 3, or 5, the ASN values in multiple sampling are generally from 80 to 85% of those in double sampling; for these plans, acceptance is possible on the second multiple sample.

It is chiefly where *c* in single sampling is 7 or more that the multiple sampling plans involve large savings in amount of inspection as compared with the double sampling plans. As previously stated, this advantage occurs chiefly where AQL values are relatively high or sample sizes are relatively large. For example, with an *n* of 32 in single sampling (code letter G), the AQL is 10% for a *c* of 7 in normal inspection. For an AQL of 1% and a *c* of 7 in normal inspection, *n* is 315 (code letter M).

The reader should note that the average sample number (ASN) is related to but not the same as the average total inspection (ATI) discussed in Chap. 11. As generally used, ASN refers to comparisons between average numbers of items inspected in sampling where other characteristics of the plans, such as their OC curves, are nearly identical. ATI applies only to cases where rejected lots are subjected to 100% inspection. This usage is adhered to consistently in this book. Such is not always the case in general literature of acceptance sampling.

12.7.3 Computing Average Sample Numbers in Double and Multiple Sampling

The following numerical examples are intended to illustrate how to calculate a point on an ASN curve. They also aim to show why each of the 15 graphs in Fig. 12.5 is applicable to the entire family of plans in the ABC standard that corresponds to a particular value of *c* in single sampling.

Consider the following plans for normal inspection from the ABC standard:

Code letter J, 0.65% AQL			Code letter P, 0.065% AQL		
Cumulative sample size	Acceptance number	Rejection number	Cumulative sample size	Acceptance number	Rejection number
Single:			Single:		
80	1	2	800	1	2
Double:			Double:		
50	0	2	500	0	2
100	1	2	1,000	1	2
Multiple:			Multiple:		
20	†	2	200	†	2
40	†	2	400	†	2
60	0	2	600	0	2
80	0	3	800	0	3
100	1	3	1,000	1	3
120	1	3	1,200	1	3
140	2	3	1,400	2	3

†Acceptance not permitted at this sample size.

Assume that product 1% defective is being submitted to the double sampling plan shown for code letter J and an AQL of 0.65%. To compute the average sample number it is first necessary to find the respective probabilities that a decision will be reached on each sample. (It was explained in Chap. 11 that such probabilities are computed in finding each point on an OC curve in double and multiple sampling.)

Assume that calculations will be based on the Poisson approximation. Then μ_{np} for each sample of 50 is 50(0.01) = 0.5. Table G gives P_0 as 0.607, P_1 as 0.303, and $P_{2 \text{ or more}}$ as 0.090. The probability of a decision on the first sample is the P_a of 0.607 plus the P_r of 0.090 = 0.697. In double sampling, the probability of a decision on the second sample is identical with the probability that a second sample will be taken, in this case 0.303. If there is no curtailment of inspection, 0.697 of the lots will require a sample of 50 and 0.303 will require a sample of 100. It follows that

$$\text{ASN} = 0.697(50) + 0.303(100) = 65.2$$

For double sampling plans in general,

$$\begin{aligned}\text{ASN} &= [P_a(n_1) + P_r(n_1)]n_1 + [P_a(n_2) + P_r(n_2)](n_1 + n_2) \\ &= n_1 + [P_a(n_2) + P_r(n_2)]n_2 = n_1 + n_2 P(n_2)\end{aligned}$$

where $P_a(n_1)$ = probability of acceptance in n_1 , and so on
$P(n_2)$ = probability of drawing n_2

Calculations regarding the ASN in multiple sampling follow the same lines, although they require considerably more arithmetic. In the multiple sampling scheme for code letter J and an AQL of 0.65%, each sample is 20. With a μ_p of 0.01, the μ_{np} for each sample is 0.2. Table G gives $P_0 = 0.819$, $P_1 = 0.163$, $P_2 = 0.017$, $P_3 = 0.001$. These figures may be used to compute probabilities of acceptance and rejection on each sample. These probabilities are as follows:

Sample	P_a	P_r	$P_a + P_r$
1st	0.000	0.018	0.018
2nd	0.000	0.045	0.045
3rd	0.549	0.060	0.609
4th	0.000	0.006	0.006
5th	0.220	0.015	0.235
6th	0.000	0.016	0.016
7th	0.058	0.013	0.071
Totals	0.827	0.173	1.000

If it is assumed that there is no curtailment of inspection, the probabilities of a decision (that is, $P_a + P_r$) on each sample may be used to compute the ASN value as follows:

$$\text{ASN} = 20(0.018) + 40(0.045) + 60(0.609) + 80(0.006) + 100(0.235) + 120(0.016) + 140(0.071) = 74.5$$

Now, let us compare the single, double, and multiple sampling plans shown for code letter P and an AQL of 0.065% with the corresponding plans shown for code letter J and an AQL of 0.65%. It is evident that the acceptance and rejection numbers are identical and that all sample sizes in the plans for code letter P are 10 times those for code letter J. If we assume a μ_p of 0.001, one-tenth of the μ_p value used in our calculations for the plan from code letter J, the μ_{np} values for the sample sizes of 500 and 200 with code letter P will be the same as those with samples of 50 and 20 with J. It follows that, if the Poisson approximation is used, the values of P_a and P_r for each sample will be identical with those calculated for a μ_p of 0.01 with code letter J. The ASN values will be 652 with double sampling and 745 with multiple sampling, exactly 10 times the values calculated for code letter J.

In Fig. 12.5 one ASN curve for double sampling and another for multiple sampling applied to the plans we have analyzed for code letters J and P and to all other plans corresponding to a c of 1 in single sampling. The use of one ASN curve to represent many different plans is possible, in part, because of the generalized scales used in plotting the curves in Fig. 12.5 and, in part, because of certain unique aspects of the ABC standard.

One relevant aspect of the ABC standard is that there is only one set of acceptance and rejection numbers in double and multiple sampling corresponding to each value of c in single sampling. Another aspect is that the ratio of double and multiple sample size to single sample size is approximately constant for all corresponding plans.

Moreover, the product of single sample size and AQL value in normal inspection is approximately constant for all plans having the same value of c in single sampling. For this reason, it is possible to show arrows indicating AQL values in normal inspection on the graphs in Fig. 12.5. (The graphs without AQL arrows apply to plans used only in tightened inspection.)

12.7.4 Effect of Curtailment on Amount of Sampling Inspection

The curves in Fig. 12.5 assume no curtailment of inspection when the rejection number is reached in the second sample in double sampling or in the second or subsequent samples in multiple sampling. (This assumption simplified the calculation of average sample numbers and made possible the plotting of one set of curves for each value of c.) However, such curtailment is a desirable practice and is commonly used. Economies due to curtailment occur particularly where quality is sufficiently worse than the AQL for a fair percentage of lots to be rejected.

The larger the sample sizes, the greater the possible savings due to curtailment. For this reason, in a comparison of corresponding double and multiple sampling plans, curtailment will cause greater economies in double sampling than in multiple sampling. It follows that, where curtailment is to be used, Fig. 12.5 exaggerates the differences in average amount of inspection between double and multiple sampling, particularly in the right-hand portion of the curves.

12.7.5 Choosing among Single, Double, and Multiple Sampling

It sometimes happens that double and multiple sampling are impracticable. There may be physical reasons why only one sample can be drawn from a lot, or it may not be feasible to provide for the variable inspection load associated with double and multiple sampling. Under such circumstances, it is necessary to use single sampling. More often, however, there is opportunity for choice. In deciding among these three types of sampling, consideration should be given to a number of factors in addition to the expected differences in average amount of inspection. Some of these factors are as follows:

1. *The psychological advantages of double sampling.* Two such advantages have played an important part in decisions to adopt double sampling. One is that borderline lots are given a "second chance" to be accepted. The other is that no lot is rejected because of a single defective article.

   There is no doubt that the idea of giving a lot a second chance has a definite appeal to practical inspectors. It seems more convincing to say that a lot was rejected after *two* samples than to say that it was rejected on the evidence of a single sample.

   Where lots are large, there is often a strong objection by the producer to the rejection of an entire lot on the basis of a single defective article found in a sample. None of the double sampling plans will reject on a single defective.

   These psychological advantages of double sampling apply also to multiple sampling. However, it is questionable whether the appeal of the third, fourth, or fifth chance adds anything to the appeal of the second chance.
2. *The expected differences in costs of administration.* These costs tend to be highest for multiple sampling and lowest for single sampling. The more complicated the acceptance plan, the greater the attention required from supervisors. Moreover, the variability of inspection load in multiple sampling and double sampling introduces extra difficulties into the scheduling of inspectors' time. In those cases where double and multiple sampling can make relatively small savings in average amount of inspection, the more complicated plans may actually increase total inspection cost.
3. *The difficulty of training inspectors to use multiple sampling correctly.* This difficulty seems to vary with the quality of the inspectors and perhaps with other matters. In some plants where inspectors have made frequent bad errors in interpreting instructions, quality control engineers refuse to specify anything but single sampling. In other plants, no serious troubles are reported in secur-

ing correct use of sampling procedures as complicated as item-by-item sequential plans (discussed in Chap. 13).

4. *The need for quick and reliable estimates of process average for various purposes including decision among normal, tightened, and reduced inspection.* In the various types of AQL systems prior to the ABC standard, the shifts among these three types of inspection depended on the estimated process average, often estimated from the samples from the most recent 10 lots. The process average was estimated from full single samples or from first samples in double and multiple sampling. The sampling error in an estimate of the recent process average is smallest with the relatively large sample size in single sampling and largest with the relatively small sample size in multiple sampling. For this reason, it may be sensible to use single sampling in the early stages of a contract even though a change to double or multiple sampling is made after there is satisfactory evidence regarding the process average.

   Even in systems (such as the ABC standard) where the process average is not formally required, it is helpful to have as good information as possible about the process average and the presence or absence of statistical control. Single sampling does the best job of providing such information.

5. *The availability of inspection personnel and facilities.* This should properly be an important factor in the choice of the type of sampling in a number of instances.

   One extreme example of this influence may be cited from the experience of the inspection department of a certain government agency. The inspectors employed by this agency typically did sampling inspection in the plants of several different producers during the course of a week, using public transportation to go to and from each plant. The length of the visit to each plant was controlled by bus schedules and train schedules. With multiple sampling, an inspector generally finished work some time before it was necessary to leave a plant. Under single sampling, time at a plant was more likely to be fully occupied. It was therefore decided to change from multiple to single sampling. This change did not increase cost, and, because of the larger single sample size, it gave much better current information regarding process quality.

   Similar cases may exist in industrial plants where it is desired to keep an inspection force intact throughout a temporary period of reduced inspection activity.

   The opposite extreme exists when the need for inspection personnel and facilities is rapidly increasing. Such a period would seem to be an appropriate one in which to introduce double or multiple sampling, particularly in those spots where such plans seem likely to make large reductions in the average inspection per lot. The alternative to reducing average inspection per lot is the hiring and training of new inspectors (presumably less competent than the old ones) and the making of investments in additional inspection facilities.

12.8 CLASSIFICATION OF DEFECTS

In AQL systems for acceptance sampling by attributes, it is common for one AQL to apply to a group of possible defects rather than to have a separate AQL value for each possible way in which product may fail to conform to specifications. Unless trivial defects are to be given the same weight as serious ones, it is essential to have a classification of defects or classification of characteristics. (These two phrases seem to be used interchangeably.) Such classifications were mentioned in Chap. 7, Sec. 7.11.

The ABC standard calls for defects to be classified as critical, major, and minor. The definitions of these are as follows:

> *Critical Defect.* A critical defect is a defect that judgment and experience indicate is likely to result in hazardous or unsafe conditions for individuals using, maintaining, or depending upon the product; or a defect that judgment and experience indicate is likely to prevent performance of the tactical function of a major end item such as a ship, aircraft, tank, missile or space vehicle.
>
> *Major Defect.* A major defect is a defect, other than critical, that is likely to result in failure, or to reduce materially the usability of the unit of product for its intended purpose.
>
> *Minor Defect.* A minor defect is a defect that is not likely to reduce materially the usability of the unit of product for its intended purpose, or is a departure from established standards having little bearing on the effective use or operation of the unit.

Product usually is subject to 100% inspection for critical defects. But there are many types of product with no defects classed as critical; all are either major or minor. It is fairly common to divide minor defects into two groups depending on their relative importance. The more important minors may be designated Minor *A*; the less important ones, Minor *B*. (In the original Army Ordnance and Army Service Forces tables, the more descriptive adjective "incidental" was used for the relatively unimportant defects of the type later classified as "Minor *B*.")

12.8.1 An Illustration of the Use of a Classification of Defects in Acceptance Sampling

Assume that the classification of defects for certain bolts includes 5 defects classified as Major, 11 classified Minor *A*, and 8 classified as Minor *B*. Bolts are to be inspected in lots of 10,000 under the ABC standard. Double sampling and inspection level II are to be used. The AQL is 1.0% for Majors, 4.0% for Minor *A*, and 6.5% for Minor *B*.

The sample size code letter L is determined by the lot size and inspection level. In this case, the acceptance criteria are:

Cumulative sample size	Major		Minor *A*		Minor *B*	
	Accept	Reject	Accept	Reject	Accept	Reject
125	2	5	7	11	11	16
250	6	7	18	19	26	27

A first sample of 125 will be drawn from the lot. This sample is inspected separately for each class of defects. The numbers of defective articles found in each class are as follows:

Major	2
Minor *A*	10
Minor *B*	7

Therefore the lot is passed for Major and Minor *B* defects on the basis of the first sample. A second sample of 125 must be drawn and inspected only for Minor *A* defects. If not more than 8 additional bolts containing such defects are found, the lot is passed. If 9 or more are found, the lot is rejected for Minor *A* defects.

In general, it is economical and convenient to inspect the same sample for all classes of defects. However, the acceptance criteria are applied separately for each class. A lot is rejected if it fails to meet the acceptance criteria for one or more classes. All characteristics checked within a class count toward acceptance or rejection within that class.

12.9 THE FORMATION OF INSPECTION LOTS†

In sampling acceptance inspection, an inspection lot is a group of articles accepted or rejected on the basis of one or more samples. An inspection lot is not necessarily identical with a production lot, a purchase lot, or a lot for other purposes.

Many practical matters such as rate of production and availability of storage space necessarily influence the formation of inspection lots. From the point of view of getting the best results from acceptance sampling, two rules should govern decisions on this matter, namely:

1. Within each lot, the factors that seem likely to cause marked variability in product quality should be as nearly constant as practicable. This may include such matters as sources of raw materials, machines, operators, and time of production.
2. Subject to the limitation of the foregoing rule, inspection lots should be as large as possible.

†For a more complete discussion of this important topic, see Freeman, Friedman, Mosteller, and Wallis, op. cit., pp. 40–43, 87–90.

The desire to have each lot come from a homogeneous source obviously conflicts with the desire to have large lots. Practical decisions usually call for a compromise between these two objectives.

The reason for the requirement of homogeneity should be evident. If most lots are relatively good and a few are relatively bad, sampling inspection can discriminate among lots and the quality of the product accepted can be much better than the average quality submitted. On the other hand, if there is a good deal of mixing of product from various sources, the percentage of defectives may not vary greatly from lot to lot; in this case the average quality of lots accepted will not be appreciably better than the average quality submitted. Where lot formation is close to production, both in time and place, there is a minimum of opportunity for mixing of product among inspection lots.

The reason for the requirement of making lots as large as possible should also be clear. It is the absolute size of the sample that governs its ability to discriminate between good and bad lots. Large lots permit larger samples than small lots. Moreover, inspection cost will be less with large lots because samples are a smaller fraction of the lot.

Just-In-Time inventory management tends to drive receiving lot sizes down and may make any form of sampling inspection impractical. In such cases the customer must rely on quality information provided by the supplier or resort to 100% inspection. An economic balance must be struck between the costs of inspection and the costs of handling and storage associated with larger inspection lot sizes.

12.9.1 Problems Arising from Resubmission of Rejected Lots

Whenever rejected lots are returned to the producer, the resubmission of these lots for another sampling inspection creates certain problems.

Obviously, it is not in the consumer's interest for lots to be submitted unchanged. This point can be illustrated by a simple probability calculation.

Assume that the probability of acceptance of a lot of given quality is 0.80. This will be the probability of its acceptance not only on the first submission but also on any subsequent submission. The probability of acceptance with a maximum of two resubmissions is as follows:

Probability of acceptance on first submission	0.80
Probability of rejection on first submission followed by acceptance on second submission = (0.20)(0.80)	0.16
Probability of rejection on first and second submission followed by acceptance on third submission = (0.20)(0.20)(0.80)	0.03
Probability of acceptance with not more than three submissions	0.99

Of course, the normal action by a reputable producer is to screen a rejected lot before resubmitting it. The ABC standard contains the following reasonable stipulation: "Lots or batches found unacceptable shall be resubmitted for reinspection only after all units are reexamined or retested and all defective units are removed or defects corrected."

Resubmitted lots that have received 100% inspection by the producer after rejection presumably are considerably better on their second submission than they were on their first submission. Therefore, the results from sampling inspection of resubmitted lots should not be viewed as representative of the process average. Criteria for determining whether normal, tightened, or reduced inspection is to be used should be based solely on the results of original inspection, and the results from any resubmitted lots should be ignored.

12.10 ACCEPTANCE BASED ON NUMBERS OF DEFECTS

A defective article is an article containing one or more defects. In our discussion up to this point, acceptance decisions have been based on the numbers of defectives contained in a sample. Decisions on lot acceptance or rejection have not been influenced by the number of defects observed in each defective article. The sampled article with eight defects has had the same influence on lot acceptance as the sampled article with only one defect; each has counted as a single defective.

Under certain conditions, it may be more reasonable to base acceptance decisions on defects rather than on defectives. For instance, some defects might be tolerable in a bolt of cloth, but it might be desired to limit the average defects per bolt in a shipment of bolts of cloth. In general, the field of application of acceptance based on defects is similar to the field for the c chart discussed in Chap. 7. With acceptance based on defects, AQL values, OC curves, and so forth, are interpreted in terms of defects per hundred units rather than as percent defective.

The entire ABC standard may be used for inspection based on defects as well as for inspection based on defectives. In general, the only changes necessary are to substitute the word *defects* wherever *defectives* is used and to interpret AQLs as defects per hundred units. Of course, the highest AQL values in the standard may be interpreted only as defects per hundred units. For example, it would be meaningless to have an AQL of 150% defective but entirely reasonable to have an AQL of 150 defects per hundred units in sampling inspection of, say, bolts of cloth (that is, 1.5 defects per bolt). Normal, tightened, and reduced inspection can be used in inspection based on defects just as in inspection based on defectives.

The binomial distribution is correct in principle for Type B OC curves based on defectives; in the ABC standard, the binomial is used for OC curves for AQLs of 10.0 or less. The Poisson distribution is correct in principle for OC curves based on defects; in the ABC standard, the Poisson is used for AQLs of 15.0 or more.

12.11 A SYSTEMATIC RECORD OF QUALITY HISTORY IS AN IMPORTANT ASPECT OF STATISTICAL ACCEPTANCE PROCEDURES

As pointed out at the start of Chap. 11, when acceptance procedures are on an informal basis, it frequently is recognized that a commonsense principle is to base the severity of acceptance criteria on the quality history of the product being sampled. An AQL system involving normal, tightened, and reduced inspection substitutes a systematic way of doing this for an informal and unsystematic way. In this systematic procedure, a written record of quality history takes the place of the inspector's memory.

It is helpful if the written record not only gives data on the current quality level but also tells whether the process is in control. Figure 12.6 illustrates a possible form for use with double sampling.[†] The scale for the control chart shown on the right can readily be adapted to any quality level by changing the value of a division on the chart.

Such quality records of sampling inspection are useful even though the acceptance criteria are not affected by past performance. They often bring out differences between the quality levels of different suppliers that had not been recognized before a formal record was kept. In this way they may bring about a decision not to do business with a supplier submitting product with an unsatisfactory process average; such a decision has an effect similar to the adoption of tightened acceptance criteria that reject a majority of the supplier's lots.

In the use of acceptance plans having large samples and an acceptance number greater than 0, some inspectors are reluctant to call the defect that requires the taking of a second sample or the rejection of a lot. Where this type of flinching exists, the situation should be evident from the written quality record. A record showing that nearly all lot acceptances have been based on the maximum number of defectives permissible suggests that inspectors have acted in this way.

[†]See H. F. Dodge and H. G. Romig, *Sampling Inspection Tables*, 2d ed., pp. 48–50, John Wiley & Sons, Inc., New York, 1959, for two forms used by the Western Electric Company for a record of sampling inspection. Many forms used by other manufacturers have been adapted from these Western Electric forms.

FIGURE 12.6 Form that combines record of sampling inspection with control chart for percent rejected.

CONTROL CHART AND DATA SHEET FOR SAMPLING ACCEPTANCE INSPECTION

Vendor	Article purchased
Inspection procedure	Acceptable quality level

Remarks

Date	Lot size	First sample		Percent defective in first sample	Second sample		Action on lot	Control chart for first samples
		Number insp'd	Defectives		Number insp'd	Defectives		

12.12 SELECTING AN ACCEPTANCE PLAN FOR AN ISOLATED LOT

The teeth in any AQL system are in the provision for tightened inspection whenever the recent quality record is bad enough. An AQL is enforced through the overall system, not through the lot-by-lot acceptance criteria used in normal inspection. As has been explained, such criteria are chosen to protect the producer by making sure that nearly all lots will be accepted as long as the process average is at the AQL value or better. Therefore, if individual lots considerably worse than the AQL are submitted, such lots have a good chance to be accepted.

Where lots are isolated or infrequent, an overall acceptance sampling system cannot be used to develop pressure for the submission of good quality. The consumer dealing with an isolated lot needs to take a good look at the OC curve of the proposed sampling plan to judge whether it gives adequate protection. If it is required to be protected against product worse than, say, 1% defective, one should not select a sampling plan indexed under an AQL of 1% in any of the various AQL tables. Such a plan will not give the required protection.

At several points we have noted that one of the serious objections to regular 100% inspection by the consumer is that such inspection does not exert enough pressure on a producer to improve the process. Of course, this objection applies only when there is a continuing relationship between producer and consumer; it is not valid when there is only one lot. In general, the consumer should be willing to screen an isolated lot whenever high quality is deemed essential and when such screening is practicable (as in the case of nondestructive tests). A possible compromise between the interests of producer and consumer might be to use an acceptance plan for isolated lots based on a stated quality standard at $p_{0.50}$, the indifference quality, with the consumer screening any *accepted* lot.

The ABC standard stipulates that, when a lot or batch is of an isolated nature, sampling plans should be chosen based on the identification of a Consumer's Risk point associated with the designated AQL. Tables for Consumer's Risks of 5 and 10%, called the *Limiting Quality* (LQ) tables, are contained in the standard. Table 12.1 reproduces the LQ table for a 10% Consumer's Risk. The term LQ, in this instance, is synonymous with the LTPD.

The user enters the table by the designated AQL and searches the column until an acceptable value of LQ ($100p_{0.10}$) is found. The sample size code letter is then read from the left-hand side. In using this table, the choice of a sampling plan is based, in effect, on the choice of two points on the OC curve, a Producer's Risk point (AQL) and a Consumer's Risk point ($100p_{0.10}$).

12.13 IMPORTANCE OF AOQL VALUES IN SAMPLING PLANS BASED ON THE AQL

Tables *U* and *V* in App. 3 are reproduced from the ABC standard. They give AOQL factors for normal and tightened inspection, respectively, for the single

TABLE 12.1 ABC-STD-105 LIMITING QUALITY TABLES FOR WHICH $P_a = 10\%$. NORMAL INSPECTION, SINGLE SAMPLING[†]

Code letter	Sample size	In percent defective ($100p_{0.10}$)															
		Acceptable quality level															
		0.010	0.015	0.025	0.040	0.065	0.10	0.15	0.25	0.40	0.65	1.0	1.5	2.5	4.0	6.5	10
A	2															68	
B	3														54		
C	5													37			58
D	8												25			41	54
E	13											16			27	36	44
F	20										11			18	25	30	42
G	32									6.9			12	16	20	27	34
H	50								4.5			7.6	10	13	18	22	29
J	80							2.8			4.8	6.5	8.2	11	14	19	24
K	125						1.8			3.1	4.3	5.4	7.4	9.4	12	16	23
L	200					1.2			2.0	2.7	3.3	4.6	5.9	7.7	10	14	
M	315				0.73			1.2	1.7	2.1	2.9	3.7	4.9	6.4	9.0		
N	500			0.46			0.78	1.1	1.3	1.9	2.4	3.1	4.0	5.6			
P	800		0.29			0.49	0.67	0.84	1.2	1.5	1.9	2.5	3.5				
Q	1,250	0.18			0.31	0.43	0.53	0.74	0.94	1.2	1.6	2.3					
R	2,000			0.20	0.27	0.33	0.46	0.59	0.77	1.0	1.4						

Code letter	Sample size	In defects per hundred units																									
		Acceptable quality level																									
		0.010	0.015	0.025	0.040	0.065	0.10	0.15	0.25	0.40	0.65	1.0	1.5	2.5	4.0	6.5	10	15	25	40	65	100	150	250	400	650	1000
A	2															120			200	270	330	460	590	700	1000	1400	1900
B	3														77			130	180	220	310	390	510	670	940	1300	1800
C	5													46			78	110	130	190	240	310	400	560	770	1100	
D	8												29			49	67	84	120	150	190	250	350	480	670		
E	13											18			30	41	51	71	91	120	160	220	300	410			
F	20										12			20	27	33	46	59	77	100	140						
G	32									7.2			12	17	21	29	37	48	63	88							
H	50								4.6			7.8	11	13	19	24	31	40	56								
J	80							2.9			4.9	6.7	8.4	12	15	19	25	35									
K	125						1.8			3.1	4.3	5.4	7.4	9.4	12	16	23										
L	200					1.2			2.0	2.7	3.3	4.6	5.9	7.7	10	14											
M	315				0.73			1.2	1.7	2.1	2.9	3.7	4.9	6.4	9.0												
N	500			0.46			0.78	1.1	1.3	1.9	2.4	3.1	4.0	5.6													
P	800		0.29			0.49	0.67	0.84	1.2	1.5	1.9	2.5	3.5														
Q	1,250	0.18			0.31	0.43	0.53	0.74	0.94	1.2	1.6	2.3															
R	2,000			0.20	0.27	0.33	0.46	0.59	0.77	1.0	1.4																

Reproduced from "Military Standard 105E. Sampling Procedures and Tables for Inspection by Attributes," pp. 28–29, Superintendant of Documents, Government Printing Office, Washington, D.C., 1989.

sampling plans in the standard. With the exception of those plans where $c = 0$, the AOQL values in tightened inspection are fairly close to the stipulated AQL values. The AOQL value is significant in an AQL plan because it is reasonable to expect that a reputable producer will screen lots that have been rejected under the sampling plan.

Some persons object to the policy of placing reliance on the quality protection promised by a computed AOQL value on the grounds that one cannot be *certain* that errors will not be made by the inspectors who screen the rejected lots. To be consistent, such persons also should object to all OC curves for sampling inspection plans because such curves assume that the inspector discovers the defective items in a sample. They also should object to *all* use of 100% inspection.

In general, there is no reason to be more concerned with the possibility of inspectors' errors in screening rejected lots than with the possibility of such errors made in other inspection activities. In fact, it usually is reasonable to expect that more care will be exercised in screening rejected lots than when 100% inspection is used in the first place.

PROBLEMS

12.1. In acceptance sampling under the ABC standard, single sampling is to be used with inspection level II, an AQL of 4%, and a lot size of 2,500. What are the acceptance criteria under *(a)* normal and *(b)* tightened inspection?
Answer: *(a)* $n = 125$, $c = 10$; *(b)* $n = 125$, $c = 8$.

12.2. In Prob. 12.1, use Table *G* to compute the approximate probabilities of acceptance of 4% nonconforming product under normal and tightened inspection.
Answer: 0.986, 0.932.

12.3. Assume that normal inspection is being used for the conditions stated in Prob. 12.1. A series of lots 8% nonconforming are submitted for acceptance. What is the approximate value of the probability that a shift to tightened inspection will be required after the first two such lots? Assume that no rejections have occurred in the preceding four lots.
Answer: 0.17.

12.4. Assume that tightened inspection is being used for the conditions stated in Prob. 12.3. After the rejection of one lot, the product quality improves to the point where lots are now only 4% nonconforming. What is the approximate probability that normal inspection will be reinstated after the next five such lots?
Answer: 0.70.

12.5. What are the acceptance criteria under reduced inspection for the conditions stated in Problem 12.1?
Answer: $n = 50$, $c = 5$; $r = 8$.

12.6. Assume that an 8% nonconforming lot is submitted under the reduced inspection plan of Prob. 12.5. Use Table *G* to compute the approximate probabilities that:

(*a*) The lot will be accepted and reduced inspection continued.
(*b*) The lot will be accepted but normal inspection will be reinstated.
(*c*) The lot will be rejected.

Answer: *(a)* 0.785; *(b)* 0.164 *(c)* 0.051.

12.7. In acceptance sampling under the ABC standard, double sampling is to be used with inspection level II, an AQL of 1%, and a lot size of 750. What are the acceptance criteria under *(a)* normal and *(b)* tightened inspection?

Answer: (a) $n_1 = 50$, $c_1 = 0$, $r_1 = 3$; $n_2 = 50$, $c_2 = 3$, $r_2 = 4$; *(b)* $n_1 = 50$, $c_1 = 0$, $r_1 = 2$; $n_2 = 50$, $c_2 = 1$, $r_2 = 2$.

12.8. In Prob. 12.7, use Table *G* to compute the approximate Type *B* probabilities of accepting lots from 1.5% nonconforming product under *(a)* normal and *(b)* tightened inspection.

Answer: (a) 0.922; *(b)* 0.640.

12.9. What are the acceptance criteria under reduced inspection for the conditions of Prob. 12.7?

Answer: $n_1 = 20$, $c_1 = 0$, $r_1 = 3$; $n_2 = 20$, $c_2 = 0$, $r_2 = 4$.

12.10. Assume that a lot from 2% nonconforming product is submitted under the reduced inspection plan of Prob. 12.9. Use Table *G* to compute the approximate probabilities that:

(a) The lot will be accepted and reduced inspection continued.
(b) The lot will be accepted but normal inspection will be reinstated.
(c) The lot will be rejected.

Answer: (a) 0.670; *(b)* 0.316; *(c)* 0.014.

12.11. Assume that the double sampling plan of Prob. 12.7 for normal inspection is being used where process averages are, respectively, 0.5, 3, and 6% nonconforming. Assuming no curtailment, compute the ASN value for each process average. Use the Poisson approximation to the binomial in your calculation. Express the respective ASN values as a percentage of the single sample size corresponding to this double sampling plan in the ABC standard.

Answer: 61.0, 79.3, 68.7; 76%, 99%, 86%.

12.12. In acceptance sampling under reduced inspection in the ABC standard, double sampling is used with code letter J and an AQL of 2.5%. Assume that a lot from product 5% nonconforming is submitted. Use Table *G* to compute the approximate probabilities that:

(a) The lot will be accepted and reduced inspection will be continued.
(b) The lot will be accepted but normal inspection will be reinstated.
(c) The lot will be rejected.

Answer: (a) 0.864; *(b)* 0.107; *(c)* 0.029.

12.13. In acceptance sampling under the ABC standard, single sampling is to be used with inspection level II, an AQL of 0.40%, and a lot size of 1,500.

(a) What plan is called for under normal inspection?
(b) What plan is called for under tightened inspection?

Answer: (a) $n = 125$, $c = 1$; *(b)* $n = 200$, $c = 1$.

12.14. For the plans called for in Prob. 12.13, assume that the inspection process has been operating on normal inspection and that the last five lots have been accepted. Suddenly the lot quality shifts to 1.6% nonconforming.

(a) Use Table *G* to calculate the approximate probability that the next lot sampled will be accepted.
(b) What is the probability that a shift from normal to tightened inspection will be required within the next five lots inspected?

Answer: (a) 0.406; *(b)* 0.908.

12.15. In the situation described in Prob. 12.13 and 12.14, tightened inspection is now in force. Corrective action has been taken on the process, and the process average has been reduced to 0.40% nonconforming product, the AQL value. The last lot inspected was rejected.

(*a*) Using Table *G*, compute the approximate probability that the next lot inspected will be accepted.

(*b*) What is the probability of a shift from tightened to normal inspection within the next five lots inspected?

Answer: (a) 0.809; *(b)* 0.346.

12.16. (*a*) Find the ABC standard normal inspection double sampling plan for inspection level II, an AQL of 1.5%, and a lot size of 3,000 units.

(*b*) Assuming no curtailment of inspection, compute the ASN values for process averages of 2.0, 4.5, and 8.0%.

(*c*) Sketch a curve of ASN as a function of process average over the range 2.0 to 8.0%.

12.17. A plan for inspecting isolated lots of 1,500 items is to be selected from the ABC standard based on a 10% probability of acceptance limiting quality (LQ) of 4.5% nonconforming product.

(*a*) Find the most appropriate single sampling plan if the AQL is 1.0%.

(*b*) If the LQ were not specified in this case, which normal inspection single sampling plan would be used under inspection level II?

Answer: (a) $n = 200$, $c = 5$; *(b)* $n = 125$, $c = 3$.

12.18. In acceptance sampling under the ABC standard, single sampling is to be used with inspection level II, an AQL of 1.0%, and a lot size of 15,000 units.

(*a*) What plans are called for under normal, tightened, and reduced inspection?

(*b*) Use Table *G* to find the approximate probability that lots with a μ_p of 0.03 will be accepted when normal inspection is in force.

(*c*) Use the results of *(b)* to find the probability that a shift to tightened inspection will be required after five such lots have been inspected on normal inspection.

12.19. Assume that rectifying inspection is to be used in the inspection process for the plans found in Prob. 12.18.

(*a*) What is the AOQL of the normal inspection plan?

(*b*) Use the results of part *(b)* of Prob. 12.18 to compute the AFI of the plan when μ_p is 0.03.

(*c*) Calculate the AOQ at a μ_p of 0.03.

12.20. In acceptance sampling under the ABC standard, single sampling is to be used with inspection level II, an AQL of 0.40%, and a lot size of 1,000 items.

(*a*) What are the single sampling plans under normal, tightened, and reduced inspection?

(*b*) Assuming that rectifying inspection is to be used, find the AOQL for the normal inspection plan.

(*c*) Use Table *G* to compute the approximate probability of acceptance under normal inspection for lots with a μ_p of 0.01.

(*d*) What is the probability of shifting to tightened inspection within the first five lots inspected for the lots described in part *(c)*?

12.21. In acceptance sampling under the ABC standard, double sampling is to be used with inspection level II, an AQL of 4.0%, and a lot size of 1,000 items.

(*a*) What plans are called for under normal and tightened inspection?

(*b*) Assuming no curtailment, use Table *G* to compute the ASN values under normal inspection for lot quality μ_p values of 0.02, 0.05, and 0.15.
(*c*) Compare the sample size for the single sampling normal inspection plan with these values of ASN, and comment on the relative efficiency of the double sampling plan.

12.22. A manufacturer ships a certain product in lots of 400 items that are subject to receiving inspection under the ABC standard, inspection level I, and an AQL of 2.5%.
(*a*) Find the appropriate double sampling plans for normal, tightened, and reduced inspection.
(*b*) Use Table *G* to find the approximate probability of acceptance for the normal inspection plan if incoming lots have a μ_p of 0.04.
(*c*) Use the results of part *(b)* to compute the ASN at a μ_p of 0.04. How does this value compare with the sample size of the corresponding single sample plan?

12.23. In acceptance sampling under the ABC standard for receiving inspection of lots of 5,000 items, a manufacturer uses inspection level II and a 0.065% AQL.
(*a*) What is the normal inspection double sampling plan?
(*b*) Use Table *G* to compute the approximate probability of acceptance of lots for which μ_p is 0.015.
(*c*) Rejection of a lot requires that the remainder of the lot be screened. Cost of screening the remaining items is charged to the supplier along with the cost of returning rejected items for replacement. What is the average number of items inspected on sampling inspection (ASN) and the average number of items screened (ATI-ASN) when μ_p is 0.015?

12.24. In acceptance sampling under the ABC standard, single sampling is used with inspection level II, an AQL of 0.10%, and a lot size of 1,000.
(*a*) What plans are called for under normal, tightened, and reduced inspection?
(*b*) Assuming that rectifying inspection is to be used, what are the AOQL values under normal and tightened inspection?

12.25. (*a*) For the acceptance sampling plans found in Prob. 12.24, use Table *G* to compute the probabilities of acceptance under normal and tightened inspection for lots with a μ_p of 0.01.
(*b*) Compute the values of the AOQ under normal and tightened inspection for lots with a μ_p of 0.01.
(*c*) What is the probability of switching to tightened inspection within the first five lots inspected on normal inspection for these lots?

12.26. In acceptance sampling under the ABC standard, double sampling is used with inspection level III, an AQL of 1.0%, and a lot size of 1,500 items.
(*a*) What plans are called for under normal, tightened, and reduced inspection?
(*b*) For the reduced sampling plan, use Table *G* to compute the approximate probabilities of accepting a lot and continuing on reduced inspection and of accepting a lot but switching back to normal inspection if incoming lots have a μ_p of 0.02.
(*c*) Use Table *G* to compute the ASN under normal inspection if the μ_p of incoming lots equals the AQL. How does this value compare with the sample size of the corresponding normal inspection single sampling plan?

12.27. Find the ABC-STD-105 double sampling plan for a lot size of 800 items, inspection level II, and an AQL of 2.5% under normal inspection.

(*a*) Calculate the approximate probability of acceptance when incoming lots contain 4% rejectable items. Use Table *G*, and ignore the relationship between the lot size and the sample sizes in making these calculations.

(*b*) Calculate the ASN for this plan when $\mu_p = 0.04$.

(*c*) Assuming that this plan is to be used in a rectifying inspection procedure, calculate the AFI and AOQ when $\mu_p = 0.04$.

12.28. In acceptance sampling under the ABC standard for receiving inspection of lots of 800 items, a manufacturer uses inspection level II and a 1.0% AQL.

(*a*) Find the appropriate single sample plans for normal, tightened, and reduced inspection.

(*b*) Use Table *G* to calculate the approximate probability of acceptance under normal and tightened inspection for lots with μ_p of 0.04.

(*c*) Assuming that normal inspection is in force, calculate the probability of switching to tightened inspection within five lots inspected when μ_p is 0.04.

(*d*) Assuming that tightened inspection is in force, calculate the probability of switching to normal inspection within five lots inspected when μ_p is 0.04.

12.29. An ABC-STD-105 double sampling system is to be used for rectifying inspection of lots of 1,000 units, inspection level II, and an AQL of 1.5%.

(*a*) What plans are called for under normal, tightened, and reduced inspection?

(*b*) Use Table *G* to calculate the probability of acceptance under normal inspection assuming that incoming lots have a μ_p of 0.04.

(*c*) What will be the AFI for the lots described in part (*b*)?

(*d*) What will be the AOQ for the lots described in part (*b*)?

12.30. The ABC-STD-105 double sampling, normal inspection plan for code letter J and an AQL of 0.65% calls for

$$n_1 = 50 \qquad c_1 = 0 \qquad r_1 = 2$$

$$n_2 = 50 \qquad c_2 = 1 \qquad r_2 = 2$$

(*a*) What are the corresponding plans under tightened and reduced inspection?

(*b*) What is the probability of acceptance of lots containing 2% rejectable items under normal inspection? Use Table *G*, and ignore the lot size in making your calculation.

(*c*) What is the ASN value at a μ_p of 0.02 under normal inspection?

(*d*) What is the probability of switching from normal to tightened inspection within the next five lots sampled under the conditions described in (*b*)?

12.31. (*a*) Find the ABC-STD-105 single sampling plans for normal, tightened, and reduced inspection for a lot size of 600 units, general inspection level II, and an AQL of 0.65%.

(*b*) What are the AOQLs for the normal and tightened inspection plans found in part (*a*) assuming rectifying inspection is to be applied?

(*c*) What are the probabilities of acceptance under normal and tightened inspection for a sequence of lots with a μ_p of 0.02? (Use Table *G*.)

(*d*) Compute the probability that a switch from normal to tightened inspection will be required after the first five lots described in part (*c*) have been inspected.

12.32. (*a*) Find the double sampling plans that correspond to the single sampling plans found in Prob. 12.31.

(*b*) For incoming lots with a μ_p of 0.02, compute the probabilities of acceptance under normal inspection on the first sample, the second sample, and overall. Use Table *G,* and ignore the lot size in making these calculations.

(*c*) Compute the ASN for the normal inspection double sampling when $\mu_p =$ 0.02. How does this compare with the corresponding single sampling plan sample size?

(*d*) Compute the AFI and AOQ for the lots described in part (*b*) where rectifying inspection is to be applied. The lot size is 600 items.

12.33. The ABC standard specifies the following single sampling plan, normal inspection, for inspection level II, a lot size of 5,000, and an AQL of 1.5%:

$$n = 200 \qquad c = 7 \qquad r = 8$$

(*a*) What are the corresponding single sampling plans under reduced and tightened inspection?

(*b*) Use Table *G* to find the probability that lots with a μ_p of 0.04 will be accepted when normal inspection is in force.

(*c*) Use the results of (*b*) to find the probability that a shift to tightened inspection will be required after five such lots have been inspected.

(*d*) Assuming that rectifying inspection is used (rejected lots are screened and defectives removed from both rejected lots and samples from accepted lots and replaced with good items), find the AFI and the AOQ based on the results obtained in (*b*).

12.34. (*a*) Find the ABC-STD-105 single sampling plans under normal, tightened, and reduced inspection for a lot size of 400 units, general inspection level III, and an AQL of 10 nonconformities per 100 units inspected.

(*b*) The corresponding normal inspection double sampling plan is

$$n_1 = n_2 = 50 \qquad c_1 = 7 \qquad r_1 = 11$$
$$c_2 = 18 \qquad r_2 = 19$$

What is the probability of acceptance of lots containing 25 nonconformities per 100 units for this double sampling plan? Ignore the depletion of nonconformities in the first sample when making this calculation.

(*c*) Rectifying inspection is used in the inspection process. Calculate the AFI and AOQ for the double sampling normal inspection plan for lots containing 25 nonconformities per 100 units.

12.35. Receiving inspection is to be performed on a nonconformities per 100 units basis using the ABC standard, inspection level I, and an AQL of 40. The lot size is 15,000 units.

(*a*) Find the single sampling plans under normal, tightened, and reduced inspection.

(*b*) Calculate the probabilities of acceptance under normal and tightened inspection if incoming lots have a μ_c of 60.

(*c*) What is the probability of switching from normal to tightened inspection within five lots if μ_c is 60?

(*d*) For the lots described in part (*c*), assume that a switch to tightened inspection has been made. What is the probability of returning to normal inspection after the next five lots are inspected assuming that no improvement is made in the process average?

12.36. An ABC standard double sampling system of plans is to be used for inspection of lots of 1,000 items at receiving inspection. A code letter K plan will be used (inspection level III) with an AQL of 0.04%.

(*a*) What plans are called for under normal, tightened, and reduced inspection?

(*b*) Find the probability of acceptance for the normal inspection plan when the incoming μ_p is 0.01.

(*c*) Calculate the ASN for the normal inspection plan for the lots described in part (*b*).

(*d*) Rectifying inspection is to be used. Calculate the AOQ and the AFI for the lots described in part (*b*).

12.37. Receiving inspection is conducted on an item on a defects-per-hundred-units basis, inspection level II, received in lots of 1,500 units. The AQL is 6.5 defects per 100 units.

(*a*) Find the appropriate single sampling plans under normal, tightened, and reduced inspection.

(*b*) Calculate the probability of acceptance under normal inspection if incoming lots have 0.15 defects per unit (15 defects per 100 units).

(*c*) What are the AOQL values under normal and tightened inspection?

(*d*) What is the probability of switching to tightened inspection after five lots containing 15 defects per 100 units have been inspected under normal inspection?

12.38. (*a*) Select from the ABC standard a system of double sampling plans for reduced, normal, and tightened inspection for a lot size of 4,000 units, inspection level III, and an AQL of 0.04%.

(*b*) Assuming that incoming lots are 0.1% defective, calculate the probability of acceptance under reduced inspection and staying on reduced inspection.

(*c*) Assuming that incoming lots are 0.1% defective, calculate the probability of acceptance of a specific lot on reduced inspection and returning to normal inspection for subsequent lots.

12.39. An ABC-STD-105 system of single sampling plans is to be used at receiving inspection of a certain item received in lots of 500 units. Sampling is to be conducted using inspection level III and an AQL of 1.5%.

(*a*) What single sampling plans are called for under normal, tightened, and reduced inspection?

(*b*) What are the AOQLs of the normal and tightened inspection plans if rectifying inspection is employed?

(*c*) An *np* chart is to be used to plot the sample data results to test for consistency of successive lots. The AQL is to be used as the target value of μ_p. What will be the central line and control limits under normal and tightened inspection?

(*d*) Use the Poisson approximation to find the probability of acceptance, AFI, and AOQ for incoming lots with a μ_p of 0.04 when on normal inspection.

(*e*) What is the probability that any one point will fall outside the control limits on the *np* chart when μ_p is 0.04? Use Table *G*.

(*f*) What is the probability of a run of 7 points between the central line and the upper control limit on the *np* chart when μ_p is 0.04?

12.40. The ABC-STD-105 single sampling plan for normal inspection, a lot size of 100 units, inspection level II, and an AQL of 0.65% is $n = 20$, $c = 0$, and $r = 1$.

(*a*) What plans are prescribed under reduced and tightened inspection?

(*b*) Use Table *G* to find the approximate probabilities of acceptance under normal and tightened inspection for lots with an incoming μ_p of 0.025.

(*c*) What is the probability of a switch from normal to tightened inspection within five lots inspected when μ_p is 0.025?

(*d*) What is the probability that a switch from tightened to normal inspection will be made within five lots inspected when μ_p is 0.025?

12.41. An ABC-STD-105 system of single sampling plans is to be used at receiving inspection of a certain part received in lots of 500 units. Sampling is to be conducted using inspection level I and an AQL of 1.5%.

(*a*) What plans are called for under normal, tightened, and reduced inspection?

(*b*) What are the AOQLs of the normal and tightened inspection plans if rectifying inspection is employed?

(*c*) Use the Poisson approximation to find the probability of acceptance, AFI, and AOQ for incoming lots with a μ_p of 0.04 when on normal inspection.

12.42. An ABC-STD-105 single sampling system of plans is to be used at receiving inspection of a part received in lots of 500 units. Sampling is to be conducted using inspection level III and an AQL of 1.5%.

(*a*) What plans are called for under normal, tightened, and reduced inspection?

(*b*) What are the AOQLs of the normal and tightened inspection plans if rectifying inspection is employed?

(*c*) Use the Poisson approximation to find the probability of acceptance, AFI, and AOQ for incoming lots with a μ_p of 0.04 when on normal inspection.

12.43. A supplier produces blank circuit cards which it ships to an electronics manufacturer in lots of 200 cards. By contract, these cards are inspected on receipt using single sampling and rectifying inspection in accordance with MIL-STD-105E (the ABC standard).

(*a*) Find the appropriate sampling plans for normal, tightened, and reduced inspection for an AQL of 0.04%.

(*b*) The supplier's process is operating in control at a μ_p of 0.03. What are the respective probabilities of acceptance on normal, tightened, and reduced inspection at this level of nonconforming product? Give the probabilities for both conditions of acceptance on reduced inspection. Use Table *G*.

(*c*) Assuming that the process described in part (*b*) is being inspected by the customer under normal inspection, what is the probability of a shift to tightened inspection within any five consecutive lots inspected?

(*d*) What are the AOQLs under normal and tightened inspection?

(*e*) What is the AFI and AOQ under normal inspection when the process is generating 3% nonconforming units?

(*f*) The contract states that the customer will provide all sampling and screening inspection, and the supplier will be invoiced for the cost of screening. For what average proportion of shipped product will the supplier be billed for this screening inspection? Assume normal inspection and a μ_p of 0.03.

13

OTHER PROCEDURES FOR ACCEPTANCE SAMPLING BY ATTRIBUTES

One trouble with 100% inspection, where it is practicable, is that the inspector merely cleans up the faults of others, sorting the good from the bad, and the production man takes it as a matter of course if just individual articles are returned to him for repair. But if a whole lot is returned to him, as when lot sampling is used, and he is required to undertake the entire corrective action, the steady outward flow of product is interrupted. If there are many lot rejections, he must get busy to find the cause and eliminate it in order to avoid further lot rejections. This is an indirect power of sampling—it forces correction of the process, where the fault lies.

—H. F. Dodge[†]

13.1 TWO USEFUL VOLUMES OF STANDARD TABLES

Anyone responsible for the choice of an acceptance sampling procedure by attributes should have two volumes available for consultation. One of these is *Sampling Inspection Tables* by H. F. Dodge and H. G. Romig.[‡] The other is *Sampling Inspection* prepared by the Statistical Research Group (SRG) of Columbia University.[§] Both of these classic volumes not only contain extensive tables of acceptance sampling plans for attributes inspection but also contain much useful material on the theory and practice of acceptance sampling.

[†]H. F. Dodge, "Administration of a Sampling Inspection Plan," *Industrial Quality Control,* vol. 5, no. 3, pp. 12–19, November 1948; and reprinted in *Journal of Quality Technology,* vol. 9, no. 3, July 1977.

[‡]H. F. Dodge and H. G. Romig, *Sampling Inspection Tables—Single and Double Sampling,* 2d ed., John Wiley & Sons, Inc., New York, 1959.

[§]H. A. Freeman, Milton Friedman, Frederick Mosteller, and W. A. Wallis (eds.), *Sampling Inspection,* McGraw-Hill Book Company, New York, 1948.

The Dodge-Romig tables were originally prepared for use within the Bell Telephone System. They were designed primarily to minimize the total amount of inspection, considering both sampling inspection and screening inspection of rejected lots. Our discussion of sampling procedures in this chapter begins with these very important tables.

13.2 THE DODGE-ROMIG TABLES

The Dodge-Romig volume contains four sets of tables, as follows:

I. Single Sampling Lot Tolerance Tables
II. Double Sampling Lot Tolerance Tables
III. Single Sampling AOQL Tables
IV. Double Sampling AOQL Tables

Sets I and II apply to the following lot tolerance percent defectives (assuming Consumer's Risk = 0.10):

0.5%	3.0%	7.0%
1.0%	4.0%	10.0%
2.0%	5.0%	

Sets III and IV apply to the following values of AOQL:

0.1%	1.5%	4.0%
0.25%	2.0%	5.0%
0.5%	2.5%	7.0%
0.75%	3.0%	10.0%
1.0%		

13.2.1 Single Sampling Lot Tolerance Tables

Table 13.1 is representative of Dodge-Romig Set I.

All the sampling plans in this table have the same lot tolerance percent defective (LTPD), namely, 5.0%. However, the plans have different values of AOQL. The tables give the AOQL figure for each plan.

The table contains six columns, each for a different value of process average percent defective. The purpose of these different columns is to indicate the plan that involves the minimum total inspection, considering both the inspection of samples and the 100% inspection of rejected lots.

TABLE 13.1 EXAMPLE OF DODGE-ROMIG SINGLE SAMPLING LOT TOLERANCE TABLES†
Lot tolerance percent defective = 5.0%. Consumer's Risk = 0.10

	Process average, %																	
	0–0.05			0.06–0.50			0.51–1.00			1.01–1.50			1.51–2.00			2.01–2.50		
Lot size	*n*	*c*	AOQL, %	*n*	*c*	AOQL, %	*n*	*c*	AOQL, %	*n*	*c*	AOQL, %	*n*	*c*	AOQL, %	*n*	*c*	AOQL, %
1–30	All	0	0	All	0	0	All	0	0	All	0	0	All	0	0	All	0	0
31–50	30	0	0.49	30	0	0.49	30	0	0.49	30	0	0.49	30	0	0.49	30	0	0.49
51–100	37	0	0.63	37	0	0.63	37	0	0.63	37	0	0.63	37	0	0.63	37	0	0.63
101–200	40	0	0.74	40	0	0.74	40	0	0.74	40	0	0.74	40	0	0.74	40	0	0.74
201–300	43	0	0.74	43	0	0.74	70	1	0.92	70	1	0.92	95	2	0.99	95	2	0.99
301–400	44	0	0.74	44	0	0.74	70	1	0.99	100	2	1.0	120	3	1.1	145	4	1.1
401–500	45	0	0.75	75	1	0.95	100	2	1.1	100	2	1.1	125	3	1.2	150	4	1.2
501–600	45	0	0.76	75	1	0.98	100	2	1.1	125	3	1.2	150	4	1.3	175	5	1.3
601–800	45	0	0.77	75	1	1.0	100	2	1.2	130	3	1.2	175	5	1.4	200	6	1.4
801–1000	45	0	0.78	75	1	1.0	105	2	1.2	155	4	1.4	180	5	1.4	225	7	1.5
1001–2000	45	0	0.80	75	1	1.0	130	3	1.4	180	5	1.6	230	7	1.7	280	9	1.8
2001–3000	75	1	1.1	105	2	1.3	135	3	1.4	210	6	1.7	280	9	1.9	370	13	2.1
3001–4000	75	1	1.1	105	2	1.3	160	4	1.5	210	6	1.7	305	10	2.0	420	15	2.2
4001–5000	75	1	1.1	105	2	1.3	160	4	1.5	235	7	1.8	330	11	2.0	440	16	2.2
5001–7000	75	1	1.1	105	2	1.3	185	5	1.7	260	8	1.9	350	12	2.2	490	18	2.4
7001–10,000	75	1	1.1	105	2	1.3	185	5	1.7	260	8	1.9	380	13	2.2	535	20	2.5
10,001–20,000	75	1	1.1	135	3	1.4	210	6	1.8	285	9	2.0	425	15	2.3	610	23	2.6
20,001–50,000	75	1	1.1	135	3	1.4	235	7	1.9	305	10	2.1	470	17	2.4	700	27	2.7
50,001–100,000	75	1	1.1	160	4	1.6	235	7	1.9	355	12	2.2	515	19	2.5	770	30	2.8

†Reprinted by permission from H. F. Dodge and H. G. Romig, *Sampling Inspection Tables—Single and Double Sampling,* 2d ed., John Wiley & Sons, Inc., New York, 1959.

For example, consider the inspection plans indicated in the table for the lot size range from 501 to 600. Six different plans, namely, $\begin{cases} n = 45 \\ c = 0 \end{cases}$ $\begin{cases} n = 75 \\ c = 1 \end{cases}$ $\begin{cases} n = 100 \\ c = 2 \end{cases}$ $\begin{cases} n = 125 \\ c = 3 \end{cases}$ $\begin{cases} n = 150 \\ c = 4 \end{cases}$ and $\begin{cases} n = 175 \\ c = 5 \end{cases}$ all have an LTPD of 0.05. But, as previously explained, if rejected lots are to be detailed, the total amount of inspection of samples and rejected lots under these plans will depend on the quality level of the product submitted for inspection. Table 13.1 simply says that this total inspection will be a minimum for $\begin{cases} n = 45 \\ c = 0 \end{cases}$ if the process average is 0–0.05% defective, that it will be a minimum for $\begin{cases} n = 75 \\ c = 1 \end{cases}$ if the process average is 0.06–0.50%, and so forth.

If there is no basis for estimating the process average, the sampling plan should be selected from the right-hand column of the table. This gives the desired quality protection and gives satisfactory lots a better chance of acceptance. Moreover, it collects data more rapidly to permit reliable estimates of the process average.

In the use of these lot tolerance tables, it should be understood that the tables contemplate the screening of rejected lots. However, the tables give the consumer the stated quality protection regardless of any provision for screening. Even though rejected lots are merely returned by the consumer to the producer with no formal provision for screening by anyone, it is reasonable to suppose that the producer will screen these lots. Under such circumstances, the consumer's use of the process average to determine the acceptance criteria may be thought of as aimed at minimizing the total inspection done by industry as a whole, considering both the consumer's sampling inspection and the screening presumably done by the producer.

13.2.2 Double Sampling Lot Tolerance Tables

Table 13.2 is representative of Dodge-Romig Set II.

Some differences between single and double sampling plans are evident from a comparison of Tables 13.1 and 13.2. These differences may be brought out to best advantage by comparing any single sampling plan with a double sampling plan for the same lot size and process average that gives the same lot quality protection. For example, consider the plans from Tables 13.1 and 13.2 for a lot size of 801–1,000 and a process average of 0.51–1.00% defective:

Single sampling	*Double sampling*	
$n = 105$	$n_1 = 55$	$n_2 = 115$
$c = 2$	$c_1 = 0$	$n_1 + n_2 = 170$
		$c_2 = 4$

The first sample in double sampling is smaller than the one sample in single sampling; in this case it is 55 instead of 105. On the other hand, the combined sample in double sampling is larger—170 compared to 105. The relative number of articles inspected in the samples in the two plans evidently depends on the quality of submitted product. If the product sampled is good enough that very few second samples have to be taken, the inspection will be substantially less with the double sampling plan. If many second samples need to be taken, the single sampling will require less sampling inspection. Dodge and Romig give a diagram that compares the inspection under single and double sampling for various lot sizes and ratios of process average to lot tolerance fraction defectives.[†] They state that over the portion of the tables most useful in practice, the saving in inspection due to double sampling is usually over 10% and may be as much as 50%.

[†]Dodge and Romig, op. cit., fig. 2-5, p. 31.

TABLE 13.2 EXAMPLE OF DODGE-ROMIG DOUBLE SAMPLING LOT TOLERANCE TABLES[†]

Lot tolerance percent defective = 5.0%. Consumer's Risk = 0.10.

Process average, %

Lot size	0–0.05 Trial 1 n_1	c_1	Trial 2 n_2	n_1+n_2	c_2	AOQL %	0.06–0.50 Trial 1 n_1	c_1	Trial 2 n_2	n_1+n_2	c_2	AOQL %
1–30	All	0	—	—	—	0	All	0	—	—	—	0
31–50	30	0	—	—	—	0.49	30	0	—	—	—	0.49
51–75	38	0	—	—	—	0.59	38	0	—	—	—	0.59
76–100	44	0	21	65	1	0.64	44	0	21	65	1	0.64
101–200	49	0	26	75	1	0.84	49	0	26	75	1	0.84
201–300	50	0	30	80	1	0.91	50	0	30	80	1	0.91
301–400	55	0	30	85	1	0.92	55	0	55	110	2	1.1
401–500	55	0	30	85	1	0.93	55	0	55	110	2	1.1
501–600	55	0	30	85	1	0.94	55	0	60	115	2	1.1
601–800	55	0	35	90	1	0.95	55	0	65	120	2	1.1
801–1000	55	0	35	90	1	0.96	55	0	65	120	2	1.1
1001–2000	55	0	35	90	1	0.98	55	0	95	150	3	1.3
2001–3000	55	0	65	120	2	1.2	55	0	95	150	3	1.3
3001–4000	55	0	65	120	2	1.2	55	0	95	150	3	1.3
4001–5000	55	0	65	120	2	1.2	55	0	95	150	3	1.4
5001–7000	55	0	65	120	2	1.2	55	0	95	150	3	1.4
7001–10,000	55	0	65	120	2	1.2	55	0	120	175	4	1.5
10,001–20,000	55	0	65	120	2	1.2	55	0	120	175	4	1.5
20,001–50,000	55	0	65	120	2	1.2	55	0	150	205	5	1.7
50,001–100,000	55	0	65	120	2	1.2	55	0	150	205	5	1.7

Lot size	0.51–1.00 Trial 1 n_1	c_1	Trial 2 n_2	n_1+n_2	c_2	AOQL %	1.01–1.50 Trial 1 n_1	c_1	Trial 2 n_2	n_1+n_2	c_2	AOQL %
1–30	All	0	—	—	—	0.	All	0	—	—	—	0
31–50	30	0	—	—	—	0.49	30	0	—	—	—	0.49
51–75	38	0	—	—	—	0.59	38	0	—	—	—	0.59
76–100	44	0	21	65	1	0.64	44	0	21	65	1	0.64
101–200	49	0	26	75	1	0.84	49	0	51	100	2	0.91
201–300	50	0	55	105	2	1.0	50	0	55	105	2	1.0
301–400	55	0	55	110	2	1.1	55	0	80	135	3	1.1
401–500	55	0	80	135	3	1.2	55	0	105	160	4	1.3
501–600	55	0	85	140	3	1.2	55	0	110	165	4	1.3
601–800	55	0	85	140	3	1.3	90	1	125	215	6	1.5
801–1000	55	0	115	170	4	1.4	90	1	150	240	7	1.5
1001–2000	55	0	120	175	4	1.4	90	1	185	275	8	1.7
2001–3000	55	0	150	205	5	1.5	120	2	180	300	9	1.9
3001–4000	90	1	140	230	6	1.6	120	2	210	330	10	2.0
4001–5000	90	1	165	255	7	1.8	120	2	255	375	12	2.1
5001–7000	90	1	165	255	7	1.8	120	2	260	380	12	2.1
7001–10,000	90	1	190	280	8	1.9	120	2	285	405	13	2.1
10,001–20,000	90	1	190	280	8	1.9	120	2	310	430	14	2.2
20,001–50,000	90	1	215	305	9	2.0	120	2	335	455	15	2.2
50,001–100,000	90	1	240	330	10	2.1	120	2	360	480	16	2.3

Lot size	1.51–2.00 Trial 1 n_1	c_1	Trial 2 n_2	n_1+n_2	c_2	AOQL %	2.01–2.50 Trial 1 n_1	c_1	Trial 2 n_2	n_1+n_2	c_2	AOQL %
1–30	All	0	—	—	—	0	All	0	—	—	—	0
31–50	30	0	—	—	—	0.49	30	0	—	—	—	0.49
51–75	38	0	—	—	—	0.59	38	0	—	—	—	0.59
76–100	44	0	21	65	1	0.64	44	0	21	65	1	0.64
101–200	49	0	51	100	2	0.91	49	0	51	100	2	0.91
201–300	50	0	80	130	3	1.1	50	0	100	150	4	1.1
301–400	55	0	100	155	4	1.2	85	1	105	190	6	1.3
401–500	85	1	120	205	6	1.4	85	1	140	225	7	1.4
501–600	85	1	145	230	7	1.4	85	1	165	250	8	1.5
601–800	90	1	170	260	8	1.5	120	2	185	305	10	1.6
801–1000	90	1	200	290	9	1.6	120	2	210	330	11	1.7
1001–2000	120	2	225	345	11	1.9	175	4	260	435	15	2.0
2001–3000	150	3	270	420	14	2.1	205	5	375	580	21	2.3
3001–4000	150	3	295	445	15	2.3	230	6	420	650	24	2.4
4001–5000	150	3	345	495	17	2.3	255	7	445	700	26	2.5
5001–7000	150	3	370	520	18	2.3	255	7	495	750	28	2.6
7001–10,000	175	4	370	545	19	2.4	280	8	540	820	31	2.7
10,001–20,000	175	4	420	595	21	2.4	280	8	660	940	36	2.8
20,001–50,000	205	5	485	690	25	2.5	305	9	745	1050	41	2.9
50,001–100,000	205	5	555	760	28	2.6	330	10	810	1140	45	3.0

[†]Reprinted by permission from H. F. Dodge and H. G. Romig, *Sampling Inspection Tables —Single and Double Sampling,* 2d ed., John Wiley & Sons, Inc., New York, 1959.

One characteristic of all the Dodge-Romig double sampling plans is that c_2 is always 1 or more. This means that no lot is ever rejected as a result of only 1 defective.

13.2.3 Single Sampling AOQL Tables

Table 13.3 is representative of Dodge-Romig Set III.

In contrast to Table 13.1 in which all the single sampling plans had the same LTPD, all the plans in Table 13.3 have the same AOQL, namely, 2%. The table gives the lot tolerance percent defective for each plan. It is of interest to note that the larger the sample size and acceptance number for a given AOQL, the lower the LTPD.

Like all the Dodge-Romig tables, columns are given for various process averages; the plan in each column is the one that gives the minimum total inspection for the process average at the head of the column. Thus all the plans on any line of the table are alike in quality protection (as measured by the AOQL) and differ only in total amount of inspection required. As remarked in the discussion of Table 13.1, if there is no basis for estimating the process average, the sampling plan should be chosen from the right-hand column of the table.

13.2.4 Double Sampling AOQL Tables

Table 13.4 is representative of Dodge-Romig Set IV.

Dodge and Romig point out that the lot tolerance concept was first developed and applied in the Bell Telephone System in 1923. The concept of the AOQL was developed and applied in 1927. Thus tables involving both concepts, and involving single and double sampling, have been available within the Bell System for application to all types of inspection for many years. It is significant that Dodge and Romig state that the double sampling AOQL tables, of which Table 13.4 is an example, have proved the most useful of all the tables.

13.2.5 Relationship between the Process Average Used in Selecting a Dodge-Romig Sampling Plan and the OC Curve of the Plan Selected

Tables 13.1 to 13.4 all have six columns, each corresponding to a stated process average. In general, the greater the process average used in entering any Dodge-Romig table, the larger the sample size for any stated lot size. The plans having larger sample sizes have steeper OC curves with resulting better discrimination between lots superior to the quality standard and lots worse than the quality standard. This point is illustrated by Figs. 13.1 and 13.2.

Figure 13.1 shows the OC curves for the six double sampling plans given in Table 13.2 (LTPD = 5%) for the lot size 4,001–5,000. All these curves naturally show that the probability of acceptance of a 5% defective lot is 0.10. At all other points, however, the OC curves differ considerably. The differences are particu-

TABLE 13.3 EXAMPLE OF DODGE-ROMIG SINGLE SAMPLING AOQL TABLES[†]

Note: Average outgoing quality limit = 2.0%

	Process average, %																	
	0–0.04			0.05–0.40			0.41–0.80			0.81–1.20			1.21–1.60			1.61–2.00		
Lot size	n	c	$100p_{0.10}$	n	c	$100p_{0.10}$	n	c	$100p_{0.10}$	n	c	$100p_{0.10}$	n	c	$100p_{0.10}$	n	c	$100p_{0.10}$
1–15	All	0	....	All	0	....	All	0	....	All	0	....	All	0	....	All	0	....
16–50	14	0	13.6	14	0	13.6	14	0	13.6	14	0	13.6	14	0	13.6	14	0	13.6
51–100	16	0	12.4	16	0	12.4	16	0	12.4	16	0	12.4	16	0	12.4	16	0	12.4
101–200	17	0	12.2	17	0	12.2	17	0	12.2	17	0	12.2	35	1	10.5	35	1	10.5
201–300	17	0	12.3	17	0	12.3	17	0	12.3	37	1	10.2	37	1	10.2	37	1	10.2
301–400	18	0	11.8	18	0	11.8	38	1	10.0	38	1	10.0	38	1	10.0	60	2	8.5
401–500	18	0	11.9	18	0	11.9	39	1	9.8	39	1	9.8	60	2	8.6	60	2	8.6
501–600	18	0	11.9	18	0	11.9	39	1	9.8	39	1	9.8	60	2	8.6	60	2	8.6
601–800	18	0	11.9	40	1	9.6	40	1	9.6	65	2	8.0	65	2	8.0	85	3	7.5
801–1000	18	0	12.0	40	1	9.6	40	1	9.6	65	2	8.1	65	2	8.1	90	3	7.4
1001–2000	18	0	12.0	41	1	9.4	65	2	8.2	65	2	8.2	95	3	7.0	120	4	6.5
2001–3000	18	0	12.0	41	1	9.4	65	2	8.2	95	3	7.0	120	4	6.5	180	6	5.8
3001–4000	18	0	12.0	42	1	9.3	65	2	8.2	95	3	7.0	155	5	6.0	210	7	5.5
4001–5000	18	0	12.0	42	1	9.3	70	2	7.5	125	4	6.4	155	5	6.0	245	8	5.3
5001–7000	18	0	12.0	42	1	9.3	95	3	7.0	125	4	6.4	185	6	5.6	280	9	5.1
7001–10,000	42	1	9.3	70	2	7.5	95	3	7.0	155	5	6.0	220	7	5.4	350	11	4.8
10,001–20,000	42	1	9.3	70	2	7.6	95	3	7.0	190	6	5.6	290	9	4.9	460	14	4.4
20,001–50,000	42	1	9.3	70	2	7.6	125	4	6.4	220	7	5.4	395	12	4.5	720	21	3.9
50,001–100,000	42	1	9.3	95	3	7.0	160	5	5.9	290	9	4.9	505	15	4.2	955	27	3.7

[†]Reprinted by permission from H. F. Dodge and H. G. Romig, *Sampling Inspection Tables—Single and Double Sampling*, 2d ed., John Wiley & Sons, Inc., New York, 1959.

TABLE 13.4 EXAMPLE OF DODGE-ROMIG DOUBLE SAMPLING AOQL TABLES[†]

Note: Average outgoing quality limit = 2.0%

Lot size	Process average, %																	
	0–0.04						0.05–0.40						0.41–0.80					
	Trial 1		Trial 2			$100p_{0.10}$	Trial 1		Trial 2			$100p_{0.10}$	Trial 1		Trial 2			$100p_{0.10}$
	n_1	c_1	n_2	n_1+n_2	c_2		n_1	c_1	n_2	n_1+n_2	c_2		n_1	c_1	n_2	n_1+n_2	c_2	
1–15	All	0	—	—	—	—	All	0	—	—	—	—	All	0	—	—	—	—
16–50	14	0	—	—	—	13.6	14	0	—	—	—	13.6	14	0	—	—	—	13.6
51–100	21	0	12	33	1	11.7	21	0	12	33	1	11.7	21	0	12	33	1	11.7
101–200	24	0	13	37	1	11.0	24	0	13	37	1	11.0	24	0	13	37	1	11.0
201–300	26	0	15	41	1	10.4	26	0	15	41	1	10.4	29	0	31	60	2	9.1
301–400	26	0	16	42	1	10.3	26	0	16	42	1	10.3	30	0	35	65	2	9.0
401–500	27	0	16	43	1	10.3	30	0	35	65	2	9.0	30	0	35	65	2	9.0
501–600	27	0	16	43	1	10.3	31	0	34	65	2	8.9	35	0	55	90	3	7.9
601–800	27	0	17	44	1	10.2	31	0	39	70	2	8.8	35	0	60	95	3	7.7
801–1000	27	0	17	44	1	10.2	32	0	38	70	2	8.7	36	0	59	95	3	7.6
1001–2000	33	0	37	70	2	8.5	33	0	37	70	2	8.5	37	0	63	100	3	7.5
2001–3000	34	0	41	75	2	8.2	34	0	41	75	2	8.2	41	0	84	125	4	7.0
3001–4000	34	0	41	75	2	8.2	38	0	62	100	3	7.3	41	0	89	130	4	6.9
4001–5000	34	0	41	75	2	8.2	38	0	62	100	3	7.3	42	0	88	130	4	6.9
5001–7000	35	0	40	75	2	8.1	38	0	62	100	3	7.3	44	0	116	160	5	6.4
7001–10,000	35	0	40	75	2	8.1	38	0	62	100	3	7.3	45	0	115	160	5	6.3
10,001–20,000	35	0	40	75	2	8.1	39	0	66	105	3	7.2	45	0	115	160	5	6.3
20,001–50,000	35	0	40	75	2	8.1	43	0	92	135	4	6.6	47	0	148	195	6	6.0
50,001–100,000	35	0	45	80	2	8.0	43	0	92	135	4	6.6	85	1	185	270	8	5.2

Lot size	Process average, %																	
	0.81–1.20						1.21–1.60						1.61–2.00					
	Trial 1		Trial 2			$100p_{0.10}$	Trial 1		Trial 2			$100p_{0.10}$	Trial 1		Trial 2			$100p_{0.10}$
	n_1	c_1	n_2	n_1+n_2	c_2		n_1	c_1	n_2	n_1+n_2	c_2		n_1	c_1	n_2	n_1+n_2	c_2	
1–15	All	0	—	—	—	—	All	0	—	—	—	—	All	0	—	—	—	—
16–50	14	0	—	—	—	13.6	14	0	—	—	—	13.6	14	0	—	—	—	13.6
51–100	21	0	12	33	1	11.7	21	0	12	33	1	11.7	23	0	23	46	2	10.9
101–200	27	0	28	55	2	9.6	27	0	28	55	2	9.6	27	0	28	55	2	9.6
201–300	29	0	31	60	2	9.1	32	0	48	80	3	8.4	32	0	48	80	3	8.4
301–400	33	0	52	85	3	8.2	33	0	52	85	3	8.2	36	0	69	105	4	7.6
401–500	34	0	56	90	3	7.9	36	0	74	110	4	7.5	60	1	90	150	6	7.0
501–600	35	0	55	90	3	7.9	37	0	78	115	4	7.4	65	1	95	160	6	6.8
601–800	38	0	82	120	4	7.3	38	0	82	120	4	7.3	70	1	120	190	7	6.4
801–1000	38	0	87	125	4	7.2	70	1	100	170	6	6.5	70	1	145	215	8	6.2
1001–2000	43	0	112	155	5	6.5	80	1	160	240	8	5.8	110	2	205	315	11	5.5
2001–3000	75	1	115	190	6	6.1	115	2	195	310	10	5.3	160	3	310	470	15	4.7
3001–4000	80	1	140	220	7	5.8	120	2	255	375	12	5.0	235	5	415	650	20	4.3
4001–5000	80	1	175	255	8	5.5	125	2	285	410	13	4.9	275	6	475	750	23	4.2
5001–7000	85	1	205	290	9	5.3	125	2	320	445	14	4.8	280	6	575	855	26	4.1
7001–10,000	85	1	210	295	9	5.2	165	3	335	500	15	4.5	320	7	645	965	29	4.0
10,001–20,000	90	1	260	350	11	5.1	170	3	425	595	18	4.4	395	9	835	1230	37	3.9
20,001–50,000	130	2	300	430	13	4.7	205	4	515	720	22	4.3	480	11	1090	1570	46	3.7
50,001–100,000	135	2	345	480	14	4.5	250	5	615	865	26	4.1	580	13	1460	2040	58	3.5

[†]Reprinted by permission from H. F. Dodge and H. G. Romig, *Sampling Inspection Tables—Single and Double Sampling*, 2d ed., John Wiley & Sons, Inc., New York, 1959.

larly striking at about half the 5% LTPD. Thus a 2.5% defective lot is almost certain to be accepted under the plan corresponding to the process average 2.01–2.50% (curve 6) but has only about a 0.50 probability of acceptance under the plans corresponding to the process averages 0–0.05% and 0.06–0.50% (curves 1 and 2). In general, where submitted lots are as good as the process average used in entering the tables, they are almost certain to be accepted.

Figure 13.2 shows the OC curves for the six double sampling plans given in Table 13.4 (AOQL = 2%). A comparison with Fig. 13.1 may be helpful in emphasizing certain similarities and certain differences between using the LTPD and the AOQL as the quality standard. Just as in the LTPD tables, any lots at the assumed

FIGURE 13.1 OC curves for the six Dodge-Romig double sampling plans for LTPD of 5% and a lot size of 4,001–5,000. *(Reproduced from H. F. Dodge, "Administration of a Sampling Inspection Plan," Industrial Quality Control, vol. 4, no. 3, pp. 12–19, November 1948.)*

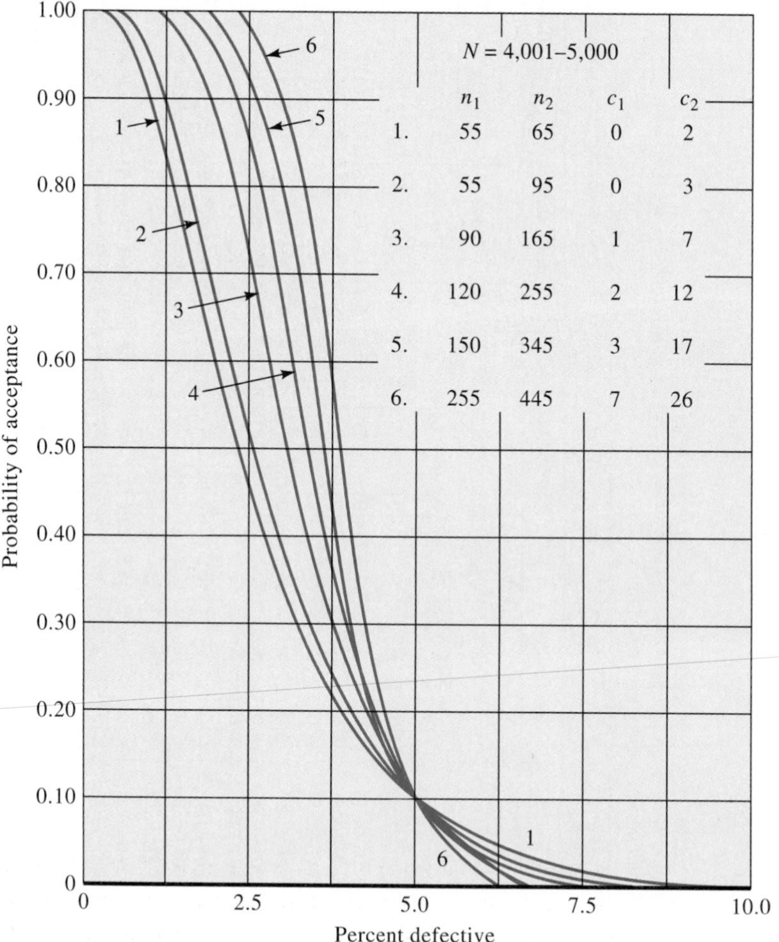

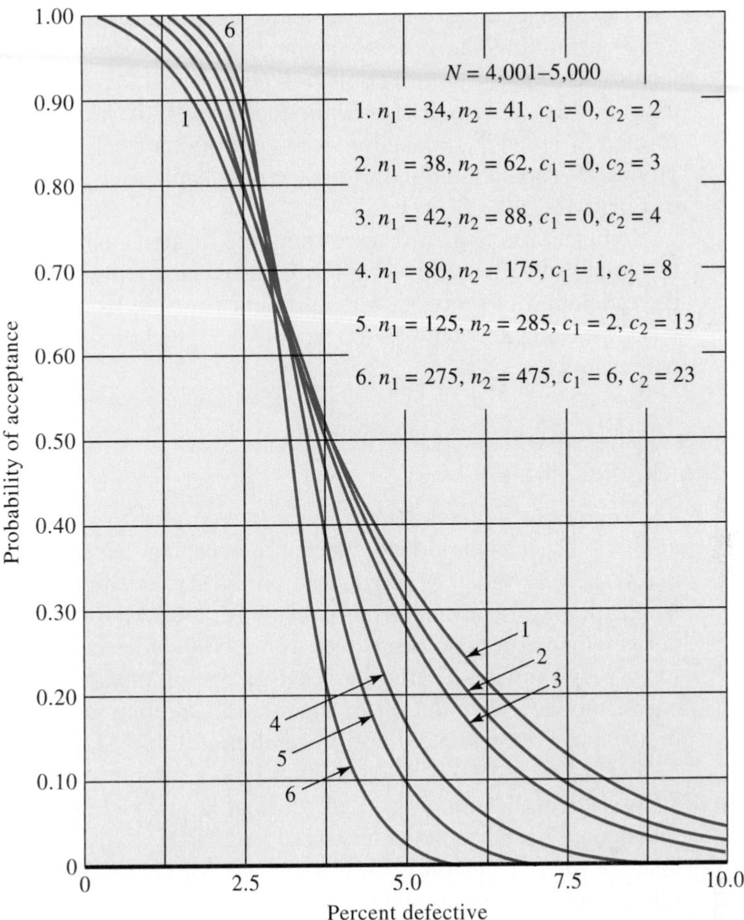

FIGURE 13.2 OC curves for the six Dodge-Romig double sampling plans for AOQL of 2% and a lot size of 4,001–5,000. *(Reproduced from H. F. Dodge, op. cit.)*

process average or better are almost certain of acceptance. The probability of acceptance of lots at exactly the AOQL value varies from about 0.84 (curve 1) to 0.99 (curve 6). The differences in the right-hand sections of the OC curves reflect different degrees of protection against accepting lots considerably worse than the stated AOQL; these differences are also brought out in Table 13.3 by the LTPD values given for each sampling plan.

13.2.6 Determining the Process Average in Dodge-Romig Inspection

The objective of minimizing total inspection depends on selecting the correct plan, which, in turn, depends on making a good estimate of the process average as a basis for the selection. It follows that in all Dodge-Romig inspection, it is advantageous to make systematic use of the results of sampling inspection to (1) determine the process average and (2) assess its stability through time. Control charts are useful tools for performing these tasks. Often it is safe to assume that lots are received in their order of production or, at least, were formed in order of production. The control chart helps to determine whether the producer is maintaining statistical control and what the long-run average level of quality is.

Complete running records of sampling results should be maintained, including the results of inspection of both samples when double sampling is used. However, the estimation of process average should be made from first samples only. Otherwise, the estimate would be too heavily weighted (biased) by the samples from the poorer lots.

13.2.7 Relationship between Lot Size and Sample Size in Dodge-Romig Tables

The Dodge-Romig tables apply to lot sizes from 1 to 100,000 and may be used for any lots that happen to be subjected to acceptance inspection. It is not necessary that inspection lot sizes be equal to production or shipping lot sizes. Where practicable, it may be advantageous to establish lot sizes for acceptance inspection purposes rather than take lots as they come. The tables help in making a decision on lot size by showing clearly the disadvantage of small lots as compared with large ones from the viewpoint of the amount of sampling necessary for a given quality protection. For example, the first column of Table 13.1 calls for a sample size of "All" (that is, 100%) of a lot of 30 or less, a sample size of 37 (that is, 37%) of a lot of 100, and a sample size of 45 (that is, 2.25%) of a lot of 2,000 all for the same quality protection (as measured by LTPD).[†] This emphasizes again the point made in Chap. 11 that it is the absolute size of the sample, much more than its size relative to the lot, that governs the quality protection.

It is evident that wherever possible it is desirable to avoid the very small lot sizes. At the same time it is apparent that the great saving in inspection for a given quality protection consists in taking lots of the order of magnitude of 1,000 rather than the conventional lots of 50 or 100. For example, in Table 13.1 a reduction from a 37% sample from lots of 100 to a 4.5% sample from lots of 1,000 is a reduction in inspection of 32.5% of the total number of articles submitted in the lots. In contrast, a reduction from a 4.5% sample from lots of 1,000 to a 2.25% sample from lots of 2,000 is a reduction of only 2.25%.

There are several possible practical objections to very large lot sizes such as those from 10,000 to 100,000. One objection is that there are often practical difficulties in bringing such large lots together for inspection purposes; the cost of doing this may more than offset the inspection savings. Another objection is that it is frequently much harder to get a random sample out of a lot of 10,000 than out

[†]The only reason the sample size jumps to 75 for lots of 2,001 to 100,000 is that $n = 75$ and $c = 1$ gives less total inspection of samples and rejected lots than $n = 45$ and $c = 0$. Even for a lot size of 100,000, the sampling plan $n = 45$ and $c = 0$ gives a 0.10 probability of acceptance for a 5% defective lot.

of a lot of 1,000. Moreover, the adverse effect on producer-consumer relationships of the rejection of very large lots is sometimes serious.

13.2.8 Making Dodge-Romig AOQL Procedures Contribute to Quality Improvement

The record of sampling inspection may contribute in various ways to the improvement of quality. The control chart maintained as part of this record will indicate out-of-control situations and will show quality trends. This gives guidance as to when to hunt for trouble and helps to concentrate executive pressure for improvement in those places where it will be most effective. The lot-by-lot record listing defects in first samples is helpful to anyone engaged in the actual hunting for the sources of trouble.

Moreover, the contribution of the sampling procedure is not limited to the supplying of useful information. The costs of carrying out screening inspection of rejected lots may be used to provide an effective financial incentive to quality improvement. Where producer and consumer are two departments of the same organization, the producer may be required to do all such screening. Or, where this is not practicable, the cost of any screening inspection may be charged against the budget of the producing department. Example 13.1 describes a case in which both these types of pressure were used.

When a supplier and purchaser are involved, an agreement can be made that whenever lots fail to pass the purchaser's sampling inspection, a stipulated deduction from the price will be made to cover the purchaser's screening costs. Under such circumstances, the AOQL to be used is a matter for prior agreement between supplier and purchaser.

Example 13.1

Use of AOQL inspection to produce quality improvement

Facts of the case In a large manufacturing plant, the practice for many years had been for the inspection department to carry out 100% inspection on numerous parts and products. On many of these items, the average percentage of defectives had been much worse than any reasonable AOQL, and it appeared that no sampling scheme would be as economical as 100% inspection.

It was decided that any required 100% inspection was properly an operating department expense and that inspection should bear only the cost of sampling inspection with sampling criteria based on reasonable quality standards. AOQL inspection was instituted with AOQL figures set at values considered to be satisfactory objectives. Many lots were rejected, and the cost of screening was charged to the operating departments. These charges caused greatly increased attention to quality among operating personnel. In a surprisingly short time, quality was brought to the desired level for the great majority of items.

This new quality consciousness in the operating departments was accompanied by a new viewpoint on the relative responsibilities of operations and inspection. Product costs were considerably reduced. Once the program reached the point where nearly all items were at qualities better than the AOQL figures adopted for them, the responsibility for carrying out all needed screening inspection was transferred to the operating departments. This transfer of responsibility applied both to the 100% inspection of any lots rejected under sampling inspection and to 100% inspection prior to sampling inspection of those few items where it was not possible to produce at an average quality equal to or better than the desired AOQL.

13.2.9 Nonproduct Applications of Acceptance Sampling

Many nonproduct applications of sampling are somewhat comparable to acceptance inspection of isolated lots of manufactured product. In such cases, relevant concepts may include OC curves, indifference qualities, Producer's Risks associated with rejection of some "good" quality, and Consumer's Risks associated with acceptance of some "bad" quality. However, in such isolated lot applications there is no chance to use the concepts of strategy and tactics that have evolved in connection with acceptance sampling systems.

Nevertheless, such concepts can be applied to advantage in those nonproduct cases where many acceptance/rectification decisions must be made. AOQL systems are of particular value in such cases.

In nonproduct activities such as clerical work, it often happens that great differences exist between the quality of work of the best individuals or teams and the quality of work of the poorest individuals or teams. It may be felt to be essential to apply 100% verification to the poorest individuals or teams, and it may be clear that it is uneconomical to apply such verification to the best ones. Example 13.2 describes a case in which it was possible to use AOQL sampling to distinguish between good and bad teams.

Example 13.2

Application of a double sampling AOQL plan to the checking of an annual merchandise inventory for a department store†

Facts of the case The top management of a large department store considered it necessary that the annual merchandise inventory be as accurate as possible. The need for accuracy existed not only because of the use of the inventory figure for accounting purposes but also because departmental inventory figures were used to provide certain useful information to management. The inventory count was in two parts. The count of reserve (warehouse) stock presented no unusual problems. However, the count of forward (on-the-floor) stock was complicated by the requirement that no merchandise ready for sale should be tied up more than a few hours.

The forward inventory count was carried out by dividing each department of the store into sections and assigning a two-person team to each section. One member of the team called information about the stock from the shelves, tables,

†When this example appeared in the second edition of this book, it was based on an unpublished paper by C. S. Brinegar. For Mr. Brinegar's own later more detailed story of this case, see his paper, "Department Store Uses of Statistical Quality Control," *Industrial Quality Control*, vol. 12, no. 2, pp. 3–5, August 1955.

bins, and the like, and the other entered the items on the inventory sheet. The stock was listed by price, description, quantity, classification within the store, and season letter. This count was carried out under considerable time pressure. In the past, a 100% check had always been made of all work performed in the forward inventory, as it was believed that such a check was necessary for reliable results. This 100% check nearly doubled the cost of the inventory and also nearly doubled the time required.

Effective use of a double sampling AOQL scheme The management of this store assigned to one individual the job of reviewing various store activities from the point of view of possible economic applications of the techniques of statistical quality control. A likely application seemed to be the substitution of an acceptance/rectification sampling scheme for 100% verification in the checking of forward inventory.

The errors that were presumably corrected in the 100% verification were those within the control of the inventory teams. These were primarily the errors caused by improper calling of the price, quantity, classification, or season letter of the item or by the improper listing of these items by the inventory writer. Fortunately, information about the types, frequency, and severity of errors that had been discovered by the check inventory teams in past years could be obtained by examining the old inventory sheets. No corrections to inventory sheets had ever been permitted by erasure; every correction required calling the inventory supervisor, crossing out the improper entry, and making the corrected entry at the bottom of the sheet. By examining the handwriting of corrected entries, it was possible to tell whether the correction had been made by the original team or by the checking team.

Some 3,000 inventory sheets from the previous year were analyzed. The analysis indicated that a few teams were usually responsible for most of the errors made in any department. A large majority of the teams produced work of quite respectable accuracy. Large errors in price or quantity were infrequent, the most common errors being over or under one unit in physical count, one dollar off in price, or listing the wrong season letter. Different types of goods seemed to differ greatly in their liability to inventory error.

As a result of this analysis, an inventory sampling plan was inaugurated for the forthcoming annual inventory. (The 100% double check was maintained in a very few spots, particularly on lines of merchandise where an error in count would prove costly.) The inventory took place during two 4-h periods from 6 to 10 P.M. following regular working days. The work of each inventory team was sampled by a "flying inspection squad." If the sample proved satisfactory, the entire night's output of the team was approved. If the sample contained too many errors, the team was immediately notified of the fact and its procedure was closely observed; all its previous work was checked 100%; subsequent work was sampled again with the possibility of a further 100% check.

In the sampling procedure, three different Dodge-Romig double sampling AOQL plans were used. The plan to be applied to each department was selected on the basis of the analysis of the previous year's inventory sheets, giving weight to the type and value of merchandise and the probable number of inventory entries. In the use of the sampling tables, one line on the inventory sheet (typically representing 10 to 125 items) corresponded to a single manufactured article subject to inspection; any error on the line caused its classification as a defective. However, a line that was correct except for an error in season letter was counted as only one-half a defective; this error affected not the count or value of merchandise but only its estimated age distribution.

Sampling plans were selected having OC curves indicating that not more than 20% of the total work would be rejected and reinspected. This decision was an essential part of the planning of the inventory as it determined the number of reinspection teams to be provided. Actually the work of only 10.4% of the teams was rejected, 48 teams out of 462.

All inventory supervisors were instructed to use idle teams in double checking even though the work checked had been approved by the sample check made by one of the flying squads. As a result of this instruction, a fair amount of approved work was double-checked. An analysis of the inventory sheets from all checked work proved that the sampling procedure had really succeeded in separating the inefficient teams from the efficient teams. In almost every case the double checking of approved teams revealed only a bare minimum of errors, one consistent with the AOQL used. The work of the rejected teams produced additional errors in over 90% of the cases examined.

The auditors and the store executives were well-pleased with this application of statistical sampling methods. The scheme was repeated in the inventory of the succeeding year, with minor administrative refinements based on the first year's experience.

13.2.10 Selecting an AOQL Value as a Quality Standard

Two figures may properly be considered in any decision on the AOQL. One is the maximum percentage of defectives that is tolerable in the outgoing product after inspection. The other is the capability of the production process, expressed as the percentage of defectives, that will be expected in most of the submitted lots. It is evident that 100% inspection by *someone* is necessary whenever the submitted quality is appreciably worse than the tolerable percentage of defectives. As illustrated in Example 13.1, the chief purpose of adopting sampling inspection in this case may be to use frequent lot rejections to force the producer either to improve quality or to carry out an adequate screening inspection.

Various considerations may enter into the decision on the maximum tolerable percentage of defectives. If the design specification for a quality characteristic contains a margin of safety, some articles that are technically classified as defectives may really be satisfactory for the purpose intended. Sometimes there may be a good chance that any defectives in the outgoing product from this inspection will be eliminated in a later manufacturing or inspection operation; in other cases this inspection may be the last chance to eliminate them. It is obviously necessary to consider the number of quality characteristics involved in the particular inspection and their importance in the functioning of the product being inspected. Comments on some economic aspects of this decision are made in Chap. 17.

It is likely to be easier to estimate the capabilities of a production process than to decide on the maximum tolerable percentage of defectives. If nearly all the submitted lots are to be accepted on the basis of the sample, the quality of these accepted lots needs to be somewhat better than the selected AOQL. An approximate rule for guidance is that, as things work out in practice, the net result of using an AOQL scheme will usually be an average outgoing percent defective not worse than half to two-thirds the AOQL value.† This rule may be viewed in reverse to judge the implications regarding process capability when a particular AOQL value is selected. For instance, suppose a 1.5% AOQL value is selected; this implies that it is believed that—except, perhaps, for occasional out-of-control bad lots—the process can ultimately be made capable of submitting lots that are not worse than about 1.0% defective.

13.3 SOME REASONS FOR NOT BASING A QUALITY STANDARD ON A PROVISION FOR SCREENING INSPECTION

The Dodge-Romig acceptance procedures described in Sec. 13.2 all gave consideration to the effect of a formal provision for screening inspection whenever the results of sampling inspection were sufficiently unfavorable. In the AOQL plans, the stated quality standard (that is, the AOQL) was dependent, in part, on this provision for screening inspection.

Where the producer and consumer are different organizations, dealing at arm's length, there may be good reasons why the consumer is unwilling for the stated quality standard to be based on the assumption that such screening inspection will occur. It may be considered impracticable or inadvisable for the consumer to carry out this screening in all cases, and the consumer may be unwilling to rely on any such screening carried out by the producer.

Screening inspection by the consumer (either of all the product or of rejected lots) may be impracticable because it is unduly costly or because of lack of sufficient inspection facilities or personnel or for other reasons. Even though practicable, it may be deemed inadvisable. As a consequence of the consumer carrying out screening inspection of all product, it may be found that inspection that should have been done by the producer is being executed by the consumer. Moreover, the pressure for quality improvement exerted by outright rejection of one or more entire lots is much stronger than the pressure exerted by the rejection of individual articles classified as defective by a consumer's 100% inspection. These points were illustrated in Example 13.1 and emphasized in Dodge's quotation at the start of this chapter.

†H. F. Dodge, "Administration of a Sampling Inspection Plan," *Industrial Quality Control,* vol. 5, no. 3, pp. 12–19, November 1948; and reprinted in *Journal of Quality Technology,* vol. 9, no. 3, July 1977. This article contains excellent concise advice regarding the development of sampling inspection procedures in a manufacturing plant.

13.4 DESIGNING SINGLE SAMPLING PLANS FOR STIPULATED PRODUCER'S AND CONSUMER'S RISKS

The topic of the choice of a sampling plan based on the assessment of both Consumer and Producer Risks was introduced in Sec. 12.12. The reader will remember that the definition of AQL, the basis of the ABC standard, related to a high quality (low value of μ_p) considered acceptable as a process average. Once such a value of the AQL is fixed, a definite point on the upper end of the OC curve may be established. The coordinates of this point are $(p_1, 1 - \alpha)$, where p_1 is the value of the AQL and α is the Producer's Risk. In the ABC standard system, α ranges from 0.01 to 0.12 with the highest values occurring for those plans where c equals 0.

In the discussion of the LQ tables, it was pointed out that, in instances where isolated lots of product are to be inspected using a sampling plan, equal attention should be directed to the Consumer's Risk. The LQ tables gave values of μ_p with probabilities of acceptance, that is, Consumer's Risk (β), of 5 and 10%. The tables were organized in such a manner that the user searched the AQL column until an acceptable Consumer's Risk point (p_2, β) was found. Thus the consumer's interest may be most closely associated with the lower end of the OC curve. While this duality in the relationship between the interests of the producer and consumer may be strongest in those cases where judgment must be passed on isolated lots of product, a strong argument could be made that the appropriate balancing of consumer and producer interests in the operating characteristics of a sampling plan is basic to the selection of a good plan.

Mathematically, the selection of two points on an operating characteristic curve uniquely defines a single sampling plan for those points. In practice, this means that the appropriate balancing of consumer and producer interests, as exemplified by the choice of a Producer's Risk point $(p_1, 1 - \alpha)$ and a Consumer's Risk point (p_2, β), leads to a unique single sampling plan. For simplicity, this fact is demonstrated using the Poisson distribution.

Since the producer wishes product with the stipulated quality p_1 accepted with probability $1 - \alpha$, the Poisson formula for probability of acceptance describing the stipulated preference is

$$P_a\,(\text{producer}) = \sum_{r=0}^{c} \frac{(np_1)^r}{r!}\, e^{-np_1} = 1 - \alpha$$

The corresponding formula describing the Consumer's Risk preference is

$$P_a\,(\text{consumer}) = \sum_{r=0}^{c} \frac{(np_2)^r}{r!}\, e^{-np_2} = \beta$$

Since p_1, p_2, $1 - \alpha$, and β are known, the two unknowns in these equations are n, the sample size, and c, the acceptance number. These two equations in two unknowns may be solved simultaneously for unique values of n and c.

It is not the intent here to present a mathematical solution to this problem but rather to introduce some work by J. M. Cameron† that simplifies the procedure of deriving sampling plans based on stipulated Producer's and Consumer's Risks. Two features of the derivation of plans are worth pointing out here. Both of these features are exemplified in the sample calculations given in the following paragraphs. First, if α and β are small and the *difference* between p_1 and p_2 is small, the resulting sample size n is likely to be very large.

Second, because of the requirement for integer values for n and c, it is virtually impossible to derive a sampling plan that passes exactly through the two points $(p_1, 1 - \alpha)$ and (p_2, β). Some compromise in the vicinity of one point or the other is almost always required.

Table 13.5, based on the Poisson distribution and adapted from two more extensive tables in the article by Cameron, can be used to design sampling plans with stated Producer's and Consumer's Risk points.‡

In the following numerical examples, we shall deal particularly with two columns of this table. The column headed $np_{0.95}$ enables us to compute a family of single sampling plans that have a common value of $P_a = 0.95$ for any stated "good" quality (expressed as a fraction nonconforming). The Producer's Risk α is 0.05. The column headed $np_{0.10}$ enables us to compute a family of plans for which the Consumer's Risk is 0.10 for any stated "bad" quality.

Assume that a family of plans is desired having a Producer's Risk of 0.05 that product 0.65% nonconforming will be rejected if submitted to sampling inspection. That is, 0.65% nonconforming is viewed as "good" quality; it is desired to protect the producer by having only 1 chance in 20 that product of such quality be rejected. Assume that it is desired to know for each plan the percentage nonconforming for which the Consumer's Risk is 0.10. The following family of plans may be computed from factors given in Table 13.5.

c	n	$100p_{0.10}$, %	c	n	$100p_{0.10}$, %	c	n	$100p_{0.10}$, %
0	8	28.8	4	303	2.6	8	722	1.80
1	55	7.1	5	402	2.3	9	835	1.70
2	126	4.2	6	506	2.1	10	949	1.62
3	210	3.2	7	612	1.9	11	1,065	1.56

The required calculations may be illustrated for the case where $c = 3$:

$$n = \frac{np_{0.95}}{p_{0.95}} = \frac{1.366}{0.0065} = 210.15 \approx 210$$

$$p_{0.10} = \frac{np_{0.10}}{n} = \frac{6.681}{210} = 0.032, \text{ or } 3.2\%$$

†J. M. Cameron, "Tables for Constructing and for Computing the Operating Characteristics of Single-Sampling Plans," *Industrial Quality Control,* vol. 9, pp. 37–39, July 1952.

‡Furthermore, more extensive tables may be developed by the use of a χ^2 table such as Table *B* of App. 3. For example, the row of values in Table 13.5 for $c = 1$ may be found by setting $\nu = 2(c + 1)$ degrees of freedom, and $2np_\gamma$ equals the column entries for the given values of γ. Thus $2np_{0.95} = 0.71$ from which $np_{0.95} = 0.355$ and $np_{0.10} = 7.78/2 = 3.890$.

TABLE 13.5 GENERALIZED TABLE OF SINGLE SAMPLING PLANS THAT HAVE CERTAIN SPECIFIED PRODUCER'S AND CONSUMER'S RISKS†

c	$np_{0.99}$	$np_{0.95}$	$np_{0.90}$	$np_{0.50}$	$np_{0.10}$	$np_{0.05}$	$np_{0.01}$	$\frac{p_{0.10}}{p_{0.95}}$
0	0.010	0.051	0.105	0.693	2.303	2.996	4.605	44.890
1	0.149	0.355	0.532	1.678	3.890	4.744	6.638	10.946
2	0.436	0.818	1.102	2.674	5.322	6.296	8.406	6.509
3	0.823	1.366	1.745	3.672	6.681	7.754	10.045	4.890
4	1.279	1.970	2.433	4.671	7.994	9.154	11.605	4.057
5	1.785	2.613	3.152	5.670	9.275	10.513	13.108	3.549
6	2.330	3.286	3.895	6.670	10.532	11.842	14.571	3.206
7	2.906	3.981	4.656	7.669	11.771	13.148	16.000	2.957
8	3.507	4.695	5.432	8.669	12.995	14.434	17.403	2.768
9	4.130	5.426	6.221	9.669	14.206	15.705	18.783	2.618
10	4.771	6.169	7.021	10.668	15.407	16.962	20.145	2.497
11	5.428	6.924	7.829	11.668	16.598	18.208	21.490	2.397
12	6.099	7.690	8.646	12.668	17.782	19.442	22.821	2.312
13	6.782	8.464	9.470	13.668	18.958	20.668	24.139	2.240
14	7.477	9.246	10.300	14.668	20.128	21.886	25.446	2.177
15	8.181	10.035	11.135	15.668	21.292	23.098	26.743	2.122
16	8.895	10.831	11.976	16.668	22.452	24.302	28.031	2.073
17	9.616	11.633	12.822	17.668	23.606	25.500	29.310	2.029
18	10.346	12.442	13.672	18.668	24.756	26.692	30.581	1.990
19	11.082	13.254	14.525	19.668	25.902	27.879	31.845	1.954
20	11.825	14.072	15.383	20.668	27.045	29.062	33.103	1.922
21	12.574	14.894	16.244	21.668	28.184	30.241	34.355	1.892
22	13.329	15.719	17.108	22.668	29.320	31.416	35.601	1.865
23	14.088	16.548	17.975	23.668	30.453	32.586	36.841	1.840
24	14.853	17.382	18.844	24.668	31.584	33.752	38.077	1.817
25	15.623	18.218	19.717	25.667	32.711	34.916	39.308	1.795
30	19.532	22.444	24.113	30.667	38.315	40.690	45.401	1.707
35	23.525	26.731	28.556	35.667	43.872	46.404	51.409	1.641
40	27.587	31.066	33.038	40.667	49.390	52.069	57.347	1.590
45	31.704	35.441	37.550	45.667	54.878	57.695	63.231	1.548
50	35.867	39.849	42.089	50.667	60.339	63.287	69.066	1.515

†This table, based on the Poisson distribution, is adapted by permission from J. M. Cameron, "Tables for Constructing and for Computing the Operating Characteristics of Single-Sampling Plans," *Industrial Quality Control,* vol. 9, p. 39, July 1952.

Now assume that a family of plans is desired having a Consumer's Risk of 0.10 that product 1.0% nonconforming will be accepted if submitted to sampling inspection. That is, 1.0% nonconforming is viewed as "bad" quality; it is desired to protect the consumer by having only 1 chance in 10 that product of such quality will be accepted. Assume also that it is desired to know for each plan the percentage nonconforming for which the Producer's Risk is 0.05. The following family of plans may be computed from factors given in Table 13.5. The calculations are similar to those illustrated in connection with the stipulated Producer's Risk.

c	n	$100p_{0.95}$, %	c	n	$100p_{0.95}$, %	c	n	$100p_{0.95}$, %
0	230	0.02	4	799	0.25	8	1,300	0.36
1	389	0.09	5	928	0.28	9	1,421	0.38
2	532	0.15	6	1,053	0.31	10	1,541	0.40
3	668	0.20	7	1,177	0.34	11	1,660	0.42

Now, assume that both stipulations are made, namely, that the Producer's Risk shall be 0.05 of rejection of product 0.65% nonconforming and that the Consumer's Risk shall be 0.10 of acceptance of product 1.0% nonconforming. It is evident that these two stipulations will not be met by any plan in either of these two tabulated families of plans. Even the tabulated plans with sample sizes above 1,000 do not come close to meeting both of these stipulations. (In fact, the sample size must be approximately 5,600 for a single sampling plan to meet these two stipulations.)

The difficulty here is that the two stipulated values of product quality, 0.65 and 1.0% nonconforming, are too close together for the stipulated values of Producer's Risk and Consumer's Risk. That is, they are so close together that the stipulations cannot be met without a sample size that is too large to be practicable in most circumstances that arise in industry.

The final column of Table 13.5 gives for each value of c the ratio of $p_{0.10}/p_{0.95}$. Assume that it is desired to have Producer's Risk (α) of 0.05 of rejection of a lot 0.40% nonconforming and a Consumer's Risk (β) of 0.10 of acceptance of a lot 1.50% nonconforming. The ratio $p_{0.10}/p_{0.95}$ is 0.015/0.004, or 3.75. The final column of Table 13.5 tells us that a ratio of 4.057 is obtained when c is 4 and 3.549 is obtained when c is 5. Because neither of these values is exactly 3.75, it is not possible to have a single sampling plan that has an OC curve passing *exactly* through the two stipulated points. A choice might be made of one of the following four plans that can be computed from Table 13.5; the OC curve of each plan passes through one of the specified points and provides a near miss to the other point.

c	n	$100p_{0.95}$, %	$100p_{0.10}$, %
4	492	0.40	1.62
5	653	0.40	1.42
4	533	0.37	1.50
5	618	0.42	1.50

It should be noted that both plans for $c = 5$ ($n = 618$ or 653) *at least* satisfy the risk levels of the two-point stipulation. The plan for which $n = 618$ will pass through the point ($p_{0.10} = 0.015$, $\beta = 0.10$) and has a risk α less than 0.05. That for which $n = 653$ will pass through the point ($p_{0.95} = 0.004$, $1 - \alpha = 0.95$) and has a risk β less than 0.10. In fact any value of n between 618 and 653 will at least satisfy the stipulated risks. This fact is useful particularly if computer programs are used to derive plans to fit two points on the OC curve. A point frequently lost in such derivations is that one of the plans for $c = 4$ may yield an OC curve with α and β risks so close to the desired value that the cost savings resulting from much reduced sample sizes outweigh the need for mathematical precision. Computers as yet are unable to make such judgments.

13.4.1 The Importance of the Acceptance Number *c* in Determining the Shape of the OC Curve

Consider our tabulated family of plans that had a Consumer's Risk of 0.10 of accepting product 1% nonconforming. If the requirement had been for a family of plans with a Consumer's Risk of 0.10 of accepting product 5% nonconforming, all values of n would have been one-fifth the figure shown in our table and all values of $100p_{0.95}$ would have been five times the figure shown. For example, the plan with $c = 8$ would have an n of 260 rather than 1,300 and a value of $100p_{0.95}$ of 1.8 rather than 0.36%.

Figure 13.3 illustrates the point that the same OC curve can be used for (1) n = 1,300, $c = 8$ and (2) $n = 260$, $c = 8$, provided the units on the horizontal scale are changed by a factor of 5. Moreover, the OC curve in Fig. 13.3 applies to all single sampling plans for which $c = 8$ if the horizontal scale is in units of np as in scale (*c*) of the figure. Such a generalized OC curve for all plans with a particular c assumes that the Poisson distribution is deemed to give a close enough approximation for the intended purpose.

This relationship between the acceptance number and the OC curve made possible the classification by acceptance number of the ASN curves contained in ABC-STD-105 and discussed in Sec. 12.7.

FIGURE 13.3 Depending on units used for the horizontal scale, this is the approximate OC curve for *(a)* n = 1,300, $c = 8$; *(b)* $n = 260$, $c = 8$; or *(c)* all plans for which $c = 8$.

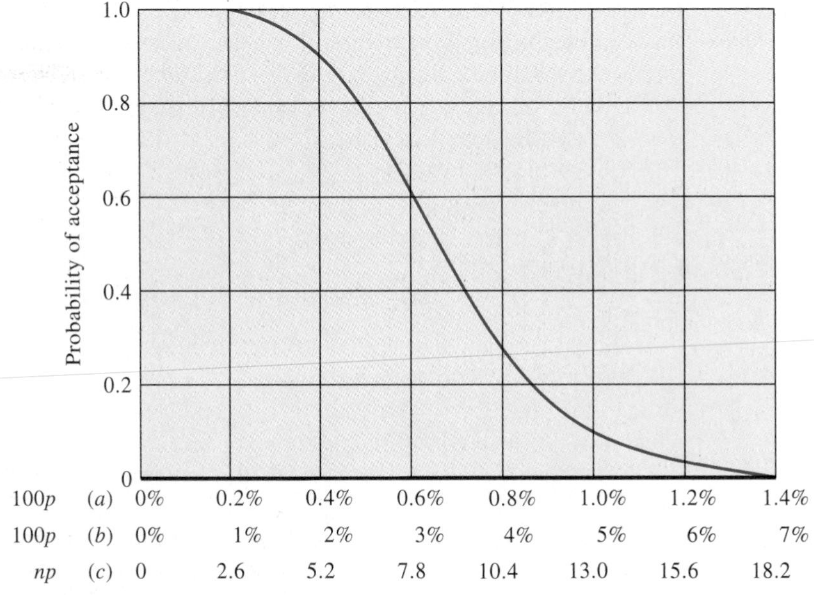

13.5 A SIMPLE AQL SYSTEM PROPOSED BY DODGE

The earliest AQL tables—Army Ordnance and Army Service Forces—were relatively simple. However, all the military standards starting with JAN-STD-105 have involved fairly extensive and complex sets of tables. The Dodge-Romig tables discussed in Sec. 13.2 also are quite extensive. The complex tables make it possible to provide many different alternative sampling plans adapted for use under many different circumstances. Nevertheless, there are often real advantages in having a system that is simple.

In a brilliant monograph written in 1959, H. F. Dodge made a number of proposals regarding the design of sampling systems based on the AQL concept.[†] One aspect of his proposals was that they made possible the design of an AQL system that could be presented on a simple one-page table. Three such systems—designated Plan 1, Plan 2, and Plan 3—are presented in the monograph. One of these, Plan 1, is reproduced here as Table 13.6.

Some of Dodge's proposals, illustrated with respect to single sampling in Table 13.6, were as follows:

1. The *same* series of preferred numbers (numbers approximating a geometric progression) is used for AQL values and sample sizes. Thus in Table 13.6 AQL values include 1.0, 1.5, 2.5, 4.0, 6.5, and 10.0%; sample sizes include 10, 15, 25, 40, 65, and 100. This gives a systematic diagonal pattern of acceptance numbers.[‡] One consequence of this pattern is that when $100\mu_p = \text{AQL}$, $n\mu_p$ is almost constant along any diagonal line of the table. (If the preferred number series used were an exact geometric progression, as would be the case if there were no rounding of numbers for more convenient use, $n\mu_p$ would be exactly constant along a diagonal.) It follows that the probability of acceptance at AQL value is almost constant along any diagonal. (This statement assumes a Type *B* OC curve and the use of the Poisson approximation.)
2. The smallest acceptance number available in normal inspection is 1. (In a number of places in this book we have noted that, when $c = 0$, the shape of the OC curve is unsatisfactory and there is generally poor discrimination between acceptable and unacceptable lots.)

[†]H. F. Dodge, "A General Procedure for Sampling Inspection by Attributes—Based on the AQL Concept," Technical Report No. 10, The Statistics Center, Rutgers—The State University, New Brunswick, N. J., Dec. 15, 1959. In January 1960, the material in this monograph was presented to the ABC Working Group and also in a public lecture at the Navy Building, Washington, D.C., sponsored by the Assistant Chief for Production and Quality Control, Bureau of Naval Weapons. The reader will note that a number of the features proposed by Dodge were incorporated into the ABC standard.

This paper, in substance, was later presented under the same title at the 1963 convention of the American Society for Quality Control. See "Annual Convention Transactions, Seventeenth Annual Convention," pp. 7–19, American Society for Quality Control, Milwaukee, Wis., 1963.

[‡]This diagonal pattern of acceptance numbers can also be observed in the ABC standard. It did not exist in the previous military standards. Although, starting with MIL-STD-105A, these standards used preferred numbers for AQL values, they did not use them for sample sizes. The numbers 1.0, 1.5, 2.5, 4.0, 6.5, and 10.0 involve rounding of the values obtained when the first number is 1.0, and each succeeding number is found by multiplying the preceding number by $\sqrt[5]{10} = 1.585$. In the ABC standard, a similar "5-series" set of preferred numbers is used for single sample sizes with the first number as 2.0.

TABLE 13.6 COMPOSITE SINGLE SAMPLING TABLE FOR PLAN 1 OF DODGE'S PROPOSED AQL SYSTEM FOR ACCEPTANCE SAMPLING BY ATTRIBUTES

Lot size, inspection level II‡	Sample size n	Acceptance number in normal inspection, c_N†										
		0.10	0.15	0.25	0.40	0.65	1.0	1.5	2.5	4.0	6.5	10.0
3–15	3											↓
16–22	4										↓	1
23–40	6									↓	1	↓
41–70	10								↓	1	↓	2
71–120	15							↓	1	↓	2	3
121–210	25						↓	1	↓	2	3	5
211–400	40					↓	1	↓	2	3	5	7
401–700	65				↓	1	↓	2	3	5	7	10
701–1,300	100			↓	1	↓	2	3	5	7	10	15
1,301–3,200	150		↓	1	↓	2	3	5	7	10	15	22
3,201–15,000	250	↓	1	↓	2	3	5	7	10	15	22	↑
15,001–80,000	400	1	↓	2	3	5	7	10	15	22	↑	
Over 80,000	650	↑	2	3	5	7	10	15	22	↑		

†For tightened inspection, acceptance number $c_T = c_N - 1$.
‡For inspection level I, use plan with next smaller sample size. For inspection level III, use plan with next larger sample size.
Source: Adapted by permission from H. F. Dodge, "A General Procedure for Sampling Inspection by Attributes—Based on the AQL Concept," Technical Report No. 10, The Statistics Center, Rutgers—The State University, New Brunswick, N.J., Dec. 15, 1959.

3. The table is designed for a probability of acceptance at the AQL value of approximately 0.95. The systematic pattern mentioned in (1) and the elimination of any plan with $c = 0$ mentioned in (2) both simplify the design of a set of plans that have a single aimed-at value of P_a at the AQL value.
4. In tightened inspection, the sample size is always the same as in normal inspection and the acceptance number is one less than in normal inspection. (This simple rule obviously would be impossible if there were any plans stipulating $c = 0$ in normal inspection.) It turns out that this rule gives an AOQL in tightened inspection that is approximately equal to the $100p_{0.95}$ in normal inspection. It follows that, if rejected lots in tightened inspection are given 100% inspection that is really effective and if such lots are then submitted and accepted, the AOQ will be no worse than the AQL.†
5. No code letters are used. The lot sizes that correspond to each sampling plan under ordinary circumstances are given in the table. These values given in the table are designated as applying to inspection level II. Where necessary, smaller sample sizes for a given lot size can be used by shifting to the next smaller sample size in the table (inspection level I); larger sample sizes can be used by shifting to the next larger sample size (inspection level III).

†In the original Army Ordnance and Army Service Forces tables, the aimed-at AOQL in tightened inspection was the stated AQL value. This feature was lost in the subsequent military standards.

A number of other features of AQL systems, not illustrated in Table 13.6, were also proposed by Dodge. One important proposal was that a shift from normal to tightened inspection be required if 2 of the last 5 (or less) lots under normal inspection were rejected. (The reader will recall that this was the rule adopted for the ABC standard.) Normal inspection was to be reinstated when 5 successive lots were accepted under tightened inspection. Dodge commented regarding these proposed rules as follows:

> The criteria given for shifting from normal to tightened inspection and from tightened back to normal inspection are very simple and easy to apply. Simple criteria have been made possible by holding to a reasonably constant probability pattern for normal inspection and a reasonably constant AOQL pattern for tightened. The rule[†] for shifting to tightened inspection "if 2 of the last 5 lots were rejected" seems appropriate as a general rule only so long as all plans under normal inspection have about the same probability of acceptance for AQL quality.
>
> To be effective and to be used, the rules for shifting from normal to tightened inspection must be reasonable, and must appear to be reasonable to inspection personnel. It is the author's observations that the penalty of shifting from normal to tightened inspection has been sufficiently drastic in certain areas of existing sampling tables and procedures based on the AQL concept, that there has been a strong tendency to avoid using tightened inspection at all. This constitutes a basic misuse of the procedures; it undermines and relegates to a position of secondary importance, the Consumer's protection against product running at a substandard level of quality. The magnitude of the change in going from normal to tightened inspection in the proposed system is relatively small, and intentionally so, yet it is adequate for the intended purpose here.
>
> The two factors, ease of application, and less drastic shifts, should be conducive to a more willing acceptance and a better following of the rules and hence a better realization of a major objective—maintenance of quality equal to or better than AQL.
>
> As indicated . . . , the maintenance of a control chart showing graphically the results of lot-by-lot inspections, while not required, is highly recommended. Such charts have been found invaluable as an aid to an understanding of the operations of the inspection system, especially when quality difficulties are encountered and shifts in inspection plans are necessary.

13.6 DESIGN OF A SEQUENTIAL PLAN HAVING AN OC CURVE PASSING THROUGH TWO DESIGNATED POINTS

A certain product is subject to lot-by-lot acceptance or rejection on the basis of a destructive test applied to a sample. The conditions of the test are considerably more severe than it is expected will be encountered in practice. All items tested are damaged to the point where they are of no further use. In order to keep the num-

[†]Dodge includes the following footnote here: "H. C. Hamaker mentions a comparable rule in a Philips specification for radio valves; Undated memo M. S. 3203, 'Attributes Sampling in Practice, Part I, General Principles,' handed to the author in Sept. 1959."

ber of items tested to a minimum consistent with the desired quality protection, it is decided to design an item-by-item sequential plan.

Such a plan may be designed so that the OC curve passes through any two points desired. In this instance, the desired points are $P_a = 0.95$, $p_1 = 0.10$ and $P_a = 0.20$, $p_2 = 0.30$. Because of the margin of safety in the test procedure, it is believed that lots are satisfactory when not more than 10% of the items would fail if subjected to this test; with a P_a of 0.95, the producer takes only 1 chance in 20 that such lots will be rejected. From the Consumer's viewpoint, there is to be 1 chance in 5 of acceptance of a lot in which 30% of the items would fail this test. In the symbols commonly used in the mathematics of sequential sampling, these desired points on the OC curve are represented by ($p_1 = 0.10$, $\alpha = 0.05$) and ($p_2 = 0.30$, $\beta = 0.20$).

Item-by-item sequential sampling was developed from the Wald sequential probability ratio test (SPRT).[†] The mathematical form that the probability ratio takes is

$$\frac{p_1^r(1 - p_1)^{n-r}}{p_2^r(1 - p_2)^{n-r}}$$

If the actual fraction nonconforming of the lot μ_p is at the level p_1, the lot should be accepted. Thus the P_a associated with the numerator of the equation is $1 - \alpha$ and that associated with the denominator is β. When the value of the probability ratio exceeds the ratio of $B = (1 - \alpha)/\beta$, the lot is accepted. Conversely, if μ_p is at the level p_2, the probabilities of rejection are α and $1 - \beta$, respectively, for the numerator and denominator of the probability ratio. Thus if the value of the ratio falls below $A = \alpha/(1 - \beta)$, the lot is rejected. B and A are associated with the acceptance and rejection lines, respectively. The solution, in terms of r and n, is found with the use of logarithms.

Figure 13.4 gives a graphical representation of an item-by-item sequential plan. The plan is fully defined by the equation of the rejection line, $Re = sn + h_2$, and the acceptance line, $Ac = sn - h_1$. To compute s, the slope of these lines, and h_1 and h_2, the intercepts, certain auxiliary symbols—g_1, g_2, a, and b—are used. The necessary computations are as follows:

$$g_1 = \log \frac{p_2}{p_1} = \log \frac{0.30}{0.10} = 0.4771$$

$$g_2 = \log \frac{1 - p_1}{1 - p_2} = \log \frac{0.90}{0.70} = 0.1091$$

$$a = \log \frac{1 - \beta}{\alpha} = \log \frac{0.80}{0.05} = 1.2041$$

[†]For a complete development of the SPRT and many of its uses, see Abraham Wald, *Sequential Analysis,* John Wiley & Sons, New York, 1947. Reprinted in 1973 by Dover Publications, Inc., New York.

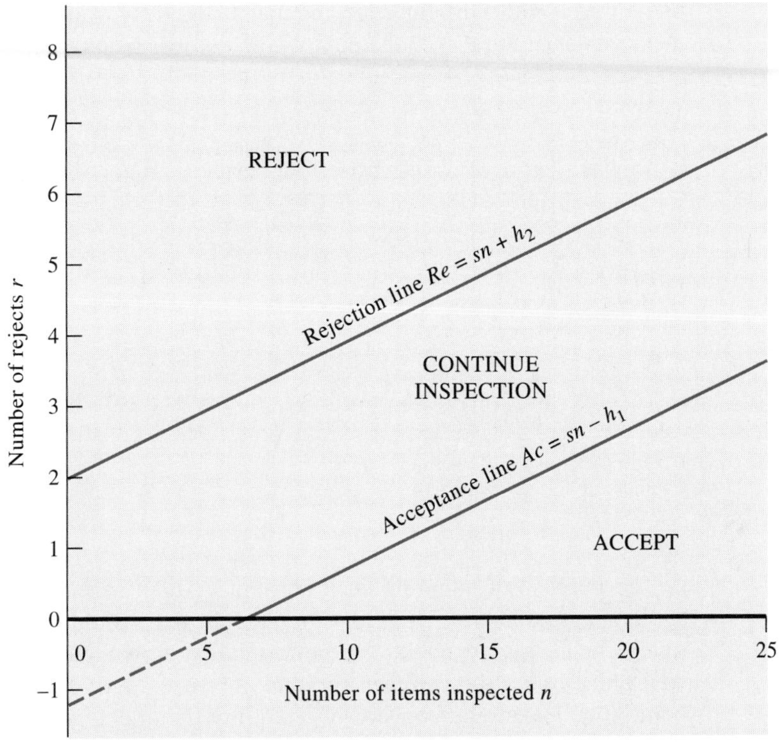

FIGURE 13.4 Graphical representation of an item-by-item sequential plan.

$$b = \log \frac{1-\alpha}{\beta} = \log \frac{0.95}{0.20} = 0.6767$$

$$h_1 = \frac{b}{g_1 + g_2} = \frac{0.6767}{0.5862} = 1.154$$

$$h_2 = \frac{a}{g_1 + g_2} = \frac{1.2041}{0.5862} = 2.054$$

$$s = \frac{g_2}{g_1 + g_2} = \frac{0.1091}{0.5862} = 0.186$$

This gives the following equations:

Rejection line: $Re = 0.186n + 2.054$

Acceptance line: $Ac = 0.186n - 1.154$

For practical use, these lines should be converted into an item-by-item table of acceptance and rejection numbers. Computed values of Ac and Re are generally not whole numbers. In the following tabulation of acceptance and rejection numbers up to $n = 24$, the rejection number is the next whole number above *Re*, and the acceptance number is the next whole number below *Ac*.

n	Acceptance number	Rejection number	n	Acceptance number	Rejection number	n	Acceptance number	Rejection number
1	†	‡	9	0	4	17	2	6
2	†	‡	10	0	4	18	2	6
3	†	3	11	0	5	19	2	6
4	†	3	12	1	5	20	2	6
5	†	3	13	1	5	21	2	6
6	†	4	14	1	5	22	2	7
7	0	4	15	1	5	23	3	7
8	0	4	16	1	6	24	3	7

†Acceptance requires a sample of at least 7 items.
‡Rejection requires a sample of at least 3 items.

The purpose of adopting such an item-by-item scheme is to reduce the average amount of inspection below the amount that would be obtained with a multiple scheme giving about the same quality protection. However, it is usually advisable to fix some upper limit to the size of *n;* otherwise, an occasional borderline lot might conceivably require the indefinite continuation of sampling.

It should be recognized that a maximum value for *n* changes the OC curve from that contemplated in the calculations to design an item-by-item sequential plan. In order to avoid a substantial effect, some writers recommend truncation at a value of *n* about three times that for a single sampling plan passing through the same two designated points on the OC curve. In the example calculations given here, the recommendation would suggest truncation at a value of *n* of about 80. In many cases, the user can derive a corresponding single sampling plan rather quickly with the aid of Table 13.5 from which an appropriate terminating value for *n* may be decided. Some users of item-by-item sequential plans accept all lots that have not been rejected at this maximum sample size; others reject all that have not been accepted.

Five points on the OC curve of an item-by-item sequential plan can be found without difficulty. Two of these are the points used in designing the plan. Two more points are established by the knowledge that when $\mu_p = 0$, $P_a = 1.00$, and when $\mu_p = 1$, $P_a = 0$. A fifth point is fixed by the relationship that when $\mu_p = s$, $P_a = h_2/(h_1 + h_2)$.

As previously mentioned, the objective in using item-by-item sequential sampling is usually to minimize the amount of sampling inspection necessary to reach a decision. The ASN function is therefore an important characteristic of the plan. Sufficiently good OC curves and ASN functions often can be sketched from the five points as follows:

Value of μ_p	OC curve P_a	ASN
0.0	1.0	$\frac{h_1}{s}$
p_1	$1 - \alpha$	$\frac{(1-\alpha)h_1 - \alpha h_2}{s - p_1}$
s	$\frac{h_2}{h_1 + h_2}$	$\frac{h_1 h_2}{s(1-s)}$
p_2	β	$\frac{(1-\beta)h_2 - \beta h_1}{p_2 - s}$
1.0	0	$\frac{h_2}{1-s}$

For the sequential plan just computed, the values would be as follows:

Value of μ_p	OC curve P_a	ASN
0.0	1.0	6.20
0.10	0.95	11.55
0.186	0.64	15.65
0.30	0.20	12.39
1.0	0.0	2.52

An explanation of the more complicated calculations of additional points on the OC curve and ASN function is contained in the basic source material on sequential analysis.†

13.7 DODGE'S CHAIN SAMPLING INSPECTION PLAN

Special problems exist in cases where there is continuing production of lots or batches but very small sample sizes are selected for each lot or batch because tests are destructive or costly. For small samples such as 4, 5, 6, or even 10, the only practicable acceptance number is 0. It has been pointed out that an acceptance number of 0 gives an unsatisfactory shape of OC curve (concave upward throughout) with poor discrimination between good and bad lots. With such small samples, the producer can be given better protection against chance rejection of lots from a satisfactory process without serious loss in consumer protection if acceptance decisions on each lot can be influenced by the results of samples from the most recent preceding lots. In presenting his chain sampling plan (designated as ChSP-1), H. F. Dodge discusses certain aspects of this type of case as follows:‡

†*Ibid.;* also Statistical Research Group, Columbia University, *Sequential Analysis of Statistical Data: Applications,* Columbia University Press, New York, 1945.

‡H. F. Dodge, "Chain Sampling Inspection Plan," *Industrial Quality Control,* vol. 11, no. 4, pp. 10–13, January 1955; and reprinted in *Journal of Quality Technology,* vol. 9, no. 3, July 1977.

Suppose we consider such a situation with a continuing supply of some processed material, such as a particular type of copper-alloy rod or sheet. Say that a single sampling plan $n = 5$, $c = 0$ is being used for tensile strength. For each lot, five standard specimens are prepared and subjected to a standard test. If all five tests meet specification, the lot is accepted. If one of the five specimen tests fails to meet specification (i.e., one defect is observed), then the lot is considered nonconforming under the sampling plan and is subject to rejection or other disposition.

At this point certain questions arise. If the immediately preceding lot, or one of the most recently submitted lots, was also found nonconforming, then it would be reasonable to consider the current lot and perhaps the whole run of recent product to be of doubtful quality and hence probably subject to rejection. On the other hand, if no tensile test failures had been observed in the samples for quite a number of preceding lots, one might reason as follows: "We can't expect perfection. A small percentage of marginal failures to meet specification is reasonable for most such products. And if some small percent defective is reasonable, this one defect that we have just observed is probably that occasional one that we must expect every now and then." Now if the current defect were in fact the occasional defect, the lot should be accepted. To be able to act as though the observed defect is merely the occasional one requires several things:

(*a*) the lot should be one of a series in a continuing supply as mentioned above.
(*b*) lots should normally be expected to be of essentially the same quality.
(*c*) the consumer should have no reason for believing that this particular lot is poorer than the immediately preceding ones, and
(*d*) the consumer must have confidence in the supplier, confidence that the supplier would not take advantage of a good record to slip in a bad lot now and then when it would have the best chance of acceptance.

Dodge describes the plan ChSP-1 as follows:

1. Conditions for Application
   (*a*) Interest centers on an individual quality characteristic that involves destructive or costly tests, such that normally only a small number of tests per lot can be justified.
   (*b*) The product to be inspected comprises a series of successive lots (of material or of individual units) produced by an essentially continuing process.
   (*c*) Under normal conditions the lots are expected to be of essentially the same quality (expressed in percent defective).
   (*d*) The product comes from a source in which the consumer has confidence.
2. Procedure
   (*a*) For each lot, select a sample of n units (or specimens) and test each unit for conformance to the specified requirement.
   (*b*) Acceptance number of defects, $c = 0$; except $c = 1$ if no defects are found in the immediately preceding i samples of n. ($i = 1, 2, 3, \ldots$)

OC curves for plans with sample sizes of 4, 5, 6, and 10 are shown in Fig. 13.5. These curves show what is expected to happen over a series of lots for any stated process average $100\mu_p$. Dodge comments further as follows:

Curves for individual ChSP-1 plans can be compared with the OC curves for basic $c = 0$ plans of single sampling. It is seen that adding the provision for using cumulative results for i preceding samples has the same effect on the characteristic curve as taking a second sample. It increases the chance of acceptance in the region of principal interest—where the product percent defective is very small. Since in addition it calls for rejection provided only that two defects are fairly close together, it modifies the basically undesirable features of the $c = 0$ single sampling plans.

FIGURE 13.5 OC curves for ChSP-1 plans with values of i from 1 to 5. For comparison, single sampling plans ($n = 4$, 5, 6, and 10, with $c = 0$) are shown. *(Reproduced from H. F. Dodge, "Chain Sampling Inspection Plan," Industrial Quality Control, vol. 11, no. 4, pp. 10–13, January 1955; and reprinted in Journal of Quality Technology, vol. 9, no. 3, July 1977.)*

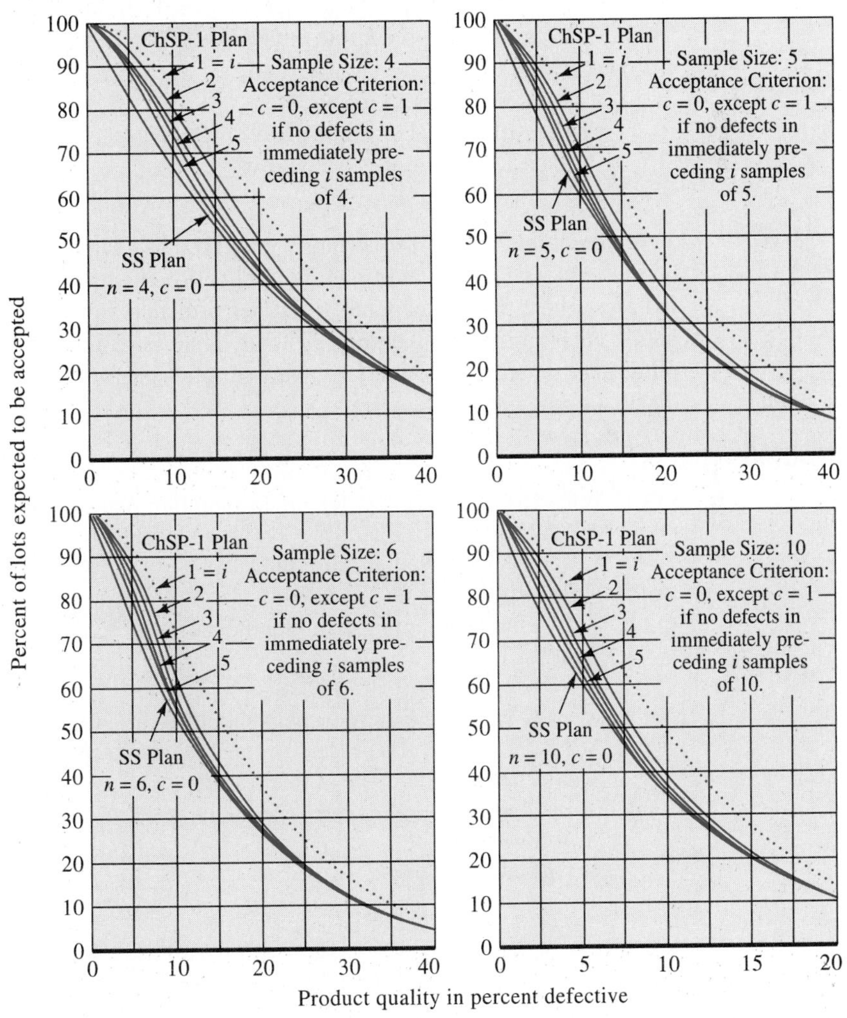

> It is believed that values of $i = 3$ or more, say three to five, will be found most helpful in practical application. The curve for $i = 1$ is shown dotted, merely to suggest that this is not a preferred choice. Theoretically, any value of i can be used. . . .
>
> With the use of this plan for measurable characteristics, for example, tensile strength and elongation of a material, it will normally be useful to keep a running chart of the measured values to provide supplementary information on the distribution of quality. Among other things, a continuing time-plot or control chart of measured values helps to identify wild points and is of assistance in problems of determining whether troubles are measurement troubles or product troubles.

The probability of acceptance in ChSP-1 is the probability of 0 defectives in a given sample plus the probability that the given sample will have exactly 1 defective and be preceded by i samples that had 0 defectives. For example, for $n = 4$, $i = 3$, and $\mu_p = 0.02$:

$$P_a = (0.98)^4 + 4(0.98)^3(0.02)(0.98)^4(0.98)^4(0.98)^4$$

$$= 0.9224 + 0.0591 = 0.9815^{\dagger}$$

It cannot be emphasized strongly enough that chain sampling plans should not be used unless all the conditions for application mentioned by Dodge are satisfied. In cases where process capability and performance indicators are provided by suppliers, such as in JIT inventory systems, chain sampling plans and skip lot sampling may be used effectively to audit supplier performance. Skip lot sampling is discussed in Chap. 14.

PROBLEMS

13.1. What Dodge-Romig (D-R) single sampling plan should be used for a lot size of 1,500 and an LTPD of 5.0% with a Consumer's Risk of 0.10 if the process average is estimated as 0.6% defective?
Answer: $n = 130$, $c = 3$.

13.2. What double sampling plan should be used for the conditions described in Prob. 13.1?
Answer: $n_1 = 55$, $c_1 = 0$; $n_2 = 120$, $c_2 = 4$.

13.3. What D-R single sampling plan should be used for a lot size of 900 and an AOQL of 2.0% if the process average is estimated as 0.9% nonconforming?
Answer: $n = 65$, $c = 2$.

13.4. What double sampling plan should be used for the conditions described in Prob. 13.3?
Answer: $n_1 = 38$, $c_1 = 0$; $n_2 = 87$, $c_2 = 4$.

† The following important paper should be studied by any readers interested in the special problems arising where small samples are necessary or desirable for one reason or another.

A. F. Cone and H. F. Dodge, "A Cumulative-Results Plan for Small-Sample Inspection," *Industrial Quality Control,* vol. 21, pp. 4–9, July 1964.

13.5. Determine from Table 13.3 the single sampling plans for an AOQL of 2%, an estimated process average of 1%, and lot sizes of 200, 1,000, and 5,000, respectively. What percentage of the product will be subject to sampling inspection with each lot size?

Answer: 8.5%, 6.5%, 2.5%.

13.6. A consumer receives lots of 2,000 items and uses a single sampling plan to accept or reject the lots. The plan calls for 200 items to be inspected and for the lot to be accepted if two or less nonconforming units are found. All rejected lots are screened by the consumer, and the resultant cost is billed to the producer. Compute the average outgoing quality (AOQ) if 0.8% nonconforming lots are submitted for inspection.

Answer: 0.564%.

13.7. In a single sampling plan, $N = 10{,}000$, $n = 300$, and $c = 1$. This is used with the stipulation that all rejected lots are to be screened. Compute the AOQ if all lots submitted are 0.5% nonconforming.

Answer: 0.271%.

13.8. In a double sampling plan, $N = 5{,}000$, $n_1 = 100$, $c_1 = 0$, $n_2 = 100$, and $c_2 = 1$.

(*a*) Use Table *G* to compute the probability of acceptance of a 1% nonconforming lot.

(*b*) Assume that a lot rejected by this sampling plan will be 100% inspected. What will be the AOQ if the submitted product is 1% nonconforming?

(*c*) Considering both the inspection of samples and inspection of rejected lots, what will be the average number of articles inspected per lot (ATI) if the submitted product is 1% nonconforming?

Answer: (a) 0.503; *(b)* 0.503%; *(c)* 2,549.

13.9. Prepare an AOQ curve for the single sampling plan $n = 100$, $c = 0$. What is the AOQL?

Answer: 0.37%.

13.10. The lot size N is 2,000 in a certain AOQL inspection procedure. The desired AOQL of 2.0% can be obtained with any one of three single sampling plans. These are $n = 65$, $c = 2$; $n = 41$, $c = 1$; and $n = 18$, $c = 0$. If a large number of lots 0.3% nonconforming are submitted for acceptance, what will be the average number of units inspected per lot (ATI) under each of these three sampling plans?

Answer: 67, 55, 123.

13.11. (*a*) Determine the D-R single sampling plan to be used for a lot size of 250 items and an LTPD of 5% with a Consumer's Risk of 10% if the process average is estimated at 1.1%.

(*b*) What percentage of the product will be subject to sampling inspection with this sampling plan?

(*c*) What is the probability of acceptance of a lot 4.0% nonconforming under this plan?

(*d*) Compute the average total inspection (ATI) at 4.0% and at 1.1% assuming that rejected lots are screened.

Answer: (a) $n = 70$, $c = 1$; *(b)* 28%; *(c)* 0.231; *(d)* 208.4, 102.4.

13.12. In Prob. 13.5, out of 100 submitted lots 1% nonconforming, how many would you expect to be subject to screening inspection for each lot size?

Answer: 16, 3, 1.

13.13. Using the answers in Probs. 13.5 and 13.12, what will be the total percentage of inspection for each of the three lot sizes, considering both sampling inspection and screening inspection of rejected lots?
Answer: 23%, 9%, 3.5%.

13.14. What D-R single sampling plan should be used for a lot size of 2,500 and an LTPD of 5% with a Consumer's Risk of 0.10 if the process average is estimated as 0.9%?.

13.15. Find the D-R 2% AOQL double sampling plan for a lot size of 300 items and a process average p of 0.005.

13.16. For the plan found in Prob. 13.15, compute the following quantities assuming that incoming lots contain 2% nonconforming items:
(*a*) The approximate probability of acceptance on the first sample using Table *G*.
(*b*) The approximate total probability of acceptance using Table *G*.
(*c*) The AOQ.
(*d*) The average fraction inspected (AFI).

13.17. A manufacturer has specified sampling in accordance with the D-R AOQL system. A 2% AOQL plan is to be used for a lot size of 3,500 items and a process average of 0.35%.
(*a*) Find the proper double sampling plan, and identify its LTPD point.
(*b*) Find the corresponding single sampling plan.
(*c*) Use Table *G* to compute the approximate probability of acceptance under the plan found in (*b*) assuming that incoming lots contain 2% nonconforming units.

13.18. (*a*) Compute the AOQ for the single sampling plan found in part (*b*) of Prob. 13.17.
(*b*) Compute the ATI for this plan.

13.19. A D-R single sampling 2% AOQL plan is to be used to sample from lots of 12,000 items.
(*a*) Find the appropriate plan for a process average of 1.5%.
(*b*) Find the corresponding double sampling plan.

13.20. (*a*) Find the D-R single sampling 5% LTPD plan for a process average of 1.3% and a lot size of 250 units.
(*b*) Find the corresponding double sampling plan.
(*c*) What are the respective AOQLs for these plans?

13.21. (*a*) For the single sampling plan found in Prob. 13.20, use Table *G* to compute the approximate probability of acceptance if incoming lots contain 3% nonconforming units.
(*b*) Compute the AOQ.
(*c*) Compute the AFI.

13.22. Compute the average number of items sampled for the double sampling plan found in Prob. 13.20 assuming incoming lots contain 3% nonconforming items. How does this average value compare with the single sample size found in part (*a*) of Prob. 13.20?

13.23. (*a*) Find the D-R 2% AOQL single sampling plan for a lot size of 450 units and a process average of 2.5%.
(*b*) Find the corresponding double sampling plan.

13.24. (*a*) For the double sampling plan found in Prob. 13.23, compute the approximate probability of acceptance on the first sample if incoming lots contain 5% nonconforming units. Use Table *G*.
(*b*) Compute the total probability of acceptance.
(*c*) Compute the AOQ.
(*d*) Compute the AFI.

13.25. (*a*) Find the D-R 5% LTPD single sampling plan for a lot size of 1,250 units and a process average of 1.0%.
(*b*) What proportion of the product will be subject to sampling inspection with this sampling plan?
(*c*) Compute the ATI at the process average.

13.26. (*a*) For the plan found in Prob. 13.25, use Table *G* to compute the approximate probability of acceptance of lots containing 3.0% nonconforming units.
(*b*) Compute the ATI.
(*c*) Compute the AOQ.

13.27. (*a*) Find the D-R 5% LTPD double sampling plan for a lot size of 3,500 units and a process average of 0.4%. What is the AOQL of this plan?
(*b*) Find the 2% AOQL double sampling plan for the same lot size and process average. What is the LTPD of this plan?

13.28. The lot size is 2,000 units in a certain D-R AOQL inspection procedure. The desired AOQL of 2.0% can be obtained with any one of three single sampling plans. These plans are $n = 65$, $c = 2$; $n = 41$, $c = 1$; and $n = 18$, $c = 0$. Answer the following questions under the assumption that incoming lots contain 1.2% nonconforming items.
(*a*) Use Table *G* to compute the approximate probabilities of acceptance for each plan.
(*b*) Compute the ATI for each plan.
(*c*) Compute the AOQ for each plan.
(*d*) Taking all three of these measures of effectiveness into account, which plan would you recommend in this case? Explain.

13.29. For a $p_{0.50}$ of 0.025, what should be the single sample sizes corresponding to acceptance numbers of 0, 1, 3, 5, and 7, respectively? (Use Table 13.5.)
Answer: 27, 67, 147, 227, 307.

13.30. Use Table *G* to determine the approximate OC curves of the plans in Prob. 13.29 that have acceptance numbers of 0, 3, and 7.

13.31. A producer and consumer agree that a single sampling plan will be used to accept and reject lots. The single sampling plan specified will pass through the indifference quality point at 2% nonconforming. (Use Table 13.5.)
(*a*) Use Table *G* to define operating characteristic curves for sampling plans with acceptance numbers of 0, 1, 2, and 3.
(*b*) Disregarding inspection costs, which plan do you feel will provide the consumer with the best protection against poor quality?
(*c*) If 200 lots of 3% nonconforming product are shipped to the consumer, how many items do you anticipate will be inspected if all rejected lots of 1,000 items are screened, according to the plan with an acceptance number of 2?

13.32. It is desired to have a single sampling plan for which the Producer's Risk of rejection of a 1% nonconforming lot is 0.05 and the Consumer's Risk of accepting a 3% nonconforming lot is 0.10. Using Table 13.5, find the plan that exactly meets the stipulation on Producer's Risk and comes as close as possible on Consumer's Risk. For this plan, what percentage nonconforming has a Consumer's Risk of 0.10%?

Answer: $n = 398$; $c = 7$; $100p_{0.10} = 2.96\%$.

13.33. It is desired that a single sampling plan for attributes data satisfy a Producer's Risk of rejection of 1.5% nonconforming lots of 0.05 and a Consumer's Risk of acceptance of 7.0% nonconforming lots of 0.10.

(*a*) Use Table 13.5 to find the plan that fits the Consumer's Risk and is as close as possible to the Producer's Risk.

(*b*) What is the percentage nonconforming that has a probability of rejection of 0.05?

13.34. A single sampling plan is to be developed to inspect large lots of a certain product based on a Producer's Risk of 0.05 of rejecting lots containing 1.0% nonconforming items. The Consumer's Risk of accepting lots containing 4% nonconforming items is to be as close as possible to 0.10.

(*a*) Use Table 13.5 to find the two plans that fit the Producer's Risk point and most nearly fit the Consumer's Risk point.

(*b*) What percent nonconforming has a 10% chance of acceptance under each plan?

13.35. A single sampling plan is to be developed based on a Producer's Risk of 0.05 of rejecting lots that contain 0.50% nonconforming items. The Consumer's Risk of accepting lots containing 3.50% nonconforming items should be as close as possible to 0.05. Use Table 13.5 to determine the plan. (*Hint:* The right-hand column in Table 13.5 contains the ratio $np_{0.10}/np_{0.95}$. You will need the ratio $np_{0.05}/np_{0.95}$.)

13.36. It is proposed that a system of single sampling acceptance sampling plans for attributes data be developed, all using an acceptance number of zero. Each plan in the system is to be designed such that the probability of acceptance at a certain unacceptable quality level is 0.05 (a specified Consumer's Risk point).

(*a*) Use Table 13.5 to compute the values of n for unacceptable quality levels of 0.01, 0.02, 0.05, 0.10, and 0.15.

(*b*) What undesirable characteristic of the OC curve is contained in the stipulation of an acceptance number of zero?

(*c*) In the application of this system of plans, it is found that all lot sizes are between 100 and 250 items. What assumptions made in the use of Table 13.5 to develop the plans do not hold? Explain your answer, and give the correct formula for the OC curve.

13.37. It is proposed to have a set of single sampling plans in which the percentage nonconforming corresponding to a Consumer's Risk of 0.10 is, as nearly as practicable, 2.0 times the percentage nonconforming corresponding to a Producer's Risk of 0.05. Use Table 13.5 to find the plan for which $100p_{0.10}$ is 1.0% and $100p_{0.95}$ is 0.4% and for which $100p_{0.10}$ is 10% and $100p_{0.95}$ is 4%. Does it seem to you that it is a good idea to stipulate that $p_{0.10}$ shall always be 2.0 times $p_{0.95}$? Why or why not?

13.38. It is desired to have a system of single sampling plans, all of which have a sample size of 60. Use Table 13.5 to find the respective values of $np_{0.99}$, $np_{0.50}$, and $np_{0.01}$ for a system of plans having acceptance numbers of 0, 1, 2, and 3. Sketch the OC curves of these plans.

13.39. A single sampling acceptance plan is to be derived to satisfy two points on an OC curve, an AQL point and an LTPD point:

$$(\text{AQL}, 1 - \alpha) = (0.005, 0.99)$$

$$(\text{LTPD}, \beta) = (0.025, 0.10)$$

(*a*) Use Table 13.5 to find the plan that comes closest to fitting the LTPD point.
(*b*) How different is this plan from the one that best fits the AQL point? What is the actual value of μ_p at $P_a = 0.99$ for the plan found in (*a*)?

13.40. Use Table 13.5 to find the single sampling plan that satisfies a Producer's Risk of rejection of 0.01 at 1.0% nonconforming product and a Consumer's Risk of 0.05 at 5.0% nonconforming product.

13.41. Three single sampling plans all fitting a Consumer's Risk point $p_{0.10}$ of 0.10 are to be developed. For each, the probability of acceptance should be approximately 0.90 when the quality level is 0.01, 0.020, and 0.04, respectively.
(*a*) Use Table 13.5 to find the three plans that best satisfy these specifications.
(*b*) What is the indifference quality level for each of these plans?
(*c*) Sketch the OC curves for these three plans, overlaying one curve on the other.

13.42. (*a*) Use Table 13.5 to find a single sampling plan that has a Consumer's Risk of 0.05 of acceptance if the incoming quality level is 0.057 and a Producer's Risk of 0.05 of rejection if the incoming quality level is 0.005.
(*b*) Find the formulas for the acceptance (*Ac*) and rejection (*Re*) lines for the item-by-item sequential sampling plan that fits the two points on the OC curve stipulated in part (*a*). What is the maximum ASN for this plan?

13.43. (*a*) Use Table 13.5 to find a single sample acceptance plan to fit the following two points on an operating characteristic curve:

$$p_{0.95} = 0.015 \quad \text{or} \quad (p_1, 1 - \alpha) = (0.015, 0.95)$$

$$p_{0.10} = 0.065 \quad \text{or} \quad (p_2, \beta) = (0.065, 0.10)$$

(*b*) Find the values of the slope and intercepts of the item-by-item sequential sampling plan that fits these two points. Find the maximum value of the ASN for this plan. (*Hint:* ASN is maximum when $\mu_p = s$.)

13.44. Determine the equations of the rejection and acceptance lines for an item-by-item sequential plan in which $p_{0.90} = 0.05$ and $p_{0.20} = 0.15$.
Answer: $Re = 0.092n + 1.719$; $Ac = 0.092n - 1.243$.

13.45. Plot an approximate OC curve for the sequential plan of Prob. 13.44. Prepare an item-by-item table of acceptance and rejection numbers for values of n from 1 to 50. In this table, make the rejection number the next whole number above *Re* and the acceptance number the next whole number below *Ac*.

13.46. (*a*) Determine the acceptance and rejection lines for an item-by-item sequential sampling plan in which $p_{0.95} = 0.010$ and $p_{0.10} = 0.050$.
(*b*) Prepare a table of values of n for acceptance numbers from 0 through 4.
(*c*) What are the corresponding values of the rejection numbers in this range of n?

13.47. Use Table 13.5 to compute the single sampling plan corresponding to the item-by-item sequential sampling plan found in Prob. 13.46.

13.48. (*a*) Determine the equations of the acceptance and rejection lines for an item-by-item sequential sampling plan in which $p_{0.95} = 0.015$ and $p_{0.10} = 0.20$.

(*b*) Sketch an ASN curve for this plan and indicate the maximum point on the curve.

(*c*) Use Table 13.5 to develop a single sampling plan satisfying the operating characteristics prescribed in part (*a*).

(*d*) Compare the value of n found in part (*c*) with the ASN values found in part (*b*).

13.49. An item-by-item sequential sampling plan is to be developed satisfying the following two points on an operating characteristic curve:

$$(p_1, 1 - \alpha) = (0.01, 0.90) \qquad (p_2, \beta) = (0.08, 0.10)$$

(*a*) Calculate the formulas for the acceptance and rejection lines for this plan.

(*b*) Approximately what number of units n separates every change in an acceptance or rejection number?

(*c*) Find values of the ASN and probability of acceptance for five values of μ_p. Indicate the maximum ASN.

(*d*) Use Table 13.5 to find a single sampling attributes plan satisfying the two points indicated in part (*a*). Which plan is more efficient and why?

13.50. (*a*) Using Dodge's simplified AQL system shown in Table 13.6, determine the acceptance criteria under normal and tightened inspection for inspection level III, 2.5% AQL, and a lot size of 500.

(*b*) What is the probability of acceptance of lots 6.0% nonconforming under both plans?

(*c*) What is the AOQL under tightened inspection for the plan derived in (*a*) assuming rejected lots are screened and nonconforming items replaced, and also that nonconforming items found in the samples from accepted lots are replaced with good items.

13.51. A manufacturer inspects all its product before shipment using Dodge's single sampling system shown in Table 13.6. Inspection level II for normal inspection is used. Lots are shipped in quantities of 1,500 under an AQL requirement of 0.65%. On receiving lots, the customer also inspects the material using a plan from the Dodge system. However, the customer uses inspection level III and tightened inspection with the same AQL.

(*a*) Find the acceptance criteria for both plans.

(*b*) Compute the probability that a lot with 1.5% nonconforming will be accepted by both plans consecutively.

13.52. A purchaser of medical supplies uses a chain sampling acceptance plan, ChSP-1, since only a limited amount of inspection is possible. The plan in use requires that five items from each lot be inspected. If no defective is found, the lot is accepted. If one defective is found, the lot is accepted only if the four previous lot samples inspected contained no defectives. Otherwise, the lot is rejected. If the lots received are 10% defective, determine the probability of accepting any given lot.

13.53. A chain sampling plan (ChSP-1) is used by a supplier to approve material for shipment. The plan in use requires that three items be inspected from each lot to be shipped. If no defectives are found in a lot, then that lot is approved. If one is found, the lot is approved only if no defectives were found in the previous five lots. Otherwise, the lot is rejected. Determine the OC curve for this plan using product qualities of 0.1, 1.0, 3.0, 5.0, 10.0, and 25.0% defective material. Is this a Type *A* or Type *B* OC curve?

13.54. A supplier has a long record of providing variable-resistance devices of consistently good quality. Receiving inspection has been performed on lots of 50 units in accordance with the ABC standard, code letter F, and an AQL of 0.65% ($n = 20$, $c = 0$). It has been decided to change to a Dodge chain sampling plan (ChSP-1) using a sample size of 10. If no device is rejected, the lot is accepted. If one device is rejected, the lot is accepted only if no rejects were found in the previous 4 lot samples.

(*a*) Calculate the probabilities of acceptance at incoming values of p of 0.01 and 0.08 for the Dodge ChSP-1 plan.

(*b*) Calculate the probabilities of acceptance at values of p of 0.01 and 0.08 for the ABC standard plan.

(*c*) Comment briefly on the relative merits of the two plans.

14

SYSTEMS FOR ACCEPTANCE SAMPLING FROM CONTINUOUS PRODUCTION

Multi-level continuous sampling plans . . . offer some important advantages. The fact that it is not necessary to accumulate a lot before making a decision is highly advantageous for large and expensive items. The production organization can deliver products more rapidly. Less storage space is required and, as long as the quality is good, both the producer and the consumer save on the inspection cost. On the other hand, as soon as the quality gets worse than the AOQL, the rate of inspection increases rapidly. Thus, there is a better chance of detecting causes of defects soon after they occur than if the product were accumulated for lot-by-lot acceptance provided corrective action is initiated as soon as the inspection rate increases. The quality control function can be of more assistance to the production organization by keeping it informed currently regarding the status of the product and by assisting it in corrective action to eliminate causes of defects. (Many of the above points are true for all types of continuous sampling plans.)[†]

14.1 DODGE'S AOQL PLAN FOR CONTINUOUS PRODUCTION—CSP-1

The earliest acceptance/rectification plan for application to continuous production was one described by H. F. Dodge in 1943. Where production is continuous, the formation of inspection lots for lot-by-lot acceptance is somewhat artificial. More-

[†]"Multi-Level Continuous Sampling Procedures and Tables for Inspection by Attributes, Inspection and Quality Control Handbook (Interim) H 106," Office of the Assistant Secretary of Defense (Supply and Logistics), Washington, D.C., 1958.

over, where conveyor lines are used, it may be impracticable or unduly costly to form inspection lots. Dodge explained his procedure (subsequently referred to as CSP-1) as follows:[†]

(*a*) At the outset, inspect 100% of the units consecutively as produced and continue such inspection until i units in succession are found clear of defects.
(*b*) When i units in succession are found clear of defects, discontinue 100% inspection, and inspect only a fraction f of the units, selecting individual sample units one at a time from the flow of product, in such a manner as to assure an unbiased sample.
(*c*) If a sample unit is found defective, revert immediately to a 100% inspection of succeeding units and continue until again i units in succession are found clear of defects, as in paragraph (*a*).
(*d*) Correct or replace, with good units, all defective units found.

Figure 14.1 gives the necessary information for the selection of such a plan for any desired AOQL. As an example, suppose the desired AOQL is 2%, and it is desired to establish a plan that calls for inspection of 1 out of every 20 pieces from the conveyor belt. Then f, the fraction inspected, is $\frac{1}{20}$, or 0.05, or 5%. In Fig. 14.1, find the value of i corresponding to an f of 5% on the curve for an AOQL of 2%. This $i = 76$. The acceptance plan is then as follows:

1. Inspect all the units consecutively as produced until 76 units in succession are free from defects.
2. As soon as 76 successive units are free from defects, inspect a sample consisting of only 1 unit out of every 20. Accept all the product as long as the sample is free from defects.
3. Whenever one of these sample units is found defective, resume 100% inspection until 76 units in succession have again been found free from defects. Then resume sampling inspection.

14.1.1 The Dodge-Torrey Modifications of Continuous Sampling Plan 1

H. F. Dodge and M. N. Torrey developed two modifications of CSP-1, referred to as CSP-2 and CSP-3, respectively.[‡] They describe CSP-2 as follows:

[†]H. F. Dodge, "A Sampling Inspection Plan for Continuous Production," *The Annals of Mathematical Statistics,* vol. 14, no. 3, pp. 264–279, September 1943. H. F. Dodge, "Sampling Plans for Continuous Production," *Industrial Quality Control,* vol. 4, no. 3, pp. 5–9, November 1947. Both these articles have been reprinted in *Journal of Quality Technology,* vol. 9, no. 3, July 1977, American Society for Quality Control, Milwaukee, Wis.

[‡]H. F. Dodge and M. N. Torrey, "Additional Continuous Sampling Inspection Plans," *Industrial Quality Control,* vol. 7, no. 5, pp. 7–12, March 1951; reprinted in *Journal of Quality Technology,* vol. 9, no. 3, July 1977. Plans CSP-1 and CSP-2 were adopted for procurement purposes by the Department of Defense and published in "Single-Level Continuous Inspection Procedures and Quality Control Handbook Tables for Inspection by Attributes, Inspection and Quality Control Handbook (Interim) H 107," Office of the Assistant Secretary of Defense (Supply and Logistics), Washington, D.C., 1959.

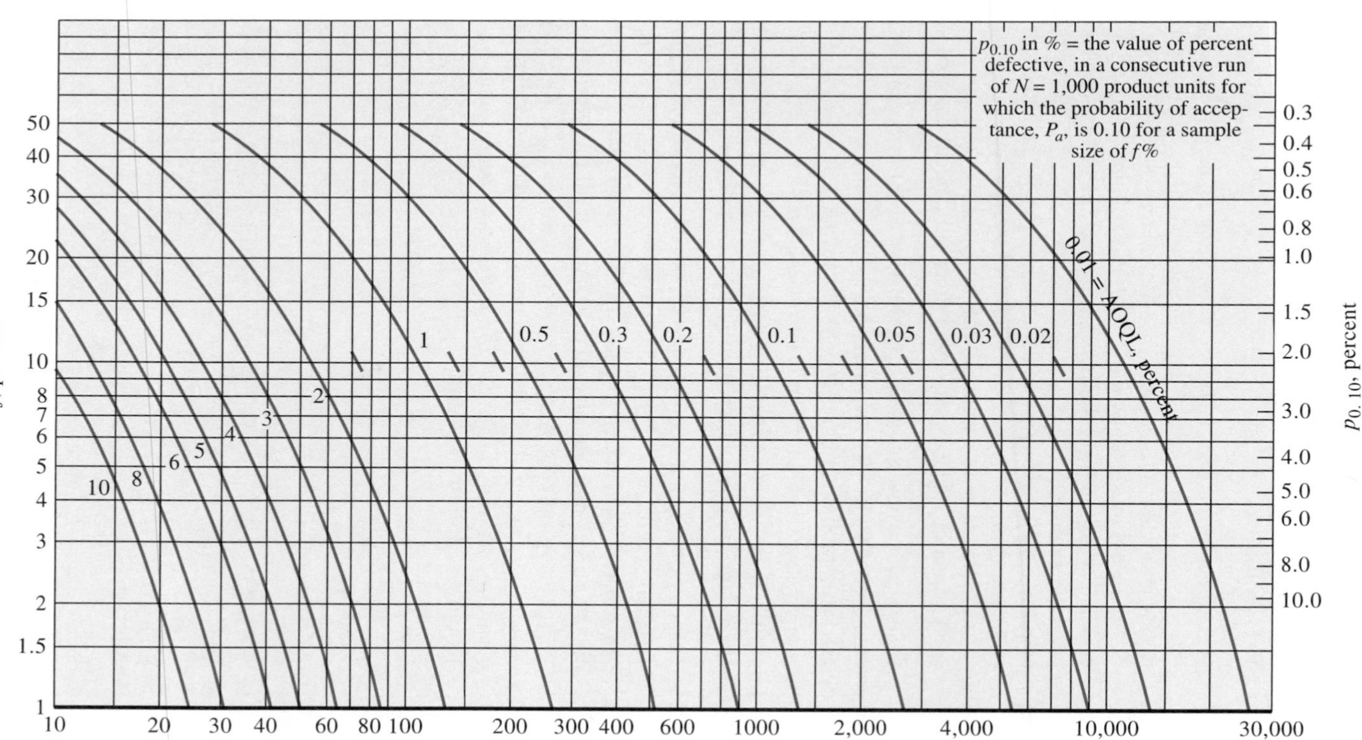

FIGURE 14.1 Curves for determining values of *f* and *i* for a given value of AOQL in Dodge's plan for continuous production CSP-1. *(Reproduced by permission from H. F. Dodge, "A Sampling Inspection Plan for Continuous Production," The Annals of Mathematical Statistics, vol. 14, pp. 264–279, September 1943.)*

Plan CSP-2 differs from Plan CSP-1 in that, once sampling inspection is started, 100% inspection is not invoked when each defect is found but is invoked only if a second defect occurs in the next k or less sample units. In other words, if two defects observed during sampling are separated by k or less good inspected units, 100% inspection is invoked. Otherwise sampling is continued.

Although the factor k might conceivably be assigned any value, the only CSP-2 plans prepared for use have been those in which $k = i$. Table 14.1 gives values of i in CSP-1 and CSP-2 using 5 and 10% samples and various AOQL values. These values for CSP-2 were obtained from a graph similar to Fig. 14.1 given in the Dodge-Torrey article.

Consider the application of CSP-2 when f is 5% and the AOQL is 2%. Table 14.1 gives $i = 96$. The acceptance plan is then as follows:

1. Inspect all the units consecutively as produced until 96 units in succession are free from defects.
2. As soon as 96 successive units are free from defects, inspect a sample consisting of only 1 out of every 20. Accept all the product as long as the sample is free from defects.
3. If one defective is found in this sampling, continue sampling inspection for the time being. However, if a second sample defective is found within the next 96 samples, resume 100% inspection immediately. Continue 100% inspection until 96 units in succession have been found free from defects. Then resume sampling under the foregoing rules.

Plan CSP-3 is a refinement of CSP-2 to provide greater protection against a sudden run of bad quality. When one sample defective is found, the next four units

TABLE 14.1 COMPARISON OF VALUE OF i FOR VARIOUS AOQLs IN CSP-1 AND CSP-2 FOR 5 AND 10% SAMPLING INSPECTION (ASSUMING $i = k$ IN CSP-2)

	f = 5%		f = 10%	
AOQL, %	i in CSP-1	i (= k) in CSP-2,3	i in CSP-1	i(= k) in CSP-2,3
0.3	510	650	370	490
0.5	305	390	220	290
1	150	195	108	147
2	76	96	55	72
3	49	64	36	48
4	37	48	27	36
5	29	38	21	29
6	24	31	17	23
8	18	23	13	17
10	14	18	10	14

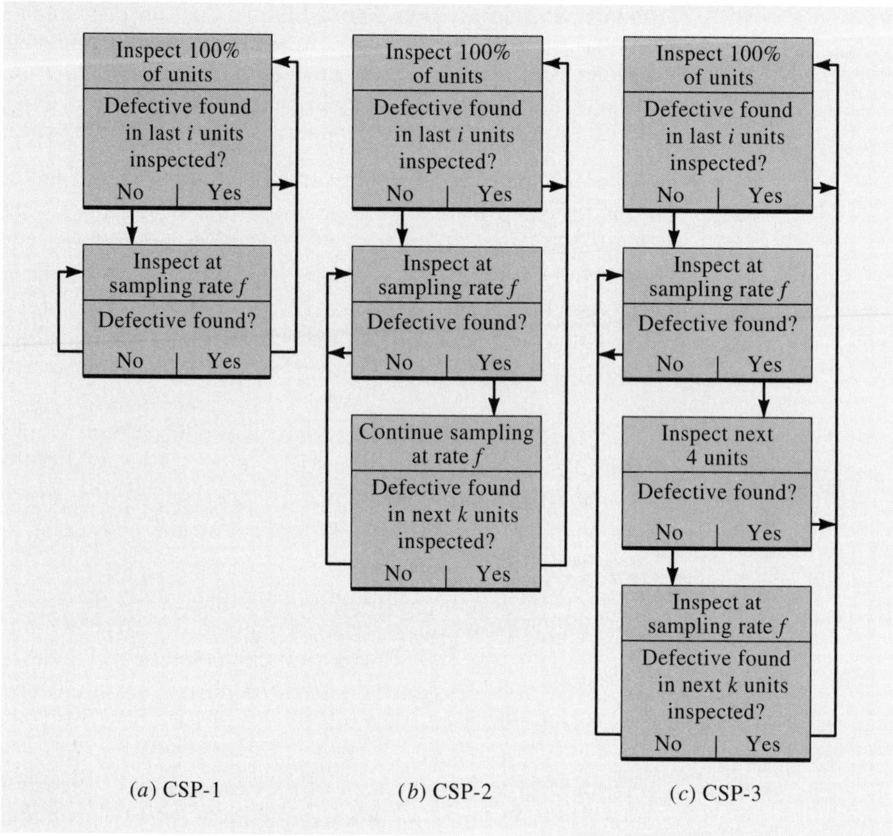

FIGURE 14.2 Flow process chart for Dodge and Dodge-Torrey continuous sampling plans CSP-1, CSP-2, and CSP-3.

from the production line are inspected. If none of these are defective, the sampling procedure is continued as in CSP-2. If one of the four units is defective, 100% inspection is resumed at once and continued under the rules of CSP-2. In CSP-3, the value of i used for a given f and AOQL is the same as in CSP-2. Figure 14.2 shows an abbreviated flow process chart of the operation of the CSP-1, CSP-2, and CSP-3 plans.

14.2 MULTILEVEL CONTINUOUS SAMPLING PLANS (CSP-M)

Sometimes it is desired to start any sampling inspection with a relatively large fraction sampled, such as $\frac{2}{3}$, $\frac{1}{2}$, or $\frac{1}{3}$. In such cases, economies can be obtained by permitting a subsequent change to a smaller sampling fraction whenever the quality observed on the initial sampling inspection turns out to be good enough. To secure these economies and to obtain certain other advantages, multilevel continuous sampling schemes have been developed that stipulate two or more different sampling fractions and give rules for shifting back and forth among the stipulated fractions.

One particular type of multilevel plan was developed and analyzed by G. J. Lieberman and Herbert Solomon.[†] Just as in the CSP plans described, the Lieberman-Solomon plans start with 100% inspection that continues until i acceptable units have been found in succession. Then sampling inspection is initiated with a fraction f inspected. If i acceptable units in succession are found under this sampling inspection, subsequent inspection is at the fraction f^2. Another i acceptable units in a row qualify for inspection at f^3, and so on. When a unit is rejected, inspection is shifted back to the next lower level. Any number of levels may be provided from two to an infinite number. The Lieberman-Solomon article provides a graph similar to Fig. 14.1 for two-level plans and gives approximate formulas for the relationship among f, i, and AOQL for plans with more than two levels. A. H. Bowker and G. J. Lieberman[‡] give graphs similar to Fig. 14.1 for two-, three-, four-, and infinite-level plans.

14.2.1 A Convenient Source for Multilevel Plans

In 1956, under a contract with the U.S. Air Force, Stanford University's Department of Industrial Engineering developed a proposed AMC Manual 74 entitled "Multi-level Continuous Sampling Acceptance Plans for Attributes." With minor additions and deletions of certain plans, this set of plans was adopted by the Air Force and later published by the U.S. Department of Defense as "Inspection and Quality Control Handbook (Interim) H 106, Multi-level Continuous Sampling Procedures and Tables for Inspection by Attributes." Table 14.2 combines the plans given in these two pamphlets. (H 106 added plans for AOQL values of 0.35, 0.15, and 0.10% and deleted those for 20 and 4%; plans for $f = \frac{2}{3}$ were shown in the proposed AMC Manual but not in H 106.)

The plans in these two pamphlets are based on the Lieberman-Solomon type of procedure combined with the special feature of the Dodge-Torrey plan CSP-3, namely, the inspection of the next four units whenever a defective is found. Figure 14.3, reproduced from H 106, illustrates the general features of all these plans by giving specific instructions for the operation of a three-level plan using an f of $\frac{1}{2}$.

14.2.2 Selection and Operation of a Multilevel Continuous Sampling Plan

Three decisions must be made in choosing a plan from Table 14.2. The AOQL and f must be selected, and it must be decided how many levels will be permitted.

[†]G. J. Lieberman and Herbert Solomon, "Multi-Level Continuous Sampling Plans," *The Annals of Mathematical Statistics,* vol. 26, pp. 686–704, 1955.

[‡]A. H. Bowker and G. J. Lieberman, *Engineering Statistics,* 2d ed., pp. 553–555, Prentice-Hall, Inc., Englewood Cliffs, N.J., 1972.

TABLE 14.2 VALUES OF i FOR MULTILEVEL CONTINUOUS SAMPLING PLANS[†]

AOQL, % defective	For $f = \frac{1}{2}$					For $f = \frac{1}{3}$					For $f = \frac{2}{3}$		
	k = 1	k = 2	k = 3	k = 4	k = 5	k = 1	k = 2	k = 3	k = 4	k = 5	k = 1	k = 2	k = 3
20.0	‡	‡	‡	‡	‡	‡	‡	4	5	6	‡	‡	‡
15.0	‡	‡	‡	4	5	‡	4	6	7	8	‡	‡	‡
10.0	‡	4	6	8	9	‡	7	10	12	13	‡	‡	‡
7.5	‡	6	9	11	13	6	11	14	16	18	‡	‡	‡
5.0	5	11	15	18	20	10	18	22	25	27	‡	5	8
4.0	7	14	19	22	25	14	23	29	32	34	‡	7	11
3.0	11	20	26	31	34	19	32	39	43	46	5	11	16
2.0	18	31	40	47	51	31	48	59	66	71	9	18	25
1.5	25	43	55	63	69	42	66	80	89	95	13	25	34
1.0	39	65	83	95	104	64	100	120	134	142	21	39	52
0.75	54	88	112	128	140	87	134	161	179	191	30	53	71
0.50	82	132	168	193	210	133	202	243	269	287	46	80	107
0.35	119	197	241	275	302	190	290	349	386	408			
0.25	167	269	337	386	422	269	406	488	540	576	96	164	217
0.15	218	446	564	636	706	450	680	815	903	960			
0.10	421	675	847	969	1,059	677	1,022	1,224	1,354	1,443			

[†]These values are taken in part from U.S. Department of Defense's "Inspection and Quality Control Handbook (Interim) H 106 (31 October 1958)" and in part from the Stanford University Department of Industrial Engineering's "Proposed AMC Manual 74 (15 June 1956)."
[‡]Sampling plans not available for values of i less than 4.

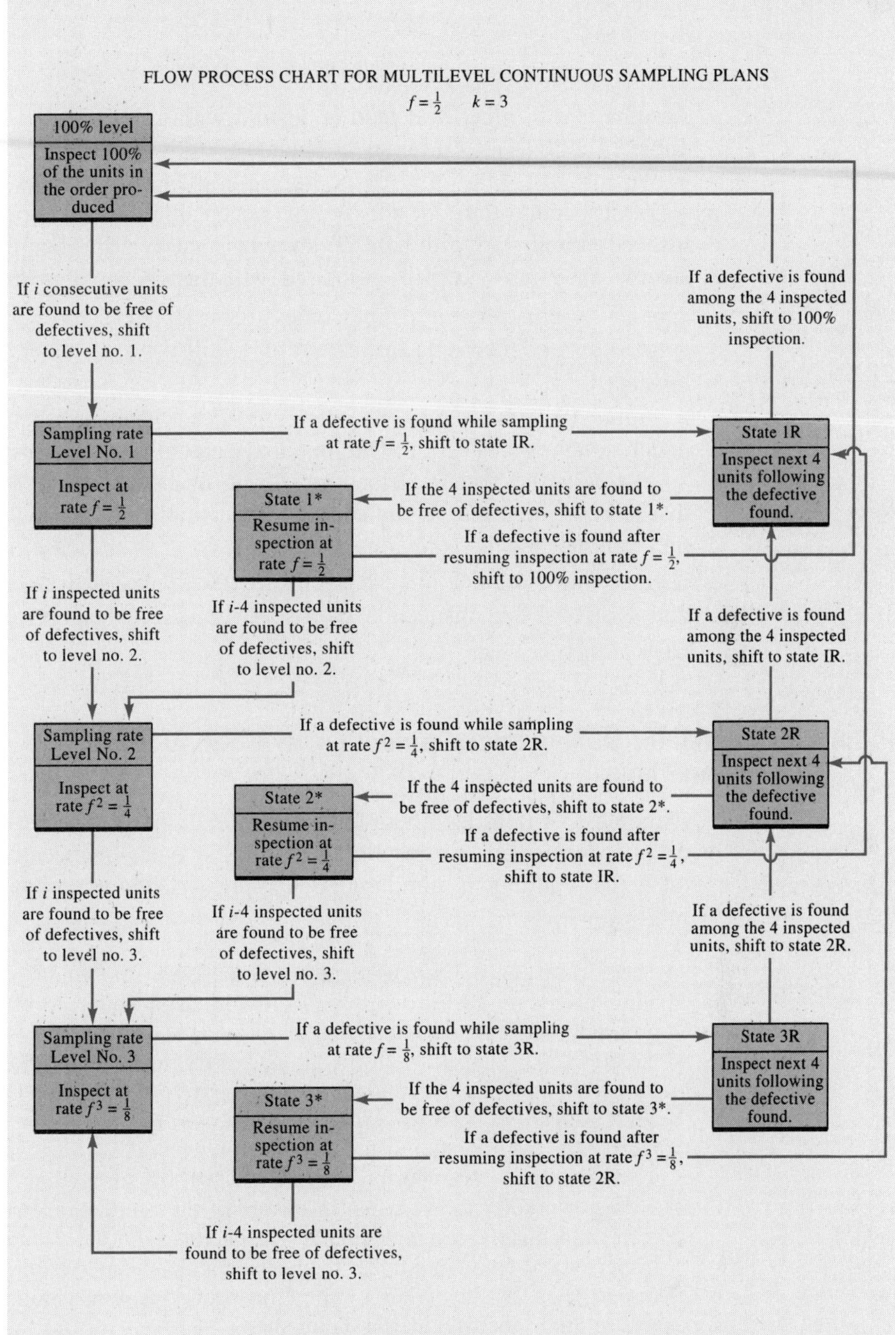

FIGURE 14.3 Example of rules for operating the multilevel continuous sampling plans of Table 14.2, assuming a three-level plan with $f = \frac{1}{2}$. *[Reproduced from "Inspection and Quality Control Handbook (Interim) H 106, Multi-Level Continuous Sampling Procedures and Tables for Inspection by Attributes," Office of the Assistant Secretary of Defense, Supply and Logistics, Washington, D.C., 1958.]*

Assume that a three-level plan is desired with an AOQL of 2.0% and an f of $\frac{1}{2}$. Table 14.2 gives $i = 40$ for such a plan.

We may use Fig. 14.3 to see how this plan will operate. There will first be 100% inspection until 40 consecutive acceptable units are found. Then sampling inspection will start with half the units inspected. If all goes well and the next 40 sample units are accepted, sampling will then be applied to $\frac{1}{4}$ of the units. If another 40 consecutive sample units are accepted, sampling inspection will be applied to $\frac{1}{8}$ of the units. Inspection of $\frac{1}{8}$ will continue indefinitely until a unit is rejected.

Figure 14.3 also shows the operating rules whenever a defective unit is found under sampling inspection. One hundred percent inspection is applied to the next 4 units, with the state described as 1*R*, 2*R*, or 3*R* depending on the sampling rate that has been in effect. If there are no rejects in these four units, sampling inspection continues at the previous rate, $\frac{1}{2}$, $\frac{1}{4}$, or $\frac{1}{8}$ as the case may be. If a unit is rejected in the 4 units, there is a shift to 100% inspection if sampling has been at rate $\frac{1}{2}$ or to 4-unit inspection with state 1*R* or 2*R*, respectively, if sampling has been at the rate $\frac{1}{4}$ or $\frac{1}{8}$.

14.2.3 Incentives for Quality Improvement under Continuous Sampling Plans

Example 13.1 described how lot-by-lot AOQL plans were used to stimulate quality improvement in a particular case. Similar types of incentive may be incorporated into continuous sampling AOQL plans. Dodge comments on this point as follows.[†]

> These plans have been used rather extensively and found most effective when administered in such a way as to provide an incentive to clear faults in process promptly. Such an incentive may be provided, for example, by requiring the production department to perform the necessary 100% inspections when defects are found. To this end, the following administrative procedure has met with good success. A regularly assigned process inspector performs all sampling inspections required; if additional assistance is needed when it becomes necessary to revert to 100% inspection or in performing the initial 100% inspections required, the process inspector notifies the foreman in charge of the production line; the foreman must then immediately assign temporary inspectors who are acceptable to the process inspector and who work under the jurisdiction and supervision of the senior process inspector on the line; when sampling inspection is reinstated the temporary inspectors return to their regular assignments.

There may be special incentive aspects of multilevel schemes owing to the fact that some producers gain satisfaction by qualifying for inspection at one of the reduced levels. This point is illustrated in Example 14.1.

[†]In his 1947 article in *Industrial Quality Control,* op cit.

Example 14.1

Quality improvement resulting from nonfinancial incentives created by a multilevel continuous sampling plan

Facts of the case under 100% inspection A manufacturer was the sole producer of a certain complex product for which there was only one customer. A few of these products were completed each working day. The contract for the manufacture of this product continued throughout several years.

At many points in the production operation, the manufacturer conducted acceptance inspection of components and subassemblies. The manufacturer's final inspection consisted of a performance test involving a check on many of the specified quality characteristics. This test was carried out on 100% of the product. In a number of cases, this test disclosed the need for adjustment of the product or replacement of certain components. After such adjustments or replacements were made, the product was given another performance test.

As soon as a product passed the manufacturer's final performance test, it was delivered to the customer. Immediately after this delivery, the customer carried out a performance test that was almost the same as the one made by the manufacturer. Although most of the products passed the customer's test, there were occasional failures that were returned to the manufacturer. The performance test was a fairly complicated and costly matter.

Occasional items of this complex product developed certain serious difficulties in their early weeks of use by the customer. It seemed evident that the difficulties were due to failure to conform to certain design specifications; the 200% final inspection had not been fully effective in removing all defects.

Consequences of adoption of a multilevel continuous sampling plan For its acceptance testing, the customer initiated a three-level sampling plan with an f of $\frac{1}{2}$. The plan was similar to the one illustrated in Fig. 14.3. Apparently the customer adopted this plan chiefly because of the high cost of the 100% inspection; it was thought that, if the manufacturer could qualify for sampling inspection, this considerable cost could be reduced. Initially, the manufacturing company made no changes in any of its production or inspection procedures.

After a short time the manufacturer qualified for customer testing of only half of the product. This event, reflecting favorably on the quality of the product, was noted with approval by the top management of the manufacturing company. When, somewhat later, the customer reverted to 100% testing because of two successive failures of product on the customer's test, the top management personnel were extremely unhappy.

The consequence of this unhappiness was a great tightening up of the manufacturer's inspection procedures on components and subassemblies, accompanied by the application of somewhat more care in the manufacturer's final performance test. Strong efforts were made to diagnose and correct certain production troubles that seemed to be causing quality variation. Soon the manufacturer qualified again for customer testing of half the product. After a while, the good record on the customer's test led to qualification for testing of one-fourth. Before the end of the contract, testing was on only one-eighth, the smallest fraction permissible under the three-level scheme.

Varying reactions to the multilevel plan The different reactions of some of the personnel in the customer's and manufacturer's organizations were interesting to outside observers of this application of multilevel continuous sampling. The persons who were most pleased seemed to be the users of the complex product in the customer's organization. There was no question that these users ultimately received better product under the sampling scheme because of the improvement in the quality of the product submitted by the manufacturer.

In contrast, some of the inspection executives in the manufacturer's organization privately expressed their displeasure over the customer's use of the multilevel plan. As far as outsiders could judge, this displeasure was due to the top management pressure on the company's inspection executives not to permit a reversion to 100% inspection or to a lower level of sampling inspection than the one in current operation. (The inspection executives felt, justifiably, that the pressure should have been directed as much to the production department as to the inspection department.)

A simple but ingenious mechanical device had been built to randomize the customer's choice of the items to be tested under sampling inspection. The customer's personnel who conducted the testing were greatly intrigued by the operation of this device. Their jobs seemed to have been made more interesting to them by this simple example of the laws of chance operating to control certain aspects of their daily activities. Moreover, it was evident that they did somewhat more thorough and careful testing under the sampling inspection than they had done under 100% inspection.

The general opinion in the manufacturer's organization seemed to be that the manufacturer's total cost on the contract had been increased a bit as a result of the customer's change from 100% inspection to multilevel sampling. However, the company's top management seemed to feel that this adverse effect on cost was more than offset by the better reputation of the manufacturer's product with the ultimate users in the customer's organization. The top management personnel seemed to feel that this better reputation put the manufacturing company in a more favorable position to receive future contracts from this customer and others.

From the viewpoint of the customer's top management, the effects of having introduced the multilevel scheme were entirely good because testing costs were reduced and incoming quality was improved. The quality improvement was somewhat of an unexpected dividend to the customer; the sole objective in introducing the scheme had been to reduce testing costs.

14.3 THE MIL-STD-1235 SYSTEM FOR SAMPLING FROM CONTINUOUS PRODUCTION

In 1959, just less than a year after publication of Handbook H 106 for multilevel continuous sampling, Handbook H 107 was published consisting of three single-level continuous sampling plans.† The set included Dodge's CSP-1 and CSP-2 and a plan developed for the U.S. Navy designated CSP-A. Most of the operating procedures of CSP-1 and CSP-2 were identical with those previously discussed with the single exception of a provision for terminating acceptance altogether when 100% (screening) inspection continues for too great a period of time. In 1962 the Handbooks H 106 and H 107 were combined into a single document designated MIL-STD-1235(ORD); however H 106 and H 107 continued in use.‡ The main procedural features of these plans have already been presented. The multilevel continuous sampling plans from H 106 were given the code designation CSP-M.

The standard was revised in 1974 at which time CSP-A and CSP-M were replaced. CSP-A was replaced by two other single-level plans designated CSP-F and CSP-V. CSP-M was replaced by CSP-T, a set of three-level plans in which the sampling rate f progresses from f to $f/2$ to $f/4$ rather than exponentially as in CSP-M. The switching procedures were simplified such that, rather than progressing through a series of intermediate states as shown in Fig. 14.3 when an item is rejected on sampling inspection, inspection reverts immediately back to 100% inspection on the rejection of an item at any level. This revised set of plans was retained in the latest revision, MIL-STD-1235B.

14.3.1 Administrative Procedures of MIL-STD-1235

Application of this standard is restricted to product flowing in assembly-line fashion, although there is no requirement that it actually be flowing on a conveyor system. Since inspection alternates between screening and sampling, ample physical facilities and personnel must be available to permit rapid 100% inspection when necessary.

In addition, very stringent homogeneity requirements are placed on the production system. All product units must be made in accordance with the same specifications and drawings under stable conditions of production that might be realized by a process in statistical control. Any unusual interruption in the production process—such as the change of material source or change of tooling or any discontinuance of manufacture beyond that resulting from the end of a shift, day, or week—is assumed to terminate a production run under homogeneous conditions.

The basic measure of effectiveness of CSPs is the AOQ function with indexing of sets of plans based on the AOQL. However, to make administration of the system consistent with the much-used ABC-STD-105, the plans were cross-referenced in terms of AQL and are indexed by both AQL and AOQL.§

†"Single-Level Continuous Sampling Procedures and Tables for Inspection by Attributes, Inspection and Quality Control Handbook (Interim) H 107," Office of the Assistant Secretary of Defense (Supply and Logistics), Washington, D.C., April 1959.

‡"Single- and Multi-Level Continuous Sampling Procedures and Tables for Inspection by Attributes," MIL-STD-1235(ORD), Department of the Army, Washington, D.C., July 1962.

§For a more complete discussion of the evolution of the plans and their indexing as contained in MIL-STD-1235(ORD), see R. A. Banzhof and R. M. Brugger, "Reviews of Standards and Specifications—MIL-STD-1235(ORD), Single- and Multi-Level Continuous Sampling Procedures and Tables for Inspection by Attributes," *Journal of Quality Technology,* vol. 2, no. 1, January 1970. AQL, in this case, relates a quality level to percentage of product accepted when on sampling inspection. This change in meaning of AQL has been the subject of much criticism.

A system of sampling frequency code letters is specified for the selection of sampling plans. Code letter designation is set by the estimated number of units in a *production interval,* for example, a day or shift. Thus the matrix format by AQL and code letter, originated in MIL-STD-105, is carried over intact to MIL-STD-1235.

14.4 AOQ FUNCTIONS OF SOME CONTINUOUS SAMPLING SCHEMES

The primary measure of effectiveness of continuous sampling schemes is the AOQ function, or, more specifically, the AOQL. The bulk of MIL-STD-1235 is composed of characteristic curves, plots of percent of product accepted on a sampling basis versus percent defective of submitted product, and AOQ and AFI functions.

Rather than attempt to reproduce all or part of these functions, approximate general formulas for the AOQ function are given for CSP-1, CSP-2, and CSP-M. Given an estimate of the process average μ_p, the following formulas may be used to determine the approximate AOQ of product:[†]

CSP-1: $$\text{AOQ} = \frac{p(1-f)q^{i-1}}{f + (1-f)q^{i-1}}$$

where p = fraction nonconforming
$q = (1 - p)$

CSP-2: $$\text{AOQ} = \frac{p(1-f)(2q^i - q^{2i})}{fq + (1-f)(2q^i - q^{2i})}$$

CSP-M: $$\text{AOQ} = p\,\frac{\sum_{j=1}^{k}\left(\frac{1}{f_j} - 1\right)\left(\frac{q^i}{1-q^i}\right)^j}{\sum_{j=0}^{k}\left(\frac{1}{f_j}\right)\left(\frac{q^i}{1-q^i}\right)^j}$$

where j = indexes over the number of inspection levels in the plan
k = maximum number of inspection levels in the plan

14.5 SKIP-LOT SAMPLING

Skip-lot sampling, developed by Harold Dodge, is a variation on continuous sampling wherein the inspection unit is a lot or batch rather than an individual item.[‡] It may be particularly useful as a quality audit procedure of supplier's

[†]R. A. Banzhof and R. M. Brugger, ibid.

[‡]H. F. Dodge, "Skip-Lot Sampling Plan," *Industrial Quality Control,* vol. 11, no. 5, pp. 3–5, 1955; and H. F. Dodge and R. L. Perry, "A System of Skip-Lot Plans for Lot by Lot Inspection," *Annual Technical Conference Transactions,* pp. 469–477, American Society for Quality Control, 1971.

material where JIT inventory management is used and receiving lots are small or where inspection is slow and costly. As is the case with chain sampling, discussed in Chap. 13, skip-lot sampling is heavily dependent on the assumption of homogeneity among lots and a good quality history. Thus it is unlikely to give satisfactory results if applied to lots coming from erratic and uncontrolled processes.

Under the plan designated SkSP-1, consecutive lots are 100% inspected until i have been found free of nonconforming items. At this point, inspection shifts to a proportion f of the lots received. As usual, the order of production of the lots must be maintained and lots inspected in that order. The values of i and f are selected from some continuous sampling system such as CSP-1, CSP-2, or CSP-M.

Under SkSP-2, rather than performing 100% inspection on each lot inspected, a *reference sampling plan* is chosen (perhaps from ABC-STD-105 or one of the Dodge-Romig systems), and each lot designated for inspection is subjected to the reference sampling plan. Under either SkSP plan the AOQL is taken to be the AOQL of the continuous sampling plan adopted. The value of the AOQL applies to the long-run average quality of lots.

14.6 FURTHER COMMENT ON CONTINUOUS SAMPLING PLANS

Various other types of continuous sampling AOQL plans may be developed. Some plans start with sampling inspection rather than with 100% inspection.[†] These plans, in effect, give a process credit for such good quality as has been shown in past samples and base the decision regarding the shift from sampling inspection to 100% inspection on cumulative evidence from past samples.

As in all acceptance/rectification schemes, a necessary condition for the successful use of continuous sampling schemes is that the submitted quality should be good enough for most product to be passed without screening. The sampling is intended to give protection against runs of bad quality. If the protection against short runs of bad quality (so-called spotty quality) is to be satisfactory, the sample percentage f should not be too low. The relative weakness of small values of f in this respect is brought out by the scale on the right-hand side of Fig. 14.1 showing values of $p_{0.10}$.

Mention should be made of one practical difference between lot-by-lot acceptance/rectification schemes and most continuous sampling schemes. In lot-by-lot schemes, an unfavorable sample from a lot results in the screening inspection of that

[†]M. A. Girshick, "A Sequential Inspection Plan for Quality Control." This was issued by the Applied Mathematics and Statistics Laboratory, Stanford University, Stanford, Calif., as Technical Report No. 16, 1954.

Abraham Wald and J. Wolfowitz, "Sampling Inspection Plans for Continuous Production Which Insure a Prescribed Limit on the Outgoing Quality," *The Annals of Mathematical Statistics,* vol. 16, pp. 30–49, 1945.

particular lot. In continuous sampling, a bad sample calls for the screening of *subsequent* production. In effect, the sample is considered as representative of the production process, and 100% inspection is applied to later articles from the same process. In some instances, where the physical conditions of production permit, a modification of continuous sampling procedures is made to require screening inspection applied to a specified number of units immediately preceding a defective sample.

14.7 UNIQUE FEATURES OF CSP PLANS AS GUIDES FOR SPC SAMPLING

Much of the data collection performed for statistical process control is done on a sampling basis. With the single exception of inspection of critical characteristics, it simply does not make economic sense to conduct 100% inspection of all dimensions and characteristics of production and service processes. This condition is reflected in many of the examples discussed in Parts One and Two of this book and presented in some detail in Chap. 17.

The procedures discussed in this chapter afford persons concerned with statistical process control a quantitative basis on which to determine sampling frequencies and procedures. That important measure of quality protection is the AOQL. Aspects of continuous sampling may be combined with sample sizes chosen from any of the attribute or variables sampling systems discussed in this book, as was discussed in the section dealing with skip-lot sampling, to provide a scientifically justifiable procedure for process control charting.

PROBLEMS

14.1. In Dodge's CSP-1, it is desired to apply sampling inspection to 1 piece out of every 15 and to maintain an AOQL of 2%. What should be the value of i?
Answer: 68.

14.2. In the Dodge-Torrey CSP-2, it is desired to apply sampling inspection to 1 piece out of every 10 and to maintain an AOQL of 3%. What should be the value of i? Assume that $i = k$.
Answer: 48.

14.3. In Dodge's CSP-1, it is desired to apply sampling inspection to 1 piece out of every 4 and to maintain an AOQL of 0.5%. What should be the value of i?

14.4. In the Dodge-Torrey CSP-3, it is desired to apply sampling inspection to 1 piece out of every 10 and to maintain an AOQL of 1%. Prepare a flowchart of the operation of this plan.

14.5. In Dodge's CSP-1, it is desired to apply sampling inspection to 1 piece out of every 30 and to maintain an AOQL of 2%. What should be the value of i? Prepare a flowchart of the operation of this plan.

14.6. In Dodge's CSP-1, it is desired to apply sampling inspection to 15% of the product and to maintain an AOQL of 0.2%. Find the value of i to be used. Prepare a flowchart of the operation of this plan.

14.7. It is desired to apply continuous sampling inspection to 1 piece out of every 10 and to maintain a AOQL of 2%.
(*a*) Find the value of i for applying Dodge's CSP-1.
(*b*) Find the values of i and k for applying Dodge's CSP-2.
(*c*) Prepare flowcharts of the operation of these plans.
Answer: (*a*) 55; (*b*) 72.

14.8. Select from Table 14.2 a four-level continuous sampling plan for an AOQL of 1.0% and an f of $\frac{1}{3}$. If a defective is found shortly after the third level is reached, describe how the plan would operate.

14.9. Select from Table 14.2 a three-level continuous sampling plan for an AOQL of 2.0% and an f of $\frac{2}{3}$. Prepare a flowchart of the operation of this plan.

14.10. Consider the question of the use of the plan selected in Prob. 14.9 after the first nonconforming unit has been found on the second inspection level.
(*a*) What action is taken if the next two consecutive units inspected are both nonconforming?
(*b*) What action is taken if the next consecutive unit is nonconforming and the following four consecutive units are all good?
(*c*) What action is taken if the next four consecutive units are all good?

14.11. Select from Table 14.2 a two-level continuous sampling plan for an AOQL of 1.5% and an f of $\frac{2}{3}$. Give detailed rules for the operation of this plan up to the time when the first nonconforming unit is found. Assume that the first reject is found shortly after the second level has been reached.

14.12. Answer questions (*a*), (*b*), and (*c*) of Prob. 14.10 for the plan selected in Prob. 14.11.

14.13. A certain manufacturing group decides to use a Dodge-Torrey CSP-2 sampling inspection plan on a line of small motors. It is decided to inspect 10% of the units when sampling and to maintain a desired AOQL of 1.0%.
(*a*) Find the appropriate values of i and k.
(*b*) Prepare a flowchart of the detailed operation of this plan.

14.14. Select from Table 14.1 a Dodge-Torrey CSP-3 plan for an AOQL of 1% and a 10% sampling rate. Determine the value of i, and prepare a flowchart of the detailed operation of the plan.

14.15. In Dodge's CSP-1, it is desired to apply sampling to 1 item out of every 20 and maintain a 1% AOQL. What should be the value of i? Prepare a flowchart of the operation of this plan.

14.16. In Dodge's CSP-1, it is desired to apply sampling to 1 item out of every 5 and maintain a 1% AOQL. What should be the value of i? Comment on the differences between this 1% AOQL plan and that found in Prob. 14.15.

14.17. Assume that the process in Prob. 14.15 and 14.16 is generating 1.5% nonconforming product. What is the actual AOQ under each of these plans?

14.18. Sketch the AOQ curve for the plan found in Prob. 14.3. Use values of p of 0, 0.005, 0.01, 0.015, and 0.02.

14.19. For the plans found in Prob. 14.7, compute the AOQ when the process is generating 3% nonconforming product.
Answer: (*a*) 1.90%; (*b*) 1.99%.

14.20. For the plan found in Prob. 14.9, compute the AOQ when the process is generating 3% nonconforming product.

14.21. (*a*) From Table 14.2, find the four-level continuous sampling plan for an AOQL of 2.0% when f is $\frac{1}{2}$.

(*b*) Assume that the inspection process has just entered State 2* ($f^2 = \frac{1}{4}$). Draw a flow diagram indicating where in the procedure the inspection process may be within the next i items inspected. Use the value of i found, not just the general statement.

14.22. In the Dodge-Torrey CSP-2 plan, it is desired to apply sampling inspection to 1 item in every 20 and to maintain an AOQL of 1%.

(*a*) What are the values of i and k?

(*b*) Calculate the AOQ when incoming lots contain 2% nonconforming units.

(*c*) Sketch the operation of the plan using the values of i and k found in part (*a*).

14.23. (*a*) From Table 14.2 choose a 2.0% AOQL multilevel continuous sampling plan with three sampling levels and a sampling proportion f of $\frac{1}{3}$.

In answering the following questions, assume that inspection is on sampling level 2 ($f^2 = \frac{1}{9}$) and that the last item inspected was rejected. Indicate the code number of the state of the system.

(*b*) What action is taken if the next item inspected is rejected?

(*c*) If the next two items inspected both are rejected?

(*d*) If the next four items inspected are accepted?

(*e*) If the next item inspected is rejected but the following four items are accepted?

14.24. A skip-lot sampling plan is to be used to audit the quality of a supplier's product. Lots of 50 units are received on a biweekly basis each accompanied by a record of C_p and C_{pk}.

(*a*) Use Fig. 14.1 to select a CSP-1 plan using an f of 50% and an AOQL of 0.2%.

(*b*) Select a four-level multilevel sampling plan for Table 14.2 with an f of 50%.

(*c*) Discuss any apparent advantages or disadvantages of these two plans as they might apply to skip-lot sampling.

15

SYSTEMS FOR ACCEPTANCE SAMPLING BY VARIABLES

I have never yet seen an inspection problem which would not benefit from the point of view that the product to be inspected was a frequency distribution.

—G. D. Edwards

15.1 SOME ADVANTAGES AND LIMITATIONS OF ACCEPTANCE SAMPLING BY VARIABLES

Most acceptance sampling is by attributes and will doubtless continue to be so. Nevertheless, the growth of knowledge of statistical quality control techniques has led to a considerable increase in the industrial use of acceptance sampling by variables. It seems likely that this tendency will continue.

One obvious limitation on the use of variables criteria in acceptance sampling is the fact that many quality characteristics are observable only as attributes. Where this is true, sampling by variables is out of the question. Nevertheless, it often turns out to be possible to devise methods of measurement in cases where at first glance it seems that inspection must be by attributes.

For those quality characteristics that can be measured, it is usually true that the cost of inspection per item is less by attributes than by variables. Often this is due to the greater economy of inspection methods using the go–not-go principle. Moreover, clerical costs are usually less with attributes inspection; it is less expensive to record merely the conformance or nonconformance to specifications than to record an actual measured value and to make computations using that value.

Perhaps the most serious limitation on the use of sampling by variables is the fact that acceptance criteria must be applied separately to each quality characteristic. This tends to increase the cost of acceptance inspection. For example, if 20 quality characteristics of a product are to be examined at a given inspection station, a single set of attributes sampling criteria can be applied to the acceptance decision. In contrast, if each characteristic is subject to variables inspection, 20 different sets of variables criteria must be used.

An additional limitation on certain types of variables criteria exists in that the computed protection against various percentages of defectives depends on an assumption regarding the form of the underlying frequency distribution of the quality characteristic. This limitation is discussed near the end of this chapter.

In spite of the foregoing limitations, acceptance sampling by variables is often preferable to acceptance sampling by attributes, particularly for those quality characteristics that are the source of troubles. Possibly only 2 of the 20 characteristics mentioned in the preceding paragraph may turn out to be troublesome. If so, it may be that variables criteria can be applied profitably to these even though attributes criteria are used for the remaining 18.

The great advantage of the use of acceptance sampling by variables is that more information is obtained about the quality characteristic in question. This may lead to a number of desirable results, as follows:

1. For a given sample size, better quality protection usually may be obtained with variables criteria than with attributes. Or, stated a little differently, for a given quality protection against various possible percentages of defectives (as reflected in the OC curve), smaller samples may be used with variables than with attributes.
2. The extent of conformance or nonconformance to the desired value of a quality characteristic is given weight where variables criteria are used. This may be important wherever there is a margin of safety in the design specifications or a twilight zone of values of the quality characteristic between clearly satisfactory and clearly unsatisfactory.
3. Variables information usually gives a better basis for guidance toward quality improvement.
4. Variables information may provide a better basis for giving weight to quality history in acceptance decisions.
5. Errors of measurement are more likely to be disclosed with variables information.

15.2 SOME DIFFERENT TYPES OF ACCEPTANCE CRITERIA INVOLVING VARIABLES

There are many different ways in which the actual measured values of quality characteristics in a sample can be used to influence decisions on acceptance of submitted product. The following general classification of types of variables criteria is intended to provide a convenient basis for discussion of the subject:

1. Criteria in which the decision depends in some way on the frequency distribution of the sample. The Shainin Lot Plot is an example of this type of plan.

2. Criteria using a control chart for variables to divide a series of consecutive inspection lots into "grand lots" with acceptance criteria applied to each grand lot.
3. Criteria in which the decision on acceptance or rejection of a lot is based on the sample average alone. Plans using such criteria may be referred to as *known-sigma* plans or plans with *variability known.*
4. Criteria in which the decision is based on the sample average in combination with a measure of sample dispersion. Such plans may be referred to as *unknown-sigma* plans or as having *variability unknown.*

Sometimes there are legal or other reasons why variables criteria should not be used for lot rejection even though such criteria are appropriate for lot acceptance. This condition may lead to some combination of variables and attributes criteria. This topic is examined following the discussion of known-sigma and unknown-sigma plans.

15.3 USING PLOTTED FREQUENCY DISTRIBUTIONS IN ACCEPTANCE SAMPLING

In the technical language of inspection, an article is *defective* if it fails to conform to specifications. A slight departure beyond specification limits makes the article defective; so, also, does a large departure beyond the limits. This definition of a defective gives no weight to the extent of the nonconformity to specifications.

Nevertheless, the extent of nonconformity may be a matter of great practical importance. Consider any measurable quality characteristic on which the designer has specified upper and lower limits. There frequently is a twilight zone of values just outside one or both specification limits within which it is satisfactory to accept a moderate percentage of "defective" articles.

In many manufacturing plants, it is common for some purchased lots to be rejected initially under the regular acceptance procedures and then finally to be accepted under some sort of material-review procedure. Sometimes there is a highly formalized procedure involving a material-review board; in other cases material review may be quite an informal matter, involving a decision by some one individual. In any event, the extent of nonconformity to specifications is usually given consideration in material-review decisions. Extra costs may be avoided and interruptions of production may be prevented by using material-review procedures to accept a product that is good enough for the purpose at hand, even though it is technically nonconforming to specifications.

Where sample frequency distributions for certain quality characteristics are plotted as part of the regular acceptance procedures, one objective usually is the provision of a more rational basis for material review. The frequency distribution is a great aid to judgment on the question of the extent of nonconformity to specifications and may yield useful results when the customer uses JIT inventory management.

Moreover, the frequency distribution has the advantage that it may disclose certain common types of departure from normality, such as skewed distributions, bimodal distributions, distributions with cutoff points, and distributions containing strays at some distance from the main distribution. The larger the sample, the better the chance to recognize such nonnormal distributions when they occur, and the less the danger that a sample really taken from a normal distribution will give a false indication of nonnormality.

Example 15.1 illustrates a common type of use of a frequency distribution on an informal basis with a good deal of judgment permitted in the acceptance decision.

Example 15.1

Acceptance decisions based on plotted frequency distributions

Facts of the case An electrical manufacturer purchased mica insulators from a number of different suppliers. Shipments generally consisted of lots of 8,000 to 10,000 pieces. All suppliers seemed to have some difficulty in meeting the dimensional specification on thickness.

The acceptance procedure involved measuring the thicknesses of a sample of 200 from each lot and plotting the frequency distribution on a check sheet similar to the one shown in Fig. 3.1. The decision on acceptance or rejection was made by the quality control engineer after examining this check sheet. No formal written criteria were established to govern this decision. Weight was given to the form of the frequency distribution, the number of items outside specification limits, and the extent to which the nonconforming items failed to meet specifications. Consideration in the acceptance decision was also given to the number of items of this product in stock at the particular moment and the hazard that returning a lot to the supplier might interfere with production schedules.

15.3.1 The Shainin Lot Plot Method

This is a more formalized method of using a frequency distribution in receiving inspection to guide decisions on acceptance or rejection. This plan, also known as the Hamilton Standard Lot Plot method, was developed by Dorian Shainin when he was chief inspector at Hamilton Standard, now a division of United Technologies. A full description was given in an article in the July 1950 issue of *Industrial Quality Control.*[†] Some later refinements were explained in a series of articles in the March 1952 issue of the same journal.[‡] The following brief discussion explains the general characteristics of the method but does not aim to reproduce the full directions for its use that are given in Shainin's 1950 article. (Reproduction of the entire 1950 article would require about 50 book-sized pages.)

[†]Dorian Shainin, "The Hamilton Standard Lot Plot Method of Acceptance Sampling by Variables," *Industrial Quality Control,* vol. 7, no. 1, pp. 15–34, July 1950. Copyright 1950, American Society for Quality Control, Inc.

[‡]Dorian Shainin, "Recent Lot Plot Experiences around the Country," *Industrial Quality Control,* vol. 8, no. 5, pp. 20–29, March 1952.

R. L. Ashley, "Modification of the Lot Plot Method of Acceptance Sampling," *Industrial Quality Control,* vol. 8, no. 5, pp. 30–31, March 1952.

Richard Wilson, "A Convenient Short Cut in the Use of Lot Plot," *Industrial Quality Control,* vol. 8, no. 5, pp. 32–33, March 1952. Copyright 1952, American Society for Quality Control, Inc.

In the Lot Plot method, the sample size always is 50. In the Hamilton Standard usage, random numbers were used in drawing the sample to ensure randomness in its selection. Figure 15.1 shows one version of the Lot Plot form. A check on the cell width used in the left-hand column of the form is obtained from the first 5 articles measured; if possible, twice the range of this set of 5 values should be from 7 to 16 cells.

The sample is divided into 10 groups of 5 articles each for the purpose of recording the measured values in the Lot Plot form. For each article in the first group, the figure 1 is entered in the appropriate cell. For example, Fig. 15.1 indicates that the 5 measured values in the first group fell in the cells 254–253, 254–253, 253–252, 252–251, and 251–250. For each article in the second group, the figure 2 is entered in the appropriate cell, and so on. As each group of 5 is entered, the range of the group in cell units is entered in the "Range" column near the right-hand side of the form.

When the 50 measured values have been recorded, the next step is to compute the $\bar{X}$ of the sample and to estimate the 3σ value of the lot from the $\bar{R}$ of the 10

FIGURE 15.1 Example of Lot Plot form. *(Reproduced from Dorian Shainin, "The Iron Age," Oct. 12, 1950.)*

HS F-958D 4/50
DATE REC'D. 8-24-50

HAMILTON STANDARD
LOT PLOT AND QUALITY REVIEW ORDER

VENDOR W. & G. Co. PART NAME Cap PART NO. 73201
P.O. NO. 321614 R.S. NO. 290667 QUANTITY 438 DATE INSP. 8-26-50
SPEC. .246 - .256 width INSPECTOR Brown SAMPLE SIZE 50

VALUE	IND	LINE No.	Plot (5, 10, 15, 20)	
		+10		
		+9		
		+8		
		+7	ULL	
.259		+6	2	
.258		+5	2 X X	
.257		+4	2 4 6 8 X Spec.	+6
.256		+3	3 4 5 6 8 8 9	+2
.255		+2	2 3 6 7 9 9 X	
.254		+1	1 1 4 4 5 7 $\bar{\bar{X}}$	
.253		0	1 2 3 3 6 6 9 X	−1
.252		−1	1 3 4 5 7 8	
.251		−2	1 5 5	
.250		−3	7 7 8	
.249		−4	9	
.248		−5		
.247		−6	LLL	
.246		−7	Spec.	
		−8		
		−9		
		−10		

PLEASE
ASK YOUR OPERATOR TO STUDY THIS DIAGRAM AND: -

1	3
2	6
3	4
4	5
5	5
6	4
7	5
8	7
9	7
10	5
11	
12	
13	
14	
15	
$\Sigma R = 51$	
$3\sigma = 6.6$	
$\Sigma x = +7$	
$\bar{\bar{X}} = +.14$	

☒ MOVE BASIC PRODUCTION SIZE (TOOL SETTING, DIE SIZE, MOLD CAVITY ETC.) .0025 IN. Low DIRECTION

☒ REDUCE PROCESS SPREAD AT LEAST .0032

☐ MAKE THE TWO BASIC PRODUCTION SIZES CENTER AT ____

☐ EXPLAIN WHY RED CIRCLED PIECES WERE LEFT IN THE LOT

IF SOME OF THE ABOVE ARE CHECKED, LOOK FOR NEXT LOT PLOT TO SEE IF YOUR ADJUSTMENTS WERE SUCCESSFUL

☐ ACCEPT OUR CONGRATULATIONS FOR A GOOD JOB! THE OTHER SIDE OF THIS SHEET TELLS YOU HOW TO GET A CHECK MARK IN THIS LAST BOX ALL THE TIME

DISPOSITION	
ACCEPT (CO. INSP.)	
Q.R. ACCEPT (HS QR)	G.H.W.
GOV'T. INSPECTOR	R.V.H

VALUE	%	EXTENT
HIGH SPEC.	15	.00425
LOW SPEC.	O.K.	

ATTRIBUTE SAMPLE DATA
SAMPLE OF ____ SHOWS ____ PIECES ____

groups of 5 in the sample. In this 1952 article, Shainin explains the shortcut method used for computing $\overline{X}$ as follows:

(*a*) Cut out the column of plus and minus line values from a blank Lot Plot form.
(*b*) Place this column to cover the line values of the Lot Plot so that the zero cell falls opposite the mode or longest horizontal row of entries.
(*c*) Note whether the +1 or −1 cell contains the most readings.
(*d*) Compare the quantities in these two cells by moving the cut column of paper horizontally to the right until the smaller of the two compared rows is just covered.
(*e*) Count the remaining uncovered squares in the partially covered cell row, multiply this count by the cell sign and value, and enter the result at the extreme right side of the grid in that cell row.
(*f*) Repeat using the +2 and −2 cells, and continue until only the zero cell remains.
(*g*) Total the results algebraically and enter the answer opposite "Totals" in the lower right-hand section of the Plot.
(*h*) Multiply this answer by 2 and point off two decimal places to the left.
(*i*) Enter this result in the ($\overline{\overline{X}}$) space.
(*j*) Point off from the middle of the mode or peak cell row the portion of a cell found by step *i*, in the direction corresponding to the sign of the *i* value, and draw a horizontal line labeled $\overline{\overline{X}}$.

The 3σ value needed to estimate the position of the lot limits is computed as 1.29 $\overline{R}$ ($[3/d_2]\overline{R}$ where $n = 5$).

In Fig. 15.1, $\overline{\overline{X}}$ (in cell units) is +0.14, and 3σ (also in cell units) is 6.6. The lot limits, designated ULL and LLL, are drawn at a distance of 3σ on either side of $\overline{\overline{X}}$. In Fig. 15.1 it is evident that the spread of the process is somewhat greater than the spread of the specifications, and that the process is not centered midway between the specification limits.

The key figures to guide the quality review are those contained in the box in the center near the bottom of the Lot Plot form. This box gives an estimate of the extent to which the extreme values of the lot fall beyond the specification limits. The computed 3-sigma lot limits supply the basis for this estimate. The box also contains an estimate of the percentage of the lot beyond each specification limit. This latter estimate is made from a diagram based on a normal-curve-area table (such as Table *A*, App. 3).

Shainin has classified typical lot plots into 11 types, as shown in Fig. 15.2. The frequency distribution in Fig. 15.1 gives no grounds for suspicion of nonnormality and is therefore treated as if the lot were normal. The rules for interpretation of nonnormal-appearing lot plots are not reproduced here. Much of Shainin's 1950 article deals with detailed instructions for handling the various types of nonnormal lot plots. For a number of the types, special rules are required for estimating lot limits; examples of this are skewed lot plots (for example, type 5), lot plots indicating screening or cutoff points (for example, type 6), and bimodal lot plots (for example, type 7). Where lot limits fall outside specification limits, the rules for action are influenced by the type of lot plot. In certain instances, doubtful lot plots lead to the use of AOQL attributes inspection. Special rules are given for the treatment of lots in which the plots indicate strays (type 11).

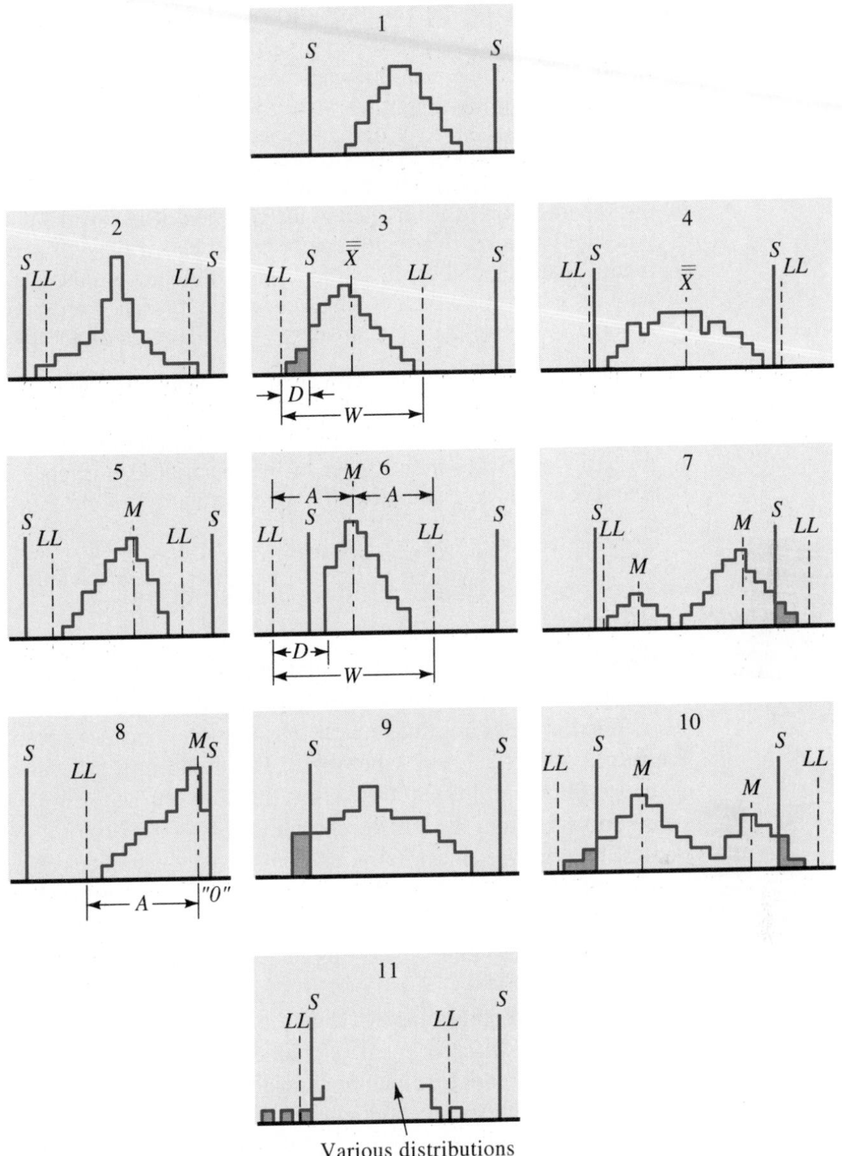

FIGURE 15.2 Eleven different types of lot plots. *(Reproduced from Dorian Shainin, "The Hamilton Standard Lot Plot Method of Acceptance Sampling by Variables," Industrial Quality Control, vol. 7, no. 1, pp. 15–34, July 1950.)*

The plotted frequency distributions in the Lot Plot method often provide a good diagnosis for quality troubles and thus lead to quality improvement. A copy of the Lot Plot form is always sent to the supplier. It will be noted that the right-hand side of the form permits the inspector to check one of several types of comments regarding quality, filling in the necessary blanks when there are quality troubles.

The reverse side of the Lot Plot form contains concise directions for a sum and range chart.

In his 1950 article Shainin made the following comment on this topic of supplier's quality improvement:

> The initial use of Lot Plots will probably reveal only a few specifications running as well as the Type 1 situation—a normal shaped distribution included well within the specification limit. If such is the case and it is surprising to your organization, you should realize that by the use of proper cell widths, of a really representative or random sample, and by the elimination of flinching on the part of the inspector, you now have a picture of the condition of the lot that corresponds to the revelations brought to your eye by a microscope of the surface of an object.
>
> The correct use by the supplier of Shewhart type control charts, of course, will bring specification after specification into line to give the inspector Type 1 Lot Plots. It is our experience, and that of several other concerns, that sending a copy of each Lot Plot to the supplier of the material also often results in an increasing frequency of Type 1 situations. By far the majority of our Plots are now of this type.

15.3.2 Some Comments on the Use of Plotted Distributions in Receiving Inspection

Example 15.1 described an informal use of frequency distributions in the inspection of purchased material. The original development of the Lot Plot technique was also aimed at receiving inspection. A common condition in such inspection is that the buyer does not know whether all the items in a lot were produced under the same essential conditions. Often, the items in an inspection lot come from two or more universes that have different values of μ or σ or both. Such a mixing of universes may be due to a mixing of items from different production sources or to a mixing of articles made at different times by a single production source that has not been in statistical control. Moreover, the lot may have been given screening inspection by the vendor, and this inspection may have been effective or ineffective.

For these and other reasons, a quality characteristic of items in a purchased lot may have many different types of frequency distribution. As indicated in Fig. 15.2, some lots are bimodal, some are skewed, some are truncated with no strays, some are truncated with strays, and so on.

Under these common circumstances in receiving inspection, there are two possible major advantages from the use of plotted frequency distributions of important quality characteristics of samples, as follows:

1. The information gained from the frequency distributions may be fed back to suppliers in a way that will cause substantial quality improvements and thus lead to the submission of lots that are satisfactory. The right-hand portion of Fig. 15.1 illustrated such a feedback; this point also was mentioned in the quotation from Shainin's 1950 article. Example 15.2 describes a case in which conditions were favorable to this type of supplier quality improvement.

2. There is better guidance to informal decisions regarding marginal lots—particularly decisions made in connection with material-review procedures. (In the original Lot Plot usage, unless a lot was accepted on the basis of the Lot Plot, it was submitted for a material-review decision. Often this decision gave weight not only to the plot for the lot under review but also to any evidence available in preceding plots from the same supplier.)

The most serious objections to the use of frequency distributions in acceptance inspection are related to the point that it is difficult to be sure about the form of the frequency distribution in a lot from inspecting a plotted frequency distribution from a moderate-sized random sample, such as 50.[†] There often are false indications of the form of the distribution in the lot. The cell interval chosen may have a considerable influence on the appearance of a frequency distribution of a sample as small as 50. Moreover, in the Lot Plot technique it is necessary for the user to decide which one of the 11 types of plot (Fig. 15.2) is applicable; this decision involves personal judgment. A particular plot may be classified in several different ways by different persons. It follows that the quality protection given by the Lot Plot or any similar procedure cannot be stated in probability terms; any computation of an OC curve must assume that the form of the frequency distribution in the lot is known.

In many of the cases where manufacturers state that they are using the Lot Plot, the actual procedures adopted differ in a number of ways from those stipulated in Shainin's articles. These manufacturers have, generally speaking, employed Shainin's basic concepts, usually with the same objectives as the original Lot Plot, even though the details of their procedures have not followed Shainin's specific rules. One interesting variation uses special templates to help in the identification of different forms of frequency distribution.[‡] Example 15.2 describes an orthodox application of the Lot Plot technique; Example 15.3 describes an unorthodox application.

Example 15.2

Successful use of advice based in part on Lot Plots to improve quality of product submitted by suppliers

Facts of the case One of the products of the *W* Company was a certain complex mechanism. Initially, all the precision parts for this mechanism were made in the *W* Company's own plant. However, when the major purchaser of this

[†]For an elaboration of this point and for criticism of Shainin's rules for nonnormal plots, see the following reports issued by the Applied Mathematics and Statistics Laboratory, Stanford University, Stanford, Calif., in connection with a contract with the Office of Naval Research:

L. E. Moses, "Some Theoretical Aspects of the Lot Plot Sampling Inspection Plan," Technical Report No. 18, 1954.

W. G. Ireson, "Some Practical Aspects of the Lot Plot Sampling Inspection Plan," Technical Report No. 19, 1954.

[‡]D. A. Hill, "The Lot Template Method of Inspection by Variables," *Proceedings of the Aircraft Quality Control Conference,* November 1953, p. 24, American Society of Quality Control, Milwaukee, Wis., 1953.

E. W. Ellis, "The Mechanical Lot Plot Template," *Industrial Quality Control,* vol. 14, no. 9, pp. 15–18, March 1958.

mechanism contracted for a greatly increased production, it was necessary to subcontract with a number of suppliers for the production of many of the precision parts. Most of these suppliers were several hundred miles from *W* Company's factory.

Lot Plots were used in receiving inspection for many important dimensions and other quality characteristics of these precision parts. Interpretation of the plots with reference to the methods used in production was helped by the fact that all these parts had previously been made by the *W* Company. Whenever a plot indicated that a supplier was having trouble in meeting a particular quality specification, the suggestions for corrective action (on the right-hand side of the Lot Plot) were written by someone who had been familiar with the particular manufacturing operation when the part had been made by the *W* Company. In this way, the advice to the suppliers could combine the *W* Company's prior knowledge gained by troubleshooting on the various operations with the knowledge gained by inspection of the Lot Plots. In most cases, this advice was successful in securing suppliers' quality improvement. Many suppliers were astonished by the fact that someone who had never visited their plants could make a correct diagnosis of the reasons for their quality troubles.

Example 15.3
Use of frequency distributions of samples for final inspection following a production operation

Facts of the case Among other products, a manufacturer made socket screws of all sizes, both standard and special. About 180 so-called Lot Plots were made daily for different production lots. Outside pitch diameter was the quality characteristic plotted. Other quality characteristics of the screws were checked using sampling inspection by attributes.

Each plot was made on a form similar in appearance to Fig. 15.1. Just as in Fig. 15.1, the end product of the analysis consisted of four sets of numbers to be inserted in the box in the center of the bottom of the form, namely, numbers indicating the estimated percent and extent beyond the high and low specification limits. However, the sampling, recording, and calculation all were different from the orthodox Lot Plot.

Fifty screws were taken by the inspector as a grab sample from a tote box containing a batch of screws. (It was felt that the screws had received so much mixing prior to inspection that it was not necessary to use random numbers to select the samples.) Typical quantities in the batch sampled varied from several hundred to several thousand. On the table by the inspector was a sheet of cardboard (or plywood) ruled into squares. The squares were marked in units from the nominal dimension. (Usually, the value of a unit was 0.0001 in.) The dial gage that showed the pitch diameter was set to show the nominal dimension as zero. When each screw was measured, it was put in the appropriate square. After 50 screws had been measured, the inspector filled in the Lot Plot form by making a check mark in the corresponding square of the form. Each square represented one item in a cell of the frequency distribution. There was no division of the 50 measurements into subgroups of 5 each as in the conventional Lot Plot form; vertical lines were drawn on the boundaries of the appropriate squares on the form so that, in effect, the completed form showed a histogram for the sample of 50. The inspector could carry out the measurements and the filling out of the form very rapidly.

The calculation, performed by office workers rather than the inspectors, was a conventional one for mean and standard deviation of a frequency distribution. Certain auxiliary tables and graphs simplified the calculation of percent and extent outside specification limits assuming a normal distribution.[†]

No attempt was made to identify the 11 types of Lot Plots given in Fig. 15.2. All plots on which the computed total percentage outside specification limits exceeded 2% were to be brought to the chief inspector's attention. So also were all plots having frequency distributions that looked peculiar. (Just as in Example 3.1, these latter were viewed as indicating that an inspection error might have been made; for this reason, the chief inspector usually required a new plot for these lots.) All other lots were accepted.

Under these instructions, an average of 15 plots per day out of the 180 were set aside for the chief inspector's attention. In effect, the chief inspector became the material-review board. The decision could be to accept, to scrap, or to screen (or to require the inspection of a new sample of 50 where inspection errors were suspected). Cost considerations usually governed the choice between scrapping and screening.

A special advantage of variables inspection for pitch diameter was the avoidance of the troubles due to gage wear that are common in attributes inspection for this quality characteristic.

15.4 USE OF CONTROL CHARTS TO IDENTIFY GRAND LOTS

In a brilliant early paper,[‡] Gen. Leslie E. Simon suggested a concept of a lot that is useful for many practical purposes. He defined a lot as "an aggregation of articles which are essentially alike." He defined "essentially alike" as meaning "that small subgroups of sample items taken from the lot in arbitrary order will respond to the Shewhart criterion of control."

Example 15.4 illustrates the application of this concept to acceptance. The control chart in Fig. 15.3 makes it evident that there were two different grand lots. The average for the first grand lot fell within the specifications; the average for the second grand lot fell outside. Each grand lot should properly have had a single acceptance/rejection decision applicable to the entire grand lot.

[†]See Wilson, op. cit., for examples of the tracing overlays and one of the auxiliary tables.

[‡]L. E. Simon, "The Industrial Lot and Its Sampling Implications," *Journal of the Franklin Institute,* vol. 237, pp. 359–370, May 1944.

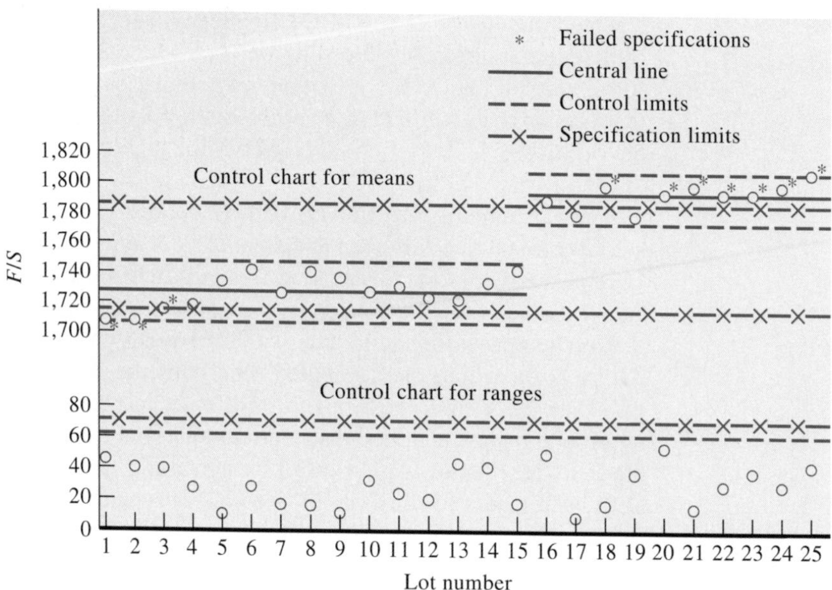

FIGURE 15.3 Control chart illustrating two different grand lots (data of Table 15.1). *(Reproduced from L. E. Simon, "The Industrial Lot and Its Sampling Implications," Journal of the Franklin Institute, vol. 237, pp. 359–370, May 1944.)*

Where this concept is used, it is necessary to put off the decision on acceptance or rejection of a sublot until the entire grand lot has been identified. This postponement of an acceptance decision is sometimes referred to as *deferred sentencing.*

Example 15.4
Grand lots are identified by control charts for $\bar{X}$ and R

Facts of the case General Simon gives an excellent example of the way in which a control chart may show the existence of grand lots, and how the neglect of grand lots in acceptance criteria may lead to accepting and rejecting the wrong lots. Table 15.1 and Fig. 15.3 are taken from his paper.

Muzzle velocities were determined for samples of 5 from 25 consecutive lots from a manufacturer of complete rounds of ammunition. Table 15.1 gives the average $\bar{X}$ and the range R from these samples. The specifications for lot acceptance or rejection were based on the values of average and range (rather than on individual values). $\bar{X}_{max}$ was 1,785 ft/s, and $\bar{X}_{min}$ was 1,715 ft/s; the maximum allowable sample range was 70. The lots rejected by this specification are marked with an asterisk (*).

Figure 15.3 shows $\bar{X}$ and R control charts for these data. General Simon's comments on the conclusions from these charts are as follows:

> It is quite evident that the first 15 lots are of essentially the same quality, and none should have been rejected. Those which were rejected (or retested until

TABLE 15.1 AVERAGES AND RANGES OF MUZZLE VELOCITY FOR SAMPLES OF 5 TAKEN FROM 25 CONSECUTIVE LOTS OF AMMUNITION†

Lot	Average muzzle velocity, ft/s	Range, ft/s	Lot	Average muzzle velocity, ft/s	Range, ft/s
1*	1710	42	16	1783	51
2*	1711	40	17	1777	9
3*	1713	39	18*	1794	15
4	1718	26	19	1773	37
5	1735	10	20*	1789	54
6	1739	25	21*	1798	15
7	1723	14	22*	1789	29
8	1741	15	23*	1788	39
9	1738	11	24*	1799	30
10	1725	31	25*	1807	44
11	1731	25			
12	1721	19			
13	1719	43			
14	1735	39			
15	1741	17			

†From L. E. Simon, "The Industrial Lot and Its Sampling Implications," *Journal of the Franklin Institute,* vol. 237, pp. 359–370, May 1944. General Simon states that the data in Table 15.1 have been altered in scale only, so as to reveal no actual military information.

they passed) represent merely so much economic loss. However, near or between Lots 15 and 16 a change occurred. If the specification criteria are really important, all the Lots 16 through 25 (not just part of them) should have been rejected. Thus, it is all too evident that Lots 1 through 15 constitute the real lot (called the grand lot) in the sense that they are *an aggregation of articles which are essentially alike.* In a like manner Lots 16 through 25 constitute another real lot, but a lot which should not pass the specification.

15.5 COMPUTING THE OC CURVE FOR A KNOWN-SIGMA VARIABLES SAMPLING PLAN BASED ON THE ASSUMPTION OF A NORMAL DISTRIBUTION

Many specifications are one-sided. That is, the specification merely states a lower limit LSL_x (L) or an upper limit USL_x (U) to apply to individual articles.† Perhaps the simplest variables test is one sometimes used with this type of specification in cases where it is believed that the standard deviation of submitted lots will remain fairly constant, that the average μ is likely to shift up and down, and that the distribution of the variable is normal.

†Much of the published material dealing with acceptance sampling by variables uses the abbreviated notation L for LSL_x, the lower specification limit, and U for USL_x, the upper specification limit. We will adopt it here for the balance of this discussion because of its simplification in repeated use in formulas.

This test requires taking a sample of size n and measuring the value of the specified quality characteristic for each item of the sample. The item values are averaged to find the $\bar{X}$ of the sample. If this $\bar{X}$ exceeds the lower specification limit L by $k'\sigma$, the lot is accepted; otherwise, it is rejected. (If the limit is an upper one U, $\bar{X}$ must be not more than $U - k'\sigma$.) The figure for σ must be estimated from past experience, possibly from a control chart that shows process dispersion to have been in statistical control. The factor k' depends on n and on the desired quality protection; the larger the value of k' for any given n, the more severe the acceptance criteria.

In the literature of variables inspection, such acceptance schemes are described as *known-sigma* plans or plans with *variability known.* In illustrating the necessary calculations for the OC curves of such plans, it may be helpful to use a numerical example. Assume that the specification for a certain product calls for a minimum tensile strength of 20,000 lb/in^2 (pounds per square inch). $\bar{X}$ and R charts have been maintained on the test results of past samples. All points have fallen within control limits on the R chart; the estimated value of σ is 1,000 lb/in^2. The $\bar{X}$ chart has shown lack of control.

Assume that it is desired to use a variables plan giving quality protection against defectives resulting from a shift in process average comparable with the protection obtainable from the attributes plan $\begin{cases} n = 75 \\ c = 2 \end{cases}$ and from the equivalent double and multiple plans shown in Fig. 11.7. In their volume on variables acceptance inspection, A. H. Bowker and H. P. Goode[†] give values of n and k' for known-sigma variables plans that have OC curves corresponding fairly well to many different attributes plans. To match the attributes plan of $\begin{cases} n = 75, \\ c = 2 \end{cases}$ n is given as 16 and k' as 1.846. In our particular example, this means that the average tensile strength of a sample of 16 test specimens should be at least

$$20{,}000 + 1.846(1{,}000) = 21{,}846 \text{ lb/in}^2$$

It is not possible to compute an OC curve for such an acceptance plan without making some assumption about the frequency distribution of the quality characteristic in question. The OC curve for this plan that was shown in Fig. 11.7 is based on the assumption of a normal distribution. Figure 15.4 illustrates the calculation of the probability of acceptance of 2% nonconforming product. It shows the frequency distribution of individual values and the frequency distribution of averages of samples of 16 when 2% of the product falls below the lower specification limit. In this calculation, two questions must be answered. (In the following explanation, μ refers to product average and $\bar{X}$ to the average of a sample of size n.)

1. *What is μ for the particular percent nonconforming?* Consult Table *A,* App. 3, to find the value of $z = (X_i - \mu)/\sigma$ corresponding to an area of 0.0200. Interpo-

[†]A. H. Bowker and H. P. Goode, *Sampling Inspection by Variables,* McGraw-Hill Book Company, New York, 1952.

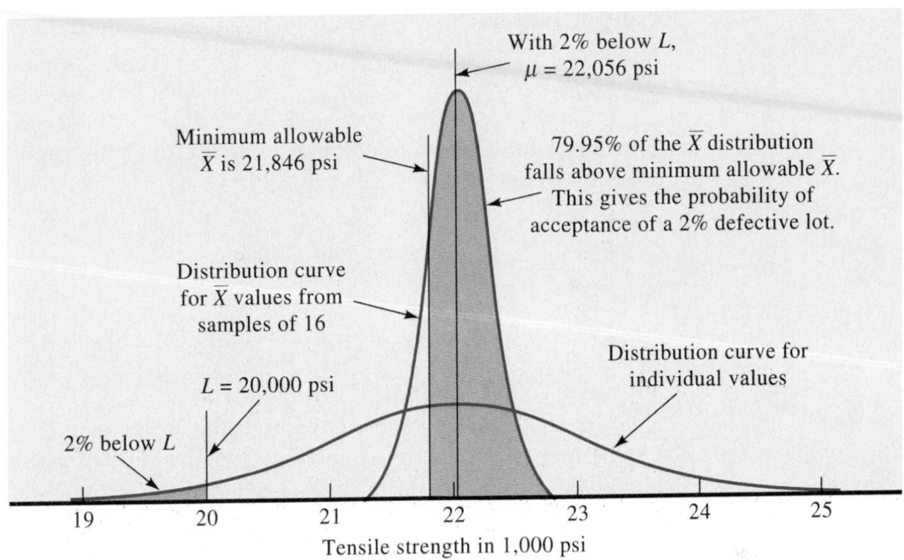

FIGURE 15.4 Diagram illustrating the calculation of the probability of accepting 2% nonconforming product under the known-sigma plan $n = 16$, $k' = 1.846$.

lation in this table shows that 2% of a normal distribution falls below the value $\mu - 2.056\sigma$. If 2% of the product falls below the lower specification limit L, then μ must be $L + 2.056\sigma$. Hence as L is 20,000 lb/in^2 and σ is 1,000 lb/in^2, $\mu = 22{,}056$ lb/in^2.

2. *With μ at this figure, what proportion of the $\bar{X}$ distribution will fall above the minimum allowable value of $\bar{X}$?* The difference between μ and the minimum $\bar{X}$ must be computed. This is 22,056 − 21,846 = 210 lb/in^2. The standard deviation of $\bar{X}$ must be computed. In this case

$$\sigma_{\bar{X}} = \frac{\sigma}{\sqrt{n}} = \frac{1{,}000}{\sqrt{16}} = 250 \text{ lb/in}^2$$

The value 210 lb/in^2 is $0.84\sigma_{\bar{X}}$. Table *A* shows that 0.7995 of a normal distribution is above the value $\mu - 0.84\sigma$. It follows that the probability of acceptance is 0.7995, or approximately 0.80.

Although the preceding calculation used numerical values of L and σ for purposes of illustration, the computed probability of acceptance is independent of these values and depends only on n and k'.

If K is the minimum allowable $\bar{X}$ value, then, in general, for a lower specification

$$K = L + k'\sigma$$

For any desired fraction nonconforming μ_p, the mean of the distribution must be

$$\mu = L + z_p\sigma$$

Normalizing these values to obtain a z_a value from which to use Table *A* to find the probability of acceptance,

$$z_a = \frac{\sqrt{n}(K - \mu)}{\sigma}$$

$$= \sqrt{n}(k' - z_p)$$

From which $P_a = 1 - F(z_a)$.

One advantage of variables sampling is the use of smaller sample sizes to obtain the same quality protection against a particular type of trouble. In the preceding example, the variables sample was 16, whereas the attributes single sample was 75.

15.5.1 Some Aspects of Unknown-Sigma Variables Sampling Plans Based on the Assumption of a Normal Distribution

The known-sigma plan illustrated in Fig. 15.4 specified a minimum allowable $\bar{X}$ of 21,846 lb/in^2 for a sample of 16. This minimum $\bar{X}$ was computed by adding 1.846σ to the L of 20,000 lb/in^2 using an estimate that σ was 1,000 lb/in^2. If it had been estimated that σ was only, say, 100 lb/in^2, the same OC curve would have been obtained by specifying a minimum $\bar{X}$ of only 20,184.6 lb/in^2 for a sample of 16. The smaller the estimate of σ, the closer $\bar{X}$ could be to L. The general relationship among the centering of a process (or lot), its dispersion, a single specification limit, and the fraction nonconforming was first discussed in Chap. 2 and illustrated in Figs. 2.4 to 2.6.

Assume that it is desired to design a variables acceptance plan for a stipulated sample size n and a specified L of 20,000 lb/in^2. Assume that nothing is known about the σ of the product sampled. Obviously it is necessary to use some measure of sample dispersion to estimate σ. Conceivably this measure might be the s or R of the sample, or if the sample is to be divided into two or more subsamples, $\bar{R}$ may be used. Because σ is unknown, it is not possible to specify any single minimum value of the sample $\bar{X}$ such as 21,846 or 20,184.6 lb/in^2. In effect, the specification has to be that $\bar{X}$ must exceed L by a stated multiple of some chosen measure of the sample dispersion.

Bowker and Goode's *Sampling Inspection by Variables,* already mentioned, is the pioneer treatise on variables sampling plans in which computed OC curves are based on the assumption of a normal distribution. This volume contains a large number of known-sigma and unknown-sigma plans having OC curves matched to those of the attributes plans given in the Columbia Statistical Research Group's volume *Sampling Inspection,* mentioned in Chaps. 11 and 12. Bowker and Goode give OC curves for all their unknown-sigma plans based on one-sided specifications.

The unknown-sigma plans in Bowker and Goode use s as the measure of sample dispersion. The reader will recall that s is the square root of an unbiased estimate of the universe variance and that it is defined as

$$s = \sqrt{\frac{\Sigma\,(X - \overline{X})^2}{n - 1}}$$

The acceptance criterion for a lower limit is

$$\overline{X} \geq L + ks$$

and for an upper limit, it is

$$\overline{X} \leq U - ks$$

MIL-STD-414 and its derivatives, discussed later in this chapter, also use s as one of the possible measures of sample dispersion. In addition, these standards permit the use of $\overline{R}$ (or, in some cases, R) as the measure of sample dispersion.

Our chapter follows Bowker and Goode in using k as the symbol for the multiplier of s in unknown-sigma plans and using k' as the multiplier for the estimated σ in known-sigma plans.

Where σ may have only one value (such as the 1,000 lb/in^2 in Fig. 15.4), there is only one value of μ that will correspond to any particular percent nonconforming in the case of a one-sided specification (such as the 2% nonconforming in Fig. 15.4). In contrast, if σ may have any value, there are an infinite number of values of μ that will correspond to, say, 2% nonconforming; each σ will have a different μ for 2% nonconforming. It follows that the mathematics of computing OC curves for unknown-sigma plans is more complicated than for known-sigma plans. To compute an OC curve for an unknown-sigma plan, it is necessary to use certain probability distributions not discussed in this book. A simple explanation of the computation of such curves along the lines of Fig. 15.4 is not possible for unknown-sigma plans.[†] Nevertheless, anyone who understands the explanation of Fig. 15.4 can doubtless also see the general reasoning that must underlie the calculation of an OC curve for an unknown-sigma plan.

15.5.2 Known-Sigma and Unknown-Sigma Plans for Two-Sided Specifications

Where both an upper limit U and a lower limit L are specified for the quality characteristic and the distribution is normal, the ability of a process to make product meeting specifications depends on the relationship between $U - L$ and the spread of the process, often assumed as 6σ. It also depends on the centering of the

[†]Some sources that explain the mathematical basis of computing such OC curves are Bowker and Goode, ibid.; A.H. Bowker and G.J. Lieberman, *Engineering Statistics*, 2d ed., Prentice-Hall, Inc., Englewood Cliffs, N.J., 1972; "Mathematical and Statistical Principles Underlying Military Standard 414," Superintendent of Documents, Government Printing Office, Washington, D.C., 1958.

process. This point was brought out by Figs. 2.1 to 2.3 and in the accompanying discussion.

In the situation illustrated in Fig. 2.1, where $U - L$ is substantially greater than the spread of the process—say, 7.5σ or more—and the distribution in a lot is normal, both specification limits will not be exceeded in a single lot. It follows that where a known-sigma plan is used with such relatively loose tolerances, the OC curve of the plan can be computed as if the specification were one-sided in the way illustrated in our discussion of Fig. 15.4. The only difference in the acceptance criterion from the case of the one-sided specification is that there are now two limits on the permissible value of the sample average; $\bar{X}$ may not exceed $U - k'\sigma$ and may not fall below $L + k'\sigma$.

The matter is more complicated in the cases illustrated in Figs. 2.2 and 2.3, where the tolerances are tight in relation to the process capability. Here some rejects are sure to be produced. For instance, with a normal distribution there will be 0.27% nonconforming even when $U - L = 6\sigma$. The minimum percentage of rejects will occur when the process is centered midway between U and L. With this centering and such tight tolerances, some rejects will be produced above the upper specification limit and others below the lower limit. It follows that in calculating the OC curve for a known-sigma variables acceptance plan for a two-sided specification with tight tolerances, it is necessary to compute percentages of rejects in both tails of the frequency distribution. Except for this need of considering both tails of the distribution, such calculations follow the lines explained in our discussion of Fig. 15.4.

Bowker and Goode give values of n and a coefficient designated as k'^* for known-sigma variables plans for two-sided specifications assuming a normal distribution. These plans are to be used where the tolerances are tight in reference to process capability. With these plans, $\bar{X}$ may not exceed $U - k'^*\sigma$ or fall below $L + k'^*\sigma$. Like the other types of variables plans in the Bowker-Goode volume, a plan is provided to match the OC curve of each of the attributes plans in *Sampling Inspection.*

Whenever unknown-sigma plans are used for two-sided specifications, the uncertainty regarding σ *always* makes it necessary to consider the possibility of rejects in both tails of the distribution. Procedures in connection with such plans are discussed briefly in our explanation of MIL-STD-414.

15.5.3 Use of the Estimated Lot Percentage Nonconforming in Acceptance Sampling by Variables

If one *knew* the μ and σ of a lot and also *knew* that the frequency distribution in the lot was normal, the exact percentage of the lot outside any specification limits could be computed. Similarly, if the μ of the lot is estimated from the $\bar{X}$ of the sample, and if it is assumed that the σ of the lot is known, or if σ is estimated from some measure of sample dispersion, and if the lot is assumed to be normal, it is possible to make an estimate of lot percentage nonconforming. Of course, any such estimate is subject to sampling errors.

We have discussed acceptance criteria in known-sigma and unknown-sigma variables plans that are expressed in terms of the relationship between the sample average and the specification limits. If a normal distribution in the lot is assumed, any such criteria may be converted into criteria expressed in terms of the maximum permissible estimated percentage nonconforming in a lot.

Example 15.3 described a case in which variables criteria were used and all lots were accepted for which the estimated percentage nonconforming was not more than 2%. This somewhat unorthodox adaptation of the Lot Plot technique was, in effect, a special type of unknown-sigma variables plan for a two-sided specification. The more orthodox Lot Plot illustrated in Fig. 15.1 also made use of estimates of percentages outside specification limits, although these estimates were to be used chiefly in an informal way in connection with material-review decisions.

The use of a variables sample to estimate the percentage nonconforming in a lot is an important feature of MIL-STD-414, "Sampling Procedures and Tables for Inspection by Variables for Percent Defective," adopted by the U.S. Department of Defense in 1957.

15.6 SOME GENERAL ASPECTS OF MILITARY STANDARD 414

In many respects this variables standard was similar to the series of successive military standards that had been used for many years for attributes sampling. Some points of similarity were that the procedures and tables were based on the concept of the AQL; that they assumed lot-by-lot acceptance inspection; that they provided for normal, tightened, or reduced inspection depending on circumstances; that sample size was greatly influenced by lot size; that several inspection levels were made available; and that all plans were identified by sample size code letter.

This variables standard was also like the attributes standards in stipulating sampling plans under normal inspection that were designed to protect the producer by making the probability of rejection of a lot having AQL percent nonconforming relatively small. Just as in the military standards for attributes, the protection to the consumer depended largely on the use of tightened inspection whenever the process average appeared to be unsatisfactory or there were other reasons to be suspicious of the process.

The plans in this standard were like those in the Bowker-Goode volume in that the OC curves given for all the plans assumed a normal distribution of any variable to which the acceptance procedures were to be applied. Tables included in the standard supplied a source for estimates of the fraction nonconforming in each inspected lot; these estimates also assumed a normal distribution of the variable.

15.6.1 Different Procedures Available in MIL-STD-414

The standard could be applied either with a single specification limit, L or U, or with two specification limits. Known-sigma plans included in the standard were designated as having "variability known." Unknown-sigma plans were designated as having "variability unknown." In the latter-type plans it was possible to use either the "standard deviation method" or the "range method" in estimating the lot variability.

Two types of computational forms were made available. In "Form 2" the decision on acceptance or rejection required the use of an auxiliary table that provided an estimate of the lot percentage nonconforming based on a "quality index" computed from certain statistics of the sample. In "Form 1," available only for one-sided specifications, this auxiliary table was not needed. The differences between the two forms applied only to the computational procedure; the two forms gave identical results as far as acceptance and rejection were concerned.

15.6.2 Exposition of MIL-STD-414

Because this standard is a relatively long one (110 pages, 8 by 10 in, with many of the pages using very fine print), it is not practicable to reproduce all its tables in this book. (Moreover, the standard itself can be obtained from The Naval Publications and Forms Center, 5801 Tabor Avenue, Philadelphia, Pa. 19120.) The tables that we have reproduced here deal chiefly with the code letter H, and all our discussions and problems relate to that code letter. Any reader who understands the use of the tables and procedures for one code letter should have no difficulty with their use for any other code letter.

15.6.3 Determining the Sample Size Code Letter

To use MIL-STD-414 it is necessary to stipulate an AQL applicable to the quality characteristic being sampled. The standard contains an AQL conversion table (A-1), not reproduced here, that must be used whenever the stipulated AQL falls between two of the AQL values used in the standard.

The standard also contains a table (A-2) similar to Table *K* of App. 3. This table gives the sample size code letter for any lot size. In contrast to the most recent military standards for attributes, which have had three general inspection levels, the variables standard provides for five such levels. Lot sizes calling for the use of code letter H are as follows:

Inspection level	Lot size
I	8,001–22,000
II	1,301–3,200
III	501–800
IV	181–300
V	66–110

It is evident from the foregoing tabulation that the decision on inspection level has an important influence on sample size. This point is perhaps shown even more strikingly in the following tabulation, which shows the sample sizes under "variability unknown—standard deviation method" for lot sizes 181–300 and an AQL of 2.5%.

Inspection level	Sample size code letter	Sample size in normal and tightened inspection
I	B	3
II	D	5
III	F	10
IV	H	20
V	J	30

The standard states that "unless otherwise specified, inspection level IV shall be used." The lower inspection levels are provided particularly for those cases where the cost of testing per unit is relatively high.

15.6.4 Tables Reproduced from MIL-STD-414

This chapter includes five tables dealing with code letter H.

Table 15.2 gives acceptance criteria in normal, tightened, and reduced inspection for all AQL values for plans based on variability unknown, standard deviation method. This table contains factors taken from Tables B-1 to B-4 of the standard. Table 15.5 contains corresponding criteria for plans based on variability unknown, range method; its factors come from Tables C-1 to C-4 of the standard. Table 15.6 has corresponding factors for plans based on known variability; these are taken from Tables D-1 to D-4 of the standard.

Table 15.4 provides a source of estimates of lot percentage nonconforming as a function of the "quality index." This incorporates information from Tables B-5, C-5, and D-5 of the standard.

Table 15.7 contains certain criteria for deciding whether to use tightened or reduced inspection rather than normal inspection for the case of variability unknown, standard deviation method. It includes factors taken from Tables B-6 and B-7 of the standard.

15.7 NUMERICAL DATA TO ILLUSTRATE NORMAL INSPECTION UNDER MIL-STD-414

Assume a one-sided specification for which $L = 20{,}000$ lb/in^2, the AQL is 2.50%, and normal inspection is to be used with code letter H. Tables 15.2, 15.5, and 15.6 indicate respective sample sizes of 20, 25, and 9.

The largest sample size, 25, is the one required for unknown variability, range method. To use the range method this sample must be divided into subsamples of 5, with ranges computed for each subsample.[†] Assume that the respective tensile strengths in pounds per square inch and the computed ranges are as follows:

[†]For the range method and sample sizes less than 10, the R value from the entire sample is used.

Subsample number	Tensile strength, lb/in^2					Range
1	22,030	21,800	20,980	20,750	21,480	1,280
2	20,570	21,110	20,270	18,970	22,110	3,140
3	22,740	21,220	21,300	21,920	21,050	1,690
4	21,780	21,800	21,580	20,990	21,740	810
5	21,170	20,390	21,300	20,760	21,440	1,050
					Sum of ranges	7,970

Assume that the first 20 of the foregoing figures are the tensile strengths obtained for the required sample under the standard deviation method with variability unknown. Assume that the first 9 of the figures are the ones obtained for the required sample with variability known.

15.7.1 Calculations with Variability Unknown—Standard Deviation Method—Form 1

From Table 15.2, it may be determined that an AQL of 2.50% with code letter H calls for an n of 20 and a k of 1.51. In effect, these criteria say that the $\bar{X}$ of a sample of 20 must exceed the L of 20,000 lb/in^2 by at least 1.51s, where s is computed from the sample.

The calculation form suggested in MIL-STD-414 is illustrated in the upper part of Table 15.3. The reader will note that the recommended computational procedure for s makes use of the identity

$$\Sigma\,(X - \bar{X})^2 = \Sigma\,X^2 - n\bar{X}^2$$

The correction factor in line 4 is, of course, equal to $n\bar{X}^2$. Obviously the method of calculation shown in Table 15.3 is designed for use on a calculator, where there is no loss of significant figures in multiplication or division.

The form refers to s as the "estimate of lot standard deviation."

$$L + ks = 20{,}000 + 1.51(804.9) = 21{,}215 \text{ lb/in}^2$$

Since the sample $\bar{X}$ of 21,309.5 lb/in^2 is not less than this figure, the lot is acceptable.

However, this minimum acceptable value for $\bar{X}$ of 21,215 lb/in^2 is not actually used in the computational procedure illustrated in Table 15.3. Instead of comparing $\bar{X}$ with $L+ks$, the same result is obtained by comparing $(\bar{X} - L)/s$ with k. In Table 15.3, $(\bar{X} - L)/s = 1.63$. Since 1.63 is not less than the stipulated k of 1.51, the lot is accepted.

15.7.2 Calculations with Variability Unknown—Standard Deviation Method—Form 2

As indicated in Table 15.3, the first 9 lines are identical in Forms 1 and 2. In effect, the tenth line is also the same except that, in Form 2, $(\bar{X} - L)/s$ is designated as Q_L, the "quality index."

TABLE 15.2 CONSTANTS REQUIRED TO DETERMINE ACCEPTANCE CRITERIA IN NORMAL, TIGHTENED, AND REDUCED INSPECTION WHEN VARIABILITY IS UNKNOWN—STANDARD DEVIATION METHOD—CODE LETTER H OF MIL-STD-414

AQL normal and reduced inspection	AQL tightened inspection	Constants in normal and tightened inspection				Constants in reduced inspection			
		Single specification limit, Form 1		Single limit, Form 2, and double limit		Single specification limit, Form 1		Single limit, Form 2, and double limit	
		n	k	n	M	n	k	n	M
0.04	0.065	20	2.69	20	0.135	15	2.53	15	0.186
0.065	0.10	20	2.58	20	0.228	15	2.42	15	0.312
0.10	0.15	20	2.47	20	0.365	10	2.24	10	0.349
0.15	0.25	20	2.36	20	0.544	7	2.00	7	0.422
0.25	0.40	20	2.24	20	0.846	7	1.88	7	1.06
0.40	0.65	20	2.11	20	1.29	7	1.75	7	2.14
0.65	1.00	20	1.96	20	2.05	7	1.62	7	3.55
1.00	1.50	20	1.82	20	2.95	7	1.50	7	5.35
1.50	2.50	20	1.69	20	4.09	7	1.33	7	8.40
2.50	4.00	20	1.51	20	6.17	7	1.15	7	12.20
4.00	6.50	20	1.33	20	8.92	7	0.955	7	17.35
6.50	10.00	20	1.12	20	12.99	7	0.755	7	23.29
10.00	15.00	20	0.917	20	18.03	7	0.536	7	30.50
15.00	. . . . .	20	0.695	20	24.53				

In Form 2, the quality index is used with an auxiliary table (B-5) to estimate the lot percentage nonconforming. Table 15.4 reproduces part of this auxiliary table. The Q_L of 1.63 gives us the estimate that the lot is 4.75% nonconforming. Since Table 15.2 states that M, the "maximum allowable percent nonconforming," is 6.17%, the lot is accepted.

As previously stated, the k values used in the Form 1 procedure are consistent with the M values used in the Form 2 scheme; therefore the decision on acceptance or rejection is not influenced by the choice between Form 1 and Form 2.

Whether or not the Form 2 procedure is used to determine the acceptance decision, it is essential to use the quality index to estimate the percentage nonconforming in each lot. These estimates are needed to compute an estimated process average and to determine whether normal, tightened, or reduced inspection should be used. This is true with the range method and with known variability as well as with the standard deviation method.

15.7.3 Calculations with Variability Unknown—Range Method

Assume normal inspection as before with an AQL of 2.5% and code letter H. Table 15.5 gives the sample size as 25. Using the Form 1 procedure, k is 0.647. For the Form 2 procedure, d_2^* is 2.358 and M is 5.98%. The following discussion illustrates the two types of computational procedure but does not include a reproduc-

TABLE 15.3 ILLUSTRATION OF CALCULATION FORMS USING MIL-STD-414—VARIABILITY UNKNOWN—STANDARD DEVIATION METHOD

Line	Information needed	Value obtained	Explanation
	Single specification limit—Form 1		
1	Sample size: n	20	
2	Sum of measurements: ΣX	426,190	
3	Sum of squared measurements: ΣX^2	9,094,204,500	
4	Correction factor (CF): $\frac{(\Sigma X)^2}{n}$	9,081,895,805	$\frac{(426{,}190)^2}{20}$
5	Corrected sum of squares (SS): $\Sigma X^2 - CF$	12,308,695	9,094,204,500 −9,081,895,805
6	Variance (V): $\frac{SS}{n-1}$	647,826	$\frac{12{,}308{,}695}{19}$
7	Estimate of lot standard deviation s: $\sqrt{V}$	804.9	$\sqrt{647{,}826}$
8	Sample mean $\overline{X}$: $\frac{\Sigma X}{n}$	21,309.5	$\frac{426{,}190}{20}$
9	Specification limit (lower): L	20,000	
10	The quantity: $\frac{\overline{X} - L}{s}$	1.63	$\frac{21{,}309.5 - 20{,}000}{804.9}$
11	Acceptability constant: k	1.51	See Table 15.2
12	Acceptability criterion: Compare $\frac{\overline{X} - L}{s}$ with k	1.63 > 1.51	
	The lot meets the acceptability criterion, since $\frac{\overline{X} - L}{s} > k$		
	Single specification limit—Form 2		
1–9	Same as in Form 1		
10	Quality index: $Q_L = \frac{\overline{X} - L}{s}$	1.63	$\frac{21{,}309.5 - 20{,}000}{804.9}$
11	Estimate of lot percent nonconforming: P_L	4.75%	See Table 15.4
12	Maximum allowable percent nonconforming: M	6.17%	See Table 15.2
13	Acceptability criterion: Compare P_L with M	4.75% < 6.17%	
	The lot meets the acceptability criterion, since $P_L < M$		

tion of the types of computational forms given in the standard (such as Table 15.3 for the standard deviation method).

The sum of the 25 tensile strengths is 531,250. The sum of the 5 ranges is 7,970. In both procedures, $\overline{X}$ and $\overline{R}$ must be computed.

$$\overline{X} = \frac{\Sigma X}{25} = \frac{531{,}250}{25} = 21{,}250 \text{ lb/in}^2$$

TABLE 15.4 ESTIMATES OF LOT PERCENTAGE NONCONFORMING FOR VARIOUS VALUES OF QUALITY INDEX AS DEFINED IN MIL-STD-414

Q_u or Q_L	Variability unknown—standard deviation method				Variability unknown—range method				Variability known
	$n = 7$	$n = 10$	$n = 15$	$n = 20$	$n = 7$	$n = 10$	$n = 15$	$n = 25$	Any n
0.00	50.00	50.00	50.00	50.00	50.00	50.00	50.00	50.00	50.000
0.10	46.26	46.16	46.10	46.08	46.29	46.20	46.13	46.08	46.017
0.20	42.54	42.35	42.24	42.19	42.60	42.42	42.29	42.19	42.074
0.30	38.87	38.60	38.44	38.37	38.95	38.70	38.51	38.38	38.209
0.35	37.06	36.75	36.57	36.49	37.15	36.87	36.65	36.50	36.317
0.40	35.26	34.93	34.73	34.65	35.36	35.05	34.82	34.66	34.458
0.45	33.49	33.13	32.92	32.84	33.60	33.27	33.02	32.85	32.636
0.50	31.74	31.37	31.15	31.06	31.85	31.51	31.25	31.07	30.854
0.55	30.01	29.64	29.41	29.32	30.13	29.78	29.52	29.33	29.116
0.60	28.32	27.94	27.72	27.63	28.44	28.08	27.82	27.64	27.425
0.65	26.66	26.28	26.07	25.98	26.78	26.42	26.17	25.99	25.785
0.70	25.03	24.67	24.46	24.38	25.14	24.80	24.56	24.39	24.196
0.75	23.44	23.10	22.90	22.83	23.55	23.22	22.99	22.84	22.663
0.80	21.88	21.57	21.40	21.33	21.98	21.69	21.48	21.34	21.186
0.85	20.37	20.10	19.94	19.89	20.46	20.20	20.01	19.89	19.766
0.90	18.90	18.67	18.54	18.50	18.98	18.75	18.60	18.50	18.406
0.95	17.48	17.29	17.20	17.17	17.54	17.36	17.24	17.17	17.106
1.00	16.10	15.97	15.91	15.89	16.14	16.02	15.94	15.89	15.866
1.05	14.77	14.71	14.68	14.67	14.79	14.73	14.69	14.67	14.686
1.10	13.49	13.50	13.51	13.52	13.50	13.49	13.50	13.52	13.567
1.15	12.27	12.34	12.39	12.42	12.25	12.31	12.37	12.42	12.507
1.20	11.10	11.24	11.34	11.38	11.05	11.19	11.29	11.38	11.507
1.25	9.98	10.21	10.34	10.40	9.91	10.12	10.27	10.39	10.565
1.30	8.93	9.22	9.40	9.48	8.83	9.11	9.32	9.47	9.680
1.35	7.92	8.30	8.52	8.61	7.80	8.16	8.41	8.60	8.851
1.40	6.98	7.44	7.69	7.80	6.83	7.27	7.57	7.79	8.076
1.45	6.10	6.63	6.92	7.04	5.93	6.44	6.78	7.03	7.353
1.50	5.28	5.87	6.20	6.34	5.08	5.66	6.05	6.33	6.681
1.55	4.52	5.18	5.54	5.69	4.30	4.94	5.37	5.68	6.057
1.60	3.83	4.54	4.92	5.09	3.58	4.28	4.74	5.08	5.480
1.65	3.19	3.95	4.36	4.53	2.93	3.68	4.17	4.52	4.947
1.70	2.62	3.41	3.84	4.02	2.35	3.13	3.64	4.00	4.457
1.75	2.11	2.93	3.37	3.56	1.83	2.63	3.16	3.54	4.006
1.80	1.65	2.49	2.94	3.13	1.38	2.19	2.73	3.11	3.593
1.85	1.26	2.09	2.56	2.75	0.99	1.79	2.34	2.73	3.216
1.90	0.93	1.75	2.21	2.40	0.67	1.45	1.99	2.38	2.872
1.95	0.65	1.44	1.90	2.09	0.42	1.15	1.68	2.07	2.559
2.00	0.43	1.17	1.62	1.81	0.23	0.89	1.41	1.79	2.275

TABLE 15.4 (*Continued*)

Q_u or Q_L	Variability unknown—standard deviation method				Variability unknown—range method				Variability known
	$n = 7$	$n = 10$	$n = 15$	$n = 20$	$n = 7$	$n = 10$	$n = 15$	$n = 25$	Any n
2.05	0.26	0.94	1.37	1.56	0.10	0.67	1.17	1.54	2.018
2.10	0.14	0.74	1.16	1.34	0.02	0.49	0.96	1.32	1.786
2.15	0.06	0.58	0.97	1.14	0.00	0.35	0.78	1.13	1.578
2.20	0.015	0.437	0.803	0.968	0.000	0.236	0.625	0.954	1.390
2.25	0.001	0.324	0.660	0.816	0.000	0.150	0.495	0.802	1.222
2.30	0.000	0.233	0.538	0.685	0.000	0.089	0.386	0.672	1.072
2.35	0.000	0.163	0.435	0.571	0.000	0.047	0.296	0.558	0.939
2.40	0.000	0.109	0.348	0.473	0.000	0.021	0.223	0.461	0.820
2.45	0.000	0.069	0.275	0.389	0.000	0.007	0.165	0.378	0.714
2.50	0.000	0.041	0.214	0.317	0.000	0.001	0.118	0.307	0.621
2.55	0.000	0.023	0.165	0.257	0.000	0.000	0.083	0.247	0.539
2.60	0.000	0.011	0.125	0.207	0.000	0.000	0.056	0.198	0.466
2.65	0.000	0.005	0.094	0.165	0.000	0.000	0.037	0.157	0.402
2.70	0.000	0.001	0.069	0.130	0.000	0.000	0.023	0.123	0.347
2.75	0.000	0.000	0.049	0.102	0.000	0.000	0.014	0.096	0.298
2.80	0.000	0.000	0.035	0.079	0.000	0.000	0.007	0.074	0.256
2.85	0.000	0.000	0.024	0.060	0.000	0.000	0.004	0.055	0.219
2.90	0.000	0.000	0.016	0.046	0.000	0.000	0.002	0.042	0.187
2.95	0.000	0.000	0.010	0.034	0.000	0.000	0.001	0.031	0.159
3.00	0.000	0.000	0.006	0.025	0.000	0.000	0.000	0.022	0.135
3.10	0.000	0.000	0.002	0.013	0.000	0.000	0.000	0.011	0.097
3.20	0.000	0.000	0.001	0.006	0.000	0.000	0.000	0.005	0.069
3.30	0.000	0.000	0.000	0.003	0.000	0.000	0.000	0.003	0.048
3.40	0.000	0.000	0.000	0.001	0.000	0.000	0.000	0.001	0.034
3.50	0.000	0.000	0.000	0.000	0.000	0.000	0.000	0.000	0.023
3.60	0.000	0.000	0.000	0.000	0.000	0.000	0.000	0.000	0.016
3.70	0.000	0.000	0.000	0.000	0.000	0.000	0.000	0.000	0.011
3.80	0.000	0.000	0.000	0.000	0.000	0.000	0.000	0.000	0.007
3.90	0.000	0.000	0.000	0.000	0.000	0.000	0.000	0.000	0.005
4.00	0.000	0.000	0.000	0.000	0.000	0.000	0.000	0.000	0.003

$$\bar{R} = \frac{\Sigma R}{5} = \frac{7{,}970}{5} = 1{,}594 \text{ lb/in}^2$$

Using the Form 1 procedure, $(\bar{X} - L)/\bar{R}$ is computed; if this computed value is equal to or greater than k, the lot is accepted; otherwise, it is rejected. In this case, $(\bar{X} - L)/\bar{R} = (21{,}250 - 20{,}000)/1{,}594 = 0.784$. Since this figure exceeds the stipulated k of 0.647, the lot is accepted. In effect, the $\bar{X}$ of the sample is required to be at least

$$L + k\bar{R} = 20{,}000 + 0.647(1{,}594) = 21{,}031$$

Under the Form 2 procedure, it is necessary to compute the quality index Q_L, now defined as $[(\overline{X} - L)d_2^*]/\overline{R}$. In this case,

$$Q_L = \frac{(21{,}250 - 20{,}000)(2.358)}{1{,}594} = 1.85$$

For this Q_L, Table 15.4 gives an estimated lot percentage defective of 2.73%. Because 2.73% is not greater than the stipulated M of 5.98%, this lot is accepted.

15.7.4 Calculation with Known Variability

Now assume normal inspection with an AQL of 2.5%, code letter H, and an estimated σ of 1,000 lb/in^2. (In fact, our sample of values for tensile strengths was taken from a table of normal random deviates, assuming that $\mu = 21{,}000$ lb/in^2 and $\sigma = 1{,}000$ lb/in^2.) Table 15.6 gives the sample size as 9. Using the Form 1 procedure, $k' = 1.49$. For the Form 2 procedure, M is 5.68% and v is 1.061. The following discussion illustrates the two types of computation procedures but does not reproduce the forms given in the standard.

The sum of the nine measurements is 187,960. (These are the values shown on page 580 as subsample 1 and the first four values in subsample 2.) Both Form 1 and Form 2 procedures call for computing the sample average.

$$\overline{X} = \frac{\Sigma X}{9} = \frac{187{,}960}{9} = 20{,}884 \text{ lb/in}^2$$

Under the Form 1 procedure, $(\overline{X} - L)/\sigma$ is computed. For our sample, this is $(20{,}884 - 20{,}000)/1{,}000 = 0.884$. If this computed value is equal to or greater than the stipulated k', the lot is accepted; otherwise, it is rejected. Since k' is stipulated as 1.49 and 0.884 is less than 1.49, this lot is rejected. In effect, this procedure stipulates a minimum value for the $\overline{X}$ of the sample of $L + k'\sigma$. In our example, the minimum $\overline{X}$ for acceptance is 21,490 lb/in^2.

Under the Form 2 procedure applicable to a single specification limit, a quality index is computed. This index Q_L is now defined as $[(\overline{X} - L)v]/\sigma$, where v (tabulated in the standard and in our Table 15.6 for various values of n) is $\sqrt{n/(n-1)}$. For the observed sample of 9,

$$Q_L = \frac{(20{,}884 - 20{,}000)(1.061)}{1.000} = 0.94$$

This quality index is used to enter a table (our Table 15.4) to determine the estimated percentage nonconforming in the lot. In this case, Table 15.4 indicates 17.4% nonconforming for a Q_L of 0.94. If the estimated lot percentage nonconforming is equal to or less than the stipulated maximum M, the lot is accepted; otherwise, it is rejected. Since M is 5.68% in this instance, the lot is rejected.

TABLE 15.5 CONSTANTS REQUIRED TO DETERMINE ACCEPTANCE CRITERIA IN NORMAL, TIGHTENED, AND REDUCED INSPECTION WHEN VARIABILITY IS UNKNOWN—RANGE METHOD—CODE LETTER H OF MIL-STD-414

AQL, normal and reduced inspection	AQL, tightened inspection	Constants in normal and tightened inspection					Constants in reduced inspection				
		Single specification limit, Form 1		Single limit, Form 2, and double limit			Single specification limit, Form 1		Single limit, Form 2, and double limit		
		n	k	d_2^*	n	M	n	k	d_2^*	n	M
0.04	0.065	25	1.14	2.358	25	0.125	15	1.04	2.379	15	0.136
0.065	0.10	25	1.10	2.358	25	0.214	15	0.999	2.379	15	0.253
0.10	0.15	25	1.05	2.358	25	0.336	10	0.916	2.405	10	0.23
0.15	0.25	25	1.01	2.358	25	0.506	7	0.702	2.830	7	0.28
0.25	0.40	25	0.951	2.358	25	0.827	7	0.659	2.830	7	0.89
0.40	0.65	25	0.896	2.358	25	1.27	7	0.613	2.830	7	1.99
0.65	1.00	25	0.835	2.358	25	1.95	7	0.569	2.830	7	3.46
1.00	1.50	25	0.779	2.358	25	2.82	7	0.525	2.830	7	5.32
1.50	2.50	25	0.723	2.358	25	3.96	7	0.465	2.830	7	8.47
2.50	4.00	25	0.647	2.358	25	5.98	7	0.405	2.830	7	12.35
4.00	6.50	25	0.571	2.358	25	8.65	7	0.336	2.830	7	17.54
6.50	10.00	25	0.484	2.358	25	12.59	7	0.266	2.830	7	23.50
10.00	15.00	25	0.398	2.358	25	17.48	7	0.189	2.830	7	30.66
15.00	. . . . .	25	0.305	2.358	25	23.79					

Note: In order to agree with symbols used in the discussion of Table 4.3 of this book as well as in other quality control literature, the symbol d_2^* is used here for the factor shown in the first column under "Single limit, Form 2, and double limit." The military standard uses the symbol c for this factor.

TABLE 15.6 CONSTANTS REQUIRED TO DETERMINE ACCEPTANCE CRITERIA IN NORMAL, TIGHTENED, AND REDUCED INSPECTION WHEN VARIABILITY IS KNOWN—CODE LETTER H OF MIL-STD-414

AQL, normal and reduced inspection	AQL, tightened inspection	Constants in normal and tightened inspection					Constants in reduced inspection				
		Single specification limit, Form 1		Single limit, Form 2, and double limit			Single specification limit, Form 1		Single limit, Form 2, and double limit		
		n	k'	n	M	v	n	k'	n	M	v
0.04	0.065	4	2.65	4	0.111	1.115	3	2.49	3	0.114	1.225
0.065	0.10	4	2.55	4	0.161	1.115	4	2.39	4	0.290	1.155
0.10	0.15	5	2.46	5	0.296	1.118	3	2.19	3	0.369	1.225
0.15	0.25	5	2.34	5	0.445	1.118	2	1.94	2	0.310	1.414
0.25	0.40	6	2.23	6	0.721	1.095	2	1.81	2	0.510	1.414
0.40	0.65	6	2.08	6	1.14	1.095	3	1.69	3	1.94	1.225
0.65	1.00	7	1.95	7	1.75	1.080	3	1.56	3	2.76	1.225
1.00	1.50	7	1.80	7	2.62	1.080	3	1.44	3	3.85	1.225
1.50	2.50	8	1.68	8	3.68	1.069	4	1.28	4	6.99	1.155
2.50	4.00	9	1.49	9	5.68	1.061	4	1.11	4	9.97	1.155
4.00	6.50	10	1.31	10	8.43	1.054	5	0.919	5	15.21	1.118
6.50	10.00	12	1.11	12	12.35	1.045	5	0.728	5	20.80	1.118
10.00	15.00	14	0.906	14	17.36	1.038	6	0.515	6	28.64	1.095
15.00	. . . . .	16	0.685	16	23.96	1.033					

Note: In order to agree with symbols used elsewhere in this book, the symbol k' is used to represent the acceptability constant in a known-sigma plan. The military standard uses k for this constant.

15.8 TIGHTENED INSPECTION IN MIL-STD-414

The standard specifies that the estimated process average at any time "is the arithmetic mean of the estimated lot percent defectives computed from the sampling inspection results of the preceding 10 lots or as may be otherwise designated." The process average is needed to determine when shifts shall be made between normal and tightened inspection and between normal and reduced inspection. Tightened inspection is required when the estimated process average is greater than the AQL with more than a certain number T of the lots used to compute the process average that have estimates of percentage nonconforming exceeding the AQL.

The standard gives values of T assuming that the process average has been computed using the preceding 5, 10, and 15 lots, respectively. The stipulated T depends on the sample size code letter and the AQL. Table 15.7 gives values of T for code letter H, assuming variability unknown and the standard deviation method. (Most T values are the same for the range method and for known standard deviation, although there are a few slight differences.)

The use of Table 15.7 may be illustrated with reference to an AQL of 2.50%, assuming that normal inspection has been in force and that the preceding 10 lots have given the following estimates of process average:

4.01%	2.02%	**4.46%**	**3.22%**	2.27%
2.56%	1.79%	**3.59%**	1.58%	**2.87%**

The estimated process average is the arithmetic mean of these 10 figures, namely, 2.84%. Thus one condition for the shift to tightened inspection is met; the estimated process average is greater than the AQL of 2.50%. However, the second condition is not met; only 6 of the lots (those in boldface) had an estimated percentage nonconforming of more than 2.50, whereas the standard (Table 15.7) stipulates that tightened inspection is not required unless more than 7 out of the 10 lots have values of estimated percentage nonconforming above the AQL. Therefore, normal inspection would be continued for the time being.

The acceptance criteria under tightened inspection are shown in Tables 15.2, 15.5, and 15.6 for code letter H. In all instances, these are the criteria under normal inspection for the next smaller AQL class. Once tightened inspection has been initiated, normal inspection is not reinstated until the estimated process average of all lots under tightened inspection is equal to or less than the AQL.

15.9 REDUCED INSPECTION IN MIL-STD-414

To initiate reduced inspection, all the preceding 10 lots (or other designated number) must have been accepted and the estimated percentage nonconforming for *each* of these preceding lots must have been less than a stated limit. Moreover, production must be "at a steady rate." For 10 lots with code letter H, unknown variability, the standard deviation method, and an AQL of 2.50%, Table 15.7 shows this limit to be 2.40%.

The acceptance criteria for code letter H under reduced inspection are given in Tables 15.2, 15.5, and 15.6. For an AQL of 2.50% with unknown variability and

TABLE 15.7 CRITERIA RELATING TO ESTABLISHMENT AND DISCONTINUANCE OF TIGHTENED AND REDUCED INSPECTION WHEN VARIABILITY IS UNKNOWN—STANDARD DEVIATION METHOD—CODE LETTER H OF MIL-STD-414

	Values of *T* for tightened inspection			Limits of estimated lot percent nonconforming for reduced inspection		
	Number of lots			Number of lots		
AQL	5	10	15	5	10	15
0.04	3	5	6	0.000	0.004	0.013
0.065	3	5	7	0.000	0.010	0.029
0.10	3	5	7	0.002	0.023	0.058
0.15	3	6	8	0.005	0.048	0.105
0.25	4	6	8	0.017	0.111	0.215
0.40	4	6	9	0.048	0.225	0.396
0.65	4	7	9	0.123	0.445	0.65
1.00	4	7	9	0.266	0.785	1.00
1.50	4	7	10	0.521	1.31	1.50
2.50	4	7	10	1.14	2.40	2.50
4.00	4	8	11	2.24	4.00	4.00
6.50	4	8	11	4.29	6.50	6.50
10.00	4	8	11	7.40	10.00	10.00
15.00	4	8	11			

the standard deviation method, n is 7 and k is 1.15; for the Form 2 procedure, M is 12.20%. Just as in the military standards for acceptance sampling by attributes, if a bad lot is submitted, it has a better chance of acceptance under reduced inspection than under normal inspection.

Reduced inspection must be discontinued and normal inspection reinstated whenever a lot is rejected, whenever the estimated process average exceeds the AQL, whenever "production becomes irregular or delayed," or whenever "other conditions" exist that "may warrant that normal inspection should be reinstated."

15.10 AN UPPER SPECIFICATION LIMIT IN MIL-STD-414

The foregoing numerical examples all have assumed a one-sided specification with a lower limit L. The only difference when there is a single upper limit U is that it is necessary to measure down from U rather than up from L; $U - \overline{X}$ replaces $\overline{X} - L$ in the analysis. This point may be illustrated by noting the definitions of the quality indexes Q_L and Q_U for the various procedures.

For unknown variability, standard deviation method, these are

$$Q_L = \frac{\overline{X} - L}{s} \qquad Q_U = \frac{U - \overline{X}}{s}$$

For unknown variability, range method, they are

$$Q_L = \frac{(\overline{X} - L)d_2^*}{\overline{R}} \qquad Q_U = \frac{(U - \overline{X})d_2^*}{\overline{R}}$$

For known variability, they are

$$Q_L = \frac{(\overline{X} - L)v}{\sigma} \qquad Q_U = \frac{(U - \overline{X})v}{\sigma}$$

The foregoing definitions of Q_L and Q_U also apply to the case of two-sided specifications.

15.11 TWO-SIDED SPECIFICATIONS IN MIL-STD-414

Only the Form 2 procedure is applicable with a double specification limit. Two types of condition are recognized. In one type, an AQL is assigned to both limits combined. In the other type, different AQLs are assigned to each specification limit.

In both types, sample size code letters are obtained just as for one-sided specifications. Moreover, in both types the same tables used for Form 2 with a single limit are used to find *n, M,* and—where applicable—d_2^* or *v*. Q_L and Q_U are computed just as for single limits, and the percentage of rejects outside each limit is estimated from a table such as our Table 15.4. These percentages are designated p_L and p_U. The estimated lot percentage nonconforming *p* is defined as $p_L + p_U$.

Where one AQL has been assigned to both limits combined, if "*p* is equal to or less than the stipulated maximum allowable percent defective *M,* the lot meets the acceptability criterion; if *p* is greater than *M* or if either Q_U or Q_L or both are negative, then the lot does not meet the acceptability criterion."

Where different AQL values have been assigned to *L* and *U,* the rules are somewhat more complicated. For example, assume code letter H, normal inspection, with unknown variability and the standard deviation method. Assume an AQL of 2.5% for the lower specification limit and 1.0% for the upper limit.

Table 15.2 tells us that *n* is 20 for both AQLs and that *M* is 6.17% for the lower limit and 2.95% for the upper. These values are designated as M_L and M_U, respectively.

Now assume that the sample of 20 has been inspected and that Q_L and Q_U have been computed following the procedure illustrated in Table 15.3. Assume that Q_L is 1.60 and Q_U is 2.10. We may enter Table 15.4 to find $p_L = 5.09\%$ and $p_U = 1.34\%$.

For the lot to be accepted:

1. p_U must be equal to or less than M_U. Since 1.34% is less than 2.95%, this condition is met.
2. p_L must be equal to or less than M_L. Since 5.09% is less than 6.17%, this condition also is met.
3. *p* must be equal to or less than the larger of M_L and M_U. Here

$$p = 1.34\% + 5.09\% = 6.43\%$$

Since 6.43% is greater than 6.17%, this condition is not met and the lot is rejected.

The general rules regarding tightened and reduced inspection are the same for two-sided and one-sided specifications.

MIL-STD-414 contains two auxiliary tables (B-8 and C-8) not reproduced here. These tables provide an aid in judging whether the dispersion of a quality characteristic is too great to meet a stated quality standard even under the most favorable conditions of centering between *U* and *L*.

15.12 THE RELATIONSHIP AMONG SAMPLE SIZES UNDER THE STANDARD DEVIATION METHOD, THE RANGE METHOD, AND THE PROCEDURES ASSUMING KNOWN VARIABILITY

There is no code letter A in MIL-STD-414. For code letters B to G with variability unknown, the standard deviation method and the range method have identical sample sizes under normal inspection, namely, 3, 4, 5, 7, 10, and 15, respectively. For code letter H and beyond, sample sizes in normal inspection are from about 12 to 20% smaller in the standard deviation method than in the range method.

For code letters of H and beyond, the choice of the standard deviation method over the range method will cause some saving in inspection cost; choice of the range method presumably will cause some saving in the cost of calculations to apply the acceptance criteria. The relative importance of these cost differences will depend chiefly on the costs per item associated with inspection. Where tests are destructive or otherwise costly, it will usually be more economical to use the standard deviation method.

For any given AQL and code letter, the OC curves for the standard deviation method and the range method are matched as closely as practicable. For this reason, the question of quality protection to the consumer should be viewed primarily as a matter of choosing the desired AQL and inspection level rather than as a matter of choice between the standard deviation method and the range method.

In the range method, all sample sizes of 10 or more are multiples of 5, and the sample must be divided into subsamples of 5 each to compute $\bar{R}$. For sample sizes of 3, 4, 5, and 7, the *R* of the entire sample is used as the measure of sample dispersion and no calculation of $\bar{R}$ is required.

The standard deviation method and the range method in MIL-STD-414 are alike in the feature that the sample size under normal inspection depends only on the code letter and is not influenced by the AQL. For example, Tables *S* and *T* in the standard for code letter H show sample sizes of 20 and 25, respectively, for all AQL values under normal inspection. In contrast, the sample size for a given code letter with known variability changes greatly with the AQL. For this reason, no simple statement applicable to all AQL values can be made about the relationship between sample sizes with known and unknown variability.

For example, Table 15.6 for code letter H shows sample sizes under normal inspection varying from 4 for an AQL of 0.04% to 16 for an AQL of 15%. This is a variation of from 20 to 80% of the required sample size with the standard deviation method, variability unknown. The paradox that sample size increases with the relaxation of the quality standard arises because the known-sigma plans in the standard were designed to match the OC curves obtained for the unknown-sigma plans.

For the popular AQL value of 1.00% and between code letters D and Q, the sample size under normal inspection for known-sigma plans varies from 40 to 32.5% of that required for unknown-sigma plans using the standard deviation method.

15.13 SOME COMMENTS ON TABLE 15.4

This table has been condensed in various ways from Tables B-5, C-5, and D-5 of the standard. Thus the quality index varies by steps of 0.05 in Table 15.4, whereas it varies by steps of 0.01 in the standard. The only values of n given in Table 15.4 for the case of variability unknown are those required for normal, tightened, and reduced inspection with code letter H; the standard, of course, gives all values of n required by any code letter.

The user of Table 15.4 needs to remember that the quality index is defined differently for each of the three parts of the standard. Given the respective definitions of quality index, Table 15.4 and Tables B-5, C-5, and D-5 of the standard are sources of unbiased minimum variance estimates of lot percentage nonconforming based on the assumption of normality. These tables may therefore be used for other purposes than the application of the acceptance criteria of MIL-STD-414.

For all three parts of the standard, Table 15.4 starts with an estimate that the lot is 50% nonconforming when the quality index is zero. Of course, a zero quality index occurs when the sample average is the same as the specification limit in question, either L or U. Whenever the sample $\bar{X}$ falls below L or above U, the quality index is negative, implying that more than 50% of the lot is nonconforming. All lots that have negative quality indexes should be rejected without further calculation.

The final column of Table 15.4, applicable to all values of n with variability known, is, in effect, merely a normal curve area table such as our Table *A*.[†]

15.14 THE CHOICE BETWEEN UNKNOWN-SIGMA AND KNOWN-SIGMA PLANS

In starting to use MIL-STD-414, the Bowker-Goode tables, or other comparable variables procedures for acceptance inspection of a new product, there may be no basis for estimating σ. Where this is true, it is necessary to start with unknown-sigma procedures.

[†]For a discussion of the mathematics underlying the estimates of lot percentage nonconforming with both known and unknown variability, see G. J. Lieberman and G. J. Resnikoff, "Sampling Plans for Inspection by Variables," *Journal of the American Statistical Association,* vol. 50, pp. 457–516, June 1955.

Nevertheless, if many lots are expected from the same production source, it always is a good idea to start gathering data to provide a basis for judgment on the question of whether or not there is statistical control of the dispersion of the quality characteristic being measured. Usually this calls for a control chart for s or R. If such a chart indicates that the lots are coming from a process that is in statistical control with respect to its dispersion, it makes sense to switch from the unknown-sigma procedure to a known-sigma one. In the initial stages of a switch to known sigma, it will be conservative to estimate σ a little on the high side. Often the estimated σ may be reduced as more evidence is accumulated.

Usually an incidental advantage of the switch to known sigma will be a reduction in sample size. In the common case where sample size is small, a more important advantage is that a much better estimate of σ is obtained from a series of samples that exhibit statistical control of dispersion than can be had from a single small sample. Even where there is not full statistical control, a conservatively chosen σ may be better than unknown sigma if it is imperative to use a very small sample size. The reader may get a feeling for the unreliability of any estimate of σ from a small sample by examining the variation of any measure of dispersion—such as R, s, or s^2—in successive samples from a statistically controlled process. For instance, in our first page of drawings from Shewhart's normal bowl (Table 3.7), R varied from 1 to 42 for samples of 4 and s varied from 0.6 to 19.1.

Of course, whenever a known-sigma plan is used, a control chart should be maintained on some measure of dispersion, R or s. Such a chart will serve to check the assumption that σ is in statistical control and, as time goes on, may provide a basis for a better estimate of the "known" value of σ that should be used.

15.15 SOME SOURCES OF OC CURVES FOR VARIABLES PLANS BASED ON THE ASSUMPTION OF NORMALITY

MIL-STD-414 contains OC curves for all code letters and all AQL values for its plans using the standard deviation method for normal inspection. These curves may be viewed as approximations to the corresponding plans based on the range method and on known variability. The standard contains such a wide variety of OC curves that it may be viewed as a convenient source from which it usually is possible to find a plan that gives an approximation to any desired OC curve.[†]

The OC curves shown in MIL-STD-414 apply to the case of a one-sided specification. A plan applicable to a two-sided specification does not have a unique OC curve because the probability of acceptance depends on the way the total percentage is divided into percentages above U and below L. Nevertheless, for any given plan in the military standard, the group of OC curves applicable to two-sided specifications is contained within a fairly narrow band. The particular OC curve applica-

[†]All the OC curves and sampling plans for MIL-STD-414 can also be found in Bowker and Lieberman, op. cit.

ble to a one-sided specification is used as an approximation to all the curves in this narrow band. The Bowker-Goode volume constitutes another convenient source of variables plans having desired OC curves.

15.16 COMMENT ON THE ASSUMPTION OF A NORMAL DISTRIBUTION IN KNOWN-SIGMA AND UNKNOWN-SIGMA PLANS

The frequency distribution of many industrial quality characteristics is roughly normal. This is particularly so where the product comes from a single source and is produced within a short period of time. For this reason, the assumption of a normal distribution is good enough for practical purposes in many instances. This assumption is most likely to be a reasonable one where inspection lots are formed close to the point of production, so that the chance for the mixing of product having different frequency distributions is held to a minimum.

Nevertheless, even though inspection lots have been produced under apparently homogeneous conditions, it is always well to view the assumption of normality with a somewhat critical eye, investigating to see whether conditions exist that are likely to cause serious departure from a normal distribution. Sometimes the underlying frequency distribution is skewed, or it may be symmetrical but either peaked or flat-topped. The percentages in the extreme tails of such distributions may differ considerably from those obtaining under a normal distribution, and the protection against stated percentages of defectives given by variables acceptance criteria may be either greater or less than the protection indicated by OC curves computed on the assumption of normality. The tighter the quality standard (for example, the smaller the AQL), the less reasonable it is to use acceptance criteria based on the assumption of normality.

One important departure from normality exists when a producer has given 100% screening inspection by attributes to a lot prior to its variables sampling inspection by the consumer. In such a case the frequency distribution in the screened lot may be truncated; one or both of the tails of the distribution may have been removed. With such truncated distributions, the variables criteria based on the assumption of normality may indicate that a lot should be rejected even though the actual nonnormal distribution in the lot may contain no defectives.

15.17 ACCEPTANCE/REJECTION PLANS MAY BE DEVISED TO ACCEPT ON VARIABLES CRITERIA BUT TO REJECT ONLY ON ATTRIBUTES CRITERIA

Two objections are sometimes raised to the use of either known-sigma or unknown-sigma plans for lot rejection. One objection relates to the point just mentioned that, for certain nonnormal distributions, the variables criteria may occasionally lead to the rejection of a lot containing no defectives.

The other objection relates to the obvious legal and psychological difficulties incident to rejecting a lot on the basis of a sample even though no defectives have been found in the sample. Consider, for example, the known-sigma plan of Fig. 15.4 in which the design specification stated that the minimum tensile strength of an individual item must be 20,000 lb/in^2, whereas the acceptance specification stated that the minimum $\overline{X}$ for a sample of 16 must be 21,846 lb/in^2. It is quite possible for a sample to fail this acceptance specification without any of the individual tensile strengths falling below 20,000 lb/in^2.[†]

These objections are sometimes met by devising double sampling plans in which variables criteria are applied to a first sample. When these criteria do not lead to acceptance, a second larger sample is taken and the final decision on acceptance or rejection is based on attributes criteria. A number of acceptance plans of this type are described in the Bowker-Goode volume.[‡]

15.18 INTERNATIONAL AND COMMERCIAL STANDARDS CORRESPONDING TO MIL-STD-414

The international standard corresponding to MIL-STD-414 is ISO/DIS 3951.[§] Promulgated in 1981, it is based on the United Kingdom Defense Standard, "Sampling Procedures and Charts for Inspection by Variables," and bears that name. The primary feature of this standard is that it uses a graphical procedure to make acceptance decisions. No procedure similar to Form 2, as discussed herein, is available. Otherwise, its coding and use of AQLs follow the pattern established by the ABC standard and MIL-STD-414.

The U.S. commercial version of MIL-STD-414 is ANSI/ASQC standard Z1.9. The original publication of this standard in 1972 was a direct copy of MIL-STD-414. However, in 1980, certain changes were made much along the lines of those adopted in the commercial version of the ABC standard, ANSI/ASQC Standard Z1.4.

The original version of MIL-STD-414 contained plans matched as nearly as possible to MIL-STD-105A(1950). However, when the ABC standard was prepared (MIL-STD-105D), a number of changes were introduced that resulted in differences in inspection levels, lot size ranges, and sample size code letter designations. As of 1994, MIL-STD-414 had not been revised to resolve these differences.

[†]A possible way to avoid this particular type of objection is to incorporate the acceptance specification as an additional requirement in the design specification. Thus two separate requirements may be made, one that individual values should be at least 20,000, and the other that averages of samples of 16 should be at least 21,846.

[‡]See also Geoffrey Gregory and G. J. Resnikoff, "Some Notes on Mixed Variables and Attributes Sampling Plans." This was issued by the Applied Mathematics and Statistic Laboratory, Stanford University, Stanford, Calif., as Technical Report No. 10, 1955.

[§]ISO 3951(1981), "Sampling Procedures and Charts for Inspection by Variables," may be obtained from the International Standards Organization, Geneva, Switzerland, or from the American National Standards Institute, New York.

The 1980 version of ANSI/ASQC Z1.9, developed by a working group chaired by E. G. Shilling, made the necessary changes to bring the commercial version into line with the commercial standard for attributes sampling.† These changes were carried forward into the latest version, ANSI/ASQC Z1.9-1993.‡

No changes were made in the basic methods and tables from MIL-STD-414 in either the 1980 or the 1993 versions. Changes were made to the code letter designations, inspection levels, and sample size code letters in 1980. Thus all the prior discussion of methodology and example values for code letter H apply to ANSI/ASQC Z1.9-1993.

The rules for switching between reduced, normal, and tightened inspection, however, were changed to those of the ABC standard. The changed rule in MIL-STD-105E regarding termination of inspection was not incorporated into Z1.9. The use of the words *defect* and *defective* was eliminated beginning with the 1980 version.

PROBLEMS

15.1. The specification on a certain dimension of a manufactured part is 1.7030 ± 0.0030 in. In a Lot Plot for this dimension, the first three digits of the measured values for all 50 items are 1.70. The final two digits are those given in Table 3.7 for drawings 101 to 150 from the Shewhart bowl, in the order stated. Prepare the Lot Plot, assuming cell boundaries of 32.5, 36.5, 40.5, and so on. Compute the upper and lower lot limits by the method described in the explanation of the Lot Plot plan. Draw these limits on your Lot Plot.

15.2. Follow the directions for Prob. 15.1, using the Shewhart bowl drawings 301–350 as shown in Table 3.7.

15.3. A sample of 50 cans of tomatoes is taken from a large lot. The drained weights of contents of these cans are to be shown on a Lot Plot. Assume that these 50 weights are the second 50 figures (samples 11 through 20) in Table 3.1, in the order given. Prepare a Lot Plot, assuming the cell boundaries as 20.25, 20.75, 21.25, and so on. Compute lot limits, and draw them on your Lot Plot.

15.4. Follow the directions for Prob. 15.3, using the third 50 measured weights (sample numbers 21 through 30) in Table 3.1.

15.5. A sample of 50 thermostatic control devices is taken from a large lot. The temperature at which each device turns "on" is to be shown in a Lot Plot. Assume that these 50 temperatures are the first 50 figures (subgroup numbers 1 through 10) in Prob. 3.1. Prepare a Lot Plot, and plot the lot limits.

15.6. A known-sigma variables plan for a one-sided specification uses $n = 9$ and $k' = 1.466$. Assuming a normal distribution and a correct estimate of σ, what is the

†For a review and clear discussion of the changes made in the standard, see Edward G. Shilling, "Revised Variables Acceptance Sampling Standards—ANSI Z1.9(1980) and ISO 3951(1980)," *Journal of Quality Technology*, vol. 13, no. 2, April 1981. This article describes as well the necessary changes to MIL-STD-414 to bring it into line with MIL-STD-105D (the ABC standard).

‡ANSI/ASQC Z1.9-1993, "Sampling Procedures and Tables for Inspection by Variables for Percent Nonconforming," is available from either the American National Standards Institute, New York, or the American Society for Quality Control, Milwaukee, Wis.

probability of acceptance of a 3.75% nonconforming lot?

Answer: 0.827.

15.7. A known-sigma variables acceptance plan for a one-sided specification uses $n = 25$ and $k' = 1.97$. Compute the probability of acceptance of a 3% nonconforming lot assuming that the frequency distribution in the lot is normal and σ is estimated correctly.

Answer: 0.326.

15.8. Assume normal inspection, MIL-STD-414, range method, variability unknown, code letter H, 1.00% AQL, single specification limit $L = 18.5$. Assume that the measured values in the sample are the first 25 drawings from Shewhart's normal bowl (Table 3.7). Use the procedures of MIL-STD-414 to estimate the lot percentage nonconforming.

15.9. Answer the question in Prob. 15.8 changing the AQL to 0.25%.

15.10. Answer the question in Prob. 15.8 using the third 25 drawings (that is, drawings 51–75) from Shewhart's normal bowl (Table 3.7). (Note that it is necessary to arrange these drawings in subgroups of 5 rather than in the subgroups of 4 shown in Table 3.7.) Will the lot be accepted or rejected? What is the actual percentage below 18.5 in Shewhart's normal bowl? (See Table 3.6.)

15.11. Assume normal inspection, MIL-STD-414, variability unknown, standard deviation method, code letter H, 2.50% AQL, single specification limit $L = 14.5$. Assume that the measured values in the sample are the first 20 drawings from Shewhart's normal bowl. Prepare a calculation using the Form 1 portion of Table 15.3 as a model. What is the value computed on line 10? Will the lot be accepted or rejected?

15.12. Change the instructions in Prob. 15.11 to call for a Form 2 calculation. What is the estimate of the lot percentage nonconforming? Will the lot be accepted or rejected?

15.13. From your solutions to Prob. 15.11 and 15.12 and from inspection of Table 15.2, determine the smallest AQL for which this lot would be accepted under normal inspection and under tightened inspection, respectively.

15.14. The single specification limit L for tensile strength of certain wire is 215 lb. MIL-STD-414 is used with normal inspection, code letter H, and an AQL of 0.65%. Variability is unknown, and the range method is to be used. A computation using the Form 2 criterion is to be made. The required sample of 25 is measured with the following results:

Subgroup 1	231, 238, 228, 231, 235
Subgroup 2	224, 245, 263, 231, 245
Subgroup 3	224, 228, 235, 238, 235
Subgroup 4	221, 242, 242, 235, 224
Subgroup 5	224, 224, 242, 252, 252

Make the necessary calculations to determine whether or not the lot should be accepted.

15.15. Assume that the same lot in Prob. 15.14 is to be inspected with variability unknown and the standard deviation method. Assume that a sample of 20 is observed and that the measured values are the ones shown in subgroups 1, 2,

3, and 4 of Problem 15.14. Prepare a calculation using Form 2 to determine whether or not the lot will be accepted. Use Table 15.3 as a model.

(*Hint:* The computations will be simplified if 200 is subtracted from each measurement before starting your calculations.)

15.16. Assume normal inspection, MIL-STD-414, variability known, code letter H, 1.50% AQL, single specification limit $L = 60{,}000$ lb/in^2 (tensile strength). The assumed value of σ is 2,000 lb/in^2. The required sample of eight is tested with the following tensile strengths: 65,060; 66,260; 65,240; 61,550; 65,760; 64,850; 63,880; 60,830. Use the Form 2 procedures of MIL-STD-414 to estimate the lot percentage nonconforming. Will the lot be accepted or rejected?

Answer: 1.29%; accepted.

15.17. Assume normal inspection, MIL-STD-414, variability known, code letter H, 2.50% AQL, single specification limit. Compute the probability of acceptance of a normally distributed lot containing 5% of nonconforming product if the σ of the lot is estimated correctly.

Answer: 0.68.

15.18. Figure 15.4 was used to illustrate the calculation of the probability of acceptance of a 2% nonconforming lot under a certain known-sigma acceptance plan. For the same plan, compute the probability of acceptance of a lot that is 8% nonconforming. Continue to assume normality and a correct estimate of σ.

Answer: 0.04.

15.19. The single specification limit U for the elongation of a certain yarn fiber is 6.86 mm per gram weight applied. MIL-STD-414 is used with normal inspection, code letter H, and an AQL of 2.5%. Variability is unknown, and the standard deviation method is used. The required sample measurements to apply the Form 2 procedure are as follows:

6.73, 7.24, 6.10, 6.05, 6.40
6.32, 6.40, 6.88, 5.82, 6.38
6.91, 6.73, 6.32, 6.38, 6.91
6.63, 6.91, 6.81, 6.32, 6.45

Make the necessary calculations to determine whether or not the lot should be accepted.

Answer: $p = 0.179$; reject.

15.20. A MIL-STD-414 variables acceptance sampling plan uses code letter H and an AQL of 0.25%. Variability is unknown, and there is a single upper specification of 6.850.

(*a*) What are the required critical numbers using the range method for tightened inspection under both Form 1 and Form 2 procedures?

(*b*) What are the critical numbers using the standard deviation method for normal inspection under both Form 1 and Form 2 procedures?

(*c*) Assuming $s = 0.125$ for a sample (and is the best estimate of σ), compute the location of the process mean for incoming lots with a μ_p of 0.04.

(*d*) Sketch the distribution of product with respect to U, the distribution of $\overline{X}$ with respect to the acceptance criterion, and compute the probability of acceptance for the lots described in (*c*). Assume the normal distribution applies.

15.21. The single lower specification limit L for tensile strength for a certain wire is 62 lb. MIL-STD-414 is used with normal inspection, code letter H, and an AQL of 1.50%. Variability is unknown, and the range method and Form 2 procedure are to be used.

(*a*) What critical numbers are required for the plan?

(*b*) A sample from a lot yields $\Sigma X = 1{,}681$ and $\Sigma R = 32$. Make the necessary calculations to determine whether or not the lot should be accepted.

15.22. The single lower specification limit on a certain item is 14.5 units. MIL-STD-414 is used with normal inspection, code letter H, and an AQL of 1.5%. Variability is unknown, and the range method and Form 2 procedure are to be used.

(*a*) What factors from the standard are required to estimate the lot percentage nonconforming and to decide whether or not to accept a lot?

(*b*) The required sample of 25 is measured with the following results:

Subgroup 1	45, 26, 37, 33, 12
Subgroup 2	29, 43, 25, 22, 37
Subgroup 3	33, 29, 32, 30, 13
Subgroup 4	40, 18, 30, 11, 21
Subgroup 5	18, 36, 34, 26, 35

Make the necessary calculations to determine whether or not the lot should be accepted.

(*c*) What is the probability of acceptance of lots having the proportion of nonconforming product found in (*b*)? Use $\bar{R}/d^*_2$ to estimate σ and the value of k listed for the Form 1 procedure to find P_a.

Answer: (b) P = 0.082, reject; (c) 0.0666.

15.23. An ABC standard single sampling plan calls for $n = 50$ and $c = 2$ for code letter H, an AQL of 1.5%, and normal inspection.

(*a*) What Form 1 MIL-STD-414 variables sampling plan corresponds to this ABC standard attributes plan? Assume σ is known and that there is a single specification limit for this characteristic.

(*b*) Find the probability of acceptance of a lot containing 5% nonconforming items under the plan found in (*a*).

15.24. Develop the formulas for P_a of a known-sigma variables acceptance sampling plan for the case where a single upper specification is given. The formulas for the lower specification case are given on pages 573 and 574.

15.25. A known-sigma variables acceptance sampling plan for a one-sided upper specification limit calls for using $n = 9$ and $k' = 1.709$. Compute the probability of acceptance of a lot containing 5% rejectable items assuming that the frequency distribution in the lot is normal and σ is estimated correctly.

15.26. A known-sigma variables acceptance sampling plan has $n = 5$ and $k' = 1.522$. When measured against a single upper specification of 950, incoming lots contain 7% nonconforming items. Distribution of product is approximately normal with a σ of 100.

(*a*) At approximately what level is this process centered?

(*b*) What is the probability of acceptance for these incoming lots?

15.27. The plan described in Prob. 15.26 has a Producer's Risk of 0.05 of rejecting lots with a μ_p of 0.012 and a Consumer's Risk of 0.10 of accepting lots with a μ_p of 0.172.

(*a*) What plan is recommended in MIL-STD-414 for code letter H, an AQL of 1.0%, and normal inspection using the Form 1 procedure when σ is known?

(*b*) What is the probability of acceptance of lots containing 7% nonconforming items under this plan?

(*c*) What percentage of nonconforming items has a 10% probability of acceptance (Consumer's Risk point) under this plan? Comment on the reasons for the difference between this value and that stated for the plan described in Prob. 15.26.

15.28. The $\overline{X}$ control-chart reject limits discussed on pages 398 to 403 constitute a special case of known-sigma variables acceptance criteria. Table 10.7 gives a V factor of 2.25 for an n of 16. Assume that these values of V and n are being used for an $\overline{X}$ chart that shows only reject limits. Assume that the process is centered so that 2% of the product is outside one specification limit and no product is outside the other specification limit. What is the probability that the first $\overline{X}$ value from a process so centered will fall within the reject limits? Assume a normal distribution and a correct estimate of σ.

Answer: 0.22.

15.29. A mixed variables attributes sampling plan is used on receiving inspection of a certain automobile subassembly because the test procedure is expensive and time-consuming. When a new lot is received, a sample of 16 units is tested for a critical voltage characteristic. If the average $\overline{X}$ is less than or equal to $U - k'\sigma$ and greater than or equal to $L + k'\sigma$, the lot is accepted. If $\overline{X}$ is outside this range, a second sample of 50 units is drawn and subjected to an automated accelerated test procedure. The lot is accepted on the second sample only if no more than one item fails during this test.

(*a*) Calculate the probability of acceptance on the first sample if $\mu = 13.0$, $\sigma = 0.2$, $U = 13.4$, $L = 10.2$, and $k' = 1.8$. What proportion (μ_p) of nonconforming units is being produced?

(*b*) What is the probability of acceptance on the second sample and in total? Use the Poisson approximation.

(*c*) Is the use of the two-sided variables acceptance procedure called for an appropriate procedure in this case? Why or why not?

15.30. A mixed variables attributes acceptance sampling plan requires the following procedures. A sample of 12 items is drawn from each lot, and a certain critical measurement is made. If the average of the sample is greater than or equal to $25.00 + 1.75\sigma$, the lot is accepted. If it is not, a second sample of 50 units is drawn and tested on an attributes basis. The lot is then accepted if the number of rejects is less than 3. Assume that σ is 0.50, that the lower specification is 25.00, and that the process follows the normal distribution and is generating 5% nonconforming product.

(*a*) What is the centering μ of this process?

(*b*) What is the probability of acceptance on the first sample?

(*c*) What is the total probability of acceptance? (Use Table *G* to find the probability of acceptance on the second sample.)

(*d*) Control charts for $\overline{X}$ and s are maintained on the results of first samples. Calculate these control limits at an assumed centering of 26.5.

15.31. To avoid the possibility of rejecting lots containing no nonconforming product, a company uses a combined variables attributes sampling plan at receiving inspection rather than a pure variables plan. The plan operates as follows: (1) Draw a random sample of 12 units from the lot, and accept the lot if $(U - \overline{X})/\sigma \geq 1.40$; otherwise, (2) draw a second random sample of 30 units, and accept the lot if no more than one unit is rejected; otherwise reject the lot.

A lot containing 2% nonconforming product is received. U is 40.00 units, and σ is known to be 2.50 units.

(*a*) What is the probability of acceptance of the lot on the first sample?
(*b*) What is the total probability of acceptance?
(*c*) Comment on the merits of this plan.

15.32. A mixed variables attributes acceptance sampling plan is to be used at receiving inspection of lots of 1,500 units in which a single characteristic is critical in assembly of the unit. Both the variables and the attributes plans are to be chosen from sample size code letter H from MIL-STD-414 and MIL-STD-105D, respectively, using an AQL of 1%.

(*a*) What is the variables sampling plan under normal inspection, Form 1 procedure, where there is a single lower specification limit L of 4.50 units, standard deviation unknown, s method?
(*b*) What is the single sampling attributes plan under normal inspection?

The variables sample is drawn from a lot, the critical dimension is measured, and $\overline{X}$ and s are calculated. If the criterion in part *(a)* is satisfied, the lot is accepted. If not, the second sample is drawn and tested on a go–not-go basis. If the number rejected r is less than or equal to c, the lot is accepted. Otherwise, it is rejected and returned to the supplier. In making the following calculations, assume that a certain lot contains 4% nonconforming product, that the standard deviation of the process is known to be 0.12 units, and that s is an unbiased best estimator of σ.

(*c*) Calculate the probability of acceptance on the first sample.
(*d*) What are the probabilities of acceptance on the second sample and the total for the lot?

16

SOME ASPECTS OF LIFE TESTING AND RELIABILITY

A word is not a crystal, transparent and unchanged; it is the skin of a living thought, and may vary greatly in color and content according to the circumstances and the time in which it is used.

—Justice Oliver Wendell Holmes†

16.1 PURPOSE OF THIS CHAPTER

For a number of types of manufactured products, both components and complex systems, life may be one of the quality characteristics specified by the designer. Acceptance sampling relative to life testing has many elements in common with acceptance sampling for the testing of other quality characteristics. Nevertheless, there are certain important points of difference. The objective of this chapter is to help the reader to recognize these similarities and differences.

The subject can be explained to best advantage with reference to certain common sampling plans based on a relatively simple assumption, namely, that throughout the period of time of interest to the life tester, the probability of failure on a life test is constant. Certain limitations of this simple assumption and some possible alternative assumptions that are more complex are discussed briefly near the end of the chapter.

†In *Towne v. Eisner,* 245 U.S. 418 at 425 (1918).

16.2 A CONVENTIONAL MODEL OF THE PROBABILITY OF EQUIPMENT FAILURE

Figure 16.1, reproduced from the AGREE Report (named from the initials of the committee that prepared it), shows a common set of assumptions that seem to give a fairly accurate description of the pattern of failures of certain types of electronic components as well as certain kinds of complex systems. Because of its shape, a failure-rate curve of the type shown in Fig. 16.1 has been called a "bathtub" curve. The initial early failure period *OA* is sometimes called the *infant-mortality* or the *burn-in* or the *debugging* period. The AGREE Report describes this failure-rate curve, in part, as follows.†

> The early failure period . . . begins at the first point during manufacture that total equipment operation is possible and continues for such a period of time as permits (through maintenance and repair) the elimination of marginal parts, initially defective though not inoperative, and unrecognizable as such until premature failure. Upon replacement of all such prematurely failing items, the failure rate will have reached a lower value (point *C*) which will remain fairly constant and which defines the beginning of the normal operating period. Because customary curve smoothing

†"Reliability of Military Electronic Equipment," p. 121, Report by Advisory Group on Reliability of Electronic Equipment, Office of the Assistant Secretary of Defense (Research and Engineering), Superintendent of Documents, Government Printing Office, Washington, D.C., 1957.

FIGURE 16.1 A common set of assumptions about equipment failure rate as a function of time under constant environmental operating conditions. *(Reproduced from "Reliability of Military Electronics Equipment," p. 120, Report by Advisory Group on Reliability of Electronic Equipment, Office of the Assistant Secretary of Defense, Research and Engineering, Superintendent of Documents, Government Printing Office, Washington, D.C., 1957.)*

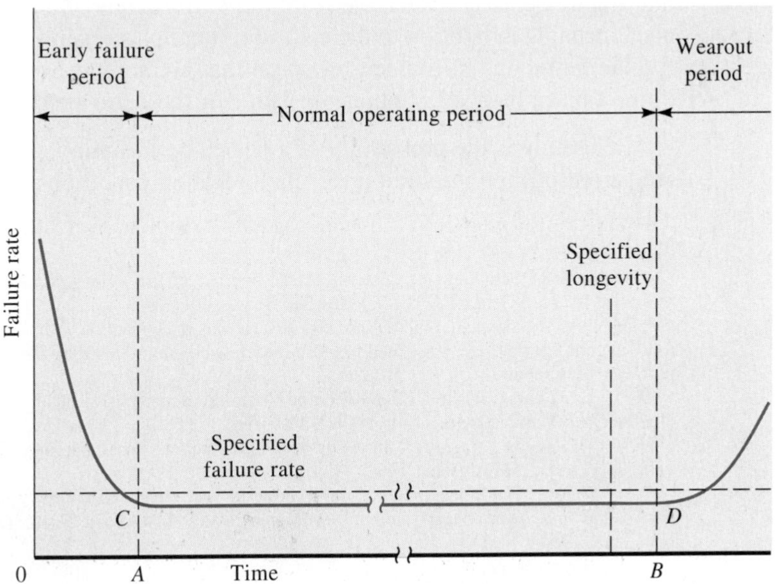

techniques, necessary to develop an average from random data points, markedly reduce the accuracy with which a point of inflection can be located, it is probable that some difficulty may generally be encountered in determining the abscissa (time) location of point *C*. . . .

The normal operating period (*A* to *B*) is that period in terms of equipment operating time in which the average failure rate is and remains essentially constant.

Consider the fairly common cases where it is believed that Fig. 16.1 gives a satisfactory picture of the failure pattern of the component or system in question and where acceptance sampling plans are based on the assumption that the failure rate is constant throughout life. In effect, the use of sampling plans based on a constant failure rate implies that only the region from *A* to *B* on Fig. 16.1 needs to be considered. Presumably, acceptance sampling will not start until all the early failures in region *OA* have occurred. Conceivably, the early failures in this region could be eliminated by a 100% "burn-in" (for components) or by an initial "debugging" (of a complex system). Therefore, the "life" subject to test by acceptance sampling is viewed as starting either from point *A* of Fig. 16.1 or from some point to the right of *A*. Presumably, also, service will terminate before the start of the "wearout period"; it therefore is good enough for practical purposes to interpret acceptance criteria on the basis of an assumption that some unknown constant failure rate will continue indefinitely for any particular lot of items subject to acceptance sampling.

16.3 SOME MODERN DEFINITIONS OF RELIABILITY

Starting in the early 1950s, the word *reliability* acquired a highly specialized technical meaning in relation to the control of quality of manufactured product.[†] Many formal definitions have been proposed that are similar in their general intent but differ a bit in their exact phrasing. Three of these are as follows:

"Reliability is the probability of a device performing its purpose adequately for the period of time intended under the operating conditions encountered."[‡]

[†]As nearly as the authors can determine, the first use of the word *reliability* in its "modern" technical sense was by Maj. Gen. Leslie E. Simon when he was director of ordnance research and development and assistant chief of ordnance of the U.S. Army; this use was in connection with the Nike missile program. The following three articles are suggested to those readers who are interested in the semantic and organizational problems of reliability:

E. L. Grant and L. F. Bell, "Some Comments on the Semantics of Quality and Reliability," *Industrial Quality Control,* vol. 17, no. 11, pp. 14–17, May 1961.

E. G. D. Paterson, "Quality Control vs. Quality Assurance vs. Reliability," *Industrial Quality Control,* vol. 19, no. 4, pp. 5–9, October 1962.

L. E. Simon, "The Relationship of Engineering to Very High Reliability," *Proceedings—Tenth National Symposium on Reliability and Quality Control,* pp. 226–232, Institute of Electrical and Electronic Engineers, New York, 1964.

[‡]This is the official definition of the Electronics Industries Association (EIA), quoted by S. R. Calabro in *Reliability Principles and Practices,* p. 1, McGraw-Hill Book Company, New York, 1962.

"The reliability of a (system, device, etc.) is the probability that it will give satisfactory performance for a specified period of time under specified operating conditions."[†]

"(a) Failure: the inability of an equipment to perform its required function.
(b) Reliability: the probability of no failure throughout a prescribed operating period."[‡]

One obvious point in common among these definitions is that reliability is defined as a probability. Another important point is that all the definitions imply the need for an exact statement of what constitutes failure (that is, inadequate or unsatisfactory performance).

Bazovsky states the modern concept of reliability in popular language as follows:[§] "Stated simply, reliability is the capability of an equipment not to break down in operation. When an equipment works well, and works whenever called upon to do the job for which it was designed, such equipment is said to be reliable."

Reliability, therefore, provides a numerical measure of "degree of excellence" through time. It is a facet of quality that, ideally, works at the interface between design and specification. Quality and reliability are not separate matters. Experts in reliability are engaged in giving advice on improving the "degree of excellence" by preventing design errors. To the extent that they are successful, they are engaged in breaking down the barriers between design and specification.

16.3.1 The Interpretation of a "Failure" with Reference to Life Testing

In studies of human mortality, usually there is no question of the moment of death of an individual. Similarly, in the types of mortality studies of physical property made in connection with the requirements of depreciation accounting, the moment of retirement of, say, a telephone pole would be evident to anyone who observed the pole being removed from service.

In contrast, where the "life" of a manufactured product is tested for purposes of acceptance inspection, the moment of termination of life may not be evident to a superficial observer. Elaborate test apparatus may be required to find the exact moment at which the performance ceases to be satisfactory. Moreover, specifications on what constitutes satisfactory performance naturally depend on the use to which the product is to be put. For example, a particular electronic device might have a short life under one set of specifications and a long life under another set.

Although there are exceptions, when a simple product fails (in the sense that its performance no longer conforms to specifications), the product usually cannot be

[†]The Bureau of Naval Weapons, U.S. Department of the Navy, "Reliability—Fundamental Concepts," Part I, A Brochure of the Material Presented in Film MN 8770a, p. 3, Government Printing Office, Washington, D.C., 1962.

[‡]AGREE Report, op. cit., p. 30.

[§]Igor Basovsky, *Reliability Theory and Practice,* p. 3, Prentice-Hall, Inc., Englewood Cliffs, N.J., 1961.

restored to its original condition of giving satisfactory performance. Generally speaking, the failure of a component is viewed as terminating its life.

When a complex manufactured device fails, it often may be possible to restore the device to its original satisfactory performance, possibly by the replacement of one or more components. Even though "life" may be renewed in this way, a relevant question regarding any such complex device is how long a time may be expected to elapse between successive failures.

16.3.2 Possible Confusion between Technical and Popular Meanings of the Word *Failure*

In Chap. 1 we pointed out that when the words *defective* and *defect* were used in their technical senses, they dealt with nonconformity to specifications. Because of the common practice of incorporating a margin of safety into specifications, a product that is defective in the technical meaning of the word is not necessarily defective when the word is used in its popular meaning. This difference in meanings has been a source of confusion in litigation about product liability. To avoid the bad effects of such confusion, Arthur Bender, Jr., and others have pointed out that it is desirable for industry to abandon the use of *defective* and *defect* in their technical senses. The reader will doubtless recall that the authors tried to follow this good advice through most of this book, substituting such words or phrases as *nonconformities, nonconforming product,* and *percent rejected* in places where earlier editions had mentioned defects, defectives, and percent defective.

The word *failure* is like *defective* and *defect* in having a technical meaning that often differs from its popular meaning. The event that denotes a failure in the sense of nonconformity to specifications may not be the same as the event that would be recognized as a failure in the intended use of a product. For example, some quality characteristic may deteriorate with the amount of use of a product; a failure for test purposes may be defined as taking place when the value of this quality characteristic falls below some stipulated figure. Moreover, the test conditions used in accelerated life testing often are more severe than the conditions that will occur in the actual use of a product.

Using the word *failure* in its technical meaning may lead to confusion in product liability litigation similar to the confusion caused when *defective* and *defect* are used in their technical meanings. It follows that it would be desirable to substitute other words or phrases where the word *failure* has been used in connection with life testing and reliability. For example, persons engaged in life testing might refer to the test *life* of a component rather than to its *time to failure.*

Unfortunately, authors who write about published acceptance procedures and tables need to adopt the language of the published documents. For this reason, even though it was possible to avoid saying *defective* and *defect* in discussing control charts and related matters in Part Two of this book, the authors have made use of these words in their discussion of acceptance sampling in Part Three. In Chap. 11, the reader was warned that these words should be interpreted in their technical senses as referring to nonconformity to specifications.

For the same reason, our discussion in the present chapter needs to use the word *failure*. Here, also, it is necessary to warn the reader that here is a word being used in its restricted technical meaning.

16.4 THE RELATIONSHIP BETWEEN A CONSTANT FAILURE RATE AND MEAN LIFE OR MEAN TIME BETWEEN FAILURES

Assume that the probability of a failure of a component or complex device is constant per unit of time. That is, the failure rate is independent of the number of hours the component or device has been operated. This is the case in region *AB* of Fig. 16.1. As explained earlier in this chapter, in those cases where the failure pattern of Fig. 16.1 is believed to apply, the assumption of a constant failure rate implies that the measurement of life will not start before time *A*. This assumption also implies that the uniform failure rate will continue for a longer time than the intended period of use; for this reason, the final wearout period can be disregarded in the design and interpretation of acceptance sampling plans.

Consider that, say, 100 components are being tested with a new component being substituted whenever one fails. Even though the probability of a failure is constant for any stated time interval, the actual numbers of failures in successive equal time intervals will be subject to chance fluctuations. When we use the mathematics of probability to compute OC curves for acceptance sampling plans used in life testing, we are recognizing the inevitability of these chance fluctuations.

Two related quantities that we shall use in computing OC curves are as follows.[†]

λ failure rate, that is, the probability of a failure in a stated unit of time. Conceivably, any unit of time could be adopted. However, for consistency throughout this chapter, we shall arbitrarily select 1 h as the time unit in all instances.

θ the arithmetic mean or average of the lives (of components), usually referred to in the literature of reliability as the *mean life*, or *mean time to failure* (MTTF). As applied to complex devices, θ is interpreted as the *mean time between failures* (sometimes designated as mtbf or MTBF). θ should be expressed in the same time units that are used for λ.

For our special case of the constant failure rate, $\theta = 1/\lambda$. (It is pointed out later in this chapter that this reciprocal relationship between θ and λ depends on the use of a time unit that is a relatively small fraction of the mean life.)

Although our discussion in this chapter is based on failure rates and mean lives expressed in units of time, it should be mentioned that other units than time may

[†]There is considerable variability in the symbols chosen by different writers on the mathematics of reliability. The symbols λ and θ adopted here are used by some writers. For the purpose of our discussion here, they have the advantage that they are not used elsewhere in this book with some other meaning. λ is the lowercase form of the Greek letter *lambda*. θ is the lowercase form of the Greek letter *theta*.

be used in certain instances. For example, the number of operating cycles of certain switching devices is a more appropriate unit for measuring life than the number of operating hours.

16.4.1 Computing an OC Curve for an Acceptance Sampling Plan Based on a Stipulated Maximum Number of Test Hours

Consider the following lot-by-lot acceptance sampling plan (taken from "Handbook H 108," Example 2C-4[†]):

Select 22 items at random from a lot. Place these items on test. Whenever an item fails, replace it with another item selected at random from the lot. If the test continues for 500 h with not more than 2 failures, accept the lot. If 3 failures occur before the 500 h of testing, reject the lot and terminate the test.

Acceptance under this plan requires 22(500) = 11,000 item hours of test with an acceptance number of 2. Under the assumption that the probability of a failure is the same for every item hour, the calculation of the OC curve is the same as if we had an ordinary single sampling attributes plan with an n of 11,000 and a c of 2. Table 16.1 illustrates the use of Table G to calculate the OC curve of this plan. The method of calculation is identical with the method first illustrated in Table 11.1. However, the failure rate λ takes the place of the fraction nonconforming μ_p. Moreover, because it is customary to plot OC curves of life testing plans in terms of mean life rather than in terms of failure rate, Table 16.1 shows the value of θ corresponding to each value of λ. Figure 16.2 shows the OC curve plotted in the usual manner for such curves.

Table 16.1 and the OC curve of Fig. 16.2 are valid for all plans requiring 11,000 item hours for acceptance and having an acceptance number of 2. They apply to the stated plan from H 108 that called for 22 items to be tested for 500 h with replacement of any items failing during the test. They also apply if 44 items are to be tested with replacement for 250 h or if 110 items are to be tested with replacement for 100 h, provided an acceptance number of 2 is used. Or, with this acceptance number, they apply to any other set of specified items and hours that have a product of 11,000. Under the assumption that the probability an item will fail is independent of the number of hours the item has been tested, one OC curve applies to all plans that have the same stipulated item hours for acceptance and the same acceptance number. OC curves for life testing plans are frequently plotted as a function of θ/θ_0, rather than as a function of θ, where θ_0 is the *acceptable value* of mean life for the component or system.

[†]The full name of this handbook is "Quality Control and Reliability Handbook (Interim) H 108—Sampling Procedures and Tables for Life and Reliability Testing (Based on Exponential Distribution)," Office of the Assistant Secretary of Defense (Supply and Logistics), Government Printing Office, Washington, D.C., 1960. In subsequent references in this chapter, the name of this document is abbreviated to "H 108."

TABLE 16.1 CALCULATION OF OC CURVE FOR ACCEPTANCE SAMPLING PLAN REQUIRING 11,000 ITEM HOURS OF LIFE TESTING WITH AN ACCEPTANCE NUMBER OF 2

Calculation assumes that the failure rate λ is independent of the age of the item tested

Failure rate per hour, λ	Mean life $\theta = 1/\lambda$ h	Expected average number of failures in 11,000 test hours (11,000 λ)	Probability of acceptance (probability of 2 or less failures, read from Table *G*)
0.00002	50,000	0.22	0.999
0.00005	20,000	0.55	0.982
0.00006	16,667	0.66	0.971
0.00008	12,500	0.88	0.939
0.00010	10,000	1.1	0.900
0.000125	8,000	1.375	0.839
0.00015	6,667	1.65	0.770
0.00020	5,000	2.2	0.623
0.00025	4,000	2.75	0.480
0.00030	3,333	3.3	0.360
0.00040	2,500	4.4	0.185
0.00050	2,000	5.5	0.088
0.00060	1,667	6.6	0.040
0.00080	1,250	8.8	0.007

FIGURE 16.2 OC curve for a test of 11,000 item hours with an acceptance number of two (based on assumption of a constant failure rate).

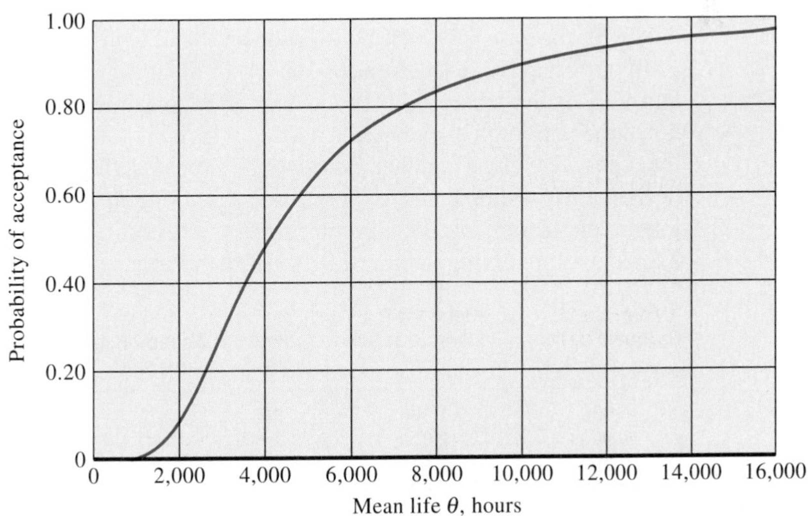

16.4.2 Discontinuance of Life Testing When the Rejection Number Is Reached

Our sampling plan from H 108 stipulated a maximum of 500 h of testing for the 22 items. However, it stipulated that testing should end when the rejection number of 3 was reached. If the third failure occurs after, say, 140 h, no more testing takes place; the decision on the lot is reached with only 3,080 item hours of test rather than with the stipulated maximum of 11,000 item hours.

In the Dodge-Romig systems described in Chap. 13 and in some of the military standards referred to in Chap. 12, it has been customary to inspect an entire single sample regardless of the number of defects found; such inspection gives a better basis for the required estimate of process average. In such systems, curtailment of inspection ordinarily occurs only in double and multiple sampling. However, there is nothing inherent in single sampling that prevents curtailing inspection as soon as a rejection decision is reached. In the usual nondestructive acceptance sampling by attributes, there may be only a small incremental cost of inspecting the remainder of a single sample after a rejection decision; the value of the information gained may be deemed to be considerably greater than this small cost. In life testing, this incremental cost usually is fairly high, and it is customary to curtail single sampling.

16.4.3 Producer's and Consumer's Risks in Acceptance Sampling Plans Used in Life Testing

OC curves for life testing plans, like OC curves for other types of acceptance sampling plans, may be interpreted in terms of stated Producer's and Consumer's Risks. The plan for which the OC curve was computed in Table 16.1 was stipulated in H 108 for a Producer's Risk of 0.10 of rejecting a lot that has a mean life of 10,000 h and a stipulated approximate value of 0.10 for the Consumer's Risk of accepting a lot that has a mean life of 2,000 h. (In Table 16.1, the Consumer's Risk for a 2,000-h mean life was computed to be 0.088.) Several of the tables in H 108 deal with the case in which both Producer's and Consumer's Risks are stipulated.

In our Chap. 13 discussion (Sec. 13.4) of the design of ordinary single sampling attributes plans where both Producer's and Consumer's Risks are stipulated, we noted the very high sample sizes needed where the lot qualities associated with the two risks were close together. As might be expected, the same difficulty arises in the design of life testing plans for both risks. This difficulty can be illustrated by the following plans from H 108, all requiring 500 h of testing with replacement to accept a lot and all involving values of 0.10 for both Producer's and Consumer's Risk.

Mean life for Producer's Risk, h	Mean life for Consumer's Risk, h	Items tested	Acceptance number	Item hours required for acceptance
10,000	1,000	10	1	5,000
10,000	2,000	22	2	11,000
10,000	3,333	63	5	31,500
10,000	5,000	205	14	102,500
10,000	6,667	660	40	330,000

It is evident that, if one starts with the premise that the producer should be protected against chance rejections of lots having mean lives of 10,000 h or more, it is going to be quite expensive to protect the consumer against chance acceptance of lots with mean lives of, say, 5,000 h or less.

16.5 AN EXPERIMENT TO ILLUSTRATE CERTAIN ASPECTS OF A CONSTANT FAILURE RATE

In Chap. 3, we used the first 400 drawings from Shewhart's normal bowl as an aid in the explanation of certain principles. There were a number of advantages gained by such use of data from bowl drawings rather than data from some industrial process. With bowl drawings, we knew the shape of the universe and its parameters. Moreover, we knew that, because the same bowl had been used throughout and because the chips had been stirred thoroughly between drawings, it was reasonable to assume that a constant system of chance causes had been operating.

In a similar way, it should be helpful to have a set of numbers of trials to failure (that is, lives) generated by a series of experiments in which the failure rate is known and is kept constant. We might generate such numbers in various ways, such as by drawing chips from a bowl (see Prob. 16.11), by throwing dice, or by the use of tables of random numbers or of random permutations.

Assume that we are interested in a failure rate of 0.02 and want to generate our data by throwing two icosahedral Japanese random dice. (These were described in Chap. 11.) A failure could be designated as a throw of, say, either 00 or 99. Successive throws could be made until either 00 or 99 occurred. The number of successive throws made to obtain a failure could be counted and recorded. This number might be viewed as the time to failure (that is, life) of one item in a sample. The experiment might be repeated as often as desired to obtain a distribution of times to failure.

The same sort of experiment might be made using a table of random numbers such as Table *X*, App. 3 (see Prob. 16.12). Or it might be done using a set of tables of random permutations of the integers 1 to 50.

Table 16.2 was generated from the Moses-Oakford tables. These tables include 400 different permutations of the integers 1 to 50. For each of the 120 experiments recorded in Table 16.2, two integers from 1 to 50 were picked from a table of random numbers. The first integer determined the position to be examined in each random permutation; the appearance of the second integer in this position constituted a failure.

Consider an industrial sampling situation in which 5 items are selected at random from each lot and are tested to determine the number of hours to failure. The averages and ranges in Table 16.2 show the considerable variation that might occur from sample to sample even though all the 24 lots had identical values of 0.02 for the failure rate λ.

The 120 lives of Table 16.2 have been rearranged in order of increasing magnitude in Table 16.3. Much of our subsequent discussion will deal with this latter

TABLE 16.2 LENGTH OF LIFE, MEASURED IN NUMBERS OF SUCCESSIVE TRIALS TO FAILURE, OBTAINED WHEN 120 SETS OF OBSERVATIONS WERE MADE WITH A CONSTANT PROBABILITY OF FAILURE OF 0.02[†]

Sets of observations	Length of life					Average of 5 values	Range of 5 values
1–5	115	105	34	120	30	80.8	90
6–10	23	14	59	10	18	24.8	49
11–15	4	28	55	11	95	38.6	91
16–20	15	54	45	5	17	27.2	49
21–25	13	50	165	5	40	54.6	160
26–30	4	46	27	30	52	31.8	48
31–35	17	20	7	67	9	24.0	60
36–40	114	132	18	39	37	68.0	114
41–45	147	3	5	23	5	36.6	144
46–50	44	140	125	11	119	87.8	129
51–55	53	86	32	148	1	64.0	147
56–60	78	8	16	14	54	34.0	70
61–65	25	118	10	9	62	44.8	109
66–70	75	2	64	171	17	65.8	169
71–75	2	1	101	26	35	33.0	100
76–80	140	93	11	148	31	84.6	137
81–85	34	52	1	33	36	31.2	51
86–90	105	56	2	5	21	37.8	103
91–95	19	1	111	28	140	59.8	139
96–100	37	98	63	73	43	62.8	61
101–105	43	17	108	35	147	70.0	30
106–110	31	14	94	5	39	36.6	89
111–115	72	127	25	25	22	54.2	105
116–120	4	33	68	20	23	29.6	64
Totals						1,182.4	2,408

[†]From L. E. Moses and R. V. Oakford, *Tables of Random Permutations,* Stanford University Press, Stanford, Calif., 1963.

table. The lives in this table are listed in the order in which they would have been observed if the 120 life tests had all started at the same moment. The arrangement of the data in this table is useful in bringing out certain aspects of acceptance sampling procedures often used in life testing.

The reader should note that Table 16.3 is similar to a tabulation of the results of a life test carried out without replacement of an item that fails. It will be recalled that our OC curve of Fig. 16.1 dealt with a life test carried out with replacement. We shall see that acceptance sampling plans involving life testing may be designed either with or without replacement.

16.5.1 Estimating Mean Life at Various Stages in a Life Test

Assume the existence in a lot or process of an unknown constant failure rate per item hour λ. We wish to estimate this probability of failure in an item hour. On the

TABLE 16.3 DATA OF TABLE 16.2 REARRANGED IN ORDER OF INCREASING MAGNITUDE
This may be viewed as representing a distribution of lives in hours that might have been found if simultaneous tests to failure had been made of a random sample of 120 items from a universe that had a constant failure rate of 0.02 per hour

1	4	9	15	20	27	34	43	54	73	105	132
1	5	10	16	21	28	34	43	55	75	108	140
1	5	10	17	22	28	35	44	56	78	111	140
1	5	11	17	23	30	35	45	59	86	114	140
2	5	11	17	23	30	36	46	62	93	115	147
2	5	11	17	23	31	37	50	63	94	118	147
2	5	13	18	25	31	37	52	64	95	119	148
3	7	14	18	25	32	39	52	67	98	120	148
4	8	14	19	25	33	39	53	68	101	125	165
4	9	14	20	26	33	40	54	72	105	127	171
Total											5,912

basis of all the information from life tests from the lot or process available at any moment, our best estimate is

$$\text{Estimated } \lambda = \frac{\text{number of failures observed}}{\text{number of item hours of test}}$$

This equation simply applies the frequency definition of probability and assumes that all trials of the event in question (that is, failure) are made "under the same essential conditions." If the probability of a failure does not depend on the age of the item tested, each item hour of test should be viewed as providing the same amount of information as any other item hour. For example, 1,000 item hours obtained from testing 1,000 items each for 1 h give just as good a basis for estimating λ as if only 10 items were tested (with replacement) for 100 h each.

The greater the number of item hours, the more confidence it is reasonable to have in the resulting estimate of λ. However, in this connection, the reader should recall the point brought out in the discussion of Table 5.1 that a small number of trials may occasionally, by chance, give an estimate closer to the true probability than will be obtained from a larger number of trials.

Because mean life is the reciprocal of failure rate, an estimate of θ in hours may be made at any time during a life test as follows:

$$\text{Estimated } \theta = \frac{\text{number of item hours of test up to the time of estimate}}{\text{number of failures up to the time of estimate}}$$

Table 16.4 shows a series of estimates of mean life such as might have been made at various times during a life test that had the results shown in Table 16.3. It should be noted that, in such a test made without replacement, the number of item

TABLE 16.4 SOME ESTIMATES OF MEAN LIFE THAT MIGHT HAVE BEEN MADE FROM THE DATA OF TABLE 16.3

Hours of test at moment of estimate	Failures up to moment of estimate	Item hours of test up to moment of estimate	Estimate of mean life, h
4	11	461	41.9
5	17	570	33.5
6	17	673	39.7
8	19	878	46.2
10	23	1,078	46.9
15	31	1,556	50.2
20	41	1,970	48.0
30	55	2,696	49.0
171	120	5,912	49.3

hours of test at any time can be computed by adding the lives in hours t_i of the items that have failed (r in all) up to the moment of estimate to the number of item hours of life observed t for the items that have not yet failed $(n - r)$. For example, the estimate after 4 h could be computed as follows:

$$\text{Estimated } \theta_r = \frac{\sum_{i=1}^{r} t_i + (n - r)}{r}$$

$$= \frac{1 + 1 + 1 + 1 + 2 + 2 + 2 + 3 + 4 + 4 + 4 + 4(120 - 11)}{11}$$

$$= \frac{25 + 436}{11} = \frac{461}{11} = 41.9 \text{ h}$$

In examining the successive estimates of mean life shown in Table 16.4, the reader should keep in mind that the data were generated using a failure rate of 0.02 per hour. This failure rate corresponds to a "true" mean life of 50 h.

16.5.2 Some Economic Considerations in Designing Acceptance Sampling Procedures under the Assumption of a Constant Failure Rate

If we were conducting a simultaneous life test on the 120 items that had the distribution of lives shown in Table 16.3, there are good reasons why we might want to stop the test before the end of the 171st hour when the 120th item had failed. If a decision to accept or reject a lot depended on the results of the test, there would be costs of storage of the lot that would be increased by a delay in reaching a decision. If the conclusions from the test were to be fed back to designers or production people to influence design or production methods on future lots, it would be desirable to have this feedback take place as soon as some dependable conclusions could be obtained. Moreover, if the probability of failure really is independent of age, every item not destroyed by the test may be assumed to end the test as good as new. A substantial part of the cost of a life test may be the cost of the items destroyed; the sooner a life test is terminated, the less will be this element of cost.

Generally speaking, the longer the lives of the items to be tested, the greater the economic advantage in using an acceptance sampling plan that will reach a decision with a maximum test period considerably shorter than the mean life. To illustrate this point, let us assume that the failure rate of 0.02 used in generating Tables 16.2 and 16.3 is interpreted as a rate per 100 h rather than per hour. (The reader will recognize that our choice of a time unit of 1 h was purely arbitrary.) The mean life would then be 5,000 h rather than 50 h; it would have taken 17,100 h before the failure of the 120th item. In such a case, there would doubtless be compelling reasons for terminating the test in a period not to exceed, say, 1,000 h.

Economic considerations therefore dictate that many acceptance sampling plans that involve life testing call for terminating a test at a time considerably short of the mean life. It therefore becomes important to estimate mean life by the methods illustrated in Table 16.4 at a time when most of the items tested are still surviving.

In choosing among alternative possible acceptance sampling plans involving life testing, the item hours of test to give the desired protection may be obtained by testing a relatively few items for a long time or more items for a shorter time. Under the assumption that the failure rate is independent of the age of the items tested, it has been pointed out that all plans with the same item hours and acceptance numbers involve identical OC curves. Therefore, the choice of a plan may involve balancing the extra costs associated with a long delay in reaching a decision against the extra costs associated with placing a larger number of items on test.

16.5.3 Use of Acceptance Criteria Based on Estimated Mean Life

In Table 16.1, we showed the calculation of the OC curve of a plan requiring 11,000 item hours of life testing with not more than 2 failures permitted. The acceptance criterion for this plan could have been expressed in terms of minimum allowable mean life rather than maximum allowable number of failures.

Suppose the 3d failure—the one causing rejection—took place at the last possible moment, namely, at the 11,000th item hour. The estimated θ would then have been 11,000/3 = 3,667 h. Any earlier occurrence of the 3d failure would have given an estimate of less than 3,667 h. Therefore, the same acceptance decision would have been reached if it had been stipulated that the estimated mean life should exceed 3,667 h.

(Incidentally, it is of interest to note that, if there had been 2 failures at the end of 11,000 item hours of test, the estimated mean life would have been 11,000/2 = 5,500 h. With this particular plan of testing, there can be no estimated mean life between 3,667 and 5,500 h.)

It may be impracticable or inconvenient to carry out a life test with replacement in the way illustrated by the acceptance plan that was the subject of Table 16.1. One possible alternative is to require a test of a stipulated number of items for a stipulated number of hours without replacement; in plans of this type, the acceptance criterion usually is stated in terms of minimum acceptable estimated mean life. Other plans involve testing a stipulated number of items until a stipulated number of failures have occurred; these plans also base the acceptance decision on the estimated mean life. In all such plans that involve a uniform failure rate, mean lives are estimated in the way that was illustrated in Table 16.4.

16.5.4 Acceptance Sampling Plans Involving the Combination of Samples from a Series of Lots

Consider the case of, say, an electronic component to be used in a service where it is desired to have a failure rate of not more than 0.00001 per hour. (Sometimes it is desired to avoid the use of very small decimal fractions; if so, a rate of 0.00001 per hour might be stated as 1% per 1,000 h.) Such a specification tells us that the mean life must be at least 100,000 h.

The need to combine the evidence of the samples from a series of successive lots is particularly great when one is dealing with a requirement for such an extremely low failure rate. Otherwise, if the decision on each lot is to be made solely on the evidence of the sample from the lot itself, the required item hours of test from each lot may be extremely high and the test may be extremely costly.

The following plan illustrates one possible method of combining the results of samples from successive lots. The plan is one of a set that has been suggested for use in acceptance sampling for certain long-lived electronic components. The sample sizes and acceptance numbers in this particular plan are related to a desired mean life of at least 100,000 h.

1. The plan starts with a qualification sample requiring 418,000 item hours of test (say, 936 items tested for 500 h). Not more than 3 failures are allowed in this sample. The sample size and acceptance number are chosen to give a Consumer's Risk of 0.40 of accepting product having a mean life of 100,000 h. (In some of the literature of reliability, such a Consumer's Risk is referred to as a "confidence level" of 0.60.)
2. After qualification has been obtained, a sample of 44 items from each lot is tested for 250 h, a total of 11,000 item hours. Not more than 1 failure is allowable to pass the lot.
3. The item hours and failures of the samples from each set of 10 successive lots are to be cumulated. The total item hours from the 10 lots will be 10(11,000) = 110,000. If a total of more than 2 failures is found at any time during the

110,000 item hours, reduced inspection is discontinued and a new qualification sample is required. If lots are held and accepted or rejected in sets of 10, these criteria give a Producer's Risk of 0.10 of rejecting product that has a mean life of 100,000 h.

4. Item hours and failures from successive samples are cumulated until 418,000 item hours, the number used in the qualification test, have been observed. If a total of more than 3 failures is found, reduced inspection is discontinued and a new qualification sample is required.

The reader will note that three different OC curves are relevant in judging how a scheme of this type affects the producer and consumer (see Prob. 16.14 to 16.19). Nevertheless, because the criteria of paragraph (1) are also applied cumulatively in (4), the OC curve of the qualification sample is the one of greatest interest to the producer. Unless the producer can make product that has a mean life considerably longer than 100,000 h, there is a substantial risk that the producer cannot qualify in the first place. (The producer has 6 chances out of 10 that its product will not qualify if its λ is exactly 0.00001.) The producer also takes the same risk that, if its product qualifies, it will be disqualified before one cycle of procedure (paragraph 4) has run its course.

Although the OC curve of the acceptance plan in paragraph (1) measures the consumer's protection relative to qualification, the consumer needs to look at the OC curve of the plan in (2) to evaluate the worst condition faced in the short run (that is, protection against accepting an individual lot after a producer achieves qualification). All things considered, it is evident that the consumer takes substantial risks of accepting product that has a failure rate much worse than the desired 0.00001.

The foregoing type of acceptance procedure has been included in this chapter to emphasize the difficulties that inevitably arise in devising acceptance sampling schemes when tolerable failure rates are extremely low. In meeting these difficulties, it is desirable that all parties concerned recognize the need for quite large sample sizes and for a substantial sharing of risks between producer and consumer.

16.6 THE "EXPONENTIAL" RELIABILITY FUNCTION THAT RESULTS FROM THE ASSUMPTION OF A CONSTANT FAILURE RATE

In generating Tables 16.1 and 16.3, we made successive trials of an event that had a constant probability 0.02 that a failure would occur on any single trial. The probability that any one trial will not result in a failure is $1 - 0.02 = 0.98$. If trials are to be continued until a failure occurs, we can apply the theorem of conditional probabilities to determine the probability of survival (that is, of no failure) after any stipulated number of trials. For example, the probability of survival after 10 trials is $(0.98)^{10} = 0.817$. In general, if λ represents a constant probability of failure per hour, the probability of an item surviving for H hours is $(1 - \lambda)^H$.

Early in this chapter, several formal definitions of reliability were quoted. Although the phrasing differed slightly, all the definitions viewed reliability as being the probability of no failure throughout some specified period of time. In this sense, the third column of Table 16.5 gives reliability values corresponding to a λ of 0.02 and to the various possible "specified periods of time" from 2 to 200 h that are given in the first column of the table. In examining Table 16.5, it is of interest to note the fairly close agreement between the computed probabilities of survival $(0.98)^H$ given in the third column and the observed fractions surviving given in the fifth column based on the experiment recorded in Table 16.3.

Nevertheless, although our experiment employed a constant failure rate of 0.02 per trial, it failed in one minor respect to simulate the results that would occur in an actual life test where the failure rate was constant. A constraint in our test was that we could obtain only integral values of lives; no fractional values were possi-

TABLE 16.5 AN ILLUSTRATION OF THE VALIDITY OF THE NEGATIVE EXPONENTIAL FORMULA TO ESTIMATE RELIABILITY (THAT IS, PROBABILITY OF SURVIVAL) UNDER THE ASSUMPTION OF A CONSTANT FAILURE RATE

The assumed failure rate λ is 0.02 per hour, the rate that was used in generating Tables 16.2 and 16.3. The values of the negative exponential can be read from Table *G*, App. 3, using the column headed $c = 0$

		Probability of survival		
Age H, h	Ratio of age to mean life, $H/\theta = H\lambda$	By theorem of conditional probabilities $(1-\lambda)^H$	By negative exponential $e^{-H\lambda}$	Actual fraction surviving among the 120 items of Table 16.3
2	0.04	0.960	0.961	0.94
4	0.08	0.922	0.923	0.91
6	0.12	0.886	0.887	0.86
8	0.16	0.851	0.852	0.84
10	0.20	0.817	0.819	0.81
15	0.30	0.739	0.741	0.74
20	0.40	0.668	0.670	0.66
30	0.60	0.546	0.549	0.54
40	0.80	0.446	0.449	0.42
50	1.00	0.364	0.368	0.37
60	1.2	0.298	0.301	0.30
70	1.4	0.243	0.247	0.26
80	1.6	0.199	0.202	0.23
90	1.8	0.162	0.165	0.22
100	2.0	0.133	0.135	0.18
120	2.4	0.089	0.091	0.10
140	2.8	0.059	0.061	0.05
160	3.2	0.039	0.041	0.02
180	3.6	0.026	0.027	0.00
200	4.0	0.018	0.018	0.00

ble. That is, we could find a life of 5 h (5 trials with each trial interpreted as an hour) or 6 h but not, say, 5.37 h. Of course, in an actual life test, failure may occur at any moment, not merely at the end of an hour. This point bears on the relative merits of the methods of computing probability of survival illustrated in the third and fourth columns of Table 16.5.

The fourth column of Table 16.5 gives values of $e^{-H\lambda}$ for our λ of 0.02. The reader will recall from the discussion of the binomial and Poisson distributions in Chap. 5 that e^{-np} is a limit of $(1-p)^n$ as n increased and p decreased with the product np held constant. Or, stated a little differently, e^{-np} gives a satisfactory approximation to $(1-p)^n$ provided p is small enough and n is large enough. Of course, the symbols that we employ are purely arbitrary; it is equally appropriate to view $e^{-H\lambda}$ as an approximation to $(1-\lambda)^H$. By comparing the values in the third and fourth columns of Table 16.5, we can see that this approximation is fairly close with a λ of 0.02. For smaller values of λ, the approximation would be even better. The success chart, discussed in Sec. 10.10, was developed from these arguments.

The theorem of conditional probabilities used in the manner illustrated in Table 16.5 is correct in principle for an artificial experiment in which lives can be only integers. However, the concept of a constant failure rate applied to actual lives presumably means a constant instantaneous failure rate; for such a rate, the negative exponential formula for probability of survival is correct in principle.

Although the expression $e^{-H\lambda}$ is more precisely described as negative exponential, the literature of life testing and reliability usually refers to it simply as "exponential." With one exception (the discussion of a graph reproduced from a source that uses "negative exponential"), the word *exponential* is the one used throughout the remainder of this chapter.

16.6.1 The Relationship between the Reliability Function and the Probability Density Function in the Case of a Constant Failure Rate

Consider any desired unit of time t (which might be, say, 10, 500, 1,000 h, and so on). Assume that there is a constant probability of 0.9 that a unit entering a time interval of length t will survive to the end of the interval. The probability of failure during the interval will, of course, be $1-0.9=0.1$. The second column of Table 16.6 gives the respective probabilities that a unit starting at time zero will survive for 1, 2, 3, 4, 5, and 6 time intervals. Of course, each figure in this column is obtained by multiplying the preceding figure by 0.9. A mathematical function (such as the exponential) that gives probability of survival is called a *reliability* function.

Each figure for probability density in the third column of Table 16.6 might have been computed in either of two ways. The probability of failure during each time interval (0, 1) might have been multiplied by the probability that the item would have survived until the start of the time interval. Or the probability of survival to the end of the time interval could have been subtracted from the probability of survival to the beginning of the interval. These two mathematical operations obviously must give identical results.

TABLE 16.6 ILLUSTRATION OF THE DISTINCTION BETWEEN A RELIABILITY FUNCTION AND A PROBABILITY DENSITY FUNCTION
The constant probability that an item will survive a time interval is 0.9

Number of time intervals from time zero	Probability of survival (reliability)	Probability of item failing during time interval (probability)
0	1.000000	
		0.100000
1	0.900000	
		0.090000
2	0.810000	
		0.081000
3	0.729000	
		0.072900
4	0.656100	
		0.065610
5	0.590490	
		0.059049
6	0.531441	

The reader will note that the probability densities in column 3 of Table 16.6 are proportional to the reliabilities in column 2. The use of a constant failure rate causes this relationship. In general, if the reliability function is exponential, the probability density function must also be exponential. However, it is important that the student of reliability understand that the reliability curve (survivor curve) and the probability density curve will not have the same shape except in the special case where probability of failure is independent of age (that is, where there is a constant failure rate).

16.6.2 Some Aspects of the Exponential Reliability Function

The numbers in the fourth column of Table 16.5 were read or interpolated from the column headed $c = 0$ in Table *G*, App. 3. This column for the probability of 0 occurrences in a family of Poisson distributions gives the value of e^{-np}. (See the original discussion of the Poisson law in Chap. 5.) Because this one column of Table *G* gives values of *e* with various negative exponents, it happens to describe the shape of all exponential reliability functions regardless of the value of λ. That is, we can observe from this column of Table *G* that, with a constant failure rate, approximately 60.7% of the items in a group will survive to half the average life of the group, approximately 36.8% will survive to the average life, approximately 13.5% will survive to twice the average life, and so on.

The figures in the fifth column of Table 16.5 give the actual fractions surviving after various numbers of trials in our experiment in which we made 120 series of trials with a known constant probability of failure of 0.02. All such observed values naturally are subject to sampling fluctuations and therefore will not agree exactly with the probabilities of survival computed from a known constant failure rate. (The observed values in the fifth column of Table 16.5 are all well within 3-sigma limits based on the known rate, using n as 120 in all cases and p in each case as the computed probability of survival shown in the same line of the third column of the table.)

Actual life testing differs from our experiment in that we never can *know* that we have a constant failure rate. If it is proposed to base engineering designs and acceptance sampling plans on the assumption that certain failure rates are constant, and if failure data are available, it is appropriate to apply various statistical tests to judge whether the data are consistent with the hypothesis of a constant failure rate. A number of such tests were summarized in a paper by Benjamin Epstein that appeared (in two parts) in the February and May 1960 issues of *Technometrics.* Dr. Epstein's useful paper was reprinted as a U.S. Department of Defense Technical Report.†

16.6.3 Some Limitations on the Use of the Reciprocal Relationship between Failure Rate and Mean Life

When the useful formula $\theta = 1/\lambda$ was introduced earlier in this chapter for the special cases of the constant failure rate, it was mentioned that the validity of this formula depended on the use of a time unit that is a relatively small fraction of the average life. We are now in a position to explain that statement.

Consider our experiment of 120 sets of trials to failure that was summarized in Tables 16.2 and 16.3. The reciprocal relationship between θ and λ is exact for the conditions of this experiment. Our distribution of failures was not a continuous one; we could have a failure on trial 22 or trial 23 but not on, say, trial 22.57. Our probability of failure applied to a trial that was an identifiable single operation rather than to an arbitrarily chosen interval of time throughout which there was a continuing chance for failure.

Now imagine that we are making life tests of a type of item for which the probability of failure *sometime* within any hour is 0.02. An experiment testing 120 such items would result in the same type of variability observed in Tables 16.2 and 16.3. However, if the moment of failure should be recorded to the nearest 0.01 h, three failures during the 23d hour would not be recorded as 23, 23, and 23 but rather as numbers such as 22.15, 22.57, and 22.89. It is intuitively evident that a continuous failure rate that would cause the probability of failure during an hour to be 0.02 would result in a slightly shorter mean life than would occur if the probability were 0.02 that a failure would occur only at the *end* of an hour.

†"Tests for the Validity of the Assumption That the Underlying Distribution of Life Is Exponential," Superintendent of Documents, Government Printing Office, Washington, D.C. 20402, 1960.

The true mean life is approximately 99% of the life given by the reciprocal formula in the case of a continuous failure rate that results in a probability of failure of 0.02 during a chosen time period. Therefore, the reciprocal formula usually may be accepted as good enough for practical purposes in the common case where the chosen time period is short enough for the failure rate per unit time interval to be 2% or less.[†]

The most important practical limitation of the reciprocal formula that always should be kept in mind is the assumption that there really is a λ, a failure rate that is independent of the age of an item. Where the assumption of a constant failure rate is used for purposes of acceptance sampling and the bathtub curve (Fig. 16.1) is believed to describe the expected failure rate pattern, two further limitations of the reciprocal formula should be kept in mind:

1. Time zero from which life is being measured is assumed to start after the infant mortality period has ended.
2. The wearing-out period is disregarded. It is obvious that, if the wearing-out period is considered, the mean life will be less than the figure given by the reciprocal formula.

Some of the acceptance sampling plans in H 108 (the U.S. Department of Defense handbook mentioned earlier in this chapter) are based on what is called a "two-parameter" exponential. These plans assume a zero failure rate for some initial period of time, followed by a constant failure rate thereafter. (The second parameter is the duration of the failure-free period.) Because the reciprocal formula $\theta = 1/\lambda$ assumes that the failure rate starts with time zero, it is necessary to add the assumed length of the failure-free period to the θ computed from the formula in order to find the mean life for this two-parameter exponential.

16.7 PRINCIPAL U.S. GOVERNMENT DOCUMENTS THAT TREAT LIFE TESTING UNDER THE ASSUMPTION OF A CONSTANT FAILURE RATE

Three principal documents have been issued by the U.S. Department of Defense dealing with sampling plans and procedures for life testing in those cases where the constant failure rate assumption is appropriate. They are H 108, which has been discussed in part in earlier articles of this chapter, MIL-STD-690B, and MIL-STD-781C.

Handbook H 108, the earliest of the three, contains three sets of plans. The first is a set of life test plans that terminate on the occurrence of a preassigned number

[†]If h represents the chosen unit time period in hours and f is the probability of failure during period h, the exponential reliability function yields the following formula for mean life in hours:

$$\theta = \frac{h \log e}{\log(1 - f)}$$

of failures. In the second set, termination occurs at a preassigned time H, or when r (the rejection number) failures have been detected, whichever occurs first. The third set contains plans for sequential life testing, based on Wald's sequential probability ratio test, in which both the acceptance/rejection numbers and total unit time on test vary. All three sets stipulate procedures for life testing either with or without replacement of failed units. In addition, the choice of individual plans may be made based on the selection of a range of stipulated Producer's Risk points (θ_0, α), and Consumer's Risk points (θ_1, β).

Handbook H 108 and the Military Standard 690 and 781, along with many other standards and specifications, resulted directly from the 1957 AGREE report. Whereas H 108, first published in 1960, emphasized the use of standard tables of life test plans, the military standards emphasize the procedural aspects of analysis of design and production results. MIL-STD-690B, "Failure Rate Sampling Plans and Procedures,"[†] establishes detailed procedures for demonstration (qualification) testing, production run tests, failure rate test records, and quality history requirements, and also provides guidance to specification writers.

Following several military directives in the early 1960s, it was determined that procedures were needed whereby manufacturers could qualify their product against established standards. Such standard data on basic electronic components were necessary in the development of military hardware systems design as well as in the prediction of reliability of these systems. MIL-STD-690B provides four inspection procedures as follows:

I. Qualification testing at an initial failure rate
II. Extension of qualification to a lower failure rate level
III. Maintenance of failure rate level qualification
IV. Lot conformance failure rate inspection

MIL-STD-781C,[‡] while equally applicable to component parts testing, is particularly useful for testing assemblies and systems. A number of the test plans included in the standard employ the principles of sequential sampling based on Wald's sequential probability ratio test. The standard emphasizes requirements for preproduction reliability qualification tests, production reliability acceptance tests employing both sampling and all-units testing, and combined environmental test conditions.

Some proposed plans specify a fixed-length test for acceptance but use a sequential-type rejection line. The sloping rejection line of the sequential procedure is intended to reject infant mortality, thus reducing the necessity for lengthy burn-in time.

[†]Failure Rate Sampling Plans and Procedures," MIL-STD-690, was first published in 1963. Revision A followed in 1965; and Revision B, which is discussed in this book, was published in April 1968. A discussion of the history of its development along with the changes that have taken place since first publication may be found in S. Grubman, C. A. Martin, and W. R. Pabst, Jr., "MIL-STD-690B, Failure Rate Sampling Plans and Procedures." *Journal of Quality Technology,* vol. 1, no. 3, pp. 205–216, July 1969.

[‡]MIL-STD-781C, "Reliability Design Qualification and Production Acceptance Tests. Exponential Distribution," U.S. Department of Defense, Washington, D.C., 21 October 1977. Revision C is substantially a complete revision, including change in title, of MIL-STD-781B, dated 15 November 1967.

16.8 THE BROAD GENERAL USEFULNESS OF SAMPLING AND TESTING STANDARDS

During the last several chapters, many documents and standards published by the U.S. government have been discussed in varying degrees of depth. In part the discussion has been directed at particular statistical methodologies, and in part it has been directed at familiarizing the reader with the wealth of developed material available. In a book such as this, which is devoted to the statistical aspects of quality control, it is natural that methodology should take precedence. Nevertheless, the reader should not lose sight of a point that we have emphasized at various places in our exposition of acceptance sampling, namely, that the improvement of the quality of product submitted for inspection often is the most important consequence of the adoption and use of an acceptance sampling system.

The authors of a review of the previous MIL-STD-781B, published in the *Journal of Quality Technology,* in 1969, made some significant observations about the impact of such documents.[†] They pointed out that these standards make possible a better framework of communication among designers and manufacturers. A primary consequence of the use of the standards ought to be an improvement in quality. The plans contained in such standards should not be viewed merely as the source for decisions on acceptance or rejection of product. Any discussion that emphasizes only the statistical aspects of the standards tends to neglect the more important effects that generally occur because of their use.

PROBLEMS

16.1. In one of the plans quoted from H 108 (page 610), 10 items were to be tested for 500 h with replacement and with an acceptance number of 1. Using values of λ that seem to you to be appropriate, make a calculation similar to Table 16.1 to find enough points to define the OC curve of this plan. Plot your OC curve showing probability of acceptance as a function of mean life.

16.2. The acceptance sampling plan of Prob. 16.1 was stipulated for an approximate value of 0.10 for the Producer's Risk of rejection of a lot having a mean life of 10,000 h and for an approximate value of 0.10 for the Consumer's Risk of acceptance of a lot having a mean life of 1,000 h. Use Table *G* to compute the respective values of these two risks.
Answer: Producer's, 0.09; Consumer's, 0.04.

16.3. In one of the plans quoted from H 108 (page 610), 63 items were to be tested for 500 h with replacement and with an acceptance number of 5. Using values of λ that seem to you to be appropriate, make a calculation similar to Table 16.1 to find enough points to define the OC curve of this plan. Plot your OC curve showing probability of acceptance as a function of mean life.

16.4. The acceptance sampling plan of Prob. 16.3 was stipulated for an approximate value of 0.10 for the Producer's Risk of rejection of a lot having a mean life of

[†]R. D. Neathammer, W. R. Pabst, Jr., and C. G. Wigginton, "MIL-STD-781B, Reliability Tests: Exponential Distribution," Reviews of Standards and Specifications department, *Journal of Quality Technology,* vol. 1, no. 1 p. 59, January 1969.

10,000 h and for an approximate value of 0.10 for the Consumer's Risk of acceptance of a lot having a mean life of 3,333 h. Use Table *G* to compute the respective values of these two risks.

Answer: Producer's, 0.10; Consumer's, 0.092.

16.5. The following table indicates the time a lot of electronic parts operated until failure. The times are presented in the order in which the trials were measured and include 30 measurements made with a constant failure rate of 0.0005 per h.

(*a*) Rearrange the data values in order of increasing magnitude in the manner illustrated in Table 16.3.

(*b*) Compute the respective estimates of mean life in hours that would have been appropriate after 1,000, 3,000, 5,000, and 7,713 hours of testing.

Set of observations	Time until failure, h				
	1	2	3	4	5
1	2,737	1,281	1,855	1,472	2,638
2	2,147	4,522	428	3,727	7,713
3	961	617	6,715	714	3,303
4	714	4,025	5,618	1,474	1,567
5	1,285	3,132	501	1,661	2,148
6	456	3,913	1,846	2,114	2,713

16.6. For the rearranged data of Prob. 16.5, compute to the nearest two decimal places the fractions surviving after 500, 1,000, 1,500, 2,000, 3,000, 4,000, 5,000, and 7,000 h, respectively. Compare these observed values with the appropriate values for probability of survival that you obtain from the column in Table *G* headed $c = 0$.

16.7. The following lengths of lives, measured in numbers of successive trials to failure, were obtained when 60 sets of observations were made with a constant probability of failure of 0.01.

Sets of observations	Lengths of lives				
1–5	45	431	16	388	34
6–10	59	27	107	233	88
11–15	276	84	7	40	13
16–20	1	106	121	24	223
21–25	277	156	261	35	81
26–30	2	24	42	56	131
31–35	31	47	6	179	2
36–40	36	128	20	78	112
41–45	301	22	4	121	20
46–50	44	7	163	245	74
51–55	51	23	107	3	10
56–60	31	62	142	296	101

Rearrange the foregoing 60 numbers in order of increasing magnitude in the manner illustrated by the rearrangement of the numbers of Table 16.2 into the numbers of Table 16.3.

16.8. Your rearranged table in Prob. 16.7 may be viewed as representing a distribution of lives in hours that might have been found if simultaneous tests to failure had been made of a random sample of 60 items from a universe that had a constant failure rate of 0.01 per hour. Compute the respective estimates of mean life in hours that would have been appropriate after the following numbers of hours of test:

(*a*) 30 (*b*) 60 (*c*) 80 (*d*) 100 (*e*) 431

Answer: (*a*) 82.8; (*b*) 81.4; (*c*) 90.5; (*d*) 95.9; (*e*) 97.6.

16.9. Rearrange the first 60 numbers of Table 16.2 in order of increasing magnitude. Assume that your rearranged table represents the lives in hours found when simultaneous tests to failure were made of a sample of 60 items out of a large lot. Compute the respective estimates of mean life in hours that would have been appropriate after the following numbers of hours of test:

(*a*) 4 (*b*) 5 (*c*) 10 (*d*) 15 (*e*) 20 (*f*) 165

16.10. An acceptance sampling plan for life testing requires that a sample of 40 items be tested with replacement for 500 h. If no more than 5 of these items fail, the lot is accepted. Otherwise it is rejected.

(*a*) Assuming that the failure rate is constant, compute the mean life for which the Producer's Risk of lot rejection is 0.05.

(*b*) Compute the mean life for which the Consumer's Risk of lot acceptance is 0.10.

16.11. Assume that a failure is designated as the drawing of a chip marked 0 to 9, inclusive, from Shewhart's normal bowl. Examine the 400 drawings given in Table 3.7 to determine values of the number of trials to failure. (Eight such values can be obtained from the first 384 drawings.) Assume that the observed number of trials to failure correspond to the lives in hours of a sample of eight electronic components subject to life tests.

(*a*) What is λ (see Table 3.6)?

(*b*) What is θ?

(*c*) What estimate of mean life would be obtained from the test to failure of the sample of eight components?

(*d*) What estimate of mean life would have been obtained after the first 10 h of test, assuming that the eight tests were carried out concurrently?

(*e*) After the first 20 h?

Answer: (*a*) $\frac{20}{998}$; (*b*) 49.9 h; (*c*) 48 h; (*d*) 38.5 h; (*e*) 31.8 h.

16.12. Assume that a failure is designated as the occurrence of 00 or 99 in one of the pairs of random numbers shown in Table *X*, App. 3. Start counting in the upper left-hand corner of the table, and count downward in the left-hand column of pairs of numbers. From the bottom of the left-hand column, go to the top of the second column from the left, and so on. Record the numbers of successive trials to failure in tabular form in the manner illustrated in Table 16.2. (Your first 4 numbers should be 4, 11, 17, and 100.) Continue until 20 values have been obtained. Rearrange these values in order of increasing magnitude in the manner illustrated in Table 16.3. (Your first 4 numbers should be 3, 4, 6, and 11.)

16.13. Assume that the values in your second table of Prob. 16.12 represent the hours to failure obtained when a sample of 20 electronic components is being tested until each component fails. Compute the respective estimates of mean life in hours that would have been appropriate after the following numbers of hours of test:

(*a*) 10 (*b*) 20 (*c*) 30 (*d*) 184

Answer: (*a*) 61.0; (*b*) 55.5; (*c*) 50.8; (*d*) 57.4.

16.14. In the qualification test (1) mentioned on page 616, 418,000 item hours of test are required with an acceptance number of 3.

(*a*) Approximately what mean life in hours corresponds to a Producer's Risk of 0.05?

(*b*) To a Consumer's Risk of 0.10?

Answer: (*a*) 307,000; (*b*) 62,600.

16.15. Plot the OC curve of the qualification test (1) mentioned in Prob. 16.14. Show the probability of qualification as a function of θ.

16.16. In the acceptance test (2) for individual lots mentioned on page 616, 11,000 item hours of test are required with an acceptance number of 1.

(*a*) Approximately what mean life in hours corresponds to a Producer's Risk of 0.05?

(*b*) To a Consumer's Risk of 0.10?

Answer: (*a*) 30,942; (*b*) 2,828.

16.17. Plot the OC curve of the lot acceptance test (2) mentioned in Prob. 16.16. Show the probability of acceptance as a function of θ.

16.18. In the test for continued qualification (3) applied to a series of 10 lots (page 616), 110,000 item hours of test are required with an acceptance number of 2.

(*a*) Approximately what mean life in hours corresponds to a Producer's Risk of 0.05?

(*b*) To a Consumer's Risk of 0.10?

Answer: (*a*) 134,500; (*b*) 20,600.

16.19. Plot the OC curve of the test for continued qualification mentioned in Prob. 16.18. Show the probability of acceptance as a function of θ.

16.20. On pages 616 to 617, there was given an illustration of a proposed type of acceptance procedure to be used in life testing where the tolerable failure rates are extremely low. With reference to the qualification criteria illustrated in (1) and applied again in (4), it has been suggested that the principle of the indifference quality should be used. In other words, the Consumer's Risk (and Producer's Risk) associated with some stipulated failure rate should be 0.50.

The specific numerical example of a qualification test given on page 616 called for a test of 418,000 item hours with an acceptance number of 3. This gives a probability of qualification of 0.40 of product having a mean life of 100,000 h.

(*a*) If the probability of qualification for such product is to be 0.50 and the acceptance number is 3, how many item hours should be tested to qualify?

Answer: 367,000.

(*b*) With your revised qualification criteria, approximately what mean life in hours corresponds to a Producer's Risk of 0.05?

Answer: 270,000.

(*c*) With your revised qualification criteria, approximately what mean life in hours corresponds to a Consumer's Risk of 0.10?
Answer: 55,000.

16.21. In connection with the matters discussed in Prob. 16.20, it has also been suggested that the qualification criteria are too lenient from the point of view of the consumer. Answer questions (*a*), (*b*), and (*c*), changing the probability of qualification to 0.10 for product having a mean life of 100,000 h.
Answer: (*a*) 668,000 item hours; (*b*) 490,000 h; (*c*) 100,000 h.

16.22. For your rearranged data of Prob. 16.7, compute to the nearest two decimal places the fractions surviving after 10, 20, 40, 60, 80, 100, 150, 200, 300, and 400 h, respectively. Compare these observed values from this sample of 60 items with the appropriate values for probability of survival that you obtain from the column in Table *G* headed $c = 0$.

16.23. A certain type of electronic component has a uniform failure rate of 0.00001 per hour. What is its reliability for a specified period of service of 10,000 h? Of 2,000 h?
Answer: 0.905; 0.980.

16.24. Although an acceptance procedure is being used to ensure the failure rate of not more than 0.00001 mentioned in Prob. 16.23, it happens that a producer qualifies who has a λ of 0.0000125. What now will be the respective values of reliability for services of 10,000 and 2,000 h?

16.25. Given a θ of 5,000 h and a constant failure rate, what is the reliability associated with a specified service period of 400 h?

16.26. Two hundred solid state electronic devices were tested to determine the failure rate of these units. Testing was conducted for 1,000 h with four units failing after 425, 575, 650, 920 h, respectively.

(*a*) Assuming a constant failure rate, calculate the total unit hours on test, $V(t)$, where failed units are immediately replaced.

(*b*) Where failed units are not replaced.

(*c*) What is the probability that one of these units will survive for a required 500 h of operation?

Answer: (*a*) 200,000 h; (*b*) 198,570 h; (*c*) 0.990.

16.27. After an initial burn-in period of 50 h, one hundred electrical units are placed on test for 600 h. Three units failed at 230, 300, and 420 h, respectively.

(*a*) Estimate the mean life θ for this unit, where failed units are not replaced, assuming a constant failure rate.

(*b*) What is the probability that one of these units will survive for a required 200, 300, and 400 h of operation?

16.28. An item has a mean life θ of 15,000 h. Assuming a uniform failure rate, what is the probability that one of these units will survive for a required 1,000 h of operation?

16.29. An item is required to have a failure rate no greater than 0.10% per 1,000 h of operation.

(*a*) Assuming a constant failure rate, what is the probability that one of these units will survive for a required 2,000 h of service.

(*b*) Determine the maximum acceptable failure rate where the probability of survival for a required 2,000 h of operation is 0.999.

Answer: (*a*) 0.998; (*b*) 0.05%/1,000 h.

PART FOUR

SOME RELATED TOPICS

17

SOME ECONOMIC ASPECTS OF QUALITY DECISIONS

Finding the correct balance between cost of quality and value of quality is not so easy, since the facts are widely scattered throughout the various company departments, the distribution chain, the customers, the vendors, and still other locations. . . . The balance to be struck (between cost and value) is not as to quality generally; it applies to each quality characteristic.

—J. M. Juran and F. M. Gryna, Jr.†

17.1 PROBLEMS OF BUSINESS ALTERNATIVES ARE PROBLEMS IN ECONOMY

Decisions of many sorts are always being called for in the management of any productive enterprise. All decisions are between alternatives—either express or implied. In business enterprises operated for profit, the real basis of a choice between alternatives is the prospective effect of each alternative on the costs and revenues of the business in both the short and the long term. Where technical considerations are involved in the alternatives, a study comparing specific money estimates of the differences between the alternatives as well as other estimated differences not readily expressible in money terms is called an *engineering economy study.*

Engineering economy studies dealing with quality matters are often more difficult than studies dealing with such matters as proposed investments in business assets, primarily because of the difficulty of expressing in money terms the probable effect of particular quality decisions. In some enterprises, this difficulty may

†J. M. Juran and F. M. Gryna, Jr., *Quality Planning and Analysis,* p. 38, McGraw-Hill Book Company, New York, 1970.

be due in part to the fact that the system of accounts does not identify certain types of costs, such as costs of spoilage and rework, in any satisfactory way. Nevertheless, to a large extent the difficulty is an inherent one in quality decisions; certain elements in such decisions are extremely hard to measure in money terms.

Perhaps for this reason, many quality decisions seem to be made on an intuitive basis without any conscious attempt to evaluate the elements of each decision in the only units that can make all the elements commensurable, namely, in money units. Intuition may be a good enough guide in many circumstances, but it may occasionally prove to be a very costly one. Moreover, it is common to find that different individuals in an organization will reach opposite conclusions on quality matters whenever the basis for such decisions depends entirely on intuition.

The brief discussion in the remainder of this chapter is intended to suggest the controlling elements in certain types of quality decisions. A recognition of these elements should be helpful in improving the rational basis for quality decisions even in the numerous cases where a money evaluation of all the elements in a particular decision turns out to be impracticable.

17.2 SOME BASIC CONCEPTS IN ENGINEERING ECONOMY

The authors of this book, in collaboration with W. G. Ireson, have suggested the following conceptual framework for economy studies.[†]

1. Decisions are among alternatives; it is desirable that alternatives be clearly defined and that the merits of all appropriate alternatives be evaluated.
2. Decisions should be based on the expected consequences of the various alternatives. All such consequences will occur in the future.
3. Before establishing procedures for project formulation and project evaluation, it is essential to decide whose viewpoint is to be adopted.
4. In comparing alternatives, it is desirable to make consequences commensurate with one another insofar as practicable. That is, consequences should be expressed in numbers and the same units should apply to all the numbers. In economic decisions, money units are the only units that meet the foregoing specification. Money units at different times are not commensurable without calculations that somehow reflect the time value of money; in some cases they are not commensurable without adjustment for prospective changes in the purchasing power of the monetary unit.
5. Only the differences among alternatives are relevant in their comparison.
6. Insofar as practicable, separable decisions should be made separately.
7. It is desirable to have a criterion for decision making, or possibly several criteria. The primary criterion to be applied in a choice among alternative proposed investments in physical assets should be selected with the objective of making the best use of limited resources.

[†]E. L. Grant, W. G. Ireson, and R. S. Leavenworth, *Principles of Engineering Economy,* 8th ed., chap. 1, John Wiley & Sons, Inc., New York, 1990.

8. Even the most careful estimates of the monetary consequences of choosing different alternatives almost certainly will turn out to be incorrect. It often is helpful to a decision maker to employ secondary criteria that reflect in some way the lack of certainty associated with all estimates of the future.
9. Decisions among investment alternatives should give weight to any expected differences in consequences that have not been reduced to money terms as well as to the consequences that have been expressed in terms of money.
10. Often there are side effects that tend to be disregarded when individual decisions are made. To consider such side effects adequately, it may be necessary to examine the interrelationships among a number of decisions before any of the individual decisions can be made.

Even though the remainder of this chapter does not refer specifically to the foregoing concepts, the discussion and the examples illustrate the application of many of them. For instance, Example 17.4 illustrates the desirability of defining the appropriate alternatives and basing a decision on their expected consequences expressed in terms of money; Example 17.5 illustrates the need to decide whose viewpoint is to be adopted and the need to recognize that only the prospective differences among alternatives are relevant in their comparison.

17.2.1 A Comment on the Semantics of Quality

In using the word *quality,* it is helpful to recognize the distinction between *quality of design* and *quality of conformance.* In the sense that a Lincoln is considered to be a better quality automobile than a Ford, or a Cadillac a better quality car than a Chevrolet, the word is used in the sense of quality of design. The designers of the higher-priced automobiles have included certain more costly features aimed to secure greater comfort, better appearance, better performance, and the like.

When we discussed specifications and tolerances in Chap. 9, Sec. 9.3, we noted that, with respect to specific quality characteristics, the designer's specifications do not always seem to be consistent with the design objectives. Some instances where this occurred were cited in Examples 9.3 to 9.5.

In the sense used in this chapter, quality of conformance relates to whether or not the quality characteristics of a product or service correspond to those really needed to secure the results intended by the designer. Used in this sense, margins of safety written into design specifications are often aimed chiefly at securing quality of conformance. Where such margins of safety are used with this objective, design specifications and acceptance specifications are properly viewed as interrelated matters.

17.3 THREE GENERAL CLASSES OF CONSEQUENCES THAT SHOULD BE RECOGNIZED IN MAKING CERTAIN QUALITY DECISIONS

Economy studies involving quality of conformance may relate to the amount and type of inspection, to production methods and objectives, and to margins of safety used in design specifications. In studies to guide decisions on these matters, it is helpful to divide the expected economic consequences of the decisions into three general classes. These consequences may be somewhat loosely referred to as (1) production costs, (2) acceptance costs, and (3) unsatisfactory-product costs.

In this usage, the expression *production costs* is intended to refer to those costs involved in the production of the article under consideration. Different design specifications may require different materials, different labor skills, different amounts of labor time, and different machines. For example, increased strength requirements for a part may change the material to be used; closer required tolerances on dimensions may call for the use of newer or different machines. This general class of costs properly includes *spoilage costs,* that is, the production expenses on all product discarded as not meeting specifications minus any receipts from the disposal of this discarded product. It also includes *rework costs* necessary to make product acceptable and screening costs, if any, on rejected lots.

The *acceptance costs* include not only testing and inspection costs but also the costs of administering the acceptance program.

The expression *unsatisfactory-product costs* is intended to refer to those costs resulting from the acceptance of product that turns out to be unsatisfactory for the purpose intended. In this sense the word *cost* should be interpreted as including a reduction in revenue as well as an increase in expense. It should be recognized that some or all of the product that is technically defective in the sense of failing to meet design specifications is not necessarily unsatisfactory for the purpose intended whenever the common practice is followed of including a margin of safety in the specifications. This distinction between product that is really unsatisfactory and product that is satisfactory even though nonconforming to specifications is an important one in any discussion of the economics of quality decisions.

Of these three classes of costs affected by quality decisions, unsatisfactory-product costs are inherently the most difficult to evaluate. Doubtless the greatest difficulty occurs in the consumers' goods industries, where the product goes to a great many different customers who make no formal acceptance tests. It is hard to predict the consequences to the manufacturer of consumers' goods when some stated percentage of its product fails to give satisfactory service to its purchasers, and it is even more difficult to place a money value on these consequences. Where the consumers' product carries a guarantee, past customer service costs can be used as a guide to judgment, and changes in these costs can be carefully watched and related to changes in design specifications and in inspection and acceptance procedures.

The most favorable circumstances exist for securing reliable information on unsatisfactory-product costs when all the product goes to one user. This user may be another department in the producer's organization, or it may be a single purchaser who is responsible for the design specification.

17.3.1 ASQC Categories of Quality Costs

The American Society for Quality Control breaks down quality costs into two broad areas: those costs related to conformance to specification and those related to nonconformance.[†] The *cost of conformance* includes costs associated with the *prevention* of nonconformance, which, in turn, includes designing, implementing, maintaining, conducting training, and auditing the quality system. Second, it includes inspection costs associated with *appraisal,* including measuring, evaluating, and auditing products, services, components, materials, and processes. The *cost of nonconformance* includes costs associated with the generation of *internal failures*—which includes all rework and losses due to nonconforming products, components, materials, or services—and costs associated with *external failures*—which includes all costs due to failure of a product in the customer's hands or service errors and deficiencies occurring in the field.

The argument is made that costs associated with conformance, especially prevention costs, usually start out low in the initial stages of implementing quality cost analysis and increase over time, at least in the short run. Costs associated with nonconformance, especially external failure costs, usually start out high and decline and hopefully disappear over time. The total of these costs normally will decline as operation after operation is brought under control and suitably within specifications. One criticism of the ASQC quality cost approach is that it pays little attention to alternative product/service design and alternative methods of production/operation aspects of the economics of quality.

Several companies have developed computer software packages to assist in preparing so-called cost of quality studies. One package with which one of the coauthors is familiar contains a very helpful checklist of some 93 cost elements broken down among prevention, appraisal, internal failure, and external failure categories.[‡] Costs can be recorded through time—say, on a monthly basis—in each of the four categories with reports and graphs plotted through time.

17.3.2 Reasons for Margins of Safety in Design Specifications

In specifying a quality characteristic such as a dimension, strength, resistance, and the like, there will often be a twilight zone of uncertainty within which the product will be satisfactory under most conditions of use but not under all conceivable conditions. Design specifications may be drawn in a way that classifies as defective all articles falling in this twilight zone. Moreover, designers often seem to believe that they cannot secure the quality characteristics that are really needed without requiring a margin of safety. This common practice has not been changed by the advent of statistical quality control techniques in industry. Some reasons for continuing to require a margin of safety are clearly stated by Wyatt H. Lewis, as follows.[§]

[†]See, for example, "Principles of Quality Costs: Principles, Implementation, and Use," 2d ed., ASQC Quality Cost Committee, J. Campanella, Ed., ASQC Quality Press, Milwaukee, Wis., 1990.

[‡]"The Cost of Quality Utility," The Harrington Group, Orlando, Fla., 1992.

[§]W. H. Lewis, "Discussion of E. L. Grant, The Economic Relationship between Design and Acceptance Specifications," *Special Technical Publication* No. 103, Symposium on Application of Statistics, American Society for Testing and Materials, Philadelphia, Pa., 1950.

1. Acceptance sampling plans, although they may guarantee a long range average quality level of no worse than say 2% defective, may accept occasional lots with as much as 8% defective. Such a high percent defective may cause dislocation of manufacturing operations at considerable cost due to lost time, excessive rework, special handling to make up for delays, etc.
2. Even a control chart using actual measurements may not detect slight shifts in the $\bar{X}$ or σ values for a matter of several samples and you might be in trouble before you realized it.
3. The vendor may have a process operating in control but there are times when causes arise to disturb the state of control. In such cases the margin of safety is very handy for acceptance of material on the basis of deviation from the specification in order to keep assembly lines going and to avoid sending operators home and cancelling orders. The cost of the latter cannot be ignored.
4. Laboratory instruments and inspection gages, although given periodic checks, sometimes drift and such drift is not detected until the next periodic check.
5. There is also the human element to consider: lack of experience, acceptance of borderline cases, etc.

In many industrial plants, it is a fairly common experience for parts first to be rejected by the inspection department because they fail to conform to specifications and later to be accepted by a plant salvage committee or material review board primarily on the grounds that their lack of conformity falls within the margin of safety included in the specifications.

17.3.3 Some Economic Aspects of Decisions on the Amount and Type of Inspection

Sometimes it is economical to do no inspection at all, sometimes 100% inspection is the most economical, and sometimes sampling inspection of one type or another is better than either. *The objective should be to select that amount and type of inspection that will minimize the sum of the production costs, acceptance costs, and unsatisfactory-product costs influenced by the decision regarding inspection.* Once this viewpoint has been adopted, certain conditions are evident that are favorable, respectively, to no inspection, to 100% inspection, and to sampling inspection.

Where submitted product is consistently satisfactory for the purpose intended, it is likely to be most economical to have no inspection whatever. In this case, there are no unsatisfactory-product costs to be reduced by inspection. Neither do there appear to be production costs such as spoilage and rework to be reduced through diagnosis by control charts or through the pressure for process improvement exerted when product is rejected. Sometimes, however, as in the case of overfill of containers, concealed opportunities may exist for reducing production costs; such opportunities might be disclosed by variables sampling inspection using control charts.

Low unsatisfactory-product costs per unit of such product may also make it economical to do no inspection whatever. For example, where unsatisfactory product is readily discovered and eliminated in a subsequent production operation, it may be cheaper to tolerate a moderate percentage of such product than to eliminate it by inspection.

Where submitted product is consistent in quality but nearly always contains a substantial percentage of unsatisfactory product, 100% inspection may be the most economical alternative. Here the choice is likely to be between 100% inspection and no inspection for acceptance purposes; with a statistically controlled product, sampling inspection cannot be expected to separate the relatively good lots from the relatively bad ones. The higher the percentage of unsatisfactory product submitted and the higher the unsatisfactory-product cost per unit of such product, the more favorable the conditions for 100% inspection as compared with no inspection. The higher the unit cost of inspection and the less the effectiveness of 100% inspection in eliminating unsatisfactory product, the more favorable the conditions for no inspection.

W. Edwards Deming, in developing what has been called his k_p rule, concluded that, when a process is operating in statistical control at a known mean μ_p, the ratio of the cost of inspecting an item c_1 divided by the cost of passing a nonconforming item (unsatisfactory-product cost) c_2 determines whether 100% inspection or no inspection should be performed. Thus we calculate $k_p = c_1/c_2$. The criterion is to perform 100% inspection if $\mu_p > k_p$. If $\mu_p \leq k_p$, no inspection is economical, and thus items are passed without inspection. This conclusion was arrived at independently by D. Guthrie and M. V. Johns in connection with their development of economic models for sampling inspection.[†] Their work shows, however, that sampling inspection is economically justified when μ_p in unknown and must be surmised.

In making economy studies regarding the amount and type of inspection, it should be recognized that sampling inspection schemes may possibly reduce unsatisfactory-product costs in two ways. One way is by the rejection or rectification of the relatively bad lots of product, thereby making the proportion of unsatisfactory product approved less than the proportion submitted. The other way is by reducing the proportion of unsatisfactory product submitted; sampling inspection may improve product quality through diagnosis of causes of quality troubles and through the exertion of effective pressure for process improvement. This improvement of product quality may also reduce production costs, particularly costs of spoilage and rework.

If this possible contribution of sampling inspection to the improvement of product quality is neglected, the following general statement may be made: The economic field for sampling inspection is where submitted product is usually good enough for no inspection to be more economical than 100% inspection and where submitted product is occasionally bad enough for 100% inspection to be more economical than no inspection.

A set of generalizations about economy studies regarding acceptance inspection that is somewhat more complete and realistic is as follows:

[†]D. Guthrie and M. V. Johns, "Bayes' Acceptance Sampling Procedures for Large Lots," *Annals of Mathematical Statistics,* vol. 30, pp. 896–925, 1959.

1. No economic comparison of alternative plans for acceptance inspection is possible without making assumptions regarding the quality of the product submitted for acceptance. The conclusions of any economy study will depend on these assumptions.
2. In general, as the level of submitted quality is improved and as its consistency is improved, it becomes economical to use acceptance schemes involving less inspection. This is one of the reasons for keeping a record of quality history and for making periodic reviews of acceptance procedures in the light of that history.
3. In any proposal for a change in acceptance procedures, it is insufficient to consider merely the expected change in acceptance costs. Attention should also be given to the probable influence of the proposed change on production costs and on unsatisfactory-product costs.
4. The most important point favorable to certain acceptance sampling systems is their prospective contribution to the improvement of submitted quality in some cases through diagnosis of quality troubles and in others through effective pressure for quality improvement. This improvement of quality decreases production costs by reducing the cost of spoilage and rework. It also tends to decrease unsatisfactory-product costs. In the long run it permits a reduction in acceptance costs.

17.3.4 Will It Pay to Use a Control Chart for Variables?

The greatest opportunities for cost reduction from statistical quality control often arise out of applications of the Shewhart control chart for variables. These savings are sometimes spectacular. They come from many sources—from reduction in cost of spoilage and rework, from reduction in inspection cost, from better control over the quality of purchased items, from the use of more economical materials or methods because of their greater reliability under statistical control, from better decisions on proposed investments in plant and equipment.

On the other hand, each control chart for variables involves some costs. The measurements of the variable must be made and recorded. Clerical labor may be required for plotting the charts and computing the averages and limits. The time of people who have good technical ability is required for interpreting charts as a basis for action. Troubleshooting based on the evidence of the charts may sometimes be a costly matter.

Each set of $\bar{X}$ and R charts may be thought of as a gamble that the resulting savings will be greater than the cost of keeping the charts. In the introduction of the control chart for variables in any mass-production industry, experience indicates that from the overall viewpoint this gamble almost amounts to betting on a sure thing; there are certain to be *some* opportunities for substantial cost savings. However, there are bound to be many quality characteristics for which $\bar{X}$ and R charts will not pay their way.

Before applying $\bar{X}$ and R charts to any given quality characteristic, it is seldom possible to be *certain* that the resulting cost savings will more than pay for the charts. It is, however, often possible to eliminate many quality characteristics from consideration by observing that sufficient opportunities for cost savings do not seem to exist, and to observe that for other quality characteristics there seem to be excellent opportunities for savings. This calls for an examination of costs that might be reduced—such as spoilage, rework, and inspection costs—and for a consideration of the possibilities of using the control-chart information as a basis for changes in design, specifications, or production or service methods.

Sometimes the question of whether it is a good gamble to use $\bar{X}$ and R charts for a given quality characteristic cannot be answered without the evidence of the charts themselves. Fortunately this is not a serious obstacle, as the cost of maintaining charts for a short period usually is small. It is always possible to discontinue control charts whenever it is clear that they are no longer justified.

Example 17.1 deals with the question of whether prospective savings in spoilage and rework will justify the expense of a control chart. Example 17.2 illustrates possible economies in the control of product weight when a control chart is used. Example 17.3 deals with a decision as to whether it is likely to be worthwhile to use a control chart for this purpose.

Example 17.1
Analysis of spoilage reports

Facts of the case In the introduction of statistical quality control in one manufacturing plant, many p charts were initiated. In most cases, these charts made use of inspection records that already were maintained for 100% inspection by attributes. The p charts indicated the quality level of various parts and products and the presence or absence of statistical control. In this way, they suggested possible places for the use of $\bar{X}$ and R charts to diagnose the causes of trouble.

The chances of reducing costs by reducing the amount of spoilage and rework on any given part or product depend on the percentage nonconforming, on the subdivision of rejects into spoilage and rework, on the unit cost of a spoiled part or product and on the average rework cost, and on the prospective future production of the part or product.

From the standpoint of possible $\bar{X}$ and R chart applications, it was evidently necessary to break down the rejects by reasons for rejection and to note particularly those rejects that resulted from failure to meet specifications on some quality characteristic that might readily be measured (dimension, weight, resistance, tensile strength, and the like). For rejections due to each such characteristic, defective work reports or other cost records should be examined to estimate the average net cost of a spoiled unit (usually the manufacturing costs up to the point of spoilage minus scrap value of the unit) and average rework cost.

In manufacturing operations on a job that has a definite foreseeable termination and that is unlikely to be repeated, the possible saving from elimination of spoilage and rework is limited by the total amount still to be produced. For instance, assume 20,000 units are still to be produced on a given job. Past spoilage due to failure to meet specifications on a certain dimension has averaged 2%; rework has averaged 4%. The net cost of a spoiled unit is \$50, and the average rework cost is \$12. The total possible saving from the complete elimination of spoilage and rework from all remaining product is therefore

$$\begin{aligned} \text{Spoilage} &= 20{,}000(\$50)(0.02) = \$20{,}000 \\ \text{Rework} &= 20{,}000(\$12)(0.04) = \underline{\quad 9{,}600} \\ &\qquad\qquad\qquad \text{Total} = \$29{,}600 \end{aligned}$$

This estimate neglects the fact that anything learned in troubleshooting on this particular job may also prove helpful in reducing spoilage and rework on other similar operations.

Where manufacturing operations on the given product are expected to continue for an indefinite period, it is desirable to estimate expected annual production. The maximum possible annual saving from complete elimination of spoilage and rework may then be estimated.

Such estimates guide the selection of quality characteristics for the application of $\bar{X}$ and R charts so that each chart has definite possibilities for substantial cost savings.

Example 17.2
Control of product weight

Facts of the case Figure 17.1, taken from an article by O. P. Beckwith, illustrates the effect of successive changes in a textile manufacturing operation. The purchase specification for this fabric stated that the average weight in ounces per square yard of a sample of five from a lot of material should not be less than

FIGURE 17.1 $\bar{X}$ chart illustrating the effect of successive changes working closer to a minimum specification limit on weight—Example 17.2. *(Reproduced from O. P. Beckwith, "A Fresh Approach to Quality Control," Textile World, vol. 94, pp. 79–81, January 1944.)*

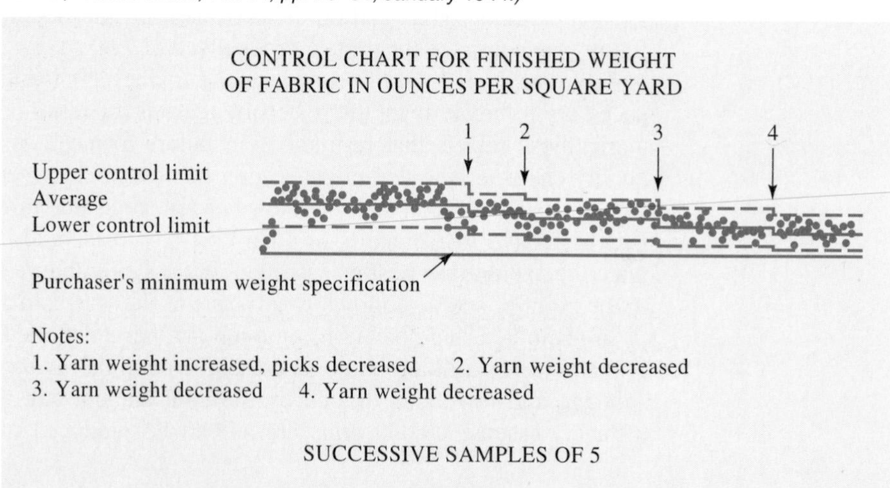

a stated value. As time went on, it proved possible to work closer and closer to this specification limit with safety.

The first section of the chart shows a few points close to the specification minimum, even though there was a relatively high average value. At this time the operation was not in statistical control. Moreover, trouble was experienced with yarn strength. Beckwith explains the successive changes as follows:[†]

> In the first change the weight of the yarn was increased, which, under the particular conditions of manufacture, made a more uniform yarn. At the same time the yarn-weight increase was compensated for by a decrease in fabric picks per inch. Meanwhile considerable investigation of the spinning process showed how yarn could be spun at required strength and uniformity, but at reduced weight. The yarn weights were therefore lowered progressively, as is reflected by the third, fourth, and fifth changes in limits.

Because the specification applied to *averages* of samples of five rather than to individual values, it was appropriate to show the specification limit on this $\bar{X}$ chart.

Example 17.3
Possible reduction in overfill of containers

Facts of the case In the packaging of a cereal product in 10-lb sacks, it was suggested that a study be made to determine the advisability of using $\bar{X}$ and R charts.

It was possible to make preliminary guesses as to possible savings prior to the use of $\bar{X}$ and R charts for an experimental period. The value of the material being sacked per operator per hour was \$500. Thus it was evident that a 2% overfill would cost \$10/h. This estimate showed that if there were really an overfill and if it could be reduced without falling below specifications, there was a chance for a good saving.

A trial period of control-chart operation was required to provide a basis for judgment as to the average amount of overfill and the possibility of improvement. This showed that the filling process was not in statistical control. If the process could be brought into control, it was evident that the average overfill could be substantially reduced. An estimate indicated that the annual cost of overfill on this particular item was nearly \$40,000. Although it was apparent that all overfill could not be eliminated, it was clear that even a small reduction would more than pay the cost of a control-chart program.

17.4 TAGUCHI'S LOSS FUNCTION

Dr. Genichi Taguchi, whose experimental designs in Japanese manufacturing date back to the mid-1950s, has developed an interesting quality cost model that today is called Taguchi's loss function. The basic philosophy is that, whenever something is produced, a cost is imposed on society. Part of that cost is borne by the

[†]O. P. Beckwith, "A Fresh Approach to Quality Control," *Textile World,* vol. 94, pp. 79–81, January 1944.

producer, and part is borne by the customer as the result of using the product or service. If these costs are plotted as a function of "quality," producer costs tend to increase with increased quality; customer costs tend to decrease because of greater efficiency, less breakdowns, and so forth. The total cost, or loss to society, is the sum of these two cost functions. It tends to decrease over some range of increasing "quality" until a minimum is reached and then increase beyond that point.

With regard to individual process operations, Taguchi focuses on their *parameters,* or the nominal dimensions, and their *tolerances,* the allowable deviation from nominal, as established by the design engineers. Where production people might be satisfied that any item produced within specifications is as good as any other, Taguchi argues that best quality is achieved by minimizing the deviation from the target or nominal dimension. The cost of quality, then, can be measured as a function of a process's deviation from target and by the inherent variability of the process. If optimally set, items at the nominal dimension will be perfect for their intended use and the loss to society will be minimal. Thus attention is focused on minimizing deviations in parameter settings ($\overline{X}_0$) from targets and reducing process variability.

The parabolic loss function equation proposed by Taguchi for two-sided specifications is

$$L = k(\sigma^2 + \delta^2) = k(\text{MSD})$$

where σ = the current best estimate of process standard deviation
$\delta = (\mu - T)$
μ = the current best estimate of process centering
T = the target or nominal dimension
k = a constant used to convert MSD = $(\sigma^2 + \delta^2)$ into monetary units
MSD = the mean-square deviation

In deriving the value of k, it is necessary to know the allowable extreme variation in the measurement S as set by the design engineers. Thus specification limits are $USL_X = (T + S)$ and $LSL_X = (T - S)$. The economic loss that occurs when a unit is produced at either USL_X or LSL_X, designated L_0, must be determined in order to put a value on k. This is probably the most difficult number to compute. Fortunately, if two competing processes are being compared, the dollar value of L_0 will be the same for each. Thus, when the loss functions for the two processes are equated, the constant k will cancel out and the problem becomes one of selecting the process that is the most likely to yield the minimum MSD.

Figure 17.2 illustrates a Taguchi loss function superimposed over an assumed normally distributed process. For this process, T is 52.5 and S is 12.5. The loss function is

$$L = k[3.2^2 + (52.99 - 52.5)^2] = k[10.24 + 0.24] = 10.48k$$

Only a very small contribution to loss is made by the deviation from target in this instance.

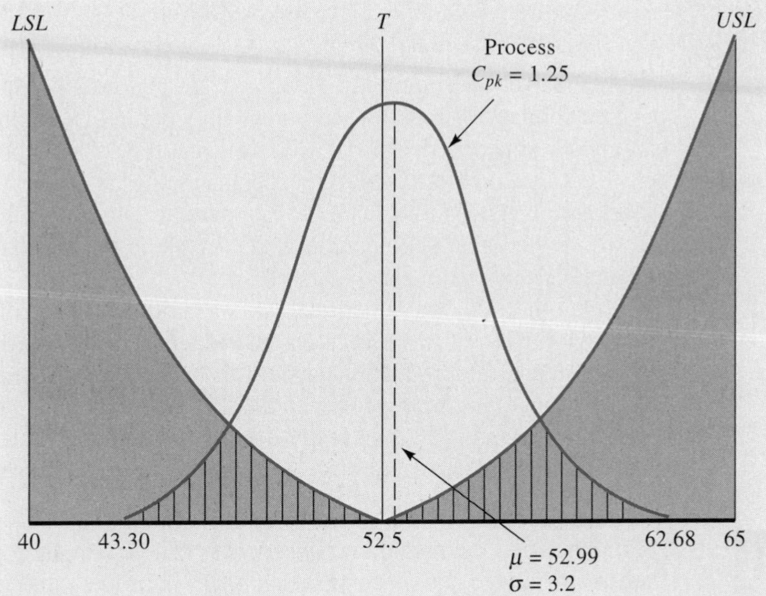

FIGURE 17.2 Illustration of a normally distributed process superimposed over a parabolic Taguchi loss function.

17.4.1 A Related Capability Index

Taguchi and others have developed an additional process capability index based on the loss function. The reader will recall that, in Chap. 9, Sec. 9.2, we developed two indexes, a pure process capability index C_p and a process performance index C_{pk}. Applying the formulas for each from Sec. 9.2, their values for the process illustrated in Fig. 17.2 are

$$C_p = (65 - 40)/19.2 = 1.30 \qquad C_{pk} = (65 - 52.99)/9.6 = 1.25$$

The new index, designated C_{pm}, uses the square root of the mean-square deviation (MSD) of the loss function, $\sqrt{(\sigma^2 + \delta^2)}$, to replace σ in the formula for C_p. Thus

$$C_{pm} = (USL_X - LSL_X)/(6\sqrt{\text{MSD}})$$

For the process illustrated in Fig. 17.2, $C_{pm} = (65 - 40)/(6\sqrt{10.48}) = 25/19.42 = 1.29$. Numerically, the value will always lie somewhere between C_p and C_{pk} if the latter are within acceptable ranges. The value of C_{pm} will become less than 1.0 when the corresponding value of either of the other indexes becomes less than 1.0. It would appear that C_{pm} is intended to combine the information given by the other two indexes perhaps in an attempt to replace them. We do not believe that it fulfills that goal.

17.5 SOME ECONOMIC ASPECTS OF THE MARGIN OF SAFETY IN DESIGN SPECIFICATIONS[†]

Designers have not commonly thought of the margin of safety to be included in specified tolerance limits as a problem in economy. Nevertheless, because such margins are aimed at quality of conformance, they should properly be viewed in the same way as the selection of acceptance criteria.

Serious adverse consequences of accepting unsatisfactory product and high unit costs of inspection are two conditions favorable to the economic use of large margins of safety in design specifications.

On the other hand, an uncritical use of large margins of safety under all conditions may turn out to be very costly. Generally speaking, the design specification determines the production method to be employed. Unnecessary margins of safety may require more costly materials, more precise machines, and more expensive workmanship than really needed to secure satisfactory product. Conditions favorable to small margins of safety in the design specification are as follows:

1. Evidence that the manufacturing process can be kept in good statistical control with an average and dispersion that result in product that is satisfactory for the purpose intended with very little margin of safety
2. Evidence that a costly change in the manufacturing process will be required in order to increase this margin
3. Either low unsatisfactory-product cost per unit of such product or relatively low cost of applying acceptance criteria to give adequate protection

17.5.1 Quality Decisions in Choosing a Production Method

The choice of a production method is influenced by the quality objective. The quality objective, in turn, is influenced by the design and acceptance criteria. Where acceptance criteria are such that lots with moderate percentages of defectives have only a small probability of rejection, it may be more economical to have an occasional lot rejected than to increase production costs in order to prevent any rejections. This seems particularly likely to happen where the producer has some degree of statistical sophistication and believes that the design specifications contain enough margin of safety so that moderately defective lots are really good enough for the purpose intended.

In some instances it may be economical to adopt production methods in which a certain amount of rework or spoilage seems inevitable. This point is illustrated in Example 17.4.

[†]For additional comment on this topic, see E. L. Grant, "The Economic Relationship between Design and Acceptance Specifications," *Special Technical Publication* No. 103, Symposium on Application of Statistics, American Society for Testing and Materials, Philadelphia, Pa., 1950. See also E. L. Grant, "Some Possible Contributions of Statistical Quality Control to Engineering Economy," Paper no. 1, Fourth National Convention ASQC, American Society for Quality Control, New York, 1950.

Example 17.4

Reducing a high percentage of defectives will not necessarily reduce costs

Facts of the case In the manufacture of a certain type of pressure gage, many different types of nonconformities were responsible for rejection at final inspection. The nonconformities causing each rejection were shown on the inspection record. More than half the rejections were shown to be made for one reason, associated with the zero registration of the gage. On the average, about 20% of the gages produced had this problem. A p chart plotted for this type of nonconformity indicated that the process was in control at this level; the day-to-day fluctuations above and below 20% nonconforming were such as might be attributable to chance. This nonconformity was always correctable by a rework operation, which, for most gages, was relatively simple.

Analysis and action An engineering study indicated that this type of defect could be almost completely eliminated by an extra operation on one of the parts of the gage. The costs of doing this for 100% of the gages were compared with the costs of carrying out the rework operation on 20% of the gages. This cost comparison showed it to be more economical to rework 20% of the gages than to carry out the extra operation on 100% of them. Consequently it was decided that, until a less costly method of eliminating this problem was discovered, the 20% level for this nonconformity and attendant rework would be accepted as normal.

17.6 SOME SPECIAL DIFFICULTIES OF ESTIMATING INDIRECT COSTS IN ECONOMY STUDIES

Most cost accounting systems recognize three classes of manufacturing costs, namely, direct material cost, direct labor cost, and indirect manufacturing expense, often called *burden* or *overhead.* The classification *indirect manufacturing expense* includes a great variety of costs, such as supervision, inspection, factory transportation, janitors, repairs and maintenance on machines and buildings, heat, light and power, manufacturing supplies, factory clerical expense, accident compensation, tools and dies, factory insurance and taxes, and depreciation on buildings and machinery. Rework and spoilage are sometimes charged directly to a job or process, and are sometimes included as part of overhead. Indirect manufacturing expense is apportioned among cost centers and applied to each job or process in proportion to something directly measurable such as direct labor cost, direct labor hours, or machine hours. The most common basis is direct labor cost. If, for example, the machine shop carried a burden rate of 150% of direct labor cost, this would mean that for every dollar of direct labor cost charged against a job in the machine shop, there would also be charged $1.50 of indirect manufacturing expense. This charge would be, in effect, an apportionment against the job of the many different indirect expenses, that is, supervision, janitors, heat, light and power, and factory taxes.

Although such an apportionment serves many useful purposes—such as valuation of inventories of goods in process and finished goods, determination of cost of goods sold, and sometimes determination of selling price—it does not follow that the saving of a dollar of direct labor in the machine shop will save $1.50 of disbursements for indirect manufacturing expense. Neither does it follow that an increase of a dollar of direct labor cost in the machine shop increases disbursements for indirect manufacturing expense by $1.50. From the standpoint of an economy study, the real question is what receipts and disbursements are likely to be influenced by the choice between alternatives.[†]

It is a general principle of all economy studies that indirect costs should *not* be included in cost comparisons simply by the uncritical application of burden rates. The relevant question regarding indirect expenses is always the probable effect of the choice between the given alternatives on each separate indirect expense. If no disbursements for indirect expense seem likely to be influenced by the choice, there is no need to consider indirect expenses in the economy study. In many economy studies relative to quality control, it may be good enough for practical purposes to limit the estimates to the direct costs involved in the alternatives being compared. These costs are not relevant to cost of quality studies as described previously.

17.7 DID IT PAY TO USE STATISTICAL QUALITY CONTROL?

Up to this point the discussion in this chapter has dealt with *prospective* uses of statistical quality control techniques. *Each* use of these techniques should be based on a favorable prospect that the particular use in question will more than pay its costs.

It sometimes happens that when statistical quality control has been newly introduced into an organization and has been used for a few months, a continuation of the *overall* use of the techniques requires evidence satisfactory to management that their use has paid on the whole. Management is inclined to look to the financial accounts and cost accounts for this evidence, and often wants before-and-after figures for comparison. Although in some cases a fairly satisfactory picture of the effect of statistical quality control may be had by examining costs before the use of the techniques and comparing them with costs after the techniques have been used for some time, this point of view has definite limitations that should be understood.

Where there have been reductions in the amount of spoilage and rework over a short period and these reductions are clearly due to statistical quality control, there is a good measure of cost saving. Also, where the amount of destructive testing has been reduced because of the use of the control chart for variables as a basis for acceptance, it is possible to arrive at a definite figure for the saving.

[†]For a more extended presentation of this point of view, see E. L. Grant, W. G. Ireson, and R. S. Leavenworth, op. cit., particularly chap. 10.

But in most cases the situation is sufficiently complicated that a fair before-and-after picture cannot be obtained solely from accounting figures. It is seldom as simple as being able to say, for example, that the charges to account 317-B-1 were $1,427 in January before statistical quality control and $955 in May after statistical quality control. (Logically, of course, such a comparison is never appropriate. The correct comparison is between the actual costs in May with statistical quality control and the costs in May as they would have been without statistical quality control; the practical difficulty is that there is no way to be certain of the latter figure.) Too many changes are constantly taking place to make it reasonable to ascribe all cost changes over a period of time to a single influence. Any before-and-after study needs to combine all the relevant evidence; the figures in the accounts constitute only one part of this evidence.

For example, the use of the control chart may reduce the number of machine shutdowns for readjustment by 75% and thus increase productive machine time and reduce costs. In any given period, this saving might be neutralized by other causes that were responsible for idle machine time.

Or an improvement in the outgoing quality from the forge shop might reduce the number of nonconforming forgings transmitted to the machine shop. This, in turn, might reduce excessive machining costs and tool breakage in the machine shop. These savings, however, would be only two of many possible influences on machine shop costs. If the unit costs in the machine shop actually decreased, this improvement might be attributed by the machine shop superintendent to causes within the machine shop, even though better forgings were really responsible for most of the savings.

Or an improvement in the quality of purchased parts might reduce the delays due to nonconforming parts on an assembly line. But there might be enough other variations in assembly costs that the effect of this improvement could not be isolated in the accounts.

Or improvement in the quality level of a final product or service might tend to reduce the number of customer complaints. But the time lag between actions and complaints might be such that this would be evident only after a long period of time.

Moreover, it should be recognized that the chances for before-and-after studies showing spectacular cost savings come chiefly in the early days of statistical quality control. As time goes on, the function of statistical quality control becomes more and more one of holding the line of better quality and improved quality assurance, and there is no "before" figure from a few months ago that can be contrasted with current performance.

Sometimes a misinterpretation of accounting figures may lead to an unsound conclusion regarding the merits of past changes. Example 17.5 is a case in point.

Example 17.5

A misinterpretation of the ratio of indirect manufacturing expense to direct labor cost

Facts of the case One of the important raw materials in a certain manufacturing operation was subject to considerable inherent variability. For this reason, it had always been assumed that satisfactory product was not possible without 100% inspection following the manufacturing operation. The percentage of spoilage disclosed by this inspection was always fairly high. Nearly half the direct labor cost charged on this operation was for this 100% inspection, which was carried out by production personnel. All the different sizes and designs of the plant's basic product required this manufacturing operation.

The plant quality control director introduced $\bar{X}$ and R charts on this operation. As often happens, it proved possible to diagnose the causes of a number of the quality troubles and to eliminate them. For most of the sizes and designs of the product, it was possible to bring the fraction rejected to a low enough figure so that an AOQL sampling plan could be substituted for the mandatory 100% inspection.

In the cost accounts of this company, factory burden (that is, indirect manufacturing expense) was apportioned to product in each department in proportion to direct labor cost. In effect, the burden rate was a fraction in which the numerator was departmental indirect manufacturing expense and the denominator was departmental direct labor cost.

The effect of the quality improvement was to make a large increase in this fraction for the particular department affected. The elimination of the direct labor for screening inspection on most of the product cut the denominator to some 60% of its previous figure. The AOQL sampling inspection was carried out by inspection personnel rather than production personnel and was therefore classified as an indirect manufacturing expense; it increased the value of the numerator of the fraction. This increase in the numerator and decrease in the denominator nearly doubled the burden rate.

The controller of this company was located at company headquarters several hundred miles from this manufacturing plant. The ratio of indirect manufacturing expense to direct labor cost was viewed as a figure that should be kept as low as possible in every department. When this ratio was greatly increased as a result of the activities of the plant quality control director, the controller was extremely critical of both the director and the head of the manufacturing department and, in fact, proposed to top management that the previous inspection scheme should be reinstated and that the plant quality control director's responsibilities should be severely curtailed. Fortunately, after a review of the matter, top management did not accept the controller's suggestions.

Comment on the case Of course, the changes brought about by these quality control activities had really caused a substantial decrease in *total* manufacturing costs. There had been a substantial decrease in spoilage. (The accounts in this company had never isolated the costs of spoilage.) The costs incident to the sampling inspection were much less than the cost of the previous 100% inspection.

It seems hardly conceivable that this controller, trained in accounting, could have been so unaware of the limitations of conclusions that may be drawn from an uncritical examination of ratios taken from the accounts. However, the controller had become preoccupied with the need to control all costs that were classified as indirect manufacturing expense and, moreover, had come to accept the ratio of such indirect expense to direct labor cost as the best measure of success in controlling indirect expense. These attitudes seemed to blind the controller to the realities of the situation and to lead to the making of recommendations that were adverse to the company's best interests.

17.7.1 Who Should Get Credit for Savings from Quality Improvement?

It is futile to argue the question, "Who was responsible for making this saving?" Any quality improvements made on the basis of a diagnosis of troubles by the control chart are necessarily based on teamwork. Inspection personnel, production personnel, methods engineers, tool engineers, design engineers, and others may all participate.

From the standpoint of managerial decisions, the important question is not "Who gets the credit?" but rather "What methods and policies shall we adopt?" In judging whether or not to use statistical quality control techniques, the question for management is, "What savings will be made with these techniques that will not be made without them?" All such savings are relevant in the decision whether to use statistical quality control, even though the statistical techniques are only one of the necessary links in the chain by means of which the savings are accomplished.

17.7.2 Some Problems Connected with Budgetary Control

The reduction of costs and the prevention of cost increases call for the continuous exercise of executive pressure. In modern industries, this pressure is exerted by the use of budgets. A principle of budgetary control is that each comparison between actual performance and the budgeted figure must follow the lines of individual responsibility. Pressure, to be effective, must be exerted on individuals—on works managers, department heads, supervisors, and operators.

Because department heads and supervisors are judged by their superiors in terms of conformance to their own budgets, they are sometimes inclined to view all decisions and proposed actions solely in terms of the effect on those budgets. This point of view often is adverse to the best interests of the business enterprise as a whole. It frequently happens that a small increase in expenditure in one department will bring about a much larger saving in another department. Or a decrease in cost in one department may cause a much larger increase in cost elsewhere.

This concentration of attention on individual budgets may, in some organizations, constitute a serious obstacle to the effective use of statistical quality control. Control charts may cause increases in inspection costs (at least temporarily) even though they may result in much larger decreases in production costs. Quality improvements in the manufacture of parts may require increased fabrication costs, even though they result in much larger decreases in assembly costs, and so forth. Nearly every application of statistical quality control and, for that matter, nearly all policies related to product quality will cross the lines of departmental responsibility.

For this reason, many decisions related to quality ought to be made at a management level higher than that of the supervisors who seem to be immediately concerned.

17.8 THE INCREASED IMPORTANCE OF COSTS RELATED TO PRODUCT LIABILITY

The 1960s and 1970s were decades in which there was an extraordinary increase in the number of lawsuits involving product liability. A distinguished authority on this subject, Prof. John Mihalasky, estimated in 1977 that the annual number of product liability lawsuits in the United States had increased to more than 500,000 from a figure of less than 5,000 some 10 to 15 years earlier. The dollar amount of representative settlements of such suits had also increased very rapidly.[†]

The increased litigation about product liability resulted in part from drastic changes in court-made law and in part from specific legislation and the establishment of new government regulatory agencies. There was, in effect, a shift from the older legal doctrine that has been described as "Let the buyer beware" to a newer doctrine that might be called "Let the producer or marketer beware." The legal basis for increased product liability litigation naturally varied from one country to another and, in the United States, from one state to another. The United States had certain special aspects of its legal system that contributed to the rapid increase in the volume and costs of such litigation, as follows:[‡]

1. The amount of the award in the United States has not been limited to the actual damages claimed to have been sustained; in addition, it has been possible to award unlimited so-called exemplary or punitive damages.
2. Such cases may be tried by juries. Experience has indicated that decisions by juries in cases of this type seem to be capricious and unpredictable and verge on becoming a national lottery. The extremely high awards that sometimes occur have encouraged producers and their insurance companies to make an out-of-court settlement in many instances rather than to let a case go to trial. (Of course such out-of-court settlements also eliminate the defendant's expenses incident to the actual trial of a case.)
3. Lawyers in the United States have been permitted to take suits for damages with the payment of any fee by the client contingent on the success of the case. If the case is lost, the client owes nothing. This feature of the U.S. legal system is quite unique compared to most other countries in the world. If the case is won, the sum awarded for damages is split between the lawyer and the client according to some agreed-on percentage (possibly 50% to each). The chance

[†]John Mihalasky, "The Status of Product Liability in the U.S.A.," *Reports, First European Seminar on Product Liability,* vol. 1, pp. 119–130, April 28/29, 1977. This seminar was held jointly by the European Organization for Quality Control and Associazione Italiana Controllo Qualità. The two volumes of reports on the seminar include papers on the status of product liability in a dozen or so countries. These reports are copyrighted by the European Organization for Quality Control, P.O. Box 2613–CH 3001, Bern, Switzerland.

[‡]See J. G. Cowell, "Products Liability—European Insurance Perspectives," *Reports, First European Seminar on Product Liability,* op. cit., vol. 2, pp. 25–30. Our list is adapted from a similar one in the Cowell paper that explains some of the differences between Europe and the United States in this matter.

for good out-of-court settlements of many cases has made such contingency-fee agreements attractive to a number of lawyers.

4. It sometimes has been possible to hold a producer liable for damages resulting from the use of that producer's product regardless of the number of years that may have elapsed between the date of the purchase of the product new and the date of the use of the product that is alleged to have caused the injuries or other damages in question.

Product liability litigation is only one of many types of litigation that has been influenced by the foregoing differences between the United States and other countries. For example, during the same years when product liability litigation was increasing so rapidly, there was also a great increase in the number of malpractice suits against physicians, hospitals, and medical supply and pharmaceutical companies and in the size of the monetary awards in such suits.[†]

17.8.1 Statistical Quality Control Techniques in Relation to Certain Economic Aspects of Product Liability

Where a manufacturer carries insurance against product liability, the cost of this insurance is one element of product liability cost. Other elements include legal expenses and other expenses in connection with the defense of lawsuits, judgments for damages in suits that go to trial and are lost, and payments in suits that are settled out of court rather than going to trial, to the extent that all of these are not covered by insurance. It needs to be recognized that although current insurance costs may give a good enough estimate of a major part of product liability costs for the immediate future, the long-run costs of insurance will be greatly influenced by a producer's experience in having to pay damages and in making out-of-court settlements. In our classification of three general classes of costs that need to be recognized in decisions about quality, product liability costs clearly should be classified as unsatisfactory-product costs.

We started this chapter with a quotation from J. M. Juran and F. M. Gryna. It stated a general principle, namely, that the economic balance to be struck between the cost of quality and the value of quality applies to each quality characteristic. This principle applies both to decisions on quality of design (such as a choice among alternative possible materials) and to decisions on quality of conformance (such as the classification of a particular nonconformity as critical, major, or minor, or the choice of an LTPD or AOQL). Rarely, if ever, is it possible to make a satisfactory estimate of the product liability cost associated with each alternative that is being considered either in decision making about quality of design or in deciding about quality of conformance.

One would not expect to find the phrases "quality of design" and "quality of conformance" used in legal actions involving product liability. Nevertheless, it is the authors' impression that a large proportion of the alleged deficiencies in man-

[†]"Why Product-Liability and Medical-Malpractice Lawsuits Are So Numerous in the United States," by E. L. Grant as told to Theodore E. Lang, *Quality Progress,* vol. 27, no. 12, Dec. 1994, pp 63–65.

ufactured product and service that give rise to such actions are really deficiencies in quality of design. Such deficiencies cannot be eliminated by the better use of the techniques of statistical quality control. What is needed is a careful design review of materials, methods, and specifications dealing with quality characteristics that seem likely to be the source of lawsuits involving product or service liability.

To the extent that awards in product liability suits depend on alleged unsatisfactory quality of conformance to specifications, the use of statistical quality control techniques ought to help. That is, the techniques may help wherever it is possible to identify those quality characteristics for which nonconformity to specifications seems to be a likely source of such litigation.

The techniques of statistical quality control may be useful here both in process control and in manufacturers' inspection of incoming parts and of outgoing product. Control charts may help to diagnose and eliminate certain causes of failure to meet specifications. If a classification of quality characteristics (a so-called classification of defects) is used in a factory's inspection of its own product, characteristics should be classed as critical if nonconformities are likely to cause product liability suits. If acceptance sampling is used for such quality characteristics, the quality standard (AOQL, LTPD, or indifference quality) should be tight. Equivalent arguments can be made relative to service industries.

PROBLEMS

17.1. A manufacturer is faced with the following choices in meeting military contract requirements: (*a*) it can inspect all product 100% under monitor by a military QC representative, or (*b*) it can employ one of two single sampling AOQL plans as follows: (1) a 1.06% AOQL plan with $n = 80$ and $c = 1$, or (2) a 0.87% AOQL plan with $n = 200$ and $c = 3$. Rejected lots are screened under either sampling plan, a process that is assumed to be 100% effective. Defective items are always found at a later assembly stage and are returned to the manufacturer at a contract penalty cost of \$5.00 per unit. Output is in lots of 2,000, and monthly production is 10,000 units. Inspection costs are as follows: \$0.10 per unit to inspect 100%, that is, to screen the lots; and \$0.25 per unit to sample from a lot. If the process is operating properly, lots are produced at 0.50% defective; if the process is not operating properly, the defect rate is 3.0%.

(*a*) What is the cost per lot for 100% inspection?

(*b*) What is the average cost per lot under sampling plan (1) if lots are 0.5% defective? If they are 3.0% defective?

(*c*) What is the average cost per lot under sampling plan (2) if lots are 0.5% defective? If they are 3.0% defective?

(*d*) Sketch the cost per lot for the three actions as a function of the proportion of the time that lots are produced at 0.5% defective and at 3.0% defective. Which alternatives yield the lowest cost per lot over the range? What is/are the break point/s in terms of the proportion of time that the process is operating properly?

17.2. For a certain process operating in statistical control, T is 0.300, S is 0.100, and σ is 0.025. If the process mean shifts to the level of either specification limit, economic analysis indicates that the loss L_0 is \$12.00 per unit.

(*a*) Calculate the values of the upper and lower specification limits, and sketch a curve similar to Fig. 17.2 representing the process correctly centered between the specification limits.

(*b*) Calculate the value of the constant k in the Taguchi loss function, and compute enough points in the specification range to properly scale your graph.

17.3. An alternative to the process shown in Fig. 17.2 has a mean μ of 49.75 and a standard deviation σ of 1.25. At the present time, the parameters of these processes may not be adjusted to recenter them at T. Which process is preferred in terms of its loss? Calculate the values of the indexes C_p, C_{pk}, and C_{pm} for this alternative process. Is there anything about the comparative values of these indexes that would lead you to the same conclusion reached by evaluating the loss function?

17.4. As an assistant to a consulting quality control engineer, you encounter the following situation in a client's plant:

A certain operation involves the packaging of a dry cereal product in 10-lb bags of heavy paper. The bags are filled by an automatic weighing and filling machine that cuts off the flow of the product when the weight of the bag and its contents has reached a certain predetermined setting. After the flow has been stopped by this automatic cutoff, there is a small amount of the cereal product that has already passed the cutoff point and therefore goes into the bag. When operating without interruption, the machine fills 110 bags per hour. The practice of the operator appears to have been to aim at an overfill of 3 oz. (This value presumably was determined on the basis of past experience that any less overfill caused trouble with the government inspector.) The weight of an empty bag has been assumed to be 7 oz. This has given an aimed-at weight of 170 oz (160 + 3 + 7) for the filled package. Every hour or so the operator has weighed a single filled package on a spring scale that weighs to the nearest half ounce. If this measured weight is above 170 oz, the operator changes the machine setting to cut off at a lower weight; if the measured weight is below 170 oz, the operator changes the machine setting to cut off at a higher weight.

The government inspector at this plant has the job of checking weights and other quality characteristics that are controlled by legislation and government regulations in the interests of the ultimate consumer. The inspector's time is kept fully occupied as the job carries the responsibility for checking many products. The inspector's practice in checking weights on this particular product has been to select 5 bags at random from each day's production. (Daily production has averaged about 1,500 bags from two 8-h shifts.) Each bag has been emptied and the contents weighed on a small platform scale weighing to 0.1 oz. If the contents of each of the 5 bags weigh 160 oz or more, the inspector approves the day's output as to weight. If 2 or more of the bags have contents weighing less than 160 oz, the entire day's output is required to be stamped "Substandard Weight." If just 1 of the bags has contents below 160 oz, an additional 5 bags are taken for inspection. If all this second sample is satisfactory, the day's production is considered to be satisfactory. If 1 or more of the second sample are below 160 oz, the day's output is considered to be all substandard weight. This particular acceptance procedure is not specified in any written instructions that have been given to the government inspector but appears to have been the traditional one at this plant. At least the present government inspector, who has worked on this job for the past 2 years, is simply following the acceptance procedure explained by the previous inspector.

You talk with the government inspector and find a conscientious and cooperative person, anxious to carry out the functions of protecting the general public and at the same time having no desire to make unreasonable demands on the manufacturer. You obtain from the inspector the record of the measurements of weights of contents of bags for the past 50 days. (No earlier records are available.) You plot $\bar{X}$ and R charts from the first samples of 5 for these days. $\bar{\bar{X}}$ is 164.1; $\bar{R}$ is 3.9. The $\bar{X}$ chart shows definite evidence of lack of statistical control with 5 points above the upper control limit and 3 points below the lower control limit. Two points are slightly above the upper control limit on the R chart; a recomputed $\bar{R}$ with these two points eliminated is 3.6. On 5 of the 50 days a second sample of 5 was taken because one of the weights in the first sample fell below 160 oz. Two of the days requiring second samples were days on which $\bar{X}$ was below the lower control limit; on the other 3 days $\bar{X}$ was within limits. Two of the 5 second samples contained one sack with contents below 160 oz. The entire day's product for these 2 days was stamped "Substandard Weight." Packages so stamped are accepted by the trade only at a 5% reduction from the regular factory price of \$11.70 for a 10-lb bag.

(*a*) Write a general discussion analyzing the statistical aspects and other related aspects of this situation. In preparing this discussion, assume that the reader is familiar with the facts as given in the preceding statement and is familiar with the terminology and general principles of statistical quality control. This discussion might be one that you would prepare for your chief or for a colleague in your own organization.

(*b*) Draft a letter to the plant superintendent giving specific proposals for any action that you recommend should be taken. Explain in as definite terms as possible the advantages that you expect will be gained by following your proposals. Insofar as possible, this letter should not assume that the plant superintendent is familiar with the terminology and concepts of statistical quality control.

(*c*) Assume that the government bureau carrying out the inspection is your client rather than the manufacturer. Draft a report to the bureau chief discussing the problems involved in setting and enforcing specifications on filling weights with a view to consumer protection. Illustrate any general statements you make by reference to this specific case. Assume that the bureau chief (or a subordinate who reads the report) is familiar with the terminology and concepts of statistical quality control; for example, he or she will understand the meaning of an OC curve if one is included in your report.

18

SOME SIGNIFICANT EVENTS IN THE DEVELOPMENT OF STATISTICAL QUALITY CONTROL

Thus in many directions the engineer of the future, in my judgment, must of necessity deal with a much more certain and more intimate knowledge of the materials with which he works than we have been wont to deal with in the past. As a result of this more intimate knowledge his structures will be more refined and his factors of safety in many directions are bound to be less because the old elements of uncertainty will have in large measure disappeared.

—Frank B. Jewett, president, Bell Telephone Laboratories, Inc.[†]

18.1 QUALITY AND STANDARDIZATION

Scientific methods for judging manufactured product did not emerge until near the end of the eighteenth century. The concept of interchangeable parts, the very foundation of mass production, did not emerge until about 1787. From that time until about 1840 the emphasis was on artisan skill, and every effort was made to produce exact parts in order to ensure interchangeability. Gradually it became apparent that for most parts used in manufacturing there was some small range (tolerance) of a dimension over which a part would remain interchangeable and function as intended. The gaging of parts on a "go" and "no-go" basis did not appear until about 1870, with the concept of the minimum requirement ("go" gage) coming first in 1840.

The next event in the evolution of manufacturing came in 1900 with the introduction of standardization. The perceived need was to establish nominal

[†]"Problems of the Engineer," *Science,* V. 75, pp. 251–256, 1932.

dimensions for items commonly used by many industries in building their products such as screws and nuts and bolts (fasteners). Great Britain was the first to establish a standards organization (1901). Next came the Netherlands and Germany. The real need for national and then international standards organizations was accelerated during and immediately after World War I. By 1920 Belgium, Canada, France, Switzerland, and the United States had established organizations. Such organizations existed in most industrialized nations by the mid-1930s.

18.2 THE TWO ARCHITECTS OF STATISTICAL QUALITY CONTROL

The two major divisions of the material contained in this book are *statistical process control* and *scientific sampling.* It is both interesting and significant that these technologies, along with so many others, were developed initially within AT&T's Bell Laboratories. Walter A. Shewhart is responsible for the development of statistical process control and Harold F. Dodge for scientific sampling for acceptance purposes.

Any explanation of their contributions ought to begin with the structure of the Bell Telephone System during the 1920s and 1930s when they were doing their pioneer work. The American Telephone and Telegraph Company (AT&T) owned both the manufacturing company, Western Electric, and the subsidiary operating companies as well as Bell Labs, which was largely the research company. The operating companies bought much of their equipment from Western Electric. Originally, each operating company did its own inspection of purchased product. Then it was concluded, no doubt by someone in AT&T headquarters, that it would be more economical for Bell Labs to represent all the various operating companies in the buyers' inspection of the product made by Western Electric. Shewhart and Dodge, who worked for G. D. Edwards at Western Electric Engineering, were all transferred to Bell Labs to assist in the effort. Suddenly there was a big and very important inspection job with a lot of bright people assigned to it. Edwards, Shewhart, and Dodge were three of these bright people. It was because of this corporate structure—where design, manufacture, and usage were all part of the same management system—that inspection and quality assurance could develop a true systems approach.

Walter A. Shewhart, Ph.D. (1891–1967), the architect of statistical process control, developed the concept of statistical control and introduced the control chart around 1924. His experiences with and teachings about the theory and application of control charts carried on during the 1920s resulted in publication of his now famous book, *Economic Control of Quality of Manufactured Product,* in 1931.[†] In it he describes statistical control as follows: "A phenomenon will be said to be

[†]W. A. Shewhart, *Economic Control of Quality of Manufactured Product,* D. Van Nostrand Company, Inc., Princeton, N.J., 1931. After being out of print for nearly a generation, it was republished in 1980 as a Fiftieth Anniversary Commemorative Reissue by ASQC Quality Press, 611 E. Wisconsin Ave., Milwaukee, Wis.

controlled when, through the use of past experience, we can predict, at least within limits, how the phenomenon may be expected to vary in the future. Here it is understood that prediction within limits means that we can state, at least approximately, the probability that the observed phenomenon will fall within the given limits."[†] It is worth noting that Shewhart intended this definition of a *constant* or *stable system of chance causes* to apply not just to the results of manufacturing operations but to *any* phenomenon. Many of his examples dealt with problems in physics and chemistry. Also he was setting the stage for using the control-chart technique to *predict* outcomes scientifically after a state of statistical control has been established. Part Two of this book deals with the many applications of Shewhart control charts.

If Walter Shewhart is the architect of statistical process control, Harold F. Dodge (1893–1976) is the architect of scientific sampling for acceptance purposes. Research in acceptance sampling began at Bell Laboratories in the 1920s. It had been put into the hands of the quality group established within Bell Labs to design, manage, and perform final inspection of Western Electric equipment on behalf of the operating companies at the production sight. Dodge and a team of statisticians and engineers were engaged in the development of methods for scientifically sampling from large numbers of essentially identical items. The resulting tables of acceptance sampling plans were intended for use in surveillance of final products as they were to be shipped to the operating units of the Bell System, and also after shipment to ensure that logistic movements and shelf life problems had not emerged.

A number of new concepts resulted from Dodge's research including those of Consumer's Risk and Producer's Risk, when 100% inspection is replaced by sampling inspection, and the calculation of probability-of-acceptance curves for specific sampling plans. These curves are known today as *operating characteristic (OC) curves.* The first application of sampling plans based on probability calculations to inspection for customers' lots was made in 1923. The plans based on Consumer's Risk, the lot tolerance percent defective (LTPD) plans, were developed first and saw their first shop use in 1926. They were first published in the *Bell System Technical Journal* in 1929.[‡] The concept of *average outgoing quality limit* (AOQL) and tables of sampling plans were developed in 1927. Harry G. Romig supervised the enormous calculating effort necessary to produce the tables. (There were no computers in those days.) Thus they were given the name "Dodge-Romig tables." The complete set of LTPD and AOQL sampling inspection tables as we know them today was published in the *Bell System Technical Journal* in January 1941.[§] The concepts and examples of the tables are discussed in detail in Chap. 13.

Dodge and others were involved in the development of the Army Service Forces (ASF) tables, which were used throughout World War II by the U.S. Army

[†]Op. cit., p. 6.

[‡]H. F. Dodge and H. G. Romig, "A Method of Sampling Inspection," *Bell System Technical Journal,* V. 8, pp. 613–631, October 1929.

[§]H. F. Dodge and H. G. Romig, "Single Sampling and Double Sampling Tables," *Bell System Technical Journal,* V. 20, N. 1, January 1941. Later they were published as *Sampling Inspection Tables,* John Wiley and Sons, Inc., New York, 1944; 2d ed., 1959.

for the acceptance of ordnance and supplies. These tables were based on a different philosophy of the inspection problem. They emphasized producer protection (the Producer's Risk) rather than Consumer's Risk and did not focus on rectifying inspection as did the AOQL plans. What they had in common was that they involved the mathematics of probability and they were calculated in a way that something was consistent throughout each set. In all of these systems, the original concept came from Harold Dodge. The impact of the philosophy of these tables still influences acceptance decision today in the descendant standards ABC-STD-105, MIL-STD-105E, ANSI/ASQC Standard Z1.4, and a number of international standards in Europe and the Orient. Much of Part Three of this book is devoted to an explanation of these concepts, most of which were developed by Dodge.

It should be pointed out that Dodge and his colleagues did not intend that sampling for acceptance purposes should stand alone or be used as a substitute for statistical process control as has been implied by some practitioners and consultants. The intent in using an acceptance sampling procedure is to verify that quality standards are being met or that time and logistic movements have not caused quality to deteriorate to an unacceptable level. In no published Bell Laboratories literature of the time is there the claim or implication that one statistical procedure was intended to replace another.

18.3 QUALITY ASSURANCE SCIENCE AND WORLD WAR II†

The entry of the United States into World War II marked the beginning of wide deployment of the tools of statistical quality control in American industry. It was at this point in history that the name W. Edwards Deming became important in association with statistical theory, sampling, and quality assurance.

W. Edwards Deming, Ph.D. (1900–1993), born in Iowa, studied engineering at the University of Wyoming. He received a master's degree in mathematics and physics from the University of Colorado and, in 1927, a Ph.D. in physics from Yale University. It was in that year, while employed by the U.S. Department of Agriculture (USDA) as a physicist and mathematician, that he met and became intrigued with Shewhart's concepts. In the mid-1930s, he took leave from the USDA to journey to England and study for a year with Sir Ronald A. Fisher in statistical design and sampling. By 1938, when, at Deming's invitation, Shewhart delivered his famous 4-day lecture, both Deming and the USDA Graduate School had gained considerable recognition for their statistical capabilities and training efforts. A short time later Deming collaborated with Shewhart in securing publication of Shewhart's notes in book form under the title *Statistical Method from the*

†The reader interested in statistical quality control during the World War II years should read the following three articles: (1) Holbrook Working, "Statistical Quality Control in War Production," *Journal of the American Statistical Association,* pp. 425–447, December 1945; (2) E. L. Grant, "Shewhart Medal Address," *Industrial Quality Control,* pp. 31–35, ASQC, July 1953; and (3) E. L. Grant, as told by T. E. Lang, "Statistical Quality Control in the World War II Years," pp. 31–36, *Quality Progress,* V. 24, N. 12, December 1991, ASQC.

Viewpoint of Quality Control.† It is on page 45 of this book that we see the first reference to what today is recognized as the plan-do-check-act cycle. Shewhart likened the three steps in manufacturing of specification, production, and inspection to the three steps of the scientific method: making a hypothesis, conducting an experiment, and testing the hypothesis. He argued that the "old" view of these three steps was that they operated in a straight line. His "new" approach was to view them as a continuum spiraling upward through stages of improvement based on new knowledge until the spiral became a circle in which inspection indicated no need to change specification. "Mass production viewed in this way constitutes a continuing and self-corrective method for making the most efficient use of raw and fabricated materials."‡ Figure 18.1 illustrates Shewhart's view of this "new" model.

Having aided the Census Bureau in setting up the U.S. unemployment survey in 1937, Deming joined the Census Bureau in 1940, where his work demonstrated the accuracy of sampling results and the power of the control-chart technique to help improve output. The extremely large number of questionnaires received by the Census Bureau required a mammoth effort to transform the data onto punch cards and have their accuracy verified. Deming demonstrated that (1) sampling could replace 100% verification inspection of punch card errors with much more accurate results, (2) control charts could be used to improve accuracy by indicating when operators were performing poorly and needed retraining, and (3) control

†Walter A. Shewhart, with editorial assistance of W. Edwards Deming, *Statistical Method from the Viewpoint of Quality Control,* 1939, published by The Graduate School, U.S. Department of Agriculture, Washington, D.C.

‡Op. cit., p. 45.

FIGURE 18.1 Shewhart's model of the "new" scientific approach to improving production.

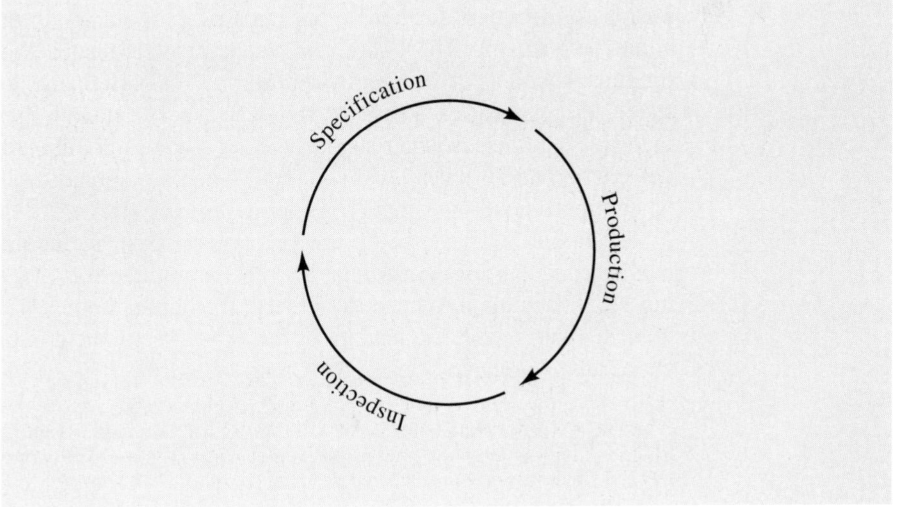

charts could help in identifying places where changes to the system might improve performance. This may be the first reported nonmanufacturing application of the control-chart technique.

Both Shewhart and Deming were involved with the application of statistical techniques to the manufacture and testing of war material during those final years before the United States entered the war. Leslie Simon, then a colonel and chief of the Ballistics Research Laboratory at the Aberdeen Proving Ground, was responsible for much of the statistical applications to ordnance development and testing. He was a prime mover along with Harold Dodge in the development of the American War Standard Z1.3—1942, "Control Chart Method of Controlling Quality during Production."[†]

Immediately after the attack on Pearl Harbor, Prof. Holbrook Working of Stanford University wrote to a number of friends and officials around the country wanting to know what the statisticians at Stanford could do to help the war effort. Deming responded with the suggestion that they establish training courses for government and industry on how to use the new control-chart and sampling techniques. Working and the senior coauthor of this book developed a 10-day (later reduced to 8-day) course.

Grant had been appointed to represent Stanford University on the Engineering, Science, and Management War Training (ESMWT) program of the U.S. Office of Education. By enlisting the aid of the Office of Production Research and Development (OPRD) of the U.S. War Production Board, he was able to obtain grant money to develop and conduct the course. It was first presented at Stanford in July and then in Los Angeles in September, 1942. Deming's aid was enlisted in presenting these first two courses. It was during these presentations that Deming illustrated the shortcomings of certain common industry sampling practices with a bead box experiment. This experiment, and its development into the now famous Red Bead experiment, are described in Chap. 20.

Shortly after the second course was presented, Grant and Working decided that a series of seminars for course participants, particularly in the Los Angeles area, would be desirable. They were designed to provide feedback on the results of SQC training as well as providing an opportunity for participants to exchange stories of successes and failures. These proved very successful from the standpoint of allowing some to showcase their successes and, at the same time, reinvigorated many of those who had not been so successful. Successes spread fastest at those locations where upper level management supported the effort.

The initial SQC course on the U.S. West Coast, including its follow-up seminars, became part of the national ESMWT training effort in 1943. Holbrook Working was called upon to head the ESMWT program from 1943-45. In 1946, shortly after the war ended, the leaders of these local seminar groups formed together to charter the American Society for Quality Control.

[†]Current versions of the entire set of ANSI/ASQC standards dealing with quality control are discussed in Chaps. 2 to 4 of this book. Dodge, who was chairman of the original writing committee, continued to serve on standards committees until his death.

18.4 JAPAN'S RECOVERY AS A WORLD-CLASS INDUSTRIAL POWER

Popular mythology in the United States likes to refer to pre-World War II Japan as a somewhat backward industrial power that produced and exported mostly trinkets and small items of dubious quality bought by Americans impoverished by the Great Depression. Few bring up the fact that, prior to the Pearl Harbor attack, Japan had conquered what are now Korea, Manchuria, Taiwan, and a large portion of China, Vietnam, and Thailand; and by the end of 1942 Japan had extended its empire to include Burma, the Philippines, Indonesia, Malaysia, Thailand, Cambodia, New Guinea, plus many strings of islands in the eastern Pacific Ocean. Its navy had moved a large armada of warships 4,000 miles across the Pacific Ocean, in secret and in silence, to attack Pearl Harbor and then returned safely home. Manufacturers capable of producing only low-grade goods don't accomplish such feats.

High-quality standards for military hardware, however, did not extend to civilian and export goods, which received very low priority during the war years. Thus the perception in the United States for a long time before and then immediately after the war had nothing to do with some inherent character flaw in Japanese culture or industrial capability. It had everything to do with Japan's national priorities and the availability of funds and materials.

Following Japan's surrender in 1945, General MacArthur was given the task of rebuilding the Japanese economy on a peaceful footing. As part of that effort an assessment of damage was to be conducted and a national census was planned for 1950. Deming was asked in 1947 to go to Japan and assist in that effort. As a result of his association with Shewhart and quality training, he was contacted by representatives from the Union of Japanese Scientists and Engineers (JUSE), and in 1950, Deming delivered his now famous series of lectures on quality control. His message to top industry leaders, whom he demanded to attend, and to JUSE was that Japan had to change its image in the United States and throughout the world. He declared that it could not succeed as an exporter of poor quality and argued that the tools of statistical quality control could help solve many quality problems. Having seen their country devastated by the war, industry and government leaders were eager to learn the new methods and to speed economic recovery. Experience was to prove to Deming and others that, without the understanding, respect, and support of management, no group of tools alone could sustain a long-term quality improvement effort.

In 1951, JUSE established the Deming Prize for Quality to recognize his efforts in support of economic recovery. Since that time, the prize has grown in stature to world recognition as a highly prestigious award. Over the years it has expanded to include several prizes for individual, companywide, factory, and small business quality improvement efforts. In 1984 its scope was expanded to allow overseas companies to participate. In 1985, the Japan division of Texas Instruments was awarded a Deming Application Prize. In 1989, just one year after companies wholly outside Japan were allowed to apply, Florida Power and Light Company was presented a prize. Their efforts at implementing the Deming philosophy had begun in 1981.

18.4.1 Three Phases of Evolution in Japanese Management

In a keynote speech for the Eighteenth Japan Symposium on Reliability and Maintainability,[†] Professor Takagi divided the development of management for quality into three phases. The first, which he called a period of "take and take," was a time during which Japan's industry directed it efforts toward absorbing as much knowledge as possible from overseas, notably from the United States, and then broadly applying it. This period extended from the end of the war until 1960. By 1960, training in the use of statistical tools was common, courses in management for quality had spread throughout industry, and the Deming Prize had taken root and was heavily influencing quality activities in industry. Perhaps most impressive of all was the fact that upper-level managers were personally taking charge of managing for quality. Annual quality audits, conducted by top management, were becoming common.

The second period, which Takagi called the phase of "give and take," was a time during which Japan learned to produce quality consumer products meeting standards normally required only by the military. This phase extended across the decade of the 1960s and perhaps into the 1970s. This was the time that many Japanese industries "arrived" in the sense that they were able to compete in world trade anywhere that high trade barriers were not raised against them. It was during this period that Japan began to "home grow" its quality experts and develop new tools and management for quality structures.

The third period, "give and give," saw many Japanese industries overtake the United States and begin to take a leadership role in the design and development of new technology.

Dr. Kaoru Ishikawa (1915–1989) is recognized worldwide as an early Japanese pioneer and leader in the quality movement. Born of a prominent Japanese family, he graduated from Tokyo University in applied chemistry in 1939. He served as an officer in the Japanese Navy from 1939 to 1941 and as a factory construction manager from 1941 to 1947. In 1947 he became a professor of engineering at Tokyo University. His interest in quality control began in 1949 when he became active in JUSE statistical methods activities. He was involved in arranging the JUSE lectures by Deming. Early in the 1960s JUSE had launched the monthly journal *Quality Control for the Foreman,* which contained many prize-winning examples of quality improvements and guidelines on how to use new tools, concepts, and methodologies.

In 1962 Ishikawa developed the quality circle concept, similar to the shop floor team concept now so popular in the United States. It spread rapidly through the more progressive export-oriented industries. The quality circle movement was directed at shop floor personnel to provide new opportunities for involvement and empowerment while generating large numbers of quality improvements. Ishikawa is credited with initiating many of the quality-related publications of JUSE and with organizing seminars, symposia, conferences, quality prizes, and consultan-

[†]Noboru Takagi, "Japan's Reliability with the Development of Electronics Industry," *Societas Qualitatis,* V. 2, N. 3, July/August 1988, Union of Japanese Scientists and Engineers, Tokyo.

cies. Many of the articles published in *Quality Control for the Foreman* became the foundation for his book *Guide to Quality Control* published in 1968.[†] One of the new tools discussed in the book—the cause and effect, or fishbone, diagram—is frequently referred to as the Ishikawa diagram. It is discussed in Chap. 9.

The "give and give" period for Japan began in the 1970s and stretched throughout the 1980s and perhaps beyond. It is during this period of material and fuel shortages, rising consumerism, and increasing demand that Japan's dominance in chosen competitive markets reached, or is reaching, its zenith. In the areas of both technology improvement and quality system improvement, Japan led the way. A decade before, upper-level management had taken charge of quality as Deming had insisted it must. Workforce involvement was reaching new heights through QC circles. The concepts that had worked so well on the factory floor were beginning to be applied in business and service areas. New quality improvements were developing at a furious pace. New tools were being invented with names like quality function deployment (QFD), hoshin planning (management by policy), total productive maintenance, just-in-time inventory management, and concurrent/simultaneous engineering. And new names of Japanese quality leaders were being added such as Yoji Akao (QFD), Genichi Taguchi, and Masaaki Imai. Some of these topics and names were mentioned at various places throughout this book. Others are in management areas beyond the scope of this text.

18.5 THE QUALITY CULTURAL REVOLUTION IN THE UNITED STATES

In the mid-1970s, many U.S. business leaders began to recognize that their ability to compete had slipped to levels that threatened corporate survival. Efforts to increase productivity took many forms. Some believed that automation would solve the problem because excessive wages paid to U.S. workers were the root cause. Some wanted the U.S. government to raise trade barriers in order to slow the flow of foreign goods. Others wanted to discover what Japan was doing that was making it such a fierce competitor.

In 1980, the National Broadcasting Company prepared and aired on TV a "white paper" entitled "If Japan Can, Why Can't We?" W. Edwards Deming was approached to present a segment on the postwar Japanese quality revolution. At the time, he was teaching at George Washington University and working on the manuscript of his book *Quality, Productivity, and Competitive Position.*[‡] Deming called for a new corporate culture based on principles of participative management; and in this new book, he presented his now famous 14 points describing the new corporate management model he envisioned. Apparently he was as persuasive in this televised

[†]Kaoru Ishikawa, ed., *Genba no QC Shuho,* JUSE, Tokyo, 1968. The first English translation was prepared by the Asian Productivity Organization in 1971 under the title *Guide to Quality Control.* Its great success led to a series of revisions and reprintings, the latest being in 1986. It is available through the American Society for Quality Control, Milwaukee, Wis.

[‡]W. Edwards Deming, *Quality, Productivity, and Competitive Position,* Cambridge, Mass., M.I.T., Center for Advanced Engineering Study, 1982.

interview as he had been at Stanford University in 1942 and in Tokyo in 1950. The reinvigorated U.S. quality movement began almost immediately.

Since their introduction to management literature in 1980, Deming's 14 points, or principles of world-competitive management, have gone through several phases of modification, some made by Deming himself and some made by others. Usually these changes have led to greater understanding of the point being made and did not change its aim or redirect its focus. One version is shown in Fig. 18.2.

It is clear that Deming's focus was on the responsibilities for quality shared by all levels of management, beginning at the top. Repeatedly he has stated that quality products and services are "made in the board room." Among other factors, he stressed the need for intimate knowledge of general business processes as well as direct production or service processes; partnering with suppliers and the elimination of "lowest bidder" thinking; the need to remove psychological and physical dangers in the system; strengthening understanding and cooperation among managers, professionals, and wage earners in a spirit of teamwork; and replacing straight-line project thinking with repeated cycles of improvement.

One of the first U.S. corporations to enlist Deming's aid was the Ford Motor Company. Quoting a December 1993 newspaper article: "In the mid-1980s, the company that had brought the world the Pinto, a subcompact notorious for exploding in rear-end collisions, unveiled the Taurus. By 1992, the Taurus unseated the Honda Accord as the top-selling model in the United States." In the decade of the 1980s the growth of quality tools, books, magazines, consultants, and membership in the American Society for Quality Control was phenomenal. U.S. industry developed a thirst for knowledge of all aspects of quality that rivaled that of the Japanese in the 1950s, and for about the same reasons.

18.6 U.S. AWARDS FOR QUALITY AND PRODUCTIVITY IMPROVEMENT

The movement to establish a quality award in the United States began in 1982 when a National Advisory Council for Quality was organized under the auspices of the American Society for Quality Control and the American Productivity Center. At the conclusion of a White House Conference on Productivity in September 1983, it was recommended that a National award be created similar to Japan's Deming Prize. In September 1985, a National Organization for U.S. Quality Award was created to determine how the award would be administered, how it would be funded, and what criteria would be used to assess applicants; to obtain the support of the White House; and to draft legislation for Congress. Top executives of a number of leading U.S. corporations, both manufacturing and service, guided the effort.

House Bill 5321 to establish a National Quality Award was introduced in August 1986. It died without hearings. It was reintroduced as House Bill 812 in January 1987, and hearings were conducted in March. In July a Senate Committee renamed it the *Malcolm Baldrige National Quality Award* after the popular secretary of commerce who had recently been killed in a tragic rodeo accident in Texas.

1. Create constancy of purpose toward improvement of products and services, with a plan to become competitive, stay in business, and provide jobs.
2. Adopt the new philosophy. We are in a new economic age. We can no longer live with commonly accepted levels of delays, mistakes, defective materials, and defective workmanship.
3. Cease dependence on mass inspection. Require instead statistical evidence that quality is built into products and services.
4. End the practice of awarding business on the basis of price tag alone. Instead depend on meaningful measures of quality, along with price. Move toward suppliers for any one item chosen for a long-term relationship based on loyalty and trust.
5. Improve constantly and forever the system of production and service to improve quality and productivity and thus constantly decrease costs.
6. Institute training and education on the job, including that for management.
7. Institute leadership. The aim of supervision should be to help people and machines do a better job.
8. Drive out fear so that everyone may work effectively for the company.
9. Break down organizational barriers between departments. Everyone must work as a team to foresee and solve problems.
10. Eliminate arbitrary numerical goals, posters, and slogans for the workforce that seek new levels of productivity and quality without providing improved methods. Such exhortations only create adversarial relationships; the bulk of the causes of low quality and low productivity belong to the system and thus lie beyond the power of the workforce.
11. Eliminate work standards that prescribe numerical quotas for the day. Replace management by numbers with helpful supervision and never ending improvement.

12a. Remove the barriers that rob hourly workers of their right to take pride in their work. Supervisors need to stress quality, not quotas.

12b. Remove the barriers that rob people in management and engineering of their right to take pride in their work. This means, *inter alia*, abolition of annual or merit rating and of management by numbers. Stress understanding of the extended process.

13. Institute a vigorous program of education and self-improvement. New skills are required for changes in techniques, materials, and service.
14. Put everyone to work in teams to accomplish the transformation. Create a corporate structure which will push the prior 13 points every day because the transformation is everybody's job.

FIGURE 18.2 W. Edwards Deming's 14 points for quality management.

In August the bill passed Congress and, as Public Law 100-107, was signed on August 20 by President Reagan.

While Congress was considering the award, committees of industrial leaders and experts in quality assurance and management were arranging for funding and developing the criteria to be used to assess applicants. An Award Office was set up within the National Institute of Standards and Technology to manage the award process.† Funding for the entire process was arranged by donations from many U.S. major corporations to a foundation set up for the purpose. Thus no federal tax money was involved. A number of leading quality experts prepared the initial set of criteria. All of this was accomplished in time to cycle through the process and present the first three awards in 1988 to Motorola, Inc., Schaumburg, Illinois; Westinghouse Commercial Nuclear Fuel Division, Pittsburgh, Pennsylvania; and Globe Metallurgical, Inc., Cleveland, Ohio.

The award criteria require that applicants respond to about thirty different items assembled under seven categories. Within each item are from two to seven or more areas that the applicant must address. Teams of examiners evaluate each application, scoring it on a 1,000-point scale. After some screening, the team, led by an experienced senior examiner, makes a site visit to clarify any questions raised by the off-site assessment and to verify the accuracy of the representations made in the application. The seven categories and their 1994 point allocations were leadership (95 points), information and analysis (75 points), strategic quality planning (60 points), human resource development and management (150 points), management of process quality (140 points), quality and operational results (180 points), and customer focus and satisfaction (300 points).‡ Two awards may be made each year in each of three economic sectors: large manufacturers, large service organizations, and small businesses, either manufacturing or service. As of 1994, plans were under way to expand the number of economic sectors.

Following the lead set by the Baldrige Award, both the National Aeronautics and Space Administration and the U.S. Department of Defense have established quality awards for their contractors. The Baldrige Award criteria have fostered the creation of similar awards in Canada, Europe, and Latin America. European and Mexican awards were made for the first time in 1992. Many states within the United States have established awards for corporations operating within their borders. Most have either adopted or slightly modified the Baldrige criteria to fit their circumstances. The award process in the state of Florida, first presented in 1993, provided for five economic sectors: private manufacturing, private service, education, health care, and public (government). Awards could be made separately to large and small employers.

†Individual copies of the criteria, application forms, profiles of winners, and examiner application forms may be obtained from Malcolm Baldrige National Quality Award, National Institute of Standards and Technology, Route 270 and Quince Orchard Road, Administration Building, Room A537, Gaithersburg, Md. 20899. Bulk orders for the criteria should be sent to the American Society for Quality Control.

‡While the seven categories have not changed since 1988 (except for minor adjustments in title), the number of items within each category, the number and definitions of areas to address, and point allocations have changed marginally. Actually, the structure of the criteria has held up very well considering the speed with which it was prepared.

18.7 QUALITY SYSTEM STANDARDS AND STANDARDIZATION

As pointed out in Sec. 18.1, the movement toward industrial standardization began in Great Britain in about 1900. Generally, the focus has been on standard weights and measures, categorizations of common fasteners, parts and assemblies, and symbols and definitions. Probably the first standard on quality control in the United States was the *Manual of Presentation of Data* published by the American Society for Testing and Materials in 1933.[†] Next to come were the American War Standards Z1.1—1941, *Guide for Quality Control;* Z1.2—1941, *Control Chart Method of Analyzing Data;* and Z1.3—1942, *Control Chart Method of Controlling Quality during Production.* While these were followed by a series of U.S. military standards dealing with various types of acceptance sampling plans and reliability test procedures, the first U.S. attempt at establishing quality system standards came in the form of a military specification, MIL-Q-9858, *Quality Program Requirements.*[‡] This specification applies to acquisitions by the armed forces of the United States, not to the consumer goods producers. It spells out requirements in quality program management, facilities and standards, control of purchases from suppliers, and manufacturing control including traceability requirements and documentation in all areas. Other special-purpose standards have been developed such as the one for regulating nuclear power plants and the one for regulating the manufacture of medical devices.

The most ambitious effort was mounted by the International Organization for Standards (ISO) in the mid-1980s in preparation for the full integration of the European Community (EC). The quality system standard is designated as ISO 9000, but in reality, it is a series of standards numbered 9000–9004. The U.S. equivalent of these standards is ANSI/ASQC Q9000 numbered Q9000–Q9004. Many nations are either adopting the ISO series or adapting it to their circumstances. In the United States, the Department of Defense has decided to abandon MIL-Q-9858A in favor of the ISO series. The documents are being updated and expanded on a continuous basis. Beginning in 1993 EC companies registered or certified under ISO 9000 may not be barred from trading within the community. Most U.S. government contractors and multinational corporations either have obtained ISO registration or are seeking it.

ISO 9000/Q9000 is entitled *Quality Management and Quality Assurance Standards: Guidelines for Selection and Use.* It is intended to guide a company toward the selection of the appropriate standard requirements. The next four standards are intended to be applied to different kinds of business operations. ISO 9001/Q9001 applies to firms that design, develop, produce, install, and maintain products. ISO

[†]The latest version of this booklet is *ASTM Manual on Presentation of Data and Control Chart Analysis,* STP 15D (Special Technical Publication 15D), American Society for Testing and Materials, 1916 Race Street, Philadelphia, Pa., 1976.

[‡]U.S. military specifications, approved by the Department of Defense, are a higher-order document than military standards. The latest revision, MIL-Q-9858A, is dated December 16, 1963. Within it, two other military specifications are incorporated by reference: MIL-I-45208, *Inspection System Requirements,* and MIL-C-45662, *Calibration System Requirements.*

9002/Q9002 applies to firms seeking accreditation for production and installation only. ISO 9003/Q9003 applies to firms seeking accreditation for final inspection and test only. ISO 9004/Q9004 contains guidelines for implementing quality management and quality system elements and is not used in third-party registration. It is being prepared in a series of parts that apply to different businesses, such as services and processed materials, or to different program elements, such as quality planning and configuration management.

A comparison of the Baldrige criteria with the ISO 9000 requirements is included in Chap. 19 where we discuss these new management models and relate them to other problem-solving models and approaches involving the use of statistical quality control. Suffice it to say here that there are substantial commonalities between the Baldrige criteria and the ISO 9000 requirements as well as substantial differences. The Baldrige criteria are not prescriptive with respect to the tools and procedures suggested for such activities as process control. Those who formulated the criteria believed that companies should be free to try whatever they believe will work. Their successes and failures will thus advance knowledge of quality improvement strategies fastest. The ISO requirements insist on the development of a "quality manual" and then expect the company to abide by what they say they are going to do. ISO 9004 is meant to serve as a guide in the preparation of that manual. It does suggest that such tools as control charts and scientific sampling plans are appropriate.

18.8 AN APOLOGY TO THOSE NOT MENTIONED

This brief retrospective on the history of statistical quality control, mainly in the United States, has left unmentioned many people and events that have had substantial impact on where the quality movement stands now, in 1995, and where it is headed as we approach the twenty-first century. Some contributors will be mentioned in association with particular techniques, philosophies, or events for which they are noted. Since this is not a book on quality management, we have attempted to limit discussion of that very complicated topic to those matters that impinge on statistical quality control just as we did with many topics in statistical analysis and experimental design. The reader interested in the history of quality should review the indexes to *Quality Progress,* the monthly magazine of the American Society for Quality Control, and its Quality Press *Publications Catalog.*

19

MODELS FOR QUALITY MANAGEMENT AND PROBLEM SOLVING

Quality management is not just a strategy. It must be a new style of working, even a new style of thinking. A dedication to quality and excellence is more than good business. It is a way of life, giving something back to society, offering your best to others.

—George Bush[†]

19.1 EVOLUTION OF THE "QUALITY WHEEL"[‡]

When Deming delivered his 1950 lectures in Japan, he presented two quality improvement models based on Shewhart's spiraling circular model, his "new" scientific approach to improving production. (See Fig. 18.1.) The model presented in the 8-day lecture to JUSE engineers is shown in Fig. 19.1*a.* It more resembles an eight-bladed rotating fan. The model presented to executives, Fig. 19.1*b,* resembles more a four-spoked wheel and illustrated a process more in line with Shewhart's model. It is Fig. 19.1*b* that Japanese industrial management referred to as the "Deming cycle," and others have referred to it as the "Deming-Shewhart cycle." These early models were the beginnings of the formulation of much more complex, and wordy, management models.

By 1959, when articles on control charts began to appear in JUSE's *Quality Control* magazine, Japanese writers began to differentiate between the Deming

[†]President George Bush, quoted from the cover of the *1992 Award Criteria, Malcolm Baldrige National Quality Award.*

[‡]Kozo Koura, "Deming Cycle to Management Cycle," *Societas Qualitatis,* Vol. 5, No. 2, JUSE, Tokyo, May/June 1991.

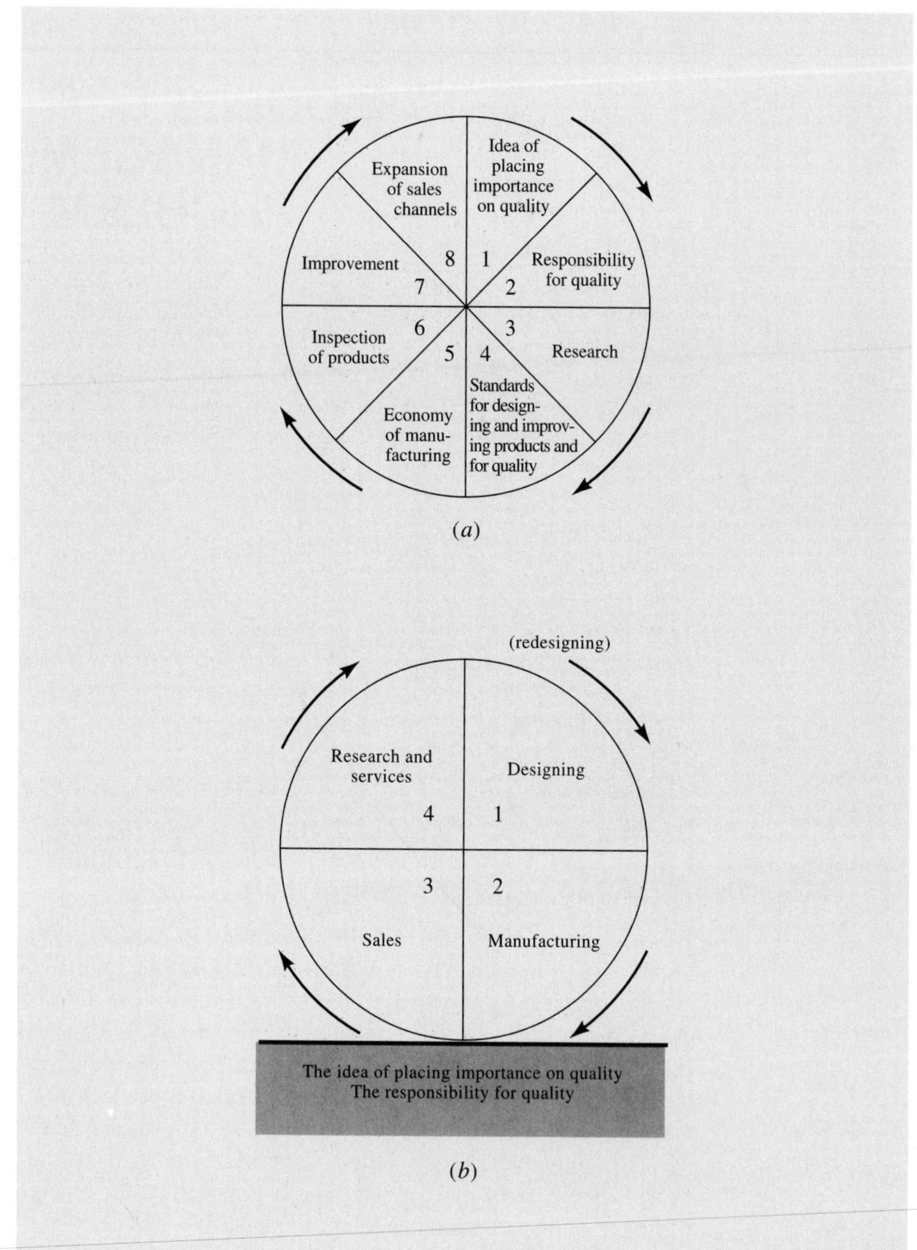

FIGURE 19.1 Deming's wheel of quality control as presented in 1950 *(a)* in the JUSE lectures to engineers and *(b)* to Japanese executives.

cycle, with its four spokes (designing, manufacturing, inspection and sales, and research and service), and the emerging management cycle, with its four spokes (planning, doing, checking results, and taking action). The most familiar form is shown in Fig. 19.2. This more generalized model, often referred to as the "PDCA cycle," has been adopted in many countries because of its broad application to business areas beyond manufacturing operations. More complex versions—some of which incorporate a standardize, deploy, verify, act (SDVA) cycle within the PDCA cycle—often are seen in the literature of total quality management. Many continuous improvement team studies use the cycle as a device around which to organize their effort. (See Fig. 19.9.)

ANSI/ASQC Standard 9004 (ISO 9004), which is designed to provide guidance for developing internal quality systems and manuals, includes a figure that is called the "quality loop." Shown here in Fig. 19.3, it resembles somewhat the wheel of quality presented by Deming to the JUSE engineers. In addition to showing somewhat greater detail, it introduces such topics as procurement, packing and storage, and disposal after use.

These simple graphics are intended to convey to the uninitiated that the quality improvement process is a continuous process, which, like the wheel of quality, must be kept continually rotating. It also acts as a *reminder* to the initiated and helps provide a focus for improvement efforts.

19.2 NAMES USED TO DESCRIBE THE QUALITY MANAGEMENT MOVEMENT

Early in the development of Japan's quality movement, the Japanese equivalent of the phrase *total quality control* (TQC) was adopted to describe the activities and

FIGURE 19.2 Illustration of the plan-do-check-act (PDCA) management cycle.

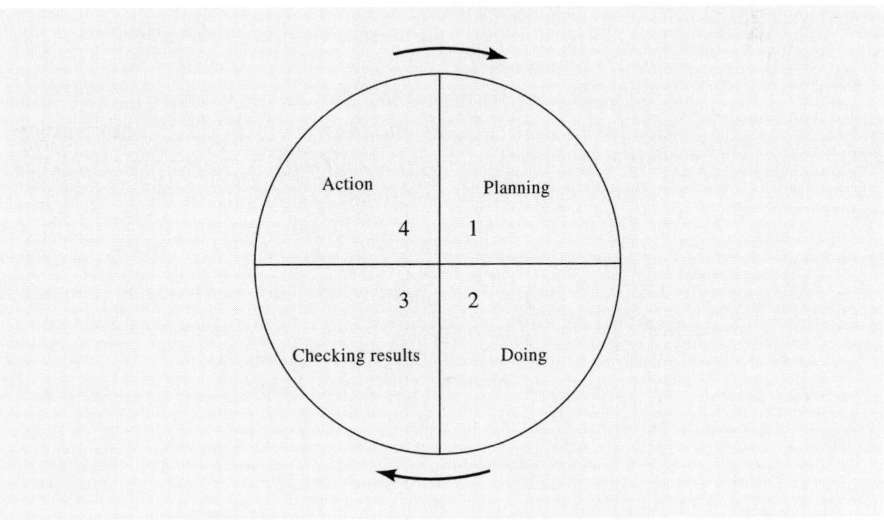

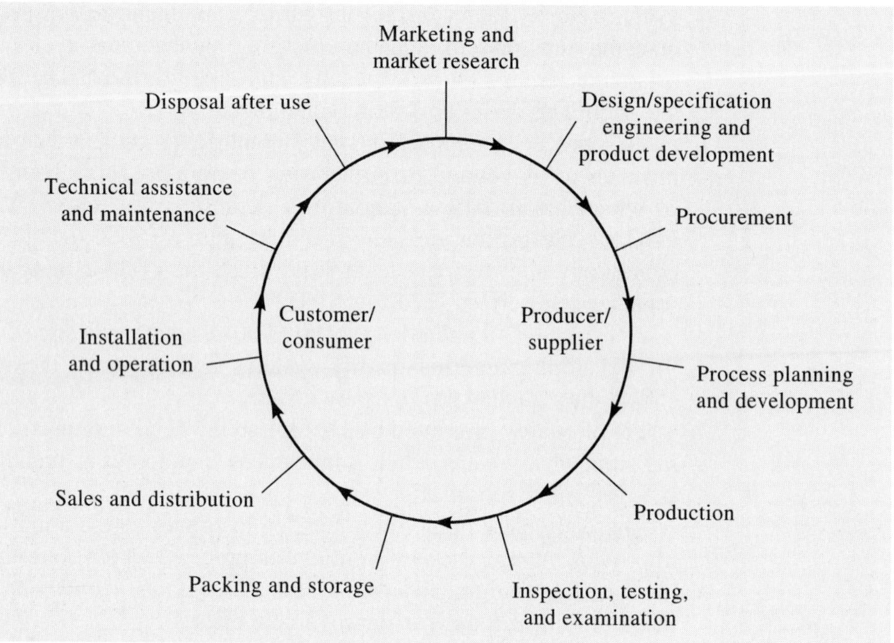

FIGURE 19.3 The quality loop as shown in ANSI/ASQC Standard Q94-1987.

the systems they were implementing. At the time, this was the title of a new book by Armand V. Feigenbaum.† That name continues in popular usage in Japan and elsewhere. Later, the Japanese coined the phrase *companywide quality control* (CWQC) to emphasize the fact that the tools and systems were to be applied to all aspects of company activities, not just to manufacturing. Some U.S. companies have adopted CWQC to describe their philosophy and systems. At least one company has expanded on that name calling their effort *customer-driven companywide quality control* (CDCWQC).

The U.S. government, as well as many other private and public concerns, uses the name *total quality management* (TQM). Other names include *total quality leadership* (TQL), *continuous quality improvement* (CQI), *continuous measurable improvement* (CMI), or simply *continuous improvement* (CI). Well-intentioned protagonists of one name or the other will argue at length over the fine points and differences. However, all involve the use of essentially the same structured approach to problem solving as well as the same management and engineering tools. The name selected by any organization seems to be more related to the predisposition of top management than to any real difference in methodology employed. In the mid-1990s the names *business process simplification, business*

†Armand V. Feigenbaum, *Total Quality Control,* 3d ed., 1983, McGraw-Hill Book Co., New York. The first edition was published in 1951. Dr. Feigenbaum, a consultant, was for 10 years manager of worldwide manufacturing and quality control for the General Electric Company.

process improvement, and *business process reengineering* were being used to describe essentially the same analytical process. Nevertheless, *total quality management* seems to be the name that is sticking in the U.S. published literature just as *total quality control* stuck in Japan.

The student of quality control needs to be aware that there can be many names for essentially the same thing. No doubt other names will become popular as time passes. An examination of the steps, processes, and procedures will reveal rather quickly whether the new name results from a real breakthrough in management technology and thinking or simply expresses the preferences of some group or some company.

19.3 THE BALDRIGE CRITERIA AS A QUALITY MANAGEMENT MODEL

By whatever name, the criteria for the U.S. *Malcolm Baldrige National Quality Award* are recognized as the road map to implementation of management for quality strategies, both nationally and internationally. If that is the case, then it is reasonable to turn for a definition to the core values on which the criteria are based. The authors suggest the following; key words and phrases have been italicized:

> *Total quality management (TQM)* is a *philosophy* and *system* that focuses on *customer satisfaction* in terms of *continuous improvement* of the quality of products and services. It embodies *leadership* and personal *commitment* by top management in fostering *team building,* as well as *employee involvement* and *empowerment.* TQM is a management system dedicated to making *decisions based on facts, data and analysis.* Improvement is driven by the twin objectives of *providing superior quality* at competitive cost and *shortening response time* to customer needs, both of which confer market advantages. Progress, measured against *quantitative performance indicators,* is assessed regularly to derive information for future *cycles of improvement.*

Each of the key words and phrases in the definition of TQM plays an important role in the Baldrige criteria themselves and in the assessment process required of an applicant. A brief description of each of these elements follows. The structure and interrelationship among the categories are illustrated in Fig. 19.4.

19.3.1 Philosophy

The philosophy is one of participative management. It is strongly influenced by Deming and many other management experts. It calls for decision making to be driven as far down into the organization as possible. Teams are used at all levels from frontline workers, what some call "natural work teams," to vertical functional interdepartmental teams, to horizontal cross-functional teams involving all levels of management. The philosophy calls for the establishment of partnerships with employees, suppliers, customers, educational institutions, and even strategic alliances with companies that otherwise might be considered competitors.

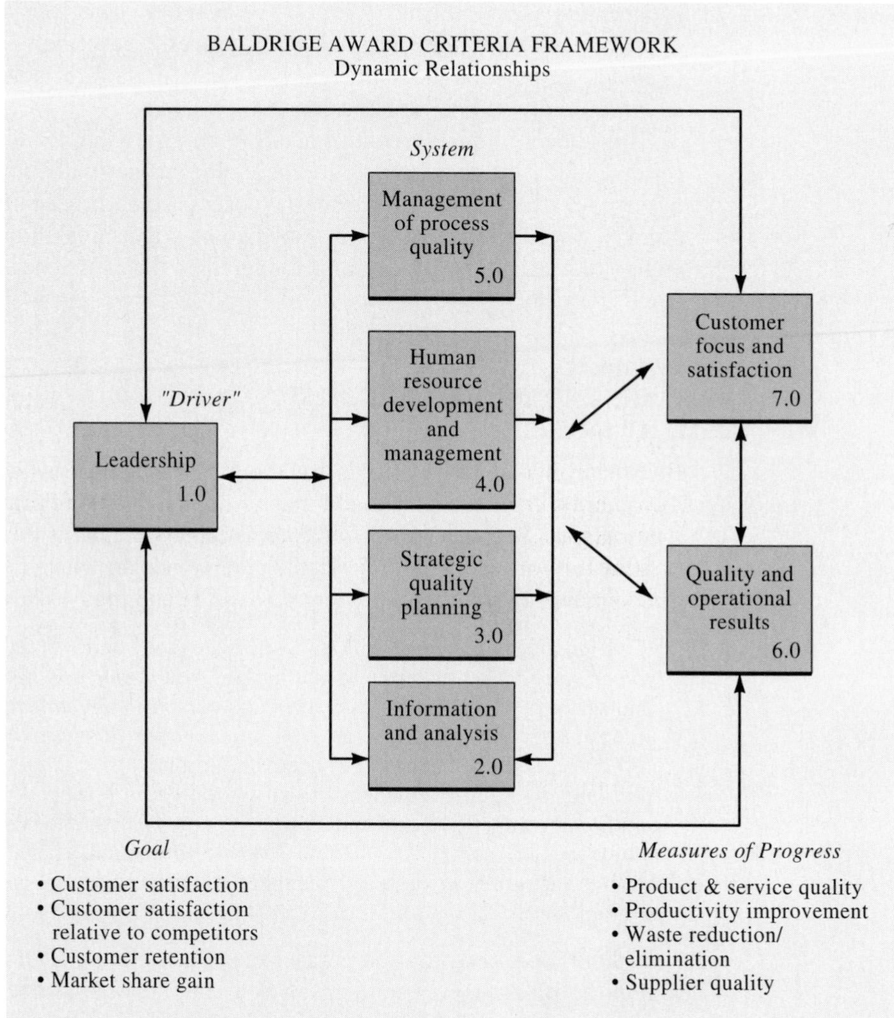

FIGURE 19.4 Structure and dynamic relationships of the Baldrige Award criteria. *(Source: 1995 Award Criteria, Malcolm Baldrige National Quality Award, p. 5, NIST, U.S. Department of Commerce.)*

19.3.2 System

The Japanese place great emphasis on the "system." It involves a number of elements including simultaneously set quality, response time, and productivity objectives; careful documentation of all processes; constant measurement and analysis; and open channels of cooperation, communication, and feedback. As Fig. 19.4 indicates, the Baldrige criteria divide the system into four parts: *information and analysis, strategic quality planning, human resource development and management,* and *management of process quality.*

19.3.3 Customer Satisfaction

Quality is judged by customers. For quality to be customer-driven, the system must be focused not only on meeting customer needs but on exceeding expectations wherever possible. This is a strategic concept and needs to address all stakeholders, including direct customers and final consumers, employees, suppliers, stockholders, the public, and the community. Customer-perceived value includes satisfaction and preference and must be measured in order to establish trends and make competitive comparisons. The Baldrige criteria recognize customer focus and satisfaction as the goal of the organization affecting all stakeholders.

19.3.4 Continuous Improvement

Increased competitiveness demands a well-designed and continuously executed approach to quality and productivity improvement. As the 1995 criteria state, it must be so embedded in the way the organization functions that "(1) improvement is part of the daily work of all work units; (2) improvement processes seek to eliminate problems at their source; and (3) improvement is driven by opportunities to do better, as well as by problems that must be corrected." It is directed at all aspects of the business in the design of products, processes, and support services. The focus is on problem prevention and relies on repeated cycles of planning, execution, and evaluation as Shewhart described.

19.3.5 Leadership

The criteria define the organization's top leadership as the driver of the entire system. (See Fig. 19.4.) Just as Deming insisted that Japanese top business executives take control of quality in 1950, the criteria recognize that it is up to management to establish the mission and to formulate the visions and values. It is through the management structure that policy is deployed throughout the organization.

19.3.6 Commitment

Building trust and confidence at all levels within an organization requires commitment by top management to the people in the system. This commitment must extend to individuals, to teams, and to partnership relationships with customers and suppliers.

19.3.7 Team Building

Deming's fourteenth point calls for management to put everyone to work in teams to accomplish the cultural transformation he envisioned. The criteria call for full participation and for reward systems that recognize both individual and team achievements at all levels in the organization. Partnering is encouraged, both internally, as among different work centers, and externally, as among suppliers and customers.

19.3.8 Employee Involvement and Empowerment

Category 4 in the criteria focuses on the management and development of the organization's human resources. Increasing skills and motivating the workforce are at the heart of improving performance. New skills require training and education in new technologies and in problem-solving methods (enlightenment). An empowered workforce must be furnished the appropriate tools (enablement) and be taught to make rational decisions. Developing trust and confidence among managers and workers also requires a concentration on safety, health, well-being, and morale.

19.3.9 Decisions Based on Facts, Data, and Analysis

Category 2 focuses on the organization's information system, on the measurement systems used to support decision making, and on how the data elements are selected, measured, and analyzed to extract greater meaning. Hunch decisions are not best, although they may be right on occasion. The simple fact is that "if you don't measure it, you can't improve it." This is the place where statistical tools provide the backbone of quality assurance.

19.3.10 Provide Superior Quality

Technically, quality begins in design. This is true whether you are dealing with products, processes, or services, and includes postdelivery support. As pointed out in Chap. 17, most product liability action stems from design error, not production or service error. In general, the costs of preventing problems and waste at the design stage are much lower than costs of corrective action downstream.

19.3.11 Shorten Response Time

How quickly we are able to react to shifting market conditions can affect competitive position for many years to come. A delay of just a few months in bringing a product or service to market can impact market share for the better part of a decade. New techniques such as concurrent or simultaneous engineering and business process simplification are aimed directly at cycle time reduction and fault prevention.

19.3.12 Quantitative Performance Indicators

These indicators must address customers, products and services, operations, markets, suppliers, employees, costs, and finances. Some indicators at the company or corporate level may derive from those at the functional or process level. For example, a large manufacturing operation may entail the control of thousands of critical dimensions using the tools of statistical quality control. Obviously, top management control mechanisms should not be expected to review thousands of charts. This information must be condensed, perhaps by charting the instances of deficiencies or the percentage of processes not meeting quality level targets.

19.3.13 Future Cycles of Improvement

The act of carrying on improvement projects is not new to business. The new ingredient, if you can call an idea suggested by Shewhart in the 1920s new, is the concept of continuous improvement or repeated cycles of improvement. Continuous improvement requires a long-range outlook and a long-term commitment to all stakeholders.

While it might appear to some readers that statistical analysis plays a very small role in this quality management model, it cannot be emphasized strongly enough how important that role is. Nor is the usefulness of the statistical concepts and tools limited to private sector producers of goods and services. They have been applied successfully in government organizations at all levels, in law offices, to health care providers, and to educational institutions. Some organizations refer to this universal approach to quality as "big Q." When the concepts and tools are applied in a very limited way to primary goods and services, they refer to it, in contrast, as "little q."

Relatively few U.S. companies apply each year for the Baldrige Award; nevertheless, distribution of the criteria numbers well over 100,000 copies annually. Many use the criteria as a model for internal self-improvement. This is true even of companies that have no intention of ever applying for the award. Many companies—including IBM, Westinghouse, and AT&T—use the criteria as a basis for internal awards presented to their best-performing operating units or divisions.

19.4 COMPARING THE BALDRIGE CRITERIA AND ISO 9000 REQUIREMENTS

The purposes of standards certification or registration and of quality awards are quite different. Certification does not guarantee the quality of any particular product or service. What it does is to ensure clients/customers that a firm's systems and procedures conform to certain minimum requirements. The ISO assessment process is focused on ensuring that a company is following its own specified quality practices.[†] Its purpose is to enhance and facilitate trade across national borders. The requirements may be very prescriptive and, in some cases, may require actions that are in conflict with company policy or management philosophy.

The objective of the Baldrige Award is to enhance U.S. competitiveness. The main purpose of quality awards is to recognize innovative activities by management and excellence in service to customers. In this regard, the Baldrige criteria endeavor to be nonprescriptive. For example, the ISO 9000 assessment process requires that the firm develop a manual describing its quality assurance process, procedures, and documentation. The Baldrige criteria place no such demand. However, the organization seeking the award would be well advised to have a quality manual handy should it receive a site visit. Certification requirements usually focus on primary products and services. Whether or not a firm has extensive

[†]For many firms, these "minimum requirements" are very demanding in comparison to the firm's current level of performance.

employee recognition programs or uses business process improvement teams in its billing department is of little or no concern. Quality awards generally focus on all aspects of an applicant's operations with a heavy emphasis on customer satisfaction, competitive quality results comparisons, executive leadership, employee involvement, and public responsibility.

Figure 19.5 shows the quality assessment elements common to both the Baldrige criteria and ISO 9000 and those that, at least in 1995, may be considered unique to each. It is not unlikely that, in the course of time, more elements will shift from a "unique" column to the "common" column. Surely certification and registration processes and quality awards will impact corporate and government activities for many years to come. Particularly in the United States, where manufacturing wages traditionally have been quite high, it is vital that corporations

FIGURE 19.5 Comparison of quality elements contained in the Baldrige criteria and those in the ISO 9000 requirements.

ISO 9000 Unique

- Contract review
- Product ID & traceability
- Inspection & testing
- Calibration
- Inspection test status
- Control of nonconforming material
- Packaging & preservation

Common

- Management responsibility
- Quality system
- Design control
- Document control
- Purchasing
- Purchased product
- Process control
- Corrective action
- Quality records
- Quality audits
- Training & education
- Field servicing
- Statistical techniques
- Quality values
- Quality management
- Quality data & analysis
- Process design
- Process control
- Assessment
- Documentation
- Supplier quality
- Quality results
- Supplier quality results

Baldrige Criteria Unique

- Executive leadership
- Public responsibility
- Competitive comparisons
- Strategic quality planning
- Human resource mgt.
- Employee involvement
- Employee recognition
- Employee well-being
- Continuous process improvement
- Business process/support service quality
- Bus. process operational & support serv. results
- Customer satisfaction

become and remain competitive. Slogans and sales promotions cannot make up the difference. As Dr. Harry G. Romig said in a 1981 interview, "There is far too much interest, time and money devoted to trickery in advertising, with a disproportionate amount of interest, time, and money given to quality. Marketing experts will tell us that their strategies pay off. But if that is true, why are we losing so much business to the foreign market?"

19.5 PROBLEM-SOLVING MODELS COMPARED TO MANAGEMENT MODELS

At the heart of all management models are the analytical tools that not only show when processes are operating as they should and the products and services meet specifications but show factually when improvements occur as well as levels and trends in meeting customer expectations. Global models for management do not provide much guidance when specific quality problems or improvement opportunities present themselves.

One of many strong points of Japanese business management in the latter half of the twentieth century has been its reliance on systematic approaches to problem solving. Thus, major producers, eager to enter the export market following World War II, adapted easily to new insights and tools. Our discussion of the Shewhart-Deming wheel of quality, and its evolution into the plan-do-check-act (PDCA) management cycle, illustrates the ease with which Japanese business leadership could absorb these concepts into their organizational culture.

While the steps in the structure may vary slightly depending on the publication or expert consulted, the nine-step model shown in Fig. 19.6 outlines the procedure. Not every step or substep may be required for every project. Application and emphasis will depend on the nature of the problem encountered.

The model applies to many different levels of quality improvement effort from the basic shop floor or frontline service level to design, new product or service introduction, and upper-level functional or cross-functional management projects. Step 9 deserves special mention because, in the fast-paced rush of our working world, it is often forgotten. It requires the team to critique its problem-solving effort to determine which actions worked well, which could be improved, and what each team member gained from the experience. Thus, as participants in the team effort move to new projects, they take with them greater knowledge and skill at problem solving.

The model outlined does not mention what tools may be appropriate at each step. It is more general in describing the problem-solving process. Indeed, a recent compilation of "continuous improvement tools" prepared as a poster by a leading U.S. electronics manufacturer listed over 100 "tools," some of which would better be described as entire tool kits. For example, control charts are called "a tool" as are cause and effect diagrams. While the construction and interpretation of the cause and effect diagram may be addressed in a page or two, the construction and interpretation of the many types of control charts that are available occupy at least half of this book. Many of the suggested tools are not statistical techniques. Nev-

1. Identify and/or define a problem
   a. List and prioritize issues as necessary
2. Organize a team
3. Study the current process
   a. Understand the process
   b. Analyze symptoms
   c. Understand customer needs
   d. Collect and analyze data
4. Determine root causes
   a. Formulate theories of causes
   b. Test the theories
   c. Set performance improvement goals
5. Identify improvement alternatives
   a. Establish immediate countermeasures
   b. Develop alternative solutions
   c. Design solutions and controls
6. Evaluate alternatives
   a. Recommend appropriate solution
   b. Set process targets, control points, measurement methods
7. Implement selected alternative
   a. Address resistance to change
   b. Conduct needed training
   c. Check performance, results versus targets
8. Establish control mechanisms to prevent retrogression
   a. Set control points and measurement methods
   b. Standardize procedures and deploy
   c. Monitor control system continually
   d. Identify unresolved issues and future plans
9. Analyze and critique the problem-solving procedure used (for improvement)

FIGURE 19.6 Illustration of steps involved in a problem-solving model.

ertheless, they do support a logical approach to problem solving and thus provide a framework for the efficient use of the statistical tools.

19.6 THE "SEVEN BASIC TOOLS OF STATISTICAL PROCESS CONTROL"

In 1971, the Asian Productivity Organization published an English translation of Dr. Kaoru Ishikawa's book *Genba no QC Shuho.* Professor Ishikawa had compiled the book as a study guide and shop floor training manual from articles published in the JUSE magazine *Quality Control for the Foreman.* Word spread quickly that this was the best book on quality control as it was practiced in Japan. The book was revised in 1976 and again in 1982, and many U.S. companies adopted *Guide to Quality Control*[†] as their training manual for supervisors and shop personnel. The tools described in its Chaps. 2 through 9 have become known in U.S. businesses as the "seven basic tools of SPC." Actually, Ishikawa's book contains 13 chapters; the other 4 are devoted to discussions on data collection, sampling principles, sampling for acceptance, and uses of binomial probability paper, with a final chapter devoted exclusively to practice problems.

Some organizations have modified the list by dropping one or more tools and adding others that they consider more "basic" to their operations. In all, one can identify at least 15 that might be called "basic tools and approaches" to facilitate successful accomplishment of quality improvement objectives. Figure 19.7 shows an augmented list that was compiled from several lists of "seven basic tools" used by various companies in the United States. Those included in "Ishikawa's basic seven" are indicated with an asterisk (*). While most have been covered extensively elsewhere in this book, a few deserve brief description at this time.

19.6.1 Process Flowcharts

Flowcharting is not a new technique, nor is it necessarily associated with quality control. Its early development is associated with work measurement, methods simplification and standardization, and computer systems analysis. Many standardized sets of symbols have been developed. Many stationers carry plastic templates of standard symbols. Special software is also available for use on personal computers. Some users prefer a simplified version that employs only boxes to indicate

[†]Kaoru Ishikawa, *Guide to Quality Control,* 2d revised ed., Asian Productivity Organization, Tokyo, 1982. This book may be obtained from the American Society for Quality Control, Milwaukee, Wis.

FIGURE 19.7 The augmented list of "seven basic tools" of statistical process control.

1. Process flowcharts
2. Cause and effect diagram*
   a. Action and effect diagram
3. Brainstorming
   a. Multivoting
   b. Consensus building
   c. Nominal group technique
4. Data collection and analysis
5. Graphs, run charts*
6. Histograms*
7. Pareto diagrams*
8. Check sheets*
   a. Process distribution
   b. Defects/defectives
   c. Defect cause
   d. Defect maps (location)
9. Scatter diagrams*
   a. Stratification
10. Control charts*
    a. For attributes
    b. For variables

operations. Often they will use some symbol such as a circle to indicate check or inspection points and the types of control mechanisms used at those points. Figure 19.8 illustrates a flowchart of a continuous process improvement model. It is based on the PDCA cycle and indicates what tools might be appropriate at various stages in an improvement project team effort.

In conducting a process quality improvement study, the flowchart is an important ingredient in "understanding the process," Step 3a in the nine-step model of Fig. 19.6. Developing the flowchart for a manufacturing or service process is necessary in order to ensure that everyone involved has the same understanding of how the process really operates. Frequently process operators have insights that are unknown to their supervisors. Often there are more storage delays and operator checks than the "official" flowchart indicates.

19.6.2 The Cause and Effect (CE) Diagram

The cause and effect diagram was developed by Ishikawa initially for use by quality circles in discovering potential causes of a problem under study. Usually these were manufacturing-related problems involving too much variation in a process. However, their use has spread to all aspects of business activity whenever it is necessary to sort out causes and organize them into mutual relationships. Cause and effect diagrams are discussed in Chap. 9, Sec. 9.2.3, and are illustrated in Fig. 9.8.

The *action and effect diagram* is a logical extension of the cause and effect diagram. Once corrective actions have been formulated, they may be entered onto the action and effect diagram. This diagram appears much like the CE diagram (see Fig. 9.8) except that the action arrows point away from the prime arrow and the actions taken and resultant effects are indicated for each contributing factor. The diagram acts as a double check to ensure that all causes have been accounted for and countermeasures planned or implemented.

19.6.3 Brainstorming

Brainstorming is a group process technique for eliciting a large number of ideas from a group of people in a relatively short period of time. It may be used to suggest problems, enumerate causes, suggest solutions and the action steps required, or find ways to implement solutions. A session may be conducted in either a very structured manner or an unstructured manner. In a structured session, ideas are offered one at a time in turn. Thus each participant is pressured to contribute. Once each person's list is exhausted, he or she "passes" until the next round. In an unstructured session, group members offer ideas as they come to mind. One advantage of an unstructured session is that the atmosphere tends to be more relaxed. A disadvantage is that one or two members of the group may tend to dominate the discussion.

Regardless of the method used, the following sequence of activities is suggested:

1. The leader writes down the topic or problem in clear view of each participant.

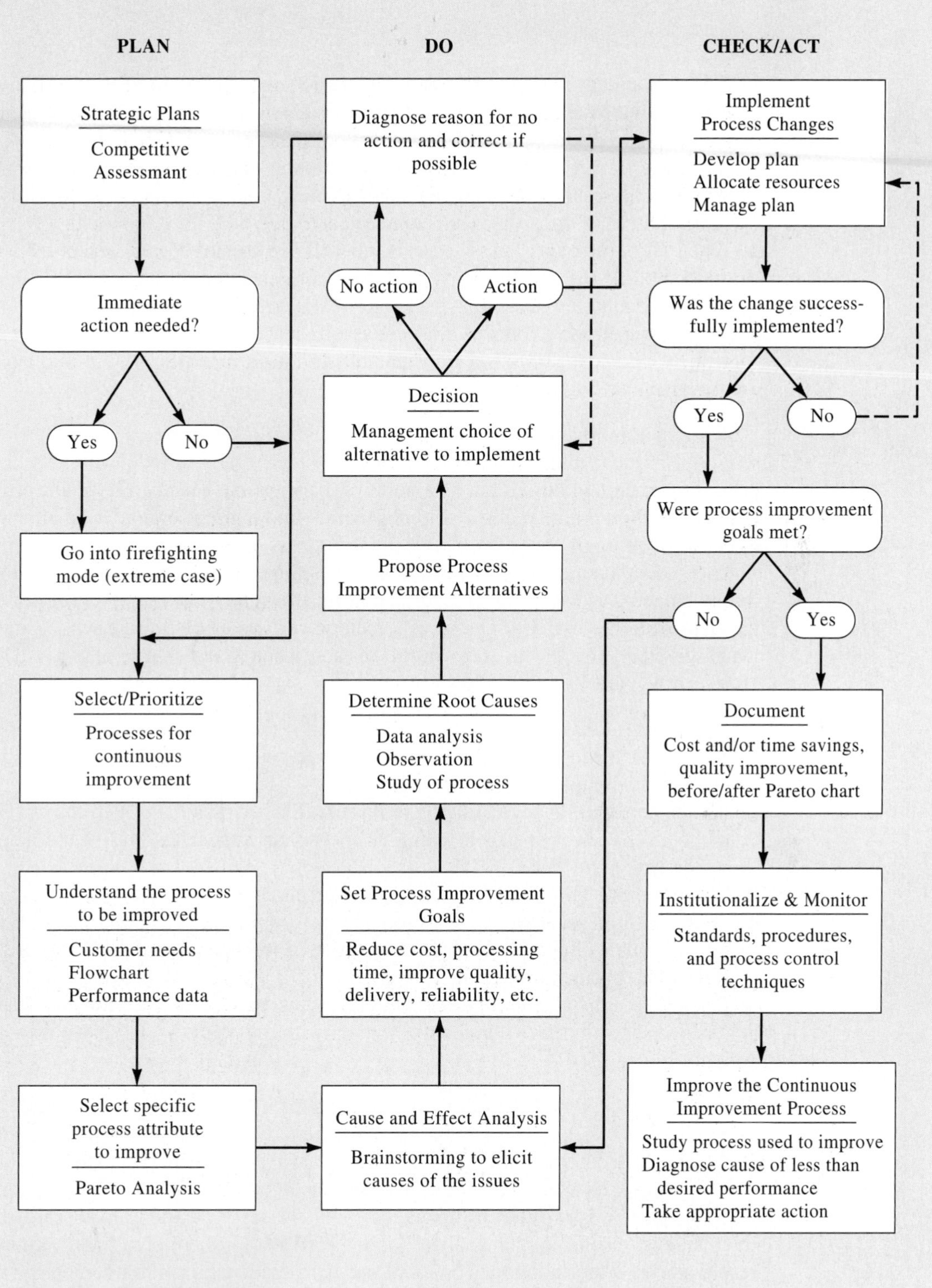

FIGURE 19.8 Flowchart of a continuous process improvement model, based on the PDCA cycle, and indicating where in the cycle certain tools might be used.

2. Each participant is asked to make a list of ideas addressing the issue. Five to fifteen minutes should be allowed for this activity. (On occasion participants may come to the session armed with lists of ideas generated beforehand.)
3. Operating in a round-robin fashion, each person presents an idea from his or her list. The session leader writes these in clear view of all. A participant may "pass" any time his or her list is exhausted.
4. When all participants "pass," that is, all lists are exhausted, the leader asks if the combined list suggests any new ideas not generated during step 2. This is called "hitchhiking" or "piggybacking."
5. Some group leaders, once a final impasse is reached, ask each person to suggest a "wild idea." Often a truly unrealistic idea stimulates a valid one from someone else.

It is important that certain rules of etiquette be followed during a brainstorming session. No one is allowed to criticize another's idea, orally or by gesture. No discussion is permitted other than as a point of clarification. Finally, the brainstorming session should not be allowed to degenerate into a gripe session or an attempt to blame; it is a part of the problem-solving process.

Once a list is completed, it is desirable to combine similar or related ideas, and it may be necessary to conduct a straw poll to select the most important or popular ideas from the list. *Multivoting* is a method for conducting a poll while limiting discussion. This is the stage in the session when discussion is allowed. The "rules of the game" are as follows:

1. Combine similar items. This is accomplished if the group agrees that the items are similar. Strong objections from the idea presenters should block the combination. The list is then renumbered.
2. Participants choose several items to discuss. Each writes his or her choices by number on a sheet of paper, listing no more than about one-third of the items on the list.
3. The votes are tallied by the session leader either by a show of hands or by secret ballot, if desired.
4. Those items with the fewest votes (say, three or less) are eliminated or set aside for future consideration.
5. Steps 2, 3, and 4 are repeated until the objective of the session has been reached. The objective may be to come up with one item (a consensus) or with a prioritized list.

Items eliminated as the result of any voting strategy should not be discarded out of hand. Particularly when attempting to identify root causes of problems or the best of a set of alternative solutions, it is necessary to "go back to the drawing board" when a cause turns out to be wrong or incomplete or when a "solution" doesn't work or is rejected by management. It is never desirable to discard the intermediate results in the analysis of a problem. Furthermore, brainstorming results derive from individual opinion, not hard evidence or measurements. The evidence, pro or con, may come later in the problem-solving process.

19.6.4 Data Collection and Analysis

Data collection and analysis are at the heart of any quality assurance, TQM, CQI, or business process reengineering strategy. Data may be generated either on a continuous measurement scale *(variables data)* or as the result of counts of one thing or another *(attributes data)*. Most of this book is devoted to the use of sampling procedures and control charts to collect and analyze data.

19.6.5 Business Graphs and Run Charts

Business graphs are the types of charts found in all business magazines. Most spreadsheet software programs provide for the automatic conversion of tables of data into a variety of graph types. The main types are bar charts, pie charts, and line charts. Variations on these charts include X-Y plots, stacked bar charts, rotated bar charts, column charts, and text charts. The ability to produce three-dimensional charts allows either a third or Z axis to be introduced onto the chart or simply makes a two-dimensional chart appear more dramatic. Shading and coloring also make presentations more dramatic. Scaling may be accomplished on one or more axes. Column charts represent the data in a single column similar to a pie chart. By mixing lines and bars on the same chart, a variety of types of charts useful for communicating quality information may be produced.

A *run chart* is a special form of line chart. It is used primarily to illustrate change, hopefully improvement, over time. Examples of the kinds of indicators particularly adaptable to run charts are percentage of incoming phone orders answered by second ring, lost time injuries per 1,000 employees per month, percentage of machining operations achieving a specified level of quality at a point in time, numbers of billing complaints from customers each month, percentage of units approved on first inspection, number of hospital emergency room admissions per day or per shift, and so on. Figure 1.1*b* is an example of a run chart for the average of sample measurements drawn in subgroups of five with the order of production maintained. Run charts are popular for reporting to management the levels and trends over time of important business operational characteristics. On occasion they are mistakenly referred to as control charts by persons unfamiliar with the Shewhart control-chart technique.

19.6.6 Check Sheets

A check sheet is a manual graphical counting mechanism for enumerating the results of sampling. Often it is the beginning of the process of translating opinions into facts, or it is used as a beginning step in the manual preparation of a histogram. Examples of check sheets used to illustrate process distributions are shown in Figs. 3.1, 3.4, and 3.5. Check sheets also may be used to (1) illustrate the distribution of nonconformities in complex assemblies or rejected units in lots of product; (2) enumerate causes of nonconformances for use in a Pareto analysis; and (3) prepare nonconformance maps, that is, diagrams of a product or form indicating the location of the more prominent nonconformances.

Computer software used in process control usually eliminates the need for preparing check sheets. Charts, histograms, and other analytical results are calculated directly from the data and may be displayed on the screen or printed on hard copy or both.

19.6.7 Histograms

The preparation and use of histograms are discussed in detail in Chap. 3. Other graphical means of presenting frequency or count data include the check sheet, frequency bar chart, and frequency polygon. A histogram, bar chart, and polygon are illustrated in Fig. 3.2*a, b,* and *c.* Examples of histograms appear in many places throughout this book.

19.6.8 Pareto Analysis

Pareto analysis is a technique to assist in determining where to begin an attack on a problem or to display the relative importance of problems or conditions. The resulting chart is used to choose a starting point for problem solving, monitor success, or identify basic causes of problems. A method for preparing a Pareto diagram is discussed in Chap. 9, Sec. 9.2.2, and is illustrated in Fig. 9.7.

19.6.9 Scatter Diagrams

If, for example, it is known or suspected that two variables move together, a scatter diagram can be used to illustrate that fact. The technical name for this phenomenon is *correlation.* If the relationship is very strong and the value of one variable tends to increase when the other increases, then the points will be close to a sloping straight line and the relationship is said to show a strong *positive* correlation. If the value of one variable appears to decrease with an increase in the value of the other, *negative* correlation is said to exist. Typical applications of scatter diagrams compare product shelf life with active ingredient content, clerical errors with overtime hours, percent fiber elongation with moisture content, and the like. Some aspects of correlation studies and scatter plots are discussed in Chaps. 5 and 11.

19.6.10 Control Charts

It was stated in Chap. 1, Sec. 1.3, that one purpose of this book is to explain the control-chart point of view in some detail. That paragraph described the concept of a "constant system of chance causes" as suggested by Shewhart. Deming referred to this as *common causes of variation,* constantly operating within a system. Variation outside this constant or stable system, called *assignable cause* variation by Shewhart and *special cause* variation by Deming, may be detected using control charts and its cause or causes eliminated through analysis and corrective action. Often these actions result in substantial improvement in the quality of products and services and reductions in spoilage, rework, and error rates. Moreover, by identifying certain of the quality variations as inevitable chance variations, the control chart tells when to leave a process alone and thus prevents unnecessarily frequent adjustments that tend to increase variability rather than decrease it. Through its disclosure of the natural capabilities of a production process, the control-chart technique permits better decisions on engineering tolerances and better comparisons between alternative designs and between alternative production or service methods. Part Two of this book discusses the many uses of control charts and provides many examples of both good and bad applications.

19.7 TRACKING AND CELEBRATING QUALITY

Figure 19.9 (on the next two pages) illustrates a device for tracking problem-solving activities that is intended to serve a number of purposes not the least of which is to show off a process improvement team's results. The name given to the two panels is "storyboard" because it traces the activities of a team throughout the time of the study. This particular effort, which took place in a hospital anesthesia care unit, follows a PDCA model. It involved data collection and analysis, control-chart construction both before and after implementing prospective improvements, a cause and effect analysis, and a Pareto analysis.

As each part of a study is completed, the new elements are added onto the storyboard, which is always prominently displayed near a workplace control center. When the project is finished, the final storyboards display examples of significant facts contained in the final report and are used to demonstrate applications of quality improvement tools. Many organizations have adopted this means of promoting and supporting continuous improvement activities and use it as a basis for individual and team awards. In one form or another, storyboards can be seen in utilities, manufacturing plants, insurance companies, service companies, government and defense installations, as well as health care facilities.

QUALITY CONNECTION STORYBOARD

Facility: Premier Anesthesia/Atlanta, GA

Project Oxygen Desaturation

Contact Dale Valentine
Director of QI
(404) 458-4842

- **Identify need for improvement:**
  Reduce oxygen desaturation upon admission to PACU.
- **Establish definition of desaturation:**
  Any patient admitted to the PACU with an oxygen level below 95% for a sustained period of 30 seconds, after the initial 30 second period following admission.

- **Organize the Team:**
  Anesthesia
  OR Nurses
  PACU Nurses
  Transport Help

Step 1: PLAN ***Plan the Improvement and Continued Data Collection***

Data collection and analysis plan:

1. Collect data for one month weekdays only
2. Use a binominal distribution; the patient is either desaturated or not desaturated
3. Record each patient on a data check sheet on the PACU wall
4. Calculate mean, upper and lower control limits
5. Display data graphically using control chart

- **Calculation of control limits:**

Upper Control Limit	Lower Control Limit
UCL = p+1.96*sqrt(p(1-p)/n)	LCL= p-1.96*sqrt(p(1-p)/n)

Control limits measure the current system's capability. A confidence level of 95% was used to establish the control limits.

- The mean of the percent of patients desaturated is 20%. This indicates that 20% of the patients who recovered from anesthesia in this recovery room were desaturated using our operational definition. The data shows that there is a large variation in the system.

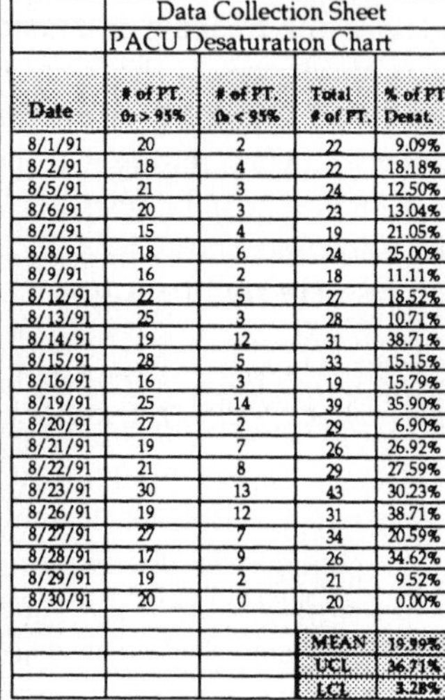

	Data Collection Sheet			
	PACU Desaturation Chart			
Date	# of PT. O_2 > 95%	# of PT. O_2 < 95%	Total # of PT.	% of PT. Desat.
8/1/91	20	2	22	9.09%
8/2/91	18	4	22	18.18%
8/5/91	21	3	24	12.50%
8/6/91	20	3	23	13.04%
8/7/91	15	4	19	21.05%
8/8/91	18	6	24	25.00%
8/9/91	16	2	18	11.11%
8/12/91	22	5	27	18.52%
8/13/91	25	3	28	10.71%
8/14/91	19	12	31	38.71%
8/15/91	28	5	33	15.15%
8/16/91	16	3	19	15.79%
8/19/91	25	14	39	35.90%
8/20/91	27	2	29	6.90%
8/21/91	19	7	26	26.92%
8/22/91	21	8	29	27.59%
8/23/91	30	13	43	30.23%
8/26/91	19	12	31	38.71%
8/27/91	27	7	34	20.59%
8/28/91	17	9	26	34.62%
8/29/91	19	2	21	9.52%
8/30/91	20	0	20	0.00%
			MEAN	19.99%
			UCL	36.71%
			LCL	3.28%

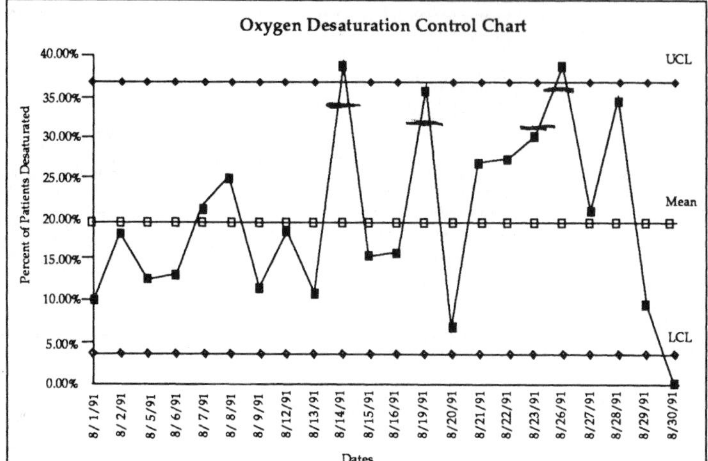

To determine the reasons for variation in the system, a cause and effect diagram is created. This technique identifies causes of problems by analyzing system components. Generally, a system can be viewed as being composed of people, processes, policies, and mechanical devices.

Cause and Effect Diagram

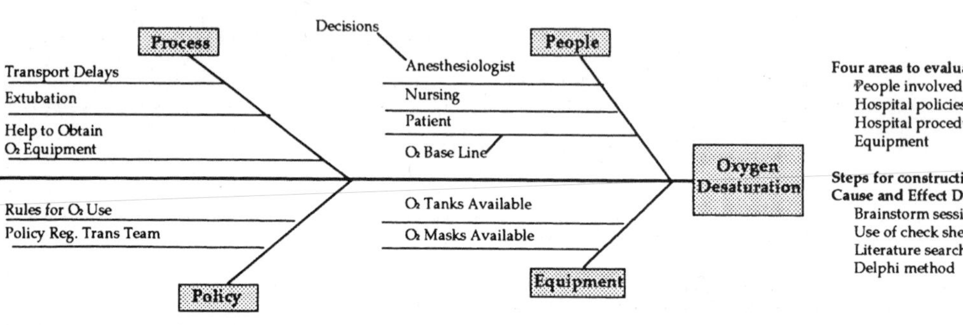

Four areas to evaluate:
People involved
Hospital policies
Hospital procedures
Equipment

Steps for construction of Cause and Effect Diagram:
Brainstorm session
Use of check sheets
Literature search
Delphi method

FIGURE 19.9 Example of a process improvement Storyboard model used to track progress and to display project results. (*Reprinted from Quality Connection, newsletter from the Institute for Healthcare Improvement, Vol. 1, No. 3, Feb. 1992, pp.8–9, 1 Exeter Plaza, Ninth Floor, Boston, MA 02116.*)

QUALITY CONNECTION STORYBOARD

Step 1: PLAN Continued..

After the cause and effect diagram has been completed, data is collected to determine which causes are the primary contributors to desaturation. A pareto diagram can identify the most common causes of desaturation.

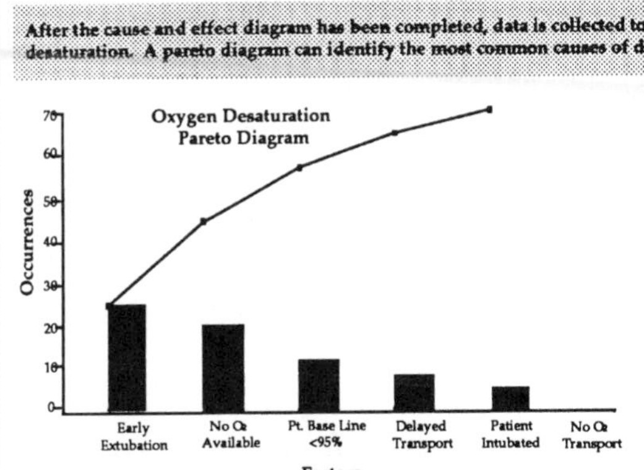

Steps for graphing the pareto diagram:

- Use check sheets to count the occurrences
- Graph using pareto graph
- Purpose is to help focus corrective actions

Conclusion:

It is determined after collecting data that early extubation and the unavailability of oxygen are the primary contributors to oxygen desaturation.

Step 2: DO *Do the Improvement, Data Collection, and Analysis*

After the primary contributors are identified, corrective action is taken.
This may involve a change in any one of the four components of the system (people, process, policy, equipment).

- Corrective actions to reduce problems of early extubation and unavailability of oxygen:
  **Check muscle relaxant level with nerve stimulator prior to extubation and establish patient's ability to support own airway by sustained neck lift*
  **Have oxygen tanks available for patient transport*
- Institute corrective actions and measure results by data collection system and graph results

Step 3: CHECK *Check and Study the Results*

The process is checked by continuing to collect data on desaturation.
If the system continues to have a large variation, Step 2 is repeated looking for other causes.

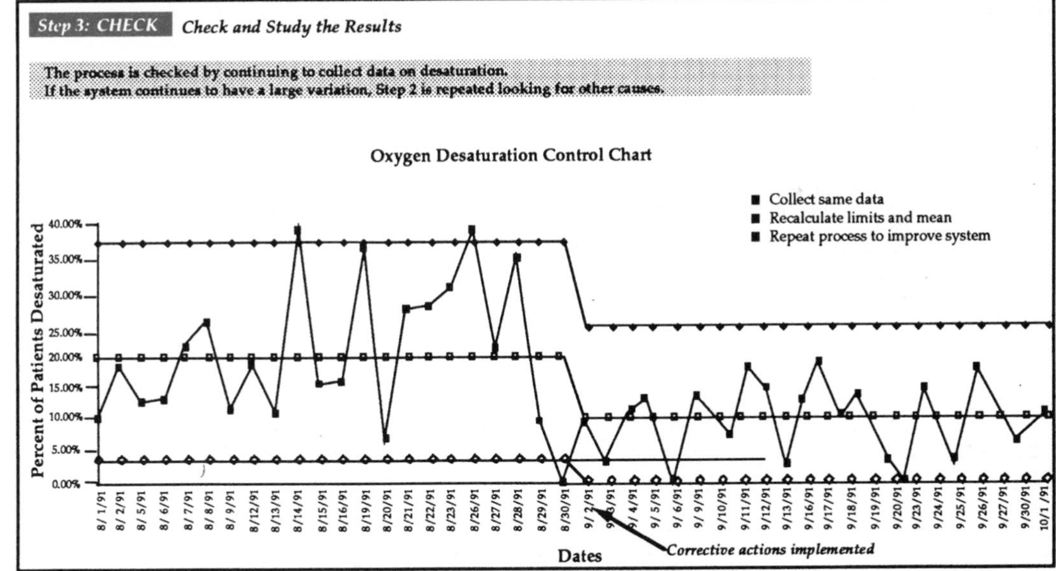

Step 4: ACT *Act to Hold the Gain and to Continue to Improve the Process*

Institute new policy and procedure for reducing oxygen desaturation.
Continue to check results of actions through sample selection of patients on a periodic basis.

FIGURE 19.9 *(continued)*

20

DEMONSTRATING THE OPERATION OF SYSTEMS OF CHANCE CAUSES

The most powerful mathematical tools are sometimes less important to the engineer than some of the simpler or less powerful tools. But often, for lack of information about either, neither is used.

—C. M. Ryerson[†]

20.1 USE OF GROUP EXPERIMENTS IN INTRODUCING THE SUBJECTS OF CONTROL CHARTS AND ACCEPTANCE SAMPLING PROCEDURES

In presenting an introductory course or seminar in statistical quality control (or, for that matter, in almost any other subject involving the operation of the laws of chance), it is helpful if 2 or 3 h is devoted to experiments that illustrate the operation of systems of chance causes. Such experiments are desirable in courses for college students as well as in their more usual setting of seminars given to supervisors, inspectors, and designers in industry. The best time for such experiments is near the start of the course or seminar.

In part, such experiments give the group a better feeling for the way in which systems of chance causes really operate. In part, also, they provide a kind of assurance of the reasonableness of certain important formulas and relationships. This is useful even to groups that do not have to take the formulas and relationships on faith. Many of us, doubtless somewhat illogically, find experimental demonstra-

[†]In the Foreword to D. M. Chorafas, *Statistical Processes and Reliability Engineering,* p. v, D. Van Nostrand Company, Inc., Princeton, N.J., 1960.

tion somewhat more satisfying than mathematical proof. Wherever practicable, the entire group should participate in each experiment (or at least enough of the group to make it obvious that the results are not being manipulated by the instructor).

A number of companies manufacture and market statistical quality control demonstration devices like those described in this chapter. Their advertisements appear in such magazines as *Quality Progress,* published by the American Society for Quality Control, Milwaukee, Wisconsin, and *Quality,* published by Hitchcock Publishing Company, Wheaton, Illinois. While computer software may be available to conduct certain demonstration experiments, physical devices usually will prove superior because of their "hands on" nature. Participants cannot get the feeling that the instructor is somehow manipulating the results.

20.2 DEMONSTRATING FREQUENCY DISTRIBUTIONS AND CONTROL CHARTS FOR VARIABLES

It is important for everyone dealing with variability of industrial product to have a feeling for the manner in which variation can occur even when a constant system of chance causes is operating. An experiment involving the drawing of chips, somewhat in the manner illustrated in our discussion of the Shewhart bowl, can help the members of a group to develop this feeling. Such an experiment can also illustrate the construction of $\bar{X}$ and R control charts and the estimation of universe parameters from relatively small samples (such as 5) and from relatively large samples (such as 125 or 150).

In their classes, the authors have used a cookie can containing several thousand chips, each with a number written on it. (See Chap. 4, Probs. 4.48 to 4.50.) First, the instructor draws a sample of five; the number on each chip is recorded on the chalkboard or on an overhead projector slide, and $\bar{X}$ and R are calculated and recorded for the subgroup. The chips are replaced, the can is closed and shaken, and one of the class now draws five chips. This drawing is repeated, with a different member of the class selecting the subgroup, until 25 subgroups have been obtained. (If there are enough chips in the universe, the timesaving practice of making replacement after each five chips rather than after each chip is good enough for practical purposes.) If matters are properly organized, a 50-min period is sufficient to conduct the sampling, compute $\bar{X}$ and R values as each subgroup is recorded on the chalkboard, compute the control limits, draw the control charts, and compute the estimated value of σ. Near the end of the period, the students are given a sheet containing the frequency distribution in the cookie can and its values of μ and σ.

Control charts and frequency distributions can also be demonstrated by the use of a *quincunx,* illustrated in Fig. 20.1. Such an apparatus was first developed by Sir Francis Galton in the 1890s. It might be thought of as an ancestor of the modern pinball machine. Beads drop through an arrangement of pins, eventually falling into any one of 25 parallel slots at the bottom of the inclined plane into which the pins are driven. The quincunx provides a visual demonstration of the effect of chance in producing variation within small subgroups, by holding sam-

FIGURE 20.1 Quality control chart demonstrator (quincunx). *(Photo courtesy Lightning Calculator Company, Troy, Mich.)*

ples of five beads above a slide located near the top of the slots, and the variation in a large frequency distribution. With the funnel located in the central position, it tends to give a distribution approximately normal (actually a binomial with $\mu_p = \frac{1}{2}$). By shifting the funnel to various positions, assignable causes may be introduced into the sampling procedure, giving rise to changed process averages and to skewed distributions.

20.3 DEMONSTRATING ACCEPTANCE SAMPLING BY ATTRIBUTES

It is advantageous to have at least one experiment devoted to variables and at least one devoted to attributes. The attributes experiment can be planned to serve several purposes. If designed to illustrate the operation of an acceptance sampling plan, it may also illustrate certain matters about *p* charts (or *np* charts) and the laws of probability.

A good way to demonstrate sampling by attributes is to use a sampling tray to scoop 50 beads out of a box containing a mixture of white and colored beads. Devices are available that operate on an open-box basis, as shown in Fig. 20.2, and on a closed-box basis, as shown in Fig. 20.3, wherein the beads cannot escape from the device. White beads are frequently designated as good articles; colored beads, as nonconforming ones. The fraction of colored beads in the box represents the μ_p of a production process. The fraction of colored beads in the tray after each scooping operation represents the fraction nonconforming in a lot of 50 from that process.

This type of tray is particularly well-adapted for use in a classroom experiment analyzing the acceptance plan $N = 50$, $n = 5$, $c = 0$ that was discussed in Example 1.3. (The authors have observed that many people who ought to know better are not aware of the serious weakness of this particular plan; this statement applies to persons of such different backgrounds as chief inspectors and Ph.D.s in mathematical statistics.) A box may be filled with, say, 1,880 white beads and 120 colored ones. The fact that the "process" actually is 6% nonconforming should not be disclosed to the class until the end of the experiment. When a member of the class scoops 50 beads out of the box, he or she has an inspection lot. The number of nonconforming units in each lot may be observed by the class and recorded. (For a process 6% nonconforming, inspection lots of 50 typically will vary from 0 to 7 or 8 rejects.) It should be emphasized that the class, which can observe the qual-

FIGURE 20.2 Open form of bead box used in quality control demonstrations. *(Photo courtesy Lightning Calculator Company, Troy, Mich.)*

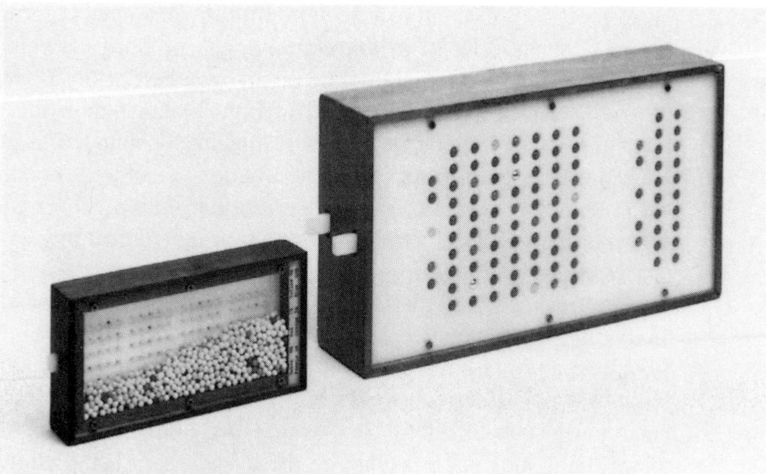

FIGURE 20.3 Closed form of bead box used in quality control demonstrations. *(Photo courtesy Lightning Calculator Company, Troy, Mich.)*

ity of the entire lot, has knowledge that would not be available to an inspector who would test only 5 of the 50 articles in the lot.

The beads from each inspection lot should be replaced in the box, and the beads in the box should be thoroughly mixed before the sampling tray is used to select the next lot. For convenience in carrying out the experiment rapidly, it is desirable that the sample be selected by choosing some one of the 10 vertical columns of 5 beads on the sampling tray. Each column on the tray may be marked with one of the 10 digits. If a 20-sided Japanese random decimal die is available, it can be thrown to select the sample from each inspection lot. Otherwise, a table of random numbers may be used. If the experiment is properly organized, a 50-min period is sufficient to arrive at acceptance or rejection decisions on 40 or 50 inspection lots, to tabulate the observations (somewhat along the lines of Table 1.3), to make the appropriate calculations, and to discuss conclusions.

Like the experiment involving drawing chips, this sampling tray experiment is a good one to use near the start of a course or seminar—considerably before any assignments dealing with acceptance sampling systems. It is helpful for the instructor to be able to make reference to such an experiment in initial discussions of various aspects of probability. Moreover, two class experiments—one involving variables and the other attributes—serve to dramatize the important difference between inspection by variables and inspection by attributes.

Even though, at the time of this class experiment, the group seems merely to be analyzing the plan $N = 50$, $n = 5$, $c = 0$, the results of the experiment may be used later to bring out a variety of other matters. For instance, the successive trays of 50 beads may be viewed as random samples from a process 6% nonconforming, and the numbers of rejects observed on successive trays may be plotted on an *np* chart; or the successive trays may be viewed as single samples with $n = 50$ from a lot or process 6% nonconforming, and the data from the experiment may be used to estimate the position of one point on each of the respective OC curves with acceptance numbers c of 0, 1, 2, 3, and so on.

While the majority of beads may be white or natural varnished oak, most commercially available bead boxes contain several colors of beads in different proportions, not just a single color. This permits the demonstration to be run at several different levels of μ_p. It also permits it to be run defining the different colors of bead as different causes for rejection. Thus Pareto diagrams may be prepared from the data if the count of each color of bead is maintained.

20.4 DEMING'S RED BEAD EXPERIMENT

This demonstration, which uses a bead box, is intended to illustrate Deming's conviction that most problems arise because of faults in the system, not simply as the result of operator error. Therefore, if the faults are to be eliminated or reduced, changes must be made in the system. The experiment gets its name from the sampling box containing white and red beads used by Deming in his many seminars to managers beginning in 1980.[†]

At the start of the exercise, class participants are told that the "X Bead Company" (where X is replaced by the dominant color of beads in the box or bowl) is in the business of manufacturing and marketing X beads. Five volunteers from the group are selected to become "production workers" in this experiment. Production of a lot of 50 beads is accomplished by using the sampling paddle to draw a sample of 50 beads from the box. A tally sheet and prescaled control chart are handed to each participant. As each worker presents a lot for approval, the count of "non-X" beads is recorded on an overhead projector foil and by each participant. For purposes of discussion, "non-X" beads will be referred to as red beads.

When only a few draws have been made, the instructor should halt "production" to point out to a worker that the sampling paddle contained some red beads. Clarify to all that the objective is to produce "X" beads and that red beads represent a loss to the company. The point being made is that, just because management makes objectives clear to the workforce, there is no assurance that they will be more able, or motivated, to meet them.

After several complete rounds of sampling, stop at a particularly high count of red beads. Advise the worker that retraining will be necessary, and have a worker with a low count demonstrate the technique employed. The point being made is that, when the difficulty is embedded in the system, retraining is powerless to correct it.

[†]According to information provided to the authors, this experiment evolved from work done by W. E. Deming for the Hewlett-Packard Company in the early 1980s. Initially it was referred to as the "Brown Bead Manufacturing Company" exercise. Deming had used the experiment as described in Sec. 20.3 in 1942 at Stanford and in 1950 in Japan to illustrate the weakness of a commonly used acceptance sampling plan. It was at Hewlett-Packard that the implications for management were introduced.

Following just one more cycle of sampling, announce a new team-based "incentive bonus" system in which each member of the team will receive a $5 bonus if, for any one cycle, the average count of red beads is less than *n,* where *n* has a less than 0.01 probability of happening. The point is that pay cannot motivate improved quality of production when the problem is in the system.

After several more draws, pause at some relatively high count and threaten to fire the individual if average performance doesn't improve by an amount that virtually ensures noncompliance. When the number is violated, fire the individual "just to make an example for others to see." This should demonstrate to the class that management pressure can be counterproductive when the problem is beyond the control of the worker.

When the fired worker is replaced and another round of sampling has taken place, ask someone in the class to predict the outcomes of the next two or three draws. The result will demonstrate that the exact outcome of any particular draw cannot be predicted with certainty.

At this point, control limits should be drawn on the chart to show that all workers had been performing "within the capability of the system" for the entire time. Variation in their outcomes was purely the result of chance. None of the management actions taken had a positive effect on the outcome because all were aimed at the workers, where the fault was assumed to lay; no action was taken to improve the system.

APPENDIX 1

GLOSSARY OF SYMBOLS

The symbols used in the literature of statistical methods are not fully standardized; the same statistical quantity may be represented by different symbols by different writers. Moreover, statistical problems are so numerous and diverse that any given symbol may be used in a number of different meanings, each meaning referring to a particular type of statistical problem. This lack of standardization complicates matters for the reader who consults a number of statistical books and articles.

The principle used in selecting symbols for this book has been to adopt for each statistical quantity that symbol which is most common in the literature of statistical quality control in the United States. In general, where applicable, the symbols used follow the standards of the American Society for Quality Control and the American National Standards Institute[†] and also agree, in most cases, with the symbols used in the publications of the American Society for Testing and Materials.[‡]

In the case of a few symbols, this conformance to common practice in the quality control literature has involved the use of one symbol with two or more meanings. However, in every case the appropriate meaning should be clear from the context. This duplication of meanings for a few symbols seemed less likely to cause confusion than the adoption for this book of symbols that are not in common use in the other literature of the subject.

[†]"American National Standard—Definitions, Symbols, Formulas, and Tables for Control Charts," ANSI/ASQC A1-1987, and "American National Standard—Terms, Symbols, and Definitions for Acceptance Sampling," ANSI/ASQC A2-1987, American Society for Quality Control, Milwaukee, Wis.

[‡]"ASTM Manual on Presentation of Data and Control Chart Analysis," STP 15D, American Society for Testing and Materials, Philadelphia, Pa., 1976.

A, B, C, etc.	in probability, used to designate events that may occur within a probability space.
A^C, etc.	the complement of the event *A*, etc. is $S - A$, that is, the occurrence of *not A*.
A	a multiplier of σ or σ_0 to determine the distance from central line to 3-sigma control limits on an $\overline{X}$ chart. It equals $3/\sqrt{n}$ and is given for values of *n* from 2 to 100 in Table *F,* App. 3.
A_1	a multiplier of σ_{RMS} to determine the distance from central line to 3-sigma control limits on an $\overline{X}$ chart. It equals $3/(c_2\sqrt{n})$ and is given for values of *n* from 2 to 100 in Table *E,* App. 3.
A_2	a multiplier of $\overline{R}$ to determine the distance from central line to 3-sigma control limits on an $\overline{X}$ chart. It equals $3/(d_2\sqrt{n})$ and is given for values of *n* from 2 to 20 in Table *D,* App. 3.
A_3	a multiplier of $\overline{s}$ to determine the distance from central line to 3-sigma control limits on an $\overline{X}$ chart. It equals $3/(c_4\sqrt{n})$ and is given for values of *n* from 2 to 100 in Table *E,* App. 3.
A_5	a multiplier of $m(R)$ to determine the distance from $\overline{\widetilde{X}}$ to 3-sigma control limits on an $\widetilde{X}$ chart (control chart for medians). See Table 10.2.
ACL	acceptance control limit. An action criterion on an acceptance control chart. For two-sided tolerances the criteria may be designated as an upper acceptance control limit, *UACL,* and a lower acceptance control limit, *LACL.*
AFI	average fraction inspected. It is the ATI divided by the lot size *N* and is generally used in the analysis of rectifying inspection plans.
AOQ	average outgoing quality.
AOQL	average outgoing quality limit.
APL	acceptable process level. The process level at which there is a risk probability of α that the statistic plotted on an acceptance control chart will fall outside the *ACL.*
AQL	acceptable quality level.
ARL	average run length. The average number of point plots on a control chart or the average number of items inspected before a shift in process level may be expected to be signaled.
ASN	average sample number. The average number of items sampled in an acceptance sampling plan.
ATI	average total inspection, that is, average number of items inspected in a given production quantity under a specified acceptance procedure.
B_1	a multiplier of σ or σ_0 to determine the 3-sigma lower control limit on a chart for σ_{RMS}. It is given for values of *n* from 2 to 100 in Table *F,* App. 3.
B_2	a multiplier of σ or σ_0 to determine the 3-sigma upper control limit on a chart for σ_{RMS}. It is given for values of *n* from 2 to 100 in Table *F,* App. 3.

B_3 a multiplier of $\bar{s}$ or $\bar{\sigma}_{RMS}$ to determine the 3-sigma lower control limit on a chart for s or σ_{RMS}. It is given for values of n from 2 to 100 in Table *E*, App. 3.

B_4 a multiplier of $\bar{s}$ or $\bar{\sigma}_{RMS}$ to determine the 3-sigma upper control limit on a chart for s or σ_{RMS}. It is given for values of n from 2 to 100 in Table *E*, App. 3.

B_5 a multiplier of σ or σ_0 to determine the 3-sigma lower control limit on a chart for s. It is given for values of n from 2 to 100 in Table *F*, App. 3.

B_6 a multiplier of σ or σ_0 to determine the 3-sigma upper control limit on a chart for s. It is given for values of n from 2 to 100 in Table *F*, App. 3.

$B_{0.001}$, $B_{0.005}$, $B_{0.025}$, $B_{0.975}$, $B_{0.995}$, $B_{0.999}$ factors for computing probability limits for control charts for s. See Table 10.3.

c number of nonconformities, usually in a sample of stated size of one or more identical units. Also, c may refer to the number of occurrences of some other chosen event in a sample of stated size.

c in single sampling acceptance schemes, the acceptance number, that is, the maximum allowable number of nonconforming pieces in a sample of size n.

$\bar{c}$ (c bar), the average number of nonconformities per sample in a series of samples of equal size.

c_0 a standard adopted for control-chart purposes for the average number of nonconformities in a sample of stated size.

c_1 in double sampling acceptance plans, the acceptance number for the first sample, that is, the maximum number of nonconforming items that will permit acceptance of the lot on the basis of the first sample.

c_2 in double sampling acceptance plans, the acceptance number for the two samples combined, that is, the maximum number of nonconforming items that will permit acceptance of the lot on the basis of the two samples.

c_2 a factor used in connection with sampling by variables that is a function of n and expresses the ratio between the expected value of σ_{RMS} from a long series of samples from a normal universe and the σ of that universe. Values of c_2 are given in Table *C*, App. 3.

c_4 a factor used in connection with sampling by variables that is a function of n and expresses the ratio between the expected value of s from a long series of samples from a normal universe and the σ of that universe. Values of c_4 are given in Table *C*, App. 3.

$C_r^n = \binom{n}{r}$ the number of combinations of n things taken r at a time.

d deviation in cells from the assumed origin of a frequency distribution.

d_i (d sub i), a numerical weighting in demerits assigned to a classification of a nonconformity or nonconforming unit to provide a means of obtaining a weighted quality score.

d_2 a factor used in connection with sampling by variables that is a function of n and expresses the ratio between the expected value of R from a long series of samples from a normal universe and the σ of that universe. Values of d_2 are given in Table C, App. 3.

d_2^* a modification of the factor d_2 that gives weight to the number of samples (subgroups) used to compute $\bar{R}$. Where more than 20 subgroups have been used, this factor is almost identical with d_2. See Table 4.3 for values of d_2^* when $n = 5$.

d_3 a factor used in connection with sampling by variables that is a function of n and expresses the ratio between the standard deviation of R from a long series of samples from a normal universe and the σ of that universe. Values of d_3 are given in Table C, App. 3.

D the number of nonconforming pieces in a lot of size N.

D_1 a multiplier of σ or σ_0 to determine the 3-sigma lower control limit on a chart for R. It is given for values of n from 2 to 20 in Table F, App. 3.

D_2 a multiplier of σ or σ_0 to determine the 3-sigma upper control limit on a chart for R. It is given for values of n from 2 to 20 in Table F, App. 3.

D_3 a multiplier of $\bar{R}$ to determine the 3-sigma lower control limit on a chart for R. It is given for values of n from 2 to 20 in Table D, App. 3.

D_4 a multiplier of $\bar{R}$ to determine the 3-sigma upper control limit on a chart for R. It is given for values of n from 2 to 20 in Table D, App. 3.

$D_{0.001}$, $D_{0.005}$, $D_{0.025}$, $D_{0.975}$, $D_{0.995}$, $D_{0.999}$ factors for computing probability limits for a control chart for R. See Table 10.4.

e napierian or natural base of logarithms = 2.71828+.

f frequency; generally, the number of observed values within a cell of a frequency distribution.

f in continuous sampling plans, the fraction of units that are inspected during the period when sampling is in effect. It is used in this sense in Dodge's CSP-1, in Dodge-Torrey CSP-2 and CSP-3, and in MIL-STD-1235. In some multilevel plans, f applies to the first level of sampling, f^2 to the second, f^3 to the third, and so forth.

i — in continuous sampling plans, the clearance number. It is the number of successive units that must be found by 100% inspection to be acceptable before acceptance may be carried out by sampling. In the multilevel continuous sampling plans, i also means the number of acceptable units in succession required under one sampling level to justify a shift to the next lower sampling level.

k — number of levels in a multilevel continuous sampling plan.

k — a factor used in connection with unknown-sigma variables acceptance plans. With a one-sided specification, the criteria are $\overline{X} \geq L + ks$ or $U - ks \geq \overline{X}$.

k' — a factor used in connection with known-sigma variables acceptance plans. With a one-sided specification, the criteria are $\overline{X} \geq L + k'\sigma$ or $U - k'\sigma \geq \overline{X}$.

k'^* — a factor used in connection with known-sigma variables acceptance plans involving two-sided specifications. The criteria are $\overline{X} \geq L + k'^*\sigma$ and $U - k'^*\sigma \geq \overline{X}$.

L — lower specification limit.

LCL — lower control limit on a control chart. $LCL_{\overline{X}}$ refers to the lower control limit on an $\overline{X}$ chart, LCL_p to the lower control limit on a p chart, and so forth.

LLL — lower lot limit. Used in the Shainin lot plot technique, it is $\overline{\overline{X}} - 3\sigma$ of a sample drawn from a lot or batch of product.

LQL — limiting quality level. In acceptance sampling, the percentage or proportion of nonconforming units in lots or batches for which, for purposes of sampling inspection, the consumer wishes the probability of acceptance to be restricted to a specified low value β. Other terms sometimes used to express this quantity are LQ (limiting quality), RQL (rejectable quality level), and UQL (unacceptable quality level).

$LRL_{\overline{X}}$ — lower reject limit for averages, the lowest possible satisfactory value of the lower control limit on a modified $\overline{X}$ chart.

LSL_X — see L and LTL.

LTL — lower tolerance limit or lower specification limit. Also see L.

LTPD — lot tolerance percent defective. See Chap. 13.

$m(R)$ — median of a set of ranges.

M — maximum allowable percent defective as used in the Form 2 calculation in MIL-STD-414. See Chap. 15.

MSD — mean-square deviation.

n — the number of pieces or observed values in any given sample or subgroup.

np number of nonconforming items in a sample of size n.

n_1 in double sampling, the number of pieces in the first sample.

n_2 in double sampling, the number of pieces in the second sample.

N the number of pieces in a given lot to be sampled for purposes of acceptance.

NPL natural process limits. Symmetrical limits about a universe mean that contain virtually all the product. Usually they are designated as an upper natural process limit (UNPL) at $\mu_X + 3\sigma$ and a lower natural process limit (LNPL) at $\mu_X - 3\sigma$.

$\binom{n}{r}$ number of combinations of n things taken r at a time.

p fraction rejected, the ratio of the number of nonconforming items to the total number inspected. Also p may refer to the ratio of the number of units in which one or more events of a given classification occur to the total number of units sampled.

$100p$ percent rejected.

$\bar{p}$ average fraction rejected, the ratio of the sum of the number of nonconforming items found in a set of samples to the sum of the number of articles inspected in the same samples.

p_l (p sub l), an estimate of the proportion of units in a lot below the lower tolerance limit.

p_u (p sub u), an estimate of the proportion of units in a lot above the upper tolerance limit.

p_0 a standard adopted for control chart or other purposes for the fraction rejected or for the probability of a nonconforming item.

p_1 (p sub 1). In process control, the proportion of nonconforming units occurring when the process is centered at the acceptable process level (APL). In acceptance sampling, the proportion of nonconforming units in a lot or a series of lots for which the probability of acceptance is high. Also see AQL and α (Type I) error.

p_2 (p sub 2). In process control, the proportion of nonconforming units occurring when the process is centered at the limiting quality level (LQL). In acceptance sampling, the proportion of nonconforming units in a lot or a series of lots for which the probability of acceptance is low. Also see LQL and β (Type II) error.

$p_{0.95}$ product fraction nonconforming having a probability of acceptance of 0.95 under given acceptance criteria.

$p_{0.50}$ product fraction nonconforming having a probability of acceptance of 0.50 under given acceptance criteria.

$p_{0.10}$ product fraction nonconforming having a probability of acceptance of 0.10 under given acceptance criteria.

P_0, P_1, P_2, etc. probabilities of exactly 0, 1, 2, etc., nonconforming items, respectively.

P_a probability of accepting any given lot or product.

P_L and P_U (P sub L and P sub U), symbols used in MIL-STD-414 to represent estimates of lot percentage defective below a lower specification limit L and above an upper specification limit U, respectively. See Chap. 15.

P_r probability of rejecting any given lot or product.

P_r^n the number of permutations of n things taken r at a time.

$P(A)$ the probability of occurrence of an event designated by A.

$P(A|B)$ the conditional probability of A occurring given that B has already occurred.

q $(1 - p)$, the probability that a particular event will not happen in a single trial.

Q quality score. A numerical indicator used to measure the relative quality of incoming materials, operations, and in-process or final products or services.

Q_L and Q_U (Q sub L and Q sub U), "quality index" as defined in MIL-STD-414. See Chap. 15.

r a constant referring to the number of occurrences of some event. Also the number of nonconforming pieces in a given sample of size n.

R the range, the difference between the largest value and the smallest value in a set of numbers.

$\bar{R}$ the average of a set of ranges.

R_0 an aimed-at or standard value of R used as the central line on a control chart for R. Usually it is derived from an assumed value of σ.

RPL rejectable process level. The process level at which there is a risk probability of β that the plotted statistic will fall inside the ACL.

s $\sqrt{\Sigma(X - \bar{X})^2/(n - 1)}$, the sample standard deviation. This is the square root of an unbiased estimate of the variance of an infinite universe as determined from a single sample of size n.

$\bar{s}$ average of a set of subgroup (sample) standard deviations.

s_0 an aimed-at or standard value of s used as the central line on a control chart for s. Usually it is derived from an assumed value of σ.

S the probability space, that is, the set of all possible outcomes of an experiment.

S_k the cumulative sum of a statistic, or of deviations of a statistic from a standard value, at time (or subgroup) k.

u nonconformities per unit, the ratio of the number of nonconformities in a sample to the total number of units in the sample; $u = c/n$.

$\bar{u}$ average nonconformities per unit; the total number of nonconformities in a set of samples divided by the total number of units inspected in the set.

u_0 the aimed-at or standard value of nonconformities per unit used for purposes of computing control limits on a control chart for u.

U upper specification limit. Also see *UTL*.

UCL upper control limit on a control chart. $UCL_{\bar{X}}$ refers to the upper control limit on an $\bar{X}$ chart, UCL_p to the upper control limit on a p chart, etc.

ULL upper lot limit. Used in the Shainin lot plot technique, it is $\bar{\bar{X}} + 3\sigma$ of a sample drawn from a lot or batch of product.

$URL_{\bar{X}}$ upper reject limit for averages, the highest possible satisfactory value of the upper control limit on a modified $\bar{X}$ chart.

USL_X see U and UTL.

UTL upper tolerance or specification limit. Also see U.

V factor for reject limits. It equals $3 - 3/\sqrt{n}$ and is given in Table 10.7.

w_i weighting factor for characteristic i used in finding a sample quality score Q.

X a number representing a value of some variable; in statistical quality control, X is usually the observed value of some quality characteristic for an individual unit. Specific observed values may be designated as $X_1, X_2, X_3, \ldots, X_i, \ldots, X_n$.

$\bar{X}$ (X bar or bar X), the average (arithmetic mean) of two or more X values. The average of n X values is the sum of the X values divided by n.

$\bar{\bar{X}}$ (X double bar), the average of a set of $\bar{X}$ values, sometimes called the grand average.

$\bar{X}_0$ (X bar sub 0), an aimed-at or standard average value of a measurable quality characteristic.

$\tilde{X}$ median of a subgroup.

$\tilde{\tilde{X}}$ median of a set of subgroup medians.

α (alpha), Producer's Risk, that is, the probability of rejecting product of some stated desirable quality under a stipulated acceptance sampling plan; also represented by $1 - P_a$. Often it is used to designate Type I error probabilities.

β (beta), Consumer's Risk, that is, the probability of accepting product of some stated undesirable quality under a stipulated acceptance sampling plan; also represented by P_c. Often it is used to designate Type II error probabilities.

δ (delta), relative deviation, used for cumulative sum control charts. The smallest shift in process level that it is desired to detect, stated in terms of the number of standard deviations of the sample average.

λ (lambda), failure rate, that is, the probability of a failure in a stated unit of time. Also used to designate the intensity parameter for the Poisson distribution.

μ (mu), the parametric value of the arithmetic mean of a distribution or universe.

μ_c (mu sub *c*), universe average number of nonconformities per sample. The parametric value.

μ_p (mu sub *p*), the probability that a particular event will happen in a single trial. In statistical quality control the event in question usually is the occurrence of a nonconforming article. The parametric value.

μ_u (mu sub *u*), the true universe average number of nonconformities per unit. The parametric value.

μ_X $\mu_{\overline{X}}$, the true universe average (variables data). The parametric value of the arithmetic mean.

θ (theta), mean life, the arithmetic mean or average of the lives of a universe of items.

σ (sigma), the known or estimated true value of universe standard deviation. The parametric value.

σ_0 a standard value of the universe standard deviation used for control-chart purposes.

σ_{RMS} (sigma sub RMS), the root-mean-square (RMS) deviation of a set of numbers about the average of the set.

$\overline{\sigma}_{RMS}$ (sigma bar sub RMS), the average of a set of σ_{RMS} values.

$\sigma_{\overline{X}}$ (sigma sub bar $\overline{X}$), the standard deviation of the expected frequency distribution of the averages $\overline{X}$ of samples of size *n*. It is equal to $\sigma/\sqrt{n}$.

σ_c, σ_{np}, σ_p, σ_R, σ_u, σ_σ the standard deviation of the sampling distribution of *c, np, p, R, u,* and σ, respectively.

! symbol for factorial; *n*! is the product of the first *n* integers.

APPENDIX 2

BIBLIOGRAPHY

The following bibliography is limited to books and pamphlets. Much additional useful material on theory and applications is available in periodical articles.

Abbott, W. H.: "Probability Charts," Wendell H. Abbott, St. Petersburg, Fla., 1962.

American Society for Quality Control: "Annual Convention Transactions." (These paper-covered volumes, starting in 1951, typically contain several hundred pages and fifty or more papers.) Published annually by American Society for Quality Control, Inc., Milwaukee, Wis.

"ASTM Manual on Presentation of Data and Control Chart Analysis," Special Technical Pub. 15D, American Society for Testing and Materials, Philadelphia, Pa., 1976.

Banks, Jerry: *Principles of Quality Control,* John Wiley & Sons, Inc., New York, 1989.

Barrentine, Larry B.: *Concepts for R&R Studies,* ASQC Quality Press, Milwaukee, Wis., 1991.

Bazovsky, Igor: *Reliability: Theory and Practice,* Prentice-Hall, Inc., Englewood Cliffs, N.J., 1961.

Besterfield, D. H.: *Quality Control,* 4th ed., Prentice-Hall, Inc., Englewood Cliffs, N.J., 1994.

Blanchard, B. S., and E. E. Lowery: *Maintainability,* McGraw-Hill Book Company, New York, 1969.

Blank, Leland: *Statistical Procedures for Engineering, Management and Science,* McGraw-Hill Book Company, New York, 1980.

Bowker, A. H., and H. P. Goode: *Sampling Inspection by Variables,* McGraw-Hill Book Company, New York, 1952.

——— and G. J. Lieberman: *Engineering Statistics,* 2d ed., Prentice-Hall, Inc., Englewood Cliffs, N.J., 1972.

Box, G. E. P., and N. R. Draper: *Evolutionary Operation,* John Wiley & Sons, Inc., New York, 1969.

———, W. G. Hunter, and J. S. Hunter: *Statistics for Experimenters,* John Wiley & Sons, Inc., New York, 1978.

Bradley, J. V.: *Distribution-Free Statistical Tests,* Prentice-Hall, Inc., Englewood Cliffs, N.J., 1968.

Brownlee, K. A.: *Statistical Theory and Methodology in Science and Engineering,* 2d ed., John Wiley & Sons, Inc., New York, 1965. Reprinted by R. E. Krieger Publishing Company, Inc., Melbourne, Fla., 1984.

Burlington, R. S., and D. C. May, Jr.: *Handbook of Probability and Statistics with Tables,* 2d ed., McGraw-Hill Book Company, New York, 1970.

Burr, I. W.: *Engineering Statistics and Quality Control,* McGraw-Hill Book Company, New York, 1953.

———: *Statistical Quality Control Methods,* Marcel Dekker, Inc., New York, 1976.

———: *Applied Statistical Methods,* Academic Press, Inc., New York, 1973.

Butterbaugh, G. I.: *A Bibliography of Statistical Quality Control,* University of Washington Press, Seattle, 1946.

———: *A Bibliography of Statistical Quality Control—Supplement,* University of Washington Press, Seattle, 1951.

Calabro, S. R.: *Reliability Principles and Practices,* McGraw-Hill Book Company, New York, 1962.

Campanella, Jack (ed.): *Principles of Quality Costs: Principles, Implementation, and Use,* 2d ed., Quality Costs Committee, ASQC Quality Press, Milwaukee, Wis., 1990.

Cavé, R.: *Le Contrôle statistique des fabrications,* Editions Eyrolles, Paris, 1953.

Cochran, W. G.: *Sampling Techniques,* 3d ed., John Wiley & Sons, Inc., New York, 1977.

——— and G. M. Cox: *Experimental Designs,* 2d ed., John Wiley & Sons, Inc., New York, 1957.

"Control Chart Method of Controlling Quality during Production, ANSI Standard Z1.3—1975," American National Standards Institute, New York, 1975.

DeBruyn, C. S. V.: *Cumulative Sum Tests: Theory and Practice,* Hafner Publishing Company, New York, 1968.

Deming, W. Edwards: *Some Theory of Sampling,* John Wiley & Sons, Inc., New York, 1966.

———: *Quality, Productivity, and Competitive Position,* MIT Center for Advanced Engineering Study (CAES), Cambridge, Mass., 1982.

———: *Out of the Crisis,* MIT CAES, Cambridge, Mass., 1986.

———: *The New Economics for Industry, Government, and Education,* MIT CAES, Cambridge, Mass., 1993.

DeVor, Richard E., T. Chang, and J. W. Sutherland: *Statistical Quality Design and Control,* Macmillan Publishing Co., New York, 1992.

Dixon, W. J., and F. J. Massey, Jr.: *Introduction to Statistical Analysis,* 4th ed., McGraw-Hill Book Company, New York, 1983.

Dodge, H. F.: "A General Procedure for Sampling Inspection by Attributes—Based on the AQL Concept," Technical Report No. 10, The Statistics Center, Rutgers—The State University, New Brunswick, N.J., 1959.

——— and H. G. Romig: *Sampling Inspection Tables—Single and Double Sampling,* 2d ed., John Wiley & Sons, Inc., New York, 1959. (The complete Dodge-Romig tables, with an explanation of how they were derived, the OC curves of all AOQL plans, and an illustration of their use at Western Electric Company.

Every inspection department in a mass-production industry should have a copy of this book.)

Draper, N. R., and H. Smith: *Applied Regression Analysis,* 2d ed., John Wiley & Sons, Inc., New York, 1981.

Duncan, A. J.: *Quality Control and Industrial Statistics,* 5th ed., Richard D. Irwin, Inc., Homewood, Ill., 1986.

Enrick, N. L.: *Quality, Reliability, and Process Improvement,* 8th ed., The Industrial Press, New York, 1985.

Farnum, Nicholas R.: *Modern Statistical Quality Control and Improvement,* Duxbury Press (Wadsworth, Inc.), Belmont, Calif., 1994.

Feigenbaum, A. V.: *Total Quality Control—Engineering and Management,* 3d ed., Fortieth Anniversary Edition, McGraw-Hill Book Company, New York, 1991.

Feller, William: *An Introduction to Probability Theory and Its Applications,* 3d ed., vol. I, John Wiley & Sons, Inc., New York, 1968.

Freeman, H. A.: *Introduction to Statistical Inference,* Addison-Wesley Book Company, Reading, Mass., 1963.

———, Milton Friedman, Frederick Mosteller, and W. A. Wallis (eds.): *Sampling Inspection,* McGraw-Hill Book Company, New York, 1948.

Fry, T. C.: *Probability and Its Engineering Uses,* 2d ed., D. Van Nostrand Reinhold Company, Inc., Princeton, N.J., 1965.

Gibbons, J. D.: *Nonparametric Methods for Quantitative Analysis,* American Sciences Press, New York, 1985.

"Glossary of Terms Used in Quality Control," European Organization for Quality Control, Rotterdam, 1965. (Includes terms in English, French, German, Dutch, and Italian.)

GOAL/QPC: *The Memory Jogger,* 2d ed., Methuen, Mass., 1988.

"Guide for Fatigue Testing and Statistical Analysis of Fatigue Data," American Society for Testing and Materials, Philadelphia, 1964.

Guttman, Irwin, and S. S. Wilks: *Introductory Engineering Statistics,* 3d ed., John Wiley & Sons, Inc., New York, 1982.

Hahn, G. J., and S. S. Shapiro: *Statistical Models in Engineering,* John Wiley & Sons, Inc., New York, 1967.

Hald, Anders: *Statistical Tables and Formulas,* John Wiley & Sons, Inc., New York, 1952.

———: *Statistical Theory with Engineering Applications,* John Wiley & Sons, Inc., New York, 1952.

———: *Statistical Theory of Sampling Inspection by Attributes,* Academic Press, 1981.

Hansen, B. L., and P. M. Ghare: *Quality Control and Application,* Prentice-Hall, Inc., Englewood Cliffs, N.J., 1987.

Harvard University: *Tables of the Cumulative Binomial Probability Distribution,* Harvard University Press, Cambridge, Mass., 1955.

Hayes, G. E., and H. G. Romig: "Modern Quality Control," Bruce Division of Benziger, Bruce, and Glencoe, Inc., Encino, Calif., 1982.

Hicks, C. R.: *Fundamental Concepts in the Design of Experiments,* 3d ed., Holt, Rinehart, and Winston, New York, 1982.

"Highway Research Board Special Report 118. Quality Assurance and Acceptance Procedures," Highway Research Board, Washington, D.C., 1971.

Hodges, J. L., Jr., and E. L. Lehmann: *Basic Concepts of Probability and Statistics,* 2d ed., Holden-Day, Inc., San Francisco, 1970.

Hoel, P. G.: *Elementary Statistics,* 4th ed., John Wiley & Sons, Inc., New York, 1976.

———: *Introduction to Mathematical Statistics,* 5th ed., John Wiley & Sons, Inc., 1984.

Holman, J. P.: *Experimental Methods for Engineers,* 4th ed., McGraw-Hill Book Company, New York, 1983.

Hopper, A. G.: *Statistical Quality Control,* McGraw-Hill Book Company, London, 1969.

Huitson, Alan, and Joan Keen: *Essentials of Quality Control,* William Heinemann, Ltd., London, 1965.

Imai, Masaaki: *Kaizen, The Key to Japan's Competitive Success,* Random House Business Division, New York, 1986.

Institute of Electrical and Electronics Engineers: "Proceedings of National Symposia on Reliability." (These paper-bound volumes, starting in 1955, are typically several hundred pages and contain many papers.) Published annually by the IEEE, Inc., New York.

Ireson, W. G., and C. F. Coombs (eds.): *Handbook of Reliability and Management,* 2d ed., McGraw-Hill Book Co., New York, 1988.

——— and E. L. Grant (eds.): *Handbook of Industrial Engineering and Management,* 2d ed., Prentice-Hall, Inc., Englewood Cliffs, N.J., 1971. (See particularly the section on "Industrial Statistics" by A. H. Bowker and G. J. Lieberman, the section on "Inspection and Quality Control" by T. C. McDermott and D. M. Cound, and the section on "Reliability Methods" by Myron Lipow and D. K. Lloyd.)

Ishikawa, Kaoru: *Guide to Quality Control,* 2d rev. ed., Asian Productivity Organization, Tokyo, 1986.

———: *What Is Total Quality Control? The Japanese Way,* Prentice-Hall, Englewood Cliffs, N.J., 1985.

Japanese Standards Association: "Standardization and Quality Control in Japan," JSA Technical Report No. 1, Japanese Standards Association, Tokyo, Japan, 1963. (This short pamphlet includes a bibliography in English of the numerous books on quality control that have been published in Japanese. It also lists more than 20 Japanese industrial standards related to quality control. All these standards have titles in English as well as in Japanese; there are a few with English translations.)

Juran, Joseph M.: *Management of Inspection and Quality Control,* Harper & Row, Publishers, Inc., New York, 1945.

———: *Juran on Planning for Quality,* The Free Press (Macmillan, Inc.), New York, 1988.

——— and F. M. Gryna, Jr. (eds.): *Juran's Quality Control Handbook,* 4th ed., McGraw-Hill Book Company, New York, 1988.

——— and ———: *Quality Planning and Analysis,* 3d ed., McGraw-Hill Book Company, New York, 1993.

Kramer, Amihud, and B. A. Twigg: *Quality Control for the Food Industry: vol. 1, Fundamentals,* 3d ed., The AVI Publishing Company, Inc., Westport, Conn., 1970.

——— and ———: *Quality Control for the Food Industry: vol. 2, Applications,* The AVI Publishing Company, Inc., Westport, Conn., 1973.

Ku, H. H. (ed.): "Precision Measurement and Calibration," Superintendent of Documents, Government Printing Office, Washington, D.C., 1969.

Lieberman, G. J., and D. B. Owen: *Tables of the Hypergeometric Probability Distribution,* Stanford University Press, Stanford, Calif., 1961.

Lindgren, B. W., and G. W. McElrath: *Introduction to Probability and Statistics,* 4th ed., The Macmillan Company, New York, 1978.

Lloyd, D. K., and Myron Lipow: *Reliability: Management, Methods, and Mathematics,* Prentice-Hall, Inc., Englewood Cliffs, N.J., 1962. Reprinted by American Society for Quality Control, Milwaukee, Wis., 1987.

Locks, Mitchell O.: *Reliability, Maintainability, and Availability Assessment,* 2d ed., ASQC Quality Press, Milwaukee, Wis., 1995.

López, D. M., and E. J. González: "Síntesis de un curso de control estadístico de calidad," American Society for Quality Control, Sección Ciudad de México, México, D. F., 1960.

Mendenhall, William and R. J. Beaver: *Introduction to Probability and Statistics,* 9th ed., Duxbury Press Division of Wadsworth Pub. Co., N. Scituate, Mass., 1994.

——— and R. L. Scheaffer: *Mathematical Statistics with Applications,* 2d ed., Duxbury Press Division of Wadsworth Pub. Co., N. Scituate, Mass., 1981.

Meyer, P. L.: *Introductory Probability and Statistical Applications,* 2d ed., Addison-Wesley Publishing Company, Inc., Reading, Mass., 1970.

"Military Standard 105E. Sampling Procedures and Tables for Inspection by Attributes," Superintendent of Documents, Government Printing Office, Washington, D.C., 1989. (Nominally, the ABC standard.)

"Military Standard 414. Sampling Procedures and Tables for Inspection by Variables for Percent Defective," Superintendent of Documents, Government Printing Office, Washington, D.C., 1957.

"Military Standard 690-B. Failure Rate Sampling Plans and Procedures," Superintendent of Documents, Government Printing Office, Washington, D.C., 1969.

"Military Standard 781-C. Reliability Design Qualification and Production Acceptance Tests, Exponential Distribution," Superintendent of Documents, Government Printing Office, Washington, D.C., 1977.

"Military Standard 1235B, Single- and Multi-Level Continuous Sampling Procedures and Tables for Inspection by Attributes," Superintendent of Documents, Government Printing Office, Washington, D.C., 1981.

Molina, E. C.: *Poisson's Exponential Binomial Limit,* Van Nostrand Reinhold Company, Inc., Princeton, N.J., 1942. (Molina's tables contain individual values and cumulative values of the Poisson, both to six decimal places.) Reprinted in 1973 by R. E. Krieger Publishing Company, Melbourne, Fla.

Montgomery, D.C.: *Introduction to Statistical Quality Control,* 2d ed., John Wiley & Sons, Inc., New York, 1991.

Moroney, M. J.: *Facts from Figures,* 3d ed., Penguin Books, Ltd., Harmondsworth, Middlesex, England; and Baltimore, Md., 1956. (Paperback; a good introduction to statistics for the layperson.)

Moses, L. E., and R. V. Oakford: *Tables of Random Permutations,* Stanford University Press, Stanford, Calif., 1963.

National Bureau of Standards: "Tables of the Binomial Probability Distribution," Applied Mathematics Series, vol. 6, Superintendent of Documents, Government Printing Office, Washington, D.C., 1949.

Ott, Ellis R., and E. G. Shilling: *Process Quality Control: Troubleshooting and Interpretation of Data,* 2d ed., McGraw-Hill Book Company, New York, 1990.

Palazzi, Aurelio: "Metodi statistici nella recerca industriale e nel controllo della produzione," Etas Kompass, Milan, 1964.

Parzen, Emanuel: *Modern Probability Theory and Its Applications,* John Wiley & Sons, Inc., New York, 1960, reprinted 1992.

Peach, Paul: *Quality Control for Management,* Prentice-Hall, Inc., Englewood Cliffs, N.J., 1964.

——— (ed.): *The ISO 9000 Handbook,* 2d ed., CEEM Information Services (available from ASQC), 1994.

Polovko, A. M.: *Fundamentals of Reliability Theory,* Academic Press, New York, 1968.

"Proceedings of the International Conference on Quality Control," Union of Japanese Scientists and Engineers, Tokyo, 1969.

"Product Quality Assurance Handbook H-57," Superintendent of Documents, Government Printing Office, Washington, D.C., 1969.

"Quality Control and Reliability Handbook (Interim) H 108. Sampling Procedures for Life and Reliability Testing (Based on Exponential Distribution)," Superintendent of Documents, Government Printing Office, Washington, D.C., 1960.

"Quality Control and Reliability Handbook (Interim) H 109. Statistical Procedures for Determining Validity of Suppliers' Attributes Inspection," Superintendent of Documents, Government Printing Office, Washington, D.C., 1960.

"Quality Control and Reliability Handbook (Interim) H 110. Evaluation of Contractor Quality Control Systems," Superintendent of Documents, Government Printing Office, Washington, D.C., 1960.

"Quality Control and Reliability Technical Report TR 3. Sampling Procedures and Tables for Life and Reliability Testing Based on the Weibull Distribution (Mean Life Criterion)," Superintendent of Documents, Government Printing Office, Washington, D.C., 1961. (This is based on a Cornell University Technical Report prepared by H. P. Goode and J. H. K. Kao.)

"Quality Control and Reliability Technical Report TR 4. Sampling Procedures and Tables for Life and Reliability Testing Based on the Weibull Distribution (Hazard Rate Criterion)," Superintendent of Documents, Government Printing Office, Washington, D.C., 1962. (This is based on a Cornell University Technical Report prepared by H. P. Goode and J. H. K. Kao.)

RAND Corporation: *A Million Random Digits with 100,000 Normal Deviates,* Glencoe Free Press Division of The Macmillan Company, New York, 1955.

Rickmers, A. D., and H. N. Todd: *Statistics, An Introduction,* McGraw-Hill Book Company, New York, 1967.

Romig, H. G.: "50–100 Binomial Tables," John Wiley & Sons, Inc., New York, 1953.

Schaafsma, A. H., and F. G. Willemze: "Modern kwaliteitsbeleid," N. V. Uitgeversmij. Centrex, Eindhoven, Holland, 1958. (This is the title of the original edition in the Dutch language. Titles of French, German, and Spanish language editions are respectively: "Gestion moderne de la qualité," "Moderne Qualitätskontrolle," and "Gestion moderna de la calidad.")

Scheaffer, R. L., and J. T. McClave: *Probability and Statistics for Engineers,* 4th ed., Duxbury Press, Boston, 1994.

Scholtes, Peter R.: "The Team Handbook," Joiner Associates, Inc., Madison, Wis., 1988.

Shewhart, W. A.: *Economic Control of Quality of Manufactured Product,* D. Van Nostrand Company, Inc., Princeton, N.J., 1931. Republished by American Society for Quality Control, Milwaukee, Wis., 1980.

——— (edited by W. E. Deming): "Statistical Method from the Viewpoint of Quality Control," The Graduate School, Department of Agriculture, Washington, D.C., 1939.

Shooman, M. L.: *Probabilistic Reliability: An Engineering Approach,* 2d ed., McGraw-Hill Book Company, New York, 1990.

Simon, L. E.: *An Engineers' Manual of Statistical Methods,* John Wiley & Sons, Inc., New York, 1941.

Slater, Roger: *Integrated Process Management: A Quality Model,* ASQC Quality Press, Milwaukee, Wis., 1991.

Statistical Research Group, Columbia University: *Sequential Analysis of Statistical Data: Applications,* Columbia University Press, New York, 1945.

Stok, T. L., *The Worker and Quality Control,* The University of Michigan, Ann Arbor, 1965.

"Supply & Logistics Handbook—Inspection H 105. Administration of Sampling Procedures for Acceptance Inspection," Superintendent of Documents, Government Printing Office, Washington, D.C., 1954.

Taguchi, G.: *Introduction to Quality Engineering: Designing Quality into Products and Processes,* Kraus International Publications, White Plains, N.Y., 1986.

——— and Y. Wu: *Introduction to Off-Line Quality Control,* Central Japan Quality Control Association, Nagoya, 1985.

"Tests for Validity of the Assumption That the Underlying Distribution of Life Is Exponential," Technical Report, Superintendent of Documents, Government Printing Office, Washington, D.C., 1960. (This government pamphlet is a reprint of two articles by Dr. Benjamin Epstein that appeared in *Technometrics* in the issues of February and May 1960.)

Tippett, L. H. C.: *Technological Applications of Statistics,* John Wiley & Sons, Inc., New York, 1950.

Wade, O. R.: *Tolerance Control in Design and Manufacturing,* Industrial Press, Inc., New York, 1967.

Wadsworth, G. P., and J. G. Bryan: *Introduction to Probability and Random Variables,* 2d ed., McGraw-Hill Book Company, New York, 1974.

Wadsworth, Harrison M. (ed.): *Handbook of Statistical Methods for Engineers and Scientists,* McGraw-Hill Book Co., New York, 1990.

———, K. S. Stevens, and A. B. Godfrey: *Modern Methods for Quality Control and Improvement,* John Wiley & Sons, Inc., New York, 1986.

Wald, Abraham: *Sequential Analysis,* John Wiley & Sons, Inc., New York, 1947. Reprinted in 1973 by Dover Publications, Inc., New York.

Weibull, Wallodi: *Fatigue Testing and Analysis of Results,* Pergamon Press, Inc., New York, 1961.

Weintraub, Sol: *Tables of the Cumulative Binomial Probability Distribution for Small Values of p,* The Macmillan Company, New York, 1963.

Western Electric Company: "Statistical Quality Control Handbook," 2d ed., Western Electric Company, Inc., New York, 1958.

Wetherill, G. B.: *Sampling Inspection and Quality Control,* 2d ed., Methuen and Company, Ltd., London, 1977.

Wilks, S. S.: *Mathematical Statistics,* 2d ed., John Wiley & Sons, Inc., New York, 1962.

Williamson, Eric, and M. H. Bretherton: *Tables of the Negative Binomial Probability Distribution,* John Wiley & Sons, Inc., New York, 1963.

Youden, W. J., *Statistical Methods for Chemists,* John Wiley & Sons, Inc., New York, 1951. Reprinted in 1977 by R. E. Krieger Publishing Company, Melbourne, Fla.

APPENDIX 3

TABLES

TABLE A AREA UNDER THE NORMAL CURVE

Proportion of the total area of the standard normal curve from $-\infty$ to z (z represents a normalized statistic)

z	0.09	0.08	0.07	0.06	0.05	0.04	0.03	0.02	0.01	0.00
−3.5	0.00017	0.00017	0.00018	0.00019	0.00019	0.00020	0.00021	0.00022	0.00022	0.00023
−3.4	0.00024	0.00025	0.00026	0.00027	0.00028	0.00029	0.00030	0.00031	0.00033	0.00034
−3.3	0.00035	0.00036	0.00038	0.00039	0.00040	0.00042	0.00043	0.00045	0.00047	0.00048
−3.2	0.00050	0.00052	0.00054	0.00056	0.00058	0.00060	0.00062	0.00064	0.00066	0.00069
−3.1	0.00071	0.00074	0.00076	0.00079	0.00082	0.00085	0.00087	0.00090	0.00094	0.00097
−3.0	0.00100	0.00104	0.00107	0.00111	0.00114	0.00118	0.00122	0.00126	0.00131	0.00135
−2.9	0.0014	0.0014	0.0015	0.0015	0.0016	0.0016	0.0017	0.0017	0.0018	0.0019
−2.8	0.0019	0.0020	0.0021	0.0021	0.0022	0.0023	0.0023	0.0024	0.0025	0.0026
−2.7	0.0026	0.0027	0.0028	0.0029	0.0030	0.0031	0.0032	0.0033	0.0034	0.0035
−2.6	0.0036	0.0037	0.0038	0.0039	0.0040	0.0041	0.0043	0.0044	0.0045	0.0047
−2.5	0.0048	0.0049	0.0051	0.0052	0.0054	0.0055	0.0057	0.0059	0.0060	0.0062
−2.4	0.0064	0.0066	0.0068	0.0069	0.0071	0.0073	0.0075	0.0078	0.0080	0.0082
−2.3	0.0084	0.0087	0.0089	0.0091	0.0094	0.0096	0.0099	0.0102	0.0104	0.0107
−2.2	0.0110	0.0113	0.0116	0.0119	0.0122	0.0125	0.0129	0.0132	0.0136	0.0139
−2.1	0.0143	0.0146	0.0150	0.0154	0.0158	0.0162	0.0166	0.0170	0.0174	0.0179
−2.0	0.0183	0.0188	0.0192	0.0197	0.0202	0.0207	0.0212	0.0217	0.0222	0.0228
−1.9	0.0233	0.0239	0.0244	0.0250	0.0256	0.0262	0.0268	0.0274	0.0281	0.0287
−1.8	0.0294	0.0301	0.0307	0.0314	0.0322	0.0329	0.0336	0.0344	0.0351	0.0359
−1.7	0.0367	0.0375	0.0384	0.0392	0.0401	0.0409	0.0418	0.0427	0.0436	0.0446
−1.6	0.0455	0.0465	0.0475	0.0485	0.0495	0.0505	0.0516	0.0526	0.0537	0.0548
−1.5	0.0559	0.0571	0.0582	0.0594	0.0606	0.0618	0.0630	0.0643	0.0655	0.0668
−1.4	0.0681	0.0694	0.0708	0.0721	0.0735	0.0749	0.0764	0.0778	0.0793	0.0808
−1.3	0.0823	0.0838	0.0853	0.0869	0.0885	0.0901	0.0918	0.0934	0.0951	0.0968
−1.2	0.0985	0.1003	0.1020	0.1038	0.1057	0.1075	0.1093	0.1112	0.1131	0.1151
−1.1	0.1170	0.1190	0.1210	0.1230	0.1251	0.1271	0.1292	0.1314	0.1335	0.1357
−1.0	0.1379	0.1401	0.1423	0.1446	0.1469	0.1492	0.1515	0.1539	0.1562	0.1587
−0.9	0.1611	0.1635	0.1660	0.1685	0.1711	0.1736	0.1762	0.1788	0.1814	0.1841
−0.8	0.1867	0.1894	0.1922	0.1949	0.1977	0.2005	0.2033	0.2061	0.2090	0.2119
−0.7	0.2148	0.2177	0.2207	0.2236	0.2266	0.2297	0.2327	0.2358	0.2389	0.2420
−0.6	0.2451	0.2483	0.2514	0.2546	0.2578	0.2611	0.2643	0.2676	0.2709	0.2743
−0.5	0.2776	0.2810	0.2843	0.2877	0.2912	0.2946	0.2981	0.3015	0.3050	0.3085
−0.4	0.3121	0.3156	0.3192	0.3228	0.3264	0.3300	0.3336	0.3372	0.3409	0.3446
−0.3	0.3483	0.3520	0.3557	0.3594	0.3632	0.3669	0.3707	0.3745	0.3783	0.3821
−0.2	0.3859	0.3897	0.3936	0.3974	0.4013	0.4052	0.4090	0.4129	0.4168	0.4207
−0.1	0.4247	0.4286	0.4325	0.4364	0.4404	0.4443	0.4483	0.4522	0.4562	0.4602
−0.0	0.4641	0.4681	0.4721	0.4761	0.4801	0.4840	0.4880	0.4920	0.4960	0.5000

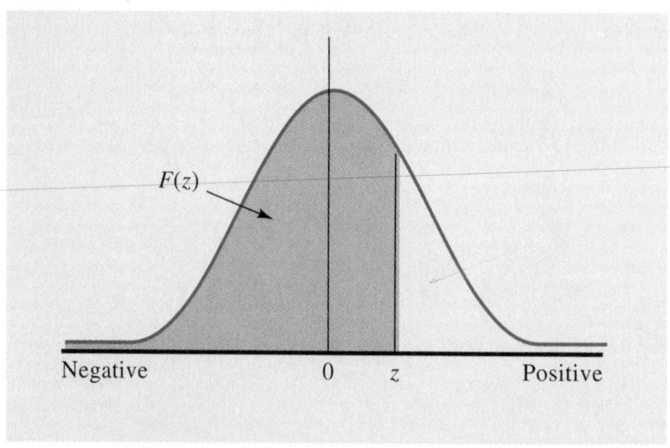

TABLE A *(continued)*

z	0.00	0.01	0.02	0.03	0.04	0.05	0.06	0.07	0.08	0.09
+0.0	0.5000	0.5040	0.5080	0.5120	0.5160	0.5199	0.5239	0.5279	0.5319	0.5359
+0.1	0.5398	0.5438	0.5478	0.5517	0.5557	0.5596	0.5636	0.5675	0.5714	0.5753
+0.2	0.5793	0.5832	0.5871	0.5910	0.5948	0.5987	0.6026	0.6064	0.6103	0.6141
+0.3	0.6179	0.6217	0.6255	0.6293	0.6331	0.6368	0.6406	0.6443	0.6480	0.6517
+0.4	0.6554	0.6591	0.6628	0.6664	0.6700	0.6736	0.6772	0.6808	0.6844	0.6879
+0.5	0.6915	0.6950	0.6985	0.7019	0.7054	0.7088	0.7123	0.7157	0.7190	0.7224
+0.6	0.7257	0.7291	0.7324	0.7357	0.7389	0.7422	0.7454	0.7486	0.7517	0.7549
+0.7	0.7580	0.7611	0.7642	0.7673	0.7704	0.7734	0.7764	0.7794	0.7823	0.7852
+0.8	0.7881	0.7910	0.7939	0.7967	0.7995	0.8023	0.8051	0.8079	0.8106	0.8133
+0.9	0.8159	0.8186	0.8212	0.8238	0.8264	0.8289	0.8315	0.8340	0.8365	0.8389
+1.0	0.8413	0.8438	0.8461	0.8485	0.8508	0.8531	0.8554	0.8577	0.8599	0.8621
+1.1	0.8643	0.8665	0.8686	0.8708	0.8729	0.8749	0.8770	0.8790	0.8810	0.8830
+1.2	0.8849	0.8869	0.8888	0.8907	0.8925	0.8944	0.8962	0.8980	0.8997	0.9015
+1.3	0.9032	0.9049	0.9066	0.9082	0.9099	0.9115	0.9131	0.9147	0.9162	0.9177
+1.4	0.9192	0.9207	0.9222	0.9236	0.9251	0.9265	0.9279	0.9292	0.9306	0.9319
+1.5	0.9332	0.9345	0.9357	0.9370	0.9382	0.9394	0.9406	0.9418	0.9429	0.9441
+1.6	0.9452	0.9463	0.9474	0.9484	0.9495	0.9505	0.9515	0.9525	0.9535	0.9545
+1.7	0.9554	0.9564	0.9573	0.9582	0.9591	0.9599	0.9608	0.9616	0.9625	0.9633
+1.8	0.9641	0.9649	0.9656	0.9664	0.9671	0.9678	0.9686	0.9693	0.9699	0.9706
+1.9	0.9713	0.9719	0.9726	0.9732	0.9738	0.9744	0.9750	0.9756	0.9761	0.9767
+2.0	0.9773	0.9778	0.9783	0.9788	0.9793	0.9798	0.9803	0.9808	0.9812	0.9817
+2.1	0.9821	0.9826	0.9830	0.9834	0.9838	0.9842	0.9846	0.9850	0.9854	0.9857
+2.2	0.9861	0.9864	0.9868	0.9871	0.9875	0.9878	0.9881	0.9884	0.9887	0.9890
+2.3	0.9893	0.9896	0.9898	0.9901	0.9904	0.9906	0.9909	0.9911	0.9913	0.9916
+2.4	0.9918	0.9920	0.9922	0.9925	0.9927	0.9929	0.9931	0.9932	0.9934	0.9936
+2.5	0.9938	0.9940	0.9941	0.9943	0.9945	0.9946	0.9948	0.9949	0.9951	0.9952
+2.6	0.9953	0.9955	0.9956	0.9957	0.9959	0.9960	0.9961	0.9962	0.9963	0.9964
+2.7	0.9965	0.9966	0.9967	0.9968	0.9969	0.9970	0.9971	0.9972	0.9973	0.9974
+2.8	0.9974	0.9975	0.9976	0.9977	0.9977	0.9978	0.9979	0.9979	0.9980	0.9981
+2.9	0.9981	0.9982	0.9983	0.9983	0.9984	0.9984	0.9985	0.9985	0.9986	0.9986
+3.0	0.99865	0.99869	0.99874	0.99878	0.99882	0.99886	0.99889	0.99893	0.99896	0.99900
+3.1	0.99903	0.99906	0.99910	0.99913	0.99915	0.99918	0.99921	0.99924	0.99926	0.99929
+3.2	0.99931	0.99934	0.99936	0.99938	0.99940	0.99942	0.99944	0.99946	0.99948	0.99950
+3.3	0.99952	0.99953	0.99955	0.99957	0.99958	0.99960	0.99961	0.99962	0.99964	0.99965
+3.4	0.99966	0.99967	0.99969	0.99970	0.99971	0.99972	0.99973	0.99974	0.99975	0.99976
+3.5	0.99977	0.99978	0.99978	0.99979	0.99980	0.99981	0.99981	0.99982	0.99983	0.99983

Characteristic	Statistic	Normalized statistic *z*
Measurement	X	$\dfrac{X - \mu_X}{\sigma}$
Subgroup average	$\bar{X}$	$\dfrac{\bar{X} - \mu_{\bar{X}}}{\sigma/\sqrt{n}}$
Binomial count	$np = c$	$\dfrac{c + 0.5 - n\mu_p}{\sqrt{n\mu_p(1 - \mu_p)}}$
Binomial fraction	$p = \dfrac{c}{n}$	$\dfrac{(c + 0.5)/n - \mu_p}{\sqrt{\mu_p(1 - \mu_p)/n}}$
Poisson count	c	$\dfrac{c + 0.5 - \mu_c}{\sqrt{\mu_c}}$

TABLE B RIGHT TAIL AREA OF THE χ^2 DISTRIBUTION†

ν	γ										
	0.995	**0.990**	**0.975**	**0.950**	**0.900**	**0.500**	**0.100**	**0.050**	**0.025**	**0.010**	**0.005**
1	0.00	0.00	0.00	0.00	0.02	0.45	2.71	3.84	5.02	6.63	7.88
2	0.01	0.02	0.05	0.10	0.21	1.39	4.61	5.99	7.38	9.21	10.60
3	0.07	0.11	0.22	0.35	0.58	2.37	6.25	7.81	9.35	11.34	12.84
4	0.21	0.30	0.48	0.71	1.06	3.36	7.78	9.49	11.14	13.28	14.86
5	0.41	0.55	0.83	1.15	1.61	4.35	9.24	11.07	12.83	15.09	16.75
6	0.68	0.87	1.24	1.64	2.20	5.35	10.65	12.59	14.45	16.81	18.55
7	0.99	1.24	1.69	2.17	2.83	6.35	12.02	14.07	16.01	18.48	20.28
8	1.34	1.65	2.18	2.73	3.49	7.34	13.36	15.51	17.53	20.09	21.96
9	1.73	2.09	2.70	3.33	4.17	8.34	14.68	16.92	19.02	21.67	23.59
10	2.16	2.56	3.25	3.94	4.87	9.34	15.99	18.31	20.48	23.21	25.19
11	2.60	3.05	3.82	4.57	5.58	10.34	17.28	19.68	21.92	24.72	26.76
12	3.07	3.57	4.40	5.23	6.30	11.34	18.55	21.03	23.34	26.22	28.30
13	3.57	4.11	5.01	5.89	7.04	12.34	19.81	22.36	24.74	27.69	29.82
14	4.07	4.66	5.63	6.57	7.79	13.34	21.06	23.68	26.12	29.14	31.32
15	4.60	5.23	6.26	7.26	8.55	14.34	22.31	25.00	27.49	30.58	32.80
16	5.14	5.81	6.91	7.96	9.31	15.34	23.54	26.30	28.85	32.00	34.27
17	5.70	6.41	7.56	8.67	10.09	16.34	24.77	27.59	30.19	33.41	35.72
18	6.26	7.01	8.23	9.39	10.87	17.34	25.99	28.87	31.53	34.81	37.16
19	6.84	7.63	8.91	10.12	11.65	18.34	27.20	30.14	32.85	36.19	38.58
20	7.43	8.26	9.59	10.85	12.44	19.34	28.41	31.41	34.17	37.57	40.00
21	8.03	8.90	10.28	11.59	13.24	20.34	29.62	32.67	35.48	38.93	41.40
22	8.64	9.54	10.98	12.34	14.04	21.34	30.81	33.92	36.78	40.29	42.80
23	9.26	10.20	11.69	13.09	14.85	22.34	32.01	35.17	38.08	41.64	44.18
24	9.89	10.86	12.40	13.85	15.66	23.34	33.20	36.42	39.36	42.98	45.56
25	10.52	11.52	13.12	14.61	16.47	24.34	34.38	37.65	40.65	44.31	46.93
26	11.16	12.20	13.84	15.38	17.29	25.34	35.56	38.89	41.92	45.64	48.29
27	11.81	12.88	14.57	16.15	18.11	26.34	36.74	40.11	43.19	46.96	49.65
28	12.46	13.57	15.31	16.93	18.94	27.34	37.92	41.34	44.46	48.28	50.99
29	13.12	14.26	16.05	17.71	19.77	28.34	39.09	42.56	45.72	49.59	52.34
30	13.79	14.95	16.79	18.49	20.60	29.34	40.26	43.77	46.98	50.89	53.67
40	20.71	22.16	24.43	26.51	29.05	39.34	51.80	55.76	59.34	63.69	66.77
50	27.99	29.71	32.36	34.76	37.69	49.33	63.17	67.50	71.42	76.15	79.49
70	43.28	45.44	48.76	51.74	55.33	69.33	85.53	90.53	95.02	100.42	104.22
100	67.33	70.06	74.22	77.93	82.36	99.33	118.50	124.34	129.56	135.81	140.17

†Taken by permission from Leland Blank, *Statistical Procedures for Engineering, Management, and Science*, McGraw-Hill Book Company, New York, 1980.

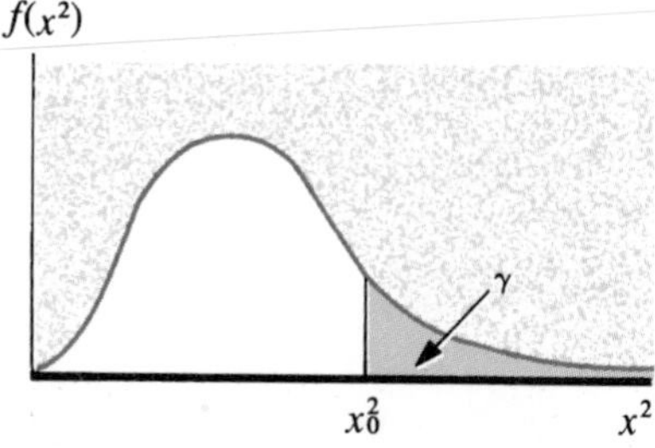

Given ν, the table gives the x_0^2 value with γ of the area above it; that is, $P(x^2 \geq x_0^2) = \gamma$

TABLE C FACTORS FOR ESTIMATING σ FROM $\overline{R}$, $\overline{s}$, OR $\overline{\sigma}_{RMS}$ AND σ_R FROM $\overline{R}$

Number of observations in subgroup, n	Factor d_2, $d_2 = \frac{\overline{R}}{\sigma}$	Factor d_3, $d_3 = \frac{\sigma_R}{\sigma}$	Factor c_2, $c_2 = \frac{\overline{\sigma}_{RMS}}{\sigma}$	Factor c_4, $c_4 = \frac{\overline{s}}{\sigma}$
2	1.128	0.8525	0.5642	0.7979
3	1.693	0.8884	0.7236	0.8862
4	2.059	0.8798	0.7979	0.9213
5	2.326	0.8641	0.8407	0.9400
6	2.534	0.8480	0.8686	0.9515
7	2.704	0.8332	0.8882	0.9594
8	2.847	0.8198	0.9027	0.9650
9	2.970	0.8078	0.9139	0.9693
10	3.078	0.7971	0.9227	0.9727
11	3.173	0.7873	0.9300	0.9754
12	3.258	0.7785	0.9359	0.9776
13	3.336	0.7704	0.9410	0.9794
14	3.407	0.7630	0.9453	0.9810
15	3.472	0.7562	0.9490	0.9823
16	3.532	0.7499	0.9523	0.9835
17	3.588	0.7441	0.9551	0.9845
18	3.640	0.7386	0.9576	0.9854
19	3.689	0.7335	0.9599	0.9862
20	3.735	0.7287	0.9619	0.9869
21	3.778	0.7242	0.9638	0.9876
22	3.819	0.7199	0.9655	0.9882
23	3.858	0.7159	0.9670	0.9887
24	3.895	0.7121	0.9684	0.9892
25	3.931	0.7084	0.9696	0.9896
30	4.086	0.6926	0.9748	0.9914
35	4.213	0.6799	0.9784	0.9927
40	4.322	0.6692	0.9811	0.9936
45	4.415	0.6601	0.9832	0.9943
50	4.498	0.6521	0.9849	0.9949
55	4.572	0.6452	0.9863	0.9954
60	4.639	0.6389	0.9874	0.9958
65	4.699	0.6337	0.9884	0.9961
70	4.755	0.6283	0.9892	0.9964
75	4.806	0.6236	0.9900	0.9966
80	4.854	0.6194	0.9906	0.9968
85	4.898	0.6154	0.9912	0.9970
90	4.939	0.6118	0.9916	0.9972
95	4.978	0.6084	0.9921	0.9973
100	5.015	0.6052	0.9925	0.9975

Estimate of $\sigma = \overline{R}/d_2$ or $\overline{s}/c_4$ or $\overline{\sigma}_{RMS}/c_2$; $\sigma_R = \overline{R}/d_3$. These factors assume sampling from a normal universe.

TABLE D FACTORS FOR DETERMINING FROM $\overline{R}$ THE 3-SIGMA CONTROL LIMITS FOR $\overline{X}$ AND R CHARTS

Number of observations in subgroup, n	Factor for $\overline{X}$ chart, A_2	Factors for R chart	
		Lower control limit D_3	Upper control limit D_4
2	1.88	0	3.27
3	1.02	0	2.57
4	0.73	0	2.28
5	0.58	0	2.11
6	0.48	0	2.00
7	0.42	0.08	1.92
8	0.37	0.14	1.86
9	0.34	0.18	1.82
10	0.31	0.22	1.78
11	0.29	0.26	1.74
12	0.27	0.28	1.72
13	0.25	0.31	1.69
14	0.24	0.33	1.67
15	0.22	0.35	1.65
16	0.21	0.36	1.64
17	0.20	0.38	1.62
18	0.19	0.39	1.61
19	0.19	0.40	1.60
20	0.18	0.41	1.59

Upper control limit for $\overline{X} = UCL_{\overline{X}} = \overline{\overline{X}} + A_2\overline{R}$

Lower control limit for $\overline{X} = LCL_{\overline{X}} = \overline{\overline{X}} - A_2\overline{R}$

(If aimed-at or standard value $\overline{X}_0$ is used rather than $\overline{\overline{X}}$ as the central line on the control chart, $\overline{X}_0$ should be substituted for $\overline{\overline{X}}$ in the preceding formulas.)

Upper control limit for $R = UCL_R = D_4\overline{R}$

Lower control limit for $R = LCL_R = D_3\overline{R}$

All factors in Table D are based on the normal distribution.

No lower control limit for R chart where n is less than 7.

TABLE *E* FACTORS FOR DETERMINING FROM $\bar{s}$ AND $\bar{\sigma}_{RMS}$ THE 3-SIGMA CONTROL LIMITS FOR $\bar{X}$ AND s OR σ_{RMS} CHARTS

Number of observations in subgroup, n	Factor for $\bar{X}$ chart using $\bar{\sigma}_{RMS}$, A_1	Factor for $\bar{X}$ chart using $\bar{s}$, A_3	Factors for s or σ_{RMS} charts	
			Lower control limit B_3	Upper control limit B_4
2	3.76	2.66	0	3.27
3	2.39	1.95	0	2.57
4	1.88	1.63	0	2.27
5	1.60	1.43	0	2.09
6	1.41	1.29	0.03	1.97
7	1.28	1.18	0.12	1.88
8	1.17	1.10	0.19	1.81
9	1.09	1.03	0.24	1.76
10	1.03	0.98	0.28	1.72
11	0.97	0.93	0.32	1.68
12	0.93	0.89	0.35	1.65
13	0.88	0.85	0.38	1.62
14	0.85	0.82	0.41	1.59
15	0.82	0.79	0.43	1.57
16	0.79	0.76	0.45	1.55
17	0.76	0.74	0.47	1.53
18	0.74	0.72	0.48	1.52
19	0.72	0.70	0.50	1.50
20	0.70	0.68	0.51	1.49
21	0.68	0.66	0.52	1.48
22	0.66	0.65	0.53	1.47
23	0.65	0.63	0.54	1.46
24	0.63	0.62	0.55	1.45
25	0.62	0.61	0.56	1.44
30	0.56	0.55	0.60	1.40
35	0.52	0.51	0.63	1.37
40	0.48	0.48	0.66	1.34
45	0.45	0.45	0.68	1.32
50	0.43	0.43	0.70	1.30
55	0.41	0.41	0.71	1.29
60	0.39	0.39	0.72	1.28
65	0.38	0.37	0.73	1.27
70	0.36	0.36	0.74	1.26
75	0.35	0.35	0.75	1.25
80	0.34	0.34	0.76	1.24
85	0.33	0.33	0.77	1.23
90	0.32	0.32	0.77	1.23
95	0.31	0.31	0.78	1.22
100	0.30	0.30	0.79	1.21

Upper control limit for $\bar{X} = UCL_{\bar{X}} = \bar{\bar{X}} + A_3\bar{s} = \bar{\bar{X}} + A_1\bar{\sigma}_{RMS}$

Lower control limit for $\bar{X} = LCL_{\bar{X}} = \bar{\bar{X}} - A_3\bar{s} = \bar{\bar{X}} - A_1\bar{\sigma}_{RMS}$

(If aimed-at or standard value $\bar{X}_0$ is used rather than $\bar{\bar{X}}$ as the central line on the control chart, $\bar{X}_0$ should be substituted for $\bar{\bar{X}}$ in the preceding formulas.)

Upper control limit for s or $\sigma_{RMS} = UCL = B_4\bar{s} = B_4\bar{\sigma}_{RMS}$

Lower control limit for s or $\sigma_{RMS} = LCL = B_3\bar{s} = B_3\bar{\sigma}_{RMS}$

All factors in Table *E* are based on the normal distribution.

No lower control limit for s or σ_{RMS} chart where n is less than 6.

TABLE *F* FACTORS FOR DETERMINING FROM σ THE 3-SIGMA CONTROL LIMITS FOR $\bar{X}$, R, AND s OR σ_{RMS} CHARTS

Number of observations in subgroup, n	Factors for $\bar{X}$ chart, A	Factors for R chart: Lower control limit D_1	Factors for R chart: Upper control limit D_2	Factors for σ_{RMS} chart: Lower control limit B_1	Factors for σ_{RMS} chart: Upper control limit B_2	Factors for s chart: Lower control limit B_5	Factors for s chart: Upper control limit B_6
2	2.12	0	3.69	0	1.84	0	2.61
3	1.73	0	4.36	0	1.86	0	2.28
4	1.50	0	4.70	0	1.81	0	2.09
5	1.34	0	4.92	0	1.76	0	1.96
6	1.22	0	5.08	0.03	1.71	0.03	1.87
7	1.13	0.20	5.20	0.10	1.67	0.11	1.81
8	1.06	0.39	5.31	0.17	1.64	0.18	1.75
9	1.00	0.55	5.39	0.22	1.61	0.23	1.71
10	0.95	0.69	5.47	0.26	1.58	0.28	1.67
11	0.90	0.81	5.53	0.30	1.56	0.31	1.64
12	0.87	0.92	5.59	0.33	1.54	0.35	1.61
13	0.83	1.03	5.65	0.36	1.52	0.37	1.59
14	0.80	1.12	5.69	0.38	1.51	0.40	1.56
15	0.77	1.21	5.74	0.41	1.49	0.42	1.54
16	0.75	1.28	5.78	0.43	1.48	0.44	1.53
17	0.73	1.36	5.82	0.44	1.47	0.46	1.51
18	0.71	1.43	5.85	0.46	1.45	0.48	1.50
19	0.69	1.49	5.89	0.48	1.44	0.49	1.48
20	0.67	1.55	5.92	0.49	1.43	0.50	1.47
21	0.65			0.50	1.42	0.52	1.46
22	0.64			0.52	1.41	0.53	1.45
23	0.63			0.53	1.41	0.54	1.44
24	0.61			0.54	1.40	0.55	1.43
25	0.60			0.55	1.39	0.56	1.42
30	0.55			0.59	1.36	0.60	1.38
35	0.51			0.62	1.33	0.63	1.36
40	0.47			0.65	1.31	0.66	1.33
45	0.45			0.67	1.30	0.68	1.31
50	0.42			0.68	1.28	0.69	1.30
55	0.40			0.70	1.27	0.71	1.28
60	0.39			0.71	1.26	0.72	1.27
65	0.37			0.72	1.25	0.73	1.26
70	0.36			0.74	1.24	0.74	1.25
75	0.35			0.75	1.23	0.75	1.24
80	0.34			0.75	1.23	0.76	1.24
85	0.33			0.76	1.22	0.77	1.23
90	0.32			0.77	1.22	0.77	1.22
95	0.31			0.77	1.21	0.78	1.22
100	0.30			0.78	1.20	0.78	1.21

$UCL_{\bar{X}} = \mu + A\sigma$ $\qquad LCL_{\bar{X}} = \mu - A\sigma$

(If actual average is to be used rather than standard or aimed-at average, $\bar{\bar{X}}$ should be substituted for μ in the preceding formulas.)

$UCL_R = D_2\sigma$ $\qquad UCL_S = B_6\sigma$ $\qquad UCL\sigma_{RMS} = B_2\sigma$

Central line$_R = d_2\sigma$ $\qquad$ Central line$_S = c_4\sigma$ $\qquad$ Central line$\sigma_{RMS} = c_2\sigma$

$LCL_R = D_1\sigma$ $\qquad LCL_S = B_5\sigma$ $\qquad LCL\sigma_{RMS} = B_1\sigma$

No lower control limit for dispersion chart where n is less than 6.

TABLE *G* SUMMATION OF TERMS OF POISSON'S EXPONENTIAL BINOMIAL LIMIT

1,000 × probability of *c* or less occurrences of event that has average number of occurrences equal to μ_c or μ_{np}

	c									
μ_c or μ_{np}	0	1	2	3	4	5	6	7	8	9
0.02	980	1,000								
0.04	961	999	1,000							
0.06	942	998	1,000							
0.08	923	997	1,000							
0.10	905	995	1,000							
0.15	861	990	999	1,000						
0.20	819	982	999	1,000						
0.25	779	974	998	1,000						
0.30	741	963	996	1,000						
0.35	705	951	994	1,000						
0.40	670	938	992	999	1,000					
0.45	638	925	989	999	1,000					
0.50	607	910	986	998	1,000					
0.55	577	894	982	998	1,000					
0.60	549	878	977	997	1,000					
0.65	522	861	972	996	999	1,000				
0.70	497	844	966	994	999	1,000				
0.75	472	827	959	993	999	1,000				
0.80	449	809	953	991	999	1,000				
0.85	427	791	945	989	998	1,000				
0.90	407	772	937	987	998	1,000				
0.95	387	754	929	984	997	1,000				
1.00	368	736	920	981	996	999	1,000			
1.1	333	699	900	974	995	999	1,000			
1.2	301	663	879	966	992	998	1,000			
1.3	273	627	857	957	989	998	1,000			
1.4	247	592	833	946	986	997	999	1,000		
1.5	223	558	809	934	981	996	999	1,000		
1.6	202	525	783	921	976	994	999	1,000		
1.7	183	493	757	907	970	992	998	1,000		
1.8	165	463	731	891	964	990	997	999	1,000	
1.9	150	434	704	875	956	987	997	999	1,000	
2.0	135	406	677	857	947	983	995	999	1,000	

TABLE G *(continued)*

μ_c or μ_{np}	c = 0	1	2	3	4	5	6	7	8	9
2.2	111	355	623	819	928	975	993	998	1,000	
2.4	091	308	570	779	904	964	988	997	999	1,000
2.6	074	267	518	736	877	951	983	995	999	1,000
2.8	061	231	469	692	848	935	976	992	998	999
3.0	050	199	423	647	815	916	966	988	996	999
3.2	041	171	380	603	781	895	955	983	994	998
3.4	033	147	340	558	744	871	942	977	992	997
3.6	027	126	303	515	706	844	927	969	988	996
3.8	022	107	269	473	668	816	909	960	984	994
4.0	018	092	238	433	629	785	889	949	979	992
4.2	015	078	210	395	590	753	867	936	972	989
4.4	012	066	185	359	551	720	844	921	964	985
4.6	010	056	163	326	513	686	818	905	955	980
4.8	008	048	143	294	476	651	791	887	944	975
5.0	007	040	125	265	440	616	762	867	932	968
5.2	006	034	109	238	406	581	732	845	918	960
5.4	005	029	095	213	373	546	702	822	903	951
5.6	004	024	082	191	342	512	670	797	886	941
5.8	003	021	072	170	313	478	638	771	867	929
6.0	002	017	062	151	285	446	606	744	847	916

μ_c or μ_{np}	10	11	12	13	14	15	16
2.8	1,000						
3.0	1,000						
3.2	1,000						
3.4	999	1,000					
3.6	999	1,000					
3.8	998	999	1,000				
4.0	997	999	1,000				
4.2	996	999	1,000				
4.4	994	998	999	1,000			
4.6	992	997	999	1,000			
4.8	990	996	999	1,000			
5.0	986	995	998	999	1,000		
5.2	982	993	997	999	1,000		
5.4	977	990	996	999	1,000		
5.6	972	988	995	998	999	1,000	
5.8	965	984	993	997	999	1,000	
6.0	957	980	991	996	999	999	1,000

TABLE G *(continued)*

μ_c or μ_{np}	c									
	0	**1**	**2**	**3**	**4**	**5**	**6**	**7**	**8**	**9**
6.2	002	015	054	134	259	414	574	716	826	902
6.4	002	012	046	119	235	384	542	687	803	886
6.6	001	010	040	105	213	355	511	658	780	869
6.8	001	009	034	093	192	327	480	628	755	850
7.0	001	007	030	082	173	301	450	599	729	830
7.2	001	006	025	072	156	276	420	569	703	810
7.4	001	005	022	063	140	253	392	539	676	788
7.6	001	004	019	055	125	231	365	510	648	765
7.8	000	004	016	048	112	210	338	481	620	741
8.0	000	003	014	042	100	191	313	453	593	717
8.5	000	002	009	030	074	150	256	386	523	653
9.0	000	001	006	021	055	116	207	324	456	587
9.5	000	001	004	015	040	089	165	269	392	522
10.0	000	000	003	010	029	067	130	220	333	458

μ_c or μ_{np}	**10**	**11**	**12**	**13**	**14**	**15**	**16**	**17**	**18**	**19**
6.2	949	975	989	995	998	999	1,000			
6.4	939	969	986	994	997	999	1,000			
6.6	927	963	982	992	997	999	999	1,000		
6.8	915	955	978	990	996	998	999	1,000		
7.0	901	947	973	987	994	998	999	1,000		
7.2	887	937	967	984	993	997	999	999	1,000	
7.4	871	926	961	980	991	996	998	999	1,000	
7.6	854	915	954	976	989	995	998	999	1,000	
7.8	835	902	945	971	986	993	997	999	1,000	
8.0	816	888	936	966	983	992	996	998	999	1,000
8.5	763	849	909	949	973	986	993	997	999	999
9.0	706	803	876	926	959	978	989	995	998	999
9.5	645	752	836	898	940	967	982	991	996	998
10.0	583	697	792	864	917	951	973	986	993	997

μ_c or μ_{np}	**20**	**21**	**22**
8.5	1,000		
9.0	1,000		
9.5	999	1,000	
10.0	998	999	1,000

TABLE G *(continued)*

μ_c or μ_{np}	c = 0	1	2	3	4	5	6	7	8	9
10.5	000	000	002	007	021	050	102	179	279	397
11.0	000	000	001	005	015	038	079	143	232	341
11.5	000	000	001	003	011	028	060	114	191	289
12.0	000	000	001	002	008	020	046	090	155	242
12.5	000	000	000	002	005	015	035	070	125	201
13.0	000	000	000	001	004	011	026	054	100	166
13.5	000	000	000	001	003	008	019	041	079	135
14.0	000	000	000	000	002	006	014	032	062	109
14.5	000	000	000	000	001	004	010	024	048	088
15.0	000	000	000	000	001	003	008	018	037	070
	10	**11**	**12**	**13**	**14**	**15**	**16**	**17**	**18**	**19**
10.5	521	639	742	825	888	932	960	978	988	994
11.0	460	579	689	781	854	907	944	968	982	991
11.5	402	520	633	733	815	878	924	954	974	986
12.0	347	462	576	682	772	844	899	937	963	979
12.5	297	406	519	628	725	806	869	916	948	969
13.0	252	353	463	573	675	764	835	890	930	957
13.5	211	304	409	518	623	718	798	861	908	942
14.0	176	260	358	464	570	669	756	827	883	923
14.5	145	220	311	413	518	619	711	790	853	901
15.0	118	185	268	363	466	568	664	749	819	875
	20	**21**	**22**	**23**	**24**	**25**	**26**	**27**	**28**	**29**
10.5	997	999	999	1,000						
11.0	995	998	999	1,000						
11.5	992	996	998	999	1,000					
12.0	988	994	997	999	999	1,000				
12.5	983	991	995	998	999	999	1,000			
13.0	975	986	992	996	998	999	1,000			
13.5	965	980	989	994	997	998	999	1,000		
14.0	952	971	983	991	995	997	999	999	1,000	
14.5	936	960	976	986	992	996	998	999	999	1,000
15.0	917	947	967	981	989	994	997	998	999	1,000

TABLE *G* (continued)

μ_c or μ_{np}	c = 4	5	6	7	8	9	10	11	12	13
16	000	001	004	010	022	043	077	127	193	275
17	000	001	002	005	013	026	049	085	135	201
18	000	000	001	003	007	015	030	055	092	143
19	000	000	001	002	004	009	018	035	061	098
20	000	000	000	001	002	005	011	021	039	066
21	000	000	000	000	001	003	006	013	025	043
22	000	000	000	000	001	002	004	008	015	028
23	000	000	000	000	000	001	002	004	009	017
24	000	000	000	000	000	000	001	003	005	011
25	000	000	000	000	000	000	001	001	003	006

μ_c or μ_{np}	14	15	16	17	18	19	20	21	22	23
16	368	467	566	659	742	812	868	911	942	963
17	281	371	468	564	655	736	805	861	905	937
18	208	287	375	469	562	651	731	799	855	899
19	150	215	292	378	469	561	647	725	793	849
20	105	157	221	297	381	470	559	644	721	787
21	072	111	163	227	302	384	471	558	640	716
22	048	077	117	169	232	306	387	472	556	637
23	031	052	082	123	175	238	310	389	472	555
24	020	034	056	087	128	180	243	314	392	473
25	012	022	038	060	092	134	185	247	318	394

μ_c or μ_{np}	24	25	26	27	28	29	30	31	32	33
16	978	987	993	996	998	999	999	1,000		
17	959	975	985	991	995	997	999	999	1,000	
18	932	955	972	983	990	994	997	998	999	1,000
19	893	927	951	969	980	988	993	996	998	999
20	843	888	922	948	966	978	987	992	995	997
21	782	838	883	917	944	963	976	985	991	994
22	712	777	832	877	913	940	959	973	983	989
23	635	708	772	827	873	908	936	956	971	981
24	554	632	704	768	823	868	904	932	953	969
25	473	553	629	700	763	818	863	900	929	950

μ_c or μ_{np}	34	35	36	37	38	39	40	41	42	43
19	999	1,000								
20	999	999	1,000							
21	997	998	999	999	1,000					
22	994	996	998	999	999	1,000				
23	988	993	996	997	999	999	1,000			
24	979	987	992	995	997	998	999	999	1,000	
25	966	978	985	991	994	997	998	999	999	1,000

TABLE *H* LOGARITHMS OF FACTORIALS

	0	1	2	3	4	5	6	7	8	9
00	0.0000	0.0000	0.3010	0.7782	1.3802	2.0792	2.8573	3.7024	4.6055	5.5598
10	6.5598	7.6012	8.6803	9.7943	10.9404	12.1165	13.3206	14.5511	15.8063	17.0851
20	18.3861	19.7083	21.0508	22.4125	23.7927	25.1906	26.6056	28.0370	29.4841	30.9465
30	32.4237	33.9150	35.4202	36.9387	38.4702	40.0142	41.5705	43.1387	44.7185	46.3096
40	47.9116	49.5244	51.1477	52.7811	54.4246	56.0778	57.7406	59.4127	61.0939	62.7841
50	64.4831	66.1906	67.9066	69.6309	71.3633	73.1037	74.8519	76.6077	78.3712	80.1420
60	81.9202	83.7055	85.4979	87.2972	89.1034	90.9163	92.7359	94.5619	96.3945	98.2333
70	100.0784	101.9297	103.7870	105.6503	107.5196	109.3946	111.2754	113.1619	115.0540	116.9516
80	118.8547	120.7632	122.6770	124.5961	126.5204	128.4498	130.3843	132.3238	134.2683	136.2177
90	138.1719	140.1310	142.0948	144.0632	146.0364	148.0141	149.9964	151.9831	153.9744	155.9700
100	157.9700	159.9743	161.9829	163.9958	166.0128	168.0340	170.0593	172.0887	174.1221	176.1595
110	178.2009	180.2462	182.2955	184.3485	186.4054	188.4661	190.5306	192.5988	194.6707	196.7462
120	198.8254	200.9082	202.9945	205.0844	207.1779	209.2748	211.3751	213.4790	215.5862	217.6967
130	219.8107	221.9280	224.0485	226.1724	228.2995	230.4298	232.5634	234.7001	236.8400	238.9830
140	241.1291	243.2783	245.4306	247.5860	249.7443	251.9057	254.0700	256.2374	258.4076	260.5808
150	262.7569	264.9359	267.1177	269.3024	271.4899	273.6803	275.8734	278.0693	280.2679	282.4693
160	284.6735	286.8803	289.0898	291.3020	293.5168	295.7343	297.9544	300.1771	302.4024	304.6303
170	306.8608	309.0938	311.3293	313.5674	315.8079	318.0509	320.2965	322.5444	324.7948	327.0477
180	329.3030	331.5606	333.8207	336.0832	338.3480	340.6152	342.8847	345.1565	347.4307	349.7071
190	351.9859	354.2669	356.5502	358.8358	361.1236	363.4136	365.7059	368.0003	370.2970	372.5959

TABLE *H* (continued)

	0	1	2	3	4	5	6	7	8	9
200	374.8969	377.2001	379.5054	381.8129	384.1226	386.4343	388.7482	391.0642	393.3822	395.7024
210	398.0246	400.3489	402.6752	405.0036	407.3340	409.6664	412.0009	414.3373	416.6758	419.0162
220	421.3587	423.7031	426.0494	428.3977	430.7480	433.1002	435.4543	437.8103	440.1682	442.5281
230	444.8898	447.2534	449.6189	451.9862	454.3555	456.7265	459.0994	461.4742	463.8508	466.2292
240	468.6094	470.9914	473.3752	475.7608	478.1482	480.5374	482.9283	485.3210	487.7154	490.1116
250	492.5096	494.9093	497.3107	499.7138	502.1186	504.5252	506.9334	509.3433	511.7549	514.1682
260	516.5832	518.9999	521.4182	523.8381	526.2597	528.6830	531.1078	533.5344	535.9625	538.3922
270	540.8236	543.2566	545.6912	548.1273	550.5651	553.0044	555.4453	557.8878	560.3318	562.7774
280	565.2246	567.6733	570.1235	572.5753	575.0287	577.4835	579.9399	582.3977	584.8571	587.3180
290	589.7804	592.2443	594.7097	597.1766	599.6449	602.1147	604.5860	607.0588	609.5330	612.0087
300	614.4858	616.9644	619.4444	621.9258	624.4087	626.8930	629.3787	631.8659	634.3544	636.8444
310	639.3357	641.8285	644.3226	646.8182	649.3151	651.8134	654.3131	656.8142	659.3166	661.8204
320	664.3255	666.8320	669.3399	671.8491	674.3596	676.8715	679.3847	681.8993	684.4152	686.9324
330	689.4509	691.9707	694.4918	697.0143	699.5380	702.0631	704.5894	707.1170	709.6460	712.1762
340	714.7076	717.2404	719.7744	722.3097	724.8463	727.3841	729.9232	732.4635	735.0051	737.5479
350	740.0920	742.6373	745.1838	747.7316	750.2806	752.8308	755.3823	757.9349	760.4888	763.0439
360	765.6002	768.1577	770.7164	773.2764	775.8375	778.3997	780.9632	783.5279	786.0937	788.6608
370	791.2290	793.7983	796.3689	798.9406	801.5135	804.0875	806.6627	809.2390	811.8165	814.3952
380	816.9749	819.5559	822.1379	824.7211	827.3055	829.8909	832.4775	835.0652	837.6540	840.2440
390	842.8351	845.4272	848.0205	850.6149	853.2104	855.8070	858.4047	861.0035	863.6034	866.2044

TABLE *H* (continued)

	0	1	2	3	4	5	6	7	8	9
400	868.8064	871.4096	874.0138	876.6191	879.2255	881.8329	884.4415	887.0510	889.6617	892.2734
410	894.8862	897.5001	900.1150	902.7309	905.3479	907.9660	910.5850	913.2052	915.8264	918.4486
420	921.0718	923.6961	926.3214	928.9478	931.5751	934.2035	936.8329	939.4633	942.0948	944.7272
430	947.3607	949.9952	952.6307	955.2672	957.9047	960.5431	963.1826	965.8231	968.4646	971.1071
440	973.7505	976.3949	979.0404	981.6868	984.3342	986.9825	989.6318	992.2822	994.9334	997.5857
450	1000.2389	1002.8931	1005.5482	1008.2043	1010.8614	1013.5194	1016.1783	1018.8383	1021.4991	1024.1609
460	1026.8237	1029.4874	1032.1520	1034.8176	1037.4841	1040.1516	1042.8200	1045.4893	1048.1595	1050.8307
470	1053.5028	1056.1758	1058.8498	1061.5246	1064.2004	1066.8771	1069.5547	1072.2332	1074.9127	1077.5930
480	1080.2742	1082.9564	1085.6394	1088.3234	1091.0082	1093.6940	1096.3806	1099.0681	1101.7565	1104.4458
490	1107.1360	1109.8271	1112.5191	1115.2119	1117.9057	1120.6003	1123.2958	1125.9921	1128.6893	1131.3874
500	1134.0864	1136.7862	1139.4869	1142.1885	1144.8909	1147.5942	1150.2984	1153.0034	1155.7093	1158.4160
510	1161.1236	1163.8320	1166.5412	1169.2514	1171.9623	1174.6741	1177.3868	1180.1003	1182.8146	1185.5298
520	1188.2458	1190.9626	1193.6803	1196.3988	1199.1181	1201.8383	1204.5593	1207.2811	1210.0037	1212.7272
530	1215.4514	1218.1765	1220.9024	1223.6292	1226.3567	1229.0851	1231.8142	1234.5442	1237.2750	1240.0066
540	1242.7390	1245.4722	1248.2062	1250.9410	1253.6766	1256.4130	1259.1501	1261.8881	1264.6269	1267.3665
550	1270.1069	1272.8480	1275.5899	1278.3327	1281.0762	1283.8205	1286.5655	1289.3114	1292.0580	1294.8054
560	1297.5536	1300.3026	1303.0523	1305.8028	1308.5541	1311.3062	1314.0590	1316.8126	1319.5669	1322.3220
570	1325.0779	1327.8345	1330.5919	1333.3501	1336.1090	1338.8687	1341.6291	1344.3903	1347.1522	1349.9149
580	1352.6783	1355.4425	1358.2074	1360.9731	1363.7395	1366.5066	1369.2745	1372.0432	1374.8126	1377.5827
590	1380.3535	1383.1251	1385.8974	1388.6705	1391.4443	1394.2188	1396.9940	1399.7700	1402.5467	1405.3241

TABLE *H* (*continued*)

	0	1	2	3	4	5	6	7	8	9
600	1408.1023	1410.8812	1413.6608	1416.4411	1419.2221	1422.0039	1424.7863	1427.5695	1430.3534	1433.1380
610	1435.9234	1438.7094	1441.4962	1444.2836	1447.0718	1449.8607	1452.6503	1455.4405	1458.2315	1461.0232
620	1463.8156	1466.6087	1469.4025	1472.1970	1474.9922	1477.7880	1480.5846	1483.3819	1486.1798	1488.9785
630	1491.7778	1494.5779	1497.3786	1500.1800	1502.9821	1505.7849	1508.5883	1511.3924	1514.1973	1517.0028
640	1519.8090	1522.6158	1525.4233	1528.2316	1531.0404	1533.8500	1536.6602	1539.4711	1542.2827	1545.0950
650	1547.9079	1550.7215	1553.5357	1556.3506	1559.1662	1561.9824	1564.7993	1567.6169	1570.4351	1573.2540
660	1576.0736	1578.8938	1581.7146	1584.5361	1587.3583	1590.1811	1593.0046	1595.8287	1598.6535	1601.4789
670	1604.3050	1607.1317	1609.9591	1612.7871	1615.6158	1618.4451	1621.2750	1624.1056	1626.9368	1629.7687
680	1632.6012	1635.4344	1638.2681	1641.1026	1643.9376	1646.7733	1649.6096	1652.4466	1655.2842	1658.1224
690	1660.9612	1663.8007	1666.6408	1669.4816	1672.3229	1675.1649	1678.0075	1680.8508	1683.6946	1686.5391
700	1689.3842	1692.2299	1695.0762	1697.9232	1700.7708	1703.6190	1706.4678	1709.3172	1712.1672	1715.0179
710	1717.8691	1720.7210	1723.5735	1726.4266	1729.2803	1732.1346	1734.9895	1737.8450	1740.7011	1743.5578
720	1746.4152	1749.2731	1752.1316	1754.9908	1757.8505	1760.7109	1763.5718	1766.4333	1769.2955	1772.1582
730	1775.0215	1777.8854	1780.7499	1783.6150	1786.4807	1789.3470	1792.2139	1795.0814	1797.9494	1800.8181
740	1803.6873	1806.5571	1809.4275	1812.2985	1815.1701	1818.0423	1820.9150	1823.7883	1826.6622	1829.5367
750	1832.4118	1835.2874	1838.1636	1841.0404	1843.9178	1846.7957	1849.6742	1852.5533	1855.4330	1858.3133
760	1861.1941	1864.0755	1866.9574	1869.8399	1872.7230	1875.6067	1878.4909	1881.3757	1884.2611	1887.1470
770	1890.0335	1892.9205	1895.8082	1898.6963	1901.5851	1904.4744	1907.3642	1910.2547	1913.1456	1916.0372
780	1918.9293	1921.8219	1924.7151	1927.6089	1930.5032	1933.3981	1936.2935	1939.1895	1942.0860	1944.9831
790	1947.8807	1950.7789	1953.6776	1956.5769	1959.4767	1962.3771	1965.2780	1968.1794	1971.0814	1973.9840

TABLE *H* (continued)

	0	1	2	3	4	5	6	7	8	9
800	1976.8871	1979.7907	1982.6949	1985.5996	1988.5049	1991.4107	1994.3170	1997.2239	2000.1313	2003.0392
810	2005.9477	2008.8567	2011.7663	2014.6764	2017.5870	2020.4982	2023.4099	2026.3221	2029.2348	2032.1481
820	2035.0619	2037.9763	2040.8911	2043.8065	2046.7225	2049.6389	2052.5559	2055.4734	2058.3914	2061.3100
830	2064.2291	2067.1487	2070.0688	2072.9894	2075.9106	2078.8323	2081.7545	2084.6772	2087.6005	2090.5242
840	2093.4485	2096.3733	2099.2986	2102.2244	2105.1508	2108.0776	2111.0050	2113.9329	2116.8613	2119.7902
850	2122.7196	2125.6495	2128.5800	2131.5109	2134.4424	2137.3744	2140.3068	2143.2398	2146.1733	2149.1073
860	2152.0418	2154.9768	2157.9123	2160.8483	2163.7848	2166.7218	2169.6594	2172.5974	2175.5359	2178.4749
870	2181.4144	2184.3545	2187.2950	2190.2360	2193.1775	2196.1195	2199.0620	2202.0050	2204.9485	2207.8925
880	2210.8370	2213.7820	2216.7274	2219.6734	2222.6198	2225.5668	2228.5142	2231.4621	2234.4106	2237.3595
890	2240.3088	2243.2587	2246.2091	2249.1599	2252.1113	2255.0631	2258.0154	2260.9682	2263.9215	2266.8752
900	2269.8295	2272.7842	2275.7394	2278.6951	2281.6513	2284.6079	2287.5650	2290.5226	2293.4807	2296.4393
910	2299.3983	2302.3579	2305.3179	2308.2783	2311.2393	2314.2007	2317.1626	2320.1250	2323.0878	2326.0511
920	2329.0149	2331.9792	2334.9439	2337.9091	2340.8748	2343.8409	2346.8075	2349.7746	2352.7421	2355.7102
930	2358.6786	2361.6476	2364.6170	2367.5869	2370.5572	2373.5281	2376.4993	2379.4711	2382.4433	2385.4159
940	2388.3891	2391.3627	2394.3367	2397.3112	2400.2862	2403.2616	2406.2375	2409.2139	2412.1907	2415.1679
950	2418.1457	2421.1238	2424.1025	2427.0816	2430.0611	2433.0411	2436.0216	2439.0025	2441.9839	2444.9657
960	2447.9479	2450.9307	2453.9138	2456.8975	2459.8815	2462.8661	2465.8511	2468.8365	2471.8224	2474.8087
970	2477.7954	2480.7827	2483.7703	2486.7584	2489.7470	2492.7360	2495.7255	2498.7154	2501.7057	2504.6965
980	2507.6877	2510.6794	2513.6715	2516.6640	2519.6570	2522.6505	2525.6443	2528.6387	2531.6334	2534.6286
990	2537.6242	2540.6203	2543.6168	2546.6138	2549.6112	2552.6090	2555.6073	2558.6059	2561.6051	2564.6046
1,000	2567.6046	2570.6051	2573.6059	2576.6072	2579.6090	2582.6111	2585.6137	2588.6168	2591.6202	2594.6241

TABLE *J* LOGARITHMS OF NUMBERS

N	0	1	2	3	4	5	6	7	8	9
10	0000	0043	0086	0128	0170	0212	0253	0294	0334	0374
11	0414	0453	0492	0531	0569	0607	0645	0682	0719	0755
12	0792	0828	0864	0899	0934	0969	1004	1038	1072	1106
13	1139	1173	1206	1239	1271	1303	1335	1367	1399	1430
14	1461	1492	1523	1553	1584	1614	1644	1673	1703	1732
15	1761	1790	1818	1847	1875	1903	1931	1959	1987	2014
16	2041	2068	2095	2122	2148	2175	2201	2227	2253	2279
17	2304	2330	2355	2380	2405	2430	2455	2480	2504	2529
18	2553	2577	2601	2625	2648	2672	2695	2718	2742	2765
19	2788	2810	2833	2856	2878	2900	2923	2945	2967	2989
20	3010	3032	3054	3075	3096	3118	3139	3160	3181	3201
21	3222	3243	3263	3284	3304	3324	3345	3365	3385	3404
22	3424	3444	3464	3483	3502	3522	3541	3560	3579	3598
23	3617	3636	3655	3674	3692	3711	3729	3747	3766	3784
24	3802	3820	3838	3856	3874	3892	3909	3927	3945	3962
25	3979	3997	4014	4031	4048	4065	4082	4099	4116	4133
26	4150	4166	4183	4200	4216	4232	4249	4265	4281	4298
27	4314	4330	4346	4362	4378	4393	4409	4425	4440	4456
28	4472	4487	4502	4518	4533	4548	4564	4579	4594	4609
29	4624	4639	4654	4669	4683	4698	4713	4728	4742	4757
30	4771	4786	4800	4814	4829	4843	4857	4871	4886	4900
31	4914	4928	4942	4955	4969	4983	4997	5011	5024	5038
32	5051	5065	5079	5092	5105	5119	5132	5145	5159	5172
33	5185	5198	5211	5224	5237	5250	5263	5276	5289	5302
34	5315	5328	5340	5353	5366	5378	5391	5403	5416	5428
35	5441	5453	5465	5478	5490	5502	5514	5527	5539	5551
36	5563	5575	5587	5599	5611	5623	5635	5647	5658	5670
37	5682	5694	5705	5717	5729	5740	5752	5763	5775	5786
38	5798	5809	5821	5832	5843	5855	5866	5877	5888	5899
39	5911	5922	5933	5944	5955	5966	5977	5988	5999	6010
40	6021	6031	6042	6053	6064	6075	6085	6096	6107	6117
41	6128	6138	6149	6160	6170	6180	6191	6201	6212	6222
42	6232	6243	6253	6263	6274	6284	6294	6304	6314	6325
43	6335	6345	6355	6365	6375	6385	6395	6405	6415	6425
44	6435	6444	6454	6464	6474	6484	6493	6503	6513	6522
45	6532	6542	6551	6561	6571	6580	6590	6599	6609	6618
46	6628	6637	6646	6656	6665	6675	6684	6693	6702	6712
47	6721	6730	6739	6749	6758	6767	6776	6785	6794	6803
48	6812	6821	6830	6839	6848	6857	6866	6875	6884	6893
49	6902	6911	6920	6928	6937	6946	6955	6964	6972	6981
50	6990	6998	7007	7016	7024	7033	7042	7050	7059	7067
51	7076	7084	7093	7101	7110	7118	7126	7135	7143	7152
52	7160	7168	7177	7185	7193	7202	7210	7218	7226	7235
53	7243	7251	7259	7267	7275	7284	7292	7300	7308	7316
54	7324	7332	7340	7348	7356	7364	7372	7380	7388	7396

| *N* | 0 | 1 | 2 | 3 | 4 | 5 | 6 | 7 | 8 | 9 |

N	0	1	2	3	4	5	6	7	8	9
55	7404	7412	7419	7427	7435	7443	7451	7459	7466	7474
56	7482	7490	7497	7505	7513	7520	7528	7536	7543	7551
57	7559	7566	7574	7582	7589	7597	7604	7612	7619	7627
58	7634	7642	7649	7657	7664	7672	7679	7686	7694	7701
59	7709	7716	7723	7731	7738	7745	7752	7760	7767	7774
60	7782	7789	7796	7803	7810	7818	7825	7832	7839	7846
61	7853	7860	7868	7875	7882	7889	7896	7903	7910	7917
62	7924	7931	7938	7945	7952	7959	7966	7973	7980	7987
63	7993	8000	8007	8014	8021	8028	8035	8041	8048	8055
64	8062	8069	8075	8082	8089	8096	8102	8109	8116	8122
65	8129	8136	8142	8149	8156	8162	8169	8176	8182	8189
66	8195	8202	8209	8215	8222	8228	8235	8241	8248	8254
67	8261	8267	8274	8280	8287	8293	8299	8306	8312	8319
68	8325	8331	8338	8344	8351	8357	8363	8370	8376	8382
69	8388	8395	8401	8407	8414	8420	8426	8432	8439	8445
70	8451	8457	8463	8470	8476	8482	8488	8494	8500	8506
71	8513	8519	8525	8531	8537	8543	8549	8555	8561	8567
72	8573	8579	8585	8591	8597	8603	8609	8615	8621	8627
73	8633	8639	8645	8651	8657	8663	8669	8675	8681	8686
74	8692	8698	8704	8710	8716	8722	8727	8733	8739	8745
75	8751	8756	8762	8768	8774	8779	8785	8791	8797	8802
76	8808	8814	8820	8825	8831	8837	8842	8848	8854	8859
77	8865	8871	8876	8882	8887	8893	8899	8904	8910	8915
78	8921	8927	8932	8938	8943	8949	8954	8960	8965	8971
79	8976	8982	8987	8993	8998	9004	9009	9015	9020	9025
80	9031	9036	9042	9047	9053	9058	9063	9069	9074	9079
81	9085	9090	9096	9101	9106	9112	9117	9122	9128	9133
82	9138	9143	9149	9154	9159	9165	9170	9175	9180	9186
83	9191	9196	9201	9206	9212	9217	9222	9227	9232	9238
84	9243	9248	9253	9258	9263	9269	9274	9279	9284	9289
85	9294	9299	9304	9309	9315	9320	9325	9330	9335	9340
86	9345	9350	9355	9360	9365	9370	9375	9380	9385	9390
87	9395	9400	9405	9410	9415	9420	9425	9430	9435	9440
88	9445	9450	9455	9460	9465	9469	9474	9479	9484	9489
89	9494	9499	9504	9509	9513	9518	9523	9528	9533	9538
90	9542	9547	9552	9557	9562	9566	9571	9576	9581	9586
91	9590	9595	9600	9605	9609	9614	9619	9624	9628	9633
92	9638	9643	9647	9652	9657	9661	9666	9671	9675	9680
93	9685	9689	9694	9699	9703	9708	9713	9717	9722	9727
94	9731	9736	9741	9745	9750	9754	9759	9763	9768	9773
95	9777	9782	9786	9791	9795	9800	9805	9809	9814	9818
96	9823	9827	9832	9836	9841	9845	9850	9854	9859	9863
97	9868	9872	9877	9881	9886	9890	9894	9899	9903	9908
98	9912	9917	9921	9926	9930	9934	9939	9943	9948	9952
99	9956	9961	9965	9969	9974	9978	9983	9987	9991	9996
N	0	1	2	3	4	5	6	7	8	9

TABLE *K* SAMPLE SIZE CODE LETTERS—MIL-STD-105E (ABC STANDARD)

Lot or batch size	Special inspection levels				General inspection levels		
	S-1	S-2	S-3	S-4	I	II	III
2–8	A	A	A	A	A	A	B
9–15	A	A	A	A	A	B	C
16–25	A	A	B	B	B	C	D
26–50	A	B	B	C	C	D	E
51–90	B	B	C	C	C	E	F
91–150	B	B	C	D	D	F	G
151–280	B	C	D	E	E	G	H
281–500	B	C	D	E	F	H	J
501–1,200	C	C	E	F	G	J	K
1,201–3,200	C	D	E	G	H	K	L
3,201–10,000	C	D	F	G	J	L	M
10,001–35,000	C	D	F	H	K	M	N
35,001–150,000	D	E	G	J	L	N	P
150,001–500,000	D	E	G	J	M	P	Q
500,001 and over	D	E	H	K	N	Q	R

TABLE *L* MASTER TABLE FOR NORMAL INSPECTION (SINGLE SAMPLING)—MIL-STD-105E (ABC STANDARD)

Sample size code letter	Sample size	Acceptable quality levels (normal inspection)																									
		0.010	0.015	0.025	0.040	0.065	0.10	0.15	0.25	0.40	0.65	1.0	1.5	2.5	4.0	6.5	10	15	25	40	65	100	150	250	400	650	1,000
		Ac Re	*Ac Re*	*Ac Re*	*Ac Re*	*Ac Re*	*Ac Re*	*Ac Re*	*Ac Re*	*Ac Re*	*Ac Re*	*Ac Re*	*Ac Re*	*Ac Re*	*Ac Re*	*Ac Re*	*Ac Re*	*Ac Re*	*Ac Re*	*Ac Re*	*Ac Re*	*Ac Re*	*Ac Re*	*Ac Re*	*Ac Re*	*Ac Re*	*Ac Re*
A	2	↓	↓	↓	↓	↓	↓	↓	↓	↓	↓	↓	↓	↓	↓	0 1	↓	↓	1 2	2 3	3 4	5 6	7 8	10 11	14 15	21 22	30 31
B	3	↓	↓	↓	↓	↓	↓	↓	↓	↓	↓	↓	↓	↓	0 1	↑	↓	1 2	2 3	3 4	5 6	7 8	10 11	14 15	21 22	30 31	44 45
C	5	↓	↓	↓	↓	↓	↓	↓	↓	↓	↓	↓	↓	0 1	↑	↓	1 2	2 3	3 4	5 6	7 8	10 11	14 15	21 22	30 31	44 45	↑
D	8	↓	↓	↓	↓	↓	↓	↓	↓	↓	↓	↓	0 1	↑	↓	1 2	2 3	3 4	5 6	7 8	10 11	14 15	21 22	30 31	44 45	↑	↑
E	13	↓	↓	↓	↓	↓	↓	↓	↓	↓	↓	0 1	↑	↓	1 2	2 3	3 4	5 6	7 8	10 11	14 15	21 22	30 31	44 45	↑	↑	↑
F	20	↓	↓	↓	↓	↓	↓	↓	↓	↓	0 1	↑	↓	1 2	2 3	3 4	5 6	7 8	10 11	14 15	21 22	↑	↑	↑	↑	↑	↑
G	32	↓	↓	↓	↓	↓	↓	↓	↓	0 1	↑	↓	1 2	2 3	3 4	5 6	7 8	10 11	14 15	21 22	↑	↑	↑	↑	↑	↑	↑
H	50	↓	↓	↓	↓	↓	↓	↓	0 1	↑	↓	1 2	2 3	3 4	5 6	7 8	10 11	14 15	21 22	↑	↑	↑	↑	↑	↑	↑	↑
J	80	↓	↓	↓	↓	↓	↓	0 1	↑	↓	1 2	2 3	3 4	5 6	7 8	10 11	14 15	21 22	↑	↑	↑	↑	↑	↑	↑	↑	↑
K	125	↓	↓	↓	↓	↓	0 1	↑	↓	1 2	2 3	3 4	5 6	7 8	10 11	14 15	21 22	↑	↑	↑	↑	↑	↑	↑	↑	↑	↑
L	200	↓	↓	↓	↓	0 1	↑	↓	1 2	2 3	3 4	5 6	7 8	10 11	14 15	21 22	↑	↑	↑	↑	↑	↑	↑	↑	↑	↑	↑
M	315	↓	↓	↓	0 1	↑	↓	1 2	2 3	3 4	5 6	7 8	10 11	14 15	21 22	↑	↑	↑	↑	↑	↑	↑	↑	↑	↑	↑	↑
N	500	↓	↓	0 1	↑	↓	1 2	2 3	3 4	5 6	7 8	10 11	14 15	21 22	↑	↑	↑	↑	↑	↑	↑	↑	↑	↑	↑	↑	↑
P	800	↓	0 1	↑	↓	1 2	2 3	3 4	5 6	7 8	10 11	14 15	21 22	↑	↑	↑	↑	↑	↑	↑	↑	↑	↑	↑	↑	↑	↑
Q	1,250	0 1	↑	↓	1 2	2 3	3 4	5 6	7 8	10 11	14 15	21 22	↑	↑	↑	↑	↑	↑	↑	↑	↑	↑	↑	↑	↑	↑	↑
R	2,000	↑	↑	1 2	2 3	3 4	5 6	7 8	10 11	14 15	21 22	↑	↑	↑	↑	↑	↑	↑	↑	↑	↑	↑	↑	↑	↑	↑	↑

↓ = use first sampling plan below arrow. If sample size equals or exceeds lot or batch size, do 100% inspection.
↑ = use first sampling plan above arrow.
Ac = acceptance number.
Re = rejection number.

TABLE *M* MASTER TABLE FOR TIGHTENED INSPECTION (SINGLE SAMPLING)—MIL-STD-105E (ABC STANDARD)

Sample size code letter	Sample size	Acceptable quality levels (tightened inspection)																									
		0.010	0.015	0.025	0.040	0.065	0.10	0.15	0.25	0.40	0.65	1.0	1.5	2.5	4.0	6.5	10	15	25	40	65	100	150	250	400	650	1,000
		Ac Re	*Ac Re*	*Ac Re*	*Ac Re*	*Ac Re*	*Ac Re*	*Ac Re*	*Ac Re*	*Ac Re*	*Ac Re*	*Ac Re*	*Ac Re*	*Ac Re*	*Ac Re*	*Ac Re*	*Ac Re*	*Ac Re*	*Ac Re*	*Ac Re*	*Ac Re*	*Ac Re*	*Ac Re*	*Ac Re*	*Ac Re*	*Ac Re*	*Ac Re*
A	2	↓	↓	↓	↓	↓	↓	↓	↓	↓	↓	↓	↓	↓	↓	↓	↓	↓	↓	1 2	2 3	3 4	5 6	8 9	12 13	18 19	27 28
B	3	↓	↓	↓	↓	↓	↓	↓	↓	↓	↓	↓	↓	↓	↓	0 1	↓	↓	1 2	2 3	3 4	5 6	8 9	12 13	18 19	27 28	41 42
C	5	↓	↓	↓	↓	↓	↓	↓	↓	↓	↓	↓	↓	↓	0 1	↓	↓	1 2	2 3	3 4	5 6	8 9	12 13	18 19	27 28	41 42	↑
D	8	↓	↓	↓	↓	↓	↓	↓	↓	↓	↓	↓	↓	0 1	↓	↓	1 2	2 3	3 4	5 6	8 9	12 13	18 19	27 28	41 42	↑	↑
E	13	↓	↓	↓	↓	↓	↓	↓	↓	↓	↓	↓	0 1	↓	↓	1 2	2 3	3 4	5 6	8 9	12 13	18 19	27 28	41 42	↑	↑	↑
F	20	↓	↓	↓	↓	↓	↓	↓	↓	↓	↓	0 1	↓	↓	1 2	2 3	3 4	5 6	8 9	12 13	18 19	↑	↑	↑	↑	↑	↑
G	32	↓	↓	↓	↓	↓	↓	↓	↓	↓	0 1	↓	↓	1 2	2 3	3 4	5 6	8 9	12 13	18 19	↑	↑	↑	↑	↑	↑	↑
H	50	↓	↓	↓	↓	↓	↓	↓	↓	0 1	↓	↓	1 2	2 3	3 4	5 6	8 9	12 13	18 19	↑	↑	↑	↑	↑	↑	↑	↑
J	80	↓	↓	↓	↓	↓	↓	↓	0 1	↓	↓	1 2	2 3	3 4	5 6	8 9	12 13	18 19	↑	↑	↑	↑	↑	↑	↑	↑	↑
K	125	↓	↓	↓	↓	↓	↓	0 1	↓	↓	1 2	2 3	3 4	5 6	8 9	12 13	18 19	↑	↑	↑	↑	↑	↑	↑	↑	↑	↑
L	200	↓	↓	↓	↓	↓	0 1	↓	↓	1 2	2 3	3 4	5 6	8 9	12 13	18 19	↑	↑	↑	↑	↑	↑	↑	↑	↑	↑	↑
M	315	↓	↓	↓	↓	0 1	↓	↓	1 2	2 3	3 4	5 6	8 9	12 13	18 19	↑	↑	↑	↑	↑	↑	↑	↑	↑	↑	↑	↑
N	500	↓	↓	↓	0 1	↓	↓	1 2	2 3	3 4	5 6	8 9	12 13	18 19	↑	↑	↑	↑	↑	↑	↑	↑	↑	↑	↑	↑	↑
P	800	↓	↓	0 1	↓	↓	1 2	2 3	3 4	5 6	8 9	12 13	18 19	↑	↑	↑	↑	↑	↑	↑	↑	↑	↑	↑	↑	↑	↑
Q	1,250	↓	0 1	↓	↓	1 2	2 3	3 4	5 6	8 9	12 13	18 19	↑	↑	↑	↑	↑	↑	↑	↑	↑	↑	↑	↑	↑	↑	↑
R	2,000	0 1	↑	↓	1 2	2 3	3 4	5 6	8 9	12 13	18 19	↑	↑	↑	↑	↑	↑	↑	↑	↑	↑	↑	↑	↑	↑	↑	↑
S	3,150			1 2																							

↓ = use first sampling plan below arrow. If sample size equals or exceeds lot or batch size, do 100% inspection.
↑ = use first sampling plan above arrow.
Ac = acceptance number.
Re = rejection number.

TABLE *N* MASTER TABLE FOR REDUCED INSPECTION (SINGLE SAMPLING)—MIL-STD-105E (ABC STANDARD)

Sample size code letter	Sample size	Acceptable quality levels (reduced inspection)†																									
		0.010	0.015	0.025	0.040	0.065	0.10	0.15	0.25	0.40	0.65	1.0	1.5	2.5	4.0	6.5	10	15	25	40	65	100	150	250	400	650	1,000
		Ac Re	*Ac Re*	*Ac Re*	*Ac Re*	*Ac Re*	*Ac Re*	*Ac Re*	*Ac Re*	*Ac Re*	*Ac Re*	*Ac Re*	*Ac Re*	*Ac Re*	*Ac Re*	*Ac Re*	*Ac Re*	*Ac Re*	*Ac Re*	*Ac Re*	*Ac Re*	*Ac Re*	*Ac Re*	*Ac Re*	*Ac Re*	*Ac Re*	*Ac Re*
A	2	↓	↓	↓	↓	↓	↓	↓	↓	↓	↓	↓	↓	↓	↓	0 1	↓	↓	1 2	2 3	3 4	5 6	7 8	10 11	14 15	21 22	30 31
B	2	↓	↓	↓	↓	↓	↓	↓	↓	↓	↓	↓	↓	↓	0 1	↑	↓	0 2	1 3	2 4	3 5	5 6	7 8	10 11	14 15	21 22	30 31
C	2	↓	↓	↓	↓	↓	↓	↓	↓	↓	↓	↓	↓	0 1	↑	↓	0 2	1 3	1 4	2 5	3 6	5 8	7 10	10 13	14 17	21 24	↑
D	3	↓	↓	↓	↓	↓	↓	↓	↓	↓	↓	↓	0 1	↑	↓	0 2	1 3	1 4	2 5	3 6	5 8	7 10	10 13	14 17	21 24	↑	↑
E	5	↓	↓	↓	↓	↓	↓	↓	↓	↓	↓	0 1	↑	↓	0 2	1 3	1 4	2 5	3 6	5 8	7 10	10 13	14 17	21 24	↑	↑	↑
F	8	↓	↓	↓	↓	↓	↓	↓	↓	↓	0 1	↑	↓	0 2	1 3	1 4	2 5	3 6	5 8	7 10	10 13	↑	↑	↑	↑	↑	↑
G	13	↓	↓	↓	↓	↓	↓	↓	↓	0 1	↑	↓	0 2	1 3	1 4	2 5	3 6	5 8	7 10	10 13	↑	↑	↑	↑	↑	↑	↑
H	20	↓	↓	↓	↓	↓	↓	↓	0 1	↑	↓	0 2	1 3	1 4	2 5	3 6	5 8	7 10	10 13	↑	↑	↑	↑	↑	↑	↑	↑
J	32	↓	↓	↓	↓	↓	↓	0 1	↑	↓	0 2	1 3	1 4	2 5	3 6	5 8	7 10	10 13	↑	↑	↑	↑	↑	↑	↑	↑	↑
K	50	↓	↓	↓	↓	↓	0 1	↑	↓	0 2	1 3	1 4	2 5	3 6	5 8	7 10	10 13	↑	↑	↑	↑	↑	↑	↑	↑	↑	↑
L	80	↓	↓	↓	↓	0 1	↑	↓	0 2	1 3	1 4	2 5	3 6	5 8	7 10	10 13	↑	↑	↑	↑	↑	↑	↑	↑	↑	↑	↑
M	125	↓	↓	↓	0 1	↑	↓	0 2	1 3	1 4	2 5	3 6	5 8	7 10	10 13	↑	↑	↑	↑	↑	↑	↑	↑	↑	↑	↑	↑
N	200	↓	↓	0 1	↑	↓	0 2	1 3	1 4	2 5	3 6	5 8	7 10	10 13	↑	↑	↑	↑	↑	↑	↑	↑	↑	↑	↑	↑	↑
P	315	↓	0 1	↑	↓	0 2	1 3	1 4	2 5	3 6	5 8	7 10	10 13	↑	↑	↑	↑	↑	↑	↑	↑	↑	↑	↑	↑	↑	↑
Q	500	0 1	↑	↓	0 2	1 3	1 4	2 5	3 6	5 8	7 10	10 13	↑	↑	↑	↑	↑	↑	↑	↑	↑	↑	↑	↑	↑	↑	↑
R	800	↑	↑	0 2	1 3	1 4	2 5	3 6	5 8	7 10	10 13	↑	↑	↑	↑	↑	↑	↑	↑	↑	↑	↑	↑	↑	↑	↑	↑

↓ = use first sampling plan below arrow. If sample size equals or exceeds lot or batch size, do 100% inspection.

↑ = use first sampling plan above arrow.

Ac = acceptance number.

Re = rejection number.

† If the acceptance number has been exceeded but the rejection number has not been reached, accept the lot but reinstate normal inspection.

TABLE *O* MASTER TABLE FOR NORMAL INSPECTION (DOUBLE SAMPLING)—MIL-STD-105E (ABC STANDARD)

Sample size code letter	Sample	Sample size	Cumulative sample size	Acceptable quality levels (normal inspection)																									
				0.010	0.015	0.025	0.040	0.065	0.10	0.15	0.25	0.40	0.65	1.0	1.5	2.5	4.0	6.5	10	15	25	40	65	100	150	250	400	650	1,000
				Ac Re	*Ac Re*	*Ac Re*	*Ac Re*	*Ac Re*	*Ac Re*	*Ac Re*	*Ac Re*	*Ac Re*	*Ac Re*	*Ac Re*	*Ac Re*	*Ac Re*	*Ac Re*	*Ac Re*	*Ac Re*	*Ac Re*	*Ac Re*	*Ac Re*	*Ac Re*	*Ac Re*	*Ac Re*	*Ac Re*	*Ac Re*	*Ac Re*	*Ac Re*
A				↓	↓	↓	↓	↓	↓	↓	↓	↓	↓	↓	↓	↓	↓	†	↓	↓	†	†	†	†	†	†	†	†	†
B	First	2	2	↓	↓	↓	↓	↓	↓	↓	↓	↓	↓	↓	↓	↓	†	↑	↓	0 2	0 3	1 4	2 5	3 7	5 9	7 11	11 16	17 22	25 31
	Second	2	4																	1 2	3 4	4 5	6 7	8 9	12 13	18 19	26 27	37 38	56 57
C	First	3	3	↓	↓	↓	↓	↓	↓	↓	↓	↓	↓	↓	↓	†	↑	↓	0 2	0 3	1 4	2 5	3 7	5 9	7 11	11 16	17 22	25 31	↑
	Second	3	6																1 2	3 4	4 5	6 7	8 9	12 13	18 19	26 27	37 38	56 57	
D	First	5	5	↓	↓	↓	↓	↓	↓	↓	↓	↓	↓	↓	†	↑	↓	0 2	0 3	1 4	2 5	3 7	5 9	7 11	11 16	17 22	25 31	↑	↑
	Second	5	10															1 2	3 4	4 5	6 7	8 9	12 13	18 19	26 27	37 38	56 57		
E	First	8	8	↓	↓	↓	↓	↓	↓	↓	↓	↓	↓	†	↑	↓	0 2	0 3	1 4	2 5	3 7	5 9	7 11	11 16	17 22	25 31	↑	↑	↑
	Second	8	16														1 2	3 4	4 5	6 7	8 9	12 13	18 19	26 27	37 38	56 57			
F	First	13	13	↓	↓	↓	↓	↓	↓	↓	↓	↓	†	↑	↓	0 2	0 3	1 4	2 5	3 7	5 9	7 11	11 16	↑	↑	↑	↑	↑	↑
	Second	13	26													1 2	3 4	4 5	6 7	8 9	12 13	18 19	26 27						
G	First	20	20	↓	↓	↓	↓	↓	↓	↓	↓	†	↑	↓	0 2	0 3	1 4	2 5	3 7	5 9	7 11	11 16	↑	↑	↑	↑	↑	↑	↑
	Second	20	40												1 2	3 4	4 5	6 7	8 9	12 13	18 19	26 27							
H	First	32	32	↓	↓	↓	↓	↓	↓	↓	†	↑	↓	0 2	0 3	1 4	2 5	3 7	5 9	7 11	11 16	↑	↑	↑	↑	↑	↑	↑	↑
	Second	32	64											1 2	3 4	4 5	6 7	8 9	12 13	18 19	26 27								
J	First	50	50	↓	↓	↓	↓	↓	↓	†	↑	↓	0 2	0 3	1 4	2 5	3 7	5 9	7 11	11 16	↑	↑	↑	↑	↑	↑	↑	↑	↑
	Second	50	100										1 2	3 4	4 5	6 7	8 9	12 13	18 19	26 27									
K	First	80	80	↓	↓	↓	↓	↓	†	↑	↓	0 2	0 3	1 4	2 5	3 7	5 9	7 11	11 16	↑	↑	↑	↑	↑	↑	↑	↑	↑	↑
	Second	80	160									1 2	3 4	4 5	6 7	8 9	12 13	18 19	26 27										
L	First	125	125	↓	↓	↓	↓	†	↑	↓	0 2	0 3	1 4	2 5	3 7	5 9	7 11	11 16	↑	↑	↑	↑	↑	↑	↑	↑	↑	↑	↑
	Second	125	250								1 2	3 4	4 5	6 7	8 9	12 13	18 19	26 27											
M	First	200	200	↓	↓	↓	†	↑	↓	0 2	0 3	1 4	2 5	3 7	5 9	7 11	11 16	↑	↑	↑	↑	↑	↑	↑	↑	↑	↑	↑	↑
	Second	200	400							1 2	3 4	4 5	6 7	8 9	12 13	18 19	26 27												
N	First	315	315	↓	↓	†	↑	↓	0 2	0 3	1 4	2 5	3 7	5 9	7 11	11 16	↑	↑	↑	↑	↑	↑	↑	↑	↑	↑	↑	↑	↑
	Second	315	630						1 2	3 4	4 5	6 7	8 9	12 13	18 19	26 27													
P	First	500	500	↓	†	↑	↓	0 2	0 3	1 4	2 5	3 7	5 9	7 11	11 16	↑	↑	↑	↑	↑	↑	↑	↑	↑	↑	↑	↑	↑	↑
	Second	500	1,000					1 2	3 4	4 5	6 7	8 9	12 13	18 19	26 27														
Q	First	800	800	†	↑	↓	0 2	0 3	1 4	2 5	3 7	5 9	7 11	11 16	↑	↑	↑	↑	↑	↑	↑	↑	↑	↑	↑	↑	↑	↑	↑
	Second	800	1,600				1 2	3 4	4 5	6 7	8 9	12 13	18 19	26 27															
R	First	1,250	1,250	↑	↑	0 2	0 3	1 4	2 5	3 7	5 9	7 11	11 16	↑	↑	↑	↑	↑	↑	↑	↑	↑	↑	↑	↑	↑	↑	↑	↑
	Second	1,250	2,500			1 2	3 4	4 5	6 7	8 9	12 13	18 19	26 27																

↓ = use first sampling plan below arrow. If sample size equals or exceeds lot or batch size, do 100% inspection.
↑ = use first sampling plan above arrow.
Ac = acceptance number.
Re = rejection number.
†Use corresponding single sampling plan (or, alternatively, use double sampling plan below, where available).

TABLE *P* MASTER TABLE FOR TIGHTENED INSPECTION (DOUBLING SAMPLING)—MIL-STD-105E (ABC STANDARD)

Sample size code letter	Sample	Sample size	Cumulative sample size	0.010	0.015	0.025	0.040	0.065	0.10	0.15	0.25	0.40	0.65	1.0	1.5	2.5	4.0	6.5	10	15	25	40	65	100	150	250	400	650	1,000
				Ac Re	Ac Re	Ac Re	Ac Re	Ac Re	Ac Re	Ac Re	Ac Re	Ac Re	Ac Re	Ac Re	Ac Re	Ac Re	Ac Re	Ac Re	Ac Re	Ac Re	Ac Re	Ac Re	Ac Re	Ac Re	Ac Re	Ac Re	Ac Re	Ac Re	Ac Re
A				↓	↓	↓	↓	↓	↓	↓	↓	↓	↓	↓	↓	↓	↓	↓	↓	↓	↓	†	†	†	†	†	†	†	†
B	First	2	2	↓	↓	↓	↓	↓	↓	↓	↓	↓	↓	↓	↓	↓	↓	†	↓	↓	0 2	0 3	1 4	2 5	3 7	6 10	9 14	15 20	23 29
	Second	2	4																		1 2	3 4	4 5	6 7	11 12	15 16	23 24	34 35	52 53
C	First	3	3	↓	↓	↓	↓	↓	↓	↓	↓	↓	↓	↓	↓	↓	†	↓	↓	0 2	0 3	1 4	2 5	3 7	6 10	9 14	15 20	23 29	↑
	Second	3	6																	1 2	3 4	4 5	6 7	11 12	15 16	23 24	34 35	52 53	
D	First	5	5	↓	↓	↓	↓	↓	↓	↓	↓	↓	↓	↓	↓	†	↓	↓	0 2	0 3	1 4	2 5	3 7	6 10	9 14	15 20	23 29	↑	↑
	Second	5	10																1 2	3 4	4 5	6 7	11 12	15 16	23 24	34 35	52 53		
E	First	8	8	↓	↓	↓	↓	↓	↓	↓	↓	↓	↓	↓	†	↓	↓	0 2	0 3	1 4	2 5	3 7	6 10	9 14	15 20	23 29	↑	↑	↑
	Second	8	16															1 2	3 4	4 5	6 7	11 12	15 16	23 24	34 35	52 53			
F	First	13	13	↓	↓	↓	↓	↓	↓	↓	↓	↓	↓	†	↓	↓	0 2	0 3	1 4	2 5	3 7	6 10	9 14	↑	↑	↑	↑	↑	↑
	Second	13	26														1 2	3 4	4 5	6 7	11 12	15 16	23 24						
G	First	20	20	↓	↓	↓	↓	↓	↓	↓	↓	↓	†	↓	↓	0 2	0 3	1 4	2 5	3 7	6 10	9 14	↑	↑	↑	↑	↑	↑	↑
	Second	20	40													1 2	3 4	4 5	6 7	11 12	15 16	23 24							
H	First	32	32	↓	↓	↓	↓	↓	↓	↓	↓	†	↓	↓	0 2	0 3	1 4	2 5	3 7	6 10	9 14	↑	↑	↑	↑	↑	↑	↑	↑
	Second	32	64												1 2	3 4	4 5	6 7	11 12	15 16	23 24								
J	First	50	50	↓	↓	↓	↓	↓	↓	↓	†	↓	↓	0 2	0 3	1 4	2 5	3 7	6 10	9 14	↑	↑	↑	↑	↑	↑	↑	↑	↑
	Second	50	100											1 2	3 4	4 5	6 7	11 12	15 16	23 24									
K	First	80	80	↓	↓	↓	↓	↓	↓	†	↓	↓	0 2	0 3	1 4	2 5	3 7	6 10	9 14	↑	↑	↑	↑	↑	↑	↑	↑	↑	↑
	Second	80	160										1 2	3 4	4 5	6 7	11 12	15 16	23 24										
L	First	125	125	↓	↓	↓	↓	↓	†	↓	↓	0 2	0 3	1 4	2 5	3 7	6 10	9 14	↑	↑	↑	↑	↑	↑	↑	↑	↑	↑	↑
	Second	125	250									1 2	3 4	4 5	6 7	11 12	15 16	23 24											
M	First	200	200	↓	↓	↓	↓	†	↓	↓	0 2	0 3	1 4	2 5	3 7	6 10	9 14	↑	↑	↑	↑	↑	↑	↑	↑	↑	↑	↑	↑
	Second	200	400								1 2	3 4	4 5	6 7	11 12	15 16	23 24												
N	First	315	315	↓	↓	↓	†	↓	↓	0 2	0 3	1 4	2 5	3 7	6 10	9 14	↑	↑	↑	↑	↑	↑	↑	↑	↑	↑	↑	↑	↑
	Second	315	630							1 2	3 4	4 5	6 7	11 12	15 16	23 24													
P	First	500	500	↓	↓	†	↓	↓	0 2	0 3	1 4	2 5	3 7	6 10	9 14	↑	↑	↑	↑	↑	↑	↑	↑	↑	↑	↑	↑	↑	↑
	Second	500	1,000						1 2	3 4	4 5	6 7	11 12	15 16	23 24														
Q	First	800	800	↓	†	↓	↓	0 2	0 3	1 4	2 5	3 7	6 10	9 14	↑	↑	↑	↑	↑	↑	↑	↑	↑	↑	↑	↑	↑	↑	↑
	Second	800	1,600					1 2	3 4	4 5	6 7	11 12	15 16	23 24															
R	First	1,250	1,250	†	↑	↓	0 2	0 3	1 4	2 5	3 7	6 10	9 14	↑	↑	↑	↑	↑	↑	↑	↑	↑	↑	↑	↑	↑	↑	↑	↑
	Second	1,250	2,500				1 2	3 4	4 5	6 7	11 12	15 16	23 24																
S	First	2,000	2,000			0 2																							
	Second	2,000	4,000			1 2																							

↓ = use first sampling plan below arrow. If sample size equals or exceeds lot or batch size, do 100% inspection.
↑ = use first sampling plan above arrow.
Ac = acceptance number.
Re = rejection number.
†Use corresponding single sampling plan (or, alternatively, use double sampling plan below, where available).

TABLE *Q* MASTER TABLE FOR REDUCED INSPECTION (DOUBLE SAMPLING)—MIL-STD-105E (ABC STANDARD)

Sample size code letter	Sample	Sample size	Cumulative sample size	Acceptable quality levels (reduced inspection)†																									
				0.010	0.015	0.025	0.040	0.065	0.10	0.15	0.25	0.40	0.65	1.0	1.5	2.5	4.0	6.5	10	15	25	40	65	100	150	250	400	650	1,000
				Ac Re	Ac Re	Ac Re	Ac Re	Ac Re	Ac Re	Ac Re	Ac Re	Ac Re	Ac Re	Ac Re	Ac Re	Ac Re	Ac Re	Ac Re	Ac Re	Ac Re	Ac Re	Ac Re	Ac Re	Ac Re	Ac Re	Ac Re	Ac Re	Ac Re	Ac Re
A				↓	↓	↓	↓	↓	↓	↓	↓	↓	↓	↓	↓	↓	↓	‡	↓	↓	‡	‡	‡	‡	‡	‡	‡	‡	‡
B				↓	↓	↓	↓	↓	↓	↓	↓	↓	↓	↓	↓	↓	‡	↓	↓	‡	‡	‡	‡	‡	‡	‡	‡	‡	‡
C				↓	↓	↓	↓	↓	↓	↓	↓	↓	↓	↓	↓	‡	↑	↓	‡	‡	‡	‡	‡	‡	‡	‡	‡	‡	↑
D	First	2	2	↓	↓	↓	↓	↓	↓	↓	↓	↓	↓	↓	‡	↑	↓	0 2	0 3	0 4	0 4	1 5	2 7	3 8	5 10	7 12	11 17	↑	↑
	Second	2	4															0 2	0 4	1 5	3 6	4 7	6 9	8 12	12 16	18 21	26 30		
E	First	3	3	↓	↓	↓	↓	↓	↓	↓	↓	↓	↓	‡	↑	↓	0 2	0 3	0 4	0 4	1 5	2 7	3 8	5 10	7 12	11 17	↑	↑	↑
	Second	3	6														0 2	0 4	1 5	3 6	4 7	6 9	8 12	12 16	18 21	26 30			
F	First	5	5	↓	↓	↓	↓	↓	↓	↓	↓	↓	‡	↑	↓	0 2	0 3	0 4	0 4	1 5	2 7	3 8	5 10	↑	↑	↑	↑	↑	↑
	Second	5	10													0 2	0 4	1 5	3 6	4 7	6 9	8 12	12 16						
G	First	8	8	↓	↓	↓	↓	↓	↓	↓	↓	‡	↑	↓	0 2	0 3	0 4	0 4	1 5	2 7	3 8	5 10	↑	↑	↑	↑	↑	↑	↑
	Second	8	16												0 2	0 4	1 5	3 6	4 7	6 9	8 12	12 16							
H	First	13	13	↓	↓	↓	↓	↓	↓	↓	‡	↑	↓	0 2	0 3	0 4	0 4	1 5	2 7	3 8	5 10	↑	↑	↑	↑	↑	↑	↑	↑
	Second	13	26											0 2	0 4	1 5	3 6	4 7	6 9	8 12	12 16								
J	First	20	20	↓	↓	↓	↓	↓	↓	‡	↑	↓	0 2	0 3	0 4	0 4	1 5	2 7	3 8	5 10	↑	↑	↑	↑	↑	↑	↑	↑	↑
	Second	20	40										0 2	0 4	1 5	3 6	4 7	6 9	8 12	12 16									
K	First	32	32	↓	↓	↓	↓	↓	‡	↑	↓	0 2	0 3	0 4	0 4	1 5	2 7	3 8	5 10	↑	↑	↑	↑	↑	↑	↑	↑	↑	↑
	Second	32	64									0 2	0 4	1 5	3 6	4 7	6 9	8 12	12 16										
L	First	50	50	↓	↓	↓	↓	‡	↑	↓	0 2	0 3	0 4	0 4	1 5	2 7	3 8	5 10	↑	↑	↑	↑	↑	↑	↑	↑	↑	↑	↑
	Second	50	100								0 2	0 4	1 5	3 6	4 7	6 9	8 12	12 16											
M	First	80	80	↓	↓	↓	‡	↑	↓	0 2	0 3	0 4	0 4	1 5	2 7	3 8	5 10	↑	↑	↑	↑	↑	↑	↑	↑	↑	↑	↑	↑
	Second	80	160							0 2	0 4	1 5	3 6	4 7	6 9	8 12	12 16												
N	First	125	125	↓	↓	‡	↑	↓	0 2	0 3	0 4	0 4	1 5	2 7	3 8	5 10	↑	↑	↑	↑	↑	↑	↑	↑	↑	↑	↑	↑	↑
	Second	125	250						0 2	0 4	1 5	3 6	4 7	6 9	8 12	12 16													
P	First	200	200	↓	‡	↑	↓	0 2	0 3	0 4	0 4	1 5	2 7	3 8	5 10	↑	↑	↑	↑	↑	↑	↑	↑	↑	↑	↑	↑	↑	↑
	Second	200	400					0 2	0 4	1 5	3 6	4 7	6 9	8 12	12 16														
Q	First	315	315	‡	↑	↓	0 2	0 3	0 4	0 4	1 5	2 7	3 8	5 10	↑	↑	↑	↑	↑	↑	↑	↑	↑	↑	↑	↑	↑	↑	↑
	Second	315	630				0 2	0 4	1 5	3 6	4 7	6 9	8 12	12 16															
R	First	500	500	↑	↑	0 2	0 3	0 4	0 4	1 5	2 7	3 8	5 10	↑	↑	↑	↑	↑	↑	↑	↑	↑	↑	↑	↑	↑	↑	↑	↑
	Second	1,000	1,500			0 2	0 4	1 5	3 6	4 7	6 9	8 12	12 16																

↓ = use first sampling plan below arrow. If sample size equals or exceeds lot or batch size, do 100% inspection.
↑ = use first sampling plan above arrow.
Ac = acceptance number.
Re = rejection number.
†If, after the second sample, the acceptance number has been exceeded but the rejection number has not been reached, accept the lot but reinstate normal inspection.
‡Use corresponding single sampling plan (or, alternatively, use double sampling plan below, where available).

TABLE *R* MASTER TABLE FOR NORMAL INSPECTION (MULTIPLE SAMPLING)—MIL-STD-105E (ABC STANDARD)

Sample size code letter	Sample	Sample size	Cumulative sample size	Acceptable quality levels (normal inspection)																									
				0.010	0.015	0.025	0.040	0.065	0.10	0.15	0.25	0.40	0.65	1.0	1.5	2.5	4.0	6.5	10	15	25	40	65	100	150	250	400	650	1,000
				Ac Re	*Ac Re*	*Ac Re*	*Ac Re*	*Ac Re*	*Ac Re*	*Ac Re*	*Ac Re*	*Ac Re*	*Ac Re*	*Ac Re*	*Ac Re*	*Ac Re*	*Ac Re*	*Ac Re*	*Ac Re*	*Ac Re*	*Ac Re*	*Ac Re*	*Ac Re*	*Ac Re*	*Ac Re*	*Ac Re*	*Ac Re*	*Ac Re*	*Ac Re*
A				↓	↓	↓	↓	↓	↓	↓	↓	↓	↓	↓	↓	↓	↓	†	↓	↓	†	†	†	†	†	†	†	†	†
B				↓	↓	↓	↓	↓	↓	↓	↓	↓	↓	↓	↓	↓	†	↑	↓	‡	‡	‡	‡	‡	‡	‡	‡	‡	‡
C				↓	↓	↓	↓	↓	↓	↓	↓	↓	↓	↓	↓	†	↑	↓	‡	‡	‡	‡	‡	‡	‡	‡	‡	‡	↑
D	First	2	2	↓	↓	↓	↓	↓	↓	↓	↓	↓	↓	↓	†	↑	↓	§ 2	§ 2	§ 3	§ 4	0 4	0 5	1 7	2 9	4 12	6 16	↑	↑
	Second	2	4															§ 2	0 3	0 3	1 5	1 6	3 8	4 10	7 14	11 19	17 27		
	Third	2	6															0 2	0 3	1 4	2 6	3 8	6 10	8 13	13 19	19 27	29 39		
	Fourth	2	8															0 3	1 4	2 5	3 7	5 10	8 13	12 17	19 25	27 34	40 49		
	Fifth	2	10															1 3	2 4	3 6	5 8	7 11	11 15	17 20	25 29	36 40	53 58		
	Sixth	2	12															1 3	3 5	4 6	7 9	10 12	14 17	21 23	31 33	45 47	65 68		
	Seventh	2	14															2 3	4 5	6 7	9 10	13 14	18 19	25 26	37 38	53 54	77 78		
E	First	3	3	↓	↓	↓	↓	↓	↓	↓	↓	↓	↓	†	↑	↓	§ 2	§ 2	§ 3	§ 4	0 4	0 5	1 7	2 9	4 12	6 16	↑	↑	↑
	Second	3	6														§ 2	0 3	0 3	1 5	1 6	3 8	4 10	7 14	11 19	17 27			
	Third	3	9														0 2	0 3	1 4	2 6	3 8	6 10	8 13	13 19	19 27	29 39			
	Fourth	3	12														0 3	1 4	2 5	3 7	5 10	8 13	12 17	19 25	27 34	40 49			
	Fifth	3	15														1 3	2 4	3 6	5 8	7 11	11 15	17 20	25 29	36 40	53 58			
	Sixth	3	18														1 3	3 5	4 6	7 9	10 12	14 17	21 23	31 33	45 47	65 68			
	Seventh	3	21														2 3	4 5	6 7	9 10	13 14	18 19	25 26	37 38	53 54	77 78			
F	First	5	5	↓	↓	↓	↓	↓	↓	↓	↓	↓	†	↑	↓	§ 2	§ 2	§ 3	§ 4	0 4	0 5	1 7	2 9	↑	↑	↑	↑	↑	↑
	Second	5	10													§ 2	0 3	0 3	1 5	1 6	3 8	4 10	7 14						
	Third	5	15													0 2	0 3	1 4	2 6	3 8	6 10	8 13	13 19						
	Fourth	5	20													0 3	1 4	2 5	3 7	5 10	8 13	12 17	19 25						
	Fifth	5	25													1 3	2 4	3 6	5 8	7 11	11 15	17 20	25 29						
	Sixth	5	30													1 3	3 5	4 6	7 9	10 12	14 17	21 23	31 33						
	Seventh	5	35													2 3	4 5	6 7	9 10	13 14	18 19	25 26	37 38						
G	First	8	8	↓	↓	↓	↓	↓	↓	↓	↓	†	↑	↓	§ 2	§ 2	§ 3	§ 4	0 4	0 5	1 7	2 9	↑	↑	↑	↑	↑	↑	↑
	Second	8	16												§ 2	0 3	0 3	1 5	1 6	3 8	4 10	7 14							
	Third	8	24												0 2	0 3	1 4	2 6	3 8	6 10	8 13	13 19							
	Fourth	8	32												0 3	1 4	2 5	3 7	5 10	8 13	12 17	19 25							
	Fifth	8	40												1 3	2 4	3 6	5 8	7 11	11 15	17 20	25 29							
	Sixth	8	48												1 3	3 5	4 6	7 9	10 12	14 17	21 23	31 33							
	Seventh	8	52												2 3	4 5	6 7	9 10	13 14	18 19	25 26	37 38							
H	First	13	13	↓	↓	↓	↓	↓	↓	↓	†	↑	↓	§ 2	§ 2	§ 3	§ 4	0 4	0 5	1 7	2 9	↑	↑	↑	↑	↑	↑	↑	↑
	Second	13	26											§ 2	0 3	0 3	1 5	1 6	3 8	4 10	7 14								
	Third	13	39											0 2	0 3	1 4	2 6	3 8	6 10	8 13	13 19								
	Fourth	13	52											0 3	1 4	2 5	3 7	5 10	8 13	12 17	19 25								
	Fifth	13	65											1 3	2 4	3 6	5 8	7 11	11 15	17 20	25 29								
	Sixth	13	78											1 3	3 5	4 6	7 9	10 12	14 17	21 23	31 33								
	Seventh	13	91											2 3	4 5	6 7	9 10	13 14	18 19	25 26	37 38								
J	First	20	20	↓	↓	↓	↓	↓	↓	†	↑	↓	§ 2	§ 2	§ 3	§ 4	0 4	0 5	1 7	2 9	↑	↑	↑	↑	↑	↑	↑	↑	↑
	Second	20	40										§ 2	0 3	0 3	1 5	1 6	3 8	4 10	7 14									
	Third	20	60										0 2	0 3	1 4	2 6	3 8	6 10	8 13	13 19									
	Fourth	20	80										0 3	1 4	2 5	3 7	5 10	8 13	12 17	19 25									
	Fifth	20	100										1 3	2 4	3 6	5 8	7 11	11 15	17 20	25 29									
	Sixth	20	120										1 3	3 5	4 6	7 9	10 12	14 17	21 23	31 33									
	Seventh	20	140										2 3	4 5	6 7	9 10	13 14	18 19	25 26	37 38									

↓ = use first sampling plan below arrow (refer to continuation of table on following page, when necessary). If sample size equals or exceeds lot or batch size, do 100% inspection.

↑ = use first sampling plan above arrow.

Ac = acceptance number.

Re = rejection number.

† Use corresponding single sampling plan (or, alternatively, use multiple sampling plan below, where available).

‡ Use corresponding double sampling plan (or, alternatively, use multiple sampling plan below, where available).

§ Acceptance not permitted at this sample size.

TABLE *R* (*continued*)

Sample size code letter	Sample	Sample size	Cumulative sample size	0.010 Ac Re	0.015 Ac Re	0.025 Ac Re	0.040 Ac Re	0.065 Ac Re	0.10 Ac Re	0.15 Ac Re	0.25 Ac Re	0.40 Ac Re	0.65 Ac Re	1.0 Ac Re	1.5 Ac Re	2.5 Ac Re	4.0 Ac Re	6.5 Ac Re	10 Ac Re	15 Ac Re	25 Ac Re	40 Ac Re	65 Ac Re	100 Ac Re	150 Ac Re	250 Ac Re	400 Ac Re	650 Ac Re	1,000 Ac Re
K	First	32	32	↓	↓	↓	↓	↓	†	↑	↓	‡ 2	‡ 2	‡ 3	‡ 4	0 4	0 5	1 7	2 9	↑	↑	↑	↑	↑	↑	↑	↑	↑	↑
	Second	32	64									‡ 2	0 3	0 3	1 5	1 6	3 8	4 10	7 14										
	Third	32	96									0 2	0 3	1 4	2 6	3 8	6 10	8 13	13 19										
	Fourth	32	128									0 3	1 4	2 5	3 7	5 10	8 13	12 17	19 25										
	Fifth	32	160									1 3	2 4	3 6	5 8	7 11	11 15	17 20	25 29										
	Sixth	32	192									1 3	3 5	4 6	7 9	10 12	14 17	21 23	31 33										
	Seventh	32	224									2 3	4 5	6 7	9 10	13 14	18 19	25 26	37 38										
L	First	50	50	↓	↓	↓	↓	†	↑	↓	‡ 2	‡ 2	‡ 3	‡ 4	0 4	0 5	1 7	2 9	↑										
	Second	50	100								‡ 2	0 3	0 3	1 5	1 6	3 8	4 10	7 14											
	Third	50	150								0 2	0 3	1 4	2 6	3 8	6 10	8 13	13 19											
	Fourth	50	200								0 3	1 4	2 5	3 7	5 10	8 13	12 17	19 25											
	Fifth	50	250								1 3	2 4	3 6	5 8	7 11	11 15	17 20	25 29											
	Sixth	50	300								1 3	3 5	4 6	7 9	10 12	14 17	21 23	31 33											
	Seventh	50	350								2 3	4 5	6 7	9 10	13 14	18 19	25 26	37 38											
M	First	80	80	↓	↓	↓	†	↑	↓	‡ 2	‡ 2	‡ 3	‡ 4	0 4	0 5	1 7	2 9	↑											
	Second	80	160							‡ 2	0 3	0 3	1 5	1 6	3 8	4 10	7 14												
	Third	80	240							0 2	0 3	1 4	2 6	3 8	6 10	8 13	13 19												
	Fourth	80	320							0 3	1 4	2 5	3 7	5 10	8 13	12 17	19 25												
	Fifth	80	400							1 3	2 4	3 6	5 8	7 11	11 15	17 20	25 29												
	Sixth	80	480							1 3	3 5	4 6	7 9	10 12	14 17	21 23	31 33												
	Seventh	80	560							2 3	4 5	6 7	9 10	13 14	18 19	25 26	37 38												
N	First	125	125	↓	↓	†	↑	↓	‡ 2	‡ 2	‡ 3	‡ 4	0 4	0 5	1 7	2 9	↑												
	Second	125	250						‡ 2	0 3	0 3	1 5	1 6	3 8	4 10	7 14													
	Third	125	375						0 2	0 3	1 4	2 6	3 8	6 10	8 13	13 19													
	Fourth	125	500						0 3	1 4	2 5	3 7	5 10	8 13	12 17	19 25													
	Fifth	125	625						1 3	2 4	3 6	5 8	7 11	11 15	17 20	25 29													
	Sixth	125	750						1 3	3 5	4 6	7 9	10 12	14 17	21 23	31 33													
	Seventh	125	875						2 3	4 5	6 7	9 10	13 14	18 19	25 26	37 38													
P	First	200	200	↓	†	↑	↓	‡ 2	‡ 2	‡ 3	‡ 4	0 4	0 5	1 7	2 9	↑													
	Second	200	400					‡ 2	0 3	0 3	1 5	1 6	3 8	4 10	7 14														
	Third	200	600					0 2	0 3	1 4	2 6	3 8	6 10	8 13	13 19														
	Fourth	200	800					0 3	1 4	2 5	3 7	5 10	8 13	12 17	19 25														
	Fifth	200	1,000					1 3	2 4	3 6	5 8	7 11	11 15	17 20	25 29														
	Sixth	200	1,200					1 3	3 5	4 6	7 9	10 12	14 17	21 23	31 33														
	Seventh	200	1,400					2 3	4 5	6 7	9 10	13 14	18 19	25 26	37 38														
Q	First	315	315	†	↑	↓	‡ 2	‡ 2	‡ 3	‡ 4	0 4	0 5	1 7	2 9	↑														
	Second	315	630				‡ 2	0 3	0 3	1 5	1 6	3 8	4 10	7 14															
	Third	315	945				0 2	0 3	1 4	2 6	3 8	6 10	8 13	13 19															
	Fourth	315	1,260				0 3	1 4	2 5	3 7	5 10	8 13	12 17	19 25															
	Fifth	315	1,575				1 3	2 4	3 6	5 8	7 11	11 15	17 20	25 29															
	Sixth	315	1,890				1 3	3 5	4 6	7 9	10 12	14 17	21 23	31 33															
	Seventh	315	2,205				2 3	4 5	6 7	9 10	13 14	18 19	25 26	37 38															
R	First	500	500	↑		‡ 2	‡ 2	‡ 3	‡ 4	0 4	0 5	1 7	2 9	↑															
	Second	500	1,000			‡ 2	0 3	0 3	1 5	1 6	3 8	4 10	7 14																
	Third	500	1,500			0 2	0 3	1 4	2 6	3 8	6 10	8 13	13 19																
	Fourth	500	2,000			0 3	1 4	2 5	3 7	5 10	8 13	12 17	19 25																
	Fifth	500	2,500			1 3	2 4	3 6	5 8	7 11	11 15	17 20	25 29																
	Sixth	500	3,000			1 3	3 5	4 6	7 9	10 12	14 17	21 23	31 33																
	Seventh	500	3,500			2 3	4 5	6 7	9 10	13 14	18 19	25 26	37 38																

↓ = use first sampling plan below arrow. If sample size equals or exceeds lot or batch size, do 100% inspection.

↑ = use first sampling plan above arrow (refer to preceding page, when necessary).

Ac = acceptance number.

Re = rejection number.

†Use corresponding single sampling plan (or, alternatively, use multiple plan below, where applicable).

‡ Acceptance not permitted at this sample size.

TABLE S MASTER TABLE FOR TIGHTENED INSPECTION (MULTIPLE SAMPLING)—MIL-STD-105E (ABC STANDARD)

Sample size code letter	Sample	Sample size	Cumulative sample size	0.010	0.015	0.025	0.040	0.065	0.10	0.15	0.25	0.40	0.65	1.0	1.5	2.5	4.0	6.5	10	15	25	40	65	100	150	250	400	650	1,000
				Acceptable quality levels (tightened inspection)																									
				Ac Re	Ac Re	Ac Re	Ac Re	Ac Re	Ac Re	Ac Re	Ac Re	Ac Re	Ac Re	Ac Re	Ac Re	Ac Re	Ac Re	Ac Re	Ac Re	Ac Re	Ac Re	Ac Re	Ac Re	Ac Re	Ac Re	Ac Re	Ac Re	Ac Re	Ac Re
A				↓	↓	↓	↓	↓	↓	↓	↓	↓	↓	↓	↓	↓	↓	↓	↓	↓	↓	†	†	†	†	†	†	†	†
B				↓	↓	↓	↓	↓	↓	↓	↓	↓	↓	↓	↓	↓	↓	†	↓	↓	‡	‡	‡	‡	‡	‡	‡	‡	‡
C				↓	↓	↓	↓	↓	↓	↓	↓	↓	↓	↓	↓	↓	†	↓	↓	‡	‡	‡	‡	‡	‡	‡	‡	‡	↑
D	First	2	2	↓	↓	↓	↓	↓	↓	↓	↓	↓	↓	↓	↓	†	↓	↓	§ 2	§ 2	§ 3	§ 4	0 4	0 6	1 8	3 10	6 15	↑	↑
	Second	2	4	↓	↓	↓	↓	↓	↓	↓	↓	↓	↓	↓	↓		↓	↓	§ 2	0 3	0 3	1 5	2 7	3 9	6 12	10 17	16 25	↑	↑
	Third	2	6	↓	↓	↓	↓	↓	↓	↓	↓	↓	↓	↓	↓		↓	↓	0 2	0 3	1 4	2 6	4 9	7 12	11 17	17 24	26 36	↑	↑
	Fourth	2	8	↓	↓	↓	↓	↓	↓	↓	↓	↓	↓	↓	↓		↓	↓	0 3	1 4	2 5	3 7	6 11	10 15	16 22	24 31	37 46	↑	↑
	Fifth	2	10	↓	↓	↓	↓	↓	↓	↓	↓	↓	↓	↓	↓		↓	↓	1 3	2 4	3 6	5 8	9 12	14 17	22 25	32 37	49 55	↑	↑
	Sixth	2	12	↓	↓	↓	↓	↓	↓	↓	↓	↓	↓	↓	↓		↓	↓	1 3	3 5	4 6	7 9	12 14	18 20	27 29	40 43	61 64	↑	↑
	Seventh	2	14	↓	↓	↓	↓	↓	↓	↓	↓	↓	↓	↓	↓		↓	↓	2 3	4 5	6 7	9 10	14 15	21 22	32 33	48 49	72 73	↑	↑
E	First	3	3	↓	↓	↓	↓	↓	↓	↓	↓	↓	↓	↓	†	↓	↓	§ 2	§ 2	§ 3	§ 4	0 4	0 6	1 8	3 10	6 15	↑	↑	↑
	Second	3	6	↓	↓	↓	↓	↓	↓	↓	↓	↓	↓	↓		↓	↓	§ 2	0 3	0 3	1 5	2 7	3 9	6 12	10 17	16 25	↑	↑	↑
	Third	3	9	↓	↓	↓	↓	↓	↓	↓	↓	↓	↓	↓		↓	↓	0 2	0 3	1 4	2 6	4 9	7 12	11 17	17 24	26 36	↑	↑	↑
	Fourth	3	12	↓	↓	↓	↓	↓	↓	↓	↓	↓	↓	↓		↓	↓	0 3	1 4	2 5	3 7	6 11	10 15	16 22	24 31	37 46	↑	↑	↑
	Fifth	3	15	↓	↓	↓	↓	↓	↓	↓	↓	↓	↓	↓		↓	↓	1 3	2 4	3 6	5 8	9 12	14 17	22 25	32 37	49 55	↑	↑	↑
	Sixth	3	18	↓	↓	↓	↓	↓	↓	↓	↓	↓	↓	↓		↓	↓	1 3	3 5	4 6	7 9	12 14	18 20	27 29	40 43	61 64	↑	↑	↑
	Seventh	3	21	↓	↓	↓	↓	↓	↓	↓	↓	↓	↓	↓		↓	↓	2 3	4 5	6 7	9 10	14 15	21 22	32 33	48 49	72 73	↑	↑	↑
F	First	5	5	↓	↓	↓	↓	↓	↓	↓	↓	↓	↓	†	↓	↓	§ 2	§ 2	§ 3	§ 4	0 4	0 6	1 8	↑	↑	↑	↑	↑	↑
	Second	5	10	↓	↓	↓	↓	↓	↓	↓	↓	↓	↓		↓	↓	§ 2	0 3	0 4	1 5	2 7	3 9	6 12	↑	↑	↑	↑	↑	↑
	Third	5	15	↓	↓	↓	↓	↓	↓	↓	↓	↓	↓		↓	↓	0 2	0 3	1 4	2 6	4 9	7 12	11 17	↑	↑	↑	↑	↑	↑
	Fourth	5	20	↓	↓	↓	↓	↓	↓	↓	↓	↓	↓		↓	↓	0 3	1 4	2 5	3 7	6 11	10 15	16 22	↑	↑	↑	↑	↑	↑
	Fifth	5	25	↓	↓	↓	↓	↓	↓	↓	↓	↓	↓		↓	↓	1 3	2 4	3 6	5 8	9 12	14 17	22 25	↑	↑	↑	↑	↑	↑
	Sixth	5	30	↓	↓	↓	↓	↓	↓	↓	↓	↓	↓		↓	↓	1 3	3 5	4 6	7 9	12 14	18 20	27 29	↑	↑	↑	↑	↑	↑
	Seventh	5	35	↓	↓	↓	↓	↓	↓	↓	↓	↓	↓		↓	↓	2 3	4 5	6 7	9 10	14 15	21. 22	32 33	↑	↑	↑	↑	↑	↑
G	First	8	8	↓	↓	↓	↓	↓	↓	↓	↓	↓	†	↓	↓	§ 2	§ 2	§ 3	§ 4	0 4	0 6	1 8	↑	↑	↑	↑	↑	↑	↑
	Second	8	16	↓	↓	↓	↓	↓	↓	↓	↓	↓		↓	↓	§ 2	0 3	0 3	1 5	2 7	3 9	6 12	↑	↑	↑	↑	↑	↑	↑
	Third	8	24	↓	↓	↓	↓	↓	↓	↓	↓	↓		↓	↓	0 2	0 3	1 4	2 6	4 9	7 12	11 17	↑	↑	↑	↑	↑	↑	↑
	Fourth	8	32	↓	↓	↓	↓	↓	↓	↓	↓	↓		↓	↓	0 3	1 4	2 5	3 7	6 11	10 15	16 22	↑	↑	↑	↑	↑	↑	↑
	Fifth	8	40	↓	↓	↓	↓	↓	↓	↓	↓	↓		↓	↓	1 3	2 4	3 6	5 8	9 12	14 17	22 25	↑	↑	↑	↑	↑	↑	↑
	Sixth	8	48	↓	↓	↓	↓	↓	↓	↓	↓	↓		↓	↓	1 3	3 5	4 6	7 9	12 14	18 20	27 29	↑	↑	↑	↑	↑	↑	↑
	Seventh	8	56	↓	↓	↓	↓	↓	↓	↓	↓	↓		↓	↓	2 3	4 5	6 7	9 10	14 15	21 22	32 33	↑	↑	↑	↑	↑	↑	↑
H	First	13	13	↓	↓	↓	↓	↓	↓	↓	↓	†	↓	↓	§ 2	§ 2	§ 3	§ 4	0 4	0 6	1 8	↑	↑	↑	↑	↑	↑	↑	↑
	Second	13	26	↓	↓	↓	↓	↓	↓	↓	↓		↓	↓	§ 2	0 3	0 3	1 5	2 7	3 9	6 12	↑	↑	↑	↑	↑	↑	↑	↑
	Third	13	39	↓	↓	↓	↓	↓	↓	↓	↓		↓	↓	0 2	0 3	1 4	2 6	4 9	7 12	11 17	↑	↑	↑	↑	↑	↑	↑	↑
	Fourth	13	52	↓	↓	↓	↓	↓	↓	↓	↓		↓	↓	0 3	1 4	2 5	3 7	6 11	10 15	16 22	↑	↑	↑	↑	↑	↑	↑	↑
	Fifth	13	65	↓	↓	↓	↓	↓	↓	↓	↓		↓	↓	1 3	2 4	3 6	5 8	9 12	14 17	22 25	↑	↑	↑	↑	↑	↑	↑	↑
	Sixth	13	78	↓	↓	↓	↓	↓	↓	↓	↓		↓	↓	1 3	3 5	4 6	7 9	12 14	18 20	27 29	↑	↑	↑	↑	↑	↑	↑	↑
	Seventh	13	91	↓	↓	↓	↓	↓	↓	↓	↓		↓	↓	2 3	4 5	6 7	9 10	14 15	21 22	32 33	↑	↑	↑	↑	↑	↑	↑	↑
J	First	20	20	↓	↓	↓	↓	↓	↓	↓	†	↓	↓	§ 2	§ 2	§ 3	§ 4	0 4	0 6	1 8	↑	↑	↑	↑	↑	↑	↑	↑	↑
	Second	20	40	↓	↓	↓	↓	↓	↓	↓		↓	↓	§ 2	0 3	0 3	1 5	2 7	3 9	6 12	↑	↑	↑	↑	↑	↑	↑	↑	↑
	Third	20	60	↓	↓	↓	↓	↓	↓	↓		↓	↓	0 2	0 3	1 4	2 6	4 9	7 12	11 17	↑	↑	↑	↑	↑	↑	↑	↑	↑
	Fourth	20	80	↓	↓	↓	↓	↓	↓	↓		↓	↓	0 3	1 4	2 5	3 7	6 11	10 15	16 22	↑	↑	↑	↑	↑	↑	↑	↑	↑
	Fifth	20	100	↓	↓	↓	↓	↓	↓	↓		↓	↓	1 3	2 4	3 6	5 8	9 12	14 17	22 25	↑	↑	↑	↑	↑	↑	↑	↑	↑
	Sixth	20	120	↓	↓	↓	↓	↓	↓	↓		↓	↓	1 3	3 5	4 6	7 9	12 14	18 20	27 29	↑	↑	↑	↑	↑	↑	↑	↑	↑
	Seventh	20	140	↓	↓	↓	↓	↓	↓	↓		↓	↓	2 3	4 5	6 7	9 10	14 15	21 22	32 33	↑	↑	↑	↑	↑	↑	↑	↑	↑

↓ = use first sampling plan below arrow (refer to continuation of table on following page, when necessary). If sample size equals or exceeds lot or batch size, do 100% inspection.

↑ = use first sampling plan above arrow.

Ac = acceptance number.

Re = rejection number.

†Use corresponding single sampling plan (or, alternatively, use multiple sampling plan below, where available).

‡ Use corresponding double sampling plan (or, alternatively, use multiple sampling plan below, where available).

§ Acceptance not permitted at this sample size.

TABLE S (continued)

Sample size code letter	Sample	Sample size	Cumulative sample size	Acceptable quality levels (tightened inspection)																																																			
				0.010		0.015		0.025		0.040		0.065		0.10		0.15		0.25		0.40		0.65		1.0		1.5		2.5		4.0		6.5		10		15		25		40		65		100		150		250		400		650		1,000	
				Ac	Re	Ac	Re	Ac	Re	Ac	Re	Ac	Re	AC	Re	Ac	Re	Ac	Re	Ac	Re	Ac	Re	Ac	Re	Ac	Re	Ac	Re	Ac	Re	Ac	Re	Ac	Re	Ac	Re	Ac	Re	Ac	Re	Ac	Re	Ac	Re	Ac	Re	Ac	Re	Ac	Re	Ac	Re	Ac	Re
K	First	32	32	↓		↓		↓		↓		↓		↓		†		↓		↓		‡	2	‡	2	‡	3	‡	4	0	4	0	6	1	8	↑		↑		↑		↑		↑		↑		↑		↑		↑		↑	
	Second	32	64																			‡	2	0	3	0	3	1	5	2	7	3	9	6	12																				
	Third	32	96																			0	2	0	3	1	4	2	6	4	9	7	12	11	17																				
	Fourth	32	128																			0	3	1	4	2	5	3	7	6	11	10	15	16	22																				
	Fifth	32	160																			1	3	2	4	3	6	5	8	9	12	14	17	22	25																				
	Sixth	32	192																			1	3	3	5	4	6	7	9	12	14	18	20	27	29																				
	Seventh	32	224																			2	3	4	5	6	7	9	10	14	15	21	22	32	33																				
L	First	50	50	↓		↓		↓		↓		↓		†		↓		↓		‡	2	‡	2	‡	3	‡	4	0	4	0	6	1	8	↑		↑		↑		↑		↑		↑		↑		↑		↑		↑		↑	
	Second	50	100																	‡	2	0	3	0	3	1	5	2	7	3	9	6	12																						
	Third	50	150																	0	2	0	3	1	4	2	6	4	9	7	12	11	17																						
	Fourth	50	200																	0	3	1	4	2	5	3	7	6	11	10	15	16	22																						
	Fifth	50	250																	1	3	2	4	3	6	5	8	9	12	14	17	22	25																						
	Sixth	50	300																	1	3	3	5	4	6	7	9	12	14	18	20	27	29																						
	Seventh	50	350																	2	3	4	5	6	7	9	10	14	15	21	22	32	33																						
M	First	80	80	↓		↓		↓		↓		†		↓		↓		‡	2	‡	2	‡	3	‡	4	0	4	0	6	1	8	↑		↑		↑		↑		↑		↑		↑		↑		↑		↑		↑		↑	
	Second	80	160															‡	2	0	3	0	3	1	5	2	7	3	9	6	12																								
	Third	80	240															0	2	0	3	1	4	2	6	4	9	7	12	11	17																								
	Fourth	80	320															0	3	1	4	2	5	3	7	6	11	10	15	16	22																								
	Fifth	80	400															1	3	2	4	3	6	5	8	9	12	14	17	22	25																								
	Sixth	80	480															1	3	3	5	4	6	7	9	12	14	18	20	27	29																								
	Seventh	80	560															2	3	4	5	6	7	9	10	14	15	21	22	32	33																								
N	First	125	125	↓		↓		↓		†		↓		↓		‡	2	‡	2	‡	3	‡	4	0	4	0	6	1	8	↑		↑		↑		↑		↑		↑		↑		↑		↑		↑		↑		↑		↑	
	Second	125	250													‡	2	0	3	0	3	1	5	2	7	3	9	6	12																										
	Third	125	375													0	2	0	3	1	4	2	6	4	9	7	12	11	17																										
	Fourth	125	500													0	3	1	4	2	5	3	7	6	11	10	15	16	22																										
	Fifth	125	625													1	3	2	4	3	6	5	8	9	12	14	17	22	25																										
	Sixth	125	750													1	3	3	5	4	6	7	9	12	14	18	20	27	29																										
	Seventh	125	875													2	3	4	5	6	7	9	10	14	15	21	22	32	33																										
P	First	200	200	↓		↓		†		↓		↓		‡	2	‡	2	‡	3	‡	4	0	4	0	6	1	8	↑		↑		↑		↑		↑		↑		↑		↑		↑		↑		↑		↑		↑		↑	
	Second	200	400											‡	2	0	3	0	3	1	5	2	7	3	9	6	12																												
	Third	200	600											0	2	0	3	1	4	2	6	4	9	7	12	11	17																												
	Fourth	200	800											0	3	1	4	2	5	3	7	6	11	10	15	16	22																												
	Fifth	200	1,000											1	3	2	4	3	6	5	8	9	12	14	17	22	25																												
	Sixth	200	1,200											1	3	3	5	4	6	7	9	12	14	18	20	27	29																												
	Seventh	200	1,400											2	3	4	5	6	7	9	10	14	15	21	22	32	33																												
Q	First	315	315	↓		†		↓		↓		‡	2	‡	2	‡	3	‡	4	0	4	0	6	1	8	↑		↑		↑		↑		↑		↑		↑		↑		↑		↑		↑		↑		↑		↑		↑	
	Second	315	630									‡	2	0	3	0	3	1	5	2	7	3	9	6	12																														
	Third	315	945									0	2	0	3	1	4	2	6	4	9	7	12	11	17																														
	Fourth	315	1,260									0	3	1	4	2	5	3	7	6	11	10	15	16	22																														
	Fifth	315	1,575									1	3	2	4	3	6	5	8	9	12	14	17	22	25																														
	Sixth	315	1,890									1	3	3	5	4	6	7	9	12	14	18	20	27	29																														
	Seventh	315	2,205									2	3	4	5	6	7	9	10	14	15	21	22	32	33																														
R	First	500	500	†		↑		↓		‡	2	‡	2	‡	3	‡	4	0	4	0	6	1	8	↑		↑		↑		↑		↑		↑		↑		↑		↑		↑		↑		↑		↑		↑		↑		↑	
	Second	500	1,000							‡	2	0	3	0	3	1	5	2	7	3	9	6	12																																
	Third	500	1,500							0	2	0	3	1	4	2	6	4	9	7	12	11	17																																
	Fourth	500	2,000							0	3	1	4	2	5	3	7	6	11	10	15	16	22																																
	Fifth	500	2,500							1	3	2	4	3	6	5	8	9	12	14	17	22	25																																
	Sixth	500	3,000							1	3	3	5	4	6	7	9	12	14	18	20	27	29																																
	Seventh	500	3,500							2	3	4	5	6	7	9	10	14	15	21	22	32	33																																
S	First	800	800					‡	2																																														
	Second	800	1,600					‡	2																																														
	Third	800	2,400					0	2																																														
	Fourth	800	3,200					0	3																																														
	Fifth	800	4,000					1	3																																														
	Sixth	800	4,800					1	3																																														
	Seventh	800	5,600					2	3																																														

↓ = use first sampling plan below arrow. If sample size equals or exceeds lot or batch size, do 100% inspection.
↑ = use first sampling plan above arrow (refer to preceding page, when necessary).
Ac = acceptance number.
Re = rejection number.
†Use corresponding single sampling plan (or, alternatively, use multiple sampling plan below, where available).
‡ Acceptance not permitted at this sample size.

TABLE *T* MASTER TABLE FOR REDUCED INSPECTION (MULTIPLE SAMPLING)—MIL-STD-105E (ABC STANDARD)

Sample size code letter	Sample	Sample size	Cumulative sample size	Acceptable quality levels (reduced inspection)†																									
				0.010	0.015	0.025	0.040	0.065	0.10	0.15	0.25	0.40	0.65	1.0	1.5	2.5	4.0	6.5	10	15	25	40	65	100	150	250	400	650	1,000
				Ac Re	*Ac Re*	*Ac Re*	*Ac Re*	*Ac Re*	*Ac Re*	*Ac Re*	*Ac Re*	*Ac Re*	*Ac Re*	*Ac Re*	*Ac Re*	*Ac Re*	*Ac Re*	*Ac Re*	*Ac Re*	*Ac Re*	*Ac Re*	*Ac Re*	*Ac Re*	*Ac Re*	*Ac Re*	*Ac Re*	*Ac Re*	*Ac Re*	*Ac Re*
A				↓	↓	↓	↓	↓	↓	↓	↓	↓	↓	↓	↓	↓	↓	‡	↓	↓	‡	‡	‡	‡	‡	‡	‡	‡	‡
B				↓	↓	↓	↓	↓	↓	↓	↓	↓	↓	↓	↓	↓	‡	↑	↓	‡	‡	‡	‡	‡	‡	‡	‡	‡	‡
C				↓	↓	↓	↓	↓	↓	↓	↓	↓	↓	↓	↓	‡	↑	↓	‡	‡	‡	‡	‡	‡	‡	‡	‡	‡	↑
D				↓	↓	↓	↓	↓	↓	↓	↓	↓	↓	↓	‡	↑	↓	§	§	§	§	§	§	§	§	§	§	↑	↑
E				↓	↓	↓	↓	↓	↓	↓	↓	↓	↓	‡	↑	↓	§	§	§	§	§	§	§	§	§	§	↑	↑	↑
F	First	2	2	↓	↓	↓	↓	↓	↓	↓	↓	↓	‡	↑	↓	¶ 2	¶ 2	¶ 3	¶ 3	¶ 4	¶ 4	0 5	0 6	↑	↑	↑	↑	↑	↑
	Second	2	4													¶ 2	¶ 3	¶ 3	0 4	0 5	1 6	1 7	3 9						
	Third	2	6													0 2	0 3	0 4	0 5	1 6	2 8	3 9	6 12						
	Fourth	2	8													0 3	0 4	0 5	1 6	2 7	3 10	5 12	8 15						
	Fifth	2	10													0 3	0 4	1 6	2 7	3 8	5 11	7 13	11 17						
	Sixth	2	12													0 3	1 5	1 6	3 7	4 9	7 12	10 15	14 20						
	Seventh	2	14													1 3	1 5	2 7	4 8	6 10	9 14	13 17	18 22						
G	First	3	3	↓	↓	↓	↓	↓	↓	↓	↓	‡	↑	↓	¶ 2	¶ 2	¶ 3	¶ 3	¶ 4	¶ 4	0 5	0 6	↑	↑	↑	↑	↑	↑	↑
	Second	3	6												¶ 2	¶ 3	¶ 3	0 4	0 5	1 6	1 7	3 9							
	Third	3	9												0 2	0 3	0 4	0 5	1 6	2 8	3 9	6 12							
	Fourth	3	12												0 3	0 4	0 5	1 6	2 7	3 10	5 12	8 15							
	Fifth	3	15												0 3	0 4	1 6	2 7	3 8	5 11	7 13	11 17							
	Sixth	3	18												0 3	1 5	1 6	3 7	4 9	7 12	10 15	14 20							
	Seventh	3	21												1 3	1 5	2 7	4 8	6 10	9 14	13 17	18 22							
H	First	5	5	↓	↓	↓	↓	↓	↓	↓	‡	↑	↓	¶ 2	¶ 2	¶ 3	¶ 3	¶ 4	¶ 4	0 5	0 6	↑	↑	↑	↑	↑	↑	↑	↑
	Second	5	10											¶ 2	¶ 3	¶ 3	0 4	0 5	1 6	1 7	3 9								
	Third	5	15											0 2	0 3	0 4	0 5	1 6	2 8	3 9	6 12								
	Fourth	5	20											0 3	0 4	0 5	1 6	2 7	3 10	5 12	8 15								
	Fifth	5	25											0 3	0 4	1 6	2 7	3 8	5 11	7 13	11 17								
	Sixth	5	30											0 3	1 5	1 6	3 7	4 9	7 12	10 15	14 20								
	Seventh	5	35											1 3	1 5	2 7	4 8	6 10	9 14	13 17	18 22								
J	First	8	8	↓	↓	↓	↓	↓	↓	‡	↑	↓	¶ 2	¶ 2	¶ 3	¶ 3	¶ 4	¶ 4	0 5	0 6	↑	↑	↑	↑	↑	↑	↑	↑	↑
	Second	8	16										¶ 2	¶ 3	¶ 3	0 4	0 5	1 6	1 7	3 9									
	Third	8	24										0 2	0 3	0 4	0 5	1 6	2 8	3 9	6 12									
	Fourth	8	32										0 3	0 4	0 5	1 6	2 7	3 10	5 12	8 15									
	Fifth	8	40										0 3	0 4	1 6	2 7	3 8	5 11	7 13	11 17									
	Sixth	8	48										0 3	1 5	1 6	3 7	4 9	7 12	10 15	14 20									
	Seventh	8	56										1 3	1 5	2 7	4 8	6 10	9 14	13 17	18 22									
K	First	13	13	↓	↓	↓	↓	↓	‡	↑	↓	¶ 2	¶ 2	¶ 3	¶ 3	¶ 4	¶ 4	0 5	0 6	↑	↑	↑	↑	↑	↑	↑	↑	↑	↑
	Second	13	26									¶ 2	¶ 3	¶ 3	0 4	0 5	1 6	1 7	3 9										
	Third	13	39									0 2	0 3	0 4	0 5	1 6	2 8	3 9	6 12										
	Fourth	13	52									0 3	0 4	0 5	1 6	2 7	3 10	5 12	8 15										
	Fifth	13	65									0 3	0 4	1 6	2 7	3 8	5 11	7 13	11 17										
	Sixth	13	78									0 3	1 5	1 6	3 7	4 9	7 12	10 15	14 20										
	Seventh	13	91									1 3	1 5	2 7	4 8	6 10	9 14	13 17	18 22										

↓ = use first sampling plan below arrow (refer to continuation of table on following page, when necessary). If sample size equals or exceeds lot or batch size, do 100% inspection.

↑ = use first sampling plan above arrow.

Ac = acceptance number.

Re = rejection number.

†If, after the final sample, the acceptance number has been exceeded but the rejection number has not been reached, accept the lot but reinstate normal inspection.

‡Use corresponding single sampling plan (or, alternatively, use multiple sampling plan below, where available).

§Use corresponding double sampling plan (or, alternatively, use multiple plan below, where available).

¶Acceptance not permitted at this sample size.

TABLE T (continued)

Sample size code letter	Sample	Sample size	Cumulative sample size	0.010 Ac Re	0.015 Ac Re	0.025 Ac Re	0.040 Ac Re	0.065 Ac Re	0.10 Ac Re	0.15 Ac Re	0.25 Ac Re	0.40 Ac Re	0.65 Ac Re	1.0 Ac Re	1.5 Ac Re	2.5 Ac Re	4.0 Ac Re	6.5 Ac Re	10 Ac Re	15 Ac Re	25 Ac Re	40 Ac Re	65 Ac Re	100 Ac Re	150 Ac Re	250 Ac Re	400 Ac Re	650 Ac Re	1,000 Ac Re
L	First	20	20	↓	↓	↓	↓	‡	↑	↓	§ 2	§ 2	§ 3	§ 3	§ 4	§ 4	0 5	0 6	↑	↑	↑	↑	↑	↑	↑	↑	↑	↑	↑
	Second	20	40								§ 2	§ 3	§ 3	0 4	0 5	1 6	1 7	3 9											
	Third	20	60								0 2	0 3	0 4	0 5	1 6	2 8	3 9	6 12											
	Fourth	20	80								0 3	0 4	0 5	1 6	2 7	3 10	5 12	8 15											
	Fifth	20	100								0 3	0 4	1 6	2 7	3 8	5 11	7 13	11 17											
	Sixth	20	120								0 3	1 5	1 6	3 7	4 9	7 12	10 15	14 20											
	Seventh	20	140								1 3	1 5	2 7	4 8	6 10	9 14	13 17	18 22											
M	First	32	32	↓	↓	↓	‡	↑	↓	§ 2	§ 2	§ 3	§ 3	§ 4	§ 4	0 5	0 6	↑	↑	↑	↑	↑	↑	↑	↑	↑	↑	↑	↑
	Second	32	64							§ 2	§ 3	§ 3	0 4	0 5	1 6	1 7	3 9												
	Third	32	96							0 2	0 3	0 4	0 5	1 6	2 8	3 9	6 12												
	Fourth	32	128							0 3	0 4	0 5	1 6	2 7	3 10	5 12	8 15												
	Fifth	32	160							0 3	0 4	1 6	2 7	3 8	5 11	7 13	11 17												
	Sixth	32	192							0 3	1 5	1 6	3 7	4 9	7 12	10 15	14 20												
	Seventh	32	224							1 3	1 5	2 7	4 8	6 10	9 14	13 17	18 22												
N	First	50	50	↓	↓	‡	↑	↓	§ 2	§ 2	§ 3	§ 3	§ 4	§ 4	0 5	0 6	↑	↑	↑	↑	↑	↑	↑	↑	↑	↑	↑	↑	↑
	Second	50	100						§ 2	§ 3	§ 3	0 4	0 5	1 6	1 7	3 9													
	Third	50	150						0 2	0 3	0 4	0 5	1 6	2 8	3 9	6 12													
	Fourth	50	200						0 3	0 4	0 5	1 6	2 7	3 10	5 12	8 15													
	Fifth	50	250						0 3	0 4	1 6	2 7	3 8	5 11	7 13	11 17													
	Sixth	50	300						0 3	1 5	1 6	3 7	4 9	7 12	10 15	14 20													
	Seventh	50	350						1 3	1 5	2 7	4 8	6 10	9 14	13 17	18 22													
P	First	80	80	↓	‡	↑	↓	§ 2	§ 2	§ 3	§ 3	§ 4	§ 4	0 5	0 6	↑	↑	↑	↑	↑	↑	↑	↑	↑	↑	↑	↑	↑	↑
	Second	80	160					§ 2	§ 3	§ 3	0 4	0 5	1 6	1 7	3 9														
	Third	80	240					0 2	0 3	0 4	0 5	1 6	2 8	3 9	6 12														
	Fourth	80	320					0 3	0 4	0 5	1 6	2 7	3 10	5 12	8 15														
	Fifth	80	400					0 3	0 4	1 6	2 7	3 8	5 11	7 13	11 17														
	Sixth	80	480					0 3	1 5	1 6	3 7	4 9	7 12	10 15	14 20														
	Seventh	80	560					1 3	1 5	2 7	4 8	6 10	9 14	13 17	18 22														
Q	First	125	125	‡	↑	↓	§ 2	§ 2	§ 3	§ 3	§ 4	§ 4	0 5	0 6	↑	↑	↑	↑	↑	↑	↑	↑	↑	↑	↑	↑	↑	↑	↑
	Second	125	250				§ 2	§ 3	§ 3	0 4	0 5	1 6	1 7	3 9															
	Third	125	375				0 2	0 3	0 4	0 5	1 6	2 8	3 9	6 12															
	Fourth	125	500				0 3	0 4	0 5	1 6	2 7	3 10	5 12	8 15															
	Fifth	125	625				0 3	0 4	1 6	2 7	3 8	5 11	7 13	11 17															
	Sixth	125	750				0 3	1 5	1 6	3 7	4 9	7 12	10 15	14 20															
	Seventh	125	875				1 3	1 5	2 7	4 8	6 10	9 14	13 17	18 22															
R	First	200	200	↑	↑	§ 2	§ 2	§ 3	§ 3	§ 4	§ 4	0 5	0 6	↑	↑	↑	↑	↑	↑	↑	↑	↑	↑	↑	↑	↑	↑	↑	↑
	Second	200	400			§ 2	§ 3	§ 3	0 4	0 5	1 6	1 7	3 9																
	Third	200	600			0 2	0 3	0 4	0 5	1 6	2 8	3 9	6 12																
	Fourth	200	800			0 3	0 4	0 5	1 6	2 7	3 10	5 12	8 15																
	Fifth	200	1,000			0 3	0 4	1 6	2 7	3 8	5 11	7 13	11 17																
	Sixth	200	1,200			0 3	1 5	1 6	3 7	4 9	7 12	10 15	14 20																
	Seventh	200	1,400			1 3	1 5	2 7	4 8	6 10	9 14	13 17	18 22																

Acceptable quality levels (reduced inspection)†

↓ = use first sampling plan below arrow. If sample size equals or exceeds lot or batch size, do 100% inspection.

↑ = use first sampling plan above arrow (refer to preceding page, when necessary).

Ac = acceptance number.

Re = rejection number.

† If, after the final sample, the acceptance number has been exceeded but the rejection number has not been reached, accept the lot but reinstate normal inspection.

‡ Use corresponding single sampling plan (or, alternatively, use multiple sampling plan below, where available).

§ Acceptance not permitted at this sample size.

TABLE U AVERAGE OUTGOING QUALITY LIMIT FACTORS FOR NORMAL INSPECTION (SINGLE SAMPLING)—MIL-STD-105E (ABC STANDARD)

Code letter	Sample size	Acceptable quality level																									
		0.010	0.015	0.025	0.040	0.065	0.10	0.15	0.25	0.40	0.65	1.0	1.5	2.5	4.0	6.5	10	15	25	40	65	100	150	250	400	650	1,000
A	2															18			42	69	97	160	220	330	470	730	1,100
B	3														12			28	46	65	110	150	220	310	490	720	1,100
C	5													7.4			17	27	39	63	90	130	190	290	430	660	
D	8												4.6			11	17	24	40	56	82	120	180	270	410		
E	13											2.8			6.5	11	15	24	34	50	72	110	170	250			
F	20										1.8			4.2	6.9	9.7	16	22	33	47	73						
G	32									1.2			2.6	4.3	6.1	9.9	14	21	29	46							
H	50								0.74			1.7	2.7	3.9	6.3	9.0	13	19	29								
J	80							0.46			1.1	1.7	2.4	4.0	5.6	8.2	12	18									
K	125						0.29			0.67	1.1	1.6	2.5	3.6	5.2	7.5	12										
L	200					0.18			0.42	0.69	0.97	1.6	2.2	3.3	4.7	7.3											
M	315				0.12			0.27	0.44	0.62	1.00	1.4	2.1	3.0	4.7												
N	500			0.074			0.17	0.27	0.39	0.63	0.90	1.3	1.9	2.9													
P	800		0.046			0.11	0.17	0.24	0.40	0.56	0.82	1.2	1.8														
Q	1,250	0.029			0.067	0.11	0.16	0.25	0.36	0.52	0.75	1.2															
R	2,000			0.042	0.069	0.097	0.16	0.22	0.33	0.47	0.73																

Note: For the exact AOQL, the above values must be multiplied by $1 - \dfrac{\text{sample size}}{\text{lot or batch size}}$.

TABLE V AVERAGE OUTGOING QUALITY LIMIT FACTORS FOR TIGHTENED INSPECTION (SINGLE SAMPLING)—MIL-STD-105E (ABC STANDARD)

Code letter	Sample size	Acceptable quality level																									
		0.010	0.015	0.025	0.040	0.065	0.10	0.15	0.25	0.40	0.65	1.0	1.5	2.5	4.0	6.5	10	15	25	40	65	100	150	250	400	650	1,000
A	2																			42	69	97	160	260	400	620	970
B	3															12			28	46	65	110	170	270	410	650	1,100
C	5														7.4			17	27	39	63	100	160	250	390	610	
D	8													4.6			11	17	24	40	64	99	160	240	380		
E	13												2.8			6.5	11	15	24	40	61	95	150	240			
F	20											1.8			4.2	6.9	9.7	16	26	40	62						
G	32										1.2			2.6	4.3	6.1	9.9	16	25	39							
H	50									0.74			1.7	2.7	3.9	6.3	10	16	25								
J	80								0.46			1.1	1.7	2.4	4.0	6.4	9.9	16									
K	125							0.29			0.67	1.1	1.6	2.5	4.1	6.4	9.9										
L	200						0.18			0.42	0.69	0.97	1.6	2.6	4.0	6.2											
M	315					0.12			0.27	0.44	0.62	1.0	1.6	2.5	3.9												
N	500				0.074			0.17	0.27	0.39	0.63	1.0	1.6	2.5													
P	800			0.046			0.11	0.17	0.24	0.40	0.64	0.99	1.6														
Q	1,250		0.029			0.067	0.11	0.16	0.25	0.41	0.64	0.99															
R	2,000	0.018			0.042	0.069	0.097	0.16	0.26	0.40	0.62																
S	3,150			0.027																							

Note: For the exact AOQL, the above values must be multiplied by $1 - \frac{\text{sample size}}{\text{lot or batch size}}$.

TABLE W LIMIT NUMBERS FOR REDUCED INSPECTION—MIL-STD-105E (ABC STANDARD)

Number of sample units from last 10 lots or batches	Acceptable quality level																								
	0.010	0.015	0.025	0.040	0.065	0.10	0.15	0.25	0.40	0.65	1.0	1.5	2.5	4.0	6.5	10	15	25	40	65	100	150	250	400	650	1,000
20–29	†	†	†	†	†	†	†	†	†	†	†	†	†	†	†	0	0	2	4	8	14	22	40	68	115	181
30–49	†	†	†	†	†	†	†	†	†	†	†	†	†	†	0	0	1	3	7	13	22	36	63	105	178	277
50–79	†	†	†	†	†	†	†	†	†	†	†	†	†	0	0	2	3	7	14	25	40	63	110	181	301	
80–129	†	†	†	†	†	†	†	†	†	†	†	†	0	0	2	4	7	14	24	42	68	105	181	297		
130–199	†	†	†	†	†	†	†	†	†	†	†	0	0	2	4	7	13	25	42	72	115	177	301	490		
200–319	†	†	†	†	†	†	†	†	†	†	0	0	2	4	8	14	22	40	68	115	181	277	471			
320–499	†	†	†	†	†	†	†	†	†	0	0	1	4	8	14	24	39	68	113	189						
500–799	†	†	†	†	†	†	†	†	0	0	2	3	7	14	25	40	63	110	181							
800–1,249	†	†	†	†	†	†	†	0	0	2	4	7	14	24	42	68	105	181								
1,250–1,999	†	†	†	†	†	†	0	0	2	4	7	13	24	40	69	110	169									
2,000–3,149	†	†	†	†	†	0	0	2	4	8	14	22	40	68	115	181										
3,150–4,999	†	†	†	†	0	0	1	4	8	14	24	38	67	111	186											
5,000–7,999	†	†	†	0	0	2	3	7	14	25	40	63	110	181												
8,000–12,499	†	†	0	0	2	4	7	14	24	42	68	105	181													
12,500–19,999	†	0	0	2	4	7	13	24	40	69	110	169														
20,000–31,499	0	0	2	4	8	14	22	46	68	115	181															
31,500–49,999	0	1	4	8	14	24	38	67	111	186																
50,000 and over	2	3	7	14	25	40	63	110	181	301																

† Denotes that the number of sample units from the last 10 lots or batches is not sufficient for reduced inspection for this AQL. In this instance, more than 10 lots or batches may be used for the calculation, provided that the lots or batches used are the most recent ones in sequence, that they have all been on normal inspection, and that none has been rejected while on original inspection.

TABLE X RANDOM NUMBERS[†]

10 09 73 25 33	76 52 01 35 86	34 67 35 48 76	80 95 90 91 17	39 29 27 49 45
37 54 20 48 05	64 89 47 42 96	24 80 52 40 37	20 63 61 04 02	00 82 29 16 65
08 42 26 89 53	19 64 50 93 03	23 20 90 25 60	15 95 33 47 64	35 08 03 36 06
99 01 90 25 29	09 37 67 07 15	38 31 13 11 65	88 67 67 43 97	04 43 62 76 59
12 80 79 99 70	80 15 73 61 47	64 03 23 66 53	98 95 11 68 77	12 17 17 68 33
66 06 57 47 17	34 07 27 68 50	36 69 73 61 70	65 81 33 98 85	11 19 92 91 70
31 06 01 08 05	45 57 18 24 06	35 30 34 26 14	86 79 90 74 39	23 40 30 97 32
85 26 97 76 02	02 05 16 56 92	68 66 57 48 18	73 05 38 52 47	18 62 38 85 79
63 57 33 21 35	05 32 54 70 48	90 55 35 75 48	28 46 82 87 09	83 49 12 56 24
73 79 64 57 53	03 52 96 47 78	35 80 83 42 82	60 93 52 03 44	35 27 38 84 35
08 52 01 77 67	14 90 56 86 07	22 10 94 05 58	60 97 09 34 33	50 50 07 39 98
11 80 50 54 31	39 80 82 77 32	50 72 56 82 48	29 40 52 42 01	52 77 56 78 51
83 45 29 96 34	06 28 89 80 83	13 74 67 00 78	18 47 54 06 10	68 71 17 78 17
88 68 54 02 00	86 50 75 84 01	36 76 66 79 51	90 36 47 64 93	29 60 91 10 62
99 59 46 73 48	87 51 76 49 69	91 82 60 89 28	93 78 56 13 68	23 47 83 41 13
65 48 11 76 74	17 46 85 09 50	58 04 77 74	73 03 95 71 86	40 21 81 65 44
80 12 43 56 35	17 72 70 80 15	45 31 82 23 74	21 11 57 82 53	14 38 55 37 63
74 35 09 98 17	77 40 27 72 14	43 23 60 02 10	45 52 16 42 37	96 28 60 26 55
69 91 62 68 03	66 25 22 91 48	36 93 68 72 03	76 62 11 39 90	94 40 05 64 18
09 89 32 05 05	14 22 56 85 14	46 42 75 67 88	96 29 77 88 22	54 38 21 45 98
91 49 91 45 23	68 47 92 76 86	46 16 28 35 54	94 75 08 99 23	37 08 92 00 48
80 33 69 45 98	26 94 03 68 58	70 29 73 41 35	53 14 03 33 40	42 05 08 23 41
44 10 48 19 49	85 15 74 79 54	32 97 92 65 75	57 60 04 08 81	22 22 20 64 13
12 55 07 37 42	11 10 00 20 40	12 86 07 46 97	96 64 48 94 39	28 70 72 58 15
63 60 64 93 29	16 50 53 44 84	40 21 95 25 63	43 65 17 70 82	07 20 73 17 90
61 19 69 04 46	26 45 74 77 74	51 92 43 37 29	65 39 45 95 93	42 58 26 05 27
15 47 44 52 66	95 27 07 99 53	59 36 78 38 48	82 39 61 01 18	33 21 15 94 66
94 55 72 85 73	67 89 75 43 87	54 62 24 44 31	91 19 04 25 92	92 92 74 59 73
42 48 11 62 13	97 34 40 87 21	16 86 84 87 67	03 07 11 20 59	25 70 14 66 70
23 52 37 83 17	73 20 88 98 37	68 93 59 14 16	26 25 22 96 63	05 52 28 25 62
04 49 35 24 94	75 24 63 38 24	45 86 25 10 25	61 96 27 93 35	65 33 71 24 72
00 54 99 76 54	64 05 18 81 59	96 11 96 38 96	54 69 28 23 91	23 28 72 95 29
35 96 31 53 07	26 89 80 93 54	33 35 13 54 62	77 97 45 00 24	90 10 33 93 33
59 80 80 83 91	45 42 72 68 42	83 60 94 97 00	13 02 12 48 92	78 56 52 01 06
46 05 88 52 36	01 39 00 22 86	77 28 14 40 77	93 91 08 36 47	70 61 74 29 41
32 17 90 05 97	87 37 92 52 41	05 56 70 70 07	86 74 31 71 57	85 39 41 18 38
69 23 46 14 06	20 11 74 52 04	15 95 66 00 00	18 74 39 24 23	97 11 89 63 38
19 56 54 14 30	01 75 87 53 79	40 41 92 15 85	66 67 43 68 06	84 96 28 52 07
45 15 51 49 38	19 47 60 72 46	43 66 79 45 43	59 04 79 00 33	20 82 66 95 41
94 86 43 19 94	36 16 81 08 51	34 88 88 15 53	01 54 03 54 56	05 01 45 11 76
98 08 62 48 26	45 24 02 84 04	44 99 90 88 96	39 09 47 34 07	35 44 13 18 80
33 18 51 62 32	41 94 15 09 49	89 43 54 85 81	88 69 54 19 94	37 54 87 30 43
80 95 10 04 06	96 38 27 07 74	20 15 12 33 87	25 01 62 52 98	94 62 46 11 71
79 75 24 91 40	71 96 12 82 96	69 86 10 25 91	74 85 22 05 39	00 38 75 95 79
18 63 33 25 37	98 14 50 65 71	31 01 02 46 74	05 45 56 14 27	77 93 89 19 36
74 02 94 39 02	77 55 73 22 70	97 79 01 71 19	52 52 75 80 21	80 81 45 17 48
54 17 84 56 11	80 99 33 71 43	05 33 51 29 69	56 12 71 92 55	36 04 09 03 24
11 66 44 98 83	52 07 98 48 27	59 38 17 15 39	09 97 33 34 40	88 46 12 33 56
48 32 47 79 28	31 24 96 47 10	02 29 53 68 70	32 30 75 75 46	15 02 00 99 94
69 07 49 41 38	87 63 79 19 76	35 58 40 44 01	10 51 82 16 15	01 84 87 69 38

[†]This table is reproduced with permission from tables of the RAND Corporation published in *A Million Random Digits with 100,000 Normal Deviates.* Glencoe Free Press Division of The Macmillan Company, New York, 1955.

NAME INDEX

SUBJECT INDEX